Blue Skies

Blue Skies

A HISTORY OF CABLE TELEVISION

Patrick R. Parsons

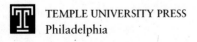

TEMPLE UNIVERSITY PRESS
Philadelphia

TEMPLE UNIVERSITY PRESS
1601 North Broad Street
Philadelphia PA 19122
www.temple.edu/tempress

⊗ The paper used in this publication meets the requirements of the American
National Standard for Information Sciences—Permanence of Paper for Printed
Library Materials, ANSI Z39.48-1992

Library of Congress Cataloging-in-Publication Data

Parsons, Patrick.
  Blue skies : a history of cable television / Patrick R. Parsons.
    p. cm.
  Includes bibliographical references and index.
  ISBN 13: 978-1-59213-287-4   ISBN 10: 1-59213-287-1 (cloth : alk. paper)
  1. Cable television—United States—History. I. Title.
HE8700.72.U6P36   2008
384.55'50973—dc22   2007021328

2 4 6 8 9 7 5 3 1

Contents

Preface

The number of books, articles and reports written on cable television is well beyond anyone's count.[1] Scholars, lawyers, engineers and journalists began a producing a torrent of paper on cable in the early and mid-1960s. Voluminous governmental reports date back to the late 1950s. The material covers every manner of cable-related issue from pressing policy and financial questions, to problems in technology, law, culture and international relations. Given the massive literature of cable television, it is reasonable to wonder why another text on the topic is necessary.

This book is, in part, the natural product of the author's own curiosity, both about the development of cable television and about the larger issue of the evolution of communications technology in society. The book also is a response to a surprising gap in the literature. Despite the thousands, perhaps tens of thousands, of studies, articles and books that have been written on cable, there have been few attempts to write a full history of the industry. Because of the nature of cable television and its role in pushing forward the horizon of communications, much of what has been written has dealt with cable television's future, how its development might impact existing telecommunications systems, the political process or even social norms and values. Far fewer works have examined cable's past.

The earliest book on the development of cable appears to have been published in 1972 by Mary Alice Mayer Phillips. *CATV: A History of*

Community Antenna Television[2] was the product of Phillips' masters' thesis and for many years was the only book on the early history of community antenna TV. It was the source of information for numerous subsequent published descriptions and offers a good description of several of the early systems and the major legal problems that confronted CATV in the 1950s and 1960s. In 1980, cable engineer Kenneth Easton published a book on the development of cable in Canada. That work, *Thirty Years in Cable TV*, drew from his personal experience in the industry and included accounts of some of the first work with community antenna systems in Great Britain.[3] An updated version of the book was released in 2000.[4] Also in 2000, pioneering cable engineer and consultant Archer Taylor published a book on the early engineering aspects of cable TV. Taylor's *History Between Their Ears* was the culmination of a special program of oral histories with cable engineers conducted by the National Cable Television Center and Museum.[5] The early 2000s brought a small surge in book publishing on specific aspects of cable development in the 1980s and 1990s, many of them featuring the industry's leading personalities. Such works included profiles or biographies of members of the cable pantheon, including John Malone,[6] Ralph and Brian Roberts,[7] and Ted Turner.[8] Because of their focus on a given industry figure or time period, these books typically provide less information on the earliest days of cable or the wider industry history.

A number of more general texts on cable television, both scholarly and popular, have been written over the past several decades. While looking at a wide range of cable topics—technology, economics, policy and social impact—their treatment of cable history usually is relegated to a chapter or only a few pages.[9] Most typical in the literature is a brief discussion of cable's overall development, generally as it relates to the specific subject of the book or article, such as the regulation of cable,[10] its social or economic implications,[11] the rise of a prominent industry figure,[12] or general reviews of cable and broadcasting.[13]

The best, perhaps the only, contemporary work to examine the full scope of cable TV history is Thomas Southwick's *Distant Signals*, published in 1999.[14] Southwick was a founder of *Cable World* magazine and reported on the industry for many years. To commemorate the 50th anniversary of the business, *Cable World* published a series of historical retrospectives by Southwick, adding them as supplements to the magazine throughout 1998. The detailed and engaging collection subsequently was printed as a full text and provides a rich and sometimes personal narrative on the rise of the industry.

Southwick's work is the exception rather than the rule, however, and with most of the population of the United States watching cable television, or programming that derives from the cable industry, there is arguably great room for additional work in the area. This book, therefore, seeks to add to the literature in several ways. It is, in contrast to some prior journalistic works, a scholarly text and attempts to provide to the extent possible full references and citations for pursuit by subsequent scholars. Unlike other works, this book begins the story of cable more deeply in the past and attempts to demonstrate the progressive evolution of the technology and the industry. Finally, this book takes a particular theoretical lens to the development of cable. It considers the complex and interacting elements of technology, economics, regulation and individual personality in the evolution of cable. Moreover, it suggests that the nature and shape of cable has been heavily influenced throughout its development not simply by the innate nature of its technology but also by the ideas or conceptions that people at various times have held about cable. It suggests that cable television is a case study in the social construction of technology. This book, therefore, is in many ways about the hopes people have for technology and how those hopes drive business decisions and public policy. It is how those visions are frequently contested by competing vested in interests and how ultimately those visions come to succeed or fail in the real world. In cable television, beginning in the 1960s, one of the dominant visions has been of a ubiquitous flexible interactive communications system capable of providing news, information, entertainment and all manner of social services. That set of utopian hopes came to be known as the "Blue Sky" vision of cable television and is taken, therefore, as the title of this book.

In writing this narrative and analysis, the author is indebted to many people who have, over the years it has taken to gather and synthesize the material, contributed greatly. First and foremost are the authors of the many accounts noted above. I have learned much from their efforts and drawn much into this text. Colleagues including Marlowe Froke, Richard Taylor, and Robert Frieden have provided a sounding board for ideas. Cable pioneers Joe Gans, Robert Tarlton, Ben Conroy, Hub Schlafly, Frank Brophy, Thomas McCauly, George Yuditsky, Jim Davidson, and younger cable hands including Joe Aman and Frank Vicente have provided a wealth of information and insight. I also am grateful to the many staff members at The Cable Center (formerly the National Cable Center and Museum) in Denver, Colorado, who have been of great assistance over the years.[15] Students who have helped in gathering information and reviewing drafts include Alan Weigmeister, Fu Yung, Katie O'Toole, and Sonya Miller. Former and current editors at Temple University Press, Peter Wissoker, who first approached the author many years ago about the possibility of publishing such a book and more recently, Micah Kleit, have been of great support. Special gratitude must be given to a man who has been a wonderful colleague and one of the great cable pioneers himself, E. Straford Smith. Obviously, any omissions or mistakes in the text (and in a work this ambitious, I am sure there will be some) are solely my responsibility. Finally, for unwavering love and support over years of writing and rewriting, unending thanks to my wife, Susan Strohm.

1

The Evolution of a Revolution
(Origins–1930s)

> *It appears to me, Miss Leete, "I said," that if we could*
> *have devised an arrangement for providing everybody*
> *with music in their homes, perfect in quality, unlimited*
> *in quantity, suited to every mood, and beginning and*
> *ceasing at will, we should have considered the limit of*
> *human felicity already attained, and cease to strive for*
> *further improvements.*

> —EDWARD BELLAMY, LOOKING BACKWARD, 1888[1]

George Gardner was young, ambitious, and willing to take a chance. He was also cold, tired, and perhaps a little frustrated. The night sky above sparkled with stars and, to him, possibilities. Still, it had been a long, sweaty, dusty day, and he had not yet found that for which he had come searching in the night sky. Gardner and three friends had begun early at the base of Jacks Mountain just north of Lewistown, Pennsylvania. They carried the heavy equipment on their backs, trekking up the rocky slopes of the mountain, through the low brush, sometimes following old trails, sometimes cutting new ones. They walked the ridges and hiked the small summits, stopping occasionally, setting up the mast and antenna, listening carefully, breaking down the equipment, and moving on to try another spot, a little higher up or a little to the west.

It was 1951 and George Gardner and his associates were on a treasure hunt of sorts. If they could capture their quarry, George thought he might be able to quit his job in the Sylvania picture tube plant at Seneca Falls, New York, and start his own business down there, in Lewistown.

Their Grail that spring night was elusive, but across the country, it was in growing demand and in Lewistown, short supply. It was, in fact, television, or in this particular case, a TV signal. To find it, Gardner had packed an electrical generator, a TV set, and a homemade antenna. The equipment was large and heavy and included a TV set with nine-inch screen housed in a sturdy wooden box. The set was needed because there were no readily available testing meters to indicate when a signal had been captured. You simply had to carry the set up the mountain, erect the antenna, fire up the generator, and hope for a picture. You had to look for the signal at night because the picture displayed on the TV screen was too faint to see in the daylight.

It took several weekends of driving down from Seneca Falls to find that signal. Gardner eventually came across the right spot on the mountain, just above Yeagertown, and the evening that he did, he and his friends watched Channel 13, WJAC-TV, out of Johnstown, Pennsylvania, about 100 miles to the west. Gardner then began to plan his next step. He would build a small stone shack to house the equipment, run an antenna wire down the mountain, boosting the signal along the way with specially made amplifiers, and hook up a demonstration TV set in town. Then, with a lot of hard work and a little luck, people would begin buying TV sets and signing up for his new service, the community antenna TV service. He would call it Lewistown Antenna Television.

George Gardner went on to successfully establish and run community antenna television (CATV) systems in towns throughout Pennsylvania. But that night on the mountain, he was one of a growing handful of entrepreneurs around the country who were simply trying to take advantage of what they saw as an expanding public need that they could turn into a modest business. In similar settings around Pennsylvania—in the coal-rich mountains of the East and the oil-soaked hills of the Northwest—and in the river valleys of Oregon and Washington; the hollows of Arkansas, Missouri, and West Virginia; in small towns literally from Maine to California, men and women were reaching out for television, setting up receiving sites often before there was even a signal to receive. George Koval in Mahanoy City, Pennsylvania, wanted to bring in a signal for his bedridden father. Abe Harter in Franklin, Pennsylvania, was tired of climbing cold, slippery roofs to install TV antennas. Ed Parsons in Astoria, Oregon, wanted to make his wife happy, and she wanted television. Jim Davidson in Tuckerman, Arkansas, already had an antenna up and lines strung when WMCT in Memphis, Tennessee, officially went on the air, just because he wanted television. Davidson and many like him knew that television was coming, and when it did, they would be ready.

Like George Gardner, those who wanted to make a business out of this idea were taking a gamble. To start their new ventures, they borrowed from friends and family, and mortgaged their existing businesses and, frequently, their homes. With the money, they bought steel and concrete for antenna sites, coaxial cable by the mile, and amplifiers by the crate. They hired construction crews, usually power and telephone company linemen who would moonlight on weekends, to install the system. Then they put an ad in the local newspaper and waited.

It is difficult—impossible, apparently—to resist the temptation to collapse into cliché and note that "the rest is history." That, however, is what this book is about. It is a history of this enterprise, one that started from many humble beginnings and evolved by the opening of the twenty-first century into a powerful and ubiquitous broadband communications network that spanned the United States and was spreading across the globe. At the start of the new millennium, the community antenna television business had grown into the cable telecommunications industry. There were nearly 8,000 cable systems in the United States, with coaxial cable running passed 99 percent of all homes with television. Subscribers in more than 65 million of those homes, about 66 percent of all TV households, received their TV programming through the cable. A typical cable TV system had well over fifty analog channels, many had 150 or more. Technologies, such as digital compression and video on demand, gave subscribers access to hundreds of channels and individual programs. More than 500 programming networks spawned by the industry supplied content to both cable and the direct broadcast satellite providers.

By 2005, cable was offering high-speed Internet access and even telephone service to millions, and it had chiseled out a place in the on-going, everyday life of the American family. MTV, CNN, the Cartoon Channel, and ESPN were increasingly replacing the old broadcast network giants, ABC, NBC, and CBS. Ted Turner was taking his place with William Paley of CBS and David Sarnoff of NBC among the legends who created modern mass media in the United States. When the United States went to war with Iraqi President Saddam Hussein in the Persian Gulf in 1991, the eyes of the country and the rest of the world turned first and most often to Turner's Cable News Network. CNN Pentagon correspondent Wolf Blitzer became a celebrity almost overnight and the face of the media changed forever.

As the industry looked into the new century, it was, as it had been at previous points in its history, in the grips of a powerful transformation. The mainspring of that transformation was captured in the term "convergence." Convergence meant the digital melding of related but previously disparate communication technologies and businesses, especially cable TV distribution channels, entertainment production enterprises, computer technologies, and the vast and wealthy telephone industry. The forging of alliances between these different interests and the concomitant integration of their technologies and businesses was, in the minds of many, to signal the start of construction of the nation's "information highway" or "telecommunications infrastructure." Over this twenty-first century digital-interactive-high-speed network, people would shop and play, conduct research and business, engage in political affairs—both local and national—work, and socialize.

In fact, the idea, or vision, of convergence was the latest manifestation of the social construction of the technology. Throughout its history, cable TV's evolution has been shaped by what people believed the technology could or should do, and the conscious attempt to mold cable by both business and political forces based on sometimes utopian and flawed ideas of its potential, has been central in influencing its development.

In the early 1990s, the telecommunications "info-structure" became not only a competitive imperative for the giant media companies maneuvering for strategic financial positions, but a point of national policy. The new communications network promised to be more than cable television, as cable television had been known for forty years. For some, the transition foreshadowed by telephone company entry into the business signaled the end of the cable TV industry. For others, it was only the next logical step in the evolution of cable television. Ultimately, the difference may only be one of semantics. For the average citizen and viewer, it will probably remain for many years, "the cable" whether it delivers MTV or baby pictures from back home in Iowa. More importantly, the technology, programming, business practices, and traditions established in the evolution of cable provide several legs upon which the new information info-structure rests. Had cable television not developed in the manner it did, the technological, business, and policy upheavals of the late twentieth century would have evolved much differently. The communications network of the future is the young creation of the communications network of the past, in large part the cable TV industry of the last half century.

It is a long road from Jacks Mountain, Pennsylvania, to the live, satellite-delivered, images of exploding rockets and antiaircraft fire over Baghdad. This book is a modest attempt to retrace the steps along that road. The story is about technology and economics. It is about geography and policy. It is about social forces that encouraged some ideas, discouraged others, and made some innovations almost obligatory and others virtually impossible. And, as much as anything else, it is about people.

The Social Context of Technological Innovation

This book is an historical narrative on the evolution of the cable TV industry, but it is also an effort to look more abstractly at the development of technology in society, using cable television as a case study. From this perspective, technical change, at least in the first instance, is seen as historically incremental and gradual, rather than radical and discontinuous.[2] Historian Abbott Usher, in the early twentieth century, was one of the first to posit technical innovation as a progressive series of small advances that, sometimes, culminated in large technical steps. Technical advances in practice or principle build upon one another and, by combining gradually, give rise to new inventions. Complex systems such as computers are the end result of a long road of prior development in sometimes widely divergent fields. The "invention" of television, for example, required the manufacture of glass, the discovery and harnessing of electricity, and a history of progress in metallurgy, among other things.

To the extent that a "revolution" in technology can be said to take place, it usually represents the culmination of such a process, or in Usher's terms, the "cumulative synthesis" of prior development.[3] It is a kind of quantum leap in the longer evolutionary path of the technology, the point, so to speak, at which the last *widget* falls into place to make the system in some way socially viable. This final piece may be a part of the technology itself or it may be a change in social context that frees or even promotes emerging technology. Television, again, is a classic example. As will be outlined, the conceptual roots of television are almost timeless,[4] but the technical, political, economic, and social conditions were not in place for a viable broadcast TV system until after World War II, when, with sufficient convergence of these elements, television took off with dramatic consequences.

The occasional debate over whether cable television constitutes an evolution or a revolution is, therefore, best answered by finding for both. In 1970, cable lawyer E. Stratford Smith penned a retrospective on the then still-young cable industry, entitling his piece, in part, "The Evolution of a Revolution."[5] The phrase very much captures the essence of many of the ideas presented here (and it is from that article that this chapter draws its title).

The social process by which a technology and an industry such as cable television arises and is shaped into such a cumulative synthesis also is an important subject of this book. Within the process of invention—whether that invention springs from the mind of an idle dreamer, an accident of industry, or a sophisticated research laboratory—comes a steady stream of new ideas and tools. For any given social or business problem, dozens and sometimes hundreds of solutions are often proposed. In the 1940s and 1950s, companies and inventors battled over transmission standards for color television. Decades later, Sony's Betamax home VCR system vied with the VHS format, and Apple computers fought for market share with the IBM-Microsoft PC standard. Frequently, market conditions, regulatory and public perception, and business decisions were more critical in determining market winners than was the technology itself.

As such fights suggest, not every technology gains a foothold, there are winners and losers. It is a well-understood axiom that technological innovation is subject to the social conditions under which it arises. As one student of technology has observed, "Not all technologies will thrive in all environments.[6] Human beings, acting in pre-established social circumstances, select the technologies that best serve their purposes under those existing conditions. The process, therefore, is local and contingent. It is frequently noted that the Chinese established the rudimentary building blocks of the printing process centuries before Gutenberg used movable metal type to print his first Bible in the mid-1400s.[7] Nonetheless, printing in Europe took hold as a technology and a business in a few short decades and became a powerful force for social change.[8] In Asia, however, social and cultural conditions that included the iconographic nature of the writing system, which required at least 5,000 Chinese characters; tight political control; and cultural traditions that prized social stability as well as the calligraphers' high art, kept printing the province of government-controlled specialists.

Moreover, the process of adopting a new technology is rarely friction free. The system typically swirls with conflict, especially as promoters of new technologies clash with those defending the old. The rise and fall of investor confidence and the availability of capital, the political machinations of ever-changing law and regulatory fashion, and the push and pull of entrenched industrial interests, all constitute, in writer Thomas Hughe's terms,[9] the integrated system within which technical change takes place. Communications scholar Brian Winston, among others, has pointed out that the social sphere is far from unidimensional. Individuals and groups of individuals with like and divergent interests and varying amounts of social power or social resources joust for position.[10]

All of which is to say that in looking at the way cable communications had evolved, one must pay close attention to the motives and resources of individuals and groups within the existing social structure. In the sphere of broadcast and telecommunications regulation, for example, analysts Erwin Krasnow, Lawrence Lognely, and Herbert Terry, have identified the key parties to the messy and often unpredictable dance of regulatory politics, including the effected industries, the Federal Communications Commission (FCC), the White House, Congress, and both last and least, the public and public interest groups.[11] Within this living system, the players compete, negotiate, threaten, lobby, and generally bring to bear whatever power they can to achieve their particular goal. Community antenna pioneers clashed not only with existing broadcasters and theater owners but even with TV antenna manufacturers, who put up a spirited political and public relations campaign against CATV operators through the 1960s.

Finally, the perception of a technology, the accepted idea of what a technology such as cable is or could become, is a critical force in shaping its development. Further, the concept or idea of the technology may not always flow directly from the inherent nature of the technology itself, and, in fact, may be at distinct odds with the actual capabilities of the device or system, setting the stage for substantial disappointment in policy and business quarters. The phrase used by some to describe this phenomenon is "the social construction of technology."[12] The social definition of cable television, the nature of the business and policy understanding of cable, has been a fundamental and repeating influence on the historical path of the technology and industry.

A particular technology, such as television, as Edward R. Murrow once put it, is in one sense nothing more than "wires and lights in a box."[13] What we think we can and should do with that box varies from group to group, and may even be quite out of keeping with the actual capabilities of the device. The early history of radio offers one of the clearest examples. Today, we think of commercial radio and television as forms of "broadcasting," the distribution of news or entertainment via the airwaves to mass audiences. But this is an idea that took some time to evolve. Prior to the advent of "radio" communication, the dominant form of electronic intercourse was wire based, the telegraph and telephone. Both were primarily point-to-point technologies, linking just two parties for personal communication. For most people, the original idea of radio, its social construction, was as a wireless form of telegraph and then telephone, hence its original name, the "wireless." Radio technology, as pure technology, was capable of a variety of functions, but the popular understanding of its nature and use determined its social application.[14]

The most detailed work on the transformation of radio technology from wireless to broadcasting has been done by historian Susan Douglas, who offers an excellent summary of the concept of the social construction of communication technology:

> Like virtually all emerging technologies, radio did not simply appear one day in its fully realized form, its components complete and its applications and significance apparent. Radio apparatus, and what all apparatus mean to a particular society at a particular time, had to be elaborately constructed. Just as individuals and institutions worked over time, to refine the invention, so did these inventors and institutions, as well as the press and the public, all interact to spin a fabric of meanings within which this technology would be wrapped.[15]

One of the central themes of this book is the manner in which the public perception of cable television, the idea of cable, has similarly shaped its development. The title, "Blue Skies," is a direct reference to the social construction of cable television. The term arose in the 1960s. It denoted a vision of cable television as something much more than just a broadcast retransmission service. Cable, according to social planners, academics, policy analysts, government regulators, and even many business people, held the potential to become a broadband communications utility, spanning the country and providing news, specialized entertainment, electronic banking, electronic mail, health care services, outlets for community expression, and a local and national forum for political debate. It could, some prophesized, advance the interests of democracy in small and large ways, and perhaps even supplant the existing system of broadcast television. Blue Skies came to suggest the unlimited, cloudless horizon of possibilities embodied in the emerging technology. It came later to carry a more pejorative connotation, one of failed promises and potential. Blue Skies came to suggest, at least in part, an unrealistic assessment of technical capacity. Blue Skies, therefore, is about the social construction of technology, generally; in this work, it is about cable television in particular. It is about the concept of cable television forged and held by industry, government, and the public, and about the role that concept plays in setting policy, crafting business plans, and making investment decisions.

Insofar as the construction of technology is equivocal, it, therefore, is contestable. An important part of the process of shaping cable has involved the political struggle to define it. The role that the ideological construction of a technology such as cable plays in evolution is twofold. First, stakeholders vary in the definition of the technology that they have coming into the business or policy arena. Some broadcasters in the 1960s saw in cable a threat to their industry; cable operators saw a legitimate business opportunity and service to the public. Further, stakeholders frequently understand well the importance of creating a definition of the situation favorable to their interests. The contemporary political terms "spin doctor" or "spinning," illustrate the point. In modern politics, the goal is often to take a politician's speech or position and color it with a particular meaning, or "spin" it in a manner intended to advance particular political goals. As will be seen, cable operators at key junctures in the industry's history worked consciously to create a specific constructed meaning for their business, designed to serve narrowly tailored goals. In molding cable television, the battleground in both industry and regulation, therefore, has been the terrain of definition and whose definition would prevail.

All of these themes—the incremental nature of technological change, the contingent nature of technical adoption within an existing social order, the importance of the social construction of technology—are woven through the history of television generally and cable television specifically. They serve as a framework for understanding and sharpening one's appreciation of the nature of cable's evolution. In keeping with this perspective then, one must look back, not just to the technical antecedents of cable television, but to the very idea of television itself.

Communicating at a Distance: Foundations

It is typically said that cable television began, or was invented, in the late 1940s or early 1950s. When and how cable television first appeared in its modern form, who invented it and whether it was "invented" at all, are issues dealt with in the next chapter. The remainder of this chapter will look at what might be called the "prehistory" of cable television, consider the historical development of television—the idea of television as well as the technology of television—especially as it relates to nonbroadcast forms of television. Insofar as cable is characterized largely, albeit not completely, by the wired transmission of electrical information, it is especially useful to look at the development of television distribution by wire.

The popular notion of the history of television typically places the birth of the medium in the early 1950s, at the start of the "Golden Era" of television. The popularity of television was exploding, with sales of TV sets accelerating annually. Programming was epitomized by serious original drama from Paddy Chayefsky, Rod Serling, and Gore Vidal, and the radio stars of the 1930s and 1940s were moving into the brightly lit sets of the TV studio.

Most of the country's TV viewers had not yet heard of community antenna television in the 1950s. Many would not come across the idea of cable until the 1960s when it spawned a flurry of utopian rhetoric; even then, it would be something most people read about rather than actually experienced themselves. Television through a wire was not a reality for most people until the 1980s. In 1955 television was, for the average viewer, a broadcast medium requiring some kind of antenna: a bent coat hanger if your means were modest and your signal strong, a fancy rooftop antenna mast if you were flush or your signal weak, and for many, a simple pair of set-top "rabbit ears."

What many outside the business or study of television may not appreciate is that the roots of modern video stretch back to before the beginning of the twentieth century. Long before scientists perfected the ability to fling TV signals through the ether, both moving and still pictures were being sent across the laboratory and around town by wire.

The history of television and cable communication, in fact, begins not with the technology, but with the dream. Like human flight, or alchemy, or transmutation, seeing at a distance is a vision that substantially predates the technical ability to realize it. For centuries people have sought methods of signaling across great distances. The Roman Emperor, Julius Caesar, built a system of high towers from which men shouted messages down the line; native Americans used puffs of smoke; and people from many cultures have flashed mirrors in the sunlight to signal distant points. One of the more famous signaling systems was built by the French in the late eighteenth century. The government erected huge semaphore signaling towers with rotating wooden arms in a system that eventually stretched across the country.

Flags, drums, trumpets, cannon fire, and flares are only a few of the many ways in which we have tried to send important information over great distances. Always one of the chief objects has been speed, to deliver the message with all haste. In this, the ultimate extension of human power was instant communication, achieved, finally and only, with the discovery of electricity, and the development of what one might arguably label the first-generation cable system, the telegraph.

The First Wire: The Telegraph

Cable television is one branch of a family tree that embraces all electronic or electrical communication and includes broadcast television and radio, telephone and even telegraph. The deepest roots of the tree, or perhaps more appropriately a "tangled vine," are traditionally traced to the founders of the science of electricity itself.

Tales of man's grasp for electrical communication often begin with early ruminations about the possibilities to be found in magnetism, especially in the form of the loadstone. The scientist and philosopher Roger Bacon considered harnessing the properties of the mysterious material for communications as early as 1267.[16] In 1569, Italian scientist Giambattista della Porta described communication between two magnetic compass needles. Around each needle were inscribed the letters of the alphabet and, as one needle pointed to a letter, the other needle, possibly miles away, would respond by pointing to the same letter. It is not known whether della Porta actually attempted to use the compass telegraph he suggested in his work *Magia Naturalis* (Natural Magic), but eventually experimenters who did build such devices would learn that "natural magic" required a greater understanding of nature than of magic.

Such understanding in the field of electricity began with the late eighteenth century experiments of Italian anatomy professor Luigi Galvani. Galvani found he could make the leg of a dead frog twitch by passing an electric current through it. This rather humble beginning opened the doors to subsequent discoveries by men like Count Alessandro Volta, Hans Christian Ohm, Michael Faraday, James C. Maxwell, and, eventually, Heinrich Hertz. Following Galvani, Volta developed the first electric battery, publishing results of his work in 1800 and stimulating interest in the field throughout Europe. Hans Christian Øersted, a Danish physicist, and André-Marie Ampère, a French physicist, demonstrated the fundamental link between magnetism and electricity, and in the early 1800s Faraday extended this connection and, working with Joseph Henry, demonstrated the ability to generate electrical current using magnetism. Exploiting this relationship, Faraday built the first electric motor, the first transformer, and the first dynamo. Faraday also discovered electromagnetic principles that later helped lay the foundation for the TV picture tube. Hertz, building on the theories and predictions of Maxwell, discovered electromagnetic waves—radio waves—in 1887, sending and detecting signals generated by a spark gap device. Hertz went on to measure the characteristics of these waves, their speed, wavelength, and their reflective and refractive qualities. The names of these pioneers are, as a result of their work, forever attached to the nature and qualities of electrical phenomenon: the ampere, the volt, the olm, and, most importantly for present purposes, the measure of the wavelength of radio communication, Hertz.

The start of communication via electricity and the beginning of cable's evolutionary trek, may be marked on the day that someone discovered that a current could be transmitted along a wire, a finding substantially predating the work of Hertz, Faraday, and even Volta. To send an electric impulse from point A to point B instantly, it seemed, was an astonishment. According to historian Alvin Harlow, an Englishman named Stephen Gray was one of the first to accomplish this in 1729.[17] In a letter to the Secretary of the Royal Society of London, dated February 8, 1731, Gray described his progress in sending electricity through lengths of wire of up to 886 feet. The development in 1745 of the Leyden jar, a precursor to the electric battery, sparked interest in this novelty, and experimental electricians throughout Europe began stretching the distance across which they could send a charge, not just through a wire, but through streams, lakes, the solid ground and, quite popularly, other people. In 1746, a clerical scientist named Abbe Nollet demonstrated this entertainment by forming a human chain of Carthusian Monks, joined together by holding links of iron wire. Nollet then sent a shock from a Leyden jar through the mile-long monkish circuit, much to the apparent amusement of the participants. Unfortunately, another explorer in this area, America's Benjamin Franklin, observed that the various experimentation seemed to lead, at the time, to little practical application, noting that he was, "chagrined a little that we have hitherto been able to produce nothing in this way of use to mankind . . ."[18]

Like so many other "firsts" in the history of communication, the first person to suggest the idea of using electricity and a wire for purposes of transmitting information is lost to us, probably irretrievably. The earliest published description of electric communication seems to have been contained in a letter appearing in *Scots Magazine*, in 1753. The letter, dated February 1, 1753, originated in Renfew, Scotland and was signed "C.M." (At least two possible authors have been credited with writing the letter). The letter,

under the caption, "An Expeditious Method of Conveying Intelligence," proposed the stringing between two distant points of as many insulated wires as there are letters in the alphabet, through which "electrical discharges should separately exhibit themselves by the diverging balls of an electroscope, or the striking of a bell by the attraction of a charged ball."[19]

A device of this type was actually built and demonstrated in 1774 by Georges Louis Le Sage. It consisted of 24 insulated wires bundled together into one cable. Each wire represented one, or in some cases two, letters of the alphabet. As the operator on one end "excited" the wire of a particular letter, its resonating partner at the far end of the device would be recorded and, in this way, a message could be spelled out.

There are several interesting points to be noted about this early device. First, it was not the only one of its kind. The early phases of technical evolution, flowing from the idea of wired communication, gave rise to a diversity of devices. Hosts of multiwire apparatus were reported throughout Europe at the end of the eighteenth century. However, all the machines were very weak. There was little power available prior to the advent of true batteries, so the distance one could signal was limited, although there were reports of experimental lines stretching several miles. Moreover, the use of multiple wires to communicate letters foreshadowed technical problems that, in their own way, were of significant concern to television generally and cable television specifically some 200 years later. One of the early goals auguring later television challenges was to build a machine that required only one wire. Again, various schemes were adopted to accomplish this. Inventors rigged clockwork mechanisms involving lettered dials that moved in sender-receiver tandem. Chemical telegraphs in which electrical pulses reacted with treated paper at the receiving site led to the development of special codes to make sense of the lines on the paper.

Advances in the general science of electricity, especially those concerning the creation of the battery and the association of magnetism with electricity, soon led to the development and widespread application of the magnetic telegraph. Following the earlier speculation of della Porta, a Russian named Paul Ludovitch Schilling may have been the first to build such a system in 1832, using magnetic needles that moved in various combinations to spell out coded messages. Following Schilling's lead, England's William Cooke and Charles Wheatstone developed a more powerful needle telegraph, introducing the device in 1837. Their magnetic telegraph, in various forms, soon came to dominate British telecommunications.

Samuel B. Morse began working in the United States on a magnetic telegraph in the early 1830s. Despite having little prior knowledge of the long history of experimentation in this area, Morse completed and demonstrated his first working device in 1837 and received a patent in 1840. As in so much of the early history of telecommunications, there is controversy over Morse's contribution. He may have taken much of the original idea for his telegraph from his shipmate, a Dr. Charles T. Jackson of Boston, on a trans-Atlantic cruise in 1932. Morse's design eventually included critical innovations developed years earlier by Joseph Henry, a widely regarded authority on electromagnetism and a professor at Princeton. The much-lauded Morse code may, in fact, have been created by Morse's business partner, Alfred Vail. Like later pioneers in the cable industry, however, it was not necessarily the technical contributions that made Morse important, but rather his ability to promote them. After much patience, some depression, and a bit of luck, Congress in 1843 approved $30,000 for Morse to build a line from Baltimore to Washington, D.C. In 1844, the line was completed, and thanks to the serendipitous convening of the Democratic presidential nominating convention in Baltimore that year, the device was successfully demonstrated, offering convention updates to Washington, D.C. The telegraph industry was booming by the mid-1850s and the first wiring of the United States well under way by 1860; the first transcontinental line was completed in 1861.[20]

There are interesting parallels between the rise of the telegraph industry and the evolution of the cable industry, one of the more intriguing but overlooked deals with a technical feat less familiar than the often told tales of Morse and Vail; it involves pictures.

Cable television can be described in many ways. It has a number of important, distinguishing characteristics. It is communication by wire, rather than through the "ether" or airwaves. It is point-to-multipoint, or mass media, rather than point-to-point. It is a visual and audio medium. It is broadband, that is, capable of transmitting large amounts of information. At certain points in its history, some of these characteristics take on greater significance than others. As simple communication by wire, its heritage arguably traces back to Morse, Le Sage, and perhaps even Stephen Gray. As a medium for transmitting pictures, as opposed to dots and dashes, however, its roots are nearly as deep, albeit less well known.

It was not long after Morse's telegraph clicked out the famous message, "What hath God wrought" in 1844, that attempts began to harness the iron wire for sending visual images. One of the earliest, if not the earliest, successful demonstrations of pictures by wire was conducted by a Scotsman, Alexander Bain, who brought several versions of his telegraph to the United States in 1848. Bain had patented a telegraph device that was, for a time, used in successful competition with Morse's apparatus. The Bain telegraphy used the reaction of chemicals to electrical impulse to etch marks on special paper. This process evolved into a facsimile prototype. The early image telegraph was strikingly similar to the facsimile machines commonly used in the twentieth century, consisting of a drum and scanning system. A specially prepared image was attached to a rotating drum; a stylus moved across the spinning drum and registered, through an electrochemical process, light and dark areas, which then were replicated at a similar receiving drum.

Frederick C. Bakewell, an Englishman, demonstrated such a machine at the London Exhibition in 1851, and in the 1860s, an Italian priest, Abbe Jean Caselli, built and patented something he called a "pantelegraph." The machine followed the scheme of Bain and Bakewell, and, with financial backing from French emperor Napoleon III, Caselli was able to establish a functioning system of visual telegraphy in the 1860s. Although the enterprise eventually failed as a business, the concept of pictures at a distance had been practically established.

Similar precursors to modern cable TV service can be seen in the early use of the next step in the development of wired communication, the telephone.

The Second Wire: The Telephone

Serious speculation about the use of electricity to carry the human voice began as early as the mid-1850s, with one pioneering German engineer, Philip Reis, popularizing the idea and developing crude devices that could transmit sounds, albeit weak, unintelligible ones. He even reportedly carved one wooden housing for a transmission device in the shape of a human ear.

The development of the first working telephone is a messy, complicated subject and, as with the telegraph, is of some controversy, with competing claims and technologies, and a resolution grounded as much in courtroom litigation as in technological prowess. The technical transition between the telephone and the telegraph—the interim evolutionary step—involved efforts to send more than one message at a time over the telegraph wire. The core objective, then as now, was to increase the power—the carrying capacity—of the existing technology. This central driving goal would find its manifestation recurring throughout the development of cable and other communications technologies. The work done by such inventors as Thomas Edison and Elisha Gray to create a "multiplexed" or "harmonic" telegraph would, in turn, set the stage for the development of a device that could carry a range of frequencies wide enough to accommodate the human voice. There are several good histories on the early telephone fight.[21] Alexander Graham Bell built on the work of many inventors before him, such as Reis, and was in stiff competition with others of his time, including Edison, to create a practical working device for the transmission of the human voice. By his own admission, the idea was not original with Bell. In fact, he initially believed (incorrectly as he later discovered) that a telephone device had already been invented and his first efforts were meant merely to replicate the feat. Whether he won the technical race is a matter of historical debate. But his success in the courtroom in defending his claims and patents would mean that in the popular histories he would be credited with the first practical telephone apparatus. Received history marks March 10, 1876, as the day he spilled acid on his pants in his laboratory and called out to his assistant, "Mr. Watson, come here, I want you," and the telephone became a reality. The first regular telephone line opened a year later between Boston and Somerville, Massachusetts, and a year after that the Bell Telephone Company was established.

The telephone soon became and remains today the world's principal point-to-point communication system. This is as opposed to a "mass medium" in the general sense of the term. The telephone allows one person to talk with another, in contrast to something like broadcasting, which is intended to link one person or programming source to thousands or even millions of listeners. This, however, has much more to do with the social construction and economic development of the telephone than it does with its inherent technical capacity. Even at its earliest stage, the telephone had the ability to provide nearly all the services now characteristic of cable television and in much the same manner. The telephone network that spread out across the United States in the late 1800s was the "second wiring" of the country, following the telegraph. And it had more flexibility to provide news and entertainment "programming" than we appreciate today. In fact, it is plausible to suggest that were it not for the development of the wireless in the early 1900s, the telephone network could have evolved directly into the principal system for electronic media today, including television.

The possibilities of a telephone-based system of news and entertainment were envisioned by futurists and realized by engineers even before Guglielmo

Marconi dreamed of harnessing Hertzian waves to broadcast information. The concept of "the wired city," the term popularized in the 1960s as a description of cable television, can be found in its basic outlines in commentary and literary fiction well before the turn of the century. Bell himself saw the potential of the telephone as more than just a dedicated line between two parties. Along with others, Bell envisioned a network of interconnected devices, any one of which could contact any other at any time (the modern telephone exchange). And even this rudimentary but critical notion of a communication network had its antecedents in then-existing technologies, as Bell's own prospectus revealed:

> At the present time we have a perfect network of gas pipes and water pipes throughout our large cities. We have main pipes laid under the streets communicating by side pipes with various dwellings, enabling the members to draw their supplies of gas and water from a common source. In a similar manner, it is conceivable that cable or telephone wires could be laid underground, or suspended overhead, communicating by branch wires with private dwellings, country houses, shops, manufactories, etc., etc., uniting them through the main cable with a central office where the wires could be connected as desired, establishing direct communication between any two places in the city.[22]

The very idea of a telephone network, therefore, draws from existing technologies. In addition, Bell's description, up to the point of person-to-person connection, is very much an outline of a classic "tree and branch" cable TV architecture. And the use of such a network for entertainment and news, as opposed to (or in addition to) one-to-one conversation, was equally on the minds of early visionaries.

While many logically saw the use of the telephone for party-to-party conversation, others saw a one-way system for distributing news and entertainment to larger audiences. This view was influenced to some extent by the initial nature of the device itself, which in its earliest form only operated in one direction (talker to listener). Telephone pioneer and Bell competitor Elisha Gray used early telephone technology in 1874 and 1875 to conduct tests of "electroharmonic" broadcasting, and Bell himself used his equipment to transmit experimental music concerts from Paris to Brantsford, Ontario, in 1876.[23] Colin Cherry noted that Bell, "at first demonstrated his telephone to audiences by playing hymns, recitations, popular songs and attempted to transmit music (not exactly hi-fi quality!), drama and news."[24] An editorial cartoon on the front page of New York City's *The Daily Graphic* in 1877 showed a sweating, wild-eyed orator shouting into a telephone receiver and sending his voice by wire to audiences in London, San Francisco, the Fiji Islands, and a several other global outposts.[25]

In one of the most popular and influential books of its day, Edward Bellamy's *Looking Backward* offered citizens of the nineteenth century a Utopian view of life in the twenty-first. The quote at the beginning of this chapter is from Bellamy's 1888 description of a citywide system of piped audio, a system that, without too much of a stretch, could describe a modern cablecast digital music service. Bellamy's system was a wire-based network of telephone lines connecting various music studios, or music halls, to the "music room" in everyone's home. As Bellamy's protagonist sits in the "music room," transfixed with wonder and pleasure at the perfect tone, quality, and volume of the music, his twenty-first century host explains:

> There is nothing in the least mysterious about the music, as you seem to imagine. It is not made by fairies or genii, but by good, honest, and exceedingly clever human hands. We simply carried the idea of labor-saving by cooperation into our musical service as into everything else. There are a number of music rooms in the city, perfectly adapted acoustically to the different sorts of music. These halls are connected by telephone with all the homes of the city whose people care to pay the small fee, and there are none, you may be sure, who do not.[26]

Bellamy's system boasted high quality music, a wide selection of choices, and as it was financed essentially by subscriber fees, there were no commercials. But Bellamy's vision was not unique. In the same year that *Looking Backward* was published, engineers and entrepreneurs began experimenting with the kind of system Bellamy described in his book. Media historians have recovered from the archives of the late 1800s a wealth of material on these efforts, including detailed descriptions of what could be characterized as the first, full-scale, protocable system ever built.

One of the first promoters of the concept of wired broadcasting, according to historian David Woods, was Hungarian native Tivadar (Theodore) Puskas. Puskas worked with Edison in the mid-1870s and, during that time, developed the idea of a central transmitting station for wired communication. Puskas later returned to Europe where he conducted "theaterphone" experiments at the Paris Exposition Internationale d'Electricite' of 1881, placing microphones on the stage of the Paris Opera House and transmitting the program to telephone booths placed in the Exhibition Pavilion. Similar experiments have been credited to French engineer Clement Ader,[27] but in either case, the idea was clearly afoot. The system was one of the more popular attractions at the Exposition, and curious fairgoers packed the exhibition houses three nights a week all summer long to see and hear live opera and theater carried electrically from the stage. Historians Robert Hilliard and Michael Keith report that the lack of amplification made the system incapable of providing reasonable reproduction of music, "because electronic amplification had not yet been developed, Ader's principal demonstration, from the Paris Opera, was a failure. The thing simply sounded bad.[28]

It sounded good enough, however, to encourage continued experimentation. Woods, as well as media historian Carolyn Marvin, have detailed the development of such systems across Europe in the 1880s and 1890s.[29] "Subscribers" to telephone media services could listen to news, musicals, stock market reports, and even Sunday morning church services from the comfort of their living rooms on these forerunners of cable.

In London, for example, a system set up for the 1892 Electrical Exhibition connected concert rooms and theaters in Birmingham, Manchester, and Liverpool to listening rooms in London.[30] Marvin remarked that, using the coin-operated "theaterphones," "it was said, anyone might listen to five minutes of theater or music for the equivalent of five or ten cents."[31] In London, Bellamy's fiction was, to some extent, realized by 1907 when a citywide system connected 600 subscribers to thirty different originating locations featuring musical performances, speeches, and church services.[32] Similar telephone-delivered plays and concerts were reported in U.S. cities, including New York, Chicago, Detroit, Rochester, Buffalo, New York; Mobile, Alabama; Wichita, Kansas; and even Oshkosh and Appleton, Wisconsin.

Perhaps the most elaborate system of this type, which was established by Puskas and his family in Budapest, Hungary, operated from about 1893 until World War I. Woods, Marvin, and others have provided detailed descriptions of "Telefon Hirmondo." Hirmondo is a Magyar word for a medieval town crier, but loosely translated the term meant "Telephonic Newseller." Telefon Hirmondo began operations in 1892, delivering at first news programming and later cultural fare, including plays, concerts, and even children's programming from the central telephone exchange. The system grew from an initial 1,000 subscribers to some 6,000 at its peak. Subscription costs were about a penny a day and installation was free. A penny, nonetheless, meant something in the 1890s and the service, according to Marvin, seemed to be restricted to the city's aristocracy.

Telefon Hirmondo assembled a full-time staff of more than 150 people and was engaged in substantial program origination, especially news. Marvin noted the strong parallel between the telephone service and radio and TV broadcasting stations of fifty years later, suggesting Telefon Hirmondo was a form of "proto-broadcasting." She also suggested, however, that the Budapest service might even more closely resemble today's cable television, pointing out that it could have evolved into the dominant form of mass media had not wireless been introduced in the early 1900s. Because broadcasting was substantially cheaper for the consumer over the long run—once a receiving set was built or purchased, service was essentially free—it had an evolutionary advantage over the wire-based systems. The lower cost also made it more accessible to all social classes, it therefore was inherently more democratic, according to Marvin.

There were attempts at replicating systems akin to the Budapest operation in the United States. In 1894, the Chicago Telephone Company used its wire network to "broadcast" local and congressional election returns to an estimated 15,000 listeners.[33] In 1911, a former *New York Herald* advertising manager, M. M. Gillam, organized the "New Jersey Telephone Herald Company." The system was modeled after the Hungarian operation, which Gillam had seen while touring Europe. Although it boasted more than a thousand subscribers, paying five cents a day, at its peak, the service became mired in a contractual dispute with the New York Telephone Company over the use of their wires. It also ran into financial difficulties and it folded within a year.[34]

Similar efforts reappeared in the 1930s. In 1931, a company called Wired Music, Inc. petitioned the New York State Public Service Commission for permission to use existing telephone lines to begin a subscription music service, noting even then that it was an old idea they were attempting to implement.[35] By 1936, the company was piping music to dozens of hotels and restaurants in New York City.[36] Another system, this one using electrical power lines to deliver a three-channel service, was attempted in Cleveland, Ohio in the mid-1930s. Incorporated locally under the name of "Muzak," it would go on to substantial success.[37]

What is apparent is that the concept and some of the technology that lie at the heart of today's cable TV systems can be found in the earliest efforts of engineers, business people, and dreamers to communicate by wire. It is in these early activities that one can see beginning of cable.

Sharing Radio Signals

The development of radio, which provided multiple channels of music, news, and entertainment to audiences without the additional costs of a monthly subscription fee, and more importantly, without the substantial cost of a wired infrastructure, meant a foreshortened life span for the wired point-to-multipoint telephone systems. Within its generous signal range, broadcasting could reach isolated country listeners as easily and cheaply as city-bound apartment dwellers. At the same time, however, radio opened up new opportunities for wired systems in the form of radio redistribution.

Community Antenna Radio

One of the chief characteristics of cable television that the earliest wired systems described above did not, and could not, anticipate was the redistribution of broadcast signals. Nonetheless, this very basic concept of the wired relay of over-the-air programming has its own heritage, one stretching back to far before the first TV set salesman erected an antenna in the mountains of Pennsylvania. The seminal purpose of community antenna television was straightforward—to pick up a weakened TV signal at the very edge of its coverage area, amplify it, and distribute it to local customers for a fee. In fact, this simple scheme was more than twenty-five years old when the first CATV systems began operation in the late 1940s.

Just as community antenna television appeared almost simultaneously alongside the first commercial TV broadcasts, community antenna radio made its first appearance shortly after the start of commercial radio broadcasting in the 1920s. One of the first reported examples of a community radio relay system came from the small farming village of Dundee, Michigan, in early 1923. "There," reported *Radio Broadcast* magazine, "the tired farmer goes in from work, closes a switch, and without any tinkering with instruments may listen to a perfectly tuned concert from almost anywhere in the country."[38] This luxury was made possible by what the magazine called "municipal radio," a system in which radio was, echoing Bell, piped in "like gas or electricity." The operation was the brainchild of the president of the local Farmer's Telephone Company. Stretching a double-mast antenna between two buildings, the company ran dedicated telephone lines to subscribers' homes throughout the town. The system required a powerful radio receiver and special batteries in the main office, but it worked well enough that people were willing to pay $1.50 a month for the service. The *Radio Broadcast* article concluded by asking, "Who will say how many Dundees, all over the country, will be adopting this system of municipal radio within the next few years?"[39] As it turned out, the answer was, not too many.

Only a handful of systems of this type were ever attempted in the United States. The telephone company in Lorain, Ohio built a dedicated wired system in the early 1930s, capable of serving 700 customers. The network used amplifiers and lines separate from the normal telephone system; it had a central retransmission station and three distribution substations that fed programs to subscribers from a variety of sources, including local live shows and radio relay programming.[40] The Muzak Corporation, firmly established in New York City by the early 1940s, ran into trouble when it used its city-wide system of leased telephone lines to retransmit the WOR broadcast of the 1941 World Series to its subscribers. Mutual Broadcasting Company brought suit to block Muzak's carriage of the games, successfully arguing that the retransmission constituted unfair competition.[41]

An excellent example of early cable radio has been provided by researcher Susan Opt in her investigation of radio relay systems in rural South Carolina in the early 1930s.[42] In several farming areas of that state, self-powered relay systems were constructed in response to the lack of regional electrification. Expansion of the nation's power grid had stalled in the Great Depression. Large rural sections of upstate South Carolina had not been wired for electricity, and the batteries necessary to run radios were an expensive luxury. Farmers, and others, were glad to pay a small monthly fee to barnyard engineers who could run a party-line styled speaker wire several miles out to their house. These local systems became a popular, if short-lived, phenomenon. As economic conditions improved and the electric power lines finally stretched out to the rural homes, the need for the radio systems vanished. While they existed however, Opt found evidence of up to a dozen such operations, providing a single channel of retransmitted radio programming to hundreds of subscribers, with many systems changing a fee of 25 cents per month.

Such systems, in the United States and around the world, illustrated the idea that technical innovation is, as noted previously, typically local and contingent. A host of specific local conditions will serve to foster or inhibit the emergence of a particular technology. In this regard, the systems of Lorain, Dundee, and South Carolina were anomalies in the United States. The declining cost of receivers, the growing familiarity with the controls, and a nearly ubiquitous national radio signal militated against the need for such technology domestically. The political–economic matrix of broadcasting in the United States led to a national system that offered almost blanket coverage of the country by several radio networks—NBC Red & Blue, CBS, DuMont— by the end of the 1930s. Variety was sufficient to make the additional cost of a subscription wire unattractive.

The same was not as true in Europe. In Great Britain public policy decisions to nationalize radio kept domestic choices to a minimum and the ethereal proximity of signals from continental Europe was such that, at least on the English coast, it was no great technical feat to pull in French radio programming. In such conditions cable radio found a market niche. One of the earliest reported cable radio pioneers was a man named Wallace Maton, of Hythe, on England's southern coast.[43] Foreshadowing the experience of later CATV pioneers, Maton, who ran an electrical shop, experimented at extending radio reception in and around his home at the urging of his wife, who wanted to hear the radio in different rooms. He soon found he could run wires even into his neighbors' homes, connecting loudspeakers there to the radio receiving set in his shop. By January of 1925 he had fashioned a commercial cable system operating throughout Hythe, charging subscribers a modest weekly fee.[44]

Historian James Welke looked in detail at early British efforts, noting that the improved signal reception, ease of operation, and low cost all contributed to the popularity of what was then known as "radio rediffusion."[45] Mirroring the experience in South Carolina, lack of electricity to many homes played a role in adoption. The Hythe system, like the one in Dundee, featured a main tuner that was located in the central receiving room or office. The subscriber was given only a speaker with one or two simple switches, controlling volume and program selection (usually offering only two or three choices). The "cable" carried amplified audio signals rather than radio frequency signals, and the wire was the twisted copper pair (or a double twisted pair, or "star quad" formation, in postwar years) of the telephone system, rather than coaxial. Moreover, the price of the self-powered service was often lower than the cost of either a purchased or rented radio set. The radio relay exchanges were especially useful in coastal towns like Hythe where ship transmissions interfered with normal radio reception. They were also popular in northeast England where the British Broadcasting Corporation (BBC) broadcast service was slow to develop.[46] By 1932, there were at least 132 such systems in Great Britain serving more than 43,000 subscribers.[47] The BBC initially saw promise in the technology as a supplement to its national broadcasting service. A 1928 BBC proposal even envisioned a national network of wireless exchanges that would constitute the primary carriage system for BBC programming, with "wire broadcasting supplemented, it is true, but in minor part, by wireless broadcasting."[48] The BBC chose not to pursue that option, however, and in time came to see the private radio relay services as a competitive threat. The BBC sought to foster a unified national system of tightly controlled broadcast programming and eventually all the radio relay systems were restricted in the number of subscribers they could serve.[49] They were also required to carry BBC programming to the near exclusion of all other signals.

Radio relay systems were common elsewhere in the world, including Sweden, Germany, Russia, Nigeria, and Malta.[50] They were exceptionally popular in the Netherlands and systems there predated those in Great Britain. By the end of the 1930s, national penetration in the Netherlands was reported at 50 percent, and in some urban areas as many as 80 percent of all homes subscribed.[51] It is worth noting then that by the time community antenna television was beginning in the United States in the late 1940s, radio rediffusion had been a reality in Europe for more than twenty years and was serving nearly 900,000 homes in England alone.

Hotel and Apartment Master Antennas

While failing on a citywide basis, cable radio did succeed in the United States in hotels and apartment buildings. Prior to development of high-quality internal antennas, radios typically required external antennas, often very long ones, and, in larger buildings, metal superstructures interfered with even those. Roof-mounted antennas were the best solution, but this gave rise to yet another dilemma, antenna population booms on the rooftops of apartment houses.

The master antenna concept, therefore, found easy application in metropolitan areas where the devices were used to service apartment houses and hotels. The famous Waldorf-Astoria Hotel in New York City, under reconstruction in 1930, installed such a system, providing six channels of radio and selectable audio service to each of its 2,000 rooms and featuring a special master antenna system for 140 apartments in the building.[52] Apartment and hotel master antenna systems were sufficiently common by the early 1930s that they engendered a copyright infringement case in which the American Society of Composers, Authors and Publishers (ASCAP) sued a Kansas City hotel for retransmitting licensed broadcast songs.[53] In picking up the signals and delivering them to the hotel's public and private rooms, the court held that the operator was illegally reproducing the protected programming. It was a case that would foreshadow similar problems for the cable TV industry.

Among the companies involved in installing apartment house and hotel master antenna radio systems in the 1930s and 1940s were a consulting firm, Amy, Aceves, and King, and the broadcasting and electronics giant, Radio Corporation of American (RCA).[54] The master antenna was seen as an added attraction for potential tenants, especially, according to the trade press, "in these days of depression, when so many apartments are empty."[55] RCA, of course, owned the National Broadcasting Company (NBC) and its two dominant radio networks, the Red and the Blue. It had a heavy vested interest in seeing that radio, in all its forms, and eventually television as well, prospered. Toward this end, RCA created its own master antenna product, called "Antenaplex." An estimated 100 RCA units were operating in New York City by 1941. Across the country, thousands of apartments and hotel rooms were using such technology when installations were curtailed by World War II.[56] Moreover, and with a prescient vision, observers in 1931 noted that these systems could, without too much difficulty, be adapted for TV reception.[57]

Television

While the telegraph and telephone systems of the 1800s and the radio relay operations of the 1930s constitute a preface to the story of cable television, cable is ultimately about wire-based video. A brief description of the development of television, with special emphasis on the role of television by wire and cable, is, therefore, fundamental to a deeper appreciation of some of the technical, regulatory, and business problems confronted in the evolution of the technology.

While "vision at a distance" had, as noted, been a dream of mankind for a thousand years, the first step in making the dream technically feasible came in 1817 when Swedish chemist Jons J. Berzelius isolated the element selenium. While he was unaware of it at the time, it turned out that selenium had the unusual property of reactance to light.

Many technical discoveries that ultimately led to wholesale changes in social practice begin accidentally, and so it was with selenium. The British Telegraph Construction and Maintenance Co., in 1873, was attempting to use selenium rods to conduct performance checks on the transatlantic cable at its terminal station at Valentia Island, Ireland. But the measurements provided by the material were highly unstable. An electrical engineer named Joseph May, looked for the cause of the problem and discovered that crystalline selenium gave off different readings of cable performance, depending on its exposure to light The greater the amount of light, the less the resistance to electricity in the selenium rod. May's supervising engineer, Willoughby Smith, investigated further and was soon proposing to harness this unique property to create a form of "visual telegraphy."[58] Smith published his early findings in the widely read *Journal of the Society of Telegraph Engineers* in 1873,[59] and shortly thereafter, schemes for adopting selenium in the wire transmission of images began attracting public attention. There was a flurry of proposals for these "electric telescopes," or "telectroscopes" In 1878, the magazine *Punch* printed a cartoon of what it proclaimed could be a new Edison invention, the "telephonoscope," coupling the emerging idea of television with the telephone line. The cartoon depicted a two-way, wide-screen TV display hung over a drawing room mantel, with parents in London talking to their daughter in Ceylon by means of this "electric camera-obscura" and associated telephone connection.[60]

According to historian, Albert Abramson, the first use of the word "television" appeared in a paper presented on August 25, 1900, at the International Electricity Congress in Paris.[61] The first reports of anyone building and demonstrating such an apparatus did not occur, however, until 1906,[62] and even then, the results were disappointing. The first wave of television-like devices typically consisted of a surface of light-sensitive selenium cells. From each receptor cell a single wire carried a charge corresponding to the amount of light received. The wires from the plate were bundled together and run to a receiving mechanism that, through one technique or another, would illuminate a corresponding cell. May himself built one of these mosaic sets. Beginning as early as 1875,[63] and continuing through the early 1900s, most television proposals were based on the bundled or multiwire scheme of linking receptor to receiver. One, shown in 1909, relied on a selenium mosaic format using only twenty-five cells in by a five by five square. Selenium proved insufficiently responsive to light and generated too weak a signal to provide the basis for a practical system, but the promise of distant vision spurred continued investigation.

Transmission for all the experimental systems, of course, was by wire because the broadcasting of images was still several years away and, for some, nothing more than a pipe dream. One observer, Marcuse Martin, writing in the June 1915 edition of *Wireless World* scoffed at the notion of broadcast television, arguing that some form of wire-based television might be made to work, but that wireless TV was "absurdly impossible."[64] It would probably be stretching the spirit of the early multiwire systems to suggest that these proposals gave cable any historical preeminence over broadcast television; a modern TV studio has a nest of cables snaking across the floor, and these early prototypes feel more like a workshop or experimental laboratory than a fully conceived distribution system. Still, it is important to remember that the distribution of TV signals by wire is a concept that arose with television itself, and finds its roots in the telegraph and telephone as much as in the radio. Moreover, it is also worth remembering that by the time broadcasting burst forth in the first decade of the twentieth century, wire-based communication had been a routine part of social life for more than fifty years. The term "wireless" itself underscored the established tradition of wired communication by the late 1800s.

Even more importantly, one could see in these initial ideas the first struggle with a problem that would challenge television engineers—both cable and broadcast—throughout the development of the cable industry, and continues to challenge them today. That problem, put in contemporary terms, is one of bandwidth. The visual image captured by the eye is composed of millions of points of data captured by the rods and cones of the retina. Mechanical systems of visual reproduction are, in comparison, substantially restricted in the amount of pictorial information they can capture, transmit, and recompose. Early systems were severely limited in their power and the pictures were correspondingly primitive. Part of the problem resided in the nature of selenium. It was quickly discovered that, while the material had a quick initial reaction to light, it recovered its resistance more slowly. Without nearly instantaneous reaction to changes in all types of lighting, transmission of moving images was impossible.

A second problem with the selenium mosaic design lay in the amount of picture detail it could transmit. Modern picture tubes have 200,000 or more picture elements—several hundred horizontal lines of dots or small squares.[65] A comparable pre-1900 multiwire selenium system would have required 200,000 selenium cells and 200,000 separate wires. Even at that time, this was recognized to be impractical. In June 1908, Shelford Bidwell wrote in *Nature* that a reasonable, two-inch square picture would require 90,000 selenium cells, a receiving device that occupied 4,000 square feet of space and a multiwire cable eight to ten inches thick. He estimated the price tag at 1.25 million British pounds, three times that much if you wanted color.[66]

An alternative to the mosaic, multiwire system involved sending picture information sequentially, in a serial format. The idea was to send one picture element at a time, scanning a picture line by line, but doing it quickly enough that the eye's natural persistence of vision perceived only the fully assembled image.

The first reported scanning proposal appeared in the November 2, 1880, edition of *LaLumiere Electrique*. Maurice LeBlanc proposed using a pair of vibrating mirrors to scan the picture area and even suggested a means for producing color images.[67]

The more famous articulation of the scanning principle arrived several years later, when the German scientist, Paul Nipkow, filed a patent for a device to dissect the picture into its constituent parts using a disk perforated by a series of twenty-four holes set in a spiral pattern. The ability to scan a picture in this manner and send the electronic pulses sequentially, meant that images could be transmitted over a single wire. The Nipkow disk was, in Abramson's words, the "master television patent."[68]

Television-related patents were issued and ideas published throughout the nineteenth century, but in keeping with the constraints of incremental technical evolution, much of the foundational technical knowledge needed to make the ideas practical remained undeveloped. There was, for example, no way to adequately amplify the feeble picture signals generated by the various image detection schemes. Experimentation nonetheless continued with Marconi and others were breaking new ground in electronics and broadcasting and paving the way for modern television by wire and airwaves. In 1904, John Ambrose Fleming patented an electronic "valve" capable of controlling the flow of electrons, the first electronic rectifier, and two years later, Lee DeForest added to it an important regulating "grid." DeForest's "Audion" tube would soon become the basis for a number of important advances in electronic technology, not the least of which was the simple amplifier tube, capable of boosting the weak radio signals generated and received by the early radio equipment and, eventually, TV signals as well.

Experimental television struggled through the first two decade of the twentieth century with halting incremental steps. Only in the early 1920s did television begin to take longer strides toward practicality. Through the 1920s and 1930s, the evolution of television was highlighted by the quantum technical transition from mechanical to electronic picture reproduction and by the economic battle for industrial control of the emerging system.

In the first decades of the twentieth century, the Nipkow disk was the standard in the mechanical reproduction of TV images. The quality of the picture was measured in terms of the number of lines that could be scanned and the number of total frames transmitted every second; the higher the number of lines and frames per second, the better the picture. No picture was very good, however, at least by today's standards. The first scanned TV pictures were small, dim, and fuzzy; a typical early system generated about thirty lines of picture information at twelve frames per second, and display screens were only an inch or two square. Nonetheless, the public was enthusiastic about the possibilities of television, and the excitement of a practical system for distant vision overcame the shortcomings in the minds of many.

The leaders in the effort to improve picture quality, and the first major figures in mechanical TV development, were John Baird in Great Britain and Charles Francis Jenkins in the United States. By the mid-1920s, both men had demonstrated enough technical success to attract private investors and had set up their own small companies. Baird, interestingly enough, secured his initial financial support simply by advertising for investors in the London Times.

Jenkins, who applied for a patent on the wireless transmission of images as early as March of 1922, is credited with the first reported broadcast of a still image in 1923. Baird reportedly sent a still picture by wire in 1923 and produced a broadcast moving image in 1925.

It was soon apparent that visual information could be transmitted by wire and through the airwaves and, for the next several years, experimentation utilized both media for TV transmission. However, the question of whether television would ultimately be distributed broadly by wire or broadcast was not a pressing one at the time; the critical issue was whether a mechanical or an electronic system of image capture would provide the superior picture.

Baird used a version of the Nipkow disk and Jenkins an arrangement of crescent-shaped revolving prisms. Some of the better Nipkow systems in the 1930s attained sixty lines of resolution. But the system of scanning a picture mechanically had inherent limitations reminiscent of those confronting the multiwire mosaic arrangements. Increasing the size and improving the resolution of the picture required increasing the number of scanning lines. Increasing the number of scanning lines, in turn, meant increasing the number of holes in the whirling disk and ultimately the size of the disk itself. The same principle held true for disks that used revolving mirrors or prisms. One observer speculated, in 1931, that a sixty-line, six-inch square picture would require a disk ten-feet wide, and that a high-resolution picture of twelve inches by twelve inches would require a Nipkow disk 400 feet across. While these projections turned out to be inflated, it was nonetheless clear to many working in the field at the time that an all-electronic system, one that scanned the image with an electron beam, would be necessary to increase both speed and picture clarity.

Such a system was, in fact, outlined very early in the history of the medium. The British scientist, Alan Archibald Campbell Swinton, proposed, in 1908, a camera and receiver based on the cathode ray tube. He offered a detailed technical explanation of his ideas in a paper presentation in 1911, but never built his system and in fact indicated he had no great interest in doing so insofar as he saw no financial future for the design. Moreover, he realized as early as 1924 that the production of such a system would require the research and development resources of a major corporation. In most of his predictions, Swinton proved prescient, and his proposals served as inspiration and direction for most of the major advances in TV technology in the 1920s and 1930s.

Unfortunately for Jenkins and Baird, they never made the requisite transition to an electronic format. In Great Britain the BBC chose an electronic system designed jointly by EMI, Ltd., and British Marconi over Baird's more primitive mechanical apparatus. Jenkins tried unsuccessfully to sell his patents to Westinghouse in 1924. Ultimately, his company, Jenkins Television, could not compete with the larger players Swinton had correctly prophesized would enter the field. Jenkins died in 1934 at age sixty-seven, his business failed. Taking over the work of these entrepreneur engineers were the major communications corporations of the day, General Electric, Westinghouse, RCA, and American Telephone and Telegraph (AT&T). In fact, these companies could be grouped into two major camps for purposes of television: AT&T and everyone else.

Zworkin, Sarnoff, and Farnsworth

The industrial structure of modern American broadcasting was forged just after World War I. Corporate giant General Electric (GE) seized upon U.S. fears of foreign domination of domestic radio to force, with the help of the U.S. Navy, the British-based corporation, American Marconi, out of the U.S. market. In 1919, General Electric created RCA to operate the radio stations it had purchased under duress from American Marconi and, within a few years, GE and RCA had entered into a patent-sharing agreement with the other major broadcasters, Westinghouse and AT&T, and together they dominated the business of electronic communications. Their interests naturally embraced television.

Westinghouse was the first with an opportunity to take a lead in the field, but it failed to recognize the future and the future literally slipped away. The missed chance came in the form of a young immigrant engineer named Vladimir Zworykin. Zworykin had spent his formative years in the St. Petersburg laboratory of Russian scientist, Boris Rosing, a pioneer in cathode ray receiver technology. In 1911, Rosing built one of the first operating electronic TV systems and was credited with inspiring some of Swinton's thinking. Fleeing the Bolshevik Revolution, Zworykin emigrated to the United States and joined Westinghouse in 1920. Building on his experience with Rosing, he took out several early patents on an all-electronic system and sometime in 1924 or 1925 demonstrated his ideas to Westinghouse executives. It did not go well. The pictures were poor, even by Zworykin's account, and he was admonished to "go work on something more useful."[69] Zworykin became disillusioned with Westinghouse and left the company for a short period. He returned within two years but it would be many more before Westinghouse became a serious contender in TV technology.

General Electric was more successful. Ernst Alexanderson, a pioneer in radio technology, headed the GE television effort beginning in 1924, using a variation of Jenkin's prism disk technology. By 1928, GE had developed a dependable scanning camera that delivered a forty-eight-line picture at sixteen frames per second. Throughout 1928, GE worked to improve the system, sending pictures via broadcast transmission and even attempting transmission over regular telephone lines, although the latter experiment proved unsuccessful.

In 1928, both GE and RCA received government approval to begin experimental TV station operations and, by the end of the year, there were about a dozen stations in the United States, all of them using mechanical scanning.[70] The role of the government in shaping both broadcast technology and industrial structure, both radio and television, is critical to the evolution of the field. Important government decisions made in the 1920s involving where and how television could be broadcast, and later decisions affecting how TV station licenses would be allocated around the nation, acted powerfully to shape the industry from its infant years. Discussion of the role of the government and its interaction with important industry players will be picked up in Chapter 2. The remainder of this chapter will focus on the technology and business issues that helped created the foundation for cable in the United States.

In 1930, Westinghouse and GE consolidated their radio receiver manufacturing operations within RCA. As part of the move, Zworkin left Westinghouse and, along with staff from the involved companies, set up shop in RCA's Camden, New Jersey research facility. When a Justice department antitrust suit subsequently led to the breakup of the Westinghouse-GE-RCA alliance, RCA became a separate company. Zworkin now worked exclusively for Sarnoff, and Sarnoff was keenly interested in Zworkin's ideas. Several years earlier, after the failure of the Zworykin's Westinghouse test, Zworkin had approach RCA head Sarnoff for support of his all-electronic system. Sarnoff was initially intrigued and eventually impressed. By 1932, he had instructed his research team to begin full-time work on Zworkin's electronic system. By 1933, they had produced a new camera tube capable of scanning 240 lines at twenty-four frames per second. The union of Zworkin and Sarnoff helped spark the quantum leap away from the mechanical scanning standard and to an all-electronic system. For their efforts, RCA and its broadcast subsidiary, NBC, were placed on a path that would lead them to domination of TV technology for the next two decades. A variety of technical and legal hurdles remained, however, one involving a patent dispute with Zworkin's chief competitor for the title of "inventor of television," Philo Farnsworth.

While Farnsworth has become a figure of almost legendary proportions as one of the "fathers," if not "the father," of television, his contributions were limited in some ways. He was exceptionally bright and inventive, but had difficulty working in an institutional setting. With the exception of an unhappy two-year agreement with Philco electronics, he was never employed by any of the major companies in the field. Backed by private investors, Farnsworth labored on his own. He was the first to build and demonstrate an all-electronic system in 1927 and was eventually successful in claiming important patent rights. He launched a number of legal attacks on RCA in defense of his patent claims, which he accused Sarnoff of infringing. But when Sarnoff settled the patent battle by signing a royalty agreement with Farnsworth, it was essentially to pay off the inventor and gain the use of some narrow and specific patent rights, especially Farnsworth's image dissector. RCA's main system relied on the contributions and work of Zworkin, and it was the Zworkin system that set the standard.

AT&T: The Once and Future King?

The June 29, 1998, edition of *Broadcasting & Cable* magazine featured a cover photo of two beaming businessmen, John Malone, head of Tele-Communications, Inc. (TCI), the nation's largest cable TV company, and Michael Armstrong, head of AT&T, the country's largest telephone company. The men were shaking hands over the caption "Founding the Telcom Age." The occasion was the purchase of TCI by the telephone giant.

The paving of the twenty-first century information highway was the result of the convergence of several important technologies, most importantly, cable television and telephone. And one of the important historical notes in the evolution of the cable TV industry and, ultimately, the telecommunications info-structure, has been the shadowing of cable by AT&T and its offspring since the inception of television itself. In the story of cable television, the end of the tale may be the marriage—shotgun or otherwise—of the cable and telephone business. It is all the more interesting, and even poignant perhaps, that the company that first led the field in TV research was AT&T. The earliest efforts of GE and then RCA, it can be argued, were attempts to catch up to the work being done by the Bell Laboratories in the 1920s. Furthermore, and even more directly to the very heart of cable television itself, was AT&T's development of that for which the industry was named—coaxial cable.

AT&T's interest in the use of its wires to transmit visual images dates back to the late nineteenth century. While it was recognized at that time that the technology was not yet available to do much on a commercial basis, AT&T was nonetheless pondering the delivery of both still and moving images by wire.[71] During World War I, the company began work on what would become facsimile transmission. It was interested in the market for press photographs and worked with the press associations in creating, by 1923, a reliable means of still photo transmission. In part because of its success there, the company then began experimenting on similar techniques for moving images. Analogizing from its distribution of still photographs to newspapers, AT&T thought it might be able to distribute newsreels to theaters.

One of the principal research engineers involved in the telephoto program, Dr. Herbert E. Ives, was directed to work up a plan for development of an AT&T television system. With the financial resources and technical expertise of the Bell Telephone Co., it took only a few months for Ives' team to produce an advanced prototype. Ives chose to build on the best practical operating system at the time, which was a mechanical variation of the Nipkow disk and the so-called "flying spot" technology in which a scene to be televised was scanned by a moving spot of light. This decision would eventually prove to be a mistake, but at the time, the Bell group took the technology to a state of the art far beyond what others were accomplishing.

Ives began his project early in 1925 and, by the end of the year, had successfully demonstrated both wire and broadcast transmission in the laboratory of a fifty-line scanning system. Bell Laboratories also claimed to have successfully televised, in the laboratory, the first motion pictures in June and July of 1925, reporting that the film images were of higher quality than the still pictures AT&T was transmitting at that time by fax.

AT&T, which had been conducting its research with a certain amount of secrecy, began planning a public demonstration for 1927. On April 7 of that year, AT&T invited reporters into the auditorium of the Bell Telephone Laboratories in New York City for the first major demonstration of television in the United States. AT&T linked the New York site to its Bell Laboratories' research facility in Whippany, New Jersey, and to a studio in Washington, D.C. The principal speaker in Washington was then Secretary of Commerce, Herbert Hoover, and the transmission medium between New York and the Capitol was wire. The link to the Whippany site was by radio. The device shown to reporters was booth-like. The speaker sat facing a hooded screen. The flickering blue spot of the scanner played across his face, and the figure of the party at the other end was seen on a tiny two-inch by two-and-a-half-inch receiver. The picture was composed of fifty scanned lines at eighteen frames per second. A larger public address display was added at the New York terminal. It consisted of a two-foot by two-and-a-half foot screen composed of 2,500 neon lamps.[72]

Participants at the demonstration were impressed by the quality of the picture on the smaller screen but reportedly found the image on the large display to be muddied and shredded by the enlargement. Nonetheless, the program of public speakers and vaudevillian musical acts was enough to earn AT&T a front-page story in the *New York Times*, with a subhead that proclaimed it to be "like a photo come to life," although it also cautioned, "commercial use in doubt."[73] The following month, across the Atlantic, Baird reportedly transmitted a TV image from London to Glasgow, Scotland, using telephone lines, but the image was weak compared with Ive's equipment. AT&T was a clear leader in the technology of the time.

Following the AT&T demonstration, Ives looked to the future. He saw the major applications of television in three areas: (1) individual, two-way communication as an adjunct to the telephone; (2) as a powerful public address system; and (3) the distribution of "scenic" events, such as sporting contests and theatrical productions. He thought that all three applications could be handled either over the airwaves or over the telephone lines, but felt that wire service would generally be preferable: "the very serious degradation of image quality produced by the fading phenomenon characteristic of radio indicates the practical restriction of radio television to fields where the much more reliable wire facilities are not available."[74]

AT&T's near-term plans, however, called for starting with a two-way video telephone and electronic distribution of newsreels. The early picturephone concept fit, of course, naturally into the Bell system, and the 1927 demonstration had many of the characteristics of a primitive picture-phone setup; the principal receivers were, by necessity, small and more in keeping with individual use, and the public display was an add-on for demonstration purposes. AT&T was also motivated by a desire to exploit its existing network of wires in carrying television, especially as it had, in 1926, departed the field of broadcasting itself.

In 1930 AT&T demonstrated its first picturephone system. Ives' team connected two AT&T research offices in New York City and again called in the press. The experiment was successful, and advanced the number of scanning lines from fifty to seventy-two, nearly doubling effective picture resolution. With this test and with a 1929 demonstration of color TV transmission, Bell had shown its ability in mechanical TV production. Ives, however, faced two technical difficulties. First, Farnsworth and RCA were making significant advances on electronic television. Second, AT&T had the recurring problem of trying to get high-information TV pictures through a wire-based system designed for only voice transmission. And as the TV research advanced and the technology improved, the first problem exacerbated the second.

AT&T knew at the very outset of its TV research that it had a problem mating its telephone wires to the demands of television. The chief obstacle again was bandwidth. All electronic information, whether it is carried through a wire or over the airwaves, travels in waves as part of the electromagnetic spectrum. Bandwidth, loosely speaking, involves with the amount of spectrum space necessary to carry a certain amount of information. The bandwidth of the typical AT&T voice circuit in 1929 was two to four kilocycles. Today's TV signal requires a bandwidth of six megacycles (or megahertz, MHz). The bandwidth necessary at the time of Ives' initial experiments for television was significantly less than that, closer to forty kilohertz (kHz) or less, but still appreciably more than the normal voice circuit. The challenge, from the beginning, was to try to fit the broad river of video information through the small pipe that constituted the Bell telephone wire system. As will be noted frequently throughout this book, the problem of fitting increasing amounts of information into scarce spectrum space was one of the principal technical problems facing the cable industry throughout its history.

Bell responded to the bandwidth challenge in a variety of ways. In its 1927 tests, for example, it used several different circuits to carry the unusual quantity of information. Separate circuits were used for voice, picture, and synchronizing data. Only by dividing the electronic stream and dedicating special paths for the different tributaries, could the system handle the necessary traffic. Researchers developed additional innovative techniques in an effort to compress the visual information into a smaller package. To send newsreels without tying up multiple circuits, AT&T tried a process of sending sequential frames of a picture over a single circuit at a slower rate, in a kind of accelerated version of the facsimile process. The idea was to transmit a motion picture a few frames at a time to a distant site to be recorded on film. In this way, one could transmit by telephone. Using this technique, Bell researchers estimated they could transmit about a foot of film every thirteen minutes. Later, Bell would adopt a similar approach for what was called "slow-scan" television, sending pictures at a slower rate of speed, for its picturephone technology of the 1950s and 1960s.

Had some of these technologies panned out, Bell might have been able to develop, before World War II, a nationwide network of switched, interactive cable television, and the country's eventual system of broadcast and cable television would have evolved much differently. For better or worse, however, the bandwidth problem frustrated the creation of such a system and nudged Bell in other directions, one of which was, nonetheless, to have a significant impact on cable.

Meanwhile, the accelerating development of the all-electronic TV system was compounding AT&T's troubled TV scheme. While Ives' team held the lead in TV research through much of the 1920s, that lead began to slip away as others moved to the more powerful electron tube technologies. By July of 1933, Ives recognized the problem and counseled a change in direction, proposing research directed at building an electronic system.[75]

Ultimately, Bell would enter into a cross-licensing agreement with Farnsworth and use versions of his image dissector in some of their later tests. While this may have helped Bell to keep abreast of trends in the production of images, again it only exacerbated the problem of transmitting those images over AT&T wires. As the number of lines per picture and the number of frames per second increased with every advance in the technology, the capacity of the lines to ship the additional information was increasingly taxed. It was simply beyond the ability of the standard telephone line to carry information that now needed hundreds of kilocycles of space.

Part of Bell's response was to simply abandon or modify some of its plans for television. But another part of its answer was a new kind of wire, a wire that eventually would serve as the physical foundation for the cable industry itself, which, of course, was coaxial cable.

The Golden Thread

The standard telephone cable was, and remains at least in essence today, two copper strands twisted around one another, and known as the twisted pair. Information in the form of electrons travels through the copper wire itself. The physical nature of this configuration severely limits the amount of information the twisted pair can carry, however. This was the underlying technical constraint that forced AT&T to distribute one TV signal across multiple voice circuits in its early demonstrations. Failed attempts in 1928 by GE to use standard telephone lines to carry TV signals cross country (Chicago to New York) underscored the inadequacy of this technology for the task.[76] The solution had to be a new kind of wire, coaxial cable.

Coaxial cable consists of a single copper wire that runs down the middle of a metal—copper or other material—pipe. An insulating material, such as hard rubber or polyethylene discs or spacers, holds the center wire in place within the tube. The central wire axis within the surrounding axis gives the coaxial cable its name. The advantage of this configuration is that electronic information, instead of traveling through the metal material itself, tends to ride on the inner surface of the outside tube. This is especially true at the higher frequencies that are characteristic of TV signals. This quality makes it possible for coaxial cable to carry a nearly unlimited amount of information. The carriage of the electronic information on the inside of the pipe also protects the transmission from outside electronic interference, in contrast to the twisted pair and similar cables that are more susceptible to extraneous radiation. The space between the center wire and the outer tube becomes a shielded, isolated slice of open spectrum space.

The theory behind coaxial conduction was understood very early in the history of electrical communication, and its first applications were in submarine telegraph lines. Early radio engineers used primitive forms of coaxial cable to carry high-frequency radio signals between their transmitters and antennas. Large-scale commercial exploitation of the technology began with AT&T's interest in using the technology to both enhance its ability to carry regular telephone messages and as a way to enter the potentially lucrative TV business.

Bell Laboratories began formulating plans and taking out patents on coaxial cable in the late 1920s, and conducted a small-scale test of coaxial transmission at its test site near Phoenixville, Pennsylvania, just outside Philadelphia, in 1929. What they discovered during these trials was that coaxial cable worked as well in the field as the paper theory said it was supposed to, something that is not true of every engineering experience.

The success of the coaxial project, coupled with the dawning realization of the technical problems associated with two-way picturephone service, prompted a modification in AT&T strategy. According to AT&T's self-published history: "The emphasis changed from two-way, telephone-adjunct systems . . . to work on circuits for broadband television networking. . . . The objective was to furnish channels for television networking in a manner analogous to the Bell System's established role in furnishing audio program channels to radio networks."[77]

In other words, AT&T would put on the "back burner" plans for entering the mass media market as a direct competitor and, following its success in providing network facilities for radio, become the physical backbone of network television as well. In 1932, AT&T met with NBC to discuss the possibility of providing a broadband network for TV program distribution. Bell also began planning a major demonstration of the capabilities of its coaxial technology, with special emphasis on television.

In 1935, it sought FCC permission to construct a 100-mile line between New York and Philadelphia. The 1-MHz cable would be able to carry 240 telephone conversations or one, 240-line TV signal, greater capacity than anyone had ever before seen. Western Union, along with existing radio and motion picture interests, immediately saw the coaxial line as the first step toward another Bell monopoly. As one contemporary described it, "They feared AT&T would lead off the television horse with its new coaxial wire as the halter."[78]

The FCC saw a similar possibility. In a 1938 staff report, the Commission reviewed the variety of possible paths TV development might take in the United States, one of them being the use of coaxial cable in an AT&T wire-based system of TV distribution. The TV system conceivably could, said the report: "develop into some sort of wire plant transmission utilizing the present basic distribution network of the Bell System, with the addition of coaxial cable or carrier techniques now available or likely to be developed out of the Bell System's present research on new methods of broad-band wire transmission."[79]

It took a promise from Bell to keep the line open to all competitive interests to secure the FCC's blessing for development of the new coaxial line. The company also promised not to pass along the $580,000 construction cost to customers.

Construction of the New York to Philadelphia line was completed by the end of 1936 and voice-only tests initiated. But delivering a quality TV picture was a significantly more complicated matter. While the coaxial cable itself provided a nearly perfect transmission medium, state of the art amplifier technology could only exploit a small portion of the available spectrum space. As cable TV operators would later learn, the real problem was in building amplifiers and terminal devices that could cope with larger bandwidths. AT&T engineers had to design an amplifier that could boost the high- and low-end frequencies across, for that time, a rather wide span of spectrum space. It took AT&T nearly a year to work out the problems associated with amplifying these electronically wide TV signals.

In the meantime, the company had also built a special scanning camera with a disk six feet in diameter to generate the 240-line picture. Eventually, new 1-MHz amplifiers were placed every ten miles along the run and, on November 9, 1937, the new system was unveiled. The demonstration featured newsreels and an animated program explaining how the system worked. The 1927 reception devices were replaced with an eight-inch cathode ray tube that gave pictures that were "only slightly colored" and "could be watched from a distance of ten feet without severe eye strain."[80]

AT&T, of course, realized that even the 1-MHz line would be insufficient to meet the increasing demands of the higher resolution RCA and Farnsworth signals. By the mid-1930s, electronic cameras were setting standards at 441 and 525 lines, a level of detail requiring more than 4 MHz of bandwidth. By 1939, AT&T had improved its amplifiers, increasing their capacity to 3 MHz. By 1940, it had advanced to the point where, in 1940, NBC could use the line to relay TV coverage of the Republican National Convention from Philadelphia to its station in New York City.

World War II forced the company to channel its energies into other activities. By the end of the war, however, AT&T was ready with a plan to lay up to 7,000 miles of coaxial cable, connecting major cities in the United States and eventually building a coaxial web binding the nation. In December of 1944, it announced construction of the first legs of a 7-MHz system capable of carrying one modern TV channel and 480 telephone calls.

An Overture

In broadcasting, meanwhile, the number of experimental TV stations grew as the technology advanced. By 1937, there were about seventeen broadcast TV stations operating in the United States. By the later 1930s, a 441-line, thirty frames-per-second standard had been informally adopted by the industry and, in 1941, the National Television System Committee recommended a 525-line standard, which the FCC subsequently approved. On April 30, 1939, Sarnoff proclaimed the start of the television age with a major demonstration of RCA television from the World's Fair in New York City.

The "inauguration" of television, like development of coaxial cable by AT&T, was one of the many incremental steps necessary in the technical evolution of cable television. The AT&T distribution system, the broadcast system, and cable were interlinked parts of a broader movement to bring television to American homes. A wide diversity of ideas and then devices arose logically from prior accomplishments. At each step, development was driven by the ideas that inventors and business men and women had about the technology, what it could be used for and how that use might profit themselves or others. From Morse to Sarnoff, people sought ways to extend the power of communication.

All the early progress was to come to an abrupt halt, however, with the start of World War II. The resources of the companies involved in TV research and manufacturing were redirected to the war effort and all but a half dozen TV stations left the air. It wasn't until 1944 that the slumbering giant of television began to awake, destined to change social life across the globe. Before that could happen, important issues of technology and policy had to be resolved, particularly matters involving the technical standards to which the new TV industry would adhere. Competing parties had heavy financial commitments in different technologies and standards, and the stakes were high. Those decisions, discussed in greater detail in Chapter 2, also would effect the development of a related industry, cable television. Part of the social change that television was going to produce would involve a small group of entrepreneurs who would bring television into homes through the coaxial cable developed by AT&T instead of through an antenna, with consequences even they could not imagine at the time.

2

Pioneering Efforts
(1930s–1952)

*We started with three channels, then went to five, then
to seven and then we went to thirteen. I remember when
we went to seven channels. There was no doubt in our
mind that was all we were ever going to need. I mean,
who wants more than seven?*

— CABLE PIONEER, MILTON SHAPP[1]

The young man was focused and busy. He was clinging to the top of a power pole in Shenandoah, Pennsylvania, on a sunny autumn day, working on the coaxial line for the community aerial that the Shen-Heights TV Association was running through town. He was new with the small company, just learning the tricky business of connecting cables, and he didn't need unnecessary interruptions. He was a bit frustrated, therefore, when a couple from the neighborhood began shouting at him from the street below. They were spirited and didn't look as if they would go away soon. His initial thought was that they were complaining about something, or perhaps just curious about his activities. In fact, turning his attention away from his work and toward the vocal pair, he quickly discovered the focus of their interest. It was the cable. How quickly, they asked, could he get the TV line into their home? He tried to explain that the company was installing service as quickly as it could. The couple was insistent, however. Couldn't he get them TV now? When he tried politely to disengage, they asked if perhaps they could pay him a little extra to run a line into their house; he wouldn't need to tell anyone. It was clear to the young man that they were serious about their desire for television. And he would soon discover that they were not alone. Similar scenes unfolded across town in the next few months, and across the nation over the next few years. In various forms it would become a common story told by CATV pioneers years later. People would accost CATV installers as they worked, even, in some anecdotes, chasing CATV trucks down the street. The offer of a little extra money under the table to get moved up on the installation schedule was not at all unheard of.[2]

Embedded within the tales, and those like it, are several deeper points about the evolution of cable television. In the first instance, the demand was substantial, but, importantly, it was not demand for CATV, it was demand for television. Driving adoption of cable through 50-plus years has been a seemingly insatiably hunger for television. CATV has been only one method for satisfying that hunger, albeit an important one. Many different types of technological solutions have been attempted over the years to perform the function of delivering video. In the evolutionary metaphor, CATV was one of the more successful forms; many others have faded into historical obscurity. It is additionally important to note that the Shenandoah couple would never have been standing on that street corner shouting up at the CATV installer if a reasonable broadcast signal had been available to them in 1953. CATV was not simply about demand, it was also about supply, or rather, the lack thereof. Television, at least until the 1990s, had been conceived of primarily as a broadcast technology. But despite public appetite, the supply of broadcast television, especially in the early 1950s, was severely limited, and it was this key condition that helped give the CATV industry its foothold in U.S. homes. The evolution of this state of affairs and its impact on the start of cable engages the triple themes of technology, policy, and economics and is the subject of this chapter, which traces the development of television broadcasting in the United States from the 1930s to the early 1950s and considers the consequent first efforts to extend the broadcast signal via local coaxial cable.

Setting Standards: Politics and Profits

For both community antenna and broadcast television, 1948 was a landmark year. In 1948 television figuratively exploded into popular consciousness. *Radio and Television News* hailed it as the "Year of Television." The number of television stations tripled that year. Sales of TV sets skyrocketed, and the first attempts to develop community antenna television began. One might fairly ask the question, however, why 1948? It is widely understood that World War II stalled the introduction of television, but experimental stations had been in operation since the late 1920s. Despite the war, it might be considered surprising that TV did not, therefore, break out earlier. In fact, there were several factors working to both advance and retard the introduction of television through the 1930s and 1940s. A lack of venture capital during the Depression kept entrepreneurs out and held down investment by all but a few major players. Those major players, especially RCA, had secured and were holding close key patents and technical knowledge necessary for the advancement of TV. Through the 1930s, RCA's David Sarnoff reportedly delayed deployment of TV in an effort to protect the corporation's massive investment in radio. Although, near the end of the Depression, he boldly attempted to force national adoption of an RCA TV standard and might have been successful had it not been for the beginning of the war, which put a hold on anything not directed toward Allied victory.

In other words, it was not the technology that was necessarily wanting in the late 1930s, it was, as suggested in Chapter 1, an issue of the social context within which television was incubating. The introduction of television is a case study in itself of a technology arising from a given economic and political context and subject to the demands and constraints of that system. Allowed to proceed only in a vise of conflicting social pressures created by industrialists on one side and government agencies on the other, it was squeezed into a negotiated form by powerful vested and combating social actors. It illustrates Brian Winston's notion of the social effort, usually but not always successful, to suppress any potential for technology to bring sudden dramatic change. While the prior chapter dealt with TV's technical adolescence, this chapter looks more closely at the relationship of that technology to the social setting, especially the government and, most importantly, the public. In this regard, the key point then is not just that TV debuted a bit later than it could have, but that the economic and political tug-of-war gave television, both broadcast and CATV, a particular shape and color. The structure of both industries was molded by the debates and decisions that took place before and after the war. This chapter also considers in detail how existing social conditions and the evolving structure of broadcasting gave rise to (and channeled) the efforts of the earliest CATV pioneers.

The story begins with the first notice given to television by federal officials. It involved the problem of bandwidth, spectrum space, and what the government was going to do with TV. As previously noted, television, whether it was going to be distributed by wire or over the airwaves, had a problem. It took a great deal of bandwidth to transmit the complicated, data-packed visual image. A television signal contains not just the pure video signal, but several synchronizing pulses to hold the picture steady, as well as the audio signal. The standard spectrum space allocated to an AM radio signal is today, as it was in 1928, only 10 kilocycles or kilohertz wide. Even the earliest TV images required ten times the spectrum space; modern broadcast television signals are 6,000-kHz or 6-MHz wide.

As explained in Chapter 1; to improve the picture quality, or picture resolution, one must increase the number of picture elements or scanning lines. And the greater the number of scanning lines required, the greater the amount of transmission space—or bandwidth—needed to send the additional information. While the spectrum—both the broadcast ether and the enclosed space of a coaxial cable—theoretically is unlimited, the technical capacity to utilize that unlimited resource has always been finite. In the late 1920s, interests involved in developing television began asking the Federal Radio Commission (FRC), forerunner of the FCC, to be assigned frequencies for experimental broadcasting. The FRC's dilemma was where to put the new service. The allocated portion of the electromagnetic spectrum at the time—and that portion most engineers felt comfortable using—ended near the top of the AM radio dial, around 1600 kHz. In October of 1928, the FRC set aside a number of channels in the AM radio band for use by television experimenters. The selected channels were only 10-kHz wide, the same width as the standard radio channel, but much too narrow to be of any real use to television. In December of 1928, the FRC, in consultation with TV engineers, allocated four experimental channels above the AM band, in the 2- to 3-MHz range. Each channel was 100-kHz wide. The FRC also set aside for future TV use three much wider channels in a frequency range beyond 23 MHz. At the same time, the FRC began extending authority for experimental broadcasting on the four usable channels and by 1931 had licensed at least 19 stations.[3]

The advantage of the new assignment was that the 2- to 3-MHz frequency range was very powerful. It is, today, the band used by shortwave and various international and military users. Shortwave transmissions are capable of traveling thousands of miles. This, however, means that two stations on the same channel (colocated) are more likely to interfere with one another if they are anywhere in the same geographic region. This would have future implications for the development of cable, but the more immediate problem at the time was the width rather than the spectrum location of the FRC assignments.

While the frequency reserved for TV was ten times that of AM, it was still insufficient to meet the growing technical demands of the video signal. Through the 1930s, as TV technology evolved from mechanical to all-electronic systems, the corresponding detail and complexity of picture information mushroomed. The forty-eight line mechanical standard of 1931 climbed to an electronic capability of several hundred lines by 1937. And while regions above 3 MHz were opening up with every advance in transmission technology, there was a concurrent increase in demand for spectrum space from other services.

The FCC responded in October of 1937 by expanding the exploitable broadcast spectrum from 20 to 300 MHz.[4] Television was moved out of the shortwave band and given nineteen new channels in a region known as the "very high frequency" (VHF) band. Seven of the channels were in the lower 44- to 108-MHz portion of the new allocation and twelve were in the higher, and as yet technically unexploitable, 156- to 294-MHz region. Most importantly, the new channels were 6-MHz wide, enough bandwidth to carry electronic TV signals.

By the late 1930s, the dominant electronics companies, especially RCA, began to feel confident that they had finally produced commercially viable systems. Sarnoff was particularly vigorous in pursuing approval to begin commercial broadcasting. He launched a campaign to have RCA's system selected as the national standard, pressuring the Radio Manufacturers Association (RMA) and, subsequently, the FCC to adopt the RCA format (441 lines at thirty frames per second) and to approve the commercial use of television. He was only partially successful. RMA's television committee recommended the RCA standard to the FCC, but business rivals Philco, Zenith, and Columbia Broadcasting System (CBS) protested. Faced with a disputatious industry, the Commission declined to anoint a single standard, declaring that experimentation on advanced techniques should continue and ultimately the marketplace should be permitted to select a successful benchmark. Sarnoff forged ahead, declaring the inauguration of broadcast television at the RCA exhibit at the 1939 New York World's Fair. His aim, in part, was to make RCA the de facto standard and force the Commission's hand. While some popular histories mark the start of U.S. broadcasting with Sarnoff's World's Fair promotion, the Commission did not immediately take the bait. It did authorize limited commercial operations, beginning in September of 1940, but cautioned the competitors to avoid promoting consumer investment in home technology that could become quickly obsolete. Sarnoff took advantage of this small opening and, with hopes of forcing it wider, launched an intense marketing campaign, slashing the price of RCA sets and proclaiming the start of commercial television. The FCC, especially Chairman James Lawrence Fly, was enraged. Fly scolded Sarnoff on national radio, and the FCC rescinded its order permitting the start of commercial broadcasting and clamped a lid on station ownership.

Within a few months the RMA, at the FCC's direction, had formed a new study group, the National Television System Committee (NTSC). It proposed a new television standard of 525 lines and thirty frames per second; it also recommended that FM rather than the AM transmission be used for the audio component due to FM's superior clarity and resistance to interference. With the industry in greater union, the FCC adopted the recommendations on May 3, 1941, and authorized the start of full commercial broadcasting the following July first.

Commercial operations began that summer with about thirty stations gaining approval for commercial status. But this was to be another false start for television. On December 7, 1941, the Japanese attacked Pearl Harbor and by early 1942, the United States was moving quickly to a war-time footing. The FCC halted the construction of new television stations and the War Production Board issued an order stopping the production of radio and TV receivers. The material and intellectual resources of the nation were diverted to the war effort. Television engineers turned their attention to radar and military communications technologies. Training in radio and electronics began for thousands of young men, some of whom would later take what they learned and use it to build the first CATV systems. For the moment, however, television went dark. Only six TV stations remained on the air through the war years.

By 1944, with the end of the war in sight, a host of interests and issues representing all the spheres of industry, technology, and policy were arraying themselves for the coming of postwar television. TV sets were seen as one of the several important kinds of consumer goods that would help drive a postwar economic recovery and TV set production was viewed as an important source of jobs for returning veterans. The nation's production facilities, expanded dramatically for the war effort, would be turned to domestic, peacetime purposes. The major corporate players were all keenly interested in maintaining and, if possible, expanding their technical and political positions in this dawning of the TV age. The stakes were high and the players were not afraid to bluff. As much as anything, it was the struggle between these entrenched interests that delayed the general introduction of television until three years after the end of the war.

It was clear that demand for spectrum space would increase substantially after the war, both nationally and internationally. Military use was expected to grow along with commercial use. The FCC, in concert with other federal agencies, began working out new allocation schemes for the postwar spectrum. Commerce, unsurprisingly, was eager to assist in this process. Waiting in the FCC's corridors with briefcases stuffed with legal, economic and technical proposals were the major players. Edwin Armstrong and other backers of FM radio wanted to retain and even expand its prime location in the lower portion of the VHF band. Prior to the war, Sarnoff, after failing to overturn important Farnsworth patents in court and unable to buy Farnsworth outright, reluctantly entered into a licensing arrangement with the iconoclastic inventor. RCA now had both prevailing electronic systems under its control. With millions of dollars invested in VHF band technology and in patents securing dominance of that technology, RCA wanted all television assigned to VHF. CBS had been effectively shut out of VHF and the Farnsworth-Zworkin patents. It, therefore, proposed moving television to the higher UHF band, where the company was developing technology for color and high-definition black and white transmission. Older interests, such as GE and DuMont sided with RCA, as did some set manufacturers who wished to jump-start receiver sales. Others, such as Westinghouse and Zenith, sided with CBS.

In May and June of 1945, the FCC issued its decisions. Adhering closely to its prewar standards, the Commission established thirteen TV channels in the VHF band, six in the lower portion between 44 and 108 MHz (low band) and seven in the higher portion between 174 and 216 MHz (high band). FM was relocated, along with other radio services, between the two parts of the VHF range. While this largely favored the RCA position, the Commission declined to shut the door on CBS and UHF. In fact, the Commission said that the space available in the VHF band would ultimately be insufficient for a nationwide competitive television system. It indicated that the industry must look to the UHF band, with its greater availability of bandwidth, for the future of color and high-definition TV.

The result was widespread confusion, especially on the part of TV set manufacturers who did not know which standard to prepare for and whether they should build VHF receivers that might be antiques before the first tube burned out. In the midst of this mess, half of the more than 150 post-War license applications were withdrawn and NBC began talking about scaling back its TV plans until the issues were resolved. CBS, on the other hand, seized the opportunity and petitioned the FCC to approve a new UHF color format, a move that only exacerbated the angry uncertainty permeating the industry.

In March of 1947, the FCC, under pressure, broke the political-technical logjam by denying the CBS petition. Citing the potential cost of converting all existing VHF receivers to UHF sets, the FCC committed to the maintenance of the VHF allocations. The technical uncertainty seemed to be resolved, and set manufacturers and potential station operators took it as a green light. The ruling, according to television historian William Boddy, "set off [an] explosion in television station licenses and receiver sales."[5] The TV floodgates were opened, manufacturing plants readied themselves for mass production, and applications for television station licenses began once again pouring into the offices of the FCC.

AT&T Spins a Web

At the same time, another major player was tightening its grip on television, as AT&T began spinning out its golden web of coaxial cable. In late 1944, Bell announced plans to begin laying an extensive coaxial network, projecting at first that 6,000 miles of cable, and subsequently 12,000 miles of line, would be installed by the early 1950s.[6] This network was designed not only to enhance AT&T telephone service capability, but to provide a new revenue stream, with station-to-station links for the promising TV industry. By 1948, AT&T had lines stretching from New York to Washington, D.C., and from St. Louis to Buffalo, New York. On the drawing board were plans to link its midwestern system, which also ran through Cleveland, Toledo, and Chicago, to the eastern net by constructing a line from New York to Pittsburgh, slated for completion in 1949. Longer-range plans called for a cross-country cable from Florida, through the southern states and into California. Microwave bridges connected Boston, Milwaukee, and Detroit to the nearest cable point.

It was clear to AT&T and to the broadcast networks that television would progress along the same lines, so to speak, as radio. Both the industrial structure of the broadcast system and the prohibitive cost of TV programming all pointed to TV networks that evolved out of their radio predecessors. Economics dictated that the networks would have to interconnect their owned and affiliated TV stations. In June of 1947, AT&T, which had been providing free, experimental service on its New York to Washington, D.C. line, filed its proposed tariff schedule with the FCC. The proposed $40 per mile per month charge was ten times that levied for radio use. Various companies, including the networks, responded by threatening to build their own system of network interconnection and, within a year, AT&T had dropped its price to acceptable levels. Television seemed to be on its way, and AT&T would serve as the backbone for national distribution of network programming.

Television (Almost) Arrives

The public and the popular press had watched the fencing between the FCC and the competing companies for more than a decade. There was skepticism and frustration, but also excitement at the prospect that television might finally be on its way. Since the mid-1930s and through one false start after another, television was said to be always "just around the corner." Proclamations on the coming out of television had been made in the mid-1930s and again just before the World War II. Newsweek optimistically reported in 1935 that television had moved "one step closer to commercial actuality" with the FCC's approval of AT&T's coaxial East Coast line.[7] But, as the entrenched players jockeyed for position in the late 1930s and the FCC issued on-again, off-again rulings, the public and the popular press became increasingly cynical.

A 1937 Business Week article likened the frustrated development of commercial television to William James' description of philosophy, which was "a blind man in a dark room looking for a black hat that isn't there."[8] In 1938, Business Week declared, once more, that "television came whamming around the corner last week," as RCA began its thunderous marketing campaign.[9] A year later, following the FCC's response to RCA's premature actions, it appeared that television had come around that corner only to slam into a brick wall. Business Week now reported that the "Television industry was viewing the future darkly this week."[10]

By 1941, the press was jaded but still hopeful, noting "Television has been so long around the corner that nobody is going to believe it when it makes its turn,"[11] which, of course, it again failed to do. Even before the War Production Board halted the manufacture of TV sets, important raw materials, such as aluminum and nickel, were already becoming scarce and television engineers were transferring to pressing defense contracts. Again, it would be several years of waiting for television.

With the apparent resolution in 1947 of many of the immediate regulatory problems and the warming of the postwar economy, television, then, was not something that could be characterized as a social surprise to a technically illiterate populace. The public knew about television and knew about its stuttering progress. Whatever frustration there might have been, however, was overshadowed by the excitement of finally seeing television arrive.

In September 1947 RCA, seeking again to steal a march on the competition, made public the technical details of its television receiver design, encouraging companies to begin selling sets based on their specifications. It was part of Sarnoff's unrelenting campaign to establish the RCA platform as the *de facto* industry standard. For better or worse, it worked to a large extent. By the end of 1948 seventy-eight different companies were selling nearly 200 different models of TV sets. The consumer could choose from three-inch tubes to "giant" twenty-inch screens, as well as direct view picture tubes or early "projection screen" receivers. None of the sets were cheap. Prices ranged from $99 to nearly $5,000 for custom-built models.[12] The average selling price was around $575 in 1947 and, although prices came down a bit each year, it was still a costly adventure.[13] As historians Christopher Sterling and Michael Kittross noted: "The typical 5-inch to 7-inch receiver cost from $375 to $500 in mid-1948—several weeks' pay for the average worker. In addition, the buyer paid an installation fee ranging from $45 to $300, depending upon antenna requirements, and usually invested in a one-year service contract and a roof-top antenna, especially in cities more than 30 miles from the transmitter."[14]

The additional expense of installing rooftop antennas would be of substantial importance to the near future of CATV. Further to the point of cost, however, unlike today's shoppers who can walk into a video store and simply "put it on the card," installment buying was strictly controlled until late 1947. Under war-time buying restrictions, the consumer had to put one-third down and had a limited time to pay the balance. The easing of such controls in 1948, along with increased competition and manufacturing efficiencies that lowered the price of sets,[15] helped feed the buying boom, and despite the obstacles, demand quickly outstripped supply. The primary bottleneck was in the manufacture of the picture tubes. There were only two major glass companies supplying the "blanks" for all the picture tubes in the country— Corning Glass Works and Kimble Glass Division of Owens-Illinois—and they were overwhelmed with orders.

Sales figures during this time were revealing. Shortly after the FCC first opened the airwaves to commercial broadcasting in 1939 and limited set sales began, fewer than 20,000 television sets were found in the country, about half of them in New York, and most of the rest in Philadelphia, Chicago, or Los Angeles.[16] Without parts for repairs or replacement during the war, the number of sets in operation actually decreased; the trade press estimated between 8,000 and 10,000 working sets were in homes at the end of 1945.[17] Sales began again slowly in 1946, with an estimated 6,500 sets sold.[18] By 1948, however, the "boom" was in full swing. Television had arrived, said the magazines, and "the spring (1948) boom in video is only a sample of the sensational progress to come during the summer, fall and winter."[19] Television was "like a locomotive on the loose,"[20] or, according to *Time*, a "young monster."

By the end of 1948, the affluent and the enthusiastic had brought nearly 600,000 TV sets into their homes. Nearly 75 percent of these were still in

New York, Philadelphia, Chicago, and Los Angeles, with about half still in and around Manhattan, but the rest were scattered through at least eighteen other major cities, from Baltimore to Salt Lake City. An estimated 1,000 new sets were being installed every twenty-four hours.[21]

Despite the high cost of the sets, people from nearly every income level were finding ways to get television. In one early test market outside of New York City, researchers estimated that 81 percent of the sets were owned by medium- and low-income families.[22] The National Association of Housing Officials debated whether to allow low-income families, living in subsidized public housing, to own TV sets. Many already had sets, some selling their cars to get them, and the moral implications of such families owning "luxury" goods was of some concern to the officials. (They decided to let them keep their TV's.)[23]

For those who could not get or afford sets or chose not to buy them, alternatives sprang up quickly. Enterprises of every type were capitalizing on the high-demand, low-availability circumstances. Bars were among the earliest adopters of the technology, drawing in sports fans and curiosity seekers. Soda shops, recreation centers, and even libraries found television a natural magnet, and a pacifier for the younger set. Philco put a television set in an airliner running service between Chicago and Washington.

The film industry began vigorously promoting theater television, bringing live sporting events and other attractions to the big screen. Even before the war, some film companies had eagerly signed on to the idea of showing wide-screen video in theater houses. For a short time, the theater system was seen as an alternative industrial structure for the public distribution of television, in direct competition with plans for commercial-based broadcast television to the home. Proponents argued that theater television would not require consumer expenditure for sets, and ticket receipts would provide an instant revenue stream for the development of the new medium, in contrast to advertising revenue which was seen as slow to mature.[24] Radio-Keith Orpheum (RKO) was particularly active in promoting theater television.

Progress on theater television stalled during the war, but picked up with the post-war TV boom. Historian Michelle Hilmes reported that by 1952 more than 100 theaters across the country had installed or were installing theater TV equipment, and the Theater Television Network was created to package and distribute news, sports, and entertainment events.[25] The entire system, however, relied heavily on AT&T's coaxial distribution system, and the telephone company set such high prices for carriage that the business could not be sustained. In an effort to outflank AT&T, promoters petitioned the FCC to set aside dedicated broadcast bandwidth for the distribution of TV programming to theaters, but the FCC rebuffed the request. As TV set penetration in the home rose, the viewer's inclination to find a theater with television (or even films) declined, and the theater TV movement all but flickered out.

The financial fortunes of broadcast television, on the other hand, were seen as quite promising in the late 1940s. On Wall Street, trading in television stocks took off with such velocity that one broker exclaimed, "The public has gone mad."[26] In less than a year, the price of TV stocks doubled. For one week in May 1950, television equities accounted for nearly a tenth of the total volume of the New York Stock Exchange. There were only about 100 television stations on the air in the entire country, and you could count on one hand the number that were turning a profit. But the public didn't care—television had arrived. Those who couldn't afford their own set went to the local bar, and those who couldn't or wouldn't go into the bars watched on the street in front of the display window of the local electronics or "white goods" store.

The Freeze

This booming interest in television drew a different sort of crowd to the FCC and drew attention to rising problems in the system. In October of 1947, there were fewer than thirty applications for TV licenses at the FCC; by October of 1948, the number had shot to 300.[27] Not all the applicants were quivering in their excitement to build; many were just covering their bets, seeking to reserve spectrum space in case television should take off. It was, after all, an expensive gamble. Putting together a TV station could take up to a million dollars. No station at the time was making money and not everyone believed that profits "were just around the corner." NBC was losing an estimated $13,000 a day on its television operation.[28] Those with better foresight, or simply better luck, began to build.

The FCC, prior to the war, had given commercial authorization to thirty-two stations. From the half dozen on the air at war's end, the number slowly crept up to seventeen by the end of 1947 and some fifty-five permits to build had been granted. By the end of 1948, about fifty stations were in operation.[29]

As the pressure for licenses mounted, a host of regulatory problems began to appear, most of them the result of previous FCC rulings. The Commission's policy decisions during the period 1945–1947 were critical in this regard, not just because they opened the way for the start of receiver production and station construction, but because they laid the foundation for the ultimate shape and nature of both the broadcast and cable TV industries in the United States. These proceedings were the base on which the Commission would build its more comprehensive 1952 allocation scheme, which in turn helped provide the market opening for thousands of small cable TV operations around the country.

It is important to bear in mind that most human and organizational decisions—whether they occur in government or the private sector—have both intended and unintended consequences. Because information and knowledge is always imperfect, there will always be unexpected fallout from any significant policy decision. In many cases, it is the unintended consequences, as much as the intended ones, which have long-term influence on the stated objective. In the case at hand, the unexpected consequences of the FCC's rule-making procedures from 1945 to 1952 had as great an impact on the development of the communications infrastructure in the United States as did the intended consequences. Most of the story is captured in what became known as, "The Freeze."

The term refers to the FCC's freeze on television licensing between 1948 and 1952, prompted, largely by the unintended consequences of its earlier rounds of decision-making. Before the end of 1947, the shortcomings of the FCC's policy were becoming painfully apparent. Throughout the 1945 allocation hearings, virtually all public and private witnesses testified to the need for twenty or thirty channels if there was to be a nationwide, competitive TV service. Despite nearly universal acknowledgement of this by both the industry and the FCC, the final allocation scheme provided for only thirteen VHF channels (Channel 1 was subsequently assigned to other services). As previously noted, the FCC initially suggested the VHF band would be a temporary home for television, with an intent to eventually move it to UHF. The denial of the CBS UHF proposal signaled that television would stay, for the time being, primarily in the twelve-channel VHF band, however. The result, to no one's surprise, was too many suitors for too few channels. The allocations in the larger markets were especially attractive. While licenses in thinly populated regions, such as the West, often went begging, the land rush for big city TV permits created a landslide of paperwork for the FCC staff. It also raised important policy questions about the adequacy of the VHF scheme.

The problem of channel scarcity interacted with the problem of signal interference. There had been an early technical assumption that television signals, because they operated at what were then the higher ranges of the usable spectrum, could not reach beyond the horizon. The effective range was thought to be about fifty or sixty miles. This was despite warnings from many experts that such waves could travel significantly farther, especially during periods of sunspot activity, and adjacent stations on the same frequency (colocated) might therefore interfere with one another. The FCC had set the spacing between colocated stations at 200 miles in 1945, but under increasing pressure to provide more outlets, the FCC in May 1948 reduced the permitted distance between such channels to 150 miles or less. The experts who predicted problems were quickly shown to be correct; stations that had been assigned the same channel but were well over 100 miles apart were casting their signals into the other's market. A station in Detroit, for example, was causing interference for a station in Cleveland. Viewers in Princeton, New Jersey, had nothing but horizontal bars on their Channel 4, due to the overlapping signals of NBC's New York station to the north and its Washington station to the south.[30]

It was clear that the VHF allocations alone would not be sufficient to satisfy demand or, given the interference problem, to establish a truly national system of broadcasting. At the same time, the Commission faced additional pressure to address the technical standards for color TV transmission and to consider the needs of educational broadcasting. As a consequence of the mounting concerns, the FCC, on September 30, 1948, shut down the system. It declared that it needed time to sort out the increasingly tangled issues. The Commission issued a temporary ban on the allocation of all TV station licenses. As originally conceived, this "freeze" on licensing was intended to last six months. It would turn out to last nearly four years. During this time, broadcast television would be limited to the 108 "pre-freeze" stations, and community antenna television would start to grow.

The freeze itself, and more importantly its resolution, would have lasting impact on CATV. While no new broadcast licenses were issued during the 1948 to 1952 period, there was, nonetheless, substantial development of both TV broadcasting and community antenna business through those critical years. The FCC's order did not diminish the public's appetite for television. If anything, it heightened the anxiety about the potential availability of television and, therefore, increased desire for it—a person often wants even more strongly that which is temporarily denied. Boddy suggests the term "freeze" itself was misleading.[31] Those stations which had been licensed prior to the FCC's order constructed towers and began broadcasting; the number of operational stations rose from forty-two[32] to 108 between 1948 and 1952. As new stations went on the air, the consumer TV craze continued apace. The number of TV sets in U.S. homes rose from about 850,000 at the end of 1948[33] to 15 million in 1952, and penetration levels rose from 0.4 percent to 34 percent.[34]

The freeze did have important implications for the structure of the broadcast industry, however. Those stations that were authorized to build found themselves in the attractive position of having limited competition for an audience more than eager to shower its postwar affluence on the cathode ray screen. The freeze was a great gift to the existing, entrenched broadcast business dominated by NBC and CBS. The networks took advantage of the licensing hiatus to solidify, in culture and industrial practice, a hold on the minds of TV viewers and advertisers. The networks often worked to extend the freeze in an effort to hold back competition. Those operators unable to fully enter TV broadcasting until after the freeze, including ABC and DuMont, would spend years playing catch-up and some, like DuMont, would ultimately fail in the effort. In the meantime, the combination of high consumer demand and the perception that the deployment of broadcast television would be delayed helped create a market opening for new businesses.

At the Edge of the Explosion

In 1948 consumers confronted two primary frustrations in their effort to get television, one was the lack of a TV set, typically due to insufficient funds, the other was the lack of a TV signal, a problem exacerbated by the FCC's freeze. For those without a set, there was at least the local TV-equipped bar, or, for a few, a neighborhood theater television. Those without a local broadcast signal had a more difficult time.

Interestingly, the lack of a signal did not deter some from purchasing an expensive TV set. Market researchers estimated that as of December 1, 1948, 50,000 of the more than 800,000 sets sold nationally had been installed in cities that were not yet served by a TV station.[35] The specter of tens of thousand of people buying TV sets they could not use seems curious at first, but speaks to the eagerness and expectation on the part of consumers who were sure television would be coming to their town soon. And while some sat, perhaps in front of blank screens, and waited, others took action. Efforts to pull in television signals from distant cities took several forms in the late 1940s. One of the most straightforward was simply to build a very tall antenna mast and hope to get lucky. Distant television or TV "DX" was a small fad before and shortly after the war. When experimental television was assigned to the short-wave frequencies for a brief period in the 1930s, receiving a signal from several hundred miles away was exciting but not a significant technical feat. When television was moved to the higher bands and signals were thought to be limited to the horizon, reports of hobbyists picking up Florida signals in Philadelphia or receiving Kansas City signals in the Ozarks were of sufficient novelty to draw comment in the press.[36]

Perhaps the most frustrated people among those trying to get television were the ones who lived just at the edge of a television signal. Neighbors nearer the TV tower could get good reception and those farther out possessed no idle hope for TV viewing. But those a certain middle distance from the station were constantly teased by a signal just a breath beyond their reach. These were the people just at the edge of the TV explosion, the potential viewers located on what was called "the fringe." It is important to remember that cable television did not start in the rural areas hundreds of miles away from the nearest station. CATV began and proliferated just at the edge of the receivable signal. The fringe was one of those problems that illustrated the evolutionary diversity of technology by inviting a host of technical solutions, CATV being only one. Even before the nearest TV station was scheduled to go on the air, people at the fringe, and beyond, were erecting multistory antennas and creating imaginative home-made receivers in an effort to catch the weak and distant signal. This raw supply-and-demand equation naturally attracted the attention of television entrepreneurs of various stripes. At the margin of the early TV markets, especially on the East Coast, was a ready clientele for companies that could deliver a usable picture and one of the first devices on the market aimed at this group was the antenna "booster."

The fringe was technically defined as the area in which the over-the-air signal dropped below a strength of 100 microvolts (mv). Boosters were simple devices designed to amplify such weak signals, and they began appearing about the same time set sales started to accelerate in late 1948. Boosters were typically small boxes that sat on top of the television with one or two dials for channel tuning. Leads from the antenna, which was usually mounted outside of the house, were attached to the back of the booster. Inside the box were two separate tube-driven amplification systems, one for the low-band channels (2–6) and one for high-band channels (7–13). The boosters retailed for between $30 and $40,[37] a price that would help them find their way into some of the earliest community antenna experiments.

Apartment Master Antennas

While "DXing" might have been the only service for those far beyond a TV signal and boosters the only help for those just at the fringe, apartment master antenna systems served another group: people within the area of a receivable signal, but denied reception on aesthetic or legal grounds. Smaller scale systems also were employed in electronic and department stores where retailers wished to run a signal to several display sets. Each situation again was an opportunity for entrepreneurs and corporations to develop and market technologies to overcome the problem and satisfy the hunger for television. In their aim, operational concept and basic technology, they also were the immediate predecessors of community antenna systems.

As the demand for television blossomed in the late 1940s, especially in New York where TV broadcasting continued even during the war, the question of wiring apartment houses evolved naturally from radio to television, although the evolution was not a particularly smooth one. The initial problems were not just technical, but aesthetic and legal. In New York City in 1943 it was estimated that 71 percent of all households were apartments in multiple-dwelling houses or apartment buildings. That accounted for about 1.57 million households and upward of 5 million people. In the borough of Manhattan alone, 98 percent of all housing was multiple dwelling.[38]

It was, therefore, with some dismay that the fledgling television industry, centered in New York, must have greeted the news on February 1, 1947, that the real estate management firm of Wood, Dolson Company was henceforth prohibiting the tenants of 100 apartment buildings under its management from installing TV antennas on the apartment building roofs.[39] Simultaneously, the trade periodical *Real Estate News* advised all New York City landlords to withhold permission to erect TV aerials.

The complaints of the apartment owners were powerful and unambiguous. If every tenant in the building wished to erect a TV antenna on the roof, there would soon be a hideous forest of metal dominating the New York skyline. In fact, such a trend already was apparent in the late 1940s.[40] The associated cables snaking across the roof and down the walls and corridors of the apartments also posed aesthetic and safety problems. There was concern about liability if a tenant tripped over an improperly installed cable or if an incorrectly installed aerial fell off a building. Owners were additionally concerned about the potential for "antenna wars" among tenants. They feared that some positions on the roof might prove more advantageous than others for signal reception, that a forest of antennas would create widespread interference causing ghosting and snow on tenants' screens. Either situation could prompt nasty tenant disputes.[41] Finally, and from the owners' perspective most importantly, all of these problems could quickly develop into lawsuits over safety, aesthetics, and renter unhappiness, hence the ban on antennas.

According to press reports at the time,[42] initial resentment and hostility on the part of the television industry gave way to a mutual effort to solve the problem. A committee of the Television Broadcasters Association met with the Real Estate Board of New York early in 1947 to discuss the issues. It was clear from the beginning that the apparent solution to the difficulty was a single master antenna on each apartment building feeding the TV sets throughout the structure. The evolutionary task then was to extend to TV the technology of the existing and widely used apartment master radio antennas.

Despite the long familiarity with radio master systems, however, several technical and economic difficulties needed to be overcome. The FCC's decision to split TV channel allocations into two separate bands created a small but vexing problem for everyone involved in TV reception, from the home viewer to, eventually, CATV operators. The unforeseen difficulty involved the separation of the so-called "low-band" Channels 2 through 6 (54–88 MHz) and the "high-band" Channels 7 through 13 (174–216 MHz). Because of the spacing, each band required its own specialized reception and amplification equipment.

A single "low gain," omnidirectional antenna could pick up signals in both bands, but was susceptible to ghost images from "echo" signals reflected off nearby buildings and it did not adequately receive weak signals. The early television technicians, therefore, recommended that separate antennas be used for at least the two bands and often suggested a separate antenna for each television signal. Each antenna could be "cut" to the length appropriate for a given signal and pointed to increase its efficiency and prevent ghosting. And because the apartment house systems of the late 1940s typically had access to only three to five stations, the arrangement didn't pose serious physical problem.

With an adequate antenna system alone, a company could construct a "passive" cable system for the distribution of several TV signals through a building. The passive system was one in which the television signals, as received off the air, were sent directly to the TV sets without any amplification, although the distribution systems nonetheless had to be properly balanced to avoid impedance mismatches and appropriately shielded to prevent interference between sets. The signals generally were distributed by coaxial cable. Even with relatively strong signals in the New York area, however, such systems could only serve a handful of apartments.

To increase the number of possible units on the system, without simply adding a new set of wires and an extra antenna array, it was necessary to amplify the signal coming from the antenna. Single strength within the normal range of a TV station can be 3,000 mv or more and 300 mv was, at this time, considered as low as one could go without amplifying the signal at the receiving antenna. The technical difficulty is that the signal strength provided by a single TV antenna is sufficient to provide a clear picture to, at most, five or six sets. And this is possible only where there is a strong initial television signal, usually within the town or city of the signal's origin, and where the signal being fed can be appropriately "balanced" or matched to the TV sets being used. Any number of receivers beyond that necessitates an amplification of the signal or signals being received.

RCA's Antenaplex used a system of amplifiers in its radio master antenna and, along with others, sought to extend the concept into television distribution. The wide band signal nature of television, however, made the task difficult. As of mid-1947, there was no amplifier capable of covering the full range of the high and low TV bands. Even if such a "broadband" amplifier had been available, the differing signal strengths of the local stations suggested that one amplifier for all signals was not the best solution. The early master antenna systems, therefore, used a separate set of boosters for each station, each booster specifically tuned to the particular strength and frequency of its signal.

At least three companies—RCA, Amy, Aceves and King (AAK), and Telicon Corp.—were developing these amplified TV master antenna systems in early 1947.[43] In 1946, RCA had adapted its master radio antenna system to distribute TV signals through its own New York City headquarters. Only Telicon, however, actually had a system on the market in 1947. By 1948, however, AAK, RCA, and another company, Intra-Video Corp. of America, were all producing and installing such systems nationally.[44] And similar systems were being manufactured and installed by other companies in hotels, hospitals, schools, and department stores.[45] Companies, such as Industrial Television Inc., were promoting their master antenna distribution systems to hotel operators by the end of 1948,[46] and many hotels were adopting the technology.

In Great Britain, apartment master antennas had been harnessed for TV use even earlier. According to Kenneth Easton,[47] a radio master antenna system was introduced in 1934 by a company called "Radio Furniture and Fittings." The company initially sought "to sell expensive radio equipment to the affluent occupants of high-class and expensive apartments in the West

End of London," but poor reception in the buildings frustrated their plans.[48] They began designing and installing apartment house master antenna systems to overcome the problem. These were early but true cable distribution systems composed of rooftop antennas and amplifiers that fed coax lines. By 1939, some thirty or forty London buildings had been wired. When the BBC officially began its first public TV broadcasting in 1936, the company quickly realized its radio master antenna system could be converted to TV distribution as well. By 1937, the firm had installed TV antennas and amplifiers and was operating apartment house cable TV systems.[49] London's single TV station was closed in 1939 at the outbreak of the World War II, but began operation again in June of 1946 and the master antenna systems were reactivated.

In the United States, there was some resistance to the adoption of the systems, largely from landlords and primarily on the basis of cost. Telicon estimated in 1947 that installation could run up to $80 an outlet, or about $4,000 for a fifty-unit building.[50] The realty industry balked at bearing the cost of these systems, arguing that TV sets eventually would have efficient internal antennas that would preclude the necessity of an outside antenna. But the master antennas nonetheless became an entrenched part of the explosion of television. Through 1948, 1949, and 1950, the popular and electronics trade press published periodic reports of the installation of master antenna systems, as well as detailed descriptions of how to make them.[51] In addition to hotel and apartment houses, they were seen as useful in department stores, schools, and hospitals. By November of 1950, an entire block of apartment buildings in Chicago was being served by one such master antenna system, which by then was composed of coaxial cables and amplified circuits.[52]

By the end of 1947, both the concept and much of the early technology of master antenna TV systems were in place, deployed ironically in the largest and most saturated television markets. Finally, and importantly, at least some of the early CATV pioneers knew about these systems.

Community Antenna Television Takes Seed

On January 11, 1949, AT&T made front-page news by joining its East Coast coaxial network with its Midwest network. An electronic "golden spike" was driven in the line joining New York to Pittsburgh. This was also the day that Pittsburgh's WDTV—later to become KDKA—officially went on the air.

Some 120 miles to the north, in the rolling oil fields of northwestern Pennsylvania, the Oliver brothers, Stephen and Frank, were, they hoped, ready. They had installed a ninety-foot antenna tower on top of their fifty-foot building in downtown Oil City, hoping to be among the first to receive the inauguration signal from WDTV. Unfortunately, the scheme failed. They tried flying large balloons from the tower to increase the height of the antenna line, but that didn't work either. They could get reception at their home in the hills overlooking the city, but signals passed without noticing over the town in the narrow river valley below.[53]

There and elsewhere across the country, would-be TV fans, salesmen, and technicians just at the edge of the signal were doing the same thing, most of them without knowing of the activities of the others. Months before Frank and Stephen Oliver sent their balloons aloft, bar owners in Mahanoy City, Pennsylvania, were showing off television from Philadelphia on sets connected to makeshift antenna lines strung on poles and trees up the adjacent mountain. In the nearby vacation area of Big Creek, Pennsylvania, radio and TV dealer Bob Tarlton installed a few thousand feet of twin-lead antenna wire and hooked up the TV set of a local labor leader, running the line to some of the neighboring cabins.

In Arkansas, an enterprising young engineer and entrepreneur, Jim Davidson, built a receiving antenna on top of his shop in the small town of Tuckerman. There wasn't an operating TV station for hundreds of miles around, but Davidson knew that WMCT, in Memphis, Tennessee, was scheduled to begin broadcasting in November of 1948. Davidson was going to be prepared, and even worked with some of the station engineers as they experimented with their signal in the weeks prior to the official sign-on.

In the summer of 1948, Oregon radio station operator L. E. Parsons listened intently while his wife, Grace, explained that, having seen television demonstrated in Seattle, she wanted it in their home in Astoria. At the time, the nearest operating TV station was in Los Angeles, but Parsons knew that on Thanksgiving Day KRSC-TV in Seattle would begin commercial broadcasting. He went to work designing what he hoped would be a way to locate the signal, 125 miles to the north, and get it into the living room.

Parsons, Tarlton, Davidson, and a few of the customers in the Mahanoy City bars would soon begin a business that would evolve into the CATV industry. And while they may have been geographically separated, their activities had a common goal and a similar technology. Along with apartment master antennas, they were part of a much wider effort to bring this new phenomenon of television into the home. The perceived problem—no television—cut across several large categories of situations, and solutions had to be adapted to the peculiarities of those differing situations. But as the core concern remained the same, the techniques and principles applied in solving the problem necessarily had shared characteristics. The technical ability to capture, amplify, and distribute TV signals moved to increasingly higher levels of sophistication as the systems advanced from store-front windows to apartment buildings to villages and towns. All of which is to underscore the notion that cable television was a technology that was not, in any meaningful way, "invented." Rather, it was a logical extension of both concepts and equipment in existence at the time. They were the incremental steps in the evolutionary process. CATV invoked and modified older forms of technology to meet a new need, drawing especially on the early television "boosters" and the apartment TV master antenna systems of the day.

In its application, the seminal problem for community antenna television was that of the fringe. The signal either reached a house, but was too weak to be of any use, or it was blocked completely by some nearby physical obstruction. In the first instance, a television booster would probably solve the problem. The second difficulty required placing an antenna away from the house in a position that allowed it to receive the signal. These often related problems and their relatively easy solutions provided the initial steps in the development of CATV.

Often, the first people to worry about the problems of the fringe were the local radio or appliance dealers who sought to expand into the TV business. According to Tarlton, it was common practice, as early as 1947, for area radio-TV shop owners to run long lines from mountain-top antennas into their shops so they could demonstrate their freshly unpacked TV sets. Davidson, in Tuckerman, Arkansas, ran a radio sales and repair shop, as did John Walsonovich Sr. in Mahanoy City. To promote TV sales, and later their CATV businesses, these pioneers invariably placed demonstration sets in their store windows, drawing curiosity seekers and prospective customers. By the early 1950s, Walsonovich had several such sets in the window of his store, each tuned to a different channel. People reportedly crowded, on a regular basis, in the street outside the shop, watching until the last station signed off.

Sometimes local appliance dealers pooled their efforts. In Shenandoah, Pennsylvania, not too far from Mahanoy City, local dealers formed an ad hoc group in early 1950 for the express purpose of building a citywide CATV system, calling in a consulting engineer and hiring a lawyer to help gain local government permission to erect a tower. Across the state in Franklin, three or four television dealers, each of whom began running their own lines in 1950 and 1951, also consolidated their neighborhood systems into a joint, city-wide company by 1953.

The appliance dealers, however, were not the only people engaged by the promise of the new medium. Lost in the shorthand history is the variety of people interested in (and involved with) television. Easton reports that Rediffusion, Inc., the radio relay company, began planning a citywide CATV system for Montreal, Canada, as early as 1949, although delays in the inauguration of broadcast held back actual operation for several years.[54] In the United States, among the first people to experiment with CATV were local radio station operators, such as Parsons, who owned and operated KAST. Pursuit of the new TV signal would be a natural outgrowth of the expertise and interest of radio operators. As early as 1947, radio station KVOS in Bellingham, Washington, reportedly began considering plans for a community aerial to distribute the signal of Seattle's KRSC-TV when that station went on the air.[55] The Bellingham station began building its system, with some advice from Parsons in late 1949, and operated it through the mid-1950s. The list of people writing to Parsons as early as 1949 to request information about community antenna equipment included industrial arts high school teachers, the owners of automobile dealerships, bankers, and officials from gas and lumber companies.

The Proto-Systems

Using equipment available in town or creating their own, the radio operators and appliance salesmen began erecting hill-top antennas and running the wires into their shops and homes and local bars, sometimes tacking on an extension to the home of a friend. Such efforts resulted in what could be called cable television "protosystems," not full community antenna networks and certainly not full-fledged businesses, but important steps in the evolution toward such systems. There are many reports, some confirmed and more unconfirmed, about such efforts. Some of these homemade mini-networks would go on to become operational cable systems, some would not. In a number of cases, the creators would continue on to construct some of the first commercial cable systems in the country.

The mountainous coal-mining region of eastern Pennsylvania was a seed bed for this kind of activity, pointing out the importance of geography to the historical development of the industry. Television broadcasting in Philadelphia, in the southeast corner of the state, had been going for nearly as long as experimentation in New York City. Three stations were licensed there before World War II, including WPTZ (Channel 3) of the Philadelphia Storage Battery Company (Philco), WFIL-TV (Channel 6), and WCAU-TV (Channel 10). WPTZ was operating again by the end of World War II and was joined by in 1947 by WFIL-TV, and in 1948 by WCAU-TV.

Moving northwest from Philadelphia, a broad rolling plain stretches about 100 miles inland until it meets the eastern slopes of the Allegheny Mountains. Small towns dot the mountaintops and hug the base of these ancient hills. Ridge after flat-topped ridge continues on into northeastern and central Pennsylvania, and tucked in the crevices between those ridges are the small towns that lived for decades off of coal mining towns, such as Jim Thorpe, Coaldale, Nesquehoning, Lansford, Tamaqua, Pottsville, Mahanoy City, and Girardville.

Reasonable reception of all three Philadelphia stations could be had in the towns that capped the mountain ridges. Frackville is one, sitting astride the ridge of Broad Mountain. In the narrow mining valley below, however, the homes of Mahanoy City are shielded from the signals as if by a huge rocky wall. The Philadelphia TV signals even today pass silently overhead.

Years before CATV was a taken-for-granted part of life in Mahanoy City, local hobbyists and businessmen were experimenting with television. The fringe problem was particularly frustrating to some of the citizens of

Mahanoy City and similarly situated towns. A reasonable TV picture could be had five minutes up the road in Frackville, but not at home in the valley below.

George Yuditsky was a technician for one of the first community antenna systems in Mahanoy City. He recalls, perhaps as early as 1946 and long before CATV came to the area, trying to string a wire from a tall antenna placed well up the side of the mountain in nearby Gilberton. He also recalls that it didn't work.[56] Stephen Koval, whose family owned the local Oldsmobile dealership, had better luck, according to one report. A tinkerer by nature, he managed to run an antenna wire down the mountain and into his house, extending it into the dealership garage. His brother, George, would later go on with Yuditsky to form one of the first CATV systems in the area.

Unheralded in early television history are the tavern owners who led the way in adopting television in many communities. The first person with a television on display for the general public in Mahanoy City, according to townspeople, was Ed "Peanuts" Trusky. Perhaps as early as 1947, "Peanuts" erected a tower partway up Broad Mountain and brought television into his poolroom and, thus, into Mahanoy City, for the first time. Three doors down, the clerics at the neighborhood church were sufficiently enchanted with the idea that they convinced Peanuts to extend the antenna line into the rectory. A friend of Peanuts, a block in the other direction, reportedly did the same. According to one version of the story, one of the many clients at the poolroom, John Walsonovich, got his idea for a CATV business in the TV glow at Trusky's poolroom. Bringing television "down the mountain" was an idea that caught on quickly.

By 1948, Joseph Kayes, owner of the West Center Street Café, sought to duplicate Peanuts' success and erected his own antenna, bringing the line down into the lounge.[57] A few blocks away, another long-time resident, Luther Holt, had opened a school for radio and TV engineers, and one of his first orders of business was to put up an antenna on the hill and run a line into the classroom. There are reports of even more people trying to get television down off the mountain in Mahanoy City, some successfully, some not.

While the geography of the area made the hills of Schuylkill and Carbon counties prime regions for early CATV development, but they were by no means unique, and the stories of Mahanoy City were probably played out in other towns and villages as TV stations began turning on their towers across the country. Tarlton, for example, reports having built the small twin-lead system, mentioned previously, in mid-1948 in Big Creek, Pennsylvania, at the request of a local official of the United Mine Workers union who said he was going to host a visit by union chief John L. Lewis and wanted television when the prominent visitor arrived. He also wanted to watch the fights. Tarlton later recalled how he and his father spending many hours scrambling up and down the steep forest banks lugging rolls of antenna wire, much of which later fell victim to disgruntled hunters who would cut the cable just to get it out of their way.[58] When they were done, however, the little network connected a handful of vacation cabins around the summer cottage.

Davidson, in Tuckerman, Arkansas, built a 100-foot receiving antenna on top of the building adjacent to his radio and appliance repair shop, Auto-Electric, while WMCT in Memphis, Tennessee, experimented with its transmitter in the fall of 1948. In addition to running a line into his shop to feed the display TV sets, Davidson ran an extension across the street, over the roof of the Tuckerman Theater, down through a cherry tree, and into the living room of Carl Toler, the local telegraph operator. Davidson snaked another line into the nearby American Legion Hall. When WMCT, still in a testing phase, broadcast a college football game on November 13, 1948, a good portion of Tuckerman's citizens were gathered around TV sets in the Legion Hall, on the sidewalk outside Auto-Electric, and in Carl Toler's living room watching Tennessee battle "Ole Miss."[59] Like many such extended antennas, Davidson's early effort did not turn into a community system. While he reportedly charged the Toler's a hookup and service fee, Davidson deemed the town too small for a community antenna business and turned his attention to larger cities (although years later he did bring cable to Tuckerman).

Distant TV reception, set-top boosters, apartment master antennas, and informal neighborhood antenna systems illustrate the variety of solutions that arose in answer to the perceived problem of a lack of television. Building on existing concepts and technologies, each attempted to resolve the dilemma within a context of specific local constraints and requirements, be they high mountains, congested apartment buildings, or the availability of spare weather balloons. And each was a response to a broader problem created in part by policy decisions made at a national level. From this diversity of technical innovations and over time, some solutions would grow and prosper; others would wither away as conditions changed and superior solutions evolved. DXing would never amount to more than a niched hobby, but neighborhood antenna systems had a future.

First Efforts

The issues of technical diversity and evolutionary process raise the question of historical primacy. A number of people have staked claims to being the "first" to build and operate a community antenna television system in the United States. At least a half dozen companies list their operational start in 1950, a few others even earlier. Some claims are off-handed and appear to be of no great consequence to the people who made them; Parsons is one example. Other claims seem to deeply engage the sentiments, even the egos, of the people involved. For many, the question is not a particularly relevant one, for others it has been intensely personal.

The thorny issue of determining the efficacy of such claims is made particularly difficult by the twin problems of documentation and definition. As to the first, there has been little focused effort to retrieve material that would illuminate the question. It has been said that the origins of cable or the identity of the first person or persons to build a CATV system are "lost in the mists of time." In fact, it is not so much that the information is lost as it is that few have ever bothered to look very closely for it. Some work on the problem has been done by the author, but a great deal has yet to be done.

This is not to say that published material is completely lacking. Existing evidence does point in certain directions, although sometimes it suggests either faulty memories or over-extended hopes on the part of some pioneers. For example, a CATV system in Emlenton, Pennsylvania, at one time listed 1946 as the year it began operations.[60] If true, this would have made it, quite clearly, the first such operation in the country, predating similar business by as much as two years. A closer look, however, reveals that the nearest likely TV station to feed the system (from Pittsburgh) did not go on the air until 1949. Even the geographic stretch to pick up a Cleveland station would not have netted a signal until 1947 when WEWS went on the air. A CATV system might, one supposes, have been built in Emlenton in 1946, but it is very unlikely that the system was carrying any TV signals.

That anecdote serves to illustrate the ease with which claims can be made, as well as the occasional ability to hold them up to the light of evidence. But for many more systems, known and unknown, the matter is less clear-cut. The historical record simply has not been sufficiently developed to confidently describe all the activities of this period.

More vexing and important is the issue of definition, especially as it relates to the evolutionary analog. If, as suggested here, cable television is in some sense an evolutionary phenomenon with antecedents that stretch back more than 100 years, how one characterizes the birth of CATV depends very much on how one defines CATV. As has been shown, wire-based information networks were in operation before the turn of the century and community antenna radio stations before World War II. Small-scale apartment master antennas were common soon after the war as were the efforts of hobbyists and small business people to extend their antennas on the fringe, sometimes allowing friends and neighbors to tap off of their makeshift masts. Therefore, does a CATV system, for example, have to connect two or more buildings in order to distinguish it from an apartment master antenna? Does it have to use coaxial cable, or will a twin lead do? Does it have to charge a monthly fee or will a one-time installation cost suffice? All these points have been used in formal and casual debate about "who was first." For better or worse, however, there simply is no objective standard by which to measure such things, no proclamation from the mount declaring the innate and intransigent qualities of CATV, nor need there be.

While work should continue on the collection of documentary evidence on the early development of the industry, the aims should not be to select and anoint an inventor, for there really isn't one. The aim should be to describe the contributions of the many people who continually nudged community antenna television along, adding in important increments to the formation and growth of both the technology and the business components of the industry. For if earlier comments are correct and cable television is an evolutionary rather than a revolutionary medium, then these concerns are best captured not in the question of "who was first," but rather simply, "who did what when."

What becomes apparent with increasing investigation into the early history of the medium is that various people were working on different forms of cable technology throughout the late 1940s and that they were, in time, aware of each other's efforts. Through personal contact and through the trade press, the early pioneers learned of each other and from each other. And through their collective trial and error achievements, the parameters of the typical 1950s CATV system began to take shape. It was through this process that twin lead and modified consumer-grade TV boosters gave way to enhanced hotel and apartment master antenna TV (MATV) systems and finally to the production and installation of dedicated CATV equipment.

Finally, and at the risk of appearing contradictory, there were several important systems established at the very start of the industry that in many ways helped set the tenor for future CATV growth. A great deal about these early systems is known and can be profitably described.

L. E. Parsons

In April of 1949, FCC Chairman Wayne Coy addressed the annual convention of the National Association of Broadcasters in Chicago. His topic was television. His prediction was that television was soon to be the dominant medium in the country. His proposal was for a "railroad system" of television distribution similar to the network established by the rail industry. This national TV system would be dominated by stations in large markets that would produce and distribute programming to a transcontinental web of simple, low-cost, repeater stations serving outlying communities and rural areas. These small town satellites would be as fully automatic as possible, simple relays for the programming supplied by the metropolitan stations.[61]

This particular vision of a national distribution system for television would eventually be displaced by a philosophy of television localism. But that would not happen for several years. On this particular Monday, at the luncheon speech before the assembled 1,200 broadcasters, one of those in the audience, Leroy Edward (L. E. "Ed") Parsons of Astoria, Oregon, was listening with more than average attentiveness.[62] And what Coy was describing sounded very much in concept like something he had been thinking about for months.

According to Parsons,[63] it all began after his first wife, Grace, visited Seattle in the fall of 1948 and saw a demonstration of television during a test conducted by KRSC-TV. She told her husband, "I want pictures with my radio." And although he told her that it was technically impossible, she persisted and he began experimenting. It was not to be a simple feat. Seattle is about 125 miles from Astoria and between the two cities is a 3,000-foot mountain range. Bringing in the signal would require the talents of someone who knew something about radio electronics. Fortunately, Parsons did. His father was an electrician for the railroad and Ed grew up in a world of crystal sets and radio tubes. By the time he was in high school, Parsons was building radio sets for the local retailer—two a week on contract. By the end of World War II, he had purchased an AM radio station in Astoria, KAST. He also operated a communications business serving the fleet of fishing boats in the small Pacific port, and with that running in the black, he went on to open his own flying school at the Astoria airport, which he also managed.

While flying was a great love throughout his life, it was his electronics background that allowed him to fashion the equipment he used throughout the summer of 1948 to locate the experimental signals of the Seattle TV station as it prepared for a formal fall opening. It helped that the manager of the TV station, Robert Priebe, was a friend willing enough to let Parsons know when the station was testing its signal (paralleling the experience of Davidson in the South).

Parsons eventually located a usable signal on the rooftop of the Astoria Hotel, not too far from his own apartment house. With permission of the hotel manager, Guyon Blisset, he erected a small antenna and set up a transformer to change the signal from Channel 5 to Channel 2. He ran a wire from the hotel to his apartment house and connected it to the very new, very expensive, and very large combination radio, phonograph, and television console he had ordered from Chicago. At nine inches, it was the largest screen in Astoria. On Thanksgiving, 1948, when KRSC officially went on the air, Mr. and Mrs. Parsons were watching television with friends. Word soon got out, of course, and the visitors began arriving regularly. As Parsons later explained: "The first problem was too many people coming into our apartment or penthouse. We literally lost our home. People would drive for hundreds of miles to see television. And when people drove down from Portland or came from The Dalles or from Klamath Falls to see television, you couldn't tell them no."[64]

So, he approached the hotel manager and suggested running a cable down the elevator shaft and into the hotel lobby. But again, the idea was more successful than anyone had bargained for. Soon, there were too many people in the hotel lobby watching television, so guests had a hard time getting in and out. Parsons next approached Cliff Poole, who owned the music store one street down from the hotel. Pool purchased a TV set and some of the necessary equipment from Parsons, who then split the cable line and ran a feed down to the store. Poole thought that having television would be a good idea for a music store and, repeating a nationwide pattern, he put the set in his shop window, added a loudspeaker for the street and turned it on when he locked up for the evening. Once more, the audience followed. In a story that echoes similar tales from the hills of Pennsylvania and elsewhere around the country, people developed a habit of gathering in the street in front of the shop to watch the weak shadows dancing on the phosphorescent screen.

According to Parsons, the Astoria police chief came to him and said that he had to do something about the TV set in Cliff Poole's window; the crowds were starting to block the street. Parsons said he asked the chief if he had any ideas. The chief suggested Parsons use the underground service channels, the "servitor," that fed telephone and power cables throughout the downtown. "Why don't you go down into the servitor and put the sets in the bars," he suggested. A short time later, every bar in downtown Astoria had a television set fed by coaxial cable from the rooftop antenna of the Astoria Hotel.[65]

As with much of the rest of the country, the bars were the first public establishments in town with television. But public hunger for TV demanded more. People wanted to bring the experience into their own homes, and they were asking, almost demanding, that Parsons extend the line down the block and into their neighborhoods. He had no city franchise and, initially, could not get permission from the telephone company or the power and light company to run the line along their poles, so he tacked the cable to rooftops and trees. To cross a street he created low-power transmitting and receiving stations and actually broadcast the signal from one block to another. Explained Parsons, "The people had amplifiers in their attics and in their upstairs rooms. Each person supplied the power for the amplifier. We ended up covering practically the whole town with cable."[66]

Eventually, Parsons moved the central receiving antenna from the top of the hotel. By cruising around town with his homemade signal testing devices he discovered what would turn out to be a common phenomenon in television reception. The topography of the region and the propagation characteristics of the television waves were such that the strongest signals were not necessarily found at the highest points in town, but rather in "fingers" of the TV signal poking through the gaps in the mountains. He reported finding one of his strongest signal sources, for example, underneath a highway bridge. Eventually he established two antenna sites and was running separate systems to feed the town.

Within a few months, the DX antenna that Parsons had set up for his wife morphed into a demand-driven, patchwork system of low-powered transmitters and coaxial cable running through underground tunnels and across rooftops. He built small wooden enclosures that looked like birdhouses to protect the tube amplifiers. By the summer of 1949 Parsons had about thirty homes and business on his system. The wired-based network was expensive and troublesome to maintain, however; in fact, Parsons really wanted to abandon it in favor of a broadcast technology. At least through 1949, Parsons harbored an ambition to move beyond wire-based distribution and build a broadcast repeater station for KRSC, along the lines suggested by FCC Chairman Coy.[67] In a May 18, 1949, letter from KRSC, Parsons was given permission to "rebroadcast" signals on an experimental, noncommercial basis.[68] And, in an August 14, 1949, letter to the FCC explaining how his cable system worked, Parsons made it clear that as soon as the Commission granted him permission to begin operating a broadcast relay station "the use of coax feed is to be discontinued as being unnecessary."[69] He stated, in fact, that the coaxial system itself was merely "a means of testing" a reception system "in preparation for the requesting of a construction permit" for a UHF repeater.

As with many similar inquiries, however, Parsons never received FCC permission to rebroadcast KRSC.[70] What he received, instead, was publicity about his community antenna scheme. Beginning in the late spring and early summer of 1949, with reports in the local newspapers that were then picked up by the Associated Press (AP), Parsons' system captured a small bit of national attention. Following the AP reports, both *Broadcasting-Telecasting*[71] and *Television Digest*[72] ran short articles on the Parsons' system in August of 1949. By the end of the year, *Sylvania News*, widely read by TV and radio trade, did a story, in April of 1950 the national magazine *Popular Mechanics* ran an article.[73] Parsons began receiving letters, dozens and dozens of letters.

Sensing a new business opportunity, Parsons, with his business partner, Byron Roman, established the Radio & Electronics Company of Astoria to design, manufacture, and sell CATV equipment. Throughout late 1949 and well into 1951, they responded to requests for information and orders for equipment. Letters came from across the country: from Salinas, California; Portland, Maine; Sioux City, Iowa; Brigham City, Utah; Winona, Minnesota; Treverton, Pennsylvania; Jefferson City, Missouri; and Berlin, Maryland. Letters came from Texas, North Dakota, Louisiana, Nevada, Illinois, Wisconsin, New York, Kentucky, South Carolina, New Jersey, and elsewhere.[74] One of the more interesting came from an appliance dealer in Pottsville, Pennsylvania, where a young man named Marty Malarkey would, a year or two later, start one of the more prominent early systems.

Many of those who wrote were simply individuals hoping that Parsons could help them get a better picture, or any picture at all, on their home set. They were early adopters of the technology and DXers, stretching out their antennas, and seeking others who did the same. Some were appliance dealers looking to improve signal reception for their showroom displays. Some were entrepreneurs intrigued by the business possibilities of community antennas.

The publicity also attracted the attention of people in towns near Astoria, and Parsons soon found himself designing and constructing systems in nearby communities, including Aberdeen, Centralia, Pasco, Kennewick, and Lewiston. Parsons, as noted previously, lent some assistance to the developers of the system in Bellingham, Washington. The Bellingham system, in fact, may have begun formal operations as early as August 1950,[75] making it, along with Parsons, one of the very first community aerials in the country.

Parsons' success had a price, however. By mid-1953 the demands of the many CATV activities had pushed him nearly to the point of a physical breakdown. By one account, he had not kept a close eye on the books and was forced into bankruptcy, leaving the company assets to his major supplier, a local heating and electrical company.[76] At his wife's insistence, Parsons took a break and pointed his small airplane north toward what he thought would be a short vacation with relatives in Canada. Halfway there he changed his mind, however, and several days later found himself in Fairbanks, Alaska. He fell in love with the open country and, for all practical purposes, never came back. He soon after sold the Astoria system to a group of local businessmen, and in 1964 it was acquired by Cox Cablevision Corp.[77]

During his relatively brief association with community television in Oregon, Parsons did not see himself as a "service" provider, but rather as a person in the CATV manufacturing and construction business. In the beginning, he did not charge a monthly fee, as was typical in subsequent community antenna operations; he charged only for the installation cost of the system, initially $100 (although by mid-1951 he was charging $3 a month for the service). He anticipated making money by building community antenna systems and, more importantly, having helped establish a market for CATV, manufacturing and marketing the equipment necessary for such systems.

More interestingly, perhaps, he illustrated the transition from protosystems to community antennas as a business. Like many of the earliest efforts, he started by stringing wire around the neighborhood, in response to requests by friends and neighbors. He was one of many "DXers" who attached any extra lead or two to his antenna for the folks across the street. His line, however, expanded far beyond what any of the previous neighborhood systems had managed and, importantly, he saw the possibility of making a profit from his work. And, through national exposure in the trade and popular press, it helped stimulate others to consider such a business.

Abe Harter

In northeastern Pennsylvania, the Oliver brothers of Oil City were not the only TV enthusiasts. Other appliance dealers, as well as the public, in both Oil City and its nearby neighbor, Franklin, longed for television and, in their pursuit of it, had the local geography working both for them and against them. Both towns are nestled deeply into the narrow Allegheny River valley, with small suburbs sitting atop adjacent bluffs. Aerials on the hilltops could pick up TV signals from cities such as Pittsburgh and Erie, but downtown Franklin and Oil City were television starved. Jack and Abe Harter, ran the Harter Brothers, appliance store in Franklin, selling, among other things, TV sets, and installing aerials for people on the ridges above the valley town. As early as the summer of 1949 and at least by the spring of 1950, the Harters had begun stringing open line wire down the mountain from an antenna tower five miles outside of town. Harter's intent was to run the line into the shop, set up a demonstration set, and declare the operation open for business. They never got the chance. As the Harters rolled their wire off the mountain-top tower and toward town, people came out of their homes and asked him to connect them to the line as he went. By the time the cable made it to the shop, the Harters were already collecting monthly service fees.[78] They soon formed Haren Corp. to run the new community antenna, and a small article in the September 1950 issue of *Electronics* magazine, hailed their novel endeavor: "An example of the improvement the installation has offered is evidenced at the local Moose Club. Here, an investment of $1,500 in a 140-foot tower, antenna, booster amplifiers and rotators produced no usable pictures. Connection to the community line resulted in good reception on four sets in the club."[79]

By the late summer of 1950, the system had a five-mile line into town and forty customers paying $9.25 a month for service; taverns paid an extra $5 a month for each additional outlet. The system used both a coaxial line and "open" line, although the latter made for interference problems during bad weather. Haren Corp. itself was a short-lived company. When a third partner and financial backer left town, the Harters joined forces with several other local TV dealers who had been working on community antennas in other sections town and together they formed Coaxial Cable Corp. in 1953.[80]

Bob Tarlton

One of the people to visit Harter and inspect his operation, probably in the summer of 1950, was Robert Tarlton. Tarlton, in many ways, was most typical of the early cable entrepreneur. Unlike Parsons and the Harters, his system was not an accident, driven block-by-block by the continuous press of public demand. It was well thought out, both technically and financially, and construction was preceded by a detailed business plan. His technology, while home-crafted initially, nonetheless drew upon extant designs in set-top boosters and apartment master antennas. And, unlike Parsons, Tarlton stayed in the business.

Moreover, Tarlton's work generated more national publicity than even Parson's operation, and through his connection with Milton Shapp, Tarlton was instrumental in establishing what would become the dominant firm in the CATV industry of the 1950s. Although Tarlton's system may not have been the first in the country, it was arguably the most influential.

Shortly after the end of World War II, Bob Tarlton, recently released from the service, joined his father in their radio shop at Chestnut and Patterson Streets in Lansford. The town of Lansford, Pennsylvania, rests at the base and runs partway up the side of Pisgah Mountain in the eastern Allegheny range, about ninety miles northwest of Philadelphia. Just above it, toward the top of the mountain, is the adjacent city of Summit Hill. As with Frackville and Mahanoy City, TV reception was available in Summit Hill, but not in Lansford. The Tarltons began selling TV sets, mostly to barrooms, commercial establishments, and a few people who lived higher up the mountain in Summit Hill.[81] But even though their shop was only three blocks down from Summit Hill, it was enough to prevent reception in the store. The solution, of course, was to place an antenna up the mountain and run twin lead into the showroom, which Tarlton recalls doing around 1947. The technology was not complex and the appliance dealers regularly worked with antennas, boosters, and related equipment in the installation of home sets. A convenient site for an appropriate antenna was located and line strung down the hill, sometimes tacked to trees, fence posts, and the eves of houses.

Tarlton was not necessarily the first in town to run an antenna wire down from the hilltop. The bar owners on Summit Hill had been doing quite well for themselves, bringing in the patrons by showing boxing matches and other sporting events. In was not long before taverns in Lansford, seeking to meet the competition, were stringing their own antenna wires up the hill.[82] Tarlton was one of several appliance dealers who helped install such systems. It did not take long under such conditions for Tarlton to begin conceiving of a community-wide antenna business for Lansford.

As he approached the problem of how to bring the signals down the mountain and into scores of Lansford homes, Tarlton was one of the first to encounter technical challenges that would soon engage other pioneers. For example, the signals transmitted through the early makeshift proto-system were typically quite weak, even more so than signals distributed by apartment master antennas because they were captured at the edge of the effective transmission range, and because the resistance in the wire itself degraded the signal over the longer course of the community-wide run. In systems that began with a twin-lead line, the TV signal would often disappear during rainstorms because of the susceptibility of the wire to interference by such weather.

As with apartment antenna systems, amplification was necessary to overcome the problem of signal degradation. Some operators, like Parsons and Walson, designed and built their own equipment; others modified existing set-top boosters. The set-top boosters were not created for such applications, however. They could not be adjusted to the different signal strengths that might exist among two or three channels and they were not built to run in series, or cascades, reamplifying the same signal every few hundred feet. The solution was to craft systems run completely on coaxial cable and amplified by equipment closely designed for this new purpose. Tarlton approached the problem not by attempting to build equipment from scratch, but by modifying existing technology.

While RCA's Antenaplex system and the Aces-Kings systems may have been leaders at the time in the apartment system business, Tarlton worked with equipment supplied by a relative newcomer to the field, Milton Jerrold Shapp. Shapp's story is detailed in Chapter 3, but for the moment it is enough to note that the young businessman had, by 1950, successfully established a small firm specializing in set-top boosters and apartment master antennas. Moreover his Philadelphia company, Jerrold Electronics, was conveniently located for the CATV pioneers of the Alleghenies. It was a relatively short trip to Philadelphia to pick up the amplifiers, and one made by Tarlton frequently through the summer of 1950 while he tinkered away, trying to shape the Jerrold equipment to meet the more rugged demands of a community antenna system. He was familiar with the Jerrold brand; he had purchased several amplifiers for a small apartment house master antenna system he installed in Lansford. The project he envisioned now was substantially larger, however, and placed a much heavier demand on the amplifiers.

When he was finally confidant that he could make the system run, he carefully began soliciting financial backing, working through friends and business acquaintances to secure the necessary resources. He needed money, but not competition, so he went to the other TV dealers in town to see if they were interested in joining the venture. He brought in two local electronics and appliance dealers, Rudy Dubosky and Bill McDonald. He recruited George Bright, the young grandson of the founder of the local department store, Bright's, and he added a close friend, attorney and state assemblyman, William Scott. Together they founded Panther Valley Television Company, Inc. (PV-TV). The original concept was to operate as a nonprofit organization, supplying service at cost and making their money through increased sales of TV sets. The local bank, however, would provide financing only if it appeared the business could generate revenue on its own, so PV-TV became a for-profit enterprise.[83] By September of 1950 the group had located a usable site for a receiving antenna on Summit Hill and had erected an eighty-five-foot tower,[84] at the same time mapping out pathways for the cable and buying supplies.

It was the unusually large number of Jerrold amplifiers that Tarlton was purchasing through the summer and fall that brought his activities to the attention of Shapp. At the Jerrold offices in Philadelphia, the curious sales agent, Bud Green, eventually asked Tarlton what he was up to, and Tarlton explained the Lansford scheme. One of Green's first responses reportedly was to asked Tarlton to sign a release pledging not to return the amplifiers he'd purchased; Green apparently did not think the amplifiers would do what Tarlton had in mind. Green then passed the word along to Shapp.

Shapp, always looking for a business opportunity, wanted to know more and called Tarlton. Shapp then drove from his offices in Philadelphia to visit the fledging operation the day before Thanksgiving in November of 1950, as lines were being strung from the newly minted tower. Shapp and Tarlton already knew each other. As a manufacturing representative, Shapp had visited Tarlton's store on several occasions. In talking over the problems of adapting the Jerrold components to the new demands of a larger system, Shapp determined he could manufacture and sell equipment specifically designed for CATV use.

Tarlton soon was being aided by Jerrold Electronics and by consulting engineers from Philco, which had a strong interest in the development of television, locally and nationally, and distributed Jerrold equipment. As Jerrold began creating specialty equipment for a new industry, PV-TV began stringing cable. The Lansford system used individual, six-tube amplifier strips (strip-band amplifiers) for each of the three channels it carried. Lines were strung by moonlighting crews from Lehigh Navigation Coal Company, using poles from three different power and telephone companies. The cable was run down Lansford's main thoroughfare, with taps at each of the town's seven cross streets. Customers were charged $100 for installation; residential homes were charged a $3 monthly service fee, commercial establishments $5.

As the system grew, Shapp and Tarlton began working together, using Lansford as a test bed for the creation of dedicated CATV equipment. (Tarlton later would go to work for Shapp, helping design and construct systems around the country). Lansford also became the focal point of a national publicity blitz, promoted by PV-TV, Shapp, and Philco. Less than a month after Shapp's visit, the *New York Times* ran a small story on the Lansford community television system, encouraged in part by a visit to the newspaper by George Bright.[85] About a week after that, a front page *Wall Street Journal* article, datelined Philadelphia and stimulated in all probability by the Philco public relations people, hailed the coming of the "community aerial."[86] The *Journal* article featured two systems known to be using Jerrold equipment, and highlighted the Lansford operation. It also mentioned Jerrold Electronics Corp. as the only company making dedicated CATV equipment at the time. Two weeks later, the Panther Valley Television company and Jerrold company were the subjects of a half-page article and illustration in *Newsweek* magazine. Jerrold subsequently used both the *Newsweek* and the *Times* articles in its publicity campaigns.

The earlier press coverage of Parsons' system notwithstanding, this was the first real burst of national press for the business of CATV. Both the new equipment that Jerrold was producing and the press, which engendered a rush of consumer and financial interest, sparked a sudden acceleration in CATV construction and helped launch the start of cable as an industry in the United States.

John Walson

In nearby Mahanoy City, for example, evidence suggests that Tarlton was indirectly influential in spawning two CATV systems. Interviews with many of the principles indicate that George Koval, who had been dallying in antenna work, asked his friend George Yuditsky to take a trip with him to Lansford to visit Tarlton's shop in 1950. With some confidence then that a CATV system could be made to work, Koval and Yuditsky began experimenting, constructing several antenna sites outside of town. Eventually, his contact with Tarlton led to an association with Shapp and, by the end of the year, engineers from Jerrold were helping Koval build a new community aerial system under the company name of City Television Corp.

At the same time, the local appliance dealer, John Walsonovich, was planning his own commercial system. Walsonovich's appliance store was a part-time venture; his full-time position was as a lineman with Pennsylvania Power and Light. In both capacities, he had witnessed, along with others, the earlier attempts to bring television into town, including the one at Trusky's. According to several accounts, Walsonovich had installed his mountain-top antenna at about the same time as the City Television Corp. tower, and by early 1951, residents of Mahanoy City were privileged to what was most likely the first ever direct competition between community television companies.

Years later, Walsonovich, who changed his name to Walson, claimed he was the first person in the country to start a CATV system, a claim that was sometimes repeated in the popular press. Historical evidence, however, suggests that Walson's story may have been the product of a memory more hopeful or romantic than accurate. Beginning in the late 1960s, Walson began telling people he had run a line down from Broad Mountain as early as the summer of 1948, and that by 1950 had a coaxial system serving 1,500 homes in Mahanoy City with three channels from Philadelphia.[87] Others around at the time have offered different memories, however. Yuditsky believed Walson started about the same time as City Television Corp. And, more importantly, the electrical engineer and instructor, Luther Holt, recalls the same. Holt's version carries special weight because he was the engineer who designed and built Walson's first set of CATV amplifiers. Holt, in fact, soon went into business as one of several small suppliers of CATV equipment in the 1950s and 1960s.

Evidence to reconcile the differing versions of events comes in several forms. *Television Factbook*, for example, has, since 1952, listed all the CATV installations they could locate in the country and has included information on the start date of each system. The information has been supplied by the companies themselves and for years, through the 1950s and 1960s, Walson listed the start date for his firm, Service Electric, as 1950. He changed that date, however, in 1967 to report a 1948 start for Service Electric. Evidence more to the point comes from newspaper stories and advertising from that early period. The local paper, the *Record American* was silent on the topic of television until 1950, when an advertisement announced plans for the construction of a community aerial system in the town, a system being developed by George Koval and company and build by Jerrold Electronic.[88] The first newspaper notice of Service Electric came in July of 1951. In interviews later in life, Walson recalled operating a three-channel system in 1948 or 1949 and expanding it to five channels by 1950. The July 1951 article, however, reports on the expansion in that month of Service Electric carriage from two channels to three, thanks to technical innovations by Holt. The newspaper article is reinforced by an advertisement taken out by Service Electric also proclaiming the additional station.[89]

Generally, contemporaneous reports suggest that community service probably began in earnest in Mahanoy City in November and December of 1950; City Television Corp. reported fifty customers in early January of 1951, with Koval stating they had been wiring about five homes a day.[90] Census figures from 1950 are also illuminating on this point. Earlier that year, the government conducted its Decennial Census and one of the questions asked was whether or not the homeowner had a television set. In mid-1950, the census takers counted about 25 TV sets in Mahanoy City, well below that needed to suggest the existence of a widely deployed community aerial.[91]

Despite this small dispute over dates, Walson was clearly an astute and energetic businessman and soon built Service Electric into the dominant CATV firm in the region. He recognized, for example, almost from the start, the desire of the public for ever greater TV variety. He placed heavy emphasis on gathering more signals for his system and he always managed to stay just a step ahead of the competition. With Holt's technology, his was the first system to expand to three channels, carrying all the Philadelphia stations by the summer of 1951. He was the first to bring New York City television into Mahanoy City via microwave and, in later years, the first to import a brand new kind of cable service, Home Box Office. It is also worth noting that Walson eventually purchased City Television Corp., absorbing its client base into his growing cable enterprise.

The Sixth Report and Order

By the end of 1950 and the beginning of 1951, community aerial television had taken root. As many as ten systems were in operation by the end of 1950, another thirty to forty by the end of 1951. As new TV stations went on the air across the country, cable systems lit up at the edges of their signals. Some systems were the work of local business people. In Franklin, Honesdale, and Pine Grove, appliance dealers followed the lead of Tarlton, designing, building, and modifying their own equipment. In other towns, cable was an idea imported by representatives of Shapp as Jerrold Electronics fanned out across the country looking for likely investment opportunities. There was little interest from established media companies. One of the few significant exceptions was RCA, which saw, as did Shapp, an opportunity to expand its apartment master antenna into a new market. For many, however, community aerials were perceived as a temporary phenomenon. According to this view or construction of the social role of CATV, once the FCC's freeze was lifted, broadcast TV would come to town and obviate the need for community antennas.

But while the freeze was a reason for some to stay away from CATV, many community antenna pioneers saw it as a window of opportunity. Their small systems, by this view, would fill the television gap by importing distant signals until broadcasters were permitted to open stations in or near their towns. As described by *Television Digest* in 1951, community antennas were an "antidote to the freeze."[92] In this way, the FCC's action therefore helped jump start community aerial construction, encouraging small businesses but filtering out larger financial interest with long-term investment horizons.

The perception of CATV as an interim service also had an impact on federal regulation. As will be seen in Chapter 4, regulators pushed aside both internal debate and public inquiries about the legal status of community antennas, in part, on the assumption that the local TV phenomenon would be short-lived.

This widespread assumption that the eventual end of the freeze would signal a concomitant end to CATV proved to be substantially mistaken, however. The licensing hiatus stimulated the growth of community antenna technology, but as has been illustrated, was not its root cause. Moreover, decisions by the FCC ending the freeze helped assure a place in the national TV system for services like community television.

On April 14, 1952, the FCC closed its moratorium on license allocation and unveiled a new plan for the national distribution of broadcast television in the United States, the Commission's historic Sixth Report and Order.[93] The Sixth Report and Order is notable on many counts. In the context of an evolutionary view of technology, it provided the last piece of a social and technical jigsaw puzzle, completing the cumulative syntheses that led to the rapid diffusion of domestic broadcasting television in the 1950s. By stabilizing technical standards and reopening the spigot of licensing, it helped coalesce decades of engineering, economic, and regulatory activity. Television penetration in the United States rose from about 34 percent (of all households) to more than 87 percent by the end of the decade. The Sixth Report and Order was, additionally, an exercise in applied political ideology. The Order was a focused effort to create a television system with a specific philosophical and social goal. This social construction of television, in turn, fixed the mold for both the national TV network system and the cable TV industry.

The Report and the industrial configurations that arose from it were grounded in a pointed social philosophy of TV access. From television's inception, the FCC was tied to a congressional mandate to provide "equality of radio broadcasting service" to "each of the states and the District of Columbia."[94] In practice, this meant forging regulatory policy to assure equitable access to radio across the nation. As it approached the question of television, the Commission extended this long-standing policy goal. The FCC's aim in creating the new allocation scheme was to strive for television service for every home in the country, and to provide at least one and preferably two TV channels to as many towns as possible. This emphasis on local service came to be known as "localism." The term was meant to suggest that television should serve as an outlet for community expression and a forum for the airing local concerns. As many towns as possible, therefore, should have their own electronic town forum provided by the new TV service. The downside to this philosophy was that technology and economics severely limited the number of stations any given town could accommodate. This meant that programming diversity was sacrificed on the altar of local expression.

The Commission did recognize the value in diversity, and sought to provide as many stations as possible to each town. It had an appreciation of the value of additional channels of service, and the new allocation plan called especially for expanding the number of possible TV outlets in larger markets. The technical nature of the broadcast spectrum, as portioned out under federal law, however, limits the number of stations in a given geographic area. Stations broadcasting on adjacent frequencies, Channels 8 and 9, for example, will interfere with one another,[95] and stations on the same frequency must be kept far enough apart geographically to avoid the kind of interference described earlier in this chapter. The FCC's pre-freeze allocation plan created twelve VHF channels and provided for TV service using those channels in 345 cities. To increase the number of potential stations, the Sixth Report and Order added seventy UHF channels and distributed television licenses across 1,291 communities, creating a total of 2,053 television assignments (242 of which were reserved for noncommercial, educational status).[96] But despite this expansion, many parts of the country found themselves with little or no television service.[97]

It was said in the 1960s that a television license was a license to print money. But this applied only to large stations and broadcast networks. In the country's smallest towns in the early and mid-1950s, it was clear to many that an FCC license was a license to lose money. For although a small town might have been accorded an FCC TV allocation, in many cases such a town could not provide the economic base necessary to support an expensive TV business. The financial problems of small town broadcasters, brought about by the 1952 Report, are treated in greater detail in Chapter 4. Here, it is sufficient to note that a certain audience and advertising base is necessary to sustain the operation of a local TV station, and many of the rural areas given FCC allocations were unable to generate such a base. To the extent this was true with allocations in the established and powerful VHF band, the situation was even worse for the new, untested UHF licenses. The technological and economic entrenchment of the VHF tuner meant that few investors were willing to take a chance on UHF and, as will be discussed later, those who did usually failed.

One of the consequences of the allocation scheme was a structure that, in practice, prohibited the creation of more than three national TV networks. As Congress explained decades later:

Under the 1952 allocation scheme, only 7 of the top fifty markets received 4 or more VHF assignments, 20 received 3 VHF assignments, 16 received 2 VHF assignments, and 2 markets received only 1 VHF assignment. Five of the top fifty markets had only UHF assignments. Consequently, one network could reach 45 of the top 50 markets with VHF stations and the second could reach 43, while a third network would be able to reach only 27 and a fourth network would have access to VHF stations in only 7 of the top fifty markets.[98]

The national audience base created through this plan was sufficient to attract the national advertising dollars necessary to produce high-quality, high-cost programming for, just barely, three networks. As late as the 1970s, analyses showed that the existing three networks reached nearly every TV home in the United States, 85 percent of them with VHF signals. Under the FCC's allocation plan, a fourth network would only reach about 60 percent of all TV homes and only about a third with a VHF signal.[99] As station owners across the country looked to affiliate with established networks capable of providing a steady stream of programming, they naturally went first to NBC and CBS, which had the network stars, the corporate name, and the production facilities. In towns with only one or two stations, the local broadcasters quickly aligned themselves with these dominant players. ABC was successful in capturing much of the remainders. Companies looking to start a fourth network, however, found themselves at a serious competitive disadvantage. The DuMont network appreciated this situation early on and challenged the FCC's allocation plan, proposing instead that the Commission create a regional system of television that would permit four networks. The FCC rejected the proposal, however, maintaining its philosophy of localism. The decision helped seal of the fate of the DuMont network, which folded in 1955.

The Sixth Report and Order, in short, helped established the template for the three-network structure that characterized U.S. broadcasting through the 1980s. At the same time, it also created a substantial market for anyone who could provide additional television choice to what would become known as unserved or under-served communities.

It was already understood that the American public had a voracious appetite for television. The hunger for additional channels and additional programming was unsated, and perhaps insatiable. The FCC grossly underestimated this demand and, while its goal of localism was noble, it failed to take into account the nature and scope of the public's interest. It was clear to many, therefore, that community television had a significant opening under the new FCC scheme. The trade press almost immediately noted that the allocation plan, when fully implemented would, "leave more than 200 communities without television service, either VHF or UHF. Many set owners in communities served by only one local station will also welcome a community master antenna system capable of pulling in signals from network stations."[100]

In February of 1953, only a few months after the FCC began once again issuing licenses, *Broadcasting-Telecasting* magazine sagaciously noted:

> The coming of a thousand or even two thousand free (television) broadcast stations does not mean the 70–80,000 subscribers to community TV systems won't continue to desire additional signals. Basically, most of the TV-locked areas are small valley towns which may have one—or perhaps two—TV stations of their own. But, community TV men feel local residents want at least a choice of the four networks and will thus still be willing to pay for the extra service.[101]

Despite the rush of station construction and the accelerating availability of broadcast signals, the peculiarities of the FCC allocation scheme left plenty of space for the nascent CATV industry. Some businesspeople, in fact, saw CATV as a more attractive investment than a broadcast outlet because it required significantly less capital to launch. As the lucrative nature of the community television business became apparent in the early 1950s, many were convinced that it was a better bet than broadcasting, especially if the only available license was UHF.

Finally, the boom in station construction itself, after the freeze, only acted to enable and encourage additional CATV growth. Community aerials could come into existence only where a TV signal existed, or rather at the fringes of such a signal. Even during the freeze, the number of stations rose steadily, providing the necessary signals to get the industry started. But as station inaugurations multiplied after 1952, the opportunities for CATV operation on the fringes of the new signals grew exponentially. Every new TV station provided at least the potential for a new community TV system keen on extending those signals into the under-served markets.

As a result, CATV, which had caught on during the freeze, grew even more furiously once the it was lifted, and community antenna systems even briefly outnumbered broadcast stations. The table below indicates that the development of cable began about a year or so after the 1948 surge of consumer equipment into the market, with the first systems beginning in 1949 and additional aerials coming on line in latter 1950. Then, as new equipment and financing became available, CATV activity began to increase in late 1951 and early 1952. Because it takes from several months to a year to organize, finance, and build a community TV system, this is the pattern one might expect, along with similar lagged development following the expansion of new broadcast outlets.

If the FCC had established a regional system of broadcasting that provided multiple signals to most of the population, along the lines of the DuMont proposal, the situation might have been different and cable's development could have been retarded. Even then, however, the wave propagation characteristics of broadcasting are such that towns nestled in the shadows

TABLE 2-1: TELEVISION AND CABLE DEVELOPMENT, 1945–1955

	COMMERCIAL TV STATIONS	TV SET OWNERSHIP (IN THOUSANDS)	PERCENT OF TV HOUSEHOLDS	CATV SYSTEMS (ESTIMATED)
1946	6	8	.02	
1947	12	14	.04	
1948	16	172	0.4	1–3
1949	51	940	2.3	3–5
1950	98	3,875	9.0	10 (approx.)
1951	107	10,320	23.5	30–50
1952	108	15,300	34.2	70
1953	126	20,400	44.7	150
1954	354	26,000	55.7	300
1955	411	30,700	64.5	400

Cable system figures, 1948–1951, are author's estimates. All other figures are from *Television Factbook*. Washington, DC: Warren Publishing, annual). Figures for 1949, 1950, and 1951 are up to December of those years; figures for 1952 and later are from January 1.

of local mountain ranges still would have needed community aerials. And communities on the fringes of secondary coverage would also have sought methods of adding signals to their sets. Only under a national system that provided multiple channels of nearly identical programming, would some form of cable not have been likely develop. But if at any time, channel allocations were distributed such that there was attractive programming available in one part of the country that was not available in another, demand for cable would have developed. Of course, such a plan was never seriously considered; it is perhaps too easy, with the benefit of historical hindsight, to fault the FCC for failing to appreciate the emerging lust for telecast programming. In fact, Shapp himself, who as much as anyone saw the need for (and promoted the vision of) CATV, seriously under-estimated the nature and extent of the public's appetite for television. Later in life, Shapp acknowledged that when he started building cable systems he never dreamed they would need more than a handful of channels. As noted at the head of this chapter, there was no doubt in his mind that seven channels were all anyone would ever need. "I mean, who wants more than seven?" he asked.[102] But one channel turned to three, three became five, five became seven, seven turned to twenty, and twenty to fifty, in a continuing upward spiral. As CATV pioneer Marty Malarkey observed in 1953, "If we can give the people a wider choice of TV fare, then there will always be a place for us."[103]

3
The Mom 'n' Pop Business
(1951–1958[1])

We're not out to make a million bucks.

—CABLE PIONEER, MARTY MALARKEY, 1951[2]

arty Malarkey was a young man managing a family business, Malarkey Music Co., in Pottsville, Pennsylvania. He sold musical instruments, sheet music, radios, and when he could, which was not often, TV sets. Pottsville is not far from Lansford and Mahanoy City, and the same hills that blocked TV signals from reaching those towns also stymied reception in Pottsville. In 1949, Malarkey took a business trip to New York City and happened to check into the Waldorf-Astoria Hotel. The hotel, of course, had been equipped with a sophisticated radio master antenna system in 1930, and that system had subsequently been updated to carry television signals. Malarkey was fascinated by the technology and called down to the front desk to see if someone could show him how it worked. A peacock proud hotel engineer took him on a tour of the facilities.[3] A few days later, Malarkey was asking a friend, over lunch back home, if such a system could work for an entire town. Malarkey was an authorized RCA TV set dealer and, encouraged by his luncheon companion, took the idea to RCA headquarters in Camden, New Jersey. The engineers there agreed to help him build a city-wide system using RCA equipment and, by early 1951, RCA technicians led by Malarkey were surveying the area for likely antenna sites. Within months, a system based on RCA technology was up and operating in Pottsville.[4] Marty Malarkey would go on to establish, along with master engineer Archer Taylor, one of the country's first CATV consulting firms, Malarkey-Taylor & Associates. And, despite his stated lack of intent, Malarkey would in time become, like many other cable pioneers, very successful.

The anecdote illustrates several important themes in the development of the industry, including the clear link between 1950s CATV systems and master antenna systems of the 1930s and 1940s, which is to say, the quantum-evolutionary nature of its development and, in equally important measure, the role of the individual in the process. As suggested in the prior two chapters, the development of technology for extending the distribution of radio and TV signals proceeded incrementally, first, through the ability to send a crude pulse of radio magnetic energy through a wire, then to do the same for voice and later video. The power of the technology was expressed in several ways, including the capacity of the distribution system, the distance over which signals could be delivered, and the clarity of the information. Each step involved the solution to an existing puzzle, and as a particular solution was adopted within the social, economic, and political context of the time, alternative solutions tended to fall by the wayside.

In the history of cable television, a variety of technical, economic, and regulatory factors were in play in the early 1950s, which suggested some means of extending broadcast television to under-served homes was likely to have evolved. Demand clearly existed, as did a number of appropriate technologies. The question was how interested businesspeople could make a sustainable living by exploiting the situation. In fact, for potential community aerial operators, the demand for television meant that as long as they could provide viewing opportunities beyond those available over the air, people would line up for the service. And if one or two channels of television were good, three or four channels were better. Not everyone, however, had a sharp sensitivity to that demand. Milton Shapp's comment in the previous chapter concerning the need for only a limited number of channels is illustrative. Many pioneers in CATV and broadcasting would have been puzzled by the suggestion that anyone would want more than even three or four channels of television. Others, however, saw the future more clearly, accurately assessing the nearly insatiable thirst for choice in information and entertainment.

This chapter looks more closely at the work of the 1950s cable pioneers as they struggled to find the financing and equipment necessary to build their first systems, negotiate agreements with local municipalities and telephone and power companies, and keep their systems running and customers satisfied. It also examines cable's first contact with federal and state authorities, the earliest efforts at pay-TV, and the start of the cable multiple system operating companies. It begins by looking at one of the industry's first national leaders.

Milton Shapp and the Founding
of the CATV Industry

Cable television in the 1950s was a slow-moving locomotive. It was not hurtling down the tracks, but neither was it going to be stopped. It was only a question of who was going to see the train and decide to jump on it. In that process, many people grabbed a handle and swung aboard. Most stepped into the cabin cars, so to speak, and began building systems and doing the best they could in their hometowns. A few, however, went straight to the engine and began the serious business of running the train. While anyone might have gone to the controls, only a handful did, among them a young engineer and businessman from Philadelphia, named Milton Jerrold Shapp. In the lore of the cable TV industry, several people have been nominated for the title of "The Father of Cable TV." If anyone can, in fact, lay a powerful claim to that honorific, it would be Milton Shapp; at the very least, he would be first in a small line for the title.

Shapp was born June 25, 1912, in Cleveland, Ohio, the son of Aaron and Eva Shapiro. He developed an early interest in Ham radio and graduated from Case Western Reserve University in 1933 with a degree in electrical engineering. His first job was driving a coal truck for $1.50 a day, but he soon found a position that more fully exploited his interest and education. He became a salesman for a Cleveland firm called "Radiart," which specialized in wholesaling replacement parts for car radios. One of the few industries that did well in the depression was radio, and Shapp and Radiart began to prosper. When the company opened an office in Philadelphia to sell directly to Philco Corp. and other manufacturers on the east coast, they asked Shapp to relocate. He did and, in 1938, he apparently struck out on his own, founding his own manufacturing rep firm, M. J. Shapp & Co.[5] As with many young men and women at the time, his career was interrupted in 1942 when he joined the Army Signal Corps and went overseas. In returning to civilian life, he picked up where he left off in 1946, reestablishing his sales network, but in 1948, Shapp met another World War II veteran whose idea would change his life.

According to cable pioneer, Archer Taylor, Shapp was on a sales trip to Baltimore when, through a local radio parts dealer, he heard about US Naval Academy graduate student named Don Kirk.[6] For his master's project, Kirk had built a simplified TV set, a so-called "gutless wonder," stripped down to its barest components. Shapp, interested in possible do-it-yourself TV kits, called on Kirk. The receiver was a low-powered device and Kirk had designed a simple preamplifier to boost the signal and improve the picture.[7] The first time Kirk showed Shapp the TV set, he could get no picture, but when Kirk added the external amplifier, an early TV booster, the signal snapped on and Shapp, according to Kirk, got very excited. "And, when I was showing this kit to Milt Shapp, it wouldn't play for us. I went out to the car and got a little gain box and put it in, and we had a good picture. Milt immediately dropped all interest in the TV set and wanted to build a booster."[8]

Shapp said he could add the amplifier to his line of goods and "peddle" any boxes Kirk built in his basement shop.[9] The amplifiers sold well and it dawned on Shapp that the TV booster business could be lucrative. He soon invested in production equipment, hired technicians, and opened a manufacturing line. In 1948, he took $500 and, along with Kirk and another engineer, Ken Simmons, founded Jerrold Electronics Corp.

Shapp quickly realized that a well-built amplifier that could improve the strength of a signal for one TV set might also, with the proper modifications, be set up to service several sets, and it was about this time that a market for multiple set antenna systems was developing. The Jerrold booster evolved into the "Mul-TV" master antenna system. The system, designed initially to be used by TV dealerships to demonstrate their equipment, was introduced to the industry in May of 1949 at the National Electronics Distributors Convention in Chicago.[10] In 1950, this product evolved again into the Mul-TV master antenna system for apartment houses, and Shapp began signing contracts. He installed his first apartment house system in early 1950 in Collingswood, New Jersey,[11] and was soon wiring buildings in Atlantic City and elsewhere. Shapp's efforts were heavily publicized, both by the industry press and by himself. By all accounts, he was a master of promotion and publicity, traits that would be critical in the expansion of his business and the industry later.

As noted in Chapter 2, Shapp's first contact with community television was through Tarlton. When Tarlton, a few hours away in Lansford, was asked to wire a local apartment building for television, he used the Mul-TV system. For Tarlton, Shapp's company was almost in the backyard, so obtaining materials was fairly easy. In addition, Jerrold equipment enjoyed a reputation for performance and reliability. It was also smaller and lighter than some competing products, making it especially attractive for CATV pioneers who would need it in great quantity.

When Shapp learned of Tarlton's efforts in Lansford and went for his historic visit, he took along with him Caywood Cooley, Jerrold's chief field engineer. Shapp, deeply taken by the idea of community TV, instructed Cooley to lend whatever assistance he could to Tarlton and made the development of specialized equipment for the operation a priority for his engineers. Said Shapp later in life, "I knew there was a pot of gold out there for Jerrold if we did things right."[12]

The challenge and the opportunity for Shapp was to be the first builder of equipment dedicated to community antenna television. Until this time, all the technology used by the early system operators was homemade or home-modified. As previously noted, Parsons and others built amplifiers largely from scratch; some converted Electro-Voice set-top boosters and Tarlton began by modifying Jerrold's equipment. In creating their protosystems, the basement and garage-based technicians used whatever materials they had handy. Often, the amplifiers would be constructed inside wooden or tin boxes hammered together from scraps found around the house. There was no standardization, no mass production, and no economies of scale. As a result, the homemade equipment often had questionable reliability.

The prospective demands of a commercial community antenna system were far beyond those of even a large apartment antenna complex, however. The lines of a CATV system would have to run for miles instead of yards and the amplification of the signal had to be stronger and cleaner than anything yet required in then-existing antenna technology. One of the principal problems, although certainly not the only one, was that the amplifiers then in use simply were not built for sustained cascading.

The problem of amplification was outlined in the previous chapter in reference to roof-top apartment systems, but for community systems, the problem was compounded several fold. Because the television signal deteriorates as it passes through the line, it needs to be amplified at regular intervals. The typical run between amplifiers in the early systems was about 1,000 feet. But the price of amplification is increased interference. Every amplifier along the line, then and now, introduces noise in the system. Noise is a loss of quality in the TV signal and is seen on the screen as a weak or fading picture, TV "snow," or rolling images. The early boosters and apartment amplifiers also acted to squeeze useable bandwidth. In many cases, the audio portion of the signal, which resides, in a sense, at the edge of the picture component, would be cut off after passing through only two or three amplifiers. And the ever-narrowing channel would soon collapse to a point where the picture itself was lost. Jerrold's task, therefore, was to design stronger, cleaner, and more reliable amplifiers for the increased demands of an outdoor, community system, and to produce them at a reasonable price in mass quantities.

One interesting historical side note was the lack of any serious attempt by the pioneering businessmen and technicians, including Shapp, to patent plans for the concept of a community aerial system. While patents could (and were) taken out on specific pieces of equipment, such as amplifiers and taps, it was generally agreed that the *idea* of a community aerial was so general and easy as to be unpatentable. In fact, a patent was taken out in the late 1930s on a system that generally described the community antenna system. It was filed in November 1937 by Louis H. Crook, a professor of aeronautical engineering at Catholic University.[13] The Crook patent, granted on November 26, 1940, described a system for delivering TV signals from a central distribution station to numerous homes by way of shielded telephone lines.[14] There is no record of any attempt to build the system or enforce the patent and, in fact, it probably would have failed. Success came from creating reliable equipment and installing and running systems.

It was the ability of the Jerrold engineers, particularly Kirk and Simmons, to produce equipment that performed well and reliably that provided the backbone of the company's success. Other companies soon followed Jerrold, including the industrial giant RCA, and smaller firms such as Entron, Spencer-Kennedy Laboratories, Ameco, and Blonder-Tongue.[15] But for years Jerrold remained the dominant supplier of equipment to the young industry.

The performance of Jerrold equipment was only part of the reason for Shapp's success and eventual influence, however. More than just a peddler of amplifiers and taps, Shapp was a fearless and aggressive promoter and salesman. It was, in part, this characteristic that was responsible for the widespread use of the Jerrold nameplate. Shapp seemed to understand that promoting his equipment went hand-in-hand with promoting the industry. Philco had contracted to sell Jerrold equipment and neither the large and established company nor Shapp himself were shy about the press, or about extolling the virtues of CATV or the company. The press blitz of late 1950 and early 1951, noted in Chapter 2, featured Tarlton's system and gave prominence to the Jerrold equipment. While the *New York Times* mentioned Jerrold only in passing, the *Wall Street Journal*, noted that Jerrold was the only producer and installer of community aerials: "Three years ago the system of bringing television to dead spots was only an engineer's dream. The engineer, Milton J. Shapp, is now president of Jerrold Electronics. His dream and an initial investment of $500 got the company started."[16]

Sensing the potential of this business—yet another avenue for tapping into the TV craze—Shapp wasted no time in seeking out new opportunities for system development. For Shapp and Jerrold, working in the neophyte industry was a little like mining for gold. One of the first steps was to identify a likely geographic location, one that lent itself to the problems and opportunities posed by the FCC's allocation scheme. He established a community operations division to look for such sites. One of the women in his office hung a national map on the wall and stuck in colored pins to designate towns where TV reception was weak, but available. It was a miner's map, in a sense, an atlas of natural resources. Shapp worked hard to discover these pockets of opportunity, then he went looking for the capital to fund their exploitation and for one or more local businessmen who might be interested in setting up a CATV system.

An important part of his enthusiasm and savvy was soon manifest in Shapp's ability to find financing for community television operations. There was very little money available to finance CATV in the early 1950s, at least at the national level. It was a risky venture in the eyes of most major banks. The people entering it were often untried and the technology was unknown. Some potential lenders shared in the popular assumption that once the FCC freeze was lifted, demand for CATV would dry up. As Shapp explained it in 1959:

At the time we started, which was late 1950 and early 1951 and 1952, this business was an extreme risk. You can't even call it a normal risk. You had, for example, the fact that everybody you talked to assumed that just as soon as the freeze was lifted as it finally was in 1952, community systems wouldn't be necessary, there would be 2,000 television stations all over America.... We couldn't go to a bank and finance a nickel's worth of this. We couldn't induce investors to come into this new business. I broke my neck for a year and a half until I got people with capital ready to come into the new business.[17]

The money Shapp eventually found came from venture capital firms in New York, and he was one of the first to develop this kind of financing for community antenna television. Seeking investment for a system he was working on in Williamsport, Pennsylvania, Shapp traveled to New York City in 1952 or 1953. There, he convinced venture capital firm J. H. Whitney Co. and later Fox-Wells to take a chance on community television. The Williamsport project was one of the first attempts to move beyond the very small towns in the east or west and into a city of some size, and that may have helped attract the interest of the financiers. Soon, Shapp had formed a working relationship with J. H. Whitney, Fox-Wells, and Goldman-Sachs. In the typical arrangement, the ownership of the finished system would be shared, with the investment houses retaining primary ownership and, occasionally, a local business partner taking a small percentage. Jerrold itself retained an equity interest of anywhere from 10 to 35 percent. "It was a beautiful deal," said Shapp.[18] More importantly, for the industry, it provided encouragement for subsequent capital investment and helped prime the financial pump for systems nationwide.

Meanwhile, the publicity generated by the Panther Valley system in Lansford and the City TV Corp. system in Mahanoy City was feeding a growing frenzy about community antennas. So many potential CATV operators were visiting those towns and then coming down to see Jerrold's operation in Philadelphia, that Shapp joked he could have run a successful shuttle bus service between the communities. At the same time, he reported holding off on new contracts after the stories broke. Despite the heavy demand, Shapp said he wanted to proceed slowly. He was concerned that any misstep, any mistakes in equipment design or failures in installation due to inexperience, could only damage the reputation of his young firm. "We were the only 'expert' then in the cable TV field," he wrote, "and we didn't know half the problems, let alone the solutions."[19]

In the spring of 1951, Shapp began negotiations with a growing list of people interested in his equipment. In what would eventually lead to bad blood between Shapp and many CATV operators and also cause Jerrold substantial grief with the Justice Department, Shapp announced a new sales policy. In order to obtain Jerrold equipment, the buyer had to enter into a mandatory service contract with the company. For Shapp, this was intended, in part, to preserve the integrity of his equipment and, consequently, his reputation. It would mean that qualified Jerrold staff would train local operators, and Jerrold engineers would be available to fix and maintain the system and the service on a regular basis. The contract also required the system owner to purchase any additional equipment only from Shapp. Finally, the contract required every system operator to pay Jerrold $5 for each new installation and twenty-five cents per subscriber per month. It was controversial within the industry and Shapp was defensive about it. CATV operators, and Jerrold's only real competitor at the time, RCA, were not reluctant to point out to potential customers the increased price of the Shapp service agreement. And Shapp was sensitive to the criticism. "In those early days politically," he once said, "it was Jerrold against most of the operators of the other systems."[20] Ultimately, this policy would run afoul of antitrust laws. In the meantime, however, Shapp continued to aggressively seek franchising opportunities and venture capital for construction.

Through his vigorous promotion of the industry, his dominance of the technology, and his growing political activities, Shapp influenced the development of cable television for a decade, and his company, Jerrold Electronics, would serve as the foundation for one of the country's first CATV system chains.

CATV as a Local Business

The period from 1952 to 1955 was one of rapid development in the new industry. The concept of community television, once it received its initial dose of national publicity, took root and was made real by local entrepreneurs in a matter of months, sometimes a matter of weeks. In sociological terms, the curve of adoption of the new technology was, in some ways, very steep. News of the business broke across the country in the newspaper reports of December 1950 and January 1951. By the late spring of 1951, systems were under construction in dozens of communities around the country and, by the fall, a coalition of CATV operators, primarily from eastern Pennsylvania, had already formed a fledgling trade association. From as few as a half dozen operating systems in late 1950, the number of firms grew to about fifty by early 1952, and to an estimated 400 by 1955.

Even though CATV had received a measure of national publicity and launched what eventually would become a national lobbying organization, the industry during this period was still intensely local. This was the age of the true "Mom and Pop" community television company. It was fueled by small-town entrepreneurs and small-town entrepreneurial spirit. Jerrold and RCA were the only significant national companies in the field and they provided only equipment and consulting; Jerrold offered construction and system maintenance, but the price was more than some operators were willing to pay. The problems of the local operators were the problems of any small business venture: finding the capital to finance the system, locating reliable equipment and suppliers, securing permission to use public and private rights of way, putting up an antenna, running wires, soliciting customers, and somehow keeping the whole thing from falling apart once it was built. For some, the business was, at least initially, a sideline, an adjunct to an existing occupation. This was especially true of the appliance salesman and broadcasters. For others, CATV was the single basket into which they placed their eggs. Most of these economic adventurers were young, in their twenties, married, recently out of the armed forces, willing to take a risk, and looking for an opportunity. All of them would find long, hard hours ahead, more problems than they bargained for, and, in some cases, rewards beyond their wildest imagination.

Getting In

George Gardner planted his aerial in a town that, somewhat unusually, had not demonstrated any real enthusiasm for television. He quit his steady job with Sylvania, moved to Lewistown, and began the project with the comment to his father, "Well, this looks like it might be something that might be useful around here."[21] He enlisted the backing of the owner of the local newspaper and the local radio station and began work.

The pioneers entered the business through a variety of doors. The earliest experiences of Parsons, Tarlton, Walson, Malarkey, and a few others have already been mentioned. Others worth noting, by way of illustrating the early trials of the community aerial operators, include: the founder of Adelphia Communications, John Rigas, and lesser-known operators, Bark Lee Yee in Pennsylvania and Roy Bliss and Tom Mitchell in Wyoming.

A few hours drive to the northwest of Malarkey's shop, a young man named John Rigas, the son of Greek immigrants, had borrowed from family and friends to purchase the local movie theater in Couldersport, Pennsylvania. An engineer by training, he worked during the day at the Sylvania Electric plant some thirty miles away in Emporium, Pennsylvania, the same plant where George Gardner had started working. Regis commuted to Couldersport to run the movie house in the evenings and it was through one of the movie studio salesmen, who traveled the region promoting new RKO films to the local theaters, that Rigas first learned about CATV. The sales rep had seen CATV in other towns he visited and encouraged Rigas to look into the business as a way of protecting his theater assets from the coming incursion of television. Rigas said he only did so to appease the rep, who regularly chided him about it. For the surprisingly low price of $100, he bought the city franchise from a local appliance dealer who had secured it initially but was unable to exploit it. As the new holder of the CATV permit, Rigas soon was approached by better-financed investors who wanted to build a system. It was the beginning of what would one day become Adelphia Cable, one of the nation's largest multiple systems operators (MSOs). In 1953, however, Rigas like many other pioneers, was wondering what he had gotten himself into.

Farther west, Roy Bliss sold Culligan soft water systems in Worland, Wyoming. A new TV station had begun operation in Billings, about 125 miles way, and Bliss, along with a friend John Huff, loaded reception gear into a small airplane and went looking for the signal. "I had to climb up to 8,000 feet before we could even get a weak signal. So we more or less gave it up," Bliss recalled.[22] But they left the equipment turned on and during a low pass over a wildcat oil well they were examining, Huff called out that they'd picked up a signal. They parked the plane back at the airport and called in a third friend, Tom Mitchell, who ran a local electronics store. Like Parsons, the men found the finger of a usable signal only a few feet off the ground, about five miles outside of Worland, and began plans to build a CATV system.

One of the more interesting pioneers was Bark Lee Yee, an immigrant from Taiwan. He came to the United State in 1939 at the age of fourteen. He traveled alone with $20 in his pocket and joined his father who ran a restaurant in Easton, Pennsylvania. He learned English in the local grade school, and eventually he took training in aeronautical engineering, spending the last year of the war in the service working on planes at LaGuardia airport in New York City. A few years later, he found himself back in Easton, servicing planes at the local airport and giving flying lessons. One weekend, a friend and neighbor, who also happened to run a local appliance store, offered Yee $150 dollars to install a TV antenna for an impatient couple who had to have television immediately. Yee made a $100 profit on the assignment and saw the long-term opportunity. Installing antennas paid significantly more than the airport job. As part of his new-found weekend work, Yee set up a small shop in his apartment to test TV reception from an antenna he installed on a nearby roof. Within weeks, everyone else in the building knew about his antenna and he soon had his apartment house and the adjacent apartment building wired.[23] Extending the idea to the rest of the town was a short, but not effortless, step from there.

Financing

Whether they were motivated by a vision and a gambling spirit, by a modest desire to goose TV set sales, or fell into it as a consequence of the quirks and turns of life, the pioneers all faced similar hurdles in getting their businesses off the ground. The first, as always, was the money.

The cost of constructing a CATV system in the early 1950s was estimated at anywhere from $30,000 to $130,000, or more, depending on the size of the system. The bulk of the cost was in the amplifiers and the coaxial cable.

One early RCA estimate placed the cost of installing cable at ninety-six cents a foot, which amounted to nearly $5,000 a mile.[24] Other experienced builders estimated construction costs at up to $7,000 a mile.[25] The large receiving tower necessary to pick up the signal was one of the cheaper items, running an estimated $1,500 to $5,000, depending on size and site.[26] Building the system in Williamsport, Pennsylvania, for example, cost about $31,000 in 1954, while a larger system in Wilkes-Barre, Pennsylvania, came to around $133,000.

The money to fund such an untried enterprise came from a variety of sources. Shapp's successful effort to develop venture capital outside the local community was the exception in the early 1950s. It would be several years before the young industry became attractive to large-scale venture capitalists or institutional investors. The local operator more typically started with a portion—sometimes large, sometimes small—of his own money, frequently borrowing from friends and relatives or taking a second mortgage on the house. It was common for operators to seek local partners, each partner then approaching the hometown bank for a loan. There is a mixed historical record on the availability of local commercial money. In some cases, bankers were skeptical of the untried and untested business. In other cases, the hometown lenders seemed willing to take a chance, not necessarily on the business idea but rather on the businessman behind it. Many CATV operators had existing businesses and bank loans were secured with personal property or through the established store or service. The fact that most of these communities were small and the social network tight-knit no doubt played a role. The five founding partners of Tarlton's Panther Valley group each contributed an initial $500 to the enterprise and together they obtained $20,000 from local banks. Gardner secured local financial support through his partnership with the town's newspaper and radio station owners. The investors in City Television Corp. in Mahanoy City included a local appliance dealer, the son of the town's Oldsmobile dealer, and the financially well-connected chief of police. The seed money covered the initial cost of building a tower and running a line into town, but operators quickly had to begin living on whatever revenue they could generate. The typical installation charge was $100 to $150, a figure competitively based on the fee to put up a roof-mounted antenna. Over the course of the 1950s, this installation fee was progressively reduced—from $150 down to as low as $25 and, in some cases, operators offered free installation in an effort to attract more price-sensitive customers. Monthly service charges ran $3 to $6.[27]

An operator would build a small section of the system, sign up as many new customers as possible, and immediately take that money to buy coaxial cable and amplifiers for the next section of the system, where the process would be repeated. In this way, many systems "bootstrapped" their way into existence, extending the cable only as funds came available from new customers.

Construction

Once an operator had money to begin construction, simply finding the right materials in sufficient quantity could be a challenge. In the mid- and late 1940s, television signals typically were carried through wire in one of two ways, coaxial cable similar to what exists in cable systems today, and "twin lead." Twin lead is the flat, plastic ribbon of wire many consumers are familiar with as the wire that connects the back of the set to the rooftop or set-top TV antenna. Twin lead is susceptible to rain fading and similar climatic disturbances. At the end of World War II, however, twin lead was relatively cheap and readily available, coax was not.

From 1950 through 1953, much of the material necessary for CATV construction was subject to federal rationing due to the Korean War. The National Production Authority (NPA), founded during World War II, operated through the Korean Conflict, exercising oversight on the production and distribution of all manner of materials. In March of 1952, for example, George Bright reported to the newly established National Community Television Association that the NPA had approved material acquisitions for one CATV system, but denied similar requests from several others.[28] According to one story, Shapp used his political connections in the Democratic Party to help ease restrictions on coaxial cable and increase the flow of coax to the industry in the early 1950s.[29] Even then, however, the cable that was available was almost exclusively war surplus. It had been used in radar and communications facilities, but was not manufactured for the special demands of community antenna television. Noted Shapp: "The cable we bought— it was during the Korean War—and what we got in many cases wouldn't even be suitable for clothesline, let alone coaxial cable for a community antenna system, because we got what was left over from what the Army took."[30]

The problem of securing the necessary materials to begin construction often led to imaginative solutions. Coaxial cable, for example, stretches and cannot support its own weight in long runs on utility poles. It has to be lashed to a stronger wire, called "strand," which is a simple length of twisted steel. As with coax, strand was a scarce commodity at the time. Recalls Gardner: "The Korean War was on and the federal government didn't really feel that the cable industry was a priority item and we couldn't get a lot of types of equipment. I can remember one of the problems that I had was getting strand. The only strand that I could find that was available was barrage balloon cable. This apparently had been used in England during the Second World War to tether balloons above London."[31]

On rare occasions, cable operators in competitive markets would even accuse each other of stealing precious coax, sneaking into the woods at night and taking down hundreds of feet of cable strung by a competitor the previous day. While such stories may be apocryphal, they illustrate much about the nature of the business at the time. Building a CATV system was not for the faint of heart.

Once cable and strand were acquired, the coax had to be pieced together and a tap created for each subscriber's home. Specially made fittings used to connect cable to associated equipment or to extend cable lines were at best poorly made and often simply unavailable. The surplus coax typically came in lengths of 100 feet, or in various leftover odd lots. CATV operators, however, needed significantly longer runs and spent much of their construction effort in these early years splicing together the surplus bits of coax. Splicing coaxial cable was a skill all its own, and decades later, Bob Tarlton recalled teaching John Walson the technique in the earliest days of CATV.[32]

The war surplus cable came in two basic thicknesses, a larger RG-11/U and smaller RG-59/U. The heavier RG-11/U was required for the long runs through town, but no simple taps were available in the early 1950s. Taps are the small junction devices that tap the TV signal off the main line and feed it into an individual home. Tap-like equipment used in apartment master antennas was available for the smaller RG-59/U, but it didn't fit the TG-11/U. Operators, therefore, would insert a section of RG-59/U in the larger cable and tap off of it for each subscriber drop. The solid copper center wires were soldered together, but the outside braided wire shields of the two cables were simply overlapped and wrapped with tape.

Every such slapdash fitting and splice was an opportunity for signal loss, or leakage, and most of these early patch quilt systems literally spewed low-level electrical radiation. The earliest systems were, as a result, notoriously inefficient carriers. The picture at the subscriber's home was sometimes so poor that operators were embarrassed by it. Eventually, manufacturers began producing usable cable specifically for the industry, in 1,000- and 1,500-foot lengths. And by 1954, specially designed and manufactured taps were being produced by Jerrold and by a new cable equipment company, Entron.[33] The designs differed a bit, and Entron for some time had a more popular product, but both provided a faster and easier means of tapping off the main line without cutting and splicing cable. By the mid- and late 1950s, other firms began supplying coaxial cable specifically for the industry, including Viking Cable Manufacturing Co. and Times Wire and Cable, Inc.

For the first few years, Jerrold Electronics and RCA were the only large firms manufacturing amplifiers specifically for CATV. There were a few, smaller scattered firms, such as Holt Electronics, supplying a limited amount of equipment. But to buy the Jerrold amplifiers meant buying the Jerrold service contract, which some operators balked at, and RCA had a reputation for selling equipment that worked well but was too large and bulky for the tastes of many of the early operators. While Jerrold soon captured the largest portion of the market, the earliest efforts often were characterized by the kind of home-built equipment discussed previously. Examples still exist of "coffee can" amplifiers, constructed in someone's kitchen or garage out of surplus or recycled electronic components and soldered together in the bottom of an empty coffee can.

Once the materials were acquired, operators had to locate a strong signal and erect the receiving antenna, often in difficult-to-reach areas. Parsons' low-lying antenna site was atypical. Antenna sites could be on heavily wooded mountaintops far beyond the reach of any existing road. Trails had to be cut, then gravel paths constructed, and concrete poured for antenna foundation sites. Operators often had to use explosives to blast a hole for the antenna base. One cable consultant told would-be operators, "If your mountaintop is anything like ours, it's 90 percent rock and 10 percent wind."[34] A small house or shack at the base of the antenna was common. Gardner built his out of rock, gathering the stones and boulders from the surrounding hillside and stacking them one by one. (By the mid-1960s, operators were often using aluminum tractor-trailer bodies. They were cheap, secure, and light enough that a head-end could be assembled in an easily accessible location then dropped into the antenna site by helicopter.) In some cases, the early antenna sites and headends were far away from any electrical lines and had to be powered by generator with fuel carried in on a regular basis. In places where poles did not exist, such as the run from the antenna site to the edge of town, they had to be installed. (While the earliest builders did lash wires to fence posts, trees, and the eaves of roofs, such jerry rigging did not last long.)

Those with the financial resources would hire trained crews from the local utility companies to run the wires along the poles. The crews would work on the weekends and in the evenings. Less well-off operators would have to string the wire themselves. To pull the cable to its proper tautness and prevent dangerous sagging, Bark Lee Yee tied the end of the cable to the bumper of his truck, pulled the wire tight, set the parking brake, and secured the cable to the pole. No one had ever really done anything like this before and many operators, drawing on a background in general electronics, engineering, business, or pure inventiveness, simply made it up as they went along. "I knew more than anybody else in our group," recalled Gardner, "and I didn't know anything."[35]

Franchising

Relationships with the local authorities were informal and, in some instances, nonexistent, at least initially. A few of the earliest operators admitted, later in life, to stringing wire and setting up makeshift systems without actually discussing their activity with municipal authorities, figuring someone would let them know if there was a problem. The press in 1950 reported that Franklin, Pennsylvania, did not have any regulations governing access to streets or utility poles, so "the line runs from house to house where permission can be obtained. This has necessitated running a line around a block to avoid certain properties whose owners refuse to cooperate."[36]

Most operators, however, sought some kind of license or permit from the city. Franchise agreements per se and even the concept of a franchise for a CATV operation did not exist, and the permit was normally a simple affair. It provided the CATV company municipal right of way in exchange for maintaining full liability or liability insurance. The city simply wanted to indemnify itself. As Tarlton explained it, "the permit was more or less a 'we don't have any objection' type of thing."[37] In 1954, *Colorado Municipalities*, a magazine for city officials and planners, conducted a survey of towns that had CATV systems, with an eye toward determining municipal obligations and practices. "Each municipal official's answer to the question, 'What are the municipal responsibilities,' was, 'none,'" reported the magazine.[38] Occasionally, a city would issue a permit saying, in effect, "We don't have authority to grant a permit for such an operation, but if we had such authority, we would authorize the system." This somewhat backhanded effort to protect the township from liability was sometimes used to leverage access to the local utility poles. The CATV operator would take the city's nonpermit to the local telephone company officials who demanded some form of authorization from the local authorities before they would allow anyone on their poles. There was no significant cost for the permits beyond the one-time, nominal license fee; which is not to say there were not occasional problems with local officials.

Near Mahanoy City is the town of Shenandoah Heights, which, like Frackville, is situated on top of a mountain. Its sister city, Shenandoah, sits at the bottom of the mountain. Television dealers in Shenandoah, looking to construct their own community antenna, asked city officials in Shenandoah Heights for permission to raise a receiving antenna in the mountaintop town. The request was denied. The town folk of Shenandoah Heights could already get off-air television, and they were concerned that the proposed CATV tower for their neighbors down the hill would cause interference with their own reception. "We have nothing to win and everything to lose," declared the township's board of supervisors.[39]

In Hazelton, Pennsylvania, John Walson was prevented from building a microwave repeater station on a hilltop by restrictive zoning laws that allowed the construction of only single family dwellings. Always ready with an imaginative solution, Walson built a two-story family home, set his microwave dishes on the second floor, and pointed them out the windows.

To guard against such problems with local authorities, some CATV operators enlisted the active support of town fathers and local power brokers. It was a practice that was to become an industry standard in some of the problematic franchising activities of later decades. But in the early 1950s, it seemed like simply good business sense. One sought out local investors who could be counted on to speak for the company when the need arose.

Marty Malarkey found this to be an almost universal solution for the problems he faced in getting his system started. "I went to an attorney I'd found whose father was chairman of the city council and I said, 'Would you like to invest in my new company?" The attorney did and a short time later Malarkey got his permit from the city. He went to the regional manager of the local Bell Telephone Co. and asked, "Would you like to invest in my new company?" The manager did, and a short time later Malarkey had permission to use Bell poles. The vice president of the Pennsylvania Power & Light (PP&L) Co. was similarly interested in the system, as was an engineer with experience with PP&L lines. Each investor purchased 1.5 percent of the company and, with RCA's technical backing, Malarkey was stringing wire in short order.[40]

Telephone Company Relations

Throughout its history cable television has both battled and wooed, and mostly battled, the telephone industry. It has been a continually contentious yet symbiotic relationship, usually a sparring match, sometimes a wedding march, but never a thing of social or political beauty. Its tone was set from the beginning of the industry, as pioneers looked for poles upon which to hang their wire. The question of whether the telephone companies would allow CATV lines on their poles, and at what price, was a problem for operators as soon as they tried to bring the coax line down from the mountain. Tacking cable to trees and fence posts was adequate for the first few months of operation, but clearly was not the foundation for a real business. The cable industry soon would come to regard the telephone companies, or "telcos" as a necessary evil, a lurking behemoth that, could, if left unguarded, swallow the cable industry whole, as a small snack.

Despite this, for the first few years the CATV pioneers had few problems mounting their lines on some kind of pole, at least no problems that a little ingenuity and, in some cases, a bit of small-time skullduggery, couldn't overcome. In the first flushes of excitement over the prospect of bringing television into town, some of the earliest experimenters did not think twice about using any pole they needed. City Televison Corp.'s builders in Mahanoy City simply began tacking line to poles for the first runs into the Koval's auto shop. The question of asking permission did not seem pressing at the time.

In Arizona, cable pioneer Bruce Merrill employed a similar, albeit more leveraged, strategy:

> The first couple systems that we built, we built without pole attachment contracts. When they declined to give us the pole attachment contract we simply went out and attached to their poles and put customers on as rapidly as we could. When they protested, we said, "Well, if you want us to disassemble this, we'll do it but there's a couple thousand people who are really going to be up in arms."

Fortunately utility companies and telephone companies reacted very slowly so we were able to pressure them through public opinion into really giving us the first pole attachment contracts, which they had refused to give us initially. I think in the first couple systems we built which were in Globe and Safford, if we'd waited until they gave us pole attachment contracts, we may never have built them.[41]

Bark Lee Yee acknowledged that he did not think much about seeking authority from the utilities when he began building his system; he just started stringing the wire. If poles did not exist, as in the run from the mountaintop down into town, they were honed from the local forest, purchased from the local utility, or, in a few cases, "otherwise obtained." The son of one pioneer once was asked whether the local utility company was reluctant to sell his father the poles he needed; the son looked away with a sheepish smile and remarked, "Well, I wouldn't exactly say we bought them" and then he changed the subject.[42]

Such a state of affairs lasted only briefly, of course. Within a short period of time, the local utilities were knocking on Bark Lee Yee's door and asking about the wire he had strung on their poles. The arrangements worked out in these earliest days varied. In Yee's case, as in some others, the community was small, it wanted television and the utilities officials were as eager as anyone to help make it happen. Yee secured permission from PP&L, in part, because the officials supported the project (and received free service). Walson, in Mahanoy City, had worked as a PP&L lineman and was a close friend of the local manager. Malarkey, of course, enlisted both the PP&L and telephone officials as investors, as did Tarlton. And, according to Tarlton, there were few difficulties in obtaining permission when dealing with either the power company or a locally owned or managed telephone company. The power companies especially encouraged any business they thought would increase the consumption of electricity. Cable's status as a good customer and substantial consumer of power also helped keep rates for power company pole rental relatively stable for many years. The problems arose when CATV operators had to deal with Bell Telephone Co.

At the time, Bell's treatment of CATV varied from benign neglect to open hostility. Depending on the area and the individuals involved, CATV was seen as anything from a nuisance to an open threat. There was, in the first instance, a basic proprietary instinct on the part of the Bells that made working with them difficult.

Some regional officials viewed community antennas in much the same way as did others, a temporary aberration brought on by the freeze. Once "real" television came to town, the wires would come down; in the meantime, the telephone company might be able to make a little money on the side. In other towns, the local Bell took an even more aggressive posture, offering construction services for CATV and leaseback arrangements for potential operators.

In the Pacific Northwest, Parsons had a very difficult time gaining access to Bell Co. poles, as he sought contracts to build systems in towns near Astoria. "Ma Bell absolutely refused to let us on the poles," he said.[43] The situation changed when city officials from Kelso and Longview, Washington, asked him to build systems in their adjacent towns. He explained the pole access problem and the authorities told him that the municipalities, in fact, owned the poles. When the telephone company protested the CATV hookup, the town threatened to revoke the telephone company's pole access. A short time after that, telephone company officials arranged a region-wide meeting with Parsons, themselves, and the power company, Pacific Power and Light. They negotiated a blanket agreement for the leasing of pole space by Parsons in the Northwest, renting space for a $1.50 a pole.

In the East, Shapp worked to encourage Bell officials in their belief that CATV would be short-lived, and that the telephone company could buy some good will with the public by supporting everyone's access to television. He also tried to convince them that individual home TV aerials were a threat to telephone company lines because they could be blown down in heavy storms. In fact, according to Shapp, a severe winter storm did knock down antennas and snarl telephone lines across the East Coast after one of his conversations with AT&T officials. Whether the disruption in service had any impact on Bell Co. officials is unknown, but most CATV operators did eventually come to an accord with the telephone and power companies, albeit at a price. It was not uncommon, for example, to have telcos and power companies demand cash bonds of $25,000 or $50,000 to make sure that if the CATV business failed, the coax would be stripped from their poles. "And they wanted cash," explained Shapp, "they wouldn't take collateral."[44] Like many others, AT&T was not entirely confident that the CATV business would be around for very long.

The typical pole attachment contract involved a $1.00 to $1.50 per-pole, per-year lease. An average system required hundreds of poles, so this amounted to a substantial operating cost. In addition to the access charges, there were, and continue to be, costs associated with preparing and maintaining the space on the poles. Often, the wires on a telephone or power pole would have to be rearranged, to provide sufficient safety space between high-voltage lines and the new cable lines. The cable company had to bear the cost of making the poles ready for the new lines, which could, in some cases, even require replacing the pole. Rearrangement costs were about $9 a pole; and the telephone company could charge as much as $150 to set a new pole.[45] Either way, it could be an expensive and time-consuming process.

There is little evidence that AT&T or the smaller telephone providers actively pursued full entry into CATV in the early 1950s, although the potential of AT&T getting into the business was enough to make many operators hesitate about the long-term potential of the business. But as it became clear that the community antenna business had legs, a variety of parties, including some in the investment community, began speculating about potential AT&T interest in CATV. And by the mid-1950s, the telephone company appeared to thinking along similar lines. It could easily capitalize on its tremendous technical and marketing muscle and on its established physical conduit into millions of American homes. The small entrepreneurs who built the CATV business, according to this scenario, would be easily shunted aside. More than one early operator thought the business would last in their town only until the local Bell company decided to buy the system, or simply build one of their own. AT&T could very well have moved into the CATV business in the late 1950s and established a monopoly position there in the same way it had built a monopoly in national telephone service. That threat was blocked, however, by action from the government. For years, the U.S. Justice Department had been investigating AT&T's alleged monopolization of telecommunications equipment and services. One of the possible outcomes of the government's case against AT&T was the breakup of the company into smaller units, a possibility Bell sought desperately to avoid. In 1956, AT&T and the U.S. Justice Department reached a negotiated settlement. It left the company intact, but, under terms of the agreement, AT&T promised to stay out of any business other than common carrier communications.[46] This included the provision of community antenna service.

The agreement not only saved small operators from the threat of an AT&T takeover, but it also assured potential investors that the industry had long-term promise. It allowed venture capital to flow somewhat more easily into the industry.

But while the 1956 Consent Decree prevented AT&T from engaging directly in the provision of cable service, it left several loopholes for the Bell system and other local telephone companies to exploit. They were, for example, permitted to build community antenna systems, then lease them back to CATV operators who controlled the content. Within a year or two of the antitrust settlement, AT&T was vigorously developing these lease-back arrangements. The terms so heavily favored the economic interests of the telephone company, however, that, according to CATV operators, they did not offer a viable business opportunity. Under some lease-back plans, construction costs for the local operator would have doubled and the leasing charges would have exceeded normal operating costs. "The economics of this was just out of the world. Nobody could make any money," Shapp explained to Congress in 1959.[47] Faced with resistance from CATV operators, AT&T pressured operators to accept a lease-back scheme, either by denying pole access under any other conditions or by raising pole rates to unacceptable levels, boosting them from $1.50 to as much as $5.00 per pole.[48] With such high fees for pole attachment, some operators were forced into the telephone lease-back option. In late September of 1958, the industry took legal action in New York, appealing both to the State Public Service Commission (PSC) and in court about AT&T's actions. The PSC ordered AT&T to rent space to CATV operators, but rates often remained very high.[49]

Marketing & Sales

For the pioneers, actually selling the service, signing up new subscribers, was less difficult, but not without its own challenges. As already noted, consumer demand in some cases was overwhelming. "The moment you opened your door," according to Malarkey, "you were overwhelmed with people who wanted to hook up to the cable You didn't have to have any marketing expertise or special talents to promote a cable system, you just began."[50] Most of the business-oriented operators nonetheless set up public demonstrations, offering free service to the home or shop that would put a TV set in a window or public place. Yee set up demonstration sets in Easton barbershops. Gardner had a set installed in the local fire station because it was the first building his line reached at the edge of town. Most of the operators would try to place free service in the radio and appliance stores around town. In some cases, the stores would offer a year's free service or free CATV hookup with the purchase of a new set (although this arrangement raised questions about anticompetitive business practices and was later discouraged by the National Cable Television Association [NCTA]). Some cable operators reportedly provided a small fee to the TV dealer whenever the dealer signed a new customer up for cable, and some operators offered new customers a discount if they could convince their neighbors to sign up as well.

Beyond the demonstrations and, perhaps a one-shot newspaper announcement that community television was now available, word of mouth was the primary publicity vehicle; extensive marketing was unnecessary. "You didn't have to advertise," said Tarlton, "You had to keep your door locked because the people were clamoring for the service. They wanted cable service."[51] Of course, anytime a new product or service is introduced into a community, a certain segment of the population will have its fears and worries. Cable pioneer Edward Allen recalls complaints from townspeople that the cable wires were "sucking all the television signals out of the air" and weakening their normal home reception.[52]

Operators sometimes had to find ways to help customers pay the $100-plus installation charge, a steep fee at the time. Some families were willing and able to save the money or pull it out of cash reserves, but they were the minority. It was not so much a question of willingness to pay as it was ability. The CATV operator, therefore, occasionally would help by underwriting a customer loan at a local bank, guaranteeing the amount of the installation fee and helping set up a payment plan. The default rate was low, according to Gardner. Also common was a simple installment plan. Yee reported having initial difficulty selling the system because of the high installation charge, until he hired a salesman who promised potential customers they could pay it off over several months. According to Yee, the salesman returned from his first day on the job with three single-spaced pages of new subscribers.[53]

The problem of high demand and limited ability to pay also led very early in the industry's history to problems of signal theft. Customers unable or unwilling to subscribe found creative methods of acquiring the signal. Often the solution was as simple as placing an antenna wire near the coaxial cable. The earliest twin-lead and open-wire systems radiated enough signal from their trunk lines that any nearby antenna could pick up the escaping signal. Even after the introduction of coax, some systems leaked so much radiation, especially at amplifier and junction boxes, that the practice was easy and common. Where systems were more tightly engineered, customers created their own leaks. Pioneer Ray V. Schneider was running a Williamsport, Pennsylvania system in the mid-1950s when he discovered people sticking pins in the coaxial cable. The pins would siphon off the passing signal, radiating it out into the neighborhood and creating a mini-broadcast station that people could receive on their normal antenna. Explained Schneider, "If the pin did not short out against the center conductor and the outer conductor, they had no problem. But once they'd short out, that'll knock everybody's system out. We found a lot of that."[54]

Programming

Programming for community antenna television was, for the most part, no more complicated or extensive than what the name implied. The vast majority of community aerial operators were content to find a good clear signal or two and get it to subscribers' homes without serious mishap. A few, however, experimented with the creation of their own programming. Usually these were the systems that had been in operation the longest and had largely saturated their market. Malarkey, in Pottsville, was offering origination service, as were Service Electric in Mahanoy City and the system in Oil City. The technology was known at the time as a "closed circuit channel," following the concept of closed circuit television available in other institutional settings.

Efforts beyond such home-produced fare were quite limited, however. A system in Cumberland, Maryland, attempted to program kinescoped broadcast network programming on its closed circuit channel. The idea was to provide a channel of network reruns in an early form of time shifting for subscribers. The operator also hoped to make some money selling advertising on the channel. The networks, however, were unwilling to provide material and advertisers were scarce.[55] The NBC television network had briefly considered selling programming to community antenna operators for origination purposes, but only when the system would not directly compete with a network broadcaster. Even that mild form of program access was abandoned within a few years, however, and, by 1958, none of the national networks, including NBC, were willing to provide programming to CATV operators.

Operators nonetheless began to use excess channel capacity for a variety of subscriber services. By 1957, Malarkey was offering twenty FM radio stations to Potttsville subscribers; others were doing local news, weather, and sports. National equipment and programming companies offered educational material. Musak, along with similar audio services, offered their content, and automated slide chains were popular for community announcements. By the late 1950s, systems were using weather scan devices, in which a camera was pointed at, or in a fancy set-up panned across, a set of time, temperature, and other weather dials.

While such live or automated origination was not widespread, it was well known and the practice made many in the industry nervous. As early as the 1953 convention of the NCTA (see below), closed circuit programming was hotly debated, especially as it might evolve into a commercial medium, a channel that could be used to sell local advertising. The trade association was wary of such use because it drew the attention and ire of local broadcasters, and the fear was that it could lead to eventual regulation under the jurisdiction of state public utilities boards. By the mid-1950s, the organization was actively discouraging local origination. Where an operator did provide origination service, the NCTA counseled the system to do so within the legal isolation of a separate company. CATV programming, however, would not soon go away and it would prove to be a source of considerable discussion and dispute for the industry in the decades to come.

System Maintenance

Beyond the challenges of obtaining equipment and permissions, building the system, and selling the product, perhaps the single most difficult job of all was keeping it all running. Especially in the earliest years, systems were a makeshift assemblage of leaky cables, untested amplifiers, and uncertain power supplies. They went dark on a semiregular basis and operators spent a significant portion of their days and nights in repairs and maintenance. Throughout its history, customer service has been the "Achilles' heel" of cable television. As much as anything else, the simple failure to answer the telephone, show up on time for an installation, or handle a customer complaint with grace and efficiency have been the bane of the local cable operator. Ineptitude in this area has driven customers to madness, local franchising authorities to distraction, and federal lawmakers to punitive legislation. Since its inception, NCTA has struggled with the problem of improving customer relations. At the very first national meeting of the NCTA in June of 1952, then-president Malarkey noted in an address to the assembly that "certain community television companies could do much to improve their handling of consumer complaints."[56] He warned that the industry would only be as successful as the service it provided its customers and that some unsavory business practices, such as tying the price of service to the purchase of a TV set, could eventually lead to government regulation. He urged operators to maintain clean offices, regular business hours, business-like billing procedures, and emergency standby arrangements.[57]

This was a large bill of fare for the 1950s small town operator, however. The local reality was that when the system went down with only one person in the office, that person had to go out and discover why. Explained Gardner:

> Back then, when the phone started ringing we didn't even bother trying to answer it because you didn't know what was wrong in the first place and you had to spend your time out there trying to fix it. It was difficult to get anyone to sit and answer a phone and say "yes, we know there is a problem." Because, everyone was yelling at us so loud it was difficult to get someone to answer the phone. The equipment was not always good enough to be called commercial even though it might have been stamped commercial.[58]

Operators were constantly battling the elements and their own equipment. One of the primary reasons CATV abandoned the so-called twin-lead or ribbon wire in favor of coaxial cable at the start of the industry was the susceptibility of twin lead to the elements. It was notorious for "rain fade," the propensity of the signal to disappear in rainstorms. Coax resisted rain fade, but had its own problems. Its ability to carry a signal was influenced by temperature. During the day, as the temperature increases, so does the resistance in the cable, creating weaker or snowy pictures. At night, the temperature and cable attenuation decrease, restoring the picture, but also creating a situation in which amplifiers can become overloaded, distorting the picture. Left alone, the picture received by the customer would grow stronger or weaker throughout the day. Cable operators, therefore, had to "tune" the system every day, turning up the power in the morning and turning it down at night, a process that involved every amplifier in the system. Some operators, such as Walson, designed their systems with amplifiers mounted low on the poles to make access and system tuning easier. Peter Walson, John's brother, was said to drive through town twice a day, stopping at each pole, and tuning the amplifiers from his car window.[59] Eventually, Jerrold and others developed gain control devices that would automatically adjust the system to temperature shifts, and these were largely implemented by the early 1950s.

The weather could also cause problems with the receiving antenna. Winter storms could coat the antenna arrays with snow and ice, weakening and even killing the signal. Operators would have to drive out to antenna sites in the midnight hours to knock the antennas free of ice. In one case, an operator, unable to scale an ice-encrusted antenna tower, used a shotgun to try to blast the ice clear.

We started out with number 8 buckshot. We got some of the ice off. A little bit of sound came back, but not enough. This was the middle of the night and we were doing anything possible to get this going. Sure enough, Wally Masada said, "Instead of a number 9 shot, let's try a number 6." So he did. We got more ice off. I said, "Wally that's enough. At least we can get back tonight. He said, "I'll bet if I gave it two more hits, it would clean it off. Sure enough, we took buckshot and away goes the antenna and he shot the whole thing down.[60]

The early tube amplifiers were an even greater maintenance problem. They had to operate outside in a protective enclosure, subject to constant vibration from the pole and the problems of varying power line voltage and temperature fluctuations and moisture from rain and snow. The early wooden enclosures offered little protection from the elements and one of the most typical problems was a burned out tube. In the earliest stripband amplifiers, each channel had its own amplifier and each amplifier had its own set of tubes. A typical pole-mounted housing might contain three strip band amps, each about an inch wide, six inches long, and five inches high. (By way of illustration, a 1990s, 55-channel cable system using this technology would need amplifier housings about four or five feet wide.) A full 1953 system would contain hundreds of tubes. When one or two tubes burned out, that particular channel went black, and tubes were always burning out. "The vacuum tubes—we'd buy them by the gross," according to pioneer Roy Bliss.[61] More than one early cable technician can recall constantly burned fingers from replacing the hot tubes.

One of the important incremental innovations of the time was the development of the "distributed gain amplifier," a system for distributing the TV signal over several tubes in an amplifier so if one tube failed the others could still carry a signal. The design also significantly increased amplifier bandwidth. The earliest distributed amplifiers were designed and manufactured by emerging CATV equipment companies such as Spencer Kennedy Laboratories (SKL), which had its roots as a supplier of advanced equipment for military use. Holding the license for a key British patent in distributed gain technology, SKL would later become a pioneer in the development of broadband equipment.[62]

Important progress was also made by a firm called Blonder-Tongue laboratories, which introduced low-cost "split-band" amplifiers in 1954. The split-band amplifier reduced the number of amplifier circuits to two, one dedicated to channels in the low VHF band and a second used for channels in the high VHF band. Five-channel broadband amplifiers, standard industry equipment by the mid-1950s, reduced the size and cost of the equipment while increasing its channel capacity. Community Engineering Corp., later C-COR, of State College, Pennsylvania,[63] also introduced some of the first amplifiers that could be mounted on the wires, rather than attached to the poles, around 1953 and 1954.

Amplifiers, of course, need power. The earliest systems relied on the subscriber to provide electricity. Alternate current (AC) lines ran from the amplifier just outside the home into an attic, a basement, even a living room, anywhere there was an outlet. In some instances, the amplifier itself was placed inside the home. While the subscribers providing the electricity received free or discounted CATV service, the arrangement left much to be desired for the operator. Far too often, a customer would accidentally kick a plug out of a wall socket, or a fuse would blow in the house. Sometimes a youngster playing where he or she should not have been playing would, literally, pull the plug on the system. Before they understood the nature of the system, some customers, when they turned in for the night, routinely pulled all the plugs in the house. In all these cases, because the systems ran in series, with each amplifier dependent upon the previous one for reception of a signal, the community antenna would be effectively turned off from the point of the power disruption. As calls from irritated customers came into the cable operator, he would have to backtrack through the system to locate the source of the outage. Operators soon abandoned their dependency on subscribers for electricity and tapped directly into the power grid on or near the poles, using a metered, dedicated system to feed the amplifiers. Even this somewhat crude technique was eventually abandoned with the introduction of cable-powered amplifiers by companies such as Community Engineering in 1953.

The improvements in reliability, picture quality, and bandwidth progressed, quickly from some perspectives, slowly from others; in all cases, following an evolutionary trajectory based on solving the problems at hand. For most subscribers, dependent upon CATV for whatever television they could get, tolerance and patience was the norm. While many complained about the outages or the rates, many more were simply happy to have television, no matter how snowy or intermittent the picture. Explained one pioneer: "We had our problems and the customers would bear with us. If we could maintain a signal of any sort they were happy. They would rather have a picture that wasn't right than no picture at all. The customer standards that we try to adhere to today, there's no way they would have make any sense back then." [64]

CATV was a small town business. Its problems were centered primarily on keeping the system operating, extending and improving the local service, and keeping the revenue coming in. It would not, however, stay parochial for long. A long line of national and international interests was forming on the industry's front porch. It was composed of a wide range of people with ideas to promote, money to invest, regulations to impose, and competition to offer. By the mid-1950s, they would be bringing the industry, for better or worse, into the national arena. Unfortunately, from the perspective of CATV operators, the first caller to come knocking on the door was the Internal Revenue Service (IRS).

Community TV, the IRS, and the Birth of the NCTA

Bob Tarlton's Panther Valley system may have been one of the first systems expressly designed for business purposes, but the drawback to that accomplishment was to be one of the first community television businesses to draw the attention of the taxing powers of the state. Tarlton's first encounter was with the township of Summit Hill. While Tarlton's system may have excited the television audiences of Lansford, the local tavern owners in Summit Hill were very unhappy. For some time they had had the monopoly on local television, and Tarlton's system meant a threat to the steady TV-driven patronage they enjoyed. According to Tarlton, the influence of these tavern owners led to Summit Hill's decision to impose a 15 percent tax on his system's gross receipts. Panther Valley challenged the tax in court and eventually won. In the meantime, however, a larger and more ominous tax burden was threatening, one that would have far-reaching consequences for the industry and its future.

Bob Tarlton said he called Marty Malarkey with the bad news. Malarkey remembered it a little differently, but they agreed completely that they shared a very real problem. It was early in 1951 and Panther Valley had been in business for only a few months when an IRS agent from Wilkes-Barre walked into Tarlton's office and told him that he was subject to an 8 percent federal excise tax. According to the IRS, Tarlton was operating a taxable, leased-wire service. Tarlton, after conferring with his lawyer, declined the IRS invitation to contribute, arguing that his operation did not constitute such a service. By the end of the summer, the IRS was threatening legal action, including a lien on the business, and Tarlton was getting nervous. A trip to the IRS offices in Washington, D.C., failed to dislodge the agency from its position and Tarlton reluctantly agreed to pay the tax if the IRS would halt its "saber rattling." On his return from the Capitol, however, he said he immediately placed a phone call to Malarkey. "Marty," said Tarlton, "we've got a problem."[65]

Under section 3465(a)(2)(b) of the IRS Code, all "wire and equipment service" used in commercial communication was subject to the levy. Under the provisions of the Code, this included businesses offering services such as stock quotations, power or watering metering, and burglar alarms. It also included, according to the IRS, income from community television systems. The tax was part of a more inclusive section that imposed a fee on various telephone, telegraph, and cable or radio communications. At a time when cable operators were living hand-to-mouth and using next week's subscriber fees to pay for last week's purchase of cable, an 8 percent revenue handicap was a serious concern. It would amount, according to one estimate at the time, to about $3.5 million annually in revenues lost to taxes.

Malarkey and Tarlton went to the telephones and began calling other operators. Soon, a small group of area CATV operators were assembled in the dining room of the Necho Allen Hotel in Pottsville, Pennsylvania, to consider their dilemma. It was Tuesday, September 18, 1951, when representatives from eight systems met: Bill Scott and George Bright came down from Lansford; Michael Sheridan was there from Wilkes-Barre; Ken Chapman drove down from Honesdale in the northeastern corner of the state; Hubert Strunk came in from Ashland; and others came in from Harrisburg, Palmerton, and Mahanoy City. The focus was on Lansford and on Chapman's system in Honesdale, which is where the IRS had first knocked on the door. Everyone knew, however, that left unchecked, the excise tax soon would be levied against every operator. The group concluded that they should organize in order to represent their mutual interests in this, as well as other, somewhat less pressing, issues, including relations with the local telephone companies. They agreed to form several committees, one to draw up a constitution and by-laws, one to investigate the tax problem, and a third to contact community TV operators around the country to solicit interest in the organization. Malarkey was elected temporary chairman of the tentatively named "National Community Television Council." All parties agreed to contribute $100 to a startup fund and to meet again in a week's time to report their progress.[66]

On September 26 they gathered again at the Necho Allen; the second meeting of what was to eventually become the National Cable Television Association, the lead trade association of the cable industry.[67] By-laws for the new corporation were adopted and Pottsville selected as the site for its headquarters. Malarkey appointed a board of directors to oversee operations until the first organizational meeting of the corporation. The board included Malarkey, George Bright, Clyde Davis of Wilkes-Barre, Claude Reinhard of Palmerton, and Hubert Strunk of Ashland.

They then turned to the immediate problem of the excise tax. Sheridan suggested that it was a legal problem more than an accounting one, and Edward Mallon, of William E. Howe & Co. (one of the accounting firms working with the operators) recommended the retention of a tax lawyer, Thomas Egan.

The group met again on October 10, 1951, and on January 3, 1952. During these meetings they worked out details of the constitution, by-laws, and dues and they decided to change the organization's name, from the National Community Television Council to the National Community Television Association, Inc. The first formal meeting of the new NCTA was held at the Necho Allen on January 16, 1952. The board was elected, Malarkey named president, and Mallon reviewed the tax problem.[68]

It took the NCTA board through the first part of 1952 to decide to retain Howe & Co. and authorize proceedings. In October, Howe reported that during the process of searching for a willing test case, the IRS had solved the problem for them by sending a bill to Pottsville Trans-Video. Malarkey's company had been paying taxes on the monthly service fees, but not on the installation charges. The IRS requested they do so and, at Howe's counsel, Trans-Video declined, stating the money constituted a contribution-in-aid to construction. The IRS agent in charge sent the material to Washington, D.C., and there it disappeared into the black hole of bureaucracy for several months. The agency eventually ruled against Malarkey, but his would not become the industry's test case.

While the IRS was still pondering the question in mid-1953, the NCTA, by September of that year, was moving on several other tracks. It determined to attempt to seek a legislative solution to the problem and obtain an amendment to the Code exempting CATV operations. This effort did not bear fruit. It did, however, lead the association to the conclusion that they would need legal counsel in Washington, D.C., and, in October 1953, they voted to retain the Washington law firm of Welch, Mott and Morgan. Welch, Mott and Morgan took over the tax case and also offered the association a room in its Washington law offices for a national headquarters, an offer quickly accepted. The selection of the firm was no accident. Former FCC staff attorney E. Stratford Smith had been assigned, while still at the Commission, to look into the new community antenna service as soon as it came to the FCC's attention in 1949. Smith was now with Welch, Mott and Morgan, but had been following the progress of the community antenna group and went to its first New York City meeting in June 1953 with the specific intent of lobbying for the association's business. It was a trip that turned out to be quite successful for the firm. By the end of 1954, the NCTA had hired Smith to work, in addition to his duties as counsel, as the association's new executive secretary.

One of the firm's first tasks was to deal with the IRS. The taxation problem eventually split into two separate but related issues. There was the large problem of the applicability of the excise tax itself, which covered all revenues generated by the community antenna company, and there was a narrow issue of the tax status of the revenue generated by the installation charge (as distinguished from the monthly service fee). Eventually, these were broken into separate court cases, with cable winning one legal battle and losing the other.

The installation charges that early operators used to bootstrap their systems into existence constituted the largest part of the operators' initial revenues. As early as 1950, the industry was considering a method of sheltering this particular revenue stream from the IRS. According to Tarlton, Panther Valley TV obtained the services of the Ernst & Ernst accounting firm because it was doing the books for partner George Bright's department store. The firm suggested that Panther Valley declare the installation fee to be a "contribution-in-aid of construction" and, therefore, immune from tax liability. The contribution-in-aid of construction concept dated back to the 1920s.[69] Rural homeowners beyond the reach of a municipality's power and light grid could have a line extended to their house if they put up some of the capital cost for its construction. This was considered to be a gift or a contribution-in-aid of the utility company's construction cost rather than revenue subject to normal taxation. There were potential dangers for CATV in claiming this protection. By invoking the utility company accounting scheme, the industry was comparing itself, in a legal sense, to other public utilities. This, in turn, opened the door for an argument that CATV companies could be more broadly regulated like public utilities and subject to similar federal, state, and local rate regulation and service obligations. In addition, if the physical structure—the wires and amplifiers—of a community antenna system was, in essence, a gift from the subscribers, the business could not benefit from the normal tax depreciation schedule associated with a firm's capital costs. It was, in several ways, lucky, therefore, that the courts found in favor of the government on the issue of contributions-in-aid of construction. Both the Tax Court[70] and the Federal Appellate Court[71] in *Teleservice Company of Wyoming Valley v. Commissioner of Internal Revenue* held that the money collected for CATV installation constituted gross revenue subject to normal taxation. The initial payments, said the Court, were simply the price of the service. No general community benefit accrued from the subscriber-by-subscriber construction of the system (an element in a contribution finding) because, said the Court "only he who bought a ticket could 'go for the ride.'"[72] An interesting side note was the industry's plea to the Court for sensitivity to the "pioneer and hazardous nature of the enterprise." The Court was unmoved, however, responding that such pioneering spirit was "of no critical value" in the case.[73]

For many systems, the loss of the *Teleservice* case meant payment of a sizable back tax bill. However, the case was not finally decided until 1958, which gave operators years of tax-free revenue. Many pioneers later argued that this grace period was critical in the initial development of the industry by freeing up the money necessary to buy the amplifiers and string the wire. By the time the tax bill finally came due, most of the systems were sufficiently established to permit payment without having to close up shop. They may not have been happy about it, but they could afford to pay it.

In the meantime, the broader tax case was proceeding along a separate path. The NCTA had found a system for its test case in Meadville, Pennsylvania, where George Barco ran the Meadville Master Antenna company. Barco and his daughter, Yolanda, were also attorneys and determined to take up the IRS challenge. They found a willing subscriber, Gus Pahoulis, and together they sued for recovery of the $70.40 Pahoulis had paid in excise taxes on the service in 1953 and 1954, and for a declaration that the tax itself was illegal. Meanwhile, a similar case was proceeding in the small town of

Mullens, West Virginia, where DeForest Lilly and Robert Jones, owners of Mullens TV Cable Service, were also contesting the IRS bill. The district court decision in the *Lilly* case was issued first, upholding the government's position, and the NCTA's law firm asked Mullens TV if they would care to have the NCTA assist them in the appeal. Lilly agreed and while the two cases were never consolidated in the courts, Welch Mott worked with Mullens TV's attorney on both cases, using the same logic and with the same results.

In the district courts in West Virginia and Pennsylvania, the court held for the IRS, finding that the community television operation was a communications "service" within the meaning of the code. This not only left the industry subject to the excise tax, it additionally put it into the same class of utilities as telephone and telegraph companies and thereby threatened to open the door to additional public utility type regulation. The NCTA needed a strategy that would attack both the narrow issue of the tax and the broader issue of cable's regulatory standing.

NCTA legal counsel Smith developed an ingenious and far-reaching strategy. Under his guidance, the cable operators chose to argue that community television was not subject to the tax because it was not, in fact, a "service" within the meaning of the clause. They positioned community aerial television as a completely passive technology. It was not, they argued, a process or a service or an activity. Rather, it was merely a piece of equipment they provided to their subscribers. They postured CATV as a passive antenna, a simple but necessary extension of the customer's receiver. Operators were not, in any legal sense, "carrying" a signal or engaging in any communicative activity. Smith went so far as to suggest the Association change its name from National Community Television Association to National Community Antenna Television Association, adding the word "antenna" to underscore the legal defense.

The critical, even historic, importance of this argument cannot be overstated. It was as powerful as it was simple and became the spine of the cable industry's legal strategy for the next decade. It was to have a major impact on judicial proceedings in the 1960s. It could be persuasively argued that without the legal concept of the passive antenna, cable television would have developed over the next several decades in a drastically different form, if it had developed at all. It was, in addition, one of the first and most important examples of the industry creating its own social definition. The passive antenna strategy was both an inventive and ultimately successful political strategy and a bold example of the social construction of technology. It furthermore illustrated the clash of social definitions, as the government sought to impose an altogether different kind of social construction on CATV.

No small part of the practical power of the industry's argument, therefore, was its success in the courts. The government had sought to define community TV as a service involved in the gathering, processing, and distributing of TV signals. It reasonably argued that CATV operators modified TV signals, strengthened them, and channeled programming through a complex system of wires and amplifiers. It further noted that community TV operators did not sell equipment, per se, to subscribers. In no way did it lease or furnish receiving equipment to the home. CATV, the government held, was in the business of providing TV signals or, in today's terms, information and entertainment. Its fungible good was programming. The industry countered, claiming to provide only equipment, an extension of the viewer's reception equipment. Overturning the decisions in their respective district court jurisdictions, the appeals courts in the two cases found the industry's self-definition more compelling.

In *Lilly*, the U.S. Court of Appeals for the Fourth Circuit concluded that stock tickers, fire alarms, and burglary alarms provide a communications service similar to those of a telephone or telegraph message, but that CATV "merely furnishes an attachment to a television receiving set."[74] The Court repeatedly stated that CATV was not a communications facility or service: "We think it clear that this community antenna service was a mere adjunct of the television receiving sets with which it was connected and was in no sense a communication service or facility such as it was the purpose of the statute to tax."[75]

The decision in *Lilly* was handed down in November of 1956 and, a few months later, in March of 1957, the Third Circuit Court reached a similar conclusion in *Pahoulis*.[76] Citing *Lilly*, the court stated, "without some reasonable limitation of the underlined phrase 'any wire and equipment service'," a host of other electronics devices including radios, record players and public address systems "would be swept under the levy."[77] CATV, said the Court, is "an aid in reception only." The Court continued: "The wire and equipment have nothing to do with the origin of the electronic signal which occurs at the television station many miles away and wholly separate from the service supplied by the community television antenna company."[78]

In fending off the IRS, the industry also created a legal and, in many ways, social definition for itself. The social meaning implicit in that legal position would subsequently color its relationship to broadcasting, the government, and the public for another ten years, when it would give way to a new and even larger social construction of cable television. The *Lilly* and *Pahoulis* cases, however, were the end of the government's attempt to impose the excise tax. Beginning in 1957, cable subscribers were given an opportunity to file for refunds from the pool of money that the industry had collected for the IRS over the intervening years, estimated at around $16 million.[79]

State's Rights

The IRS was not the only governmental body interested in CATV. In addition to the local franchising agencies, state governments and state public utilities commissions were beginning to recognize the existence of community an-tenna television. Through most of cable's early history, the states exercised minimal regulatory control, although they exhibited enough curiosity to keep CATV lawyers worried. CATV was seen by the states, for the most part, as either a local problem involving the stringing of wire and the regulation of local business or, alternatively, as a federal issue better handled by the FCC. The earliest pioneers, in addition to seeking FCC approval for their systems, sometimes contacted state authorities. In Rice Lake, Wisconsin, a broad-cast executive looking to set up the town's first community TV system in 1951 went to the state public utilities commission for approval. In what was probably the first such decision of its kind, the state commission turned him down, ruling that it had no authority in the field of community antennas; the utilities board recommend he go talk to the FCC.[80]

Despite this general disinterest in community television, the NCTA was worrying as early as 1952 about the prospect of state regulation, and not without cause. While most states avoided control, a few pursued it. While the Pennsylvania and Utah public utilities commissions followed Wisconsin's lead, declining authority, but Maine had assumed jurisdiction over commu-nity antenna installations and California was considering it. Through the mid-1950s, attorneys general in several states, including Arizona,[81] New Mexico,[82] and Washington,[83] ruled that cable was not a public utility sub-ject to state control. Different conclusions were reached, however, in Cali-fornia and Wyoming. In California, the utilities commission found that the attachment of the cable to utility poles provided them the lever to regulate the business and the Public Utility Commission (PUC) looked to control rates and services. The NCTA spent several years and a great deal of money fighting and eventually defeating repeated attempts by the California PUC to regulate cable.[84]

A similar lengthy and expensive legal battle was waged in Wyoming where the state Public Service Commission (PSC) in 1954 asserted jurisdic-tion over CATV, issuing a certificate of convenience and necessity to Cokeville Radio & Electric Co. to build a system in Cokeville.[85] When the PSC subse-quently attempted to extend its control to all the community antenna systems it could find in the state, about 127, some of them filed suit.[86] The courts eventually ruled against the Wyoming PSC, holding CATV regulation to be outside its preserve.[87] Either explicit or implicit in all these cases was the question of whether CATV was, by definition, a public utility. As with the tax situation, the legal and social definition of CATV was in play, in state regulatory arenas and in state legislatures.

Beginning in the mid-1950s, legislative proposals to declare CATV a pub-lic utility, while not widespread, were sufficiently frequent to engage NCTA's attention. Pennsylvania considered CATV regulation in 1955. Arizona, Mon-tana, and Washington did the same in 1957; Oregon in 1961 and Arkansas in 1963. In most cases, the efforts were defeated, however. In Montana, for example, the governor vetoed a bill in the late 1950s that would have given the state control over CATV.

The issue of cable's status as a public utility would become more pronounced as its social definition and perceived role in society changed in the 1960s, and is discussed in greater detail later in the text. As far as 1950s legal restraints were concerned, the decade would prove to be good times for community antenna operators. Little was required at the local level and, as will be seen in Chapter 4, the industry had a philosophical friend at the FCC. It was not a picture that would remain stable into the 1960s; for its first decade, CATV could worry about other issues.

Pay-TV Comes to Bartlesville

Community antenna television, by anyone's definition, was television for a fee. The subscribers paid the CATV operator a monthly sum and received one or more channels of programming that they would not have otherwise been able to obtain. But the label "pay-TV" has always meant something more, it has meant charging for commercial-free programming, typically first-run movies and sporting events not available on broadcast network television. Fees for commercial-free programming have been assessed either on a per-program basis or as a flat monthly charge. Pay-TV has also provided one of the most contentious issues and hard-fought battlegrounds in the evolution of cable television.

The idea of subscriber or pay-TV dates back at least to AT&T's 1920s vision of wired distribution of programming to theaters. In the 1930s, Zenith Corp. began developing what it called the "Phonevision" pay system. The system was designed to broadcast a scrambled TV signal. Subscribers would call their local telephone company which, in turn, would send a signal through the telephone line to the viewer's TV set, triggering a device to descramble the broadcast picture. Customers could purchase three films a day for $1. Zenith actually tested the idea in the 1950s in Chicago.

International Telemeter Corp. (ITC), a division of Paramount Pictures, experimented with a coin box service in the 1950s. For Paramount, as well as for other Hollywood film studios, pay-TV was an attempt to harness the emerging technology of television to their own business plan, something of a fitting irony insofar as the new medium of television was siphoning movie goers away from the theaters. Hollywood would, over the next several decades, carry on a pronounced love-hate relationship with pay-TV, first in its broadcast form and later in cable.

A third company, Skiatron, was also active in promoting a pay-TV system in the 1950s and 1960s. Skiatron used a set-top decoder with an IMB punch card system to authorize and track viewing.

All three companies, Zenith, ITC, and Skiatron, had to adapt or invent new equipment and processes, illustrating again the broader principle of diversifying technological alternatives. Additionally to the point of the evolutionary metaphor, the promoters had to battle competing entrenched and emerging interests and technologies in an effort to secure a place for themselves in the market. Aligned against them were powerful broadcasters and theater owners who accurately saw the service as an economic threat. Some of the contests were waged in the marketplace, others in the regulatory arena.

The FCC approved limited experimental pay-broadcasting for Skiatron and Zenith in 1950 and 1951, but after that progress stalled in the face of mounting opposition. The Commission held lengthy hearings on pay-TV throughout the1950s. It was one of the agencies most fiercely contested proceedings.[88] The request for comments in 1955, for example, drew more responses than anything the Commission had previously experienced; more than 20,000 individuals filed papers.[89] In 1959, after years of FCC and congressional review, the Commission finally approved service on an experimental basis.[90] In the interim, however, the inability to conduct pay-TV tests over the airwaves provided a market opening for CATV.

Pay-TV promoters saw CATV as a means of testing the service with a technology that offered greater security against home piracy of their programming and did not require government approval. As a result, the use of community antenna systems to deliver pay service started very early in the industry's history. In November of 1953, with only about 300 CATV systems in operation around the country, the first pay-TV CATV service went into operation in the well-heeled vacation community of Palm Springs, California. The desert resort community seemed like a perfect site to test pay-TV. An occasional home to Hollywood executives and movie stars like Bob Hope, Palm Springs was cut off from regular Los Angeles TV reception by a beautiful but TV-signal-opaque ring of mountains. The prospective subscribers were affluent to rich and had a professional interest in entertainment. In 1953, International Telemeter installed its pay-cable system in town. Normal community antenna service, which gave subscribers the major Los Angeles broadcast stations, cost about $5.40 a month. But, with a special scrambling device, installed for $21.75, customers could also get current films and sporting events for about $1 a program. The home-based descrambling unit was International Telemeter's coin box system. Customers paid for the programming by inserting nickels, dimes, and quarters into the slot. By early 1954, Paramount's Telemeter subscription system had signed up 200 pay-TV subscribers out of a total of 700 community antenna customers.[91] Motion picture companies supplying the films for the pay service soon came under pressure, however, from area theater owners and film distributors who were unhappy about the competition. Following a threatened lawsuit from a Palm Springs drive-in theater owner, the supply of films all but dried up, and Paramount found it could not satisfy customer expectations playing only Paramount films. In April of 1954, it shut down the service.[92]

Even more widely publicized, likely through the efforts of Milton Shapp, was the trial of pay-TV conducted on a small CATV system in Bartlesville, Oklahoma, in 1957 and 1958. Shapp had written about the potential for CATV as a vehicle for pay programming as early as February of 1953,[93] and told the FCC in 1955 that CATV was the best technical avenue for pay service. The Bartlesville project was a joint venture of Shapp and Henry Griffing's Video Independent Theaters. As with Palm Springs, Bartlesville was an affluent community, the oil-rich home base of corporate giant Phillips Petroleum. Griffing, owner of a chain of regular film houses, saw theater attendance declining with the rising popularity of television and decided that if he couldn't get people to go to the movies, he would try to bring the movies to them. Unlike the Palm Springs experiment, the system was designed and built specifically to deliver pay programming, which was described by Griffing as an extended theater. Customers would still get local off-air TV service using a standard antenna. The physical plant was built by Griffing and Jerrold, but formally owned and maintained by Southwestern Bell Telephone Co. in a lease-back arrangement. Where Palm Springs charged on a per-program basis, Griffing thought a monthly fee would be more acceptable to subscribers. So, for a flat rate of $9.50 a month, customers received two channels, featuring thirteen first-run films each month and the other used for frequent reruns of the movies.

The service was inaugurated in September of 1957 with the film "Pajama Game." The CATV industry was guardedly enthusiastic about the project, sensing the potential bonanza of added monthly income. Operators worried, however, about the reaction of economically and politically powerful broadcasters and theaters owners and about the possibility that the telephone companies might get involved in the business. The Bell companies were looking at the possibility of getting into pay CATV and broadcasters and theater owners were working furiously on a variety of fronts to nip the pay-cable threat in the bud. Bartlesville, and similar pay-TV proposals, met with fierce opposition from the broadcast television and the motion picture theater industries. Film exhibitors, already financially damaged by the rise of commercial television after World War II, saw pay-TV as yet another competitive foe. Broadcasters had similar feelings. They argued that movies and sporting events which otherwise would be available for over-the-air broadcasting could be "siphoned" away by pay services willing to pay more for the programming. Only those viewers able to buy the premium service would be able to see that programming. Federal regulators worried that such a system would disadvantage less affluent viewers and harm over-the-air broadcasting.

In Bartlesville, broadcasters responded to Griffing's pay-movie scheme by airing scores of films in an effort to hold on to the movie-hungry audience. Customers also seemed unhappy with the monthly service charge, according to Griffin. He had hoped for 2,000 subscribers, but never signed up more than 800, even after lowering the subscription charge to $4.95 a month and adding carriage of the local TV stations to the system. Because he owned the town's theaters, no opposition arose from that quarter, but the costly lease-back contract with AT&T inflated overhead, and in the end, the scheme could not break even. The experiment in Bartlesville lasted about a year, closing in May of 1958.

Two other pay experiments of note were attempted in the 1960s. In 1962, Zenith ran a broadcast trial out of RKO General's WHCT-TV in Hartford, Connecticut. Subscription never attained levels sufficient to justify costs, however. The single-channel black and white service faced substantial over-the-air commercial competition and was handicapped by its failure to make the move to color television. The service was closed by 1969. In 1963, a company called Subscription Television (STV), Inc. founded by a former Hollywood film executive, Matthew Fox, and run by Pat Weaver, previously the president of NBC television, took a run at pay-TV in California. Weaver licensed the Skiatron system and used both wire and broadcast to distribute baseball games and movies to subscribers in San Francisco and Los Angeles. The effort was again opposed by theater owners, who launched a heavily funded publicity campaign to outlaw the service. Flying a "Save Free TV" political banner, the theater owners, supported by broadcasters, gained enough signatures to place a referendum on the 1964 California ballot prohibiting any fee for the delivery of a TV signal to the home (referendum supporters were careful about the language because closed-circuit television events such as prize fights were shown in theaters). The measure passed. The courts overturned the law a year later, but by then STV had filed for bankruptcy.[94]

Ultimately, the development of pay-TV would require a change in the technology and a change in the political climate. Massive political opposition, coupled with unwieldy technology and consumer apathy, killed its early development. FCC regulations adopted in 1970 would further restrict the potential of pay programming on cable. Only with the development of the satellite distribution system and the relaxation of regulatory controls would pay-TV start to play an important role in the business.

MSO Seed

Shapp and Griffing shared more than an interest in pay television. Both were prominent as two of the earliest holders of multiple CATV properties. The mom-and-pop business typified the fledgling community antenna business for a decade or more, but as Chapter 5 will detail, the forces of industrial concentration and amalgamation would begin working in cable in earnest in the 1960s. What would become known as MSOs, or multiple system operators, eventually would evolve into global entertainment and information giants, with hundreds of systems and millions of subscribers. Their lineage, however, can be traced back to the small systems that began operations in the 1950s, and the handful of businessmen who began collecting them, initially in small numbers. Following the notion of proto-systems in Chapter 2, these small chains of CATV systems could be considered proto-MSOs. The proto-MSOs were tiny in comparison to twenty-first century communications conglomerates such as Comcast and Time Warner. The neophyte MSOs of the 1950s boasted five to ten systems and only a few thousand subscribers, sometimes less, but from them sprang corporate giants. Along with Shapp and Griffing, the earliest MSOs included pioneers such as Bruce Merrill in Arizona; Charles A. Sammons of Texas; Bob Magness, founder of what would become the largest MSO in the United States; and Bill Daniels, who would be financial midwife and spiritual step-father to many of the top MSOs in the country.

An examination of the evolution of the economics and eventual social and political power of the cable industry is greatly aided by a closer look at the introduction of new capital into the CATV industry and the establishment of these first MSOs.

Jerrold Electronics

The industry's first, and for many years leading, multiple system operator was also the chief equipment supplier to the young business. Early in the 1950s, as word of the potential of community television spread, more and more business people found their way to Shapp for equipment and advice. In most cases, Jerrold would assist when possible in securing outside funding, help plan and construct the system with Jerrold equipment, and, of course, sign them to the company's long-term service agreement. It was only a matter of months after he met Tarlton that Shapp was building systems in nearby, and similarly TV-deprived, communities in the mining mountains of eastern Pennsylvania. Shapp had completed initial construction of Mahanoy City's City Television Corp. by the early winter of 1950 and, eyeing the colored pins on his U.S. map, was soon looking at possible CATV sites across the country. By 1952, his reach extended to the West Coast where he contracted to develop systems in Southern California's San Bernardino and Laguna Beach areas.

Shapp promoted turnkey systems. Jerrold would provide all the equipment and construct the physical plant. While the company did not make coaxial cable, Shapp had entered into a "gentleman's agreement" with Larry deGeorge of the cable company Times Wire and Cable. Times Wire would not make electronic equipment in competition with Jerrold and, in return, Jerrold would use and recommend Times Wire coax exclusively. Jerrold would also often help find financing for the system. Businessmen who purchased Jerrold's package also signed on for the Jerrold service agreement. By 1953, Shapp claimed to have installed 132 systems around the country. The 1954 *Television Factbook* listed about 280 CATV systems in the country—190 of them had been installed by Jerrold. *Fortune* magazine in October of 1954 estimated the company's gross revenues at about $4.5 million.[95]

In addition to his construction activities, Shapp was an indefatigable promoter of the industry, the fortunes of which were tied to his own. He wrote op-ed pieces for the trade press, lauding the accomplishments and future of CATV and defending it from its critics.[96] While his controversial service agreement caused for a mixed relationship with CATV operators, he rarely missed an opportunity to get the accomplishments of community television into the press, going so far as to call his own news conferences during the annual industry conventions.

By March of 1955, Jerrold's activity in system development and operations had become sufficiently extensive that the company formed a separate division devoted exclusively to its community television activities. Jerrold began promoting management services for those who simply wanted to own or invest in community systems. Entry into ownership of the systems themselves was, therefore, a next logical step for the company and, in November of 1955, Jerrold announced the purchase of its first, in Key West, Florida. Two months later it announced the acquisition of a second system, in Ukiah, California, noting also that it now had management responsibilities for eleven other CATV operations. Jerrold continued buying over the next few years and by 1958 owned ten systems around the country, the largest string of community TV properties at the time. They included systems in Washington, California, Arizona, Idaho, Iowa, New Jersey, and Alabama, most with 2,000 to 4,000 subscribers each and accounting in total for about 33,000 CATV customers.[97]

Jerrold's growing prominence, both through ownership and equipment supply, had its downside, however, in attracting the attention of the U.S. Justice Department. The concern was Jerrold's controversial service contract. In early 1957, the U.S. Justice Department brought suit against Jerrold, charging that the exclusive equipment and services agreements violated antitrust laws as restraints of trade. The government's case went to trial in late 1959, with one of its objectives being the removal of Jerrold from cable system ownership. While denying any relationship to the antitrust suit, Jerrold began selling off its cable properties in early 1959, selling its minority interest in the Key West system for $800,000, about $250 per subscriber. In March of 1960, Shapp additionally began stepping away from leadership of his company, selling more than 80 percent of his stock to businessmen Carl M. Loeb and Jack Wrather. That summer, the government won most of its suit against Jerrold.[98] The company was forbidden to acquire any new CATV systems without approval of the Court and was ordered to cease its exclusive and restrictive service and equipment practices. While the Court did not require Jerrold to divest its existing systems as the government had sought, the company nonetheless announced the following month the sale of its remaining nine systems. The purchaser was a new face in the community antenna business, a company based in Los Angeles, called H&B American Corp. With the Jerrold systems as a nucleus, H&B would soon become the country's largest and, arguably first, formal "MSO."

Proceeds from the sale of the cable systems were used, in part, to purchase the hi-fi audio component maker Harman-Kardon. Cofounder Sidney Harman became president of Jerrold in 1961 while Shapp settled into the role of Chairman of the Board. Shapp then began a two-year foray into politics, working as a special consultant to the U.S. Commerce Department, assisting in the development of the Peace Corps and making a short-lived bid for the Democratic nomination for U.S. Senator in Pennsylvania. It was the beginning of a new political career for Shapp, one that would lead to the Governor's mansion in Pennsylvania.

Henry Griffing

A different set of MSO roots can be found planted deep in the flat, dusty plains of Texas and Oklahoma. The sower of this seed was Shapp's occasional partner in business and politics, Henry Griffing. The Bartlesville pay-TV experiment, promoted by Shapp but run chiefly by Griffing, was only a small part of Griffing's regional media empire. Through his company, Video Independent Theaters (VIT), Griffing owned, by the mid-1950s, more than 150 conventional and sixty drive-in theaters spread across Texas, Oklahoma, and New Mexico. He was part owner of television station KXTV in Oklahoma City and held construction permits for TV stations in Hot Springs, Arkansas, and Santa Fe, New Mexico.[99] A subsidiary of VIT, Vumore, sprang from Griffing's first CATV system, established in Ardmore, Oklahoma, in 1952. By 1959, however, Vumore controlled community antenna businesses in thirteen southwestern communities, along with a microwave transport company, Mesa Microwave, serving some of those cable systems.

Griffing's corporate size and influence were measured by his participation in CATV politics nationally. In addition to playing a major role alongside Shapp in the defeat of S. 2653, detailed in the next chapter, Griffing was one of a select number of industry representatives called to testify before Congress during its regular reviews of CATV issues in the late 1950s. By June of 1960, Vumore owned thirteen CATV systems and its president, Larry Boggs, was selected as vice-president of NCTA.

Unfortunately, in August of 1960, Griffing, his wife, and their two young children, were all killed in the crash of their single-engine Cessna on a flight from New Jersey back home to Oklahoma.[100] The plane, piloted by Griffing, left the Teterborough, New Jersey, airfield and soon after disappeared. The wreckage was found in the Pennsylvania mountains about seventy miles south of Pittsburgh.[101]

Griffing's CATV holdings would provide the entry point for a much larger corporation, however. In 1961, the family's estate sold VIT, including the Vumore subsidiary and its CATV systems, to the corporate giant RKO General.

Bruce Merrill

Another small chain began its story in nearby Arizona. Like many cable pioneers, Bruce Merrill left military service in 1946 looking for a postwar career. One of ten children in a conservative Mormon family, Bruce's older brother, Paul, operated a small string of radio stations in their home state of Arizona. Through the late 1940s, Bruce Merrill fostered a new CPA practice, but would also have long talks with his broadcast-business brother about the potential for television in Arizona, a state limited to only one station during the freeze years. Around 1950, the brothers began seeing popular press reports of the emerging community systems in Pennsylvania and Oregon and, at Paul's urging, Bruce decided to give CATV a try. With loan money secured in part by a steel pipe firm in which the Merrills had an equity interest, they opened their first CATV system in Globe, Arizona, Bruce's birthplace, in 1952. The single-channel system carried Channel 5 out of Phoenix. It was the start of Merrill's Antennavision, Corp., a company that would begin building, collecting, and then selling systems over the next two decades.

Merrill's interests were not confined to just system operation, however. From the start, the brothers were dissatisfied with the available CATV equipment, much of which was still being converted from apartment master antenna supplies. Merrill recalled that their initial RCA equipment did not perform satisfactorily, and the Spencer-Kennedy amplifiers were of good quality but demanded high maintenance. Merrill knew that Jerrold made good equipment, but he was not interested in the contractual tying arrangements required by Shapp. So, they decided to make their own equipment, hiring a University of Arizona engineering graduate, Earl Hickman, who had been working as a radio engineer for Paul Merrill. The endeavor worked so well that by 1956 Merrill formed AMECO, (Antennavision Manufacturing and Engineering Co.) and began selling his own equipment to the industry.[102]

Building their own microwave links to import Phoenix signals, Merrill started several more systems in the next few years, including one in Silver City, New Mexico. In 1957, he combined his various CATV and microwave properties into Antennavision, Inc. of Phoenix. By 1959, Merrill's Antennavision, headquartered in Phoenix, had ten systems operating in Arizona with about 9,000 subscribers. The systems ranged in size from 150 subscribers to 2,500.

Merrill continued his franchising and acquisition activity, expanding into California's Imperial Valley (Valley Telecasting), Texas, Kentucky and Tennessee, and eventually up the eastern seaboard into New York and Connecticut. By the mid-1960s, Merrill, who was one of many who frequently bought and sold systems, had interests in more than thirty CATV operations, most of them in small towns. He became NCTA Chairman in 1964.

Bill Daniels

It was noted earlier that if anyone deserved the accolade, "Father of Cable Television," it was probably Milton Shapp. Some would disagree with that assessment, and award the honor to Bill Daniels, and it must be acknowledged that at the very least Daniels was a close second. Bill Daniels was a navy fighter pilot and aircraft carrier flight director during World War II. Following the war, he joined his father's by-then successful insurance business in Hobbs, New Mexico, but he was recalled to service during the Korean Conflict and did not fully end his military experience until 1952. When he attempted to re-enter the insurance firm, his brother had taken over control of the office and Daniels, not wanting to take a second place, decided to look for new opportunities, intending to open his own insurance firm in another area. He landed in Casper, Wyoming, and began a successful insurance business there, but during his search for a new place to call home, he had passed through Denver and in a bar there saw, for the first time in his life, television. He was 32 years old at the time. He later recalled thinking, "My God, what an invention this is."[103] Daniels continued thinking about television in Casper, a city then outside the reach of any TV signal. At the same time, Jerrold's people were on their national hunt for CATV opportunities. The mutual interests of Shapp and Daniels met in the person of Jerrold marketing chief, Zal Garfield, who discovered Daniels' interest and suggested the insurance man travel east to learn the CATV business from Marty Malarkey, who was already moving into the CATV consulting field. Malarkey charged Daniels $500 a day for two days of consultation, a price Daniels willingly paid. Daniels returned to Casper, obtained a one-page permit from the city and called in Jerrold to build the system in 1953.

Of the many systems being built at the time, the Casper operation was unusual because the nearest available signal was in Denver, far outside the reach of a typical mountaintop receiving and coaxial distribution architecture. Jerrold and Daniels developed a scheme to pick up the Denver signal on a mountaintop near Laramie, Wyoming, and then lease AT&T (Mountain States Telephone and Telegraph) microwave facilities to retransmit the signal 200 miles to Casper. Daniels, thereby, helped pioneer this method of distant signal importation, which became particularly necessary and soon politically controversial, in the widely dispersed towns of the West and Northwest.

Daniels put only $5,000 of his own money into the business; the real capital for construction and initial operations came from a group of local oilmen he assembled (Daniels, as had his father, specialized in insurance for the oil industry) along with a prominent lawyer. These men secured banks loans amounting to $375,000 to start the system.

Daniels went on to build systems in Rawlins, Wyoming, and Farmington, New Mexico, and he would continue to build and buy systems through the rest of his career, but an additional opportunity soon presented itself, one which would place him at the very heart of the cable industry for the next several decades. Daniels became involved in the NCTA. He was the NCTA's second president following the initial tenure of his early business mentor Marty Malarkey in 1956. In his capacity as head of the national trade organization, Daniels received regular inquiries on how to start or buy a CATV system. He offered what advice he could, frequently referring callers to Jerrold for equipment and services, but it also dawned on him that there was a business opportunity in CATV consulting.[104] When he left his NCTA post he made one of those life-altering decisions. He moved to Denver, as he had wished to do for some time, and in 1958 set up shop as Daniels & Associates, entering the business as a CATV consultant, broker, and investment banker.

It was the first company of its kind in the young industry and would become one of the largest and most influential. Daniels' initial mailing announcing his start in the business got a nibble from a Pennsylvania operator who wanted to sell his system for a million dollars.[105] Daniels soon found a buyer in Charles Sammons, a successful Dallas-based businessman. According to Daniels, Sammons was guest at a business dinner in Rapid City, South Dakota, during which Daniels spoke on the potential of CATV. Intrigued, Sammons contacted Daniels the next day and the deal for the Pennsylvania system was soon completed. Sammons, however, had no experience in the business and was looking for someone to help run the operation; Daniels volunteered his own System Management Division to do the job, a division that did not, in fact, exist up until Daniels created it on the spot for that purpose. Moreover, Daniels, as part of the management agreement, took a 20 percent equity stake in the cable system, setting the pattern for subsequent deals through the next several years. Although it was created on the fly, by 1959, Daniels management company was running systems for a number of investors and held equity stakes in as many as twenty-three.

In the coming decades, the hand of Daniels and Associates would touch, sometimes lightly sometimes heavily, the development of the country's largest and most influential cable companies. Out of his offices would arise some of the industry's most successful businessmen, including cable leaders such as Monroe "Monty" Rifkin, Amos "Bud" Hostetter, and Glen Jones. For much of its history, the cable industry could be accurately described as a small club, and so by extension, Daniels and Associates was one of the chief club houses. While the title "Father of Cable Television" could therefore be made to fit, it might be more accurate to describe Daniels as the midwife, present at and in charge of the birth of many new companies. Or, if the masculine gender is preferred, he could be described as the "Father of Cable MSOs."

Charles Sammons

Charles Sammons was already a well-established businessman when he entered the community antenna business. Sammons, orphaned at an early age, nonetheless rose through the business ranks and, in 1938, founded the Reserve Life Insurance Co. of Dallas. He began diversifying in the 1950s, moving into the hotel industry, including the Jack Tarr chain of hotels, steel with the Standard Steel Works of Kansas City, and milling with company called Flour Mills of America. Ever watchful for lucrative investment opportunities, Sammons was convinced by Daniel's presentation and by the attractive financial characteristics of the CATV business. His experience with the first CATV property was such that his company determined to buy more and by the end of 1960 he had fourteen systems. They were housed under the corporate structure of the Trans Video division of Sammons Hotel Operating Co., a firm that in fact had begun as the CATV operation of Marty Malarkey in Pottsville, Pennsylvania, and which Malarkey sold to Sammons. They initially were managed by Daniels' Systems Management Division.

The Sammons' company would continue through the 1960s and into the next several decades to be an active buyer and seller of cable systems, and in itself become a sizable MSO. But while Sammons would become a top MSO over the next several decades, cable would be only a piece of the larger Sammons business empire. Sammons Enterprises was formed in 1962 to oversee the multifaceted investment interests. Reserve Life Insurance Co., by itself, was a billion dollar company by 1973. Sammons' cable company, in fact, was something of an oddity in the industry in its general lack of involvement in the political affairs of cable television. Most of the early pioneers, such as Daniels, Kahn, and Magness, would become important architects of the industry and play powerful roles in its business and political evolution. Sammons himself, and the Sammons cable interests would never come to play any key roles in the shape of the business, content to see it as largely that, one more business in a closet full of business operations and investment vehicles. At the same time, Charles Sammons and Sammons Enterprises would become important sources of philanthropy, establishing hospital and art charities in Dallas and elsewhere.

Bob Magness

Beginning in the 1970s and through the 1990s, the largest, most powerful, and often most controversial cable television company in the United States was Tele-Communications, Inc., TCI. The company that, by 1998, controlled 14 million subscribers and had interests around the world was started in a small Texas town by an otherwise unpresupposing part-time rancher named Bob Magness. Robert John Magness was born in Clinton, Oklahoma, in 1924 into a farming and railroad family. Drafted at age 18, he served in the infantry in World War II and returned to complete his bachelor's degree at Oklahoma State College in Weatherford. He went to work for Anderson Clayton, a large cotton company, and married Betsy Preston. The new Mrs. Magness came from a cattle ranching family, so the couple began raising their own herd as a sideline, while Magness continued to work in the cotton business, primarily buying seed for cotton oil products. It was in a cotton gin that he met two men who had blown a rod in their truck and needed a lift to Paducah, in North Texas, where they were setting up a CATV system. Magness gave them a lift and listened to their talk of community antenna television. "So I listened and thought a little bit about it, and about a week later, I had thought enough about it, so I went down and looked them up and talked to them some more, and about thirty days later we were stringing wire," recalled Magness.[106]

The Magnesses sold their cattle, secured loans from a local bank—Magness was a friend and hunting buddy of the mayor, who also happened to be the local banker—and from Betsy's father. They opened their first small system in Memphis, Texas, in 1956. Like many of the pioneers, Magness spent a lot of time on poles tacking up coaxial cable, while Betsy did the books on the kitchen table. Magness was soon tapped into the ways of the young CATV industry. As did the system builders in Paducah, Magness used Jerrold equipment and expertise. The system prospered and, in a few short years, Griffing's Vumore offered to buy it. Magness went in search of a new CATV opportunity and found one, through Daniels, in Bozeman, Montana. Bob and Betsy Magness moved to Bozeman in 1958. With another cable pioneer, Paul McAdam, Magness built a microwave link connecting Salt Lake City to Bozeman and began offering imported big city TV stations. It was a good start, and Magness soon was looking for more. From these often modest beginnings, pioneers like Magness and Daniels would build cable empires and become millionaires and billionaires in the process. There was much work ahead, however.

Abel Cable

Shortly after Strat Smith was named NCTA executive secretary, he called a meeting with a public relations firm. The freshman industry needed to start thinking about national advertising and promotion, and the conversation turned to the possibility of a logo or symbol. John Smith, from the PR firm, suggested something along the lines of the Electric Utility industry's "Reddy Kilowatt" mascot, a smiling stick figure drawn in lightening bolts with a light bulb head. Picking up on the idea, Strat Smith suggested a similarly smiling figure with a TV tower for a body, coaxial cable for arms, and a picture tube-shaped head. John Smith went off to sketch what would become the cable industry's new icon, Abel Cable, joined a few years later by a female companion, Ima Cable. For nearly two decades, Abel Cable would be the ubiquitous Cheshire cat-like presence of CATV, painted on the sides of CATV service trucks, shop windows, and billboards.

Through the mid 1950s, the community antenna business was, for the most part, as simple as its name implied and Abel Cable promised, a friendly community service, spreading slowly but steadily through the small towns of the United States. Hundreds of these systems were being built, but none were very large, and all told, they accounted for less than 1 percent of the nation's television viewing population. The focus of the industry was on building the hometown business, securing adequate technology, and making the customer reasonably happy. Operators had a national organization that looked after important but relatively narrow issues, such as tax relief, telephone company relations, and relations with broadcasters. In addition to adopting the Abel Cable logo, the NCTA initiated such modest measures as forming a committee in 1955 to develop a code of good business practices.

As a result, through much of the 1950s, the local cable businessmen and women enjoyed a hospitable environment locally and nationally. They also enjoyed an important absence of any significant governmental control, much to the envy of their broadcast counterparts.

At important points in this placid portrait, spots of trouble were beginning to appear, however. The slowly evolving business of CATV was gathering a small constituency of supporters and benefactors. As such, it was also catching the eye of parties with competing economic and political interests. Beyond the IRS, the industry was starting to confront the problems associated with its own success. In small ways, it was putting pressure on the existing broadcast system and the system was preparing to push back. While the events at first were isolated, they would nonetheless be the harbingers of the legal and economic challenges cable would have to deal with for the next several decades. Most of the problems began in the West, but it would be in Washington, D.C. that the real battles would be fought.

4

Abel Cable Goes to Washington (1950–1960)

> *In the development of community antenna systems... I felt that I could no more protect a television station from a community antenna service than I could protect it from a local drive-in movie house or whatever other form of attraction might develop within the service area of a television station.*
>
> —FCC CHAIRMAN JOHN DOERFER, U.S. SENATE HEARINGS, MAY 27, 1958[1]

The FCC was aware of community antenna television from the start. L. E. Parsons, of course, had written the FCC early on in his experiments, describing the technology, the results, and the prospects. He had actively, albeit unsuccessfully, sought approval for a broadcast retransmission system to replace his coaxial cable. While the agency rejected his request, it did send a field engineer to visit Parsons' system. The engineer's report ended up on the desk of a young FCC staff attorney. E. Stratford Smith would later become the NCTA's chief legal counsel and first executive director, but he got his start in the Common Carrier Bureau of the FCC. Smith recalls that with the engineer's report came a request from his superiors that he look into the question of whether or not community TV services fell within the Commission's jurisdiction.[2] It was around the fall of 1949, according to Smith, and it was the first serious consideration the FCC would give to CATV. It would, as it turned out, be years before the Commission issued an opinion on community television, but Smith began his investigation shortly after receiving the Astoria, Oregon, material. He visited Abe Harter's CATV system in Franklin, Pennsylvania, probably in the fall of 1950, and took a shorter trip in 1951 to Pottsville, Pennsylvania, where he talked with Marty Malarkey. He began drafting a memorandum and opinion on the FCC's relationship with the new broadcasting-related business.

While CATV operators worked through the technical and business challenges of the 1950s, traced in the previous chapter, they also came to confront a growing federal regulatory presence. Broadcasting and telecommunications in the United States has nearly always existed in the context of significant federal control. An extensive, detailed, and frequently changing statutory framework, enacted by Congress and executed by the FCC, guides and constrains the activities of the related industries. The evolution of cable television has been powerfully shaped by this legal environment, which itself has been subject to great changes over time. Consideration of the relationship between cable television, Congress, the FCC, and the courts is a major topic of this book, and the first encounters between cable and the government in the 1950s is the focus of this chapter.

The FCC, to begin, was feeling some modest pressure in the early 1950s to act on the legal status of community television. The first inquiries came primarily from people who wanted to get into the business. As interest in community TV grew in 1951 and 1952, the FCC began receiving letters from eager entrepreneurs wondering whether they needed Commission approval to start a system, and if they did, what application forms might be necessary and what restrictions might apply. Broadcasters, too, were curious about the legal status of CATV. Was it a service subject to FCC oversight? If so, how would it be defined and regulated? Was it a common carrier, a retransmission service, a broadcaster? Under the Federal Communications Act of 1934, the Commission was given the authority to regulate, broadly, two kinds of services, broadcasting and common carriage. The former, under Title III of the Act, included everyone who used the airwaves (outside of the military and other governmental agencies), most especially radio and TV operators. The common carrier authority, Title II of the Act, included, primarily, telephone and telegraph services. In order for the FCC to invoke authority, CATV would have to be declared, at least *de facto*, a broadcasting service or a common carrier. Many people wanted to know which, if either, it would be. To these requests, like the earliest ones sent by Parsons, Tarlton, and Walson, the FCC replied tersely and repeatedly that it had "as yet taken no position with respect to its jurisdiction over such systems, but that the matter (was) receiving attention, particularly with respect to the status of such operations under the provisions of Title II of the Communications Act, as amended, applicable to common carriers."[3]

The Commission, in short, was studying the issue. There was some hope in both the broadcasting and community TV industries that the FCC would move with reasonable dispatch in making a determination. In July of 1951 *Broadcasting-Telecasting* noted that, while the Commission had not yet taken a position on CATV, "a staff report is being prepared for the Commissioners' consideration, and it is likely that soon the FCC will decide whether community antenna systems come under its jurisdiction."[4] The article further revealed that FCC staff felt community systems probably could be regulated as common carriers, although not as broadcasters. At the end of 1951, *Broadcasting-Telecasting* hopefully reported that the "year-long study is near an end, with the staff report to be handed to the Commission before the end of January."[5] But this prediction was overly optimistic. The Commission was still several years away from a public proclamation.

The eventual disposition of Smith's report is unclear; the first internal FCC memo found to date on community television was drafted in 1952. Possibly, the result of the staff investigation alluded to in the trade press, the March 1952 interoffice report was signed by the Commission's General Counsel, the heads of the Common Carrier and Broadcast Bureaus, and the agency's Chief Engineer.[6] It offered a brief history of the industry, outlining the policy issues and recommending a more detailed investigation, one hopefully leading to the issuance of a formal policy. Among the items worth considering in such an investigation, suggested the memo, was the question of whether CATV could be regulated by virtue of any property rights held by broadcasters in their programming. The memo concluded that community TV systems could be regulated as common carriers under Title II. It also noted that, given the increasing number of inquiries, the Commission could not "defer for very long taking a position" on CATV.[7] Despite the admonition, the item never made it onto the FCC's agenda. It would wait another six years.

Initial Skirmishes: 1951–1956

In their examination of the FCC's hesitancy on CATV regulation, Roger Noll, Merton Peck, and John McGowan argued that the delays were institutional, a part of the organizational culture of the FCC:

> When faced with an issue that is very likely to involve an appeal to external judgment no matter what the decision, the agency will employ two tactics: (1) It will engage in interminably long information-collection proceedings, not only to delay the unhappy day when its decision must be defended, but also to try exhaustingly to find a conflict avoiding compromise to test the resolve of the interest groups and their willingness to spend money and political capital to support their point of view; and (2) it will attempt to shift responsibility for the decision outside the agency by appealing to Congress, the courts, other regulatory agencies, or the executive to provide direction.[8]

CATV, suggested Noll, Peck, and McGowan, was a case in point. Other, more practical forces were at work as well. The FCC was fully occupied at the time with its Sixth Report and Order, its attempt to construct a framework for the national allocation of broadcast television licenses. Further, as previously noted, there was some thinking that once the Report and Order was finally released, the natural development of the broadcasting system would make unnecessary extension services, such as CATV, and the community antenna technology would quietly whither away. Even some of the earliest participants in the industry felt their business was only a temporary bridge to widespread availability of broadcasting, although the failure of the Sixth Report and Order very quickly brought an end to such speculation.

In short, institutional inertia, the relatively small size of the issue, the press of more important work, and the conception of CATV as an annoying little problem might simply disappear, all probably contributed to the FCC's initial lack of attention. Despite this initial lack of action, however, circumstance would repeatedly bring CATV back to the door of the FCC in the early 1950s.

Belknap, the Avoided Opportunity

The FCC's clear knowledge and, at the same time, active avoidance of numerous CATV-related problems were illustrated early in the industry's development in its handling of the *Belknap* case.[9] Here, the Commission missed (or, more accurately, fled) an easy opportunity to address the policy issues posed by the new technology. As the trade press declared in October of 1951: "[The] peg on which the FCC can hang its long-pending policy decision regarding community-TV installations seems to be at hand. Application by J. E. Belknap & Assoc. of Poplar Bluff, Mo., for common carrier microwave links between Memphis and the Missouri cities of Kennett and Poplar Bluff for the purpose of relaying TV signals has been filed with the Commission."[10]

J. E. Belknap was a druggist in Poplar Bluff, Missouri. Like so many other small-town entrepreneurs at the time, he saw possibilities in CATV. Within months, and perhaps only weeks, of the nationally announced opening of the Panther Valley system, Belknap was talking to other businessmen in Popular Bluff about getting a similar system established in their town. Unlike Lansford or Mahanoy City, however, there were no signals available within a reasonable distance of Poplar Bluff. Television would have to be "imported." The nearest operating station was WMCT in Memphis, Tennessee, about 150 miles to the south. Like Tarlton, Belknap gathered a group of local businessmen: C. B. Bidewell, a hardware store owner; William Cohen, a merchant; Ernest Dunn, a hotel operator; and John Davis, an insurance salesman. The partnership proposed to set up a wired television system and bring the WMCT signal into town via microwave. As their plans developed, they envisioned a much larger network of microwave TV distribution, but in their initial application, they started in their own backyard.

About halfway between Popular Buff and Memphis is the town of Kennett, Missouri. The proposed microwave relay would require two hops, the first from a receiving tower outside Memphis, which would pick up and relay the WMCT signal to a second tower at Kennett. The second hop was from the tower at Kennett to the receiving station at Popular Bluff. Belknap and his partners additionally arranged to tap the signal at the Kennett tower and feed a CATV system planned for that town as well.

On July 1, 1951, Belknap and Associates filed applications with the FCC for two experimental microwave relay stations, one in Osceola, Arkansas, across the Mississippi River and about forty-five miles from Memphis, and another in Kennett.[11] The application stated the Belknap relay would be a common carrier feeding the Popular Bluff CATV system, also owned by the Belknap partnership, and a proposed CATV system in Kennett, operated by a separate company, the Kennett Distributing Co. The plans included a schedule of rates to be charged to Kennett and other possible "customers" based on the number of TV sets attached to the system. In addition to the immediate request for the two microwave licenses, Belknap stated in the application that, if all went well, it planned to expand its operations. First, it would extend the microwave feed east beyond Poplar Bluff to nine other communities in Missouri, Kentucky, and Illinois (most of them along U.S. Route 60). Then, it would build a second system to the north, using St. Louis station KSD-TV to feed three towns in south central Illinois, Duquoin, Benton, and West Frankfort. There was potential for eventual merger of the two relay systems to carry both the Memphis and St. Louis signals to the listed cities as well as to future CATV systems that might be built in the area.

The application posed a number of interesting problems for the Commission, and, as it turned out, for Belknap. It was, in the first instance, not clear that the proposed operation qualified as a common carrier under law. Belknap claimed common carrier status because it was willing to sell the signal it carried to other CATV systems. Unfortunately, there were no other systems in the area at the time. The FCC staff, in its initial recommendation to the Commission, noted that the Kennett system itself existed only on paper. If that system failed to materialize, Belknap Associate's only real customer would be itself, through its Poplar Bluff CATV business. A suspicious Commission staff concluded that the real purpose of the application was simply to build a relay system to serve its own community antenna company. The staff also was unhappy with Belknap's proposed rate schedule, which did not look like a typical common carrier schedule of fees. Finally, there was a vexing property rights issue involving the signals of the effected TV stations. In October of 1951, not long after the filing, WMCT and KSD-TV joined to protest the Belknap proposal. The stations charged that Belknap's use of their signals would constitute an illegal rebroadcast of their programming and a violation of their property rights. They asked the FCC for a hearing.

The most promising aspect of the case was the FCC's suggestion that it might serve as the foundation for a broader inquiry into the legal status of CATV generally. The recommendation by the staff, approved by the Commission, was that the common carrier status of Belknap's microwave relay was contingent upon the status of its Poplar Bluff CATV operation. Since that system was its only customer, the microwave system as a whole was a common carrier only if this CATV operation and, more importantly, all CATV operations were common carriers. Such a determination, however, required additional thought. Nearly a year after the filing, the Commission responded to Belknap's request with a letter stating that it could not tell whether Belknap qualified as a common carrier given the current record and would look further into the issue. Consonant with the related March 1952 memo, the staff recommended a list of important policy questions: What was the status of CATV under the Communications Act? Were they broadcasters or common carriers, and what was the nature of the Commission's jurisdiction? What was the relationship between the growth of CATV and the Commission's allocation plan for television under the, then pending, Sixth Report and Order? What effect would CATV have on the development of broadcast television in smaller communities? Did CATV's unauthorized use of broadcast signals constitute a violation of the property rights of broadcasters?

It was July of 1952. There were fewer than 100 CATV systems in operation and the NCTA was holding its very first national convention in Pottsville. *Broadcasting-Telecasting* announced the FCC's action in the Belknap case with an almost audible sigh of relief, noting that the Commission had designated a case to determine the issue of its jurisdiction over CATV.[12] Belknap, said the article, would be the CATV industry "guinea pig."

Belknap, however, had no desire to be a guinea pig, and the Commission, it appeared, had no desire to press the partnership into such a role. In response to the Commission's letter, Belknap filed an amended application on October 14, 1952. The group divested its interest in the Poplar Bluff CATV system, effectively splitting the CATV and microwave into separate companies. (Two partners retained only a 5 percent interest each in the CATV system). Belknap proposed to operate as a "specialized" common carrier, feeding TV signals to any CATV operation that wished to take the service, but the partners would avoid any significant ownership interest in the client systems. At its next consideration of the new application, in July of 1953, the Commission decided that because Belknap no longer owned a CATV system, the issue of the legal status of CATV generally was no longer relevant to the proceedings.[13] The door to a broader consideration of the FCC's authority over CATV swung quietly but firmly shut. The Commission did not have to confront the difficult list of issues outlined in the prior proceedings and took no obvious steps to find a way to address them directly.

The primary issues left in the Belknap application included the rate schedule and the serious and important question of property rights. Belknap also amended its proposed rate card to provide for a flat fee for CATV customers, and one that was substantially cheaper than the rates AT&T was charging for similar services. The Commission remained concerned about the property rights issue, but observed that the law was vague in this area and varied state to state. It requested Belknap to file a detailed legal opinion in support of its contention that it was not in violation of state or federal laws. Belknap's response, in part, was to claim insulation from any such liability.

Under the terms of its model agreement with CATV clients, the microwave system would be held harmless by the CATV operation for any infringement of copyright or other property rights. To the FCC, Belknap now looked like any other common carrier and at very competitive rates. The staff recommended approval of the revised and limited applications and, in May of 1954, they finally were granted.[14]

It was a lengthy and ultimately unfulfilling process, but it did illustrate an important historical point. That is, from the very first case and the very first years of CATV, most of the significant policy issues that would haunt the industry for the next four decades were already recognized and articulated. In keeping with the evolutionary nature of the development of the technology and the social implications of that development, the fundamental legal questions arose early on and out of preexisting social and industrial structures. Those issues, in specific, included questions of copyright, retransmission consent, general regulatory oversight, and the impact of community antenna systems on the development of broadcast television. Most of the major political battles that cable television would fight in the years to come were foreshadowed in the earliest efforts to bring television into town by wire and microwave.

The Failure of UHF

While the FCC tried not to ponder the problem of CATV, television, both broadcast and wired, began to grow. Broadcast television expanded primarily in the larger markets because, as noted previously, the FCC's allocation plan distributed licenses in part on the basis of population. The smaller markets, with little or no broadcast service, became the targets of opportunity for community antenna operators and others. Moreover, to the extent that smaller markets were accorded licenses, they were often UHF rather than VHF permits, and this quickly came to constitute a new and, as it turned out, long-term, problem. The FCC's promotion of UHF technology in the early 1950s and again in the 1960s would play a significant role in both advancing and retarding the development of CATV. It began with the Sixth Report and Order.

In the Sixth Report and Order, the Commission chose to privilege localism over diversity. In its execution of this goal, however, it relied on a technology—UHF broadcasting—that proved insufficient for the task. The Commission had simply ignored market realities that, at least through the 1950s, doomed the widespread implementation of UHF service.

The problems were in the inherent inferiority of UHF as compared with VHF transmission and in the real-world cost of running a commercial TV station. A variety of technological problems plagued UHF. Its signal is not as powerful as VHF and does not travel as far. UHF requires significantly more power than VHF to generate an equivalent signal. While the FCC permitted UHF to operate at substantially higher power levels, the more potent transmitters and greater use of electricity to run them meant greater startup and operating costs. At the consumer end, most TV sets in the 1950s came equipped from the factory with only standard VHF tuners. To receive a UHF station, a viewer had to purchase and install a special UHF tuner, for about $40, and the add-on devices seldom provided as good a picture as the built-in VHF tuner. Manufacturers, who saw little consumer demand for UHF tuners, built and marketed them poorly, if at all.

The FCC's allocation plan also called for VHF and UHF stations, in many cases, to operate in the same markets. Under provisions of the 1952 Report and Order, 110 communities would be designated VHF only, 910 would be UHF only, and 255 would be "intermixed." This "intermixture" meant that viewers who wanted to receive both services had to switch back and forth between the two tuners, assuming they were willing to purchase the extra UHF tuner in the first place. In some community antenna markets, the presence of a UHF station worked to the advantage of both the CATV and UHF company in an interesting technology symbiosis. The CATV system could convert the UHF signal to a VHF channel that every subscriber could receive without expensive modification of the home TV set. Most UHF towns did not have the CATV solution available, however.

The challenge for UHF was exacerbated by the problem of market size, especially insofar as UHF licenses were disproportionately allocated to smaller towns. By some early estimates, a local TV station had to generate $500,000 annually to stay in business. Revenues are based on ratings, and ratings are limited by available audience size. Studies in the early 1950s predicted that network affiliated stations would need an unduplicated audience of between 25,000 and 50,000 to remain financially viable.[15] There were hundreds of communities across the country substantially smaller than this. In fact, the figures showed that in some instances licenses were not economically viable even if one counted the population of the entire state. According to LeDuc, the FCC's scheme provided the average applicant for a TV license in Wyoming with an audience base of about 3,600: "In Nevada, the base was 5,000 citizens per license allocation; in Montana, 5,100; in North Dakota, 5,600. Such assignments were obviously only empty gestures, offering service on paper which could never be provided."[16]

The lack of built-in UHF tuners and the disincentives to purchase and install them cut further into the potential viewership for a UHF station, which again, had to contend with higher start up costs than VHF operations.

Finally, there was the exacerbating issue of programming. Audiences as early as the 1950s came to expect a certain level of technical sophistication in TV programming, a "professional" look. The level of production quality they expected could be achieved, generally, only by high-cost Hollywood- or New York-based production processes. This was the kind of programming created and distributed by the major TV networks through their growing web of affiliated stations. The networks, naturally, sought the strongest affiliates in each market and that was almost always a VHF outlet. In many cities, affiliate contracts had been fixed before 1952 with the pre-freeze VHF stations. In intermixed markets where UHF and VHF stations competed for network affiliation, the UHF station was nearly always the loser. In short, UHF stations rarely had network quality programming to offer viewers. And such programming could not be mimicked by a local, studio-based talk show. And so, hampered by initially small and then further dwindling audiences, and restricted technologically and geographically, the fortunes of UHF stations disproportionately assigned to smaller markets began a rapid spiral downward. The FCC had set aside more than 1,300 UHF allocations in the Sixth Report and Order. By 1954, 127 were on the air, the high water mark for the decade.[17] As the marketplace realities set in, UHF operators began returning licensees to the FCC. By 1960, there were only 75 operating UHF stations in the country. And, insofar as the FCC had relied heavily upon UHF to bring television to American homes, the failure of the technology meant the failure of the plan overall. The Commission had allocated television licenses for 1,275 U.S. communities in 1952; six years later only 308 had an operating TV station. As Noll, Peck, and McGowan noted in 1973, "Localism has obviously run aground on the hard rocks of economic reality."[18] Of course, one person's wrecking reef is another person's swimming hole.

Boosters, Translators, and Power Politics

The problem of economic viability for the pioneer broadcasters was perhaps most serious in the western states where the population base necessary to sustain a station was close to nonexistent. None of this was lost on potential western broadcasters. E. B. "Ed" Craney, owner and operator of a group of Pacific Northwest radio stations, tried to create a statewide television market in Montana in 1953. He gathered together all the applicants for TV station licenses and proposed they jointly build and operate several high-powered, mountaintop transmitters to create a unified Montana audience. The plan was eventually abandoned "for lack of cooperative interest,"[19] but Craney would go on to launch his own stations in the West and, in doing so, come to play an important role in the political life of the cable television industry.

While the small and widely separated communities of the West could not support extensive broadcast television, residents nonetheless shared the national thirst for it. The need for low-cost alternatives to local broadcasting in these areas became quickly apparent to everyone, including the business people of the young community TV industry.

CATV exploited a variety of holes in the FCC's allocation plan. Community operators felt they could sell television—in towns that were too small to support any broadcast outlet, even if a license had been allocated; in towns that had only one or two stations, where CATV could provide a third or fourth (DuMont) network; and in even towns that had a full complement of network affiliates, but had one or, preferably two, independents sufficiently near for CATV pickup and redistribution.

Broadcasters viewed such business plans with mixed emotions. Some worried about community antennas working their way into the cracks and crevices of the nation's "real" TV service. "We must not sponsor the growth of a system that will prevent the establishment of television stations in smaller markets," declared, Michael Hanna, general manager of WHCA-AM-FM, in Ithaca, New York.[20] *Broadcasting-Telecasting*, in February of 1953, reported that broadcasters such as Hanna and WMCT in Memphis, were complaining, sometimes loudly, about CATV, but other "telecasters rub their hands in glee" over the prospect of expanding their audience base into unserved communities. "The split is so pronounced," stated the magazine, "that the National Association of Radio and Television Broadcasters (NARTB), traditionally in the forefront in defense of free broadcasting, has decided to keep hands off."[21]

Broadcasting-Telecasting itself took a position more in sympathy with Hanna, suggesting in an editorial in January of 1953 that it was a good time for the broadcast industry to reassess the "usefulness" of community antennas. It worried that if CATV discouraged the development of stations in small towns it could, in the long run, "do more harm than good." The solution, said the magazine, might be satellite or boosters stations that would allow distant broadcasters to beam their pictures directly into under- or unserved communities. Such devices, unlike CATV systems, could be controlled by the broadcasting industry.[22]

In fact, it was just this vision that the CATV industry dreaded the most. After its preoccupation with the tax problem, it was the competitive threat of TV boosters that kept CATV industry owners and operators awake at night. In 1953, scores, perhaps hundreds, of tiny hamlets and villages, especially in the West and Northwest, were too small to support either a broadcast TV station or, in some cases, even a CATV system. These were the seedbeds of the repeater technology of the early 1950s, the satellites, boosters, and translators of the small-town West. These devices worked in much the same way as community antenna systems, picking up and amplifying TV signals at the edge of a broadcast station's effective range. Instead of sending the signal into town on a wire, however, the low-powered antennas rebroadcast the signal over the airwaves. "Boosters" rebroadcast the signal on the same channel on which it originated; "satellite" stations moved the signal to a different channel before repeating it, usually staying, however, in the VHF band; "translators," by FCC fiat, moved the signal from the VHF band to a UHF channel. Boosters, especially, were wildly popular in the West in the early and mid-1950s. They were cheap to install and maintain, requiring almost no operational oversight. You turned them on and left. Boosters were set up by town clubs and civic committees, by local appliance salesmen, by loosely formed nonprofit associations, and by all manner of hobbyists. They were often quite crude, housed in a tarpaper shack or in the shell of a broken-down bus. By the mid-1950s, there were hundreds of these devices providing TV service in the western states. There was only one serious drawback to most of them: they were illegal.

For the FCC, boosters constituted a threat to the orderly development of U.S. television service; they disrupted the well-planned pattern of community broadcasting. In addition, they posed a potential hazard by interfering with other forms of broadcasting, including two-way radio and aviation communication. Finally, they violated the 1934 Communications Act's prohibition on the unauthorized rebroadcasting of signals.

In one of its earliest actions against satellite stations, the FCC shut down an unlicensed repeater run by Sylvania Electric Products in Emporium, Pennsylvania, in 1950. The managers of Sylvania's television tube manufacturing plant in the forested hills of north central Pennsylvania argued the necessity of the satellite for testing its equipment. After paying a $2,500 fine, the company was granted an experimental license to continue operating the station, but this was one of only a handful of such exemptions granted by the FCC.

More typical was the experience of an appliance and TV repair shop in Bridgeport, Washington, which formed the nucleus of a group known as C.J. Community Services. Like most "repeater communities," Bridgeport was a tiny, isolated settlement in north central Washington, about midway between Seattle and Spokane. Nestled deep in the Columbia River gorge, it was too small to attract even the interest of early CATV prospectors. The price of installing a coaxial system in Bridgeport was estimated at about $28,000, while it cost less than $2,000 for the appliance store technician to build two repeaters for Channels 4 and 6 out of Spokane, about 110 mile to the east.[23] C.J. Community services was created as a nonprofit corporation. Some eighty members funded construction of the low-powered transmitters and paid $5 annually for maintenance. Their simple repeaters were cheap, reasonably effective, and quickly spread to other villages along the river. Soon the boosters of C.J. Community Services were broadcasting television to Pateros, Brewster, Nespelem, and other towns in the mountains of central Washington. According the FCC, however, they were also generating complaints about interference from aeronautical and other nonbroadcast services. The agency ordered its field agents to shut down the boosters. FCC agents went into the mountains with a measure of reluctance. "Some of the booster operators threatened to protect their repeaters with guns," recalled Smith.[24] Apparently without gunplay, FCC personnel managed to padlock and seal the small transmitter shacks. Within days, however, the seals and padlocks were broken by "unknown agents" and the power restored.[25] The Commission retaliated with new threats against the operators, but the booster supporters responded in kind, working along two fronts. First, they appealed directly to the FCC, arguing that boosters were legally akin to CATV systems, that is, outside the authority and responsibility of the agency. Second, they sought the support of their representatives in Washington, D.C.

In November of 1954, Washington state's U.S. Senators, Henry (Scoop) Jackson and Warren Magnuson, sat down with FCC Commissioner Robert E. Lee and members of the FCC staff. The senators, especially Magnuson, were concerned about the FCC's handling of the problem of TV service in the West and they suggested that the FCC invite comments in a rulemaking procedure to permit boosters, or something like them, in the underserved areas.[26] It was a request that had a larger-than-normal set of congressional teeth. The next month, Magnuson, following a democratic congressional victory in the 1954 elections and a change of majority in the Senate, announced that he would accept the chairmanship of the Senate Interstate and Foreign Commerce Committee, the committee charged with oversight of the FCC.

In February of 1955, Magnuson further announced the Committee would continue its hearings on the rapidly developing TV industry, looking specifically into the problems of UHF television. These hearings, subsequently entitled "The Television Inquiry" would run on and off until 1960 and consider a wide range of television issues, including network practices, frequency allocation, service to smaller communities, boosters, translators, and CATV.

In June of 1955, Magnuson followed up his informal suggestion to Lee with a letter requesting the FCC begin a study of the problems of TV service to the smaller communities of the West, a request the Commission was pleased to honor. Meanwhile, CATV operators were starting to worry. That same month, at the NCTA's fourth annual convention held in New York City, President Malarkey told the delegates that the growing booster phenomenon was one of the Association's chief concerns. Milton Shapp warned that the CATV industry "cannot be exposed to the satellite threat" and urged the association to find ways to protect the heavy investments people had made in their industry.[27] The NCTA responded with filings in the FCC's investigation into boosters and, specifically, the Bridgeport case.

Through the early summer, the FCC took comments on the booster problem and, in October, C.J. Community Services received good news. The FCC hearing examiner had found in favor of the Bridgeport booster, ruling that, like CATV systems, repeaters stood outside the authority of the Commission, and because the hearing officer found that boosters did not interfere with other services, he recommended that the agency take no action against them.[28] The ruling drew a howling protest from the FCC's Broadcast Bureau and, in January of 1956, the Commission reversed the decision ordering all such operations off the air.[29] The Commission remained concerned that on-channel VHF boosters, especially the kind of slap-dash, unmonitored operation typical in many places, would create unacceptable interference with other services.

Although they rejected Bridgeport plea, the Commission was nonetheless keenly aware of the service problems of the western audience, the collapsing facade of the UHF allocation plan, and the increasing pressure from Western politicians. The Commissioners, therefore, proposed an alternative—VHF-to-UHF "translators." These repeaters would convert the VHF, or other UHF signals, into UHF channels set aside specifically for that purpose and in a part of the spectrum that would minimize the chance of interference. The Commission estimated the costs of such repeaters at about $1,000.[30] Moving to a specialized UHF band, with its assorted technical drawbacks, was not what the booster clubs had in mind, however, and the illegal operations continued.

The tug of war between the FCC and western interests, including booster operators and politicians, continued for another four years, and even then was not completely resolved. At one point in 1956 a maverick, Colorado Governor, Edwin S. Johnson (D-Colorado), went so far as to grant state "licenses" for boosters in his state, an action quite outside his authority.

In March of 1956, Bridgeport won a court stay of the FCC's prohibition on boosters, pending a resolution of its appeal against the agency. By issuing a cease and desist order, the FCC had purposively guided the case into Federal Court to avoid the possibility of a state judge in a local jurisdiction giving regional businessmen an overly sympathetic ear.[31] And in May, the FCC formally implemented its UHF translator policy, giving booster operators a grace period to convert to the approved service.

The eventual appeals court decision in Bridgeport was a complicated and backhanded victory for the FCC. Technically, the agency had ruled that it was forced to issue a cease and desist order against the booster company because FCC rules, in essence, had no provision for such service. The District of Columbia Circuit Court ruled that the agency did have the authority to *not* issue the cease and desist order and, in fact, should not have issued it until it concluded its rulemaking in the case.[32] It certified the Commission's general authority over boosters, but severely chastised the FCC for failing to act expeditiously on the issue. Scolding that "throughout the past 22 years of the Commission's life, it has failed to adopt rules under which signals from stations KXLY-TV and KHQ-TV, useless in Bridgeport without the booster, may be picked up, reinvigorated and made available to the residents of the town,"[33] the Court reversed the FCC's decision and remanded it back to the Commission.

This time the Commission moved quickly. By the end of June it had re-stated its concerns about boosters and closed its proceedings on the matter by formally rejecting all proposals for on-channel booster service; it would stick with its translators.[34] Within a week, however, Colorado Governor, Stephen McNichols, asked the Commission to once again open the proceedings to give booster operators a chance to prove they could operate safely. The Commission once again granted the request and held yet another round of proceedings, coming, a year and a half later, to again the same conclusion.

Throughout the unending process, the NCTA and other CATV interests waged a continuing campaign against approval of the boosters. In the short run at least, CATV operators could not compete with the much cheaper boosters and translators, so the repeater services delayed implementation of community TV in perhaps hundreds of small towns. While CATV operators were more focused in the first instance on exploitation of the larger communities of the West, such as Laramie, Wyoming, and Great Falls, Montana, many of the smaller towns were still appealing markets. The mushroom-like growth of the low-cost delivery alternatives meant that CATV was shut out of these towns for years while the FCC hemmed and hawed about the repeater services and while full-power broadcasters worked to establish their own dominance in these areas.

During this time, the NCTA filed repeated comments with the FCC, arguing that boosters would hurt existing VHF service and the associated CATV service as well. It sponsored field studies, including one by engineer-consultant Archer Taylor. The intent of the studies was to demonstrate the shoddy, amateurish construction and maintenance typical of the western repeaters. While acknowledging that booster or translator service was appropriate for some hard-to-reach regions that even CATV could not economically serve, the trade group lobbied strenuously for restrictive controls. Shapp, at one point, proposed a hybrid system where boosters would receive distant signals then feed them into coaxial systems for delivery to the home. Other CATV equipment manufacturers also asked for extended delays in the authorization of booster service. The industry, in short, wasted huge amounts of time, money, and emotional energy on the booster problem, because, ultimately and despite the delays it caused in CATV diffusion, it was a problem that solved itself.

From the privileged position of historical hindsight, it is easy to see that all the repeater technologies had a limited practical life expectancy. The FCC's effort to push translators, for example, turned out to be every bit as effective as its UHF allocation plan, and for many of the same reasons. The UHF converters needed for the translators were scarce and expensive; UHF transmissions, which were easily blocked by mountains and ridges, were ill-suited to the rugged western geography; and, FCC estimates to the contrary, many felt the construction and maintenance costs of UHF translators were substantially more than that of the simple boosters. By mid-1958, the FCC estimated that illegal boosters outnumbered legal translators by as much as ten to one. But the boosters too contained the seeds of their own destruction.

It was clear that both boosters and translators had no hope of survival for the simple reason that there was insufficient incentive to maintain them. Neither the social nor the economic context of the time provided a market niche for this technology. A CATV system and a broadcast station, in theory at least, generated revenue. Boosters and translators did not. They were the leisure-time hobbies of committed town groups, co-ops, and nonprofit corporations, such as C.J. Community Services. While some TV stations maintained satellite transmitters to extend their own signals, most of the tiny repeaters eventually fell into disrepair as interest on the part of the volunteer operators waned. They soon realized that running the repeaters took more time and money than they anticipated. As Malarkey later recalled: "We used every trick in the book to slow down the growth of translators, unsuccessfully I might add, but it turned out they were so low powered and the frequencies they were using were so high that if a signal bumped into a venetian blind it wouldn't go through it. The signal would be affected by the leaves of a tree, or the snow or the rainfall."[35]

Then, according to Malarkey, the operators found out that they had to service them—parts, labor, repairs, tubes, technical expertise. "It just didn't work."[36] The rise and fall of the broadcast repeaters was a clear illustration of several of the elements of an evolutionary perspective. Building on existing technical knowledge, they were part of the diversity of solutions thrown at the problem of "no TV." Within a commercial context, however, they could not survive. Proponents could not generate sufficient resources to maintain them and, eventually, technical solutions with a better social fit, specifically community TV and eventually regular broadcast TV, prevailed.

The repeater services, it turned out, were more of a political problem than a long-term competitive threat to community antenna operators. Moreover, they were only a part of a larger web of political difficulties spinning out of the western states. Several important additional elements were involved. There was the general problem of television service to smaller communities, epitomized by the struggle between repeaters, CATV and broadcasters. The larger policy question was how to provide TV service to underserved communities in light of the failure of the FCC's allocation scheme. There was also the issue of an increasingly strained relationship between Congress and the FCC. The Commission's awkward postponement of action on boosters constituted an on-going irritant to western political interests, and these political interests held powerful positions on important congressional committees. (This may, in part, be why the agency declined to actually prosecute anyone after the Bridgeport case. Instead, it kept extending the deadline for booster operators to switch to the approved service.) And the rift between the Commission and Congress was not good news for CATV insofar as the Commission, by both design and neglect, had largely left the industry alone, a situation that some senators thought ought to be corrected. Finally, few political problems are without an economic base, and the deeper tectonic force at work here was the unrelenting expansion of CATV. It was moving out of noncompetitive and, therefore, nonproblematic markets and into markets with established economic-political relationships. For a variety of reasons, therefore, the first major clashes between CATV operators and broadcasters came in the West.

Storm Clouds over Big Sky Country

The best towns for early CATV had several characteristics. They had a sufficient critical mass of homes to make economic sense; wiring rural areas was inefficient because of the high cost of building a physical plant to connect a few widely scattered houses. Attractive markets also lacked their own local broadcast service, but were close enough to an available signal to make community television technically feasible. The supply of towns thusly situated was soon exhausted, however. If community television was to expand, it would have to move into areas with available over-the-air signals, and even into towns with one or two TV stations. CATV, therefore, began spreading out from communities like Lansford and into places like Casper, Wyoming.

There were a number of implications to this migration. It meant that operators had to begin offering program choice; they had to provide signals beyond those already available over the air in the given market. They did this in several ways. Technically, CATV operators kept pushing at the limits of channel capacity, finding methods to advance from three channels, to five channels, and later to 12, as described in Chapter 6. But in many cases, a standard, low-band, five-channel systems afforded more capacity than was necessary to carry signals available locally off the air. The excess capacity was used in a few isolated cases for local origination, but more commonly for the importation of distant signals. Such importation, especially in the West, implied the use of microwave transport.

While Belknap was one of the first companies to attempt microwave importation, it was by no means alone in the effort. The use of microwave to import distant signals was one of those logical next steps quickly and widely grasped by the pioneers. Belknap was important in removing a regulatory barrier for microwave use. LeDuc points out that the FCC's final decision to authorize the Belknap application was the "sounding bell" in a surge of microwave-CATV development. In part because of FCC approval and in part because of advances in microwave technology, CATV operations now were free to construct their own relay systems or use the threat of construction to drive down the charges of existing carriers, primarily AT&T. Along with Belknap, one of the first people to take seriously the possibility of long-distance importation of TV signals was Bill Daniels. In 1953, while the FCC was still pondering the request by Belknap and Associates to build their own system, Daniels, in concert with AT&T, filed his request with the FCC to import Denver television into Casper. In November, the FCC authorized the service, the first of its kind.

While the struggle over boosters and translators was staged primarily in Rocky Mountain hamlets, it was in the larger, economically attractive towns, such as Casper, that the competing forces of broadcasting and CATV met for the real fight. Most of the serious action originated in Wyoming and Montana, principally because these were the respective home states of two of CATV's most vociferous opponents, two men who, more than anyone else would galvanize and lead a large segment of the broadcasting industry into a protracted regulatory contest with the CATV industry. The men were Ed Craney and Bill Grove, both regional broadcasters. Ed Craney was mentioned previously as the man who tried unsuccessfully to launch a statewide television service in Montana. Among his properties was KXLF-TV in Butte, Montana. Undaunted by the lack of support from his fellow broadcasters, Craney hoped to create his own interlocking network of full-power and booster or translator stations throughout the northwestern territory and, perceptively, saw CATV as an obstacle. Grove was the general manager and vice-president of KFBC-TV in Cheyenne, Wyoming.

As Bill Daniels and others began building their community antenna businesses in cities strung out along the eastern base of the Rocky Mountains, the signals from the stations owned by Craney and Grove became the raw material for their CATV product. Regional broadcasters took strong proprietary interest in their signals, however. In late 1954 and early 1955, the station owners spoke. The first complaints were lodged by Grove and by television station KOA out of Denver. KOA did not seem as interested in restricting carriage of its signal as it did in controlling it. In December 1954, the station sent a letter to Daniels and to J. E. Collier, who operated the CATV systems in Laramie, Wyoming, and Sterling, Colorado. The letter asked them to sign an "affiliate" contract specifying terms and conditions for carriage of its signal, but it did not seek to halt carriage or exact any fees from the CATV operators. The KOA letter, signed by executive vice-president Don Searle, explained: "The rapid growth of community antenna systems has convinced us that it is now necessary to formalize arrangements between our stations and companies which, on a commercial basis, relay our signals to homes of subscribers."[37]

Grove took a stronger position. He sent letters to the same systems, requesting that they halt the use of his station's programming. A few months later, Craney sent a similar letter to Norman Penwell, owner of Bozeman Community Television, Inc. in Bozeman, Montana, (the system was later purchased by TCI founder Bob Magness). Craney demanded that Penwell cease carriage of his Butte TV station, KXLF, stating that his signals were intended to be free to the public and were not transmitted for the purpose of realizing a profit for someone else.[38]

In 1953, in Asheville, North Carolina, the city council had refused to permit construction of a community antenna system following pressure from local broadcasters. WBTV had denied the community antenna backers permission to relay its signals and the holder of a local UHF construction permit had threatened to abandon plans for the station if a CATV were allowed in. The Rocky Mountain broadcasters were hoping for similar victories, but they were to be disappointed. In each of the western cases, the response of the CATV operators was to inform the broadcaster that they had no intention of changing any of their programming practices. The NCTA's general counsel, Strat Smith, did fly to Denver to meet with Grove and KOA officials, but little was settled. Smith, despite Daniel's use of microwave importation, recommended that the CATV owners adopt the industry's standard defense against legal assault, the "passive antenna" shield. Penwell's response to Craney presented the position in clear, simple terms: "The function of our company is merely that of receiving on behalf of our subscribers, the signals which you broadcast and to which they are entitled. This company does not realize any profit on the programs broadcast by your station...the remuneration received...is for the antenna service...and is in no way related to the programs which may be broadcast by your station."[39]

In the trade press, the response was summarized more bluntly, with *Broadcasting-Telecasting* declaring: "For the third time in less than a year, a community television operator has told a broadcaster to go jump in the lake."[40]

While the television stations took no immediately legal action in these cases, the NARTB did form a committee to look into to the issue and to begin lobbying the FCC to take control of CATV. Bill Grove was named chairman of the committee. But broader NARTB interest in the issue waxed and waned over the next several years. While Craney, Grove, and a few others were tireless in their efforts to beat back the community TV technology, most broadcasters had more important issues to attend to. CATV at this time reached less than 1 percent of the population. In communities in the Midwest and the Southwest, broadcasters and CATV operators were on good terms and enjoyed the symbiotic relationship that characterized the first few of years of cable growth in the East. It is important to note, at the same time, that while some broadcasters enjoyed the extension of their audience to other towns via cable, they typically did not report the additional viewership or ask that it be measured by the ratings services. In fact, the added audience was not necessarily an economic boon to broadcasters. An expanded audience size meant the station owner would have had to pay increased prices for syndicated film and programming usually sold on a per-thousand home basis. And they would be unlikely to recoup those costs through increased advertising revenue because local businesses were not interested in selling to distant communities. For the most part, however, CATV was a fairly small part of the broadcaster's world, locally and nationally. The broadcast industry at large, dominated by the networks and large market stations, was itself in the throws of rapid technological, economic, and regulatory redefinition; the tiny community TV business was far down on its list of concerns.

That said, it should still be noted that, despite the relatively small size of the western anti-CATV coalition, the issues this group raised were substantive and looming; if the broadcasters of Wyoming, Montana, and Utah had not advanced them in 1955, someone else soon would have. The real world complaints of the "Big Sky" broadcasters were, of course, economic, not philosophical. Community television constituted direct competition for existing stations and an impediment to broadcast expansion. But this relatively straightforward economic core problem radiated out into a complex set of intersecting issues involving business viability, local service, property rights, and even political ideology. All of these issues (and more) were deeply intertwined and embedded between the lines of the letters exchanged by Craney and Penwell. While the issues would unfold in increasing detail, and occasional confusion, over the next several decades, the seeds of virtually all of the important issues were manifest, as in Belknap, in these very first skirmishes.

Concerned broadcasters typically began with the blunt accusation of economic harm. CATV, it was claimed, would wound local broadcasting in several ways. First, it would bleed off or divert viewership of the local station. With CATV offering one or more alternatives to the hometown station, viewership would fragment and the broadcasters would attract only a portion of the potential audience. As viewership decreased, so would advertising rates, and thereby, revenue. Advertising income, it was argued, additionally would be reduced to the extent that national retailers felt they no longer had to buy time on the local station. If a CATV system was importing a distant signal into the local market, national advertisers who purchased time on the distant station were reaching that smaller market essentially free of charge. They need not pay directly to advertise in that city. So, went the argument, began the downward spiral of the hometown broadcaster, as the injured revenue stream limited the station's ability to acquire higher-priced, higher-quality programming. Inferior programming meant fewer viewers, which meant lower revenues and so on.

One of the most obvious places to look for salvation from this economic strangulation was the FCC. The first formal broadcaster petition requesting that the Commission take charge of CATV came from a West Virginia broadcaster in August of 1954. J. Patrick Beacom of WJPB-TV in Fairmont, West Virginia, warned that the uncontrolled proliferation of CATV systems constituted "a serious economic threat to allocated and established television stations."[41]

By invoking the language of "allocated and established" stations, Beacom was suggesting to the Commission that the issue of economic harm was not one isolated only to his station and his financial well-being. The economic harm model spoke to the nature of the public good at several levels. CATV, by threatening the financial viability of a local station, threatened the philosophy of local broadcasting, community expression, and ultimately the efficacy of the nation's system of electronic communication as a vehicle for democratic self-governance, at least according to the broadcasters. At the extreme, if a CATV system competitor forced the failure of a local broadcast outlet, the community would be deprived of an important source of local expression and reflection. The primary thrust, again, of the FCC's Sixth Report and Order was to ensure that every community, when possible, had at least one television "voice." Where CATV service prevented the development of a new TV station or forced the abandonment of an existing one, the underlying philosophical goals of the FCC's national plan for television service would be thwarted.

In addition, and to add "insult to injury," the raw materials of the CATV operation were obtained from the broadcasters themselves, at zero cost. This was not just unfair, according to the broadcasters, it was illegal. Broadcasters such as Craney claimed a property right in their signals. CATV operators were taking that property without the permission of the owners and reselling it. The TV station owner was unable, thereby, to control the use of his property and furthermore was denied any part of the profits realized in its resale.

Beyond the initial claim of common law ownership, broadcasters argued that CATV pickup and redistribution of their programming constituted an unauthorized "rebroadcasting" of the TV signal, a violation of section 325(a) of the Communications Act of 1934 (paralleling the violations of the illegal boosters). CATV operators, therefore, stood accused of violating both regulatory and civil law. This, in time, would come to be more commonly expressed by the complaint that CATV owners were "pirates" or "parasites" who made their money by "stealing" broadcast television signals.

The CATV industry had several set responses to these complaints. On the issue of economic harm, they pointed out that no evidence existed to support such a claim. And for reasons to be outlined later, it was a counter-claim that for many years was easy to document, even though it would prove at times to be of little persuasive force for regulators.

On the question of property rights, the industry, as per Smith and Penwell's initial arguments, stood by its passive antenna defense. Arguable in the original nature of community aerials, the position required even greater conceptual tolerance in situations involving microwave importation. The very concept of the passive antenna system, sitting benignly at the furthermost edge of a receivable signal—the fringe area amplifier—was giving way to a new technological and economic paradigm. CATV was beginning to obtain signals, or programming, from elsewhere in the country and, quite actively, transport those signals into the local system, in the same sense that the local grocer ships fresh produce into town from distant states.

From its very first fight over tax liability to its successful defense against copyright infringement in the mid-1960s, the industry survived legally on its posturing as a passive antenna service, and for the canyon systems of 1952, this seemed a rational, defensible theory. When CATV operators began, however, to establish microwave repeater stations, identify distant TV signals as likely targets, and selectively import those signals into new markets, the ingenious legal fiction of passivity arguably began to weaken. Broadcasters could more powerfully claim that CATV operators were taking an active role in program acquisition. Microwave importation of television, suggested CATV opponents, was only a passive extension of the viewers' home antenna in approximately the same sense that AT&T long-distance lines are a passive extension of the home telephone. Despite these arguments, however, CATV would remain loyal to its passive antenna position and it would prove in the long run to be a wise choice.

The three concerns—economic harm, the subsequent injury to public interest, and the alleged violation of property rights—would become the primary engines driving cable's legal battles at the FCC, in Congress, and in the courts. While the television stations in the Belknap and West Virginia cases may have been the first to complain formally about CATV, they were not going to be the last.

Call to Battle: 1956–1960

The FCC, as noted, has been criticized for failing to be more proactive during this period, for failing to have a consistent and fully conceptualized plan for the integration of CATV into the scheme of national broadcasting, and much of the criticism is warranted. For most of the 1950s, the FCC's chief response to CATV and the complaints it gave rise to, was silence. With the exception of the Belknap case—in which the CATV issue was deftly separated—and a purely technical rule regulating electrical emissions in 1956,[42] the FCC took no substantive action on CATV from its inception around 1949 until 1958, at which time it stated it could and would do nothing. The Commission might have, during those intervening nine years, adopted rules governing CATV, in conjunction with UHF and other services, designed to increase diversity in both programming and transmission systems in the United States and to help rationalize and save a failing allocation scheme.

Disparagement of the FCC's policy of benign neglect can be overstated, however, for while the Commission may not have had a plan, it did have a philosophy, and, as illustrated in the growing regulatory controversy over CATV in the late 1950s, it applied that philosophy with some consistency and for as long as it could in the face of powerful, growing political opposition.

John Doerfer and the FCC Response

The ideological touchstone for the Commission in the latter 1950s was that of the free market. Commissioner John C. Doerfer best exemplified the free market position and was most often the Commission's spokesperson on CATV.

A case study of Doerfer's role also underscores the principle that individuals, when situated in a position of power, can and do make a difference in the forging of the relationship between technology and society. Legal scholars have noted not just the obvious power of individual FCC Commissioners but also the influence of their particular personal and ideological backgrounds. Regulatory analyst Lawrence Lichty has examined FCC decision-making through such a lens, explaining:

> [C]hanges in the direction and emphasis of the Commission's regulation of broadcasting are a function of the members serving on the Commission at those specific times. Further, the personal experience, education, occupational background, and governmental philosophy of the members of the Federal Radio Commission and Federal Communications Commission directly influence the direction and emphasis of the agency's policies.[43]

Erwin Krasnow, Lawrence Longely, and Herbert Terry have additionally pointed out the special leverage often available to the Chairman of the FCC, both in bureaucratic terms to set agenda and influence priorities and, also when possible, to use the power of the office to help forge important voting coalitions among Commissioners.[44] As an FCC Commissioner and eventually Chairman Doerfer was well placed to extend, through his regulatory position, his political philosophy, and he did not hesitate to exercise the power.

Doerfer had been Chairman of the Wisconsin State Public Service Commission when he was named to the FCC in early 1953 by recently inaugurated President Dwight D. Eisenhower. Broadcast historian Erik Barnouw noted that Doerfer was a protégé of anti-Communist crusader and Wisconsin Senator Joseph McCarthy, and Eisenhower hoped to placate McCarthy and McCarthy's sympathizers with the Doerfer appointment.[45] Eisenhower also appointed, later in 1953, Commissioner Robert E. Lee, also a close friend and confidant of McCarthy.[46] Eisenhower ultimately named six new people to the Commission between 1953 and 1958 and elevated Doerfer to the Chairmanship in July 1957.

Largely as a result of Eisenhower's selections, the tenor of the FCC changed markedly in the 1950s. The regulatory activism of the New Deal FCC under presidents Franklin D. Roosevelt and Harry S. Truman was replaced by a Commission much more inclined to let market forces play out, unfettered by government regulation. Historian James Baughman additionally notes that the new outlook was largely in keeping with public opinion in the 1950s, which privileged free enterprise as a cultural value.[47]

While this laissez-faire ideology was a crucial factor in FCC inactivity in CATV, it was not the only factor. Other forces working against regulation included the relative size of the CATV problem, a grossly inefficient administrative process, and the personal failings of the commissioners themselves. For the FCC, as for the broadcasting industry, CATV was still a very small part of the national television system and was easily shunted aside amidst the flurry of larger problems the Commission was faced with, including oversight of the quickly evolving broadcast TV business. The FCC's administrative process was hamstrung by congressionally mandated procedural restrictions. Intended to safeguard the due process rights of broadcasters and the public, these constraints limited contact between the Commissioners and the FCC staff, leading to a lack of communication and negotiation within the agency. The results were often Commission decisions that reversed or ignored staff recommendations. These problems were exacerbated by a hostile Congress that often inserted itself into the detailed affairs and cases of the Commission through both formal proceedings and informal pressure. The hostility of Congress, in turn, was partly the product of a Commission that came to be known as much for scandal and inefficiency as anything else.

Following Barnouw's unflattering characterization of the ideology of the Eisenhower Commissioners, others offered the criticism that they were also, as a group, largely incompetent and often corrupt. Lucas Powe Jr. wrote that if Doerfer was not a McCarthy man; in fact, Commissioner Robert E. Lee was. With George McConnaughey, a well-connected Republican, these three constituted the new Republican core of the Commission. Joining them, according to Powe, were: "Richard Mack of Florida, pliable and corrupt; conservative Democrat T.M. Craven, from the Storer Broadcasting Group; House Speaker Sam Rayburn's nephew Robert A. Bartley; and lifetime FCC employee Rosel Hyde."[48]

Baughman paints one of the more colorful pictures of the Commission, which had abandoned activist regulatory philosophy for a set of much closer ties to the industry under its charge: "In the 1950s the Commission entered what one former chairman dubbed 'the whorehouse era.' Another commissioner, borrowing on the same metaphor, described it as the time 'when the commission lost its virginity, and liked it so much that it turned pro.'"[49]

"By the late fifties," states Baughman, "the FCC was a regulatory rat's nest, waiting to be exposed."[50] The late 1950s, in many ways, was a difficult time for television and for the FCC. Changing programming practices by the networks, the replacement of the so-called "Golden Age" programming with cheap westerns and game shows, made the networks easy targets for popular TV critics and, more importantly, for observers in Congress. The problems were compounded by the noted "whorehouse era" of FCC regulation, a situation described somewhat more formally in the academic literature as the "capture theory" of governmental regulation, in which federal regulators and industry leaders develop a relationship too friendly to be of honorable service to the public.

The FCC has been the "bete noir" of Congress almost from its inception. Krasnow and Longely have pointed out that "throughout its nearly fifty-year history, perhaps no federal agency has been as frequent a target of vilification and prolonged investigation by Congress as the FCC."[51] At no time was that truer than the late 1950s, when congressional investigations brought to light the incestuous relationship between Commissioner Mack and some broadcasters, and cast similar aspersions on Doerfer. Congressional probes, spearheaded by Representative Oren Harris (D-Arkansas), Chair of the Commerce Committee and the Special subcommittee on investigations into the infamous network quiz show scandals of 1959, cost broadcasters dearly, but cut even deeper wounds into an already damaged FCC, as the agency was appropriately blamed for its lack of oversight. Harris's investigations in 1957 and 1958 of FCC *ex parte* contacts with industry officials led to the forced resignation of Commissioner Mack (who was later indicted by a Miami grand jury for bribery). Doerfer faced similar pressure, but resisted with the support of Eisenhower and managed to hold his position on the Commission. Finally, however, his admission of spending a week on the fishing yacht of broadcast mogul George Storer was too much for even Eisenhower, who asked him to step down in March of 1960.

Doerfer has been held out by a number of analysts as a case study of an industry stooge. He certainly was a staunch defender of the networks in their battle against most forms of regulation, speaking strongly in defense of the networks even during the infamous quiz show scandals. Some have also suggested AT&T was influential in putting Doerfer and some of his co-commissioners in the FCC.[52] He was unapologetic about his socializing with industry executives, stating it was folly not to take advantage of their expertise and hear the issues of those he was appointed to regulate and, within limits, he had a point. By looking only at his pro-industry positions, critics miss an important subtlety, for Doerfer was an antiregulatory champion, even when regulation would have benefited broadcasters.

Doerfer's touchstone was market independence and regulatory restraint, in the extreme. This only reflected the broader political agenda of the Eisenhower administration, which saw itself, according to some analysts, as an administration heavily committed to promoting the interests of business and industry generally.[53] Doerfer's philosophy, therefore, was usually consonant with those of the broadcast industry, but such was not the case in every instance, and this goes far to explain his position of CATV. From within an almost Darwinian philosophy of unrestricted free market struggle, it is easier to see how Doerfer could dismiss concerns, for example, about broadcast industry concentration of ownership: "Concentration does not bother me. . . . Somebody has to be dominant. Somebody has to be big,"[54] and at the same time, scoff at broadcaster suggestions that they ought to be protected from new competitors.

Doerfer's political position with respect to the newly emerging technology and business of CATV, therefore, would not have been difficult to predict. His views were unveiled in his first public speech on the emerging industry in November of 1954. Only a few months after WJPB-TV in Fairmount,

West Virginia, had asked the FCC to regulate CATV, Commissioner Doerfer told a Chicago convention of the National Association of Railroad and Utilities Commissioners that he saw no good reason for the government to get involved in the new television service. He stated: "In my opinion it is doubtful that the Federal Communications Commission has jurisdiction over the community antenna television system, as such, particularly with respect to the regulation of rates charged for installation or monthly charges for services rendered. Jurisdiction of the Federal Communications Commission over these functions is not only doubtful but in my opinion undesirable."[55]

Jurisdiction was doubtful, according to Doerfer, because (1) no clear cut statutory law granting such jurisdiction existed; and (2) no evidence indicated that CATV constituted broadcasting, common carriage, or interstate commerce by wire under the understood intent of the Communications Act. Jurisdiction was undesirable, in his view, for both practical and philosophical reasons. Practically, Doerfer dreaded the specter of rate regulation—the bureaucratic nightmare of Commission oversight of the rates and services of potentially thousands of small CATV systems. "If such duties are suddenly thrust upon the Commission," he declared, "it has neither the personnel or funds to undertake them at this time."[56]

The heart of Doerfer's regulatory reluctance, however, appeared to lay in his free market ideology. He acknowledged that the sharpest attack on CATV came from the possibility of the new service frustrating the Commission's cherished policy of providing television service to all parts of the country. He was keenly aware of the competitive threat of CATV in the smaller markets. For Doerfer, however, such competition was a natural and positive force in the marketplace. As he explained, a broadcaster asking for legal protection from a CATV operator was as "untenable" as "a gas company asking protection from an electric company or visa versa."[57] It was the choice of the consumer as to which service would prosper and which would fail; the community would chose between the localism provided by the broadcaster or the variety provided by the CATV system. "To date," he stated, "the Commission has never denied an application for a broadcast service upon the grounds of 'too much competition.'"[58]

In this, Doerfer and the rest of the Eisenhower Commissioners had some support in judicial interpretation of the Communications Act. The Supreme Court had ruled in a 1940 case,[59] that potential economic harm to competitors did not constitute justification for the denial of a broadcast license. In that case, an existing radio station in Dubuque, Iowa, had tried to block the construction of a new station by the local newspaper claiming there wasn't enough advertising to support one, let alone two stations. The Supreme Court held that the Communications Act "recognizes that the field of broadcasting is one of free competition."[60] Importantly, the Court qualified its ruling, stating, "We hold that resulting economic injury to a rival station is not, in and of itself, *and apart from considerations of public convenience, interest, or necessity,* an element the petitioner must weight. . .in passing on an application for a broadcasting license."(emphasis added).[61]

The fundamental philosophical bias of Doerfer, and perhaps the national political climate, was such that the Court's rule concerning harm to a competitor was more prominent and lasting than the qualification concerning the public interest. Doerfer's free market mantra was repeated, a few months after his Chicago speech, in a June 1955 address before an appreciative NCTA convention in New York: "The avowed objective of the FCC is to make possible for everyone in the U.S. at least one free television service. In my opinion it would be more consistent with American philosophy to accomplish this by providing opportunity rather than imposing artificial restraints or outright prohibition of a competing CATV service by governmental fiat."[62]

Doerfer acknowledged that CATV could possibly be subject to either state or local regulation, but that was a question that only time and the courts could answer. He declared that if CATV were to be regulated as broadcasting, existing over-the-air telecasters would have to compete evenly with CATV. "[B]roadcasters are in a free competitive field," he said, "They are not entitled to protection solely from competition of other broadcasters." Regulation of CATV as a common carrier, he added, could lead to the privileging of CATV over broadcasting, and even open the door for eventual rate regulation of broadcasters themselves.

For Baughman, the Commission's failure to more actively intervene in the UHF dilemma was consistent with the "Eisenhower majority. . .advocacy of the free market as broadcast regulator."[63] The same was clearly true for CATV. The clarity and strength of this philosophy is highlighted by the Commission's clear and detailed understanding of the complex issues posed by CATV, an understanding evidenced by the series of internal agency memos, beginning in 1952, that were discussed previously. Revealed within those memos is a sophisticated debate over the legal standing of community antennas. The discussions typically began with the attempt to define CATV within the context of the Communications Act. Section 3 of the Act defines specific terms in the law; unfortunately, those definitions are often vague and therefore subject to political contestation, what in contemporary terms would be considered "spin control," and the stakes in these sparring matches were high. At issue for CATV were sections 3(a), which defined "communication by wire"; 3(b), which defined "communication by radio"; and 3(h), which defined "common carrier."

The 1952 memo concluded that CATV could not be characterized as broadcasting under section 3(o), insofar as it did not use the airwaves and could not be fairly considered broadcasting in the common sense of the term. Conversely, the staff in 1952 did find CATV to be communication by wire, under section 3(a). CATV was, in their eyes, interstate and foreign commerce because the TV signals carried by systems typically originated from the networks and crossed, at some time, state lines. Finally, systems offered their product to any and all customers willing to pay the monthly fee, held it out for hire, and so could reasonably be classified as "common carriers" under section 3(h). The memo suggested that control over CATV would primarily involve regulation of rates and services, following the model in telephony.[64]

The record is incomplete on the internal memos concerning CATV during this period and is composed primarily of documents included or referred to in the materials submitted by the Commission during the Senate's 1958 and 1959 hearings on television. However, through August of 1954, when Fairmount filed its request that FCC take control of cable, through November of 1954 when Doerfer gave his speech in Chicago, through early and mid-1955 when tensions began running high between broadcasters and CATV operators in the West, the FCC remained deadly quiet in public. In October of 1955 Richard Solomon, an attorney in the FCC's General Counsel's office, penned a memo referencing two earlier but undated documents, on the question of FCC jurisdiction.[65]

The earlier, but undated, notes were identified as the "Smith-Neustadt" and the "Straussburg" memorandae.[66] The Smith-Neustadt memo, which may have been the original March 25, 1952, discussion paper or its progeny, argued, that CATV could and ought to be regulated as a common carrier, but did not qualify as broadcasting. The "Straussburg" memo argued that CATV could be regulated as neither and the Commission had no authority in the area. Solomon weighed into the internal debate on the side of Smith-Neustadt. Interestingly enough, Solomon leaned heavily on the concept of broadcaster property rights as the lever for prying open FCC authority. Common carriers, he argued, by definition carried some form of property, as in railroads and bus lines. Television programming, he found, could be considered property belonging to broadcasters. "The CATV systems," he concluded, "could be said to carry property (that is, a particular television program) from their receiving antenna to the homes of the consumers."[67] This property was furnished on a nondiscriminatory basis to everyone in the service area and, therefore, CATV systems could be considered common carriers. The implications of the analysis went far beyond the issue of common carrier to touch on the critical property rights issue. Despite the lengthy and sophisticated internal examination, however, the Commission did not act; it did not respond to CATV publicly.

Frontier Broadcasting

The on-going silence of the Commission ran parallel with the building tensions in the West between broadcasters like Craney and the regional CATV systems, until April 6, 1956, when a coalition of thirteen broadcasters, led by Craney and Grove, filed a complaint against 288 CATV systems in thirty-six states, all of the known systems in operation at the time.

The case was officially known as *Frontier Broadcasting v. Collier*,[68] and it would become the site of the Commission's first formal pronouncement on the legal status of CATV. Frontier Broadcasting was the company home for Grove's KFBC. Following the logic of the 1952 FCC memo, the coalition of western broadcasters played the common carrier card, arguing CATV ought to be regulated under Title II of the Communications Act and subject to regulation of charges and practices. Reciting the by-now standard list of complaints, the coalition claimed that CATV importation of large market broadcast signals into a small town would discourage advertisers from buying redundant time on the small town stations. The local broadcast station, therefore, would succumb, over time to the CATV competitor and the intent of the 1952 Report and Order would be frustrated. Furthermore, rural viewers, who could not in any event subscribe to the CATV system and whose only source of television would be the small town station, would be denied all TV service. The broadcasters requested that the FCC assume jurisdiction over all CATV systems or, at the very least, begin formal hearings on the problem.

While the Commission's attention was more heavily drawn to the issues of Pay-TV, UHF and even boosters and translators at this time, the staff eventually got around to considering the CATV problem, and on July 25, 1957, more than a year after the *Frontier* filing, it penned an internal memo to the Commission.[69] The memo did not bode well for the broadcasters. The FCC's Broadcast Bureau, Common Carrier Bureau, and General Counsel were unanimous in advising the Commission to stay out of CATV. While the staff felt that CATV could be regulated as a common carrier, it would be counter productive to do so. Such authority, argued the staff, would still not be sufficiently expansive under existing law to require CATV systems to obtain operating permits. The Commission could not, therefore, control entry into the market, which was what the broadcasters were really seeking. As to the economic argument itself, the staff found no evidence of a detrimental impact on the development of local broadcasting and no evidence that CATV posed a threat to the Sixth Report and Order. "While CATV systems undoubtedly have some impact on the development of other television services, it is clear that broadcasters are not deterred from establishing stations when the economics are deemed favorable," the memo stated.[70] Finally, the staff again voiced the concern that any attempt to saddle the Commission with jurisdiction would stretch its resources beyond the breaking point. Therefore, the staff proposed that the FCC seek congressional legislation to amend the language of the Communication Act to remove any and all possibility that the Commission would ever have to regulate CATV.

It was clear that the themes sounded by Doerfer in his speech a few years earlier were becoming the backdrop of FCC thinking on CATV. Even without his participating in the vote, the Commission in January of 1958 rebuffed the next formal complaint against CATV. A microwave service had applied for permission to carry several Montana and State of Washington TV signals to various CATV systems through Montana. A nonprofit organization operating one of the area's many TV boosters complained that importation of the signals into their town, Havre, Montana, would adversely affect voluntary contributions needed to continue operation of the booster and could inhibit the development of a broadcast station in their town. The booster club wanted the microwave license rejected.[71] The Commission said "no." The booster club's complaint was with the CATV system, not the microwave system, said the Commission. More importantly, financial injury to the club was private injury not demonstrably public injury and "we can find no reason to attempt to deprive the public a choice between competing modes of television service."[72]

The development of the Commission's philosophy on CATV was, at least in retrospect, unambiguous, and when it finally did respond formally to the *Frontier* complaint, the decision was narrow, almost terse. The Commission had reviewed the July 25, 1957, staff memo and instructed the staff to prepare a memorandum opinion and order "disclaiming jurisdiction over CATV and dismissing the (Frontier) complaint."[73] The draft order was presented to the Commission in March of 1958 and adopted formally on April 2, 1958. Despite the years of debate and discussion, it was the FCC's first official ruling on the legal status of CATV.[74]

The opinion reviewed the complaints against CATV, including the economic issue and its relationship to the Sixth Report and Order. The Commission did not responded directly to those complaints, however, and the decision was glaring by their absence. Neither was the property rights question raised. The language and logic of the decision rested solely on the narrow question of CATV's status as a common carrier. The Commission noted, but did not adopt, the NCTA's position that CATV was only a passive extension of the viewer's antenna. While the definition of what, in law, constituted a common carrier was, at best hazy, admitted the Commission, it felt that congressional intent was to define the term in its ordinary or traditional sense. Here, the staff took a half step back from its earlier opinion that CATV could arguably held to be a common carrier. While conceding that CATV shared some characteristics of a communications common carrier, it differed, said the Commission, in one crucial respect: The system owners, not the customer, chose the information to be carried. "No individual subscriber has the option nor may he compel the CATV system to receive or deliver a particular signal at a given time."[75] The CATV operator's editorial discretion separated it, for example, from a telegraph or telephone company. Echoing, literally, the 1957 staff memo, the decision noted parenthetically that even if common carrier jurisdiction could be invoked, it would not supply the regulatory tools necessary to prevent the development of new CATV systems in small towns. Finally, and very briefly, the Commission observed that CATV could not arguably be regulated as broadcasting since it used wires rather than radio transmission of signals.[76]

In one sense, *Frontier* brought to an end to several years of not-so-patient waiting by the affected industries. But if the FCC thought, as the *Broadcasting-Telecasting* report suggested, that it "had washed an old irritant out of its hair,"[77] it was painfully mistaken. The opinion and order instead presented at long last a large and solid target for various stake holders to shoot at, and the firing commenced immediately.

Within two weeks of the decision, Frank Reardon, the owner of KGEZ-TV in Kalispell, Montana, had written Senate Commerce Committee Chairman Magnuson complaining that competition from a CATV system had forced his station off the air and the FCC was culpable. The broadcast coalition in *Frontier* filed a petition requesting a reconsideration of the decision, and Commissioner Rosel Hyde faced angry western broadcasters at a meeting arranged by Senator Mike Mansfield (D-Montana) in Pocatello, Idaho. Magnuson announced he would call the Commission before his committee to explain its decision in *Frontier*, and *Broadcasting* magazine labeled CATV "a young monster." At month's end, the National Association of Broadcasters (NAB) had convened its annual national convention in Las Vegas, where one of the hot topics was the FCC and CATV. Groups of frustrated broadcasters gathered in the halls and around dinner tables to share their complaints; the long-dormant NAB Committee on Community Antenna Television was reactivated, with Grove as chairman and Craney assisting. When the FCC commissioners took their places before the convention delegates, it was clear that the policy debate over CATV was not over. As a result of his Idaho meeting, Hyde reported that the scales had fallen from his eyes, and he now perceived more clearly the plight of the broadcasters. He worked to convince his fellow commissioners to reconsider the matter and, at the convention, announced that the FCC would begin a new study of the CATV problem.

The Commission's decision to sidestep CATV regulation had, ironically, thrust CATV into the spotlight. *Broadcasting* published its most detailed and lengthy piece to date on the history and issues involved. The article began by observing that "Television's bonus baby—CATV—has suddenly turned into a problem child."[78] It noted that while the networks and large markets had historically treated CATV as a happy accident that extended their market, the "young monster" now threatened small market telecasters. It also worried that the wired networks could be used in conjunction with proposed Pay-TV systems, a thought that chilled every broadcaster. CATV could and did compete successfully with broadcasting and was unlikely to go away of its own accord, according to the industry publication. It listed a few instances in which CATV operations had stymied the development or continuation of local TV stations and it reviewed the property rights issue, noting that numerous stations had warned CATV operators not to carry their signals, and that, in each and every case, the broadcasters had been ignored. As yet, no broadcaster had been willing to take the issue to court, however.

Against this thickening background, Magnuson prepared his witness list for the Senate hearings, the FCC scrambled to open a new docket on the problem and the NCTA met behind closed doors to plot its response. Magnuson's long-range goals were clear—to protect and advance the interests of broadcasters generally and western broadcasters in particular. Even at this relatively early date in the development of broadcasting, politicians understood the powerful role radio, and now television, was playing in political campaigning; maintaining a good relationship with the broadcasters in their state was a simple matter of reelection common sense. The hearings were not designed around any legislative proposals, but rather were intended to bring more pressure to bear on the FCC and, if necessary, to provide a record for possible subsequent legislation. Hearings were scheduled to begin on May 27, 1958, and the list of witnesses for the first day was a roster of aggrieved television station owners, including Grove, Craney, and Reardon, whose now-dark KGEZ-TV station constituted the prime bit of evidence against community antenna television.

Special counsel for the committee during the hearings was Kenneth Cox, an attorney from the Pacific Northwest and former legal assistant to Magnuson. Cox, now on Magnuson's staff, had served as special counsel on earlier television hearings before the panel. He would eventually come to play a major role in the regulatory fate of cable television and this set of hearings would be his introduction to the CATV field.

In something of a preemptive strike, the Commission, on May 22, only days before the Senate hearings began, issued its "Notice of Inquiry" in the CATV problem.[79] The Commission presented a list of fourteen questions for public comment. It wanted, generally to know the dimensions of the problem, how many people were served by competing CATV and broadcast stations, and to what extent CATV service impaired local broadcasting. It was interested in comments on the possible legal rationale for asserting jurisdiction and on any proposals for possible legislative action. Among the important policy questions were whether or not economic injury served as a justification for FCC intervention, and whether the public interest was better served by a single, local broadcast voice or a multiplicity of imported CATV signals. The latter was particularly surprising in its implied questioning of the long-standing Sixth Report and Order philosophy of localism.

Meanwhile, the board of directors of the NCTA huddled in their Washington offices hammering out a strategy to meet the sudden rush of unwanted attention. According to Smith, the FCC's notice in docket 12443 of the possibility of requiring broadcaster permission for the carriage of signals is what got the NCTA board to take seriously the possibility of regulatory action. "That's what really put the fear of the Lord into the Board," said Smith.[80] The result of their deliberations would be a "bend but don't break" strategy on regulation, in which the NCTA would acquiesce on some issues in order to be able to influence the overall development of regulation in hopes of controlling and containing it.

On Tuesday, May 27, FCC Chairman Doerfer took his seat as the first witness before the Senate Interstate and Foreign Commerce Committee hearings on CATV and defended the FCC's position. In aggressive questioning by a clearly probroadcast Ken Cox, Doerfer reiterated his regulatory philosophy: "In the development of community antenna systems... I felt that I could no more protect a television station from a community antenna service than I could protect it from a local drive-in movie house or whatever other form of attraction might develop within the service area of a television station."[81]

The sparring between Cox and Doerfer lasted throughout the day and touched on a wide range of issues, including the important underlying questions of whether public service was best achieved by a number of imported signals or one local signal and the degree to which a CATV system might or might not effect economic harm on local broadcasters, questions for which no one had any real answers. Ultimately, Doerfer said the FCC was not asking for legislation and, in fact, preferred that the Committee hold off on proposing legislation until the FCC could complete its investigation in docket 12443.

Vincent Wasilewski, head of the NAB, told the Committee that CATV represented "a real and growing threat" to both the TV allocation plan and to small town broadcasters.[82] The NAB argued that the FCC did have authority to regulate microwave-fed CATV systems and ought to do so when the well-being of a local broadcast station was at stake.

The Committee turned then to the broadcasters themselves. One after another, the aggrieved western television station owners took to the microphone to list their complaints. About a dozen broadcasters, including Craney, Grove, and Reardon, recited the litany of concerns about economic hardship, broadcaster property rights, FCC inaction, and the philosophy of localism explicit in the Sixth Report and Order. CATV was accused of destroying local television, stealing broadcaster property, and generally of being a "parasite" on legitimate television. The FCC was asked, in essence, to "rope" and "brand" CATV.[83]

A small contingent of representatives of the CATV industry was scheduled to testify on the third and final day of hearings, but the long list of broadcast witnesses and the abbreviated session of the second day forced a postponement of their testimony. In fact, the appearance that the hearings were more political than substantive was underscored by the poor attendance of committee members. At times, only one or two senators were present and the session was halted completely after about three hours on the second day, to the obvious annoyance of the witnesses, when no senators could be found to attend.

Beyond the regulatory and political substance of the hearings was one interesting side note. For perhaps the first time in the written record, the term "cable television" was used to refer to the industry instead of the then accepted "community antenna television." The reference, ironically enough, was from CATV nemesis Ed Craney, who noted almost off-handedly, "Cable television is commonly called community antenna television."[84]

In mid-June, about two weeks after the first round of Senate hearings, the NCTA held its Seventh Annual National Convention in Washington, D.C. The previously scheduled panels and meetings were cancelled or postponed as members gathered to plan a strategy to meet the regulatory challenge. On the second day of the convention, delegates left the hotel and fanned out across the Capitol to lobby their senators and representatives. Meetings were held with the NAB and the FCC, and Doerfer reassured the Association that he thought the FCC would "stand firm" against the pressure.[85]

Later that month, on June 24, the Commerce Committee reconvened to listen to the CATV operators' side of the story. Witnesses included representatives of the NCTA, city officials from small towns served by CATV, local operators, and, of course, Milton Shapp. They testified that their industry served an important public role in bringing television to many who would not otherwise be able to receive it, and that any complaints came from a very small group of disgruntled western broadcasters who blamed CATV for financial problems attributable, in fact, to the nature of broadcast economics in those special situations. Furthermore, and more importantly, CATV, as a passive antenna service, was not liable in the area of broadcaster property rights, they stated.

Lloyd Calhoun, a New Mexico CATV operator and newly elected Chairman of NCTA appeared with the organization's Executive Director, Edward Whitney, and General Counsel Smith. Smith presented most of the NCTA case, arguing again that the FCC had no formal authority to regulate CATV. In an attempt, however, to appear flexible and at the same time shape the nature of future regulation, he presented NCTA's newly developed position that the industry was open to limited, reasonable control:

> ... If an appropriate area of Federal regulation can be devised which will assist in the orderly development of a nationwide competitive system and guarantee to the CATV system the rights and obligations which are inherent in regulation, as well as subject such systems to the duties and responsibilities imposed by the legislation, this industry, or at least that portion of it represented by the National Community Television Association, is willing to cooperate in drafting and sponsoring such legislation before this committee.[86]

The NCTA's position was developed, as much as anything, out of fear of an array of regulations, including possible state control, common carrier and rate regulation, and, most importantly, possible copyright liability. This was the first public sign of the industry's desire to cooperate with proponents of regulation, and, for its own purposes, help nudge the direction of rulemaking away from the most egregious possibilities. While it may have planted a hopeful seed of compromise among some of the interested parties, it later would bear bitter fruit for many of the players.

In the meantime, the message from Magnuson's hearings was not completely lost on the FCC. On the one hand, the Commission seemed to stand by some of its earlier pronouncements. In response to a list of posthearing follow-up questions by Committee Counsel Cox, the FCC reiterated its philosophy that the financial interests of individual stations had to be balanced against the public's interest in a diversity of signals and that there was nothing in existing law that clearly gave them authority over CATV or other forms of technology that extended the TV signal into underserved areas. In this, it foreshadowed much of what it would have to say several months later in its report in docket 12443.

Revealing its ambivalence and caution, however, the Commission in early October announced a modification of its policy on the granting of microwave applications to carriers providing service to CATV systems. Under the new policy, the Commission would approve applications only for towns which had no existing TV station and no chance of getting one. All other applications were placed into a "deep frost" pending outcome of the docket no. 12443 investigation. The intent was, in contrast with other FCC pronouncements, to protect the financial health of the existing local stations. The frost prompted several microwave companies, including one called Carter Mountain Transmission Co., to file suit in federal district court seeking to force the FCC to act on the applications. The Commission, did, in fact, grant a few applications in cases not involving local broadcasters, including one for Carter Mountain; for the most part, however, new attempts at importing distant signals came to a halt. This change of administrative course was important in several ways. First it was, to some extent, a response to the broader political troubles of the FCC at the time, and second, it was the first link in a chain of events that would culminate several years later in a major transition in FCC policy on CATV.

The Cox Report

On January 12, 1959, the Senate Committee's staff report, authored by Cox and immediately dubbed the "Cox Report," was released to the public (although formally dated December 26, 1958).[87] In keeping with the theme set by the western Senators, the Cox Report was seen then, as now, to be generally critical of, if not openly hostile to, the position of the FCC. Cox was broadly disparaging of most Commission activity, or lack thereof, in the area of not just CATV, but of boosters, satellites, and translators as well. His comments on the Commission's *Frontier* report, were illustrative: "However indefinite its rationale may have been, the Commission's final action in the matter made it perfectly clear that it did not intend to regulate CATV in any way whatsoever. . . . What is disturbing is the long delay on the part of the Commission in deciding the question, and the apparently haphazard manner in which it reached its final result."[88]

The report recited, sometimes with implicit and sometimes explicit approval, the list of charges against CATV: damage to the goals of the Sixth Report and Order, destruction of the aim of local service, unfair competition with local broadcasters leading in some cases to economic failure of local station, and so on. As to the latter charge, Cox used a then-recent Federal Appeals Court decision to suggest the Commission revisit its long-held philosophy of privileging market forces.

In *Carroll Broadcasting Co. v. FCC*, the Disrice of Columbia Circuit Court cast a jaundiced eye on the FCC's repeated reluctance to consider economic harm to local broadcasters.[89] The Commission had eagerly used the Supreme Court's 1940 *Sanders* guideline, rejecting the relevance of competitive harm, but had conveniently ignored the Court's qualifying consideration of the public interest element in such situations. The Appeals Court, in *Carroll*, redirected the Commission's attention to the question of how competition might affect the community. In circumstance similar to those of *Sanders*, the Court noted that, "economic injury to an existing station, while in and of itself not a matter of moment, becomes important when on the facts it spells diminution or destruction of service."[90] If the public is better served by one healthy station than two weak ones, then the FCC's choice is clear, suggested the Court. A laissez-faire philosophy was good only insofar as it advanced the broader public interest. Cox's use of this new decision in his report was direct and blunt, attacking what he called the Commission's "comfortable theory that it is not responsible for any of the consequences of the forces of competition. It is to be hoped that this refusal to accept responsibility for the direct consequences of its acts will have been somewhat tempered by the decision of the court of appeals in the *Carroll* case."[91]

Cox acknowledged only three situations in which CATV systems may have played a role in the shutting down of a local station and admitted that the facts in each case were inconclusive. He nonetheless moved to the conclusion that "there is a good deal that seems almost self-evident about some aspects of this claim that there is danger that broadcast service in smaller communities will be destroyed."[92]

He stated categorically that local service should be privileged over multiple-channel service when the two conflict and recommended that CATV systems be required to carry local signals if they did not do so voluntarily. He also argued against duplication of local broadcast programming, suggesting that regulation might be necessary in the absence of voluntary compliance. Cox concluded that Commission had to exercise authority over the entire system of radio and television, because the TV industry could not "thrive and grow" if it continued to be only "half-regulated." He urged to Commission to quickly conclude docket 12443 and, if it failed to find it had authority over CATV, to immediately seek such authority from Congress. In his proposed scale of five "television values," local stations were most valuable and CATV, at number 5, least valuable.[93] The Committee itself did not take direct action following the report and, in fact, never formally adopted it, but the message was nonetheless clear.

The 1959 Report and Order

Within a few months, the FCC issued its reply to Cox and the Senate Committee. Its April 1959 Report and Order in docket 12443 was the most expansive and definitive statement yet by the FCC on its perception of the role of CATV in the national scheme of broadcasting,[94] and it could not have made Cox, Magnuson, or the western broadcasters happy. The Commission either rejected or ignored the most of their arguments and philosophy. At the same time, the CATV operators received a probably unintended shock as well. While the Commission, for the most part, did not waiver from its steady free market ideology, it did suggest the simple requirement that CATV operators obtain retransmission consent, the permission of the broadcaster to carry the signal. While this may have seemed logical and unproblematic to the Commission, it was one of the worst possible nightmares for the CATV industry.

The Commission had taken comments on the proceedings throughout the summer of 1958. The by-now standing cast of players restated, and sometimes extended, their positions. The western broadcasters argued for protection against financial catastrophe and adherence to the doctrine of localism; the NCTA praised the history of the industry and its contribution to American enjoyment of broadcasting. The NAB weighed in heavily on the issue of retransmission consent, and all the broadcasters challenged the FCC to either assert authority or request Congress to pass legislation specifically granting it control. For its part, the NCTA repeated the position Smith had taken in the 1958 hearings that competition was preferable to regulation, but that the industry was willing to discuss and even support reasonable oversight.

In Frontier, the FCC had avoided or neglected most of the major issues, deciding the case on the relatively narrow matter of CATV's status as a common carrier. In the 1959 Report and Order, the Commission met the principal problems head on. As in its response to the Cox questionnaire, the Commission's opening paragraphs foreshadowed its conclusions by noting that the evidence of economic harm to broadcasters was not overpowering. In fact, the Commission collapsed its fourteen original questions in the docket into seven categories, including the factual evidence of economic harm to broadcasters from auxiliary services; legal questions on the possible sources of FCC authority over cable and whether economic harm, if it existed, justified regulation; the philosophical balance between localism and diversity; and legislative recommendations.[95] The heart of the report was the

Commission's concern about competitive harm. Here, despite the pleading of Cox, Craney, and Grove, the FCC simply could find none. It noted (following NCTA's lead) that of ninety-six stations that had gone dark between 1952 and 1959, only three involved CATV systems. Reviewing each (including the Kalispell, Montana, and Fairmount, West Virginia complaints) in detail, it concluded that factors other the CATV competition were (or could have been) determinative. A special note was made, it seemed, of the Kalispell case, a complaint that played some role in sparking the 1958 hearings by Magnuson. While the station did go dark in April of 1958, blaming the local CATV operation, it returned to the air that October and was still in operation at the time of the FCC's report.

The Commission recognized that CATV competition probably did (and would) continue to play a role in the financial health of local broadcasting, acknowledging that multiple TV signals no doubt fragmented local audiences and placed a competitive strain on broadcasters. But, concluded the Commission, "in what situations this impact becomes serious enough to threaten a station's continued existence...or whether these things will probably happen in any particular situation—we cannot tell from the data before us."[96] The Commission did finally acknowledge that the economic well-being of local broadcasting was tied to the public interest, following the admonishment of the District of Columbia Circuit Court in *Carroll*, but it limited the application of the rule so severely as to make it almost useless in practice. The FCC would restrict CATV or other competitive services when and only to the extent that a local broadcaster could demonstrate potential damage to the public interest, and "proof of such economic injury 'is certainly a heavy burden,'" stated the Commission.[97] Further, the FCC would consider such complaints only on a case-by-case basis.

The Commission similarly rejected complaints about the competitive injury caused by microwave importation of distant signals. Citing its earlier decision in *Intermountain*, it declared that it could not (and would not) restrict microwave importation, in part because to do so would constitute an illegal restraint based on content. Again, it would consider broadcaster complaints on a case-by-case basis; but in the meantime, it was lifting its "deep frost" on the applications for microwave authorizations to serve local CATV systems.

The Commission could find no basis in the economic harm argument to assert control over CATV and it repeated its findings from *Frontier* that CATV was neither a common carrier nor a broadcast service. And while it never formally declared its philosophical preference of diversity over localism, it required little reading between the lines to take the Commission's meaning. "[W]e do not now envision where we could find that the public interest would be disserved by affording an opportunity for choice of service and the benefits of competition and diversity of expression."[98]

The Commission's unrepentant adherence to the doctrine of free market competition was evident throughout the report: "Rather, good service is shown on this record to be on occasion a result of *competition*—the competition provided by the auxiliary services" (emphasis in the original).[99]

The Commission seemed more concerned about creating broadcaster "monopolies." Yes, said the Commission, localism was important, but it should be left in the first instance for the community to determine, in the open market, its preferences, and should there be a threat that the only local broadcasting service to the rural areas would go dark because of an urban choice, well, the Commission could take that up on a case-by-case basis as well.

The Commission flatly denied that there existed in law any clear property rights to the broadcast signal. Such an issue was one for the courts or Congress to determine and the assumption of such a right could not support a requirement that CATV systems obtain permission to use the signal. NBC, in its filing with the Commission, had argued that the legislative intent of section 325(a) of the Communications Act, which required permission for the retransmission of broadcast signals, strongly implied such a property right and, further, specifically implied protection against CATV retransmission. As to the first part, the Commission acknowledged such an original intent but, for reasons not clearly articulated, suggested that conditions had changed sufficiently from 1927 when the clause was enacted.

An interesting historical note was NBC's citation of original Senate debate on the retransmission clause, in which Senator Clarence Dill (D-Washington), one of the sponsors of the 1927 Radio Act, explained the purpose of the clause to protect the originators of broadcast material: "Otherwise we would have a broadcasting station spending a large amount of money to prepare and present a program from that station, and then under the modern methods of rebroadcasting it could be picked up and broadcast from other stations, and *particularly over the wired wireless*, and money charged for listening to it." (emphasis added) [100]

The "wired wireless," as NBC accurately argued, was an unambiguous reference to radio rediffusion, community antenna radio of the 1920s. The FCC responded that a subsequent comment by Senator Dill that rebroadcasting meant, "namely the reproduction by radio of the broadcasting waves," [101] in the FCC's words "seemed to exclude reproduction or distribution by wire as in the case of CATV." [102] In fact, in the context, Dill was distinguishing radio from printed publication of material, and the "radio" was the end device in a wired "wireless" system. In short, the Commission, by design or ignorance, misstated the intent, and perhaps force, of section 325(a). Had it been accurately represented, the Commission would have had little choice but to impose retransmission consent requirements on CATV operators at that time.

None of which is to say the Commission was opposed to the concept retransmission consent. In fact, it was one of only two substantive legislative recommendations to come from the report. The Commission recommend Congress amend section 325(a) to extend the consent requirement to CATV. The Commission also saw merit in a provision that CATV systems be required to carry all local broadcast signals, although, again, the Commission felt it needed authorizing legislation from Congress. The Commission rejected a variety of other proposed controls on CATV, including restrictions on ownership, and restrictions on CATV duplication of locally available broadcast programming. The Commission characterized the latter proposal as an "unwarranted invasion of viewers' rights" to programming.[103]

With the exception, then, of the local carriage and rebroadcasting consent provisions, the Commission stood its philosophical ground, declaring that it would not seek legislation to restrict CATV development because it could not imagine a situation in which it would use such power if it had it.[104] Broadcast industry supporters, as well as later policy historians such as LeDuc, criticized the FCC's opinion as a missed opportunity to sculpt a more inclusive and coherent plan for the national development of television. While this is true, the history of later FCC control, as will seen, provides little support for the position that national television service would have benefited more by regulation than by regulatory restraint.

The 1959 report marked something of a change in both the tenor of the debate and a change in some of the strategies of the players. The base concern for everyone remained economic viability and this continued to be a catch basin for heated discussion. But discussion about mechanisms for asserting control moved away from common carriage and CATV as broadcasting to an increasing concern for property rights, both in the form of retransmission consent and in duplication of programming. Therefore, while the Report and Order may, on the surface, have appeared to give the industry a marginal victory over the western broadcasters, it held within it the "ticking time bomb" of broadcaster property rights. It was assumed by everyone in the industry, with probable cause, that once broadcasters obtained some form of copyright control over their signals, they would begin charging CATV operators for use of those signals. This, then, was not a regional problem involving a handful of misanthropic broadcasters in Montana and Wyoming, but a national issue striking at the heart of the CATV business. It was unclear just how much broadcasters would charge for access rights, but, at best, it would mean a reduction in CATV cash flow and, at worst, it could jeopardize the existence of the local system. Such a worst-case scenario saw broadcasters using an untenable fee structure to simply force operators out of business. In retransmission consent, therefore, the very fabric of the business was placed in peril.

As a result, the NCTA accelerated its program of cooperation and co-optation. At an NCTA Board of Directors meeting in Phoenix, Homer Bergren suggested the organization draft and submit its own, mild bill, to give FCC control over CATV and, thereby prevent harsher legislation promoted by broadcast interests. Henry Griffing, now a member of the NCTA legislative committee, was politically well connected in his home state of Oklahoma and approached its Senator, A. S. "Mike" Monroney (D-Oklahoma). Griffing and Monroney were close personal friends, and Monroney was on the Senate's Interstate and Foreign Commerce Committee's Subcommittee on Communications. Two weeks after the FCC's Report and Order, the NCTA announced publicly that it was preparing a formal request to Congress to amend the Communications Act to provide for FCC control of CATV. Monroney would shepherd the legislation. Meanwhile, the broadcasters were at work preparing their own bill.

On Tuesday, June 30, Chairman John Pastore (D-Rhodes Island) gaveled the Subcommittee on Communications into session to consider the contending pieces of legislation. Before the Subcommittee were four bills, three embodied the FCC's legislative proposals in docket 12443. The first two, S.1739 and S. 1741, were directed at authorizing booster service to underserved areas. S. 1801 contained the FCC's proposal to require retransmission consent and carriage of local broadcast signals for CATV operators. The fourth bill consolidated the other three and then extended the proposals to require the FCC to issue a finding on the economic impact on local broadcasters for any proposed CATV operation. All CATV systems would be licensed and no license would be granted without an FCC finding that its existence would not injure the local broadcaster(s). This bill, S. 1886, of course was the legislation sponsored by the western broadcasters and introduced by Senators Frank E. Moss (D-Utah) and James Murray (D-Montana).

Early in the first day's hearing, Henry Griffing's ally on the Subcommittee, Monroney, received permission to have the NCTA bill, S. 2303, included in the hearings. This bill, to some, was a toothless tiger. It required the Commission to issue certificates for CATV operations consonant with the public interest, convenience, and necessity, and it empowered the FCC to issue rules in furtherance of the public interest standard. The proposal was short on specifics, however. It mentioned neither must carry nor retransmission consent. No time limit was set on the certificates. The only significant detail was the grandfathering of existing systems to allow them to extend their lines without FCC permission.

The first day of the hearings, Tuesday, June 30, was taken up by testimony on the VHF booster bills. Wednesday, July 1, opened discussion of the so-called "Moss-Murray and Monroney bills." The witnesses provided a parade of familiar faces and positions. Moss opened a morning of broadcaster attacks on CATV, stressing the need to give the FCC the tools to provide "free local television in our smaller communities."[105] Grove, Pengra, Reardon, Craney, and others followed. Concerns about the competitive endangerment of local broadcasting and the 1952 allocation plan were joined more fully by worries about broadcaster property rights. With the cloud of McCarthyism only beginning to dissipate over the country, Craney drew an oblique parallel between the cable copyright problem and Communism, noting that "Reds have no property rights."[106] Later in the hearings, Harold Fellows, President of the NAB, would add the weight of that organization to the call for protection of broadcaster property rights.

When the CATV interests got their chance a few weeks later, the industry's conciliatory position took center stage. There was no real need for legislation, according to the NCTA, but it was willing to accede to FCC regulation of a broadcasting, rather than a common carrier, nature. Cases of competitive injury to broadcasters were said to be few and far between—most relations with broadcasters were friendly and mutually enhancing. Those cases that did arise could and should be dealt with by the FCC on a case-by-case basis, with regard to the public interest standard. Carriage of local signals was, by and large, a good idea and a sensible business practice, but a blanket rule requiring it could be too restrictive to accommodate the many differing circumstances. The only point on which the industry was unwilling to move was retransmission consent. NCTA President A. J. Malin told the Subcommittee of a suggestion by a television station in Springfield, Massachusetts that the fair cost of using its signal might be around $20,000 a year. The specter of those kinds of fees for CATV use of the broadcast signal made for a deeply entrenched NCTA position on the issue. Griffing, testifying on behalf of the Monroney bill, added that confusion over who actually owned program rights—the local broadcaster, the network, or the program producers—could create chaos in the actual implementation of a property rights scheme.[107] Interestingly enough, when Milton Shapp stood before the Subcommittee it was to argue against any legislation that would impede the development of the cable industry. He attacked in detail broadcaster allegations against CATV and praised the efforts of the CATV industry to bring television to underserved areas. He did not, however, offer any specific endorsement of the Monroney bill, other than a passing, perhaps grudging, reference to FCC oversight of the industry. In a possible foreshadowing of his activities over the next year, his remarks seemed largely aimed at blocking legislation generally, although no one on the Subcommittee seemed to detect his subtle divergence from the NCTA position.

The FCC, for its part, served as something of a cross between silent guest at the feast and whipping boy. Broadcasters and some senators continued their assault on FCC inattention to the area. Doerfer raised the ire of Pastore toward the end of the hearings when he declined to speak on behalf of the entire Commission. "Well," barked Pastore, "maybe we'd better get the whole Commission up here then, or someone who can talk."[108] The FCC observed, formally, that the Moss-Murray bill could, through its vaguely worded provision protecting local broadcasters, effectively ban all CATV systems in any towns with local stations. Alternatively, the wording could be construed in such a way that no justifiable enforcement was possible. The FCC similarly critiqued the CATV-sponsored bill, suggesting it lacked sufficient detail to provide meaningful guidance in regulation and did not include a specific retransmission consent clause, which the Commission endorsed. Doerfer, himself, stuck to his free market guns. "If you are going to try to develop a...communication nationwide system within the competitive concept, then you...and these local broadcasters will have to accept the risks involved," he told the Subcommittee.[109] But some senators saw instead a program of FCC inactivity leading toward the destruction of the broadcasting system and the exchange between them and Doerfer at some points was heated, with the FCC Chairman declaring: "Don't ask me what I think if you don't want to hear it."[110]

Throughout the proceedings, Chairman Pastore tried to play the even-handed man in the middle, striving for the compromise position. In this, he was very much the seasoned pragmatist. He recognized the fight was between two large moneyed interests and wanted, as much as anything it seemed, simply to get a workable piece of legislation. "I would rather have a bill," he declared, "than just have a debate."[111] Even though he sparred with Doerfer over the lackluster performance of the FCC in protecting broadcasting, he was equally as skeptical of broadcaster's efforts to insulate themselves from local competition: "[L]et the subscriber make up his mind whether he's getting a good deal one way or another. Why wouldn't free enterprise bring out a fairer and more equitable solution to the problem," he asked his colleagues on the committee at one point.[112]

It was within this context that he pushed NCTA legal counsel Smith to accept that the Monroney bill was sufficiently flexible to accommodate some of the provisions of the Moss-Murray proposal. He managed to win from Smith the acknowledgement that competitive impact on local broadcasters could be an element in the FCC's public interest standard. Broadcasters were free to lodge complaints and present evidence for FCC consideration. Similarly, required carriage of local signals and nonduplication of signals could be considered by the FCC under the NCTA bill. Again, any broadcaster could bring a local issue to the Commission. Smith maintained, however, that such issues had to be considered individually and, in the area of competitive impact, the burden of proof was on the complaining broadcaster. Retransmission consent itself remained non-negotiable. "It could, sir, destroy the industry," he told Pastore.[113] And, as LeDuc, correctly pointed out, Pastore seemed to seize on Smith's comments as the break he needed to work out a compromise package:

> SENATOR PASTORE: I understand your position to be clearly that you have no objection to the spirit of the so-called Moss bill, with the exception of being required to obtain consent?
>
> SMITH: If we define the spirit of the Moss bill to put community antenna systems under the same general type of regulation, with the same public interests standards applying, whatever it is appropriate to apply them as broadcasting, yes.[114]

But Pastore failed, perhaps in his enthusiasm, to discern the important qualifiers attached to Smith's flexibility. There was a wide latitude in the interpretation of the Monroney bill and the NCTA position. And it was within those margins of interpretation that the next and final legislative round would be fought to a bloody draw.

S. 2653

Pastore took the various pieces of legislation and, in committee, redrafted them. The VHF booster and CATV issues were moved into separate bills. Monroney's bill, S. 2303, became the CATV bill. It was altered to reflect the Subcommittee's notion of a compromise position. By the time it was reported out of committee, in September of 1959, it had a new number, S. 2653.

S. 2653 was a melding of the Monroney and the Moss-Murray bills. It offered a definition of CATV as a service subject to broadcast-like, rather than a common carrier, controls, and exempted from regulation systems with fewer than fifty subscribers and apartment master antennas. It required that all CATV systems be licensed by the FCC under the public interest standard, with existing systems given automatic licenses. In answer to the FCC's earlier complaint that the Monroney bill lacked specificity, it directly applied a host of broadcast requirements to CATV. It included a provision that gave local broadcasters the right to ask that special conditions be applied to the local CATV system that would serve to protect the broadcaster in a one-broadcaster town. But the broadcaster had to initiate the request. Under the prior broadcaster bill, deleterious impact on local broadcasting was assumed, unless demonstrated otherwise. It also required carriage of local broadcast signals and provided for the development of rules to control duplication of programming. (This bill, as early versions, also contained language ensuring that the technical quality of the locally carried signal remain strong, but this was in everyone's interest and never controversial.) Most importantly, however, it did not require broadcaster consent—no property rights were established in the bill.

Broadcasters were rather quiet on the bill, perhaps in part because they got much of what they wanted, and what they didn't get—carriage consent—they were pursuing in the courts. The CATV industry was less muted. Pastore, seeking to gather more information and take the proposal to the people, scheduled a tour of the western states to gather comments. Through October and December, he took the committee to stops in Montana, Idaho, Utah, Colorado, and Wyoming. He heard from many local people on boosters and repeaters and the general importance of television to their lives. He also heard complaints about the legislation, mostly from CATV operators. Many of the complaints were not well focused, leading to a certain degree of frustration on the part of the Rhode Island pragmatist. In quizzing a then-young CATV operator, Bob Magness, about his criticism of the bill, Pastore pleaded, "I am asking what is wrong with the bill we reported out? That is what I can't understand, the opposition on the part of the CATV with relation to the bill."

Pastore was asking for specific "perfecting" suggestions to improve the bill, but Magness, admitting that he had not read the bill closely, could offer none.[115] A few days later, however, another CATV mogul in the making, Bill Daniels, spoke to the committee in Denver on behalf of the NCTA and he could and did offer specifics.

The industry, stated Daniels, opposed S. 2653 on a number of points: its grandfather clause failed to adequately protect existing systems from local broadcasters, the proposed definition of CATV as a program distribution service did not accurately reflect its true status as a passive antenna service; the proposal on nonduplication of signals was too restrictive; in competitive situations, local broadcasters were privileged over local CATV systems; and the local carriage clause was unnecessary.[116] "We don't feel we can live with this bill," declared Daniels.[117]

In place of the flawed legislation, Daniels offered an NCTA plan to form an ad hoc committee of CATV, broadcaster, and booster interests, under the direction of the FCC, to work out the multitude of problems that had been haunting the industry for so long. Pastore was encouraging, "we don't want to pass anymore laws than we have to," he told Daniels.[118] But the FCC was less receptive.

Shortly after Daniels' testimony, Malin sent a formal letter to the FCC proposing the establishment of the ad hoc committee. Grove, Smith, and Whitney even sat down at a meeting in Denver with the chief of the FCC's broadcast bureau, Harold Cowgill, to explore the idea. But, eventually, the Commission backed away from the idea. They would not play a role in NCTA's attempt to derail the legislation.

In the meantime, the industry's tactics appeared to be backfiring. Word was getting out that NCTA was being intransigent and there was growing fear among some operators that Senators Moss and McGee, who were leading the fight against CATV, would act quickly to bring the bill to the floor and even propose amendments strengthening the broadcasters' hand. On February 7, 1960, the NCTA board, meeting in Washington, D.C., adopted a resolution aimed at softening their image. They sought to assure Pastore and other members of the committee that they were, indeed, flexible on the legislation and only sought a bill that was fair to all parties. "It was an answer," the NCTA said, "to a mistaken impression that the CATV industry was opposed to any regulation," according to *Broadcasting* magazine.[119] The device seemed to work, as the bill was delayed in going to the floor for several months, while the various sides arrayed their lobbying forces.

The actual position and strategy of the NCTA was somewhat more complex and qualified than press reports indicated, however. The resolution itself stated:

Be it resolved that the Legislative Committee instruct its Washington Office and professional representatives to seek amendments to

> S. 2653 which would make it workable for the CATV industry and, if such amendments can be obtained, to support passage of the Bill in the Senate, with the reservation that the CATV industry shall be free to seek further improvements in the interest of the public and the industry before the House of Representatives and its committees, and further reserving the right of the industry to oppose legislation in its entirety in the House of Representatives if workable legislation appears to be unattainable.[120]

It was a politically sophisticated, multistage strategy. The industry would attempt to win some compromises, by way of friendly amendments, on the Senate floor. If successful, they would not oppose the bill there, even though it remained unsatisfactory. The NCTA felt it had more friends in the House, which had not yet begun considering a companion bill, and could gain further modifications there. As a final line of defense, they could seek changes in the conference committee. If, at any of these stages, it appeared that they would not eventually succeed, they reserved the right to announce outright opposition. Unfortunately, the plan turned out to be too subtle for some members and too risky for others. While members of the Association's professional staff began conferring with the Senate, powerful voices in the industry began mumbling in dissatisfaction.

On February 15, 1960, Whitney and Smith, along with NCTA Legislative Counsel Harold Miller, sat down with Senators Pastore and Magnuson and their staff to talk about amendments. The senators agreed to some minor alterations, including a legislative history that would help protect CATV against state regulation—another serious concern of the industry. The senators also apparently agreed to be receptive to further modification in the House-Senate Conference Committee, if the bill made it that far. Smith and Miller also began holding quiet talks with the NAB in an effort to seek further amendments that would smooth the waters for everyone, although these talks did not come to fruition.

The NCTA staff and others began expressing some optimism that the amendments would help improve the legislation, not enough to make it fully acceptable, but enough to prevent the need for a full frontal assault. In one NCTA membership bulletin, the association noted: "Senator Pastore has been advised that NCTA appreciates the time and interest he has taken in considering proposed amendments. These amendments do not meet all of the objections of the industry to S. 2653, but they are considered to be helpful."[121] The parties in Washington agreed that further skirmishes could await House consideration.

This portrait, however, was subject to repainting in other quarters and soon there was an impression on the part of some that NCTA now supported S. 2653, at least in part. By late March, a new coalition was forming within the industry and a major rift was forming in the NCTA. Leading the emerging movement against any legislation were three dominant figures in the industry: Henry Griffing, George Morrell of Midwest Video Corp., and Shapp. Griffing and Shapp, especially, were among the largest and most powerful group owners. And while Griffing had been instrumental in the development of S. 2303, Shapp, as noted, had always been lukewarm about regulation. It was Shapp's strong contention that the battle against government control of the industry should be fought and could be won, at both the state and federal level. Shapp may have been influenced in his thinking by former FCC General Counsel Harry Plotkin, who warned Shapp that any legislation that gave FCC regulatory control would only open the door for ever-increasing regulation. The political schism was exacerbated in that the legislative committee that was supposed to be overseeing NCTA activities on the bill had as members both Shapp and Griffing.

On April 5, Griffing set up a meeting with a number of key industry players, and included Zal Garfield of Jerrold Electronics, one of Shapp's chief lieutenants. Less than two weeks later, the NCTA Board convened a special session in Chicago to hear objections to the Association's strategy on S. 2653. Shapp, Griffing, and Morrell stated their case. The Board voted against supporting any organized campaign in opposition to the bill, fearing the political repercussions of a sudden change of tack. But in a concession to the powerful forces represented by the three men, it declared that it would not object to individual representations against the bill. In doing so, however, it opened the floodgates to political chaos.

Almost immediately, Griffing and Shapp launched a campaign to generate support "at the grass roots level" against the bill. They sent flyers to every system operator in the country warning them "Your business is in jeopardy" and urging them to phone or write their senators. Along with Shapp, Griffing, and Morrell, the flyers were signed by nineteen other cable operators, including Bob Magness. Hundreds of telegrams, form letters, and specially prepared pamphlets rushed toward Washington. Shapp was acting in part out of his sincere belief that any regulation was bad for the industry and that the NCTA was underestimating its political strength. He also began expressing concern that key senators and congressmen were under the impression that NCTA supported S. 2653 and passage in both houses might therefore be expedited, foreclosing an opportunity to modify the bill at other stages.

Smith and Miller meanwhile felt that some legislation was inevitable and the current bill was less toxic than many of the possible alternatives, including common carrier status or state control. They encouraged the "lighting of backfires" within the organization and soon another letter-writing campaign, favoring the existing political strategy of patience and compromise, was underway. The cable operators of the Pacific Northwest were especially vocal in their support of this position.

Within a matter of weeks, the industry was divided and confused. In early May, the Board called another special meeting. This war council, held in New York, was pivotal. Shapp and Griffing won the support of Malin and then the rest of the Board. Newly fearful, even perhaps panicked, that S. 2653 would strangle the industry in its cradle, the NCTA, on May 9, sent a letter to Monroney:

> Dear Senator Monroney: We have been asked to advise you of the official position of the National Community Television Association, Incorporated, with respect to S. 2653. The NCTA does not support, and has not supported, S. 2653 as reported by the Communications Subcommittee of the Senate Interstate and Foreign Commerce Committee. Amendments which have been discussed, in our opinion, do not make the bill acceptable to the CATV industry, nor do we believe it to be in the public interest. Respectfully, A. J. Malin, President[122]

Senate leaders were shocked. *Broadcasting* magazine described Pastore as "particularly incensed."[123] The apparent betrayal was palpable, especially to Pastore, who had worked so hard to get a piece of legislation that would work. Any hope that Magnuson, Pastore, or the others who had sought a compromise, would introduce the friendly amendments or work for the industry in conference committee was now gone. The moment that Malin's letter reached the Capitol, Shapp and his colleagues had won. The only recourse for the industry was all-out attack.

A call went out for every available CATV operator to come to Washington. On Sunday night, two days before S. 2653 was scheduled to come to the floor of the Senate, more than 200 operators rallied at a local hotel. They were warned of the dangers of the bill and set out the next day to personally lobby their senators.

CATV was not, in fact, a pressing issue for many congressmen. The Senate Communications Subcommittee was deep into a contentious and high-profile debate over whether to require broadcast networks to provide free air time for presidential debates. For some senators, the appearance of the CATV local operator was the first they'd really heard of cable television or its relatively small-scale problems. Shapp reported visiting one senator who told him he had received two letters from constituents supporting the CATV bill and one letter opposing it. But, said the Senator, since Shapp and his companion (a local cable operator) were now there to oppose the legislation, that made it three against two and so the senator would vote against.

In one respect, Shapp turned out to be right; the NCTA had more political strength than it realized. When the bill came to the floor on Tuesday, May 17, the Senate was evenly split. With the Senate gallery packed to overflowing by the visiting CATV operators, the senators began a debate that would run for two acrimonious, exhausting days. Pastore, backed by senators from the West and Northwest, led the arguments in favor of the bill. Monroney, and his sharply spoken Oklahoma colleague, Senator Robert Kerr

(D-Oklahoma), led the opposition. "Senator Kerr's dogged opposition," reported *Broadcasting*, "and Sen. Pastore's equally emphatic support touched off exchanges between the two which got as near as Senators usually ever get to name-calling."[124] The *New York Times* described it as "a bitter shouting match."[125] At one point, Monroney suggested that the bill was a plot by Washington communications lawyers to pad their wallets by entangling hundreds of cable operators in federal red tape and turning them into lucrative clients. Reports that the industry ever supported the bill were untrue, said Monroney, who turned, pointing into the gallery at NCTA Counsel Smith, and declared that Smith had never spoken for the NCTA.[126]

Minor amendments were offered; some were withdrawn. CATV operators conferred with their senators, on both sides, and on Wednesday, Monroney and Kerr moved to recommit the bill to committee. The motion passed by one vote, 39 to 38. For all purposes, S. 2653 was dead.

Years of FCC decision-making, congressional hearings, and industry politics had come to an end. Neither S. 2653 nor its recently introduced companion piece in the house (HR 11041) would make it out of committee. Neither would Congress have any interest in expending additional energy on similar bills that would be proposed over the next several years. For the moment, CATV remained free from federal regulation.

Shortly after the vote, Smith wondered in a letter to a colleague what, if any, impact the defeat would have on the future development of the industry.[127] In later years, Shapp would declare this fight to be a great victory for cable. The record suggests that, given the amount of physical and emotional energy that went into the struggle, both might have been disappointed. It is arguable that the rise and fall of S. 2653, while a dramatic passage in the political life of cable, had little real long-run impact on the industry. Many of the controls that Congress sought were assumed by the FCC a few years later without congressional mandate. And the most significant contests, over copyright and broadcaster property rights, would be fought out in the courts.

At the same time, S. 2653 was something of a political coming-of-age for the industry. CATV was now a real player, albeit a very small one, in the national policy debate, and CATV itself a problematic, but enduring, component of the national broadcasting system. The episode did nothing to dispel the image of CATV operators as mavericks, risk-takers, and perhaps even pirates in the worst and best senses of the word. But these pirates, if that they were, clearly had moved from the shoals of their home coast to deeper waters, and they were starting to amass real treasure.

5

Cable's New Frontier (1960–1966)

I've got six kids. I've got to have that cable.

—ELMIRA, NEW YORK, HOUSEWIFE, 1966[1]

I f Al Malin was a little nervous, he had a right to be. He had received the standard introduction, the biography and small accolades, and now approached the podium to address the Ninth Annual Exhibit and Symposium of the National Community Television Association. As out-going president, it was his job to help heal the still bleeding political wounds, mostly self-inflicted, that the Association had sustained in the prior six months. Bruised egos and smoldering grudges were endemic. Moving the group ahead, after what he understatedly described as his "controversial year in office," would be a tall order.

It was June 21, 1960. The 500-plus members of the NCTA were gathered at the Fontainebleau Hotel in Miami. With a different internal dynamic, it could have been handshakes and cigars all around. Only a month before, Senate bill S.2653 had been, for all intents and purposes, defeated, and it appeared that the CATV business would remain largely unregulated. The victory came at a cost, however. The political battle had badly split the small association, and the alienation among former colleagues soured what could have been an exuberant event.

Instead of celebration, formal and informal caucusing occupied the hours. The western operators continued to argue for controlled regulation; Shapp and his powerful allies lobbied against it. Legislative committees met almost around the clock to formulate a new NCTA policy, while the hallways, lounges, and bars buzzed with sometimes tense conversation about the rift.

On the first day of the convention, Malin, in his address to the membership, tried to set a conciliatory and unifying tone. He attributed the internal difficulties to the normal growth pains of a healthy, young organization. He spoke of the need to mature from a primarily volunteer association and to one with greater professionalization, and of the need for better communication among the varying constituencies. He traced the short history of the body, praising, to help appease, the northwestern Association, in particular. In the end, he asked for mutual trust among the members.[2]

The speech may have helped, but afterward the lobbying and debate continued to percolate. When it was over, Shapp and his allies had once again prevailed. The NCTA formally adopted a position opposed to any form of federal regulation. It also reorganized, adding its first paid chief executive, former American Rayon Institute president William Dalton, hired in 1961. It also added a full-time association attorney and hiked its dues to cover the new operating costs.

While formally raising battlements against federal control, the industry, at the same time, declared its intention to build bridges to the broadcasting industry, especially seeking to reach out to cable-scarred station owners such as Ed Craney and Bill Grove. A suspicious broadcaster might have justifiably worried, however, that while the cable association was reaching out in friendship with its right hand, the left hand was sneaking around to lift the broadcaster's wallet, for there was another major agenda item, in addition to politics, at the Miami convention—it was pay-TV.

The convention featured two major addresses, one from Paul MacNamara, Vice President of International Telemeter; the second from a new face in CATV, a businessman by the name of Irving Kahn. Both wanted to talk about their plans for CATV-delivered pay television. International Telemeter had been testing its coin-box, pay-TV design over a system in Toronto; Kahn's company, TelePrompTer, Inc., was building an even more advanced technology that held the promise of shopping and voting via the living room TV set.

Talk of pay-TV made many people nervous, broadcasters certainly, but also a fair number of community antenna operators listening in the hotel ballroom. From their perspective, the nascent CATV business was beginning to gain a little financial traction, drawing the attention of potential investors and, to some extent, the public at large. Loose talk of pay-TV could generate yet more unneeded antipathy from broadcasters and their allies—more self-inflicted wounds.

In many ways, Malin had been right on point when he attributed such internal problems to an industry growth spurt. The NCTA no longer was the close-knit group of small businessman it had been. It was larger and more diverse, and diversity brought conflict. CATV's success had obvious benefits, including an expanding national footprint as it moved from small town to small town. Its proven record of financial performance, as well as its potential, was drawing the attention of the larger business community, as was its agility in avoiding federal oversight. On the other hand, growth brought the young business into positions of potential conflict with entrenched business and political interests.

Industry Overview

The community antenna television industry was on the move in the early 1960s, and so was the nation. Social and cultural forces were shifting in powerful ways that, while not completely obvious at first, would come to have great significance for the development of cable. The election of a new president, John F. Kennedy, in November 1960, signaled a change in the political climate in the United States. Americans began to turn their eyes away from the aftermath of the Great Depression and World War II, looking hopefully forward at what was labeled "The New Frontier." It was the social cradle in which community antenna television began to mature from a local to a national presence. CATV, after the defeat of S. 2653, was preparing for the next stage in its evolution, taking it to a place where politics and business were played for higher stakes.

This chapter examines community antenna television in the United States from 1960 to about 1966. The first half of the chapter considers the growth of the business, paying special attention on the financial factors that drove the spread of the technology and to the expansion of the early CATV multiple system operating companies. The second part of the chapter examines the beginning of federal regulation of cable television and additionally looks at CATV relations with the telephone industry. The chapter begins with a brief overview of CATV technology, programming, and industrial growth in this period.

Evolving Technology

By 1960, the technical platform upon which CATV based its business was cost effective and more than sufficient for most local applications. CATV carrying capacity, in fact, frequently exceeded local need. Technical power often surpasses existing social necessity, or economic feasibility, and so it was with CATV. In the early 1950s, the industry relied on single-channel strip-band amplifiers—using as many as necessary to retransmit available signals—and then moved on to low-band broadband amplifiers that could carry Channels 2 through 5 (or 6), plus the FM radio band. The low-band systems gave small town operators five channels, enough to bring in three network stations and two independents, so during this period, most CATV systems were three-to-five channel affairs. Even by 1964, 85 percent of all cable systems carried three to seven channels[3]; only 11 percent carried eight or more.[4] Using either the low-band or split-band amps—depending on signal availability—CATV operators through the late 1950s and into the early and mid-1960s had plenty of channels for local needs.

Greater capacity was available; the decision not to employ it was strictly a matter of cost and need. As noted in Chapter 3, the engineering firm SKL, using proprietary methods, had created a full-channel, broadband amplifier as early as 1950. Extending work done for the military in World War II, the technicians at SKL built a 216-MHz booster capable of carrying all twelve VHF channels and FM radio. But as Archer Taylor pointed out in his history of early CATV technology, the early 1950's market was not ready for the device: "SKL amplifiers were significantly higher priced than Jerrold's. Because the industry saw no immediate need for more than three channels, they resisted SKL in favor of the more conventional, lower-priced, single-channel designs from Jerrold."[5]

SKL modified its design and produced a cheaper, low-band version of the amplifier, and other suppliers developed their own distributed gain circuits. The consequent low-band equipment and the low-high, split-band boosters manufactured by Blonder-Tongue, Jerrold, and others became industry standards by 1956 and 1957. The first recorded twelve-channel CATV system did not open until 1960 when Elmira Video in Elmira, New York, began operations using SKL equipment and carrying channels imported from nearby towns in the region, including Buffalo, Syracuse, and Rochester.

Gradual improvements were being made in head-end equipment, connectors, and the coaxial cable itself. Cable shielded in aluminum with a foam dialectic entered the market in the early 1960s, improving the price-performance equation. The most important technical advances of the period involved efforts to replace tubes with transistors. Commercial realization of transistor-based amps would not come until the mid-1960s, however, with industry-wide deployment starting in the second half of the decade. (Details of that process are treated later in the text.)

Programming

Despite the publicity given to pay-TV, CATV programming in the early 1960s was, for the most part, what it had been from the start, retransmitted off-air broadcast signals. Cable was still community antenna television. Some operators were more aggressive than others in importing broadcast service. Communities of the type mined by Jerrold in the early 1950s, where nearby signals could be picked up off air and distributed wholly by coaxial network, were becoming more difficult to find. Operators were increasingly looking at situations such as those in the Northwest where distant signals had to be imported by microwave.

A few operators did experiment with pay service. The efforts of Irving Kahn, especially, are described below. But in the early 1960s, pay-TV remained much more of a CATV dream, or if you were a broadcaster, a nightmare. The role of pay-TV as business motivator was important, but it would be years before pay became a business reality.

Despite the NCTA's internal lobbying against local origination, some operators continued to offer limited local fare. In isolated cases, this meant live, local talk or talent shows; TelePrompTer's system in Farmington, New Mexico, carried local bingo games. More typically, it meant an automated or semiautomated weather or news service, although the standard CATV weather channel was little more than an inexpensive camera pointed at a set of weather dials, showing time, temperature, and humidity. A "news" channel was, more often than not, a camera pointed at an Associated Press wire machine. Merrill's experience was typical:

> It was about that time people started making their own weather channel devices. We put one together early on, using the oscillating camera scanning about five weather instruments. We also carried a channel for a long time that consisted of a camera focused on an AP machine—one of those machines that printed out the news items as they come off the tape. That thing just sat there and clattered out and the camera looked at it so that anybody that wanted to take the time to read it could keep up with the late developments in the news. On those first five or six systems that's about all we carried.[6]

Armed with such limited, but still highly prized technology and programming, the community antenna industry began to morph toward cable television.

Growth

CATV was expanding in the early 1960s, but the precise dimensions of that growth are difficult to pin down. Accurate figures on the business through the 1950s, and even into the latter 1960s, are almost impossible to obtain, although there are rough estimates from industry observers and from the primary trade group, the NCTA. A variety of problems plagued efforts to get a clear count of systems and subscribers. Until 1967, the FCC put little effort into tracking, in any detailed fashion, CATV growth. With one important exception, noted below, the agency wasn't compelled to initiate an industry census until after it asserted general regulatory jurisdiction. Trade publications, especially *Television Digest,* and to a lesser extent, *Television Magazine,* reported regularly on the growth of cable, but even they acknowledged that reliable figures were hard to come by. *Television Digest,* arguably the best source, conducted an annual survey of CATV systems, but response was voluntary and the information not always accurate. NCTA was popular source of information, but the group has never represented all CATV operators and it acknowledged that its early figures were estimates. Emerging systems were often so small, sometimes only a few dozen customers, that they were easily overlooked, and operators had few reasons to self-report, especially when there was so much to do in getting the business started. Guessing growth also was an effort to hit a moving target. Through the early 1960s, new systems were popping up around the country every week. Even the FCC's first formal census, taken in 1966, produced numbers that were, by all observers' accounts, probably badly deflated. The problem was in part methodological, an issue of survey design, and partly political. By design, the agency did not look at systems with fewer than fifty subscribers, and there were many. Politically, insofar as the FCC's survey sprang from its rising interest in CATV regulation, a significant percentage of operators who were polled chose to boycott the survey. Some refused to concede FCC authority to poke around in their financial affairs. Others felt that reporting, at least symbolically, would constitute an acknowledgement of the agency's authority over the business. Many of those who did return the FCC forms attached a note stating that their participation was not to be considered an endorsement of FCC jurisdiction. As a result, the FCC's first CATV census[7] reported 1,449 CATV systems in operation with about 1.86 million subscribers, compared with an NCTA estimate of about 1,750 systems and 2.5 million subscribers.

One of the best sources of general information about the industry came from an FCC-sponsored study conducted in late 1964[8] as part of the Commission's investigation into whether to take control of CATV. While it was not a census, it did gather detailed financial information from several hundred CATV operators and provided a good snapshot of business conditions at the time.

Despite this fuzziness in the data, however, the broader parameters of industry evolution can be roughly traced. The figures for cable's first two decades reveal a rapid and fairly constant rate of expansion. From 1952 to 1955, the number of systems rose from about seventy-five to an estimated 400, and subscribers grew from fewer than 15,000 to more than 150,000. Expansion of CATV was dependant, of course, on the general diffusion of TV sets into American homes through the 1950s, but TV set adoption followed a fairly steep curve, from about 9 percent national penetration in 1950 to nearly 86 percent by 1959[9]; by 1959, an estimated 800-plus CATV systems were serving more than 700,000 of these new TV viewers. Cable growth continued through the early 1960s, with systems increasing to more than 1,500 by 1965, and subscribers to well over 1 million, and as many as 1.6 million by some counts. According to one report in December 1964, an estimated fifteen community antennas were being raised every month.[10] Even a conservative estimate suggests the number of CATV subscribers was doubling about every five years through this period.

Revenue figures were even shakier than system and subscriber counts during this time, but contemporaneous dollar estimates mirror customer growth, rising from between $20 and $25 million in 1957, to an estimated $50 to $70 million in 1963,[11] and $90 to $100 million by 1965.[12] While the numbers in isolation may seem impressive, they are best viewed in the broader context of the development of the national broadcast television system. (Appendix A details general growth figures for cable television from 1948 to 2005.) Even with its accelerating success, CATV still served only a small fraction of the national TV market. CATV's share of the viewing public came to only about 1.2 percent in 1958 and even by 1965 had risen to only 2.4 percent. The actual impact of CATV on the broadcast industry and the awareness of the American public was, at best, tiny.

Geographically, CATV remained a creature of the smaller, underserved towns and hamlets of the United States. Nearly 90 percent of all systems were located in towns with populations of 25,000 or less. The nearest three-station TV market was more than forty miles away for 90 percent of all systems, and 77 percent of all systems were at least forty miles from a city with only two TV stations.[13] The FCC's 1964 survey showed about half of the 1,257 systems polled had fewer than 655 subscribers. About 90 percent had fewer than 3,000 customers, and only twelve systems had more than 7,000.[14] In fact, the average size of a community antenna system hovered around 1,000 until the latter 1960s.

MSO Seedlings

Despite CATV's relatively small size and modest impact, there was a growing interest in the industry on the part of broadcasters, the government, and the press. The headlines, in fact, looked quite promising: "The Golden Antenna of CATV,"[15] "Television's Lusty Offspring,"[16] "CATVs Pass 1,400—Still Going."[17] A front-page story in the Wall Street Journal in December 1964 heralded CATV's "dizzying growth," noting that the number of subscribers had doubled over the previous five years.[18] Of course, the number of subscribers also had doubled in the five years prior to that and, to some extent, the jolt of press coverage betrayed a lack of historical perspective, but much more was at play. It was not so much that the pace of CATV's deployment had accelerated relative to the prior decade, but rather that the nature of CATV's development had begun to change in important ways. Along with key alterations in the political and regulatory climate, the shape and texture of the CATV business itself was beginning an important set of transformations.

In the first half of the 1960s, the CATV industry began attracting sophisticated corporate and private investment; small, locally owned systems began selling out to the larger groups and construction of new systems accelerated in markets with one or two existing broadcast stations.

While the industry's financial track record was sound, it was the promise, as much as the performance, of CATV that attracted this interest. The powerful force of public perception—the social construction of cable technology—was coming into play. To its credit, the same Wall Street Journal article noted, "it is CATV's potential, rather than its accomplishments so far, which is prompting most of the argument and posing most of the questions."[19] That perceived potential was one of several important factors driving new money into the industry and, in doing so, helping alter its basic structure.

Pioneers like Shapp, Daniels, Merrill, and Griffing built or bought their systems, for the most part, one by one, creating a fledgling industry as they went. Slowly, they forged their small "chainlettes," assembling the industry from the inside out. Through the 1950s, CATV remained a mom-and-pop business, while the embryos of these soon-to-be multiple system operations gestated. Then, with the arrival of a new decade, came a fresh presence in community antenna ownership, large or at least larger companies and corporations. These were businesses that had little or no prior connection to CATV, but for a host of reasons brought significant amounts of capital and power to the table. Unlike their pioneer predecessors, the new companies had an appetite for ever-larger numbers of cable properties. The first to enter from what might be described as outside of the industry were Irving Kahn's TelePrompTer Corp. and H&B American Corp., Kahn in late 1959 and H&B in 1960. The arrival of these two firms, large by CATV standards, generated significant attention in the trade press. Both would quickly grow into industry leaders and eventually they would merge. A third company, larger, older, and more powerful than either TelePrompTer or H&B, would buy into the industry in 1961. That company, RKO General, would exert its own important influence on CATV.

Irving Kahn & TelePrompTer Corporation

In the legendary line of those exercising historic influence over the evolution of cable television, Kahn stands closely behind Shapp and Daniels. He was one of several people who, given his place in the power structure and his personal vision, was able to help shape the contours of the new technology.

Irving Berlin Kahn was born in Newark, New Jersey, in 1918. His parents, Russian Jewish immigrants, named him after his famous songwriter uncle. After graduating from the University of Alabama in 1939, Kahn, like Berlin, went into show business, first as a publicist for big bands, then as a public relations agent with 20th Century Fox. He served in the Army Air Corps in World War II and then rejoined Fox, eventually becoming Vice President of its radio and television division. One fateful day an actor, Fred Barkau (Barton), brought Kahn an idea for a device involving a rolled paper script that actors could use to read their lines on camera. Shapp gave the idea to Fox engineer Hub Schlafly to develop into a viable piece of equipment. It was the prototype of what would become the television teleprompter. Fox was not interested in promoting the device, so Kahn, along with his two new partners, Schlafly and Barton, founded TelePrompTer Corp. in 1950. He presented the invention to TV executives and, within a few years, it was standard issue on many of the then-live network broadcasts.

Within a few years Kahn was nosing around for other business ideas. Speakers at public gatherings, including political conventions, found the TelePrompTer a useful device, and that success led TelePrompTer to begin staging entire meetings for business, civic, and political groups. Western Union bought an interest in TelePrompTer in 1956 with the idea of carrying the TelePrompTer-staged meetings over Western Union lines on a closed circuit basis. It was a natural evolution for Kahn. While at Fox, he tried to promote an idea for using television to distributed motion pictures to the nation's theaters in order to eliminate the need for expensive film stock. Combining television, special event entertainment, and wires remained a theme in his business career in the 1950s. He floated one proposal for a nationwide, wired system that would interconnect every radio and TV station, advertising agency, and advertising placement (rep) firm in the country, not for program delivery but for information exchange, potentially for the rapid distribution of ratings data. That idea failed to mature, but he continued to explore live, closed-circuit TV distribution of sports events to theaters. He was working on a system to bring those closed circuit events directly into the home when he encountered the idea of CATV. Kahn needed AT&T to construct the physical infrastructure of the system he was attempting to create on Long Island, but the company's terms were, in Kahn's view, unreasonable. He was explaining his problem to a friend, Marty Kodell, then editor of *Television Digest*, over lunch. According to Kahn, Kodell told him, "Hey, a guy came into my office just this week and told me he is going to start becoming a broker in a new business. If you come back to the office, I'll give you his card." The business was community antenna television and Kahn saw in it an opportunity to test his closed-circuit plan He got the business card from Kodell. "I called that guy up," recalled Kahn. "That guy was named Bill Daniels. I think I was his first customer." Daniels told Kahn he had a 750-subscriber system for sale in Silver City, New Mexico and Kahn could have it for $130,000.[20]

Kahn sent his Chief Financial Officer, Monroe "Monty" Rifkin, to investigate. Through Daniels, Kahn purchased his first system, from Merrill's Antennavision—the Silver City, New Mexico, operation—in November 1959. Soon after, Kahn bought two of Daniels' own systems, the properties in Rawlins, Wyoming, and Farmington, New Mexico, for a total of about $600,000 and 27,000 shares of TelePrompTer stock. In June, Kahn added a system in Liberal, Kansas, to his holdings, switching to a new broker, Blackburn & Co. for the transaction. Through the early 1960s, TelePrompTer continued system acquisitions, adding properties from Elmira, New York, to Honolulu, Hawaii. By the fall of 1962, Kahn had fifteen systems and more than 30,000 subscribers, and he was acquiring more.[21] TelePrompTer, while adding only a few new systems in the early 1960s, doubled its subscriber base to 60,000 in 1966. Moreover, Kahn's company was one of three, along with Sterling Manhattan and a local start-up, CATV Enterprises, Inc., to secure

CATV franchises for New York City, and the company spent large portion of its resources in developing a potentially lucrative portion of Manhattan.

TelePrompTer, through steady acquisition and merger, would become the nation's largest CATV company by the 1970s. Kahn himself would grow into a colorful, sometimes controversial, leader in the industry, a millionaire, a visionary, arguably the chief architect of cable's transition to a world force in communications, and, eventually, a convicted felon.

H&B

The second outside entrant and dominant company at the start of the new decade was a firm called H&B American Corp. The company sprang from manufacturing roots in the 1940s. Susquehanna Mills, originally a seed company, made its mark in textile production and aircraft frame manufacture during World War II, but its fortunes turned downward in the years after the war, and in 1954, it merged with H&B American Machine Co., taking that name. Under the guidance of a West Coast financier named David Bright, H&B diversified into a variety of lines, from steel-rolling mills to earth-moving equipment to backyard barbecue grills. Through the 1950s, the company expanded and prospered, but took a sudden change of direction in 1960 when it entered the community antenna business. Through its earlier acquisitions, H&B had assumed substantial debt, and the cash flow promised by the new CATV business looked like a way to service some of that obligation.

In what was then the largest deal of its kind, H&B paid $5 million for Jerrold's nine remaining CATV systems. H&B was instantly one of the largest players in the industry. It also picked up two microwave companies in the Jerrold sale. The company moved aggressively to increase its CATV holdings, buying three more community TV properties and two more microwave systems by the end of 1961. It added another ten systems and one microwave operation in 1962 and, with twenty-two systems and more than 63,000 subscribers, it was the largest operator in the United States. The move into CATV proved to be so successful for H&B that it sold of all its other interests and went full time into the community antenna business.

Starting with its nine seminal Jerrold systems, Beverly Hills-based H&B grew to thirty systems and some 90,000 subscribers by 1966, making it then the largest CATV company in the country.

RKO/Vumore

The third major player to enter the scene in 1961 was the corporate giant General Tire and Rubber Co. General Tire owned the old-line film studio RKO General. RKO General, in turn, owned theaters and broadcast stations across the country.

For about $4.5 million, RKO purchased most of the business holdings of Henry Griffing, following his unfortunate death. These holdings included all of Griffing's theater properties and his community antenna systems. The CATV company, Vumore, was maintained, with RKO operating the systems under that name. By 1962, RKO's Vumore controlled twenty-seven systems and about 30,000 subscribers. By 1966, it had grown to thirty-three systems and 44,000 subscribers.

Tele-Communications, Inc.

While the new entrants were looking over the landscape and getting their bearings, some of the early pioneers were busy building or rebuilding their own fledgling cable empires. Settling quickly into his new home in Bozeman, Montana, Bob Magness went straight to the business of expansion, opening new systems and acquiring or investing in others. By 1964, Magness had full or partial ownership of five systems in Montana and two in Texas.[22] Magness also bought into Montana-Idaho Microwave, Inc., a microwave common carrier serving the Bozeman system, as well as one in Livingston, Montana, and a TV station in Billings. In doing so, he forged a key business relationship with co-owner, A. L. Glasmann.

In addition to his holdings with Magness, Glasmann's regional media empire included KUTV-TV in Salt Lake City, KLIX radio and TV in Twin Falls, Idaho, and the Standard Examiner Publishing Co. in Ogden, Utah. Interestingly, Glasmann and his son-in-law partner, KUTV-TV Vice President George Hatch, had been active in opposing community television in the West in the late 1950s and Hatch would continue to represent broadcaster interests in negotiations with the cable industry in the 1960s. But the business of broadcasting and cable would create their own strange bedfellows and by the mid-1960s, the Glasmann-Hatch-Magness group had widespread media interests in the West, including an expanding roster of CATV properties. The cable systems were held specifically within Community TV, Inc., with Magness as President and Hatch as Vice President.

In 1965, Magness and Hatch, consolidating most of their cable holdings, relocated Community TV headquarters from Bozeman to Denver. By then they had purchased CATV and microwave systems, in whole or in part, from a several of the pioneering western operators, including J.E. Collier, Thomas Mitchell, and Roy Bliss. The growing company had, as a result, full or partial ownership of more than twenty-four CATV systems across Montana, Nevada, Utah, Nebraska, and Wyoming, as well as Western Microwave, Mountain Microwave, and Carter Mountain Microwave.

Corporate ownership was divided between Magness, Glasmann, Hatch, and their various print and broadcast companies, including The Standard Corp., Kearns-Tribune Corp., and Copper Broadcasting Co. In 1968, the community antenna and microwave companies were consolidated and the company was renamed, Tele-Communications, Inc. It was only the beginning, however, for Bob Magness.

Sammons, Daniels, Jerrold, Cooke & Friends

Following the entry of TelePrompTer and H&B, the forces of small-scale consolidation and chain acquisition became increasingly pronounced. Existing businessmen, such as Charles A. Sammons, began exploiting the cash flow potential of the CATV business, Daniels and his associates began aggressive expansion and investment. Jerrold left the business entirely, but another new face, Jack Kent Cooke, came to the table. From these various beginnings, some humble and some not, the cable industry's MSOs structure, featuring companies that would in short time come to dominate and characterize the business, began to evolve.

Sammons, for example, continued his CATV system investment, then, mentored by Daniels, took some of his profits out of the market in mid-1962, by selling eighteen properties to a new company, formed for the purpose, called Televents Corp. The sale of 43,500 subscribers for $10 million was the largest, in dollar terms, in the industry's history. The new company, while publicly represented as a group of largely unnamed investors headed by former NBC television network Vice President Alfred Stern, was the creation of Daniels, who concocted and brokered creation of the new firm. Carl Williams was a partner in Daniels and Associates and President of Daniels Systems Management division, which operated the Sammons properties. He took over as President of Televents; Daniels and Associates maintained a 31 percent ownership interest in the company. By 1965, Stern had renamed the firm Television Communications, Inc. (TVC) and had also become Chairman of the NCTA. His MSO had more than 54,000 subscribers. Daniels, meanwhile, continued to buy systems for his own company while managing many others. Sammons also continued to buy and sell systems, and would, in time, become one of the nation's largest operators.

Another CATV chainlette was assembled by Fred Lieberman, who began his CATV career as a sales representative and manager for Jerrold. Around 1956, Lieberman struck out on his own, founding Telesystems Corp. By 1964 he had built it into one of the larger MSOs in the country, with systems serving thirty-two communities. Two years later, the company held full or partial ownership in nineteen systems and 47,000 subscribers, plus an interest in another young chain, GenCoE.

By 1963, Jerrold's diversification into non-CATV fields had failed to pay off as hoped. Shapp bought back a large block of Jerrold stock from Loeb, reassumed the office of the President and began steering Jerrold back into the CATV business. Within two years Jerrold had secured equity interests or franchises for CATV systems in more than a dozen communities, and was looking for more. However, while the company would maintain is preeminent position as an equipment supplier to the industry and despite the system investments, it would not develop into a major system owner after this.

Shapp's interests turned increasingly to politics and he made an unsuccessful bid as the Democrat candidate for the governorship of Pennsylvania in 1966. Concerned that his political activities were distracting him from his corporate obligations, the Jerrold Board of Directors asked him to step down as company leader in June 1966. Shapp subsequently sold his interest in the company he founded and bought a small group of cable properties in the late 1960s. Jerrold was sold to General Instruments (GI) in 1967. GI would continue as a leader in cable television equipment manufacturing while Shapp went on to win the Pennsylvania governor's seat in 1970. He won re-election in 1974, leaving his cable legacy behind.[23]

Cooke began his CATV activities in 1964. Cooke was an already successful, some said semiretired, Canadian businessman, who had moved to posh Bel Air in southern California after making a fortune in publishing. Cooke was seeking fresh opportunities and thought CATV had a future. In October 1964 he bought his first system. In less than a year, he had spent $22 million acquiring cable properties for his newly formed American Cablevision, Inc. By mid-1966, American Cablevision owned twenty-one systems and controlled more than 77,000 subscribers.

While it would be years before it became a major industry force, another young company worth noting was American Cable Systems. The company was founded, in part, as a reaction to "sans-a-belt" pants (elastic waist slacks that did not need a belt). Ralph Joel Roberts had become a successful Philadelphia businessman in the 1950s. Born in 1920, the son of Russian emigrants, he lost his parents as a teenager, but overcame the loss and graduated from the Wharton School in 1941. He served as an officer at the Philadelphia naval shipyard during World War II and afterward began manufacturing and selling golf clubs. He then went to work for a Philadelphia advertising agency and after that, the Muzak Corp.

Roberts moved to a senior management position at the Pioneer Belt & Suspender Co., in Philadelphia and, in 1955, bought the company. By 1961, however, he thought that the sans-a-belt craze might mean trouble for his belt and suspender firm. He sold the business that year and launched a venture capital company, International Equity Corp. He was, therefore, looking for new investment opportunities when he ran into an acquaintance who was, at that time, looking to sell a community antenna company he owned in Tupelo, Mississippi. Recalled Roberts, years later, "I'd never heard of Tupelo and I didn't know what cable was, but I could see the cash flow was steady and predictable and Dan Aaron, who sold me the system, agreed to manage it."[24] Aaron had worked for Jerrold, going on to start his own cable brokerage firm; now he worked for Roberts. Roberts bought the system, and founded American Cable Systems in 1963. By the end of the decade, it would be renamed Comcast.

Pouring over the returns from the FCC's 1966 CATV census, *Television Digest*, at the end of that year, published a roster of the industry's fifteen largest companies, those with more than 25,000 subscribers.[25] H&B, American Cablevision, TVC, RKO/Vumore, and TelePrompTer led the list. Many of the leaders shared ownership situations. Of the fifteen companies reported, ten had partial, as well as full, ownership in multiple-system companies. Bruce Merrill, for example, through various holdings, including American Cable TV. Inc. and Community TV Reception Co., Inc., had 100 percent ownership of seventeen systems, 90 percent ownership of eleven systems, 80 percent ownership of two systems, and 50 percent ownership of two more. Jerrold had 100 percent of two systems, 90 percent of one, 50 percent of seven, 33.3 percent of two, 13 percent of three, and 12.35 percent of two. It became fairly common through the 1960s for individuals to have partial ownership stakes in several different CATV enterprises.

Ownership and Public Policy

The issue of concentration of ownership is an old one in media studies and has been the subject of countless books, articles, statutes, regulations, and lawsuits. The underlying economic forces are not novel to CATV. The community antenna business followed a pattern of industrial evolution classic to the onset of a new technology. It was characterized by a mushrooming of small businesses, made possible in part by reasonably low barriers to entry and an absence of dominant, entrenched enterprises. Entrepreneurs and venture capitalists frequently spring into such markets hoping to gain an early foothold. After a period of time, however, as the market begins to mature, weaker competitors fall by the wayside, either through failure of their business models or through acquisition by more aggressive or better-managed firms. Economies of scale often play a role. One of Karl Marx's important early insights, a core concept in his theory of capital and one borne out again and again in practical experience, is that over time, markets concentrate as capital collects into fewer and fewer hands. The evolution of the daily newspaper industry in the United States is a frequently cited example. It arose from its industrial revolution roots in the mid- and late 1800s in the form of the penny press and reached its competitive zenith shortly after the turn of the century. Slowly, but inexorably, from about 1920 to today, direct municipal competition diminished while chain ownership grew.

In any industry, especially in the media, if the market constricts too tightly for public policy comfort, antitrust or regulatory mechanisms may kick in, although such controls wax and wane with political fashion. Cable television followed this general pattern. The late 1950s and the first half of the 1960s was a period of rapid expansion in the number of CATV systems, but it was also a period that saw ownership configurations start to coalesce into the early makings of the industry's classic MSO structure. By mid-1965, an estimated 40 percent of CATV systems and 60 percent of subscribers were in the hands of companies or individuals who owned two or more systems,[26] as noted above. At the same time, however, the vast majority of those companies owned only two or three systems and, in many cases, the systems were situated in contiguous or nearly contiguous communities. John Walson, for example, fashioned a large subscriber base by building out to several systems in small towns near his original home in Mahanoy City. In fact, by 1965, only nine firms had more than a dozen CATV systems each, scattered around the country. Overall, the four largest firms in the CATV industry controlled about 19.6 percent of all subscribers and the eight largest firms about 28.2 percent, figures that were to remain fairly constant throughout the decade. *Television Digest*, in 1965, noted that, insofar as the largest cable operators (such as H&B American) served less than 5 percent of total cable homes, "there isn't much of a monopoly problem in this industry."[27] It added that prospects for any increasing concentration appeared slim, given the proliferation of new companies. That is, even though MSOs were increasing, the industry overall was also growing apace, keeping concentration ratios fairly stable.[28]

To the extent that regulators worried about CATV ownership at this time, therefore, the focus of concern was on the issue of the local CATV-broadcaster connection. Specifically, the concern was with companies that owned both CATV systems and broadcast stations in a given community. In fact, despite their apparent political and economic warfare, the broadcasting industry was building a sizable appetite for community antenna systems. One of those taking the lead in probing the opportunities of broadcast and CATV ownership was RKO.

Both RKO and H&B had ambitious growth plans, and the two began talking in late 1962. RKO made its first public move in the spring of 1963, acquiring 10 percent of H&B's stock, and soon after increasing its stake to 20 percent. In October, the two companies formally announced their intention to merge, causing heads to turn in the industry and at the FCC. The proposed deal would have transferred RKO's Vumore subsidiary and its CATV systems to H&B; Vumore itself would have been liquidated. At the same time, RKO would have increased its ownership stake in H&B to a controlling 56 percent, creating the largest CATV provider in the country with approximately fifty systems and 100,000 subscribers.

Because the transaction involved several broadcast properties, both microwave feeders for CATV systems and RKO's five television stations, FCC approval was necessary. But, as is discussed later, the FCC was now looking more closely at CATV. The political climate, along with the composition of the Commission, had changed. Moreover, RKO was not the only broadcaster interested in owning CATV systems. More and more broadcasters, for a variety of reasons noted below, were entering or thinking about entering the business. In March 1964, against the backdrop of a changing industrial and regulatory climate, the FCC indicated it would look into the question of whether broadcasters, such as RKO, should be permitted to own and operate CATV systems. In April, it released a formal Notice of Inquiry.[29]

The Commission had expressed mild concern some six years earlier that a company in Farmington, New Mexico, which partially owned the local CATV system (a Daniels' property), might be violating at least the spirit of the duopoly rule if it was granted the local VHF TV license.[30] That was a relatively narrow case, however. The RKO-H&B deal had a much wider policy horizon. In announcing its inquiry, the Commission noted its apprehension about the purchase of community antenna systems by broadcasters generally, and worried specifically about the impact on the public interest of an RKO-H&B marriage. The proposed merger presented the ownership problem, said the Commission, in its most "acute form. . .Even though joint-CATV ownership may serve the public interest in some special circumstances, grant of this (RKO-H&B) application would put under common control one of the largest owners of television broadcast stations and probably the largest group of CATV systems in the country."[31]

Broadcasters interested in CATV investments, including Cox and Time-Life, filed strongly worded protests against cross-ownership restrictions. And the Commission, by 1965, determined that the problem was not sufficient to require action at that time, so they closed the proceeding.[32] A five-to-two majority could find no abuses, but the Commission reserved the right to look into cross-ownership issues in the future if it thought the situation warranted it. While the inquiry did not culminate in formal ownership rules, the Commission did briefly suspend its consideration of the RKO-H&B application, along with several others, while it pondered the broader policy question. That postponement was sufficient to kill the merger and, in September 1964, the companies abandoned it, although RKO maintained its heavy equity interest in H&B.

RKO, as noted, was far from being the only broadcaster to test the CATV waters. In fact, almost all the major broadcast groups entered community antenna operations between 1960 and 1965. In July 1963, Cox paid $1.8 million for a CATV system in Aberdeen, Washington. By 1965, Cox reported full or partial ownership of CATV systems serving ten communities and, with more than 40,000 subscribers, was one of the nation's largest multiple system operators. By then, it also owned five VHF TV stations and four radio stations. Another large multimedia company, Storer, acquired nine systems in the same period, most of them in southern California. By the end of 1964, *Broadcasting* magazine declared broadcasters were "flocking into the business so fast that it is believed one in five communities where CATV applications are being pursued have one or more broadcasters knocking on their door."[33]

The FCC's 1966 CATV census revealed that 28 percent of all CATV systems were owned wholly or in part by companies or individuals with financial interests in broadcasting. It listed 175 broadcasters in CATV, with seventy-seven of them holding ownership stakes in two or more systems. These included many of broadcasting's dominant players. CBS owned an interest in a system in Vancouver, British Columbia. General Electric Broadcasting Co., entering the industry in 1967, owned five systems with more than 25,000 subscribers. NBC, Time-Life Broadcasting, Newhouse Broadcasting through its Newchannels Corp. subsidiary, Westinghouse Broadcasting Co., Scripps-Howard Broadcasting, and Avco Broadcasting, all owned community antenna property.

The FCC's ownership Inquiry slowed, but did not stop the flow of broadcast dollars into the CATV business. The Commission's chief concern was with situations involving colocation of broadcast and cable operations; and many broadcasters were participating in CATV investments outside their service area. When the Commission ended its inquiry without taking action in 1965, the buying accelerated. By 1969, broadcasters owned 32.2 percent of the nation's CATV systems.[34]

The Economic Attraction

In the early and mid-1960s, the community antenna business was moving out of the minors and into the big leagues. CATV was attracting the attention of national financial institutions, established broadcasters, and venture capitalists. Well-heeled entrepreneurs, such as Cooke, were grabbing franchises, and buying and selling systems with vigor. There were a number of reasons. The failure of S. 2653 and the lack of any significant governmental regulation increased the attraction of the business. More fundamental, however, was CATV's potential as a launching pad for pay television, its steady cash flow and favorable capital depreciation schedule, and for some broadcasters, the need for local protection.

Pay-TV Possibilities

For years, the NCTA had labored to construct a social definition of CATV that cast it as a politically and economically benign adjunct to the rooftop antenna. Not everyone in the industry was willing to sign on to the story, however. As early as the Bartlesville experiment, some operators had been "licking their chops" over the prospect of CATV-delivered pay programming. Bill Daniels, in 1957, called for a "wedding" of the community antenna and theater businesses. The combination, he said, "could form the most lucrative and pleasant partnership that has been seen or will be seen in American business."[35]

The vision of this "happy marriage" was the principal motivator for some of the earliest diversified entrants, including Kahn and RKO. In fact, the nation's second largest theater chain, National Theaters Corp. (later National

General Corp.), bought into CATV in 1959, and with a half dozen systems was briefly one of the country's largest cable owners. National Theaters would leave the business before the end of the decade as the pay promise fizzled, but others were more tenacious. As noted above, Kahn was already at work on a wired system for pay-TV when he stumbled on to CATV. It was the mutual interest of Daniels and Kahn in pay-TV programming that prompted Daniel's suggestion that TelePrompTer look into community TV.[36] "Let's be honest," Kahn later explained. "Originally, I didn't know what cable was and I wanted a place to experiment with Pay TV. That's what got me into it."[37] Kahn also later confided that he purchased the Silver City, New Mexico, system in part to shield his experiments from both antipay forces and potential competitors. "We figured if we tried out our pay-TV there and it was successful, who will know? We can hide it until we develop it. If it's a failure, who will know?"[38] And Kahn lost no time in launching trial runs. In one highly publicized test in June 1960, he fed a heavyweight championship boxing match between Ingemar Johansson and Floyd Patterson to thirteen CATV systems, as well as to 230 theaters. CATV subscribers were asked to pay $2 for the fight, although payment was on the honor system, because TelePrompTer had no method of isolating the signal to individual homes or of even determining who had watched the fight, and Kahn had larger plans for the future. He hoped to be able to buy and program first-run films, championship baseball and football games, and hit Broadway shows. All would be offered to subscribers on a pay-per-view basis.

Kahn was not at all shy about pressing the adoption of a pay-TV model on the rest of the industry, lobbying hard to get other operators to follow his lead. In his address to the 1960 NCTA convention in Miami, he unveiled a metering device for CATV home viewing, dubbed "Key-TV," that he hoped would lead to the union between pay and cable television. Old-line operators and the NCTA were chagrined. They saw pay-TV as an opening for regulators to take control of the industry, and it ran contrary to their legal posturing as a passive antenna service. Pay-TV was a tool broadcasters could use to argue for the regulation of CATV; the NCTA had been working to suppress talk of industry interest in pay programming, publicly suggesting that most operators were too small to make attractive targets for a pay-service. Pioneer Ben Conroy recalled Kahn's presentation at the convention: "A lot of our people, if not horrified, were shaking their heads because they didn't want to be associated with being a program service. We weren't a programmer; we were just a re-transmission service."[39]

The newer, larger operators were not inhibited by such concerns. RKO's designs called for more than just a retransmission service. In the mid-1940s, it had been one of several Hollywood studios promoting the possibilities of theater television of the kind that Kahn pursued in the1950s. With a decades-deep library of theatrical films, the company saw great potential in

CATV-based home delivery of its product, as well as the broadcast distribution noted earlier. Purchasing the theater and CATV holdings of Griffin, the man behind the 1957 Bartlesville pay-cable experiment, seemed only natural for the company. *Barron's* reported in 1962 that companies such as RKO and TelePrompTer saw a "gold mine" in the possibility of pay-TV, predicting, prophetically, "If the FCC finally flashes the green light for fee-TV and such major viewing fare as Broadway openings and first-run movies is made available, CATV operators believe that they will have little trouble signing up subscribers."[40]

For broadcasters, the specter of CATV as a pay-programming service was a profound and long-standing fear, and one that often evoked fiery rhetoric from TV station owners. Broadcasters, along with theater owners, had battled pay-TV since nearly the beginning of television itself. Throughout the early 1960s, the NAB cautioned eternal vigilance lest pay-TV be allowed to "come in through the back door of CATV."[41] The problems associated with the potential of pay-TV were, from the broadcaster's point of view, compounded by the additional specter of networked CATV systems. If the idea of individual systems offering local pay fare on a system-by-system basis made broadcasters nervous, the idea that cable systems would interconnect into a regional or national system for delivering movies and sports events, absolutely terrified them. A system of nationally interconnected CATV systems would, two decades later, be the key to a quantum leap in television evolution and, although that transformation was years away, the potential was already obvious to both cablecasters and broadcasters. As early as 1958, broadcasters publicly voiced concerns about the possibility of large clusters of CATV systems, especially those in the East, the Mountain states, and on the West Coast, connecting to form regional pay-TV networks. Discussion of the likelihood of CATV systems interconnecting into a national network was common in the trade and even the popular press, with one financial magazine going so far as to declare in 1964 that establishment of such a system was "merely a matter of time."[42] Broadcasters were fearful not just of the possibilities of pay-TV over national CATV networks, but of the likelihood that such networks might also sell national advertising.[43] By the mid-1960s, broadcasters assumed that that it was only a matter of time before cable operators found a way to create such a system.

As a matter of political posturing and despite the rhetoric of operators such as Kahn, the NCTA offered assurances through the mid-1960s that such fears were unfounded, even unrealistic. NCTA President Frederick Ford in early 1965 labeled the concern a red herring, "I think it's a bugaboo some opponents have tried to tie on CATV's neck," he claimed.[44] But he was being either disingenuous or naïve. The broadcasters' fears about possible cable interconnection were exceptionally well founded. According to Bruce Merrill, he, Shapp, and Griffing, along with a few others in the late 1950s, "had several meetings and had actually embarked on a project to build our own terrestrial microwave system to cover the entire United States for cable."[45] And with Western Union holding a 16 percent ownership stake in TelePrompTer at the time, such discussions were not merely flights of fancy. Cox Broadcasting, in 1964, was also mulling over the idea of a microwave link between New York City and Chicago that would feed TV stations from both cities to CATV systems in every town in between.[46] Extensive regional microwave systems, designed to feed CATV systems, were spreading throughout the country in the 1950s and 1960s. Cablecasters and broadcasters alike saw the logical extension of these regional systems into a national network capable of creating a unified cable audience base, and the idea never ceased to thread through industry thinking and policy debates. As will be discussed in some detail later, Kahn and others vigorously pursued a means of interconnection, spearheading efforts that led eventually to its creation and the radical transformation of the nation's TV system. The plans for terrestrial interconnection failed not because the market was lacking but rather because the means of interconnection were, at the time, beyond the reach of the industry. AT&T raised rates for such carriage to monopolistic heights, well beyond what made any financial sense or CATV, and FCC rules that prohibited signal importation in the major markets, as will been seen, tipped the economic equation against national interconnection. In betting on the potential for pay services, however, TelePrompTer was taking a longer and more risky view. Other new entrants to the industry had nearer-term goals.

The Broadcasters

Broadcasters initially entered cable for one of two primary reasons, either as a form of self-protection or as a legitimate business enterprise. Small town broadcasters, especially, saw operation of their own community antenna system as a form of local blocking or co-optation. Their intent was to stymie the entry into their community of a "real" CATV operator. Where the local broadcasters also owned the CATV system, they could restrict or control two of the most powerful threats to the fragmentation of the audience—importation of distant signals and pay-TV. If either were to appear on the broadcaster's local CATV system, it would be only under their auspices. Operators could thereby conceivably use the CATV as an additional programming service, doubling or tripling their programming capacity and boosting potential advertising revenue. In a one- or two-station town, the broadcast-CATV owner might import otherwise unavailable network signals, although any competing local stations would be at a distinct disadvantage. In Syracuse, New York, for example, Newhouse Broadcasting applied for a CATV franchise in 1964 primarily to protect its existing TV station, WSYR, from the possibility of a real cable operator building in town. It additionally sought to guard its flank by purchasing a CATV system forty miles from Syracuse in

Rome, New York. The company reported that it was going into the CATV business "reluctantly," and only for defensive purposes.[47]

Some broadcasters, CBS chief among them, also purchased community systems as a means of testing the waters. They were looking to gauge the possible level of competitive harm that the new medium might present, and to assess where, if at all, CATV might fit as a component of their own corporate structure. CBS explained that if pay-TV did develop as a new form of TV commerce, they wanted to be in the game, and CATV, as noted, was one avenue for exploiting such a service. NBC pursued a similar strategy, but ABC kept more than an arm's length from CATV for many years.

Finally, broadcasters such as Cox simply saw CATV as an attractive business opportunity. Storer, similarly, viewed cable as a logical and lucrative extension of the company's business plan. As President George B. Storer, Jr. observed, "Anything that brings in four, five, or six bucks a month (per customer) ad infinitum, looks like a pretty good business to me."[48]

Financials

In the early 1960s, it was fairly said, no one, or almost no one, went broke investing in community antenna television.. The single most powerful factor accounting for the success of the cable business, and the astute entry of the larger players in this period, was the fact that the systems, with amazing consistency, made money. The industry liked to boast that in its entire history, only a handful of community operators ever went out of business. And looking over the historical record, it is difficult to find evidence of any honest and reasonably intelligent early cable operator going bankrupt (although in later years, some came close). There are tales of fly-by-night con men in the early days, who would enter a town, collect advance payments for a promised system, and then flee with the receipts without stringing an inch of wire. Legitimate businessmen and women who failed in the early days, however, are indeed difficult to document.

A common driver of business expansion and acquisition are scale economies, the benefits derived from aggregating back office costs, accounting expenses, and even legal fees and then reducing them on a per-unit basis. In television nationally, audiences are accumulated for sale to advertisers and sponsors' costs-per-thousand (CPM) of viewers goes down as ratings go up. Cable, before 1975, did not enjoy any substantial scale economies. Systems were not interconnected for the development of audiences for national advertising or direct pay services. It was largely a local business with back office accounting, with sales and local legal relationships in the hands of one or two system employees. National issues were handled, for the most part, by the trade association.

Despite this, the industry offered a variety of financial attractions. While capital costs were high, operating expenses were low, cash flow was exceptional, tax laws involving depreciation were highly favorable, and there was pretty much no end to consumer demand.

Supply and Demand

Taking this last point first, the basic commodity offered by CATV was highly prized. As noted earlier, demand, especially in CATV's early days, often outstripped supply. No matter how many channels were available in a given community, the public has always been interested in more, especially when "more" meant not just more of the same, but more choice and greater variety. This has always been the foundation stone upon which cable television has built.

In the early 1960s, however, the idea that the television universe might some day encompass 100 or more special interest, advertiser-supported channels was within reach of only a few far-sighted individuals. For most of the viewing public, just getting all three networks was an eagerly anticipated event. Because of the FCC's allocation policy and the technical and financial difficulties that retarded UHF development, 30 percent of the country received only one over-the-air signal by 1961, and 60 percent of the country could receive only two broadcast TV signals.[49] Even by 1965, only 544 commercial stations were on the air; another 1,171 available licenses, most of them UHF, went wanting.[50] Nine states still lacked a full three-station market, fourteen states had only one three-station city.[51] Of the 253 TV markets in the continental United States, only ninety-three had three or more stations; 172 had one or two stations, and more than half of them just one.[52]

Strong demand meant that CATV was relatively recession proof. Out-of-work customers in economically depressed communities were observed to give up their telephone service before they gave up their CATV,[53] and historic television versus telephone penetration figures give credence to such anecdotes. As one woman from Elmira, New York, told reporters in 1966, "Look, I can go next door to phone. But I've got six kids. I've got to have that cable."[54]

The development of color television also stimulated the demand for CATV, especially in larger cities. One of cable's chief selling points was its ability to deliver a picture significantly sharper and clearer than that available over the airwaves. CATV improved black and white transmission as well, helping eliminate ghosts and fading, but these were tolerable problems on many black and white screens and ones many people were not willing to pay to resolve. Color television was much more demanding, however, and double images and snow made those pictures almost impossible to watch. Especially in places where distance, terrain, or skyscrapers degraded the color image, cable was seen as an easy solution. While the growing popularity of color sets was a factor in popular demand, CATV, nonetheless, remained a business and it was for other business reasons that CATV prospered.

Costs and Depreciation

Cable system construction was expensive. In the 1960s, tower and associated antenna gear costs ranged from $10,000 to $25,000; an adequately equipped head-end building ran between $15,000 and $20,000.[55] Costs for overhead lines ran from $3,500 to $4,000 per mile in the late 1950s and early 1960s. For an average system with forty to fifty miles of plant, final figures could range from $150,000 to $200,000, or higher,[56] and underground construction could be 25 percent more.

Because capital costs were largely fixed per mile of plant, the number of subscribers per mile was critical. Costs to add customers to an existing line were low, and every new customer meant costs were spread out over a wider subscriber base. Subscriber-per-mile penetration rates, therefore, had to reach a certain critical mass. The break-even point, by one industry estimate, was about thirty-three subscribers per mile.[57] The target, in a good residential area, was about 100 homes per mile,[58] although by the late 1960s the industry average was around forty.[59] All homes per mile added over the break-even point were, in the vernacular, gravy.

Operating costs, at the same time, were low. Cable owners paid nothing for their raw material, TV programming. Municipalities did charge for the privileging of wiring the community. A franchise fee of 2 to 3 percent of gross revenue was common, although it could go higher. In some cases, a flat fee was paid to the city, although such arrangements were becoming rare by the mid-1960s. It took very few employees to maintain the physical plant; management of the business was minimal and in the mom-and-pop cable operation, construction, maintenance, and management were typically handled by the owners and one or two employees. Wages for those employees were usually low; CATV was not unionized. Operators paid the local telephone or power company rent for pole space. In the early 1960s, the utilities were charging between $1.50 and $3.50 per pole annually, although again, some charged as a high as $5.00. (An effort by the telephone companies to colonize the cable business, beginning in the mid-1960s, led to attempted rate hikes and its traced more fully later in this text.) Those systems that used microwave to import some or all of their signals paid for the service when the microwave company itself was not owned by the cable operator. In systems that depended on importation from an unaffiliated provider, microwave costs were often a sizable bite out of the budget. According to the FCC's 1964 study, average annual microwave charges came in at $12,600, although the range was broad, from $1,200 to about $50,000 annually.[60]

In examples offered by *Broadcasting* magazine in 1958, average monthly expenses for a 6,000-subscriber system that imported at least one signal were estimated at about $16,000—$6,000 for microwave fees and $10,000 for wages, advertising, and normal expenses. Monthly subscriber fees were estimated at $8, giving the system a net pretax income of $32,000. A smaller system serving 4,000 customers at $3.50 a month with no microwave fees and expenses of $8,000 a month would see a before-tax net of $6,000.[61] The FCC's 1964 study mirrored the early examples of costs. Expenses, as a percentage of gross revenue included the following: wages, 10 to 30 percent, depending on system size; pole attachment 2 to 4 percent; advertising 1 to 5 percent; maintenance 2 to 10 percent; franchise fees 2 to 5 percent; and microwave fees 2 to 20 percent.[62] Average subscriber fees were estimated at about $5 a month and the average profit margin of the systems studied was 57 percent.[63]

A $125 installation fee was common in the early and mid-1950s, but those charges dropped dramatically through the late 1950s and early 1960s. The hook-up fees were originally based on the price for the installation of a roof-mounted aerial, and justified in part by the concern of some operators that the young business had a short life span and they would need to recoup costs and make a profit as quickly as possible. As it became clear in the latter 1950s that CATV was a business that would not be displaced by broadcasting or absorbed by the telephone company, installation prices began to decline, in part due to increasing resistance by consumers. Also helping drive down installation costs was the empirical observation that subscriber counts rose as those fees fell. While charges could vary substantially from system to system, by 1960 they were as a low as $25, and in some cases hook-up was free.

In places where plant construction was still underway, the lack of an installation charge meant that the full expense of building had to be borne by the monthly service fee. By the mid-1960s, however, the larger moneyed interests built using more easily acquired investment capital. When systems reached structural maturity, operators, in some cases, sought to upgrade their plant and install new equipment to expand channel capacity, requiring a new cycle of construction financing. For operators who did not upgrade, any charges for installation on a mature system were pure profit. And further to the next point, the end of system construction, representing the near complete wiring of the area, often signaled the time for the owner to sell out.

The Cash Flow Equation

The upshot of cable's low operating costs and steady monthly income was a terrific cash flow—$5 each month, every month from every subscriber— noted by previously by Storer. CATV, in a phrase of the early operators, "threw off a lot of cash." This had numerous advantages beyond the obvious. Operators who needed the money could (and did) use it to finance new construction without seeking outside loans or investment, the so-called "boot-strapping" method of building a system. For companies that secured building capital through other means, the heavy cash flow could be used to service debt, either the debt of the CATV system itself or debt carried by the parent company.

By the end of the 1950s, cable's ability to harness cash flow and build value was becoming well known, making the acquisition of outside financing significantly easier than it had been several years earlier. Cable's consummate financier, Bill Daniels, explained years later:

> [M]y job was to sell (bankers) and show them cash flow figures and why it was a hell of a business. Now we had then and we still have today, fast write-offs, because technically the business changes so quickly. So while you can buy a building today and get 20 years depreciation, we could get five to eight years. And we have that steadily, and our business has been a cash flow business since the Day One. And that means that we pay no taxes because we are continually replacing the equipment. But it was also high cash flow so we could service a lot of debt; well, bankers love that.[64]

And the bankers did love it, as did some insurance companies and other financial institutions looking for investment opportunities. Daniels reported getting New York bankers and then the larger insurance companies, including Travelers, Allstate, John Hancock, and Equitable, interested in CATV by the mid-1960s. TelePrompTer made industry headlines in 1964 when it secured $4 million in long-term financing from Chase Manhattan and Franklin National banks. Bruce Merrill, NCTA Chairman at the time, called it a "major milestone" in the maturation of the industry.[65] Morgan Guarantee Trust, working with Cooke, reported $15 million in CATV financing in 1965.[66] The cost of borrowing decreased substantially as well, from the early days of cable when interest was 10 to 14 percent, if it was available at all, to levels of around 5.5 to 6 percent in the mid-1960s. While some companies, such as H&B, used CATV's favorable financial numbers to pay off existing corporate debt. More often, cash flow was used to procure and justify increasing amounts of new debt, and cable soon became known as a highly leveraged industry, a situation that would create problems in future years.

But it wasn't cash flow alone that attracted investors; CATV thrived more than anything else in the sunshine of an exceptionally favorable relationship between the tax laws and technology. CATV enjoyed a very liberal depreciation schedule that allowed operators to write down a system's capital expenditures over five to seven years. Because in the first year or two capital costs were high and sales were relatively low, a gross margin of 40 percent on sales might amount to 20 percent on investment. By depreciating costs over five years, at 20 percent per year, operators could show no taxable earnings for the first several years of operation and, thereby, avoid paying a 52 percent corporate tax rate. In the meantime, the value of the system would climb along with the subscriber count. At the end of the five-year cycle, operators could, and very often did, sell out, for capital gains. The new owners would then start the process all over again. For individual investors in cable property, the annual systems losses could be allocated on a prorated basis to each and claimed as a loss on their tax returns, again while the value of the asset itself, the cable system, grew. Prior to revisions in the tax laws in 1962 and 1964, operators would pay only the 25 percent capital gains tax on the system's market value. After the change in the tax laws, operators had to pay the income tax rate, then 48 percent, on their depreciated assets, and this slowed system trading and encouraged more stability in the market. The cash flow–depreciation equation remained attractive, however, encouraging investment, in part because once the five-year cycle was completed, the operators could begin again by reinvesting in new equipment and such reinvestment was a legitimate industry need. Manufacturers, such as AMECO and Jerrold, were increasing the capacity of their amplifiers and moving the industry standard, first to five channels and then to 12. As already noted, the added capacity was not always necessary, especially in the smaller towns, but in markets with the potential for regional importation, it offered a wonderful opportunity to expand product and subscriber base and enter a new round of capital depreciation.

Because the depreciation equation encouraged cyclical system turnover, creating, buying, and selling of CATV property became a brisk business in the late 1950s and early 1960s. Sammons was a leading exponent of the economic equation, but Daniels was the reigning champion, in large part because of his brokerage activities. According to one report, Daniels & Associates brokered at least seventy-three systems sales between 1960 and 1964 for a total of $65 million.[67] Daniels, along with companies such as Jerrold, also continued to survey the country for new investment opportunities, towns where a CATV system might be viable. When they found such towns, they would pull together local and national investors, hire a turnkey construction company to build the system, and then hand operations over to a dedicate management firm. Jerrold, not surprisingly, specialized in turnkey construction, and Daniels offered his firms' management services to corporate owners. As many as half of all new systems were, by some estimates, constructed on the turnkey basis in the mid-1960s.[68]

As part of the process of system trading, CATV developed its own approach to pricing and valuation. Systems were usually measured on either a cash-flow or a per-subscriber basis and, by both measures, the value of systems rose steadily in the early 1960s. The 1960 sale of Jerrold's systems to H&B, for example, came in at about $150 per subscriber. Kahn bought the Silver City system for $214 per subscriber and the Farmington, New Mexico, system for $138 a subscriber. By 1965, systems were selling for anywhere from $150 up to $400 a subscriber.[69] Cable pioneer George Gardner even recalls selling one of his systems to Cox in 1962 for about $550 a subscriber.[70] On a cash flow basis, systems were reported selling for three to four times cash flow in 1960, seven to eight times cash flow in 1962,[71] and ten to fifteen times cash flow by 1964.[72] It was a good business, at least for a while.

The Politics of Cable Regulation

While CATV was a nice business, it was by no means the foundation of U.S. television. By 1960, "the three-headed dragon" of national broadcast TV— NBC, CBS, and ABC—had established its hegemony. As a business, television was only about twelve years old, but it was socially and economically entrenched. CATV was just a bothersome sideshow for most broadcasters; that, however, was about to change.

Frequently, when a new technology rubs up against an older one, the result is social friction. CATV pushed on existing structures at a number of points and in a variety of ways. Where the contestation was political, it brought into play not just CATV operators and broadcasters, but in keeping with the framework outlined in Chapter 1, the FCC, Congress, and eventually the White House. And while cable had largely won its important political fights in the 1950s, that was to change dramatically in the new decade. The cause was not an alteration in the inherent nature of the social or regulatory system, but rather a critical change in personnel and a consequent transformation of political ideology at the FCC.

The death of S.2653 was a significant victory for cable operators looking to avoid federal regulation. Those hoping it would signal the end of efforts to restrict CATV development would soon be disappointed, however. U.S. society writ large was in the process of a wholesale cultural transformation. The social fabric of the nation was being rewoven in the 1960s, and not always with ease. The post–World War II generation, the Baby Boomers, were coming of age. Television was now in nearly everyone's home and through its lens the nation was learning about civil rights struggles in the South and the difficulties spawned by a war in Southeast Asia. The effects of the changing social climate would ripple through the nation's political institutions, eventually rocking the FCC and, in turn, the community antenna industry.

The most dramatic initial consequences would come with the change of presidential administration and the subsequent restaffing of the Commission itself. Like the herald of a new age, however, the trumpets of change could be heard in the distance even before the arrival of the legions. For CATV, the 1950s problems in the West were only a prologue to a looming national battle, and lingering cases and conflicts from that period would help usher in the new age. As before, the harbinger issues would revolve around microwave. Operators, like Daniels and Magness, needed microwave to get their product to market, and it would be on the back of a microwave signal that, FCC regulation would ride in.

The Microwave Connection & Carter Mountain

Despite the appeal of cable's fundamental economics, its growth was still limited by the nature of its product. The marketability of cable's only real commodity, retransmitted broadcast signals, was restricted to those communities without full access to broadcast television. And cable in the early 1960s was beginning to saturate its coverage of underserved areas. Communities without adequate broadcast TV coverage were being quickly claimed and wired. The remaining isolated rural areas and farmhouses were too widely dispersed to make service economically feasible. Moreover, as the number of operational TV stations increased, the demand for cable decreased. In cities with a full complement of network affiliates, cable could attract customers only by importing regional, independent stations (those stations not affiliated with a network).

It was common for the operators of CATV systems that were especially isolated from broadcast service in the mid-and late 1950s to develop their own microwave links and associated microwave companies to bring distant signals into the television-deprived towns. Many of the larger CATV operators used microwave on a regular basis and, by the mid-1960s, there was a fairly extensive infrastructure of regional CATV microwave networks existed, especially in the western states. Merrill's microwave subsidiary, Antennavision Service Co., provided feeds for his systems in Arizona and California; Griffing had a microwave company serving his operations in Texas and Oklahoma. Microwave systems were active Montana, Wyoming, the state of Washington, Colorado, and Utah. Some of the earliest microwave towers fed CATV systems in Pennsylvania and New York. More than twenty-five of these "relay links" were estimated in operation by 1958,[73] fifty by June 1959,[74] and more than seventy by 1965.[75]

As in the seminal Belknap case, broadcasters complained, or not, based on their read of the financial implications and their sensitivity to the property rights issues. As early as Magnuson's 1958 congressional hearings, some broadcasters had argued that the FCC possessed authority to regulate cable operators who used licensed microwaves. When importation subjected the local broadcaster to competitive harm, claimed broadcasters, the public interest standard of the Communications Act, along with the FCC's doctrine of localism, were damaged. The use of microwave facilities for such purposes, therefore, could be prohibited or conditioned in such a way as to protect local stations.

Nonetheless, following the Commission's decision in Belknap and up until its short-lived "deep frost" in 1958 and 1959, as a matter of administrative routine, the FCC granted CATV operators, typically through the device of a separate microwave company, a common carrier license to engage in importation. In January 1958, for example, the FCC granted a request by Montana Microwave to import three Spokane, Washington, TV stations and one from Butte, Montana, into CATV systems in Missoula, Kalispell, and Helena, Montana. The owner of the local TV station, Channel 12, KXLJ-TV, in Helena, however, was the fiery CATV opponent Ed Craney. Craney immediately lodged a protest with the FCC, charging the imported signal would threaten his station, but the Commission rebuffed his petition for reconsideration.[76]

The Helena complaint, as it turned out, would be among a set of cases that served to bridge two distinct eras in CATV regulation: the 1950s era, characterized by the FCC's general laissez-faire attitude and the failure of federal legislation, and the era of the 1960s, marked by FCC activism and ever-tightening regulation. This set of cases, for the most part, dealt with situations involving the use of microwave facilities to import distant TV signals into local markets. They focused primarily, although not exclusively, on the issues of economic harm to local broadcasters and alleged violations of the retransmission clause of section 325(a). Broadcasters, as a side issue, sought to deny CATV operators access to microwave grants by charging that they were not bone fide common carriers under FCC rules. The most important of these cases would come to be known as *Carter Mountain*.

Carter Mountain Transmission Corp. of Cody, Wyoming, was built by pioneers Roy Bliss and Tom Mitchell.[77] In the mid-to late 1950s, Carter Mountain distributed one channel of imported TV to Bliss and Mitchell's systems in Riverton, Lander, Worland, Basin, and Greybull, Wyoming. In April 1958, Carter Mountain applied to the FCC to expand its service. Its plan was to offer a second channel, Casper, Wyoming's, KTWO-TV, to the Riverton and Lander systems, while opening a new two-channel service in Thermopolis, Wyoming.

The Commission, at that time, had just issued its opinion in *Frontier* and it appeared the FCC was going to do little to interfere in the business affairs of CATV operators. The small storm unleashed by that decision, in addition to sparking both congressional hearings and an FCC reexamination of its position, also swept up Carter Mountain's importation plans, along with the plans of several other similar situated CATV operators. The "deep frost" that accompanied the Commission's 1959 review chilled a number of microwave applications, including Carter Mountain's. Also among the applications being processed at the time were several by Griffing's company, Mesa Microwave. Griffing formally requested a thawing of the "frost" but was turned down. He responded by filing suit,[78] and was joined by Carter Mountain and others in an attempt to force the FCC to move on the microwave requests.

A federal court upheld the Commission's action, but the FCC nonetheless approved two of the pending microwave applications, including Carter Mountain's request to serve Miles City, Montana. The approved applications involved situations in which no local broadcast stations existed. Carter Mountain's request for service to the other towns, however, faced opposition.

The broadcaster in Riverton, KWRB-TV, protested that importation of the Casper signal would present a severe economic threat to the station.[79] While they had not objected to the original application, KWRB owners Joseph and Mildred Ernst had complained earlier about CATV. Mildred was one of the western broadcasters to testify against community antenna service during the Senate's 1958 hearings. They also objected to the Carter Mountain plan in comments filed in the FCC's 1959 Inquiry.

Interestingly, and in contrast, the FCC continued to support the grant of a CATV microwave license that it had made to Montana Broadcasting prior to the deep frost, doing so in the face of Ed Craney's strenuous objections and highly publicized announcement that he would shut down his TV station in Helena if the CATV system began importing the Spokane signals. In January 1959, those signals starting running through the community antenna wire in Helena and, true to his word, Craney closed down his station. He also filed suit in the District of Columbia Circuit Court seeking a reversal of the FCC's decision to permit the importation.

The deep frost ended, as noted in Chapter 4, when the Commission finally issued its CATV report in April 1959. The FCC released all of the deferred microwave grants, including Carter Mountain's. The CATV companies were free to begin construction or importation. The thaw was brief, however; the grants would not stick. Almost as quickly as they were approved, the Commission reversed itself and postponed their activation. In a replay of the 1958 *Frontier* opinion, the political heat from the Commission's 1959 decision was quick and intense.

In June, one week before the Senate convened again to discuss the FCC's position on CATV, the FCC postponed five of the microwave grants it had just issued and ordered hearings on them.[80] The Commission said it was not back-peddling on its recent decision to decline CATV regulation, but instead premised its postponement on the question of whether the microwave carriers in question were true common carriers under the law. The Commission especially wanted to look at the issue of the financial relationship between the microwave companies and the CATV systems they served, noting that there was heavy common ownership between the firms.

One result was an order for Montana Microwave to halt its importation into Helena pending the hearing. A request by Montana Microwave to the D.C. Circuit Court to stay the FCC's ordered was turned down and, in August, the Spokane signals were turned off and Craney turned his TV station back on.

The Commission's decision to suspend the grants was telling; embedded within the action was one of the earliest signs of a changing FCC policy toward CATV, a change made more manifest in July 1959, when the Commission, pursuing the microwave-CATV ownership issue, took its first real step toward full regulation of community antenna television.[81] It did so, moreover, without public hearing or public comment. Most of the microwave systems at issue existed in some form of co-ownership situation with the CATV systems they served. They were either subsidiaries of the CATV companies or subsidiaries of a corporation that owned both the microwave and CATV systems. No matter what the ownership structure, the CATV systems were often the only customer of the microwave facility. In its July action, the FCC amended section 21 of its common carrier rules, adding a new section (21.709) requiring common carrier microwave companies to report the percentage of their unaffiliated customers.[82] If the microwave company's business was more than 50 percent affiliated CATV, the new rule required a special hearing on the grant application to determine its potential impact on local broadcasters, an issue the Commission had not previously been concerned with. The Commission avoided any rulemaking proceedings on the amendment by labeling the change procedural rather than substantive. CATV companies resorted to a variety of methods to circumvent the rule, including selling the microwave company or entering into complicated sales and lease-back arrangements.

The move was an indication that the Commission was willing to begin promulgating rules to deal with at least some aspects of the CATV situation. In addition, while the Commission's 1959 Report and Order, taken overall, was a victory for the industry, when examined more closely, it revealed the fracture lines in the FCC's long-standing hands-off policy. In the Report, the FCC acknowledged the possible need to protect broadcasters' property rights and sought legislation toward that end (specifically, amending section 325(a) to require retransmission consent). The FCC itself was in a state of political shambles, and pressure from Congress was intensifying. At the strong urging of the Senate to conduct a more hands-on investigation of the CATV problem, in August 1959, the Commission embarked on its own tour of the West. That fall, Broadcast Bureau Chief Harold Cowgill capped that trip with a pregnant observation. Acknowledging that the Commission had for many years opposed regulation of community antenna television, he told reporters, "there seems no choice but to re-examine" that decision.[83]

On May 20, 1960, two months after the resignation of Chairman Doerfer and only two days after the collapse of S.2653 in the Senate, the FCC announced new action on the long-pending microwave grants.[84] It ruled in favor of the broadcasters, declaring that it would hold evidentiary hearings to examine more closely the allegation of possible economic harm. Preliminary hearings were held in October 1960, and through 1960 and 1961 several of the cases were resolved in negotiated settlements or through ownership changes. In 1961, for example, Ed Craney sold KXLJ, as the aging broadcaster looked to retirement, at last hanging up his gloves in the fight against CATV. (Ironically, the station was sold again in 1963 to a group of CATV operators headed by Magness; the call letters were changed to KBLL.) A peaceful settlement was not to be reached, however, for Carter Mountain.

In May 1961, the FCC hearing examiner released his initial decision in the case, denying the broadcaster's complaint and supporting Carter Mountain's request.[85] In his decision, the examiner maintained the Commission's 1950s laissez-faire philosophy on CATV. The examiner was also applying a standard policy of the Common Carrier Bureau that as long as the carrier was abiding by the laws, the impact of the signal conveyed by that carrier on some other business was not relevant to the grant of the license. An oral argument was held before the Commission on December 14, 1961, but apparently was more for show than substance, for shortly after the meeting, the Commission announced it was reversing the hearing examiner's findings. The vague suggestion of marginal control over CATV, which was murmured shortly after the 1959 report, now took solid form. While the Common Carrier Bureau defended its position in favor of CATV, the Broadcast Bureau spoke strongly against it. The formal decision was issued in February 1962, four years after the original application. While FCC actions of the previous year had made it clear that the Commission's philosophy on CATV was shifting, the decision in *Carter Mountain* was still a dramatic turn, both in its substantive impact and its expressed logic. It signaled a complete reversal of the Commission's position of only a few years before.[86]

In the 1959 Report and Order, the FCC, in addition to rejecting the notion that it had the responsibility to protect the economic health of local broadcasters, also rejected related arguments that it ought to restrict microwave importation of competing signals. Declared the Commission:

In essence, the broadcasters' position shakes down to the fundamental proposition that they wish us to regulate in a manner favorable toward them vis-à-vis any nonbroadcast competitive enterprise. Thus, for example, we might logically be requested to invoke a prohibition against access to common carrier facilities by such enterprises as closed circuit music and news services, closed circuit theater television operators, and possibly, even ordinary motion picture and legitimate state operators, magazine and newspaper publishers, etc. comprising all of the entities which compete with broadcasting for the time and attention of potential viewers and listeners. The logical absurdity of such a position requires no elaboration.[87]

By 1962, the Commission no longer saw the "logical absurdity" of the argument, but instead embraced the concept that its job was to protect the local broadcaster by nearly any means necessary, it seemed, including the restriction of common carrier relay of CATV signals.

We do not agree that we are powerless to prevent the demise of the local television station, and the eventual loss of service to a substantial population; nor do we agree that the Commission's expertise may not be invoked in this instance to predict this ultimate situation. Thus after weighing the public interest involved in Carter's improved facility against the loss of the local station, it must be concluded, beyond peradventure of a doubt, the need for the local outlet and the service which it would provide to outlying areas outweighs the need for the improved service which Carter would furnish.[88]

In *Carter Mountain*, the Commission concluded that the CATV system was responsible for a loss of KWRB revenue, and expansion of the CATV service probably would force the station off the air. The Commission stated it would approve the requested extension of service if the microwave operator could obtain from its CATV customers an agreement not to duplicate the local TV station's programming and to guarantee carriage of the local station if requested by the station.

In 1959, the Commission had held that the activities of the common carrier customer, so long as they were legal, were not relevant to the regulation of the carrier itself. That is, common carriers could not be restricted by the uses to which the signals were lawfully put. The 1962 Commission reversed its stance on this point as well. The Commission now embraced the long-standing argument of the western broadcasters that CATV was an economic threat to local TV stations, which could result in the loss of service to small town and rural viewers. The Commission asserted this authority in the absence of any systematic evidence that CATV would, in fact, pose such a threat to local broadcasters. The potential for harm was simply assumed. As to the prima fascia reversal of long-standing FCC policy, as embodied in the 1959 Report, the Commission declared that, "To the extent that this decision departs from our views in (the 1959 Report and Order), those views are modified."[89]

Carter Mountain signaled the FCC's intention finally to bring cable under its yoke. As Commissioner Rosel Hyde later told Congress, the objective was "to integrate the CATV operation into the national television structure in a manner which (did) not undermine the television broadcast service."[90] Any new applications for microwave use would be considered on a case-by-cases basis using the criteria of *Carter Mountain*. These were the Commission's first formal restrictions on cable TV.

The Changing Commission

The tidal shift in policy positions, an almost near reversal of ideological course between 1958 and 1962, was a dramatic and historic turnabout for the FCC and for the CATV industry. Its causes are worth examining.

In regulatory decision-making, the dominant model for political scientists of the 1950s, and even into the 1960s, was capture theory, the notion that regulatory bodies, especially the FCC, were the captive agents of the industries they were assigned to regulate. Analyses based on capture theory could be compelling and offer significant insight into the evolution of the industry and the relationships between the economic base and the regulatory structure. Even at the level of specific personal interaction, capture theory has, at times, been an appropriate lens. The decline and fall of Commissioners Mack and, perhaps to a lesser extent, Doerfer, are cases in point. Wide-sweeping structural explanations can be taken too far, however, providing too little sensitivity to the more complex and finely detailed interactions of social and political life. Businesspeople, politicians, and regulators, therefore, play out their parts in the larger context of taken-for-granted cultural values, the social water, so to speak, in which they swim. But at the same time, people are neither structural nor cultural puppets; they have pointed goals and an assortment of political tools and resources. Krasnow, Longley, and Terry's classic *Politics of Broadcast Regulation*, is a prominent example of an attempt—both in theory and in practice—to reveal the complex, often highly personal, and politically charged environment of competing players afield in the process of forging policy.[91] FCC historian James Baughman also takes issue with "champions of 'captive' theses," noting the structured political limitations within which the FCC works, specifically its dependence on the three major branches of government, any of which are capable of handcuffing the agency at any point.[92]

No political institution, furthermore, is historically static. The FCC of 1955 was not the FCC of 1965. And the primary difference was not the nature of its legislative mandate or staff bureaucracy, but rather the change in personnel, which brought with it the application of fundamental shifts in political philosophy. Explains Baughman "In contrast to some of their predecessors, (FCC Commissioners) [Newton] Minow, Henry, Cox and other Commissioners were not the servants of the industry they policed; nor apparently were those Commissioners who disagreed with them and other liberal exponents of regulation."[93]

The changing FCC position vis-à-vis CATV, therefore, had little to do with any change in the medium itself, either its basic technology or its business plan. The change was in the nature of the Commission, its outlook on the world, and its vision of the relationship between CATV and broadcasting.

As Deorfer came under increasingly harsh scrutiny in the late 1950s, his laissez-faire regulatory philosophy was forced to take a back seat to political reality. Congressional hearings, convened almost immediately after the FCC's 1959 Report, have to be viewed within the larger context of a Commission in very hot political water. Doerfer and the FCC had to bend, at least a bit, to the growing demands of Congress. The small, but quick and important, change of course taken by the FCC in its position on microwave licenses shortly after the 1959 opinion can therefore be seen as a political response to mounting pressure for more active oversight.

Even more fundamental to the evolution of the Commission and its position on CATV, however, was the presidential election of 1960. The old world, represented by Dwight D. Eisenhower, was taking its leave and the new world, the "New Frontier," of John F. Kennedy, was entering. It was not just a change of politics but a change of culture. The new decade signaled a shifting view of the relationship between the individual and society and a concomitant adjustment of philosophy on the role of government in business. The policies of state noninterference were being replaced by a more activist mood, and nowhere was that change more evident than at the FCC.

The Commission, as it turned the decade, was composed of the remnants of a fading era. Frederick Ford, a respected Republican appointee of Eisenhower, had generally taken a moderate approach to regulation and had often opposed the more freewheeling style of Doerfer. Upon his elevation to Chairman, replacing Doerfer, he began almost immediately to tighten the reigns on broadcasting, in part in an effort to reform the Commission's tarnished image. Republican Rosel Hyde, a Commissioner since 1944, remained one of the "reluctant regulators" from the Doerfer regime, believing in free speech, free markets, and free enterprise. Republican Robert E. Lee, the former J. Edgar Hoover aide and McCarthy ally, was described as a philosophical enigma whose voting record was a series of self-contradictions. Democrat Robert Bartley, a Commissioner since 1952, was nephew of Speaker of the House Sam Rayburn, a Texas populist but with a restrained view of regulation. T.A.M. Craven, was a long-time Commissioner and honest engineer, who actually brought some technical expertise to the head table of the agency. Democrat John Cross, sponsored by his political patron Representative Oren Harris, replaced Mack in 1960 and characterized himself as ethically clean if not too smart—a description no one disputed.

Following Kennedy's election, new faces and a new philosophy began to take over. The agency ushered in its own New Frontier in 1961 when Kennedy made his first FCC appointment. Newton Minow was a young lawyer with political ties to the Kennedy family. He had no real broadcast experience, but he did possess a zealous enthusiasm to clean up television. His famous introduction to the broadcast industry came May 9, 1961, in a speech to the annual convention of the NAB, in which he invited his audience to sit down and watch a full day of television. "I can assure you," he told his stunned listeners, "that you will observe a vast wasteland. You will see a procession of game shows, violence, audience participation shows, formula comedies about totally unbelievable families, blood and thunder, mayhem, violence, sadism, murder, Western badmen, Western goodmen, private eyes, gangsters, more violence and cartoons. And, endlessly, commercials.[94]

The "vast wasteland" speech set a collective grimace on the face of the broadcast crowd and raised concerns about their regulatory future. Minow's regulatory zeal lacked any real political support, however, at least in the beginning. As Baughman explained, he was, at first, largely isolated on the

Commission. The other members were more cautious in their regulatory outlook, and it took two more Kennedy appointments—William Henry in 1962 (replacing Cross) and Kenneth Cox in early 1963 (replacing Craven)—to obtain even a minority alliance for Minow. Henry, named to the Commission in part by virtue of his association with Robert Kennedy, shared Minow's outlook. Cox, with a long record of regulatory activism in broadcast matters, had already been named Chief of the Broadcast Bureau by Minow when he was appointed to the Commission itself. With this nucleus in place, however, the character of the Commission was sufficiently altered to spark a radical restructuring of CATV policy over the next several years.

Even before the new Commissioners were installed, however, the shape of CATV's regulatory future was outlined by Chairman Ford in a January 1961 address to the NCTA in Washington, D.C.[95] There, Ford revealed FCC-proposed legislation that would give the agency authority to intervene, on a case-by-case basis, in cable-broadcaster disputes. The proposal did not go so far as to give the FCC licensing authority over cable and, in fact, never came to fruition, but if the NCTA attendees listened closely, they could heard the winds of changing whispering through Ford's remarks. By the end of the year, Newton was on the Commission, Cox was Broadcast Bureau Chief, and the agency had announced its decision in *Carter Mountain*. Five Commissioners voted in favor of the policy change in *Carter Mountain*: Ford, who had warned the industry such a move was likely; Lee, Minow, Craven, and Hyde. Cross voted against the majority. His close alliance with Representative Harris, who strongly opposed FCC interference in CATV affairs, may have been influential. Bartley did not participate.

Two factors stand out as particularly important in the vote. One was the appearance of Cox as head of the Broadcast Bureau. Since the late 1950s Cox had arguably been the most vocal advocate of federal regulation of community television. He was now in a position to influence policy directly, and it was his Broadcast Bureau that forcefully intervened in the common carrier proceedings arguing the broadcaster's case. Cox was hand-picked for his new job by Minow and served as Minow's principal advisor on CATV. Minow's exposure to CATV was not substantial and Cox's opinion carried weight. Cox was smart, experienced, and articulate. The Broadcast Bureau would come to be known as "Cox's Army" and exercise its power with gusto.

A second important factor in the *Carter Mountain* vote and an influence in subsequent CATV decisions, was the Commission's troubled but hopeful relationship with UHF. UHF was one of the few issues that Commissioners seemed to be able to agree on in the early and mid-1960s. The failure of the service was as obvious as it was frustrating; by 1962, only 102 of the authorized 1,544 UHF stations were in operation. A variety of proposals to save UHF were floated. Commissioner Lee was the most adamant supporter, advocating a complete abandonment of the VHF spectrum in favor of an all-UHF system. He had campaigned tirelessly for UHF interests since the early 1950s, bestowing on himself the title of "the UHF Ogre."[96]

While less zealous about UHF, other Commissioners were nonetheless optimistic that it might be able to break the monopoly of the airwaves held by the three major networks and help promote program diversity by creating more national broadcast outlets. Ford was a strong supporter of the idea and Minow quickly came on board. Through 1961 and 1962, UHF reform was the Commission's primary goal. One solution was "de-intermixture," but that idea faced fatal opposition from Senators and Representatives in the affected cities and states. By dropping any proposals for de-intermixture, the Commission was able to sell a plan that would require all TV sets to be manufactured with UHF, as well as VHF, tuners. By September 1962, the Commission had been given congressional authority under the new All-Channel Receiver Act to set a date for manufacturer compliance, and it did so. After April 30, 1964, all TV sets would have to be able to receive both VHF and UHF signals.[97]

The FCC's commitment to UHF played an important role in its CATV decision-making insofar as community antennas were considered by some as a threat to the health of UHF. CATV policy, therefore, was to some extent less about community antenna service directly than it was about protecting, both substantively and symbolically, UHF broadcasting. More broadly then, the FCC's personnel changes led to significant revisions of regulatory thinking about UHF, commercial VHF, and CATV in the first few years of the 1960s. CATV, along with broadcasters, found themselves in a new political universe.

Business Radio Service (BRS): The Next Step

From 1960 through 1962, the industry's response to these changes was largely to sit tight, holding fast to the political standard adopted at the Miami convention. The NCTA continued to oppose any suggestion of federal control, and employed as much rhetoric as it could muster to support its claim that the FCC lacked the authority to regulate CATV, noting repeatedly that Congress had declined to provide the Commission with such power. For a brief period, the industry was able to sustain this position; although the FCC was interested in pursuing the CATV problem, the bulk of its time was absorbed by UHF matters, at least through 1962.

The Commission's decision in *Carter Mountain* in December 1961 worried CATV operators, but they maintained the hard line against even moderate regulation, and it was a reasonable position to take at the time. The FCC-backed legislation, introduced in 1961, was arguably mild. It shunned licensing, still concerned about the ability to manage the case load, and it sought the power to require local carriage and institute network nonduplication rules in markets where small broadcasters felt such protection was necessary. Most important, from the CATV perspective, passage of the bill seemed unlikely. The powerful Chairman of the House Commerce Committee, Arkansas Democrat Oren Harris, took a blunt and public position that no regulation of CATV was needed or required, going so far as to chastise the FCC in a speech at the NCTA's 1962 national convention for its *Carter Mountain* ruling.[98] In fact, the bill soon died of legislative neglect and there was little indication that Congress was interested in pursuing the CATV issue.

This state of uneasy denial could not last long for the industry, however. By the end of 1962, the Commission, with the UHF controversy settled, moved CATV to the top of its agenda. It announced a new microwave rulemaking proceeding, and made it clear to everyone that, with UHF in its wake, CATV would be its primary issue for 1963. Broadcast Bureau Chief Cox declared publicly that they would seek congressional legislation in 1963 providing jurisdiction over community antenna service, adding that he personally felt CATV posed a threat to the development of both local and UHF broadcasting.[99]

This hard rhetoric took legal form in December 1962 when the Commission announced rulemaking proceedings on its first formal carriage and program duplication proposals. The action came quietly in the guise of a proposed amendment to the rules of the microwave Business Radio Service (BRS).[100] BRS, an alternative to a microwave Common Carrier license and was designed for dedicated services supplying a limited customer base, as was the case for many CATV-microwave situations. BRS was assigned to the 12,000 mc band, as opposed to the common carrier 6,000 mc band, which meant it could not transmit a clear signal as far as the Common Carrier service, at least not without additional retransmission towers and greater cost. CATV operators, therefore, had an early preference for Common Carrier allocations. Improvements in 12,000 mc technology and the 1959 customer service restriction on common carriers prompted operators to begin looking at the BRS band for importation, however, and BRS applications began mounting in the early 1960s.

Seizing this opportunity, and following the logic of *Carter Mountain*, the FCC placed a freeze on these applications pending outcome of the new rulemaking proceedings. The proposed rules would prohibit CATV systems from carrying any BRS imported programs that duplicated the network programming of a local broadcasters for thirty days before and thirty days after the airing of a particular show. These became known as the "nonduplication rules." The proposals also called for affected CATV systems to carry all local signals requested by the station owners. These were the "must carry" rules. The Commission announced it would only approve CATV-related applications for BRS licenses if the operators voluntarily agreed to abide by the proposed rules, and several did. (As an interesting side note, the rules were strikingly similar to ones proposed to the Commission by Craney and rejected in 1960.[101]) The rules were not extended to the Common Carrier service at this point, in part because *Carter Mountain*, which raised questions about the FCC's authority vis-à-vis cable and Common Carrier microwave, was still before the courts. While limited in their application to only the few CATV systems that used the BRS service, therefore, these proposed rules nonetheless would provide the model for restrictions that would eventually engulf the industry.

In May 1963, the Washington, D.C., Circuit Court upheld the FCC in *Carter Mountain*, relying in part on the Commission's authority to exercise its expert judgment in a case that directly involved broadcast licensees.[102] The Court found the FCC's concerns about the economic health of the local TV station eminently reasonably and in keeping with the agency's responsibility to safeguard the public's interest in broadcasting. While the restrictions on the Common Carrier could have a detrimental impact on the associated CATV system, that impact, concluded the Court, was outweighed by the larger public interest, as assessed by the Commission. The decision became the foundation of the FCC's "economic impact" rationale, which would be used to restrain cable's growth over the next decade.

Freed by the Court's holding and appropriately doubtful of any congressional assistance in its CATV activities, the FCC took the next step in solving what it labeled the "CATV problem." In December 1963, it announced a new proceeding on microwave importation.[103] The good news for CATV operators was that the thirty-day nonduplication window of the proposed 1962 BRS rules was shortened to fifteen days. The bad news was that the rules now included not just the BRS band but regular Common Carrier licenses, as well. The Commission said it recognized the valuable service CATV could offer, but the development of the industry should not come at the cost of crippling local TV. The Commission, therefore, froze all microwave applications for CATV-related businesses. It would grant no new licenses or modifications until its inquiry was completed. Most ominously of all, at least from a CATV perspective, was the widely reported rumor that the FCC staff, perhaps influenced by Cox, was now of the opinion that the Commission had full authority to assert jurisdiction over all CATV operations, whether or not they were associated with microwave facilities.[104] It would not be too much longer before the world discovered that those rumors were true.

Industry leaders were finally getting the message that the new FCC was serious about CATV, and they began to change course. The FCC's activity was not, parenthetically, the only legal problem it was confronting; as detailed below; new attacks were being mounted against the industry's copyright shield. The recently installed NCTA president, William Dalton, therefore extended an olive branch to broadcasters, suggesting as early as October 1962 that the two sides begin a conversation about their mutual concerns. Community antenna operators approached the Commission, as well. By early 1963, the NCTA had formed a special committee and commenced negotiations with the FCC about working together on federal CATV regulation, clearly with an eye toward co-opting or channeling any regulatory efforts and preventing more harm to the industry than was necessary. Those talks broke down within a few months, but the Appellate Court decision in May 1963 upholding *Carter Mountain* and the FCC's subsequent proposed extension of the *Carter Mountain* rules in December helped renew the NCTA's appetite for conversation.

Reversing its vote of the 1960 Miami convention, the NCTA Board of Directors declared they would seek "simple" legislation, and they began a new round of meetings with FCC and NAB negotiators in January 1964. Through mid-1964, Committees of the NCTA and NAB worked on a legislative package that offered concessions on all sides. For their part, CATV operators were willing to promise carriage of all local signals and swear off any local advertising. They were even willing to consider a prohibition on local origination. In return, broadcasters were asked to back off on demands that CATV pay copyright or signal carriage fees. Progress was thereby made on a draft bill that would give the Commission the authority it had long sought, but would also offer CATV operators a reduction of the previously imposed nonduplication window, from fifteen days to only one day.

The NCTA was also spurred in its support for federal regulation by the hovering specter of state and local control. CATV operators were seeking franchises in larger, politically more sophisticated cities and those cities were demanding franchise fees of 5 percent or more, control over subscriber rates, and public access to channel capacity. Legal differences across jurisdictions exacerbated the problem, especially for the new MSOs, who preferred a uniform national legal environment for their far-flung operations. No one wanted to face a patch quilt of inconsistent, even conflicting, state and local rules governing rate regulation, franchises fees, and even content. At the 1965 NCTA convention, out-going NCTA head Merrill noted that twelve states had bills pending to designate CATV a public utility, and utilities commissions in another seven states were thinking about asserting jurisdiction. The National Association of Regulatory Utilities Commissioners was drafting model state legislation aimed at CATV. Until federal legislation preempted local action, Merrill told the attendees "we can expect a continuous battle in many states."[105]

While the private parties negotiated, the FCC moved ahead on several fronts. In the spring of 1964, it began work on the creation of a new microwave service expressly for CATV. The Community Antenna Relay Station, or CARS band, would include twenty frequencies in the 12.7 to 12.95 GHz band, adjacent to the BRS range, giving cable operators a service that could be more cleanly controlled and contained. The CARS allocation was adopted the following year and CATV operators urged to migrate to the new band.[106] The FCC, in 1964, also issued its Notice of Inquiry on cross-ownership of CATV and broadcast properties, and, more importantly, launched its own study (noted previously) on the potential economic impact of CATV operations on small town broadcasters. The Commission allocated $20,000 to fund the study and hired Washington economist Dr. Martin Seiden to survey the industry and prepare a report.

The NAB, meanwhile, had already hired its own expert to look at CATV, commissioning a $25,000 study by Dr. Franklin M. Fisher, associate professor of economics at MIT.[107] The NAB effort reviewed data on 500 TV stations and 1,000 CATV systems and concluded that for every 1,000 subscribers on a cable system, the local broadcaster could expect to see a reduction of anywhere between 10 and 50 percent on average net profits. The level of economic damage depended on whether CATV carried the local station and whether the town had one or two broadcasters. In situations where the local CATV operator duplicated the programming (via importation) of the local TV station in a one-station town, the station's revenues would sink to zero when the CATV system signed up 2,500 subscribers. If the CATV system declined to carry the local TV station, it would have to amass only 1,650 subscribers to drive the local station out of business. The NAB praised its own study as the first solid empirical evidence of CATV's damaging impact.

The Fisher report shows that the adverse impact of CATVs on local television is already severe. In addition, the rapid growth of CATVs is a present threat to all but the largest stations ... television broadcast stations cannot long both survive the competitive assault of CATVs and, at the same time, continue high cost public service programming.[108]

The NCTA's response was an analysis of the Fisher report by City College of New York statistician, Dr. Herbert Arkin, who predictably attacked the method and the conclusions of the NAB document. The rebuttal, however, relied less on econometrics and than it did on history. As it had for several years, the industry simply pointed to the small number of cases in which CATV operators came into economic conflict with local broadcasters and the smaller number in which any harm could be demonstrated. Comments filed with the FCC by Jerrold were illustrative. Jerrold noted that of the 107 UHF stations that went off the air between 1952 and 1964, only twenty were in CATV towns, and in two of those the broadcaster had testified that CATV had worked to their benefit. NCTA noted that representatives of nineteen local TV stations had testified before Congress in 1958 and 1959 that CATV had put them on the "brink of disaster" and, left uncontrolled, CATV would surely drive them off the air. By 1964, however, eighteen of those stations were still operating.[109]

The FCC's economist, Seiden, positioned himself in the middle ground. He was sharply critical of the NAB report, finding both its assumptions and conclusions flawed. Echoing the protests of the cable industry, Seiden could find no evidence of direct CATV-generated economic harm to local broadcasters. He noted, as few others did, that lack of information about CATV systems, their size and programming, had prevented advertisers from even taking cable into account in most of their planning and buying. Because advertiser revenue was unaffected, "CATV," concluded Seiden, "has not had a direct economic impact on broadcasters."[110] He also foresaw no deleterious impact of CATV importation on the potential health of UHF stations in larger markets. The well-being of big city UHF's would be much more heavily dependent on their ability, or inability, to obtain good quality programming at reasonable prices. On the other hand, Seiden did warn of the likely economic harm to broadcasters in the smaller one- and two-station markets due to what he called the "indirect" impact of audience fragmentation.[111]

On the evidence, the FCC was left with only conflicting and inconclusive findings. It nonetheless retained its thirst for regulation and was now being pressed by some members of Congress to adopt a coherent regulatory scheme for the emerging medium. While Representative Harris warned the FCC against unilateral action, Senator John Pastore, Chairman of the Senate Communications Subcommittee, was leaning on the FCC for a policy statement.

The FCC had hoped that the affected industries could, on their own, draft a compromise plan and prevent a brutal fight in Congress. In fact, committees from the NCTA and NAB did craft a legislative proposal that would have included provisions for "must carry," a ban on local origination, and a nonduplication clause. Unfortunately, the broadcast industry was internally fragmented on the issue. Small-town broadcasters, chain broadcasters who owned CATV systems, and the large broadcast networks all had divergent interests. Despite the 1950s track record compiled by western broadcasters, many other small-town station owners, seeing the potential for a negotiated settlement that protected their interests, tended to be open to compromise. Companies with CATV holdings, such as Cox, Storer, and CBS, also took a softer line on CATV regulation. The networks were carefully guarding their property rights and had taken public positions against CATV use of their signals and programming without permission. But given their potential interest in owning community antenna systems, CBS and NBC were resistant to what they viewed as any draconian policies.[112] CBS Vice President Richard Salant stated the network opposed efforts to restrict pay-TV by broadcast or wire, as well as any attempt to restrict original programming on cable, explaining that if pay-TV were part of the inevitable future, CBS would be a part of it. "It seems to us that the principles of free competition apply equally well here as in wired pay TV," said Salant. "And so, to the extent that CATV does not use broadcast channels, I doubt that we can demand regulation and prohibition from the government."[113]

In contrast, ABC, the smallest and most vulnerable network, took a lead in lobbying for tighter federal control. While CBS's Salant was calling for moderation, ABC, with no CATV interests, was filing motions with the FCC arguing for strict limits on signal importation. Following ABC as one the most adamant and energetic opponents of CATV, were the large market broadcasters who feared the incursion of community antennas into their cities. The Association of Maximum Service Telecasters (AMST) represented the metropolitan stations and repeatedly filed briefs in support of FCC control.

This internal factionalism would, in the end, doom hopes for a negotiated settlement. In December 1964, the NAB's Committee on the Future of Television in America (FTIA), which included among its members long-time CATV foes Grove and Clair McCollough, rejected the compromise plan agreed to (and submitted) by its own subcommittee. The public rationale for rejection was the lack of sufficient protection against the practice of "leapfrogging signals." Leapfrogging involved importing signals from distant towns while skipping over, or leapfrogging, a broadcast signal from a town closer to the CATV system. Leapfrogging occurred when the more distant station had superior programming fare or one that added diversity to the CATV lineup. CATV–broadcast operators had representation on the committee, but were not present for the vote and later suggested the anti-leapfrogging issue was only a "subterfuge" meant to torpedo the compromise.

The negotiated settlement had not been the NAB's only hope in resolving the CATV problem. Broadcasters had also been following a court case that could have given them the legal leverage they sought in their battle against community antenna operators. An appellate court decision in the case went against them and the Supreme Court declined review. The broadcast industry was forced, therefore, looked once again to the Commission for help. By the end of 1964, the *Wall Street Journal* was reporting that broadcaster pressure to regulate CATV had become so powerful that "the government and industry officials expect the agency (FCC) to begin regulation of all cable systems next year with or without enabling legislation from congress."[114] For the CATV industry, the path toward regulation now looked wide, slippery, and downhill. In early 1965, the NCTA hired a new president, the out-going Chairman of the FCC, Frederick Ford, who they hoped would have the experience and the connections necessary to steer the looming regulatory debate in a direction favorable to industry interests.

Seizing Control: The FCC's First Report and Order (1965)

With little help expected from the affected industries, Congress or the courts, the FCC found itself in a situation where it would have to act, if it acted at all, largely on its own. But the NAB was not the only organization in disarray. The FCC itself was a house divided on the question of CATV regulation. Staff recommendations ran the gamut from mild to highly restrictive controls, and it was unclear until a final vote was taken whether a majority of Commissioners could be assembled around any single regulatory plan.

In such situations, the importance of the individual, set within a given social context, has been noted. Nowhere was it more readily illustrated than in the splintered positions of the FCC Commissioners in the 1960s.[115] Frequently, the Commission split four to three, with a single vote determining the outcome of landmark policy. On significant broadcast matters, Minow was often frustrated in his short time at the agency, able to gather only the support of Henry and Cox. In 1963, Minow stepped down, replaced by Kennedy's fourth and final appointment, Lee Loevinger, and Henry was elevated to the FCC Chairman. These changes acted against the activist platform.

Loevinger was a cautious Commissioner. A one-time Minnesota Chief Justice, he was appointed head of the U.S. Justice Department's Antitrust Division in 1961, but soon clashed with Attorney General Robert Kennedy. In May 1963, "the FCC appeared the most convenient dumping ground" for Loevinger, and it reportedly required some arm-twisting to get him to take the post.[116] Moreover, Loevinger's real career goal was that of a federal judge and some observers claimed that following the assassination of John F. Kennedy in December 1963, he molded his philosophy more closely to that of the Johnson administration in hopes of securing a judiciary post,[117] although like Doerfer, his principles may have been more truly felt. In his actions and words, Loevinger looked more like Doerfer than Minow, noting once that, "If I am to err, I would rather err on the side of restraint."[118] This new "reluctant regulator" soon gathered a conservative coalition with Hyde, Ford, and Bartley.

By the time of the vote on the proposed new rules for CATV in early 1965, a Loevinger-led antiregulation faction was already in gestation. Leovinger, with Bartley, opposed the rules. The regulatory activists, Henry and Cox, supported them. Lee, by virtue of his backing of UHF, was also a supporter. He told an NAB gathering in November 1964 that he would urge the Commission to take "the giant step" of regulating of cable, and felt the agency already had sufficient authority to do so.[119] The conservative Hyde, who would frequently line up with Loevinger in the months to come, seemed to sit on the fence. If Hyde took a restrained position with Loevinger and Bartley, any decision would have to await the tie-breaking vote of James J. Wadsworth, who was shortly due to replace the retired Ford. In the end, however, Hyde moved to join the activists. In an eerie echo of a few years prior, when S.2653 fell by one vote, delaying federal control of cable, one vote again determined the fate of the new medium. This time, however, the four to two vote went against CATV.

In April 1965, only a few months after Ford left the Commission to become NCTA President, the FCC issued its First Report and Order on CATV.[120] The new rules applied to all microwave-fed community antenna systems, extending to them the local carriage and nonduplication requirements of *Carter Mountain*. If a microwave-fed system was within the Grade A contour of a TV station, about sixty miles, it would have to carry that station's signal. It would also have to black out any imported programming that duplicated the local station's materials, in a window of protection that extended fifteen days before and after the local program aired. In cases where two broadcasters within range of the CATV operator carried the same network programming, the weaker station could be bumped where there was limited channel capacity and where it was necessary, by law, to carry a nonaffiliated or noncommercial station.

The majority opinion also indicated the Commission was considering prohibitions on CATV program origination, on the ownership of a CATV system and broadcast outlet in the same town (cross ownership), and on the practice of leapfrogging, and it sought comments on the proposals. The FCC also noted that, while it could not affect copyright issues directly, the rules were fashioned with sensitivity to the property rights of broadcasters.

In a sense, the First Report and Order was simply a formal enunciation of the philosophy expressed in *Carter Mountain* and other forums. The stated intent was to protect small market broadcasters, which the NAB's Fisher report had claimed would be vulnerable to redundant CATV programming. In the end, the Commission conceded it did not have the empirical tools necessary to support a case for economic harm,[121] but proceeded to issue rules nonetheless on the unsupported presumption that economic harm was a future possibility and that it was better to act to protect broadcasters before such harm appeared.

At the same time, the Commission maintained its efforts to keep administrative costs to a minimum. To acquire protection, for example, local broadcasters had to make a formal request to the Commission; coverage was not assumed. Existing CATV-microwave operations were protected from the new rules with a grandfather clause that dissolved only if they sought a change of service.

In his dissenting opinion, Commission Loevinger was acid; one could almost see the smoke rising from the page. In one of the more memorable, and quoted, passages in CATV regulation, Loevinger wrote: "It seems to me that in its approach to the CATV problem, the Commission is doing the wrong thing for the wrong reason in the wrong manner to deal with the wrong problem."[122]

The wrong problem, for Loevinger, was the narrow concern about CATV's competitive impact on broadcasting. He would have preferred a broader inquiry into the role of cable in a coherent national scheme of telecommunications regulation, privileging neither cable nor broadcasting. Echoing Commission language of the 1950s, he also observed that the Commission's logic—extending regulatory control over a service that impacted broadcasting—in effect, gave the FCC nearly unlimited power over any segment of U.S. industry and commerce that touched on the use of the airwaves. There was nothing Loevinger could find in the facts or arguments of the situation that had changed materially since the Commission's 1959 decision.

He was also sharply critical of his fellow commissioners, whom he accused of having made up their minds about CATV well before hearing any facts and arguments in the proceedings. Underscoring the importance of individuals and ideology in the regulatory process, Loevinger declared: "The only thing that has changed since the Commission last disclaimed the jurisdiction it now asserts is the personnel of the Commission."[123]

More important and ultimately far-reaching than the First Report and Order itself, however, was a Notice of Inquiry and Proposed Rulemaking released with the Report.[124] The Notice declared the Commission had authority to regulate even non–microwave-affiliated cable systems through its plenary power over interstate communication by wire, and it would hold further hearings to determine whether it should assert that authority. The prior findings of the Commission that it had no such authority and the lack of congressional action to provide it such authority were not, explained the Commission, "determinative."[125] The Notice had one more important philosophical turn. The intended subject for protection was no longer small-market telecasters, but, in keeping with the Commission's renewed concern about UHF, now swiveled toward large-market UHF stations. The Notice erected a high barrier to cable expansion in major markets (those with four or more commercial channel assignments) by requiring any system wishing to import signals to demonstrate that the microwave grant would not threaten the development of UHF service in that area. It required, in effect, the proof of a negative proposition, a showing logically impossible to make.[126] LeDuc summarized the force of the opinion:

> It was in part 1 of the Notice that the FCC asserted for the first time its belief that the unfair element of cable competition gave the agency jurisdiction over all CATV systems, and that the only issue to be resolved by further hearings was the need to assert that authority. Part 2 had even greater immediate impact. It imposed with a single paragraph a freeze upon major-market cable expansion which would extend beyond the end of the decade.[127]

The cable industry's public reaction to the report was a mixture of anger and, at the same time, conciliation. Kahn, capturing wide if not complete industry sentiment, criticized the FCC's unilateral action, but continued to be open to milder congressional legislation. A panel of speakers at that summer's 1965 NCTA convention noted that CATV bore some responsibility for passage of the rules by failing to build better relations with broadcasters. They suggested that a broad framework of government regulation that integrated broadcasting and cable was not unreasonable and suggested that another set of compromise rules could be forged. At the same time, they were angered by what they felt were regulations that suppressed CATV operations only for the benefit of broadcasters rather than the larger public good.

In the words of *Broadcasting* magazine, the FCC was clearly "the villain" at the convention.[128]

House Commerce Committee Chairman and FCC critic Harris, for his part, was upset that the FCC did not consult with his committee in developing the package, so he introduced a bill (HR 7715) that would delay any FCC action on CATV until Congress had an opportunity for review. But on the Senate side, Pastore and Magnuson indicated they were comfortable with the FCC's action, and without any chance of Senate action, Harris's bill died in committee.

Extending Control: The Second Report and Order (1966)

Because the First Report and Order applied only to community antenna systems that used microwave facilities, only about 250 systems, or 17 percent of total, were directly effected.[129] The Notice of Inquiry issued with the First Report stated the FCC had the authority to regulate all cable operations and would consider the possibility of doing so. Extensive comments were filed and arguments made, all along predictable industry lines, through the Summer and Fall of 1965. By February 1966, the Commission was ready to bring all of the cable industry under its wing, doing so in its Second Report and Order.[130]

The pretense and limitations posed by the microwave link were abandoned. Reaffirming the language of the 1965 Notice of Inquiry, the Commission simply asserted that it had authority under the Communications Act of 1934 to regulate cable as a form of interstate communication by wire, whether or not a given system was fed by microwave, and now had reason to apply that authority.[131]

The political maneuvering inside the Commission had not abated. The vote again was close, and it was only through the efforts of Henry, who modified his own position, that an odd majority could be forged in passage of the new rules. As *Broadcasting* explained it at the time:

> Chairman Henry is credited by his colleagues for the degree of unanimity that was achieved. "It was very close," said one Commissioner in commenting on the Commission decision. "It could have failed by an eyelash." ... The Chairman moderated his own previously hard line and abandoned the even harder line advocated by the staff. This costs him the support of Commissioner Cox, who favored stricter regulation. But it won the support of Commissioner Loevinger and held the vote of the other Commissioners.[132]

The result was a decision in which economic harm from CATV was assumed. The FCC turned its gaze away from Seiden's analysis, which suggested little threat to large-market UHF, and away from NCTA data that showed no actual evidence of harm to small-market broadcasters. Some parties pointed out that, given the widespread lack of UHF tuners, cable actually promoted UHF viewing by converting the signal to the usable VHF band for many subscribers. The FCC dismissed this material with an explanation that such studies were no longer necessary. In an amazing display of chutzpah and circular logic, it declared that it would not await new studies to demonstrate economic harm to broadcasting because, insofar as the Commission had already assumed that conclusion, it would be irresponsible to delay action against CATV. Broadcasting, in short, was to be protected against even phantom danger. Local carriage and nonduplication restrictions were, therefore, extended to every CATV system, although the window of protection was reduced from fifteen days to one (in part because it would be easier to administer).

In its carriage and nonduplication regulations, the FCC was simply extending regulations already attached to microwave-affiliated cable operators. In a significant additional restriction, however, the Commission formally froze the development of cable television in the top 100 U.S. markets. The agency determined that it would not permit the importation of distant signals into any of the top 100 markets unless the cable operator could demonstrate that such importation would not threaten existing UHF broadcast outlets. The importation ban was justified, in part, because the Commission felt the nonduplication rules alone were insufficient to protect independent stations, both UHF and VHF, and the non-network syndicated programming it hoped would flourish in the larger cities. The FCC felt that UHF especially had its best chance in the country's major markets, but that the service would not prosper, or would be retarded in its efforts to do so, if it had to face the additional burden of competing with CATV imported signals. Cable operators might, after notifying the FCC, be allowed to import signals into markets under the top 100, unless the local TV station objected.

In one respect, the Report was an astonishing reversal of philosophy. For years, the chief concern and object of policy debate was the health and well-being of the small-town broadcaster. The FCC had now effectively locked CATV out of the largest markets and went so far as to suggest that cable's most profitable social role would be in the smaller communities. The Commission, on the one hand, explained that cable was a poor alternative to UHF service because it could not serve rural homes, given the high costs of wiring low-density areas, and that CATV was a form of pay-TV that put local TV service out of the financial reach of some people in the community. Furthermore, and in language that would soon come back to haunt the Commission, the FCC concluded that, "most important, CATV does not serve as an outlet for local self-expression."[133] Despite all this, the Commission went on to conclude that, "it is in the markets below 100 that there may be under-served areas where CATV can make its most valuable and traditional contributions."[134] The legacy of the years of bitter battle in the West had evaporated. The small-town broadcaster was no longer the issue—the political pressures exerted by big-city broadcasters and the longing for a UHF renaissance had joined to now keep CATV from moving into the country's major markets.

The Commission off-handedly dismissed criticisms of regulatory inconsistency. With respect to the FCC's position in the 1950s, the Commission now declared it was free to correct previous rulings that now appeared "clearly erroneous."[135] Bartley dissented, as he had in the First Report and Order, and Loevinger wrote a separate opinion in which he signed on to the substance of the rules, but questioned the FCC's general authority over CATV and recommended congressional action instead. Interestingly, Commissioner Cox wrote a partial dissent complaining that the rules did not go far enough to control CATV, and bemoaning what he saw as backsliding on the part of some of his fellow commissioners.[136] He said he thought he detected a small but growing sentiment that cable offered some diversity in TV fare and such diversity might be worthy of a measure of protection.[137] It was, in fact, a mildly prophetic observation.

The Second Report and Order did not land quietly on the steps of Capitol Hill. "The community antenna television issue exploded in Washington last week and there was fallout all over town," said *Broadcasting* magazine, under the headline "Wild Escalation in CATV Fight."[138] Congressman Harley O. Staggers (D-West Virginia), Chairman of the House Commerce Committee, and Pastore, in the Senate, while generally supportive of the FCC, promised hearings into the issue. Congressman Walter Rogers (D-Texas), Chairman of the House Communications Subcommittee, accused the Commission of bad faith for not consulting with Congress before releasing the report.

NCTA Chairman Ford, along with some larger cable operators, was diplomatic in his reaction, welcoming the relaxation of the nonduplication rule and offering support of a congressionally crafted program of telecommunications legislation implicitly more friendly to CATV. Other community antenna owners were less hesitant to excoriate the Commission. Jerrold President Robert Beisswenger called it "an unprecedented attempt to restrict the freedom of TV reception."[139]

FCC Chairman Henry saw the Second Report and Order as a good interim measure, but, along with others, longed for a more solid legislative rock on which to stand, and went back to Congress. In March 1966, the House Commerce Committee, under Staggers, held hearings on three separate cable bills.[140] One was introduced by Staggers, who generally supported the Second Report and Order, but want to provide a congressional seal of approval. His bill represented the FCC's request for legislative authority over cable.[141] A second bill, introduced by Rogers, would have prohibited FCC regulation of cable altogether. Rogers was highly critical of the Commission for usurping the authority of Congress in this area. The third bill, introduced at the request of the NCTA, was more conciliatory than the Rogers' bill, granting FCC general authority over CATV, but drastically circumscribing its control.[142]

The political rhetoric was heavy and heated from all sides during the hearings, but the process was expedited by the steady hand of Staggers, seen as a moderate on the issues. The hearings eventually led, in June of 1966, to Committee approval of a bill which largely mirrored the FCC's proposals. It confirmed FCC authority and banned most CATV origination programming, save for automated time and weather displays or public affairs programs approved by the Commission. Cable operators called Committee approval of the bill their "darkest hour" and predicted the end of industry development if it became law.[143] The NCTA orchestrated strong opposition to further action on the bill and, fortunately for the industry, no counter move was organized in support. The bill died in the House Rules Committee, and there was insufficient agreement among the contending parties to resurrect it.

The Second Report and Order still tightly bound CATV operations, however. National CATV penetration in 1965 was about 2.4 percent. The top 100 markets accounted for about 90 percent of all TV homes in the country. Under the regime of the Report and Order, CATV would have little chance of wiring those homes and growing beyond its small-town universe of service.

Legitimizing Control: Southwestern and Buckeye (1967)

Broadcasters were not happy, either. Like Cox, they felt the Second Report and Order did not go far enough to control CATV. Vincent Wasilewski, President of the NAB lashed out at alleged industry plans to wire the entire nation and then charge heavily for access to television. CATV operators, he declared:

> Hope to fatten and feed on present free television programming with the ultimate aim of displacing it with programs for which they can make a charge. They will bring free programming to the American people as long as it suits their purpose ... until they are able to tell their audiences—now captive by wire—that they must pay or have no television at all.[144]

Midwest Television, Inc was a particularly distraught broadcast company. Midwest was based in Champaign, Illinois and owned stations around the country, including KFMB-TV in San Diego, California, and had been active in arguing for strict federal control of CATV. The company felt the new rules were too anemic to protect broadcasters, so less than two weeks after the release of the Second Report and Order, Midwest filed a petition complaining to the FCC. The petition eventually was expanded to encompass five CATV companies in the San Diego area, only three of which were actually in operation at the time of the filing (two were in the planning stages). Each carried or intended to carry the signals of the Los Angeles TV stations. Midwest, arguing market fragmentation and economic damage from such importation, asked the FCC to block the cable operators from extending their importation service to any customers beyond those already served on February 15, 1966, the effective date of the Second Report and Order. In essence, it was seeking a freeze on the expansion of the three existing systems and a halt to the construction of the two under development.[145] Southwestern Cable Co., one of the targets of the action, noted in response, that the request, if granted, would lead their bank to pull its financing and mean bankruptcy for the company.

The FCC nonetheless imposed a temporary ban on importation, pending a full hearing,[146] and the case moved into the courts. Southwestern appealed the FCC's action and the Ninth Circuit Court in 1967 found for Southwestern, declaring the Commission lacked authority to promulgate such rules.[147] Recalling the Commission's own findings from the 1950s, the Court concluded that the FCC's authority was limited to broadcast licensees and did not extend to CATV systems. The Commission, of course, appealed to the Supreme Court.

Meanwhile, a similar case was also working its way through the judicial system. Buckeye Cablevision in Toledo, Ohio, was importing, along with other stations, WJIM-TV from Lansing, Michigan. The FCC issued a cease and desist order against Buckeye and the company filed suit, this time in the D.C. Circuit Court, the same jurisdiction that had supported the Commission in *Carter Mountain*. It did so again in *Buckeye*, upholding the FCC action with a logic completely in opposition to the Ninth Circuit Court ruling in *Southwestern*.[148] The arguments were standard issue: The Commission had no authority over CATV insofar as cable did not constitute broadcasting under Title III of the Communications Act. Buckeye asserted a variety of other claims as well, including its status as intra- rather than interstate commerce. But the Court would hear none of it. Cable was an "aligned activity" that could affect the regulatory structure of television, said the D.C. Circuit Court. CATV, moreover, did constitute interstate communications wire service, subject to FCC oversight, by virtue of its importation of signals from an adjoining state.[149] The Court, in footnote, also rejected the industry's passive antenna self-characterization: "It is clear that a CATV system is more than a passive recipient of television signals, indistinguishable from a rooftop antenna; CATV systems engage in commercial retransmission of the signals they receive."[150]

As in *Southwestern*, the losing party appealed. With *Buckeye* finding in support of FCC regulation and *Southwestern* against, the Supreme Court agreed to hear arguments in *Southwestern*. It would become one of two landmark decisions the high court would issue in 1968. The second decision involved another problem that had hounded operators from the start, copyright.

Property Rights in the Courts

The D.C. Circuit Court's rejection, in *Buckeye*, of the industry's passive antenna defense was indicative of a growing trend in CATV jurisprudence. Discussion and debate about what CATV was and, more importantly, what it could become was already incubating by 1966 and will be treated in detail in the next chapter. In the courts, the industry's passive antenna posture had, by 1966, begun to fall on hard times, especially in the area of copyright, and after a series of wins in the late 1950s, the property rights battle was beginning to turn against the CATV.

Opening Salvos

The legacy of case law began on familiar ground in the Northwest. Three cases, all originating in that CATV-troubled part of the country and featuring long-time rivals, were involved. The first spotlighted, once again, Ed Craney. Failing to make headway with Congress or the FCC in the late 1950s, Craney filed suit in U.S. District Court in Montana in September 1958 in an effort to assert and defend property rights in his broadcast signal. It was the first suit of its kind, but it would not be the last. It involved Craney's stations in Helena, KXLJ, and Butte, WXLF. In his suit, *Z-Bar Net v. Helena Television*, Craney charged the Helena CATV operator with pirating the Butte signals and redistributing them in Helena without his permission.[151]

The other two property rights cases were closely related and somewhat complicated in their genesis. Both rose from a local business spat between W. L. Reiher's CATV system in Twin Falls, Idaho, and the Glasmann-Hatch broadcast business. Reiher and Glasmann had clashed before, butting heads in the 1958 Senate hearings. Now they met in court. Reiher's system, Twin Falls Cable Vision, Inc. carried several channels, including the local station, KLIX-TV, owned by Glasmann. In late 1960, Reiher announced he intended to add three new channels to his system, KSL-TV, KUTV, and KTVT, all from Salt Lake City. They were the regional network affiliates representing, respectively, CBS, ABC, and NBC. Reiher planned to import them via his microwave company, Idaho Microwave, Inc., and filed the appropriate applications with the FCC. But KLIX, in an arrangement common for the period, was already rebroadcasting, on a selective basis, network programming it picked up from the three Salt Lake City affiliates, and substituting its own local commercials. KLIX clearly did not want the local CATV operator importing the same programs, especially with Salt Lake City commercials, and protested the grant of the microwave application. The application fell into the FCC's 1960 microwave freeze.

In addition, Glasmann and Hatch, who also owned KUTV, joined the other two Salt Lake City broadcasters in a suit in federal court against Reiher.[152] Meanwhile, in a separate case, Reiher brought an antitrust action against KLIX, and Glasmann-Hatch countersued. In this second suit, Reiher charged the defendants with conspiring to destroy CATV operations in that region by encouraging the construction of illegal TV boosters and fostering local ill-will against CATV operators. The broadcasters responded, in part, by charging Reiher with unfair competition.[153]

As is often the case, the immediate, albeit tangled, claims of the contending parties were pointedly local and personal. But they also harbored several important and reoccurring legal issues important to CATV and broadcasters, including questions of common law copyright protection, the doctrine of unfair use, and the problems of exclusive and nonexclusive contracts. As noted, the FCC had expressed sympathy, but nonetheless ruled against broadcasters in their contention that CATV interception and carriage of signals violated section 325(a) of the Communications Act. Broadcasters like Glasmann and Hatch, therefore, sought redress through other relevant laws.

Unfortunately for the broadcasters, the relevant laws were not naturally aligned in their favor. Local broadcasters, in the first instance, do not typically hold the copyright on network or syndicated programming, which belong to the original author or copyright owner, often the production studio that created the work. In addition, a whole host of small but tenacious problems associated with the application of copyright protection hampered broadcaster claims.

The early suits, for example, sought to invoke common law copyright protection, which adheres to unpublished works upon their creation, but is extinguished once they are published. In other words, TV station owners had to show that the broadcasting of programming did not constitute "publication"; for once published under common law, the material was considered fair game and lost its protection. In the Craney case, the Court held that by broadcasting the programs to the community, Z-Bar intentionally made them public within the meaning of state law. Craney, therefore, held no actionable property interests in the signals.

Z-Bar Net v. Helena was a victory for CATV, but it was limited to Montana state law. A larger win for CATV came out of the dispute in Twin Falls. Broadcasters had long complained of CATV's unjust enrichment through the sale of programming that was "pirated" from them and "resold" to subscribers. Broadcasters argued that the practice constituted illegal "unfair competition." In the courts, the doctrine of unfair competition had a substantial history that included a key and potentially applicable case, *International News Service v. Associated Press*,[154] in which International News Service was held to have engaged in unfair competition by taking news dispatches from Associated Press postings and distributing them as their own to their wire service clients. While it is accepted in copyright law that news itself cannot be copyrighted, Associated Press was found to have earned a "quasiproperty right" in its material by virtue of the time and money invested in its gathering and dissemination. The Salt Lake City broadcasters, in *Intermountain Broadcasting v. Idaho Microwave, Inc.*, argued that Reiher was engaged in a parallel and unfairly competitive practice.

They lost on a technicality, however. A detail in the relevant law required that any claimed quasiproperty right in the programming be gained by virtue of an exclusive contract. In *Intermountain*, the relevant contracts held by the broadcasters and the Helena station were nonexclusive. Judge William Sweigert, therefore, found that the Salt Lake City broadcasters had no actionable property rights. The judge additionally held that the community antenna operator, in distinction to *International News Service*, was not presenting the programming as his own, and that the two companies in litigation were not direct competitors, as in the wire service case—Salt Lake City broadcasters sold advertising time and CATV operators sold television reception. In this, Judge Sweigert subscribed fully to the view of a CATV system as a passive antenna service. He did leave the door open to subsequent litigation, however, by suggesting the plaintiffs might prevail if they were able to show an exclusive license arrangement for their programming locally.

Where the Salt Lake City station did not have a case, the local broadcaster, KLIX, therefore, might. Taking that advice, KLIX, in the second suit, brought before the same judge evidence of just such an exclusive local license.[155] A year after his first decision (almost to the day), Judge Sweigert, in July 1962, found that Glasmann did possess an exclusive contract for the distribution of the relevant programming in Twin Falls, and that exclusivity would have been violated by Reiher upon importation of the Salt Lake City stations. Because the local TV owner had paid for the programming and the CATV operator had not, the CATV operator was guilty of unfair competition. The Court also found that the CATV and the local TV station were, in this case, in direct competition with one another. Reiher appealed, however, and in July 1964 the Ninth Circuit Court reversed the decision of the Idaho Court. Falling back on common law copyright, the appeals court held that one could not bootstrap property interests into existence by declaring an exclusive contract. There were no actionable property rights in the first instance.[156]

The Supreme Court declined to review the case, and with the Ninth Circuit Court's ruling, broadcaster hopes of protecting claimed property rights via the unfair competition doctrine ended. They would need to seek relief through other means, such as a change of heart on the part of the FCC, action by Congress, or reliance on litigation by the copyright holders themselves.

Happily for the broadcasters, action in at least two of these venues was well underway long before the Ninth Circuit Court's 1964 opinion. As noted, the FCC had already begun to clamp down on microwave importation of signals by 1962. And, while the CATV industry was keeping a close eye on

Intermountain, in the summer of 1960, it was also keenly aware that a very sizable Hollywood corporation and the owner of extensive films holdings, United Artists, was preparing a major copyright infringement case against CATV and was looking around for an appealing target.

Fortnightly

In the Northwest property rights cases, the broadcasters had hit a legal wall. Their efforts to assert property claims through the doctrine of unfair competition and common law copyright had failed. They turned therefore to the copyright protection provided under constitutional and statutory law. The "Founding Fathers" held the protection of creative work to be sufficiently important that they included it in section one (article 8) of the Constitution. It was not intended to simply shield the interests of individual writers, inventors, and creators, but more broadly to protect the social interests in maintaining a free and continuing flow of technical and creative work. The collective public interest in advancement in the arts and sciences was the primary goal.

The statutory provisions that detailed the nature of those important rights were, for most of the first half of the twentieth century, codified in the Copyright Act of 1909. Writers and producers of creative material, including such things as television programs, found their copyright protection under this law. That protection adhered only to the creators of the material, however, not to the broadcast networks or local TV stations that simply redistributed the programming (unless, of course, they were also the program creators). Broadcasters, in short, could not, in most cases, invoke copyright law to prevent CATV operators from using their signal, although some had considered it. As early as 1954, a TV film production company had threatened to sue a Reno, Nevada, CATV operator for importing its programs from a San Francisco station, and the New York Yankees and Brooklyn Dodgers had made similar threats against Pennsylvania CATV systems retransmitting their games.[157] But if the broadcast industry hoped to establish any copyright claims against cable, it was necessary for a clearly unambiguous copyright holder to take the lead.

In 1960, United Artists Television, Inc., a major motion picture and television production company, did just that, filing suit against two West Virginia CATV systems. The industry had known for some time that the suit was in preparation, and all parties in cable and broadcasting had been preparing for the challenge. United Artists had paid more than $20 million to ZIV Television productions for a library of films it planned to syndicate in the UHF market, including stations in Pittsburgh, Pennsylvania, Wheeling, West Virginia, and Steubenville, Ohio. Each station, in turn, was being carried by CATV systems in Clarksburg and Fairmont, West Virginia, both owned at the time by Fortnightly Corp. The United Artists contract specifically limited exhibition of the films to the assigned broadcasters and made it clear that those rights could not be sublicensed to any CATV systems.

It would take many years before the case was resolved. It would become the second landmark decision issued by the Supreme Court in 1968, alongside *Southwestern*. In the interim, CBS would file its own copyright suit in December 1964, this one against TelePrompTer. CBS sought to have its action combined with the United Artists case, but the court declined to consolidate the two actions. CBS then suspended active pursuit of its claim, awaiting the outcome of the United Artists complaint. The actual trial in *United Artists v. Fortnightly* did not begin until 1966.

The *Fortnightly* case was one of cable's most critical. To lose the copyright battle would mean that operators would have to begin paying for all the programming they took freely from the airwaves, a multibillion dollar prospect. Copyright liability could spell the end of business for many and severely retard the growth of the industry as a whole. A strong defense was essential and it arose logically from past legal successes. The idea of community antenna television, its social construction, was at the heart of the industry's copyright defense. Under the law, copyright infringement required unauthorized reperformance of a licensed work. Using the passive antenna façade that it had built to ward off tax liability, the industry argued that it could not possibly infringe on any copyrights because it did not, in any sense, reperform the protected work. Reperformance was a communicative activity and the community antenna, again, was simply a passive extension of the viewer's antenna.

The FCC looked to the United Artists copyright case with hopeful anticipation. The Commission, while willing for a variety of reasons to take control of CATV, would nonetheless have preferred that Congress or the courts settle as many of the problems as possible, copyright being central among them. With the United Artists case, the Commission, like many others, had reason to believe there would be a favorable resolution of the problem in the courts. The General Counsel to the FCC during this period, Henry Geller, went so far as to suggest, many years later, that the Second Report and Order itself could best be seen as a holding action until the judicial system dealt with copyright. Geller recalled that the Commission believed the courts would ultimately find in favor of the broadcasters, and the FCC was prepared to relax or rescind the importation rules once CATV became subject to the copyright laws.[158]

The Commission additionally hoped that Congress would act to bring copyright law to bear on cable, a view that sprang from a study that had been grinding away for years and appeared to be finally coming to a close in 1966. It was widely felt that the Copyright Act of 1909 was hopelessly out of date, and the Registrar of Copyrights had been laboring over a rewrite proposal since the 1950s. Cable was only a small part of that effort, which embraced creative work in all forms. Nonetheless, the Registrar noted that CATV, "which was not even referred to in the 1961 (copyright) *Report* now promises to be one of the most hotly debated issues in the entire revision program."[159]

While arguments proceeded in a Manhattan federal district courtroom in *United Artists v. Fortnightly*, Congress also worked on legislation that would require CATV operators to pay for their content and, in May 1966, a subcommittee of the House Judiciary Committee recommended a copyright regime for cable (HR 4347) that would have insulated the industry from liability for the carriage of local signals, but would have required copyright payments for carriage of distant signals. It was a "half loaf" for the industry, but was generally greeted favorably by operators who feared any even worse proposal. In fact, the worst case scenario of full copyright liability was on its way. Within weeks of the judiciary subcommittee recommendation, the industry received word on the *Fortnightly* case. As the trade press put it, their "free ride was over."[160] The historic ruling went against CATV and came like a hammer blow to the business interests of the industry.

The key to the decision, as everyone knew, would be the courts' view of the legal, and by implication social, definition of CATV. Was it an active processor of information and entertainment, or simply an expensive set of rabbit ears? Did community antennas "reperform" protected work? Precedent in common law copyright was set in the 1931 Kansas City hotel case noted in Chapter 1. There, the Supreme Court ruled against a hotel owner who was operating a radio master antenna for his guests.[161] In retransmitting the radio signals, the Court ruled that the hotel was violating the copyright interests held in the music. Also as noted in Chapter 1, a New York state court upheld a similar principle in 1941 when it enjoined Muzak Corp. from retransmitting a radio broadcast of the 1941 World Series. In a phrase that came to haunt the cable industry, the plaintiff in that case, Mutual Broadcasting, accused Muzak of being "an interloper or parasite" by making a profit off the Mutual broadcast.[162] CATV operators, clinging fast to their passive antenna defense, attempted to distinguish the 1931 hotel case by arguing the hotel owner actively controlled the content in a way that 1960s CATV operators did not.

The federal district court's decision in *United Artists* was unambiguous, however. In a lengthy and detailed treatise, the court critically examined the legal and rhetorical claims of CATV.[163] It methodically dissected and rejected the industry's long-standing argument that community antenna systems were only, in concept, passive devices. The court found CATV to be an active form of wired communication that performed and reperformed programming within the meaning of the law. The District Court stated:

> Defendant's systems are not passive antennas. They are sophisticated, complex, extremely sensitive, highly expensive equipment, especially constructed and designed to reproduce the electronic waves received from the originating station and to propagate and

transmit the new electromagnetic waves through an elaborate network of coaxial cables.[164]

For the first time in its history, CATV was faced with the prospect of having to pay for its programming. But the contest was far from over. The industry could, and did, appeal.

The Telephone Companies

Despite the copyright setback, the legal system did advance the CATV cause in one area—its relationship with the telephone industry. CATV operators sometimes used power company poles to string their lines, but the vast majority of coaxial cable was hung from telephone poles. In 1971, Bell affiliates alone accounted for more than half the country's CATV pole access.[165] While the 1956 Consent Decree had prevented AT&T from engaging directly in the provision of CATV service, it left several other market openings for the Bell system. As noted in Chapter 3, Bell subsidiaries were still permitted to build cable TV systems, then lease them back to cable operators. The New York legal proceedings discussed earlier briefly retarded AT&T lease-back efforts, but by no means ended them because courts across the country, including those in New York, were beginning to find that pole attachment issues did not fall within state public service jurisdiction.[166] State authorities, in other words, were declining to intercede in CATV-telephone company business relations.

Moreover, while AT&T was the dominant local and long distance provider, it was not the only telephone company in the country. Numerous other small, local and regional firms offered phone service and these companies were not bound by the 1956 Consent Decree. They were as free as anyone else to construct a CATV system in their town, and as community antenna service grew in financial appeal, they began to do just that.

AT&T, however, had a special incentive to explore the CATV business. It was certainly drawn by the economic potential of what looked like a related business, but it was also looking to the future and the possibility that the high-capacity wires of CATV could be harnessed to provide advanced information services, even telephone service itself. Loose talk about CATV being able to provide sophisticated services such as electronic shopping and banking was already being heard (see Chapter 6), and Bell clearly saw this as its business territory. In this regard, CATV was seen as a potential competitor with superior technology. Bell did not want to be left in the technological dust, relegated, as AT&T management put it, to "an outmoded, voice-only equivalent to Western Union."[167]

In 1964, AT&T therefore launched a campaign to enter and to the extent possible control the industry. It did not go unnoticed. "After 15 years," observed NCTA legal counsel Strat Smith, "the telephone companies want in."[168] AT&T began by opening a special planning department to manage its CATV initiatives and, by May 1965, had filed plans to develop lease-back service in twenty-six states.[169] It also marshaled its local Bell Operating Companies (BOCs). The BOCs were ordered to increase pole rates and institute various operational restrictions on local CATV operators.[170] Rates, which averaged around $1.00 to $1.50 a pole (per year) in the late 1950s and early 1960s suddenly doubled to $2, $4, or more.[171] Some BOCs proposed fees as high as $10 a pole. Additional charges were levied for any FM radio service or local origination channels offered by the CATV operator. The telephone company restrictions frequently forbade CATV companies from offering such services as pay-TV, educational TV, closed circuit TV, and any future interactive services,[172] and the AT&T tariffs restricted CATV access to underground Bell ductwork. In one of the most extreme cases, Southern Bell, in late 1964, notified CATV operators that it would no longer rent pole space for cable lines at all. Instead, the company would only provide lease-back contracts.[173] Southern Bell eventually backed away from its draconian policy, but kept poles rates high.

Lease-backs, formally labeled "channel service," were an especially bad deal for most CATV operators because they increased the costs of installation and operation by as much as 30 percent over traditional ownership. Some operators responded by setting their own poles, but this incurred its own additional cost, anywhere from $25 to $35 per pole, and failed to become widespread practice. Despite complaints from community antenna business, AT&T's program worked. By 1967, companies in the Bell system were leasing channel service facilities in 178 communities and had increased their share of new cable construction from under 1 percent in 1964 to more than 28 percent.[174]

Meanwhile, independent telephone companies not bound by the Consent Decree set up cable subsidiaries to offer direct service in their communities. The United States Independent Telephone Association (USITA) issued a notice in February 1965 recommending its members enter the CATV business. The president of the association, Paul Henson, also suggested that poles rates for existing or would-be operators be set at a prohibitive $11 to $16 a year. General Telephone and Electronics, the largest independent, created GTE Communications to build and operate CATV systems and, by 1969, was running some twenty-five of them around the country.[175] United Utilities, the third-largest independent, created United Transmission, Inc. to develop CATV service in all of its 500 communities. For several years, both companies completely foreclosed local competition by refusing all pole attachment requests from CATV operators. By early 1969, telephone companies in total owned 6.5 percent of all the CATV systems in the country.[176]

The telephone companies' incursion into the CATV business also attracted the attention of regulators, politicians, and watchdog groups.

AT&T's near monopoly of national telephone service had been a long-standing public policy concern, and any interest shown by the company in additional lines of business raised regulatory eyebrows. One fear was that the telephone companies, especially AT&T, would use the river of revenues from their protected monopoly telephone business to deliberately under-price new competitive ventures, like the community antenna service, with an eye toward driving out the competition. There was even concern that the telephone companies would cross-subsidize their business lines, raising residential telephone rates above normal levels in order to fund construction of competitive CATV systems. Such worries were ammunition for CATV's counter-attack.

In 1965, the Merrill's Ameco filed a complaint with the FCC charging AT&T with violating terms of the 1956 Consent Decree. Commissioner Loevinger (formerly with the Justice Department's antitrust office) suggested the complaint be forwarded to the antitrust department, which then began an investigation of possible anticompetitive practices by AT&T.[177]

The FCC itself became involved, as well. The Commission by that time had determined that it could regulate CATV service, and that meant service, in any form, including any offered by the telephone companies. In August 1965, FCC staff sent letters to the state public service boards and commissions expressing concern about telephone company activity in the CATV business and reminding them of the Commission's intent to act on CATV regulation generally. On April 6, 1966, two months after releasing its Second Report and Order, and having generated little positive response from the states following its previous letter, the Commission ordered AT&T and General Telephone to file all proposed tariffs for CATV service with FCC rather than with state utilities commissions.[178]

By October 1966, the FCC initiated an investigation into its authority over telephone company-supplied CATV facilities, indicating that such service would probably be brought under the purview of section 214 of the Communications Act. That section required telephone companies to demonstrate that their lease-back agreements were in keeping with the public interest, convenience, and necessity. Within two years, the Commission would, in fact, apply the section 214 requirements, taking even more severe steps after that to push the telephone companies out of the cable business. Community antenna television, fighting economic and political battles on many fronts, appeared in the mid-1960s to be gaining ground on this one.

Taming Cable

While the cable industry sparred with federal regulators, giant broadcast companies, and AT&T through the 1960s, none of these foes was more vexing, nor more continually irritating, than an organization much smaller in scope and now nearly lost as a historical footnote. That organization was TAME, an ad hoc lobbying group of the television antenna manufacturing industry. The formal name of the group was the "Television Accessory Manufacturers Institute." The slightly inaccurate TAME acronym was created to denote the main purpose of the group, which was, as it declared on its bright orange promotional brochures, to "TAME CATV." While it was tiny in comparison to the broadcast and telephone industries, it was a public and political fireball in its assault on CATV, and for good reason. No industry was more completely threatened by CATV than was the TV antenna manufacturing industry. Every time a home hooked up to the cable, the TV antenna came down, and those with CATV would never need another aerial. Early cable operators even used TV antennas as marketing tools. According to pioneer Ed Allen,

> We would perform as a service taking down these antennas on the roofs, and keeping them, so we could destroy them in exchange for the normal installation charge. It was a free installation if you would turn in your antenna. A lot of the people wanted to get rid of these antennas because they were ugly. We ran a campaign in which we vilified the antennas. We called them "bird perches."[179]

Cable, clearly, had no friends in the TV antenna business. TAME widely distributed flyers and press releases in communities considering CATV systems. TAME newspaper advertisements warned, "Here's Why Cable TV is Bad for Our Community," and "How to Fight Community Antenna Systems."[180]

TAME worked at the federal level too, filing petitions with the FCC to control and even suspend the national development of cable television. The organization was known to file amicus curia court briefs in litigation involving CATV and, for years, the organization was a thorn in cable's side.

In retrospect, of course, there was a measure of futility in TAME's efforts. The antenna manufacturers' situation was a classic example of technological displacement. As new technologies evolve, existing devices that served similar purposes but with less power, flexibility or perhaps higher costs, are shunted aside. In this manner, entire industries are transformed or eliminated. While television did not put an end to radio—in fact radio flourished in the 1950s and 1960s—it did radically change the nature of the business. Buggy whips went out as the automobile came in, and, arguably, television antennas were the buggy whips of the cable television age. They can still be found in local hardware stores, but are limited to a substantially diminished national market. Ultimately, TAME had only a negligible and transitory impact on the evolution of cable. The antenna manufacturers were no match for the larger tectonic forces driving adoption of cable television, and the slow but steady disappearance of rooftop antennas in small towns across the country through the 1960s was a telling marker of the progress of CATV.

Another interesting marker was the changing nomenclature of CATV in the mid- and late 1960s. From the start of the business and through the early 1960s, the industry and the technology were commonly and formally know as "community antenna television," or sometimes "community television" and abbreviated CATV. CATV was the title used by the popular and trade press as well as by the industry itself. Its national association was the National Community Television Association—at least that was its name until 1967. Beginning around 1965 another term began to seep, slowly, into use. In February of that year the *New York Times* headlined a story about the industry using the term "Cable-Television."[181] A few days earlier, *Newsweek* had run a story, relying primarily on the standard CATV terminology, but dropping in the descriptors "cable TV" and even "cable industry," perhaps only for variety.[182] Through 1965 and 1966, the technology remained, for most people and most purposes, "CATV." When the NCTA formally changed its name to the National Cable television Association in 1967, however, "cable" began to replace "community antenna TV" in common parlance. Both terms were used in the popular and trade press through 1968, but as the decade drew to a close, the community antenna label began to fade from common discourse. "CATV" held on in vestigial fashion through the early 1970s, but by then, the cable nomenclature was asserting its dominance.

The transition in language was much more than just an interesting artifact of the evolving nature of the medium. It was a manifestation of one of the most dramatic and far-reaching changes in the social perception and social treatment of cable TV in its history. FCC regulation in the first half of the 1960s had stalled the development of community antenna television, in part because of the ideological lens through which the Commission viewed and defined the medium. The change of language in the second half of the 1960s signaled a new vision and, eventually, a new ideology at the Commission and beyond. Cable, for many people, would soon take center stage in a radical new approach to thinking about telecommunications and telecommunications technology. The changing view would have advantages and disadvantages for people in the business, although initially most saw only the brighter side of the change. As a result, and despite the difficulties cable operators had in the early 1960s with the federal government and the broadcast industry, some in the business thought they could just begin to see in the overcast horizons of the period some glints of blue sky through the clouds.

6
The Wired Nation (1966–1972)

As cable systems are installed in major U.S. cities and metropolitan areas, the stage is being set for a communications revolution—a revolution that some experts call "The Wired Nation." In addition to the telephone and to the radio and television programs now available, there can come into homes and into business places audio, video and facsimile transmissions that will provide newspapers, mail service, banking and shopping facilities, data from libraries and other storage centers, school curricula and other forms of information too numerous to specify.

—RALPH LEE SMITH, "THE WIRED NATION," 1970[1]

It was a new vision: television programming on demand, teleconferencing, electronic banking, shopping, and health care assistance; an electronic town square dedicated to democratic discourse, the advancement of local and national governance; instant news, information, and educational opportunity. It was an electronic cornucopia of communications goods and services. It was the utopian ideal of communications technology. It was "the wired nation," and the wired nation was the new model for cable television.

It was a revolution not in the technology, but in the social meaning of the technology, and it arose with powerful and far-reaching effects in the mid- and late 1960s. It was, in the course of cable television history, one of the most dramatic social reinventions of the medium, the archetypical example of the social construction of technology. While the technology itself was evolving in a slow but steady manner, the social conception of what CATV was and could become was taking a radical turn, a turn that would have a profound impact on pubic perception, public policy, and, ultimately, the business and nature of cable. This was the period during which the industry was to shed its original label—*community antenna television*—and take on a new moniker, *cable television.*

It was also the period that gave rise to the term "Blue Skies." Blue Sky thinking, as noted in the first chapter, was the imagining of cloudless, unlimited horizons of technical possibility. The mid-1960s marked the beginning of cable's Blue Sky period, although, in one form or another, Blue Sky thinking would shape the business through its next thirty-plus years.

The industry would come to play an active role in advancing much of the Blue Sky rhetoric, but in its early stages, from the view of the operators, the transition to a new concept of cable was not entirely voluntary. Beginning as early as the mid-1950s, and increasingly through the early and mid-1960s, cable suffered from a self-inflicted conceptual schizophrenia. Depending on with whom you talked, CATV was either a passive extension of the viewer's home television antenna, or a potentially powerful new communications system capable of generating bountiful revenue streams through pay-TV, and perhaps even interconnecting into national cable programming networks. For small operators who could not afford to think in such grand terms and for CATV lawyers who had to navigate tricky legal and political waters, CATV was best seen as a cold, dead wire; in fact, its entire copyright defense was built on it. For growing MSOs such as TelePrompTer, CATV had to be something more, something larger, something that would entice institutional and private investment.

This conceptual schism widened in the early 1960s, and the industry's ambivalence was illustrated at a 1965 NCTA convention panel discussion on the topic of origination programming. Marcus Bartlett, the head of Cox Broadcasting's CATV operations, was encouraging local origination but only for public service events such as city council meetings and high school football games. NCTA legal counsel Strat Smith was more cautious, recommending that any origination be kept separate from the primary passive antenna function. The experienced lawyer suggested that the former could arguably be regulated, but the latter should remain immune from federal control.[2] NCTA lawyers realized that no matter what cable did physically with the signal, the *perception* of what it did, could and should be molded to the benefit of the industry; they were successful in this effort for many years. By the mid-1960s, they were losing ground at the FCC and in the courts. The Commission had declared that cable systems did not and could not provide outlets for community expression, it was clearly was worried about cable's potential to offer pay-TV, and the agency supported legislation that would have banned local cable origination programming. In the courts, industry attempts to fashion a favorable (that is, passive) view of the medium had fallen on hard times. Despite earlier successes, the Federal District Court in *United Artists* and the District of Columbia Circuit in *Buckeye* had rejected the position out of hand.

It is understandable, then, that 2,000 CATV operators, equipment manufacturers, and associated parties gathered in June 1966 for the fifteenth annual NCTA convention amid loud grumbling that is was time to change their image and their practices. In the previous six months, the FCC had issued its Second Report and Order asserting jurisdiction over all CATV systems, the Federal District Judge in *Fortnightly* had held them for the first time liable for copyright, and the Staggers House Commerce Committee had approved the FCC's draft legislation giving the agency full reign over cable. Veteran CATV operator and NCTA Chairman Ben Conroy surveyed the legal landscape and declared, "It's a brand new ballgame."[3]

Whatever the motive or source, if the industry was going to be perceived and treated as an active communications medium, then it might as well start taking advantage of the situation. As industry pioneer Bruce Merrill, put it, "we may as well be hung for a sheep as for a lamb."[4] In hallway and hotel room conversations a new consensus was brewing, the passive antenna scenario had served its purpose and run its course. It was time to discard the aging facade and move on. In this case, moving on meant exploiting the potential of CATV to develop its own programming and harness that programming to generate new revenue.

It was also a shrewd political tactic and one fashioned in part as an implicit threat to broadcasters. Broadcasters had sought to control CATV, especially in the larger cities, out of fear of the competitive impact, and CATV operators had been willing to compromise in ways designed to dampen any potential harm. With CATV operators losing their legal battles on all fronts, however, they had little to lose in moving from compromise to threats, in effect telling broadcasters they were willing to start wholesale programming in direct competition with local TV stations.[5]

The industry's legal advisors were still issuing strong cautionary notes. The copyright case was in progress, and public mumbling about cable creating and distributing its own product would be seriously counterproductive to arguments in court. Although the legal staff could not completely fend off the growing grass roots sentiment, they could at least delay some of its more overt manifestations. At the 1966 convention, a few operators went so far as to suggest that it was time to change the name of the organization. If the "community antenna" philosophy was being pushed out the back door, then perhaps the sign on the front door ought to reflect it. Proposals were floated to change the organization's title from the "National Community Television Association," a label reflecting passive community antenna service, to "The National Cable Television Association." On the surface, the name change seemed incidental to the substantive evolution taking place in the proposed business operations. It was a significant symbolic move, however, one that carried substantial rhetorical and therefore political weight. It was a public declaration of a new kind of business and a new kind of technology. The lawyers were able to prevent the name change until 1967 when, from the industry's official perspective, it made the transition from community antenna television to cable television.[6] In important ways, however, both inside and outside the business, that change had been in gestation for several years.

This chapter examines cable television from about 1966 to 1972, a period in which both regulation and business plans were heavily influenced by the

Blue Sky vision. It considers first important changes in the technology of cable that, in part, were responsible for the rise of Blue Sky rhetoric. It also looks at the gestation of the Blue Sky concept and its early appearance in the policy and in the popular literature, followed by details regarding the regulatory and judicial developments of this period, subsequently, a review of the growth of the business generally and industrial structure in particular. Relations with the telephone industry are described toward the end of the chapter, which closes with a look at local franchising and a more skeptical note on the growing utopian view of cable TV.

Evolving Technology

An important part of the evolution was technical, of course. Incremental but key advances improved cable's reach and reliability. In 1970, a pair of cable engineers, Arie Zimmerman and Israel "Sruki" Switzer, introduced a method for minimizing distortion and extending the range of a cable system. The design, labeled the "harmonically related concept" or "harmonically related carriers" (HRC), locked the frequency of each cable channel to a multiple of the standard 6-MHz TV carrier, increasing signal stability. Other, more dramatic, improvements in carrying capacity helped power the technology's emerging social standing, however. In the early and mid-1960s, the industry moved from a tube-based technology to one driven by transistors, and cable engineers created something called the "set-top box." Both worked to multiply cable's channel capacity and that, in turn, help feed the new social construction of cable television.

Solid State

Tube technology had serious drawbacks. It was bulky, fragile, and burned a lot of electricity, which meant dollars. Tubes gave off a great deal of heat and amplifier boxes had to be constructed with adequate ventilation. The scorched fingers of linemen constantly having to replace spent tubes was only a small reason to look for a better technology.

Solid-state electronics arose, again in evolutionary fashion, from the earliest crystal radio sets. The ability of certain materials, such as silicon or germanium, to act on electrical currents was well known. A host of scientists and engineers, including those at Bell Laboratories, began harnessing those properties with the aim of replacing the large and inefficient electronic tube before World War II. A team of Bell researchers, including William Shockley, William Brattain, and John Bardeen, conducted much of the ground-breaking research that led to the transistor. A modest press release announcing the development of the device was released in 1948, but it would be several years before a practical version could be reliably manufactured at a cost-efficient price.[7] Even then, as transistor manufacturing got underway in the mid- and later 1950s, the new technology remained more expensive and less reliable than tubes. Despite the initial problems, many in the cable industry, as elsewhere, saw that solid-state devices would eventually displace tubes, and experimentation began.

The first use of transistors in cable television was reported around 1958 (the date is uncertain) when Dr. Henry Abajian of Westbury Electronics used them to amplify signals in a small CATV system built by his brother in Vermont. This was a limited experiment, however. Widespread adoption of solid-state technology required the participation of an established CATV equipment manufacturer. AMECO was willing to give it a try. By the late 1950s, electronics firms licensed by Bell were beginning to mass produce transistors. One of them was Motorola, which had a production plant in Phoenix, not far from Bruce Merrill's AMECO headquarters. Around 1958, Merrill, looking to the future of the company, decided to investigate solid-state technology and had a box of transistors brought over from Motorola. The early reliability problems of transistors, coupled with the special demands of cable, meant that usable units were initially difficult to find. AMECO engineers would search through the 1,000 Motorola transistors—the company sent them to AMECO in large barrels—to find thirty or forty that met their specifications. Eventually, working with the manufacturer, AMECO developed a series of dependable transistors suitable for the demands of CATV. "So," explained Merrill, "we came out with a line of transistorized amplifiers which immediately got everybody's attention. That was really what put AMECO on the map."[8]

The AMECO device, hurriedly prepared for a 1960 debut at the industry's annual western show convention, was housed in a small metal box designed to be attached to the pole. Fearing linemen would use it as a toehold for climbing, however, AMECO placed a large "NO STEP" sticker on the box. The "NO STEP," as it became known, was the first transistorized amplifier marketed for CATV.[9]

Transistors offered a cornucopia of benefits to the industry. In comparison to tube technology, transistors were smaller, lighter, and, over the long term, would become significantly cheaper to make and buy. Amplifier boxes, therefore, could be made smaller, lighter, and cheaper. Transistors consumed substantially less power than tubes, which also reduced operational costs.

Along with AMECO, Jerrold became an early leader in the manufacture and sales of transistorized equipment, although not without some growing pains. According to Archer Taylor, Jerrold engineers, comfortable with tubes, were at first "bewildered" by the transistor. "They did not know how it worked or what to do with it."[10] Hiring knowledgeable transistor engineers, Jerrold soon worked through its problems, however. In the early 1960s, it developed, first, its Model TMl and, then, its Starline One transistorized amplifier. The Starline One had an additional advance in that the unit was encased in a sealable, die-cast housing that could be hung on strand. This kind of unit soon replaced the sheet metal, pole-mounted boxes typical up to that time. Unfortunately, the RCA transistors that powered the Starline One were not quite ready for the demands of cable television, and heat-related problems caused a high failure rate. Gardner was one of the early purchasers of the Jerrold amps. "The Star Line One came out in the ... '60s. They were as much a disaster as they were a success. I don't think anyone was very proud of the first generation of solid state equipment."[11] Nonetheless, by mid-1965, most major CATV equipment manufacturers, including Jerrold, Spencer-Kennedy, Entron, and Viking Cable Co., were offering solid-state equipment.

By 1967, Jerrold had taken another significant step. Overcoming its early reliability problems, Jerrold expanded the channel capacity of its Starline series and, in doing so, helped reshape social thinking about cable television. The new model was the Starline 20. The "20" stood for the number of channels it could transmit. In fact, the amplifier could carry twenty-one channels and, according to Taylor, the designator 20 was used instead of 21 probably only because it was a round number and sounded better.[12] The Starline 20 could amplify five channels in the low band, 54–88 MHz, and sixteen channels in the high and midband (120–216 MHz).

Jerrold introduced the Starline 20 at the 1967 NCTA convention. C-Cor Electronics and others soon followed. Within a few years solid state, broadband equipment was coming into common use. Cable could now deliver twenty-plus channels of video and more of audio. A dual cable system, such as one displayed in 1970 in San Jose, California, could provide forty-two channels of programming. Ironically, the channel capacity of a cable amplifier had now surpassed that of the standard VHF television tuner. Technically, cable could deliver more channels than any consumer set could use.

Set-Top Converters

The first place where this became a practical problem was New York City. In America's small towns, a five-channel CATV system was usually sufficient to meet all local TV needs. But New York had a full compliment of broadcast stations. Seven of the twelve VHF channels in the city were occupied by full-power stations, all transmitting from the top of the Empire State Building; and there were two UHF stations. City franchise agreements required some cable channels be used for local origination or access purposes. The operators themselves, TelePrompTer was one, had plans of their own for programming. Relying on just the traditional VHF tuner, cable in 1966 was running out of channel space in Manhattan.

The solution was the set-top converter, or set-top box, a now-ubiquitous fixture in American family rooms. According to Taylor, one of the first, if not *the first*, dual heterodyne cable converter was created by Ron Mandell and George Brownstein of International Telemeter. They demonstrated the device for TelePrompTer's Hub Schlafly in 1966, as one way of dealing with the New

York capacity problem. Importantly, the converter was also seen as a way of eliminating an annoying interference problem for Manhattan customers, which arose when a customer's TV set received the normal cable signal from, for example, Channel 7, but inadvertently picked up the over-the-air Channel 7 a millisecond earlier. The double signal caused unacceptable interference. The converter solved the problem by transferring all the incoming signals onto one output channel (unused Channel 12 in New York). Today, set-top boxes typically use Channel 3 or Channel 4 as their output.

At the same time, the converter opened up a substantial new range of available cable spectrum. As noted in earlier chapters, standard VHF TV tuners receive signals in the VHF low band, (54–88 MHz or Channels 2–6) and the VHF high band (174–216 MHz or Channels 7–13).[13] The home TV set tuner, however, could not read between about 88 MHz and 174 MHz, which was now used by the Starline 20 and similar amplifiers. The UHF tuner then coming into widespread use, operated at 470–806 MHz, beyond the range of cable amplifiers (hence the practice of converting UHF signals to usable VHF frequencies). The set-top box could use the full band, and convert the signals into something the consumer set could "see." Operators wedged about nine channels into the new midband range.

The expanded capacity gave rise to a small problem in technical nomenclature because the amplifier harnessed bandwidth not used by broadcasters. The midband channels, therefore, were labeled with letters, A–I. As bandwidth continued to expand over the years, the industry moved up the alphabet, with channels J–W assigned to the "superband," and double lettering adopted with development of 400-MHz amplifiers in the late 1970s. When capacity reached 500 MHz, in the early and mid-1980s, the alphabetical designations were dropped and numerical designations implemented.

Combined with the evolving, transistorized amplifiers, the set-top converter helped break the ceiling on channel capacity set by the VHF tuner. The box would also be harnessed for other uses, primarily security. It could and would be used to filter pay-TV signals, although that was still a few years away.

Blue Skies: The Social Construction of Cable TV

Milton Shapp never shared the passive antenna party line. In 1964, he said he saw a day "not too far distant" when virtually the entire country would be wired for television, and CATV, "along with low-priced home video recorders, will revolutionize concepts for television programming."[14] By the end of 1965, the idea of cable as a communications utility, something far beyond a simple broadcast retransmission service, was catching on in the trade and popular press. *Television Magazine* reported on cable's potential to create special "circuits," or channels, for stock market information, newspaper facsimile, and even home shopping.[15] In May 1966, the editor *Electronics World* echoed Shapp, proclaiming: "Someday every hamlet, town and city will be interconnected like a massive spider web. CATV will become as common as the telephone connection and each home will receive, on one or more sets, dozens of channels providing education, cultural material, and entertainment."[16]

U.S. News and World Report, in April of 1966, offered its own summary of the emerging vision:

It is suggested that homes will be linked by cables not only to the TV outlets, but to stores and banks. Merchants will use extra channels to display their wares more fully than they can on the usual spot commercial ... The housewife may be able to do much of her shopping without leaving the home—select a dress from the television screen, electronically place her order for the dress, and direct her bank to make the payment.[17]

Life magazine noted cable's potential to create its own local programming, develop nationally interlinked networks, and negotiate for network-quality shows. As did others, it saw cable supplanting the print media, again with home-delivered facsimile newspapers, magazines, and books.[18] It was an idea adopted by some in the newspaper industry itself. The president of the American Newspaper Publishers Association (ANPA) recommended its members buy into cable television; they wanted to participant in, rather than be a victim of, the changing technology.

Broadcasters who saw even the passive community antenna as a danger to their monopoly hold on television were doubly troubled by this kind of talk. Broadcaster and cable operator interests were primarily economic, however and the new social construction of cable had implications far beyond the limited sphere of business and commerce.

The mid- and late 1960s were time a great social and political ferment in the United States. The election of John F. Kennedy, as already noted, marked a cultural turning point in the nation's history, a self-aware transition between the values and authorities of World War II, the Eisenhower and McCarthy eras, and a fresh ideal of youth and energy that the young president and a blossoming economy promised. The old rarely falls lightly before the new, however, and the social world of the United States in the decade of the 1960s was rife with conflict. The clash of cultures and ideologies that epitomized the era would have profound consequences for the nation's idea of itself and the nature of its political and social role in the world. The assassinations of John F. Kennedy, Robert Kennedy, and Martin Luther King shocked the country. Clashes in the South over civil rights and the nationwide conflict over our involvement in Vietnam divided Americans, black and white, young and old.

Protestors took to the streets in the suburbs and the cities, and sometimes the peaceful demonstrations turned violent.

Through it all, television brought the spectacle, the tragedies, and the conflicts, for the first time, live and in living color, into nearly every living room, every night. Problems of race, international relations, poverty, and eventually sexism that had lived just beneath the surface in the 1950s were breaking into the national lexicon and the national debate. The forum for that debate, as often as not, was television. Despite the social turmoil, the 1960s also was a period of hope, of looking ahead, of harnessing technologies to solve social problems. Society generally, and a growing contingent of influential progressive activists in particular, were looking for solutions to the problems of the age and, in their search, they found, among other things, cable television. Pronouncements, proposals, and prophecies for the expansion of cable television into a twenty-first century communications utility circulated within in communities of policy makers and soon spread out into the public at large. In this manner, the cultural values of the period helped forge a new social definition of cable TV.

By the late 1960s, the idea of cable television as something more than just a system for retransmitting broadcast game shows and sitcoms was a dominant theme in the popular and scholarly press. Cable scholar Thomas Streeter analyzed the phenomenon in insightful detail, noting a variety of important and repeating ideas in the discourse of the times.[19] Cable, first of all, was coming to be viewed in near utopian terms, presenting the medium "as embodying the potential for solving numerous social problems and dilemmas."[20] By this perspective, cable could help ameliorate social conflict by bridging the gap between divergent and antagonistic social groups. It could serve as the local voice of the people, in a way that constricted commercial broadcasting could not and would not. It could expand educational opportunities, improve health care, offer specialized programming to every subculture in the nation, and provide a forum for discussion and collective action on the important local and national issues of the day. Cable television, in the most pronounced versions of this imagining, held the power to help fully restructure social relations and social life. At points, it had an eerie resemblance to the utopian musings of Edward Bellamy's *Looking Backward*, noted in Chapter 1 of this book. As it was with Bellamy, so it was with many of the proponents of this new "cable cure." According to Streeter, such views:

> [P]romoted a revolutionary vision of technical progress in the field of telecommunications: the growing use of communication satellites, the increasing involvement of computers in data transmission, and the increasing capacity of broadband coaxial cable transmission techniques were not isolated developments or mere continuations in the technological revolution of communications systems but were all part

of a revolutionary development comparable to that brought about by print, or by the industrial revolution.[21]

Streeter suggests the impetus for this developing vision came from five general interest centers: the industry itself, influential policy makers centered around Eugene Rostow, a group of economists concerned with regulatory issues, liberal elites, and organized progressive groups seeking to foster new democratic forms of communication.

Starting with people from within the cable industry itself, Jerrold's 1964 prediction of a "Blue Sky" future was being dressed up in work clothes and paraded in front of the press well before the 1966 NCTA convention. In 1964, Irving Kahn saw the possibility of extending his Key-TV pay TV system into an interactive home shopping service. Operators by late 1965 and early 1966 were talking about cable's potential for burglary, fire alarm, and facsimile news services. By 1966, both Associated Press (AP) and United Press International (UPI) had signed deals with cable equipment vendors to develop news text services. Activities at the 1966 NCTA convention, noted above, further help propel the new concept of cable. Going beyond just the loose talk of the convention hallways, NCTA President Frederick Ford called for a new cable programming initiative. Turning against the passive antenna concept and previous shyness about cable-originated programming, Ford used his keynote address to trumpet a call for cable-originated material. Leonard Reinsch, president of Cox broadcasting, heavily invested in cable, observed that "broadcaster opposition to CATV is forcing CATV into origination and maybe the sale of time."[22]

Ford's rhetoric was not just economic in nature. Embedded in the language was a dart aimed at the heart of the 1952 Sixth Report and Order. As an emerging communications resource, cable television could now be positioned as a First Amendment speaker and an outlet for local expression, in service to the FCC's fourteen-year-old mantra of localism. Ford described cable as every community's small-town voice, and used the device to attack FCC-supported legislation (HR-13286) that would have banned cable-originated programming.

Following the convention, the NCTA executive board directed Ford to begin a national campaign to promote local public-service programming. By the fall, the NCTA was putting together press kits and discussing a programming "code." By the end of 1966, the idea of cable TV as an information utility was beginning to ripple out to broader constituencies, and the second half of the decade saw a flurry of public policy reports on cable.

The Carnegie Commission Report on Educational Television was a two-year research effort published in January of 1967.[23] Its aim was to develop strategies for improving and extending instructional and public television, with a special emphasis on the latter (the Commission drew a clear distinction between the two). The report set forth a dozen specific proposals, most having to do with increased federal funding, and included a keystone rec-ommendation to establish a "Corporation for Public Television" to receive and administer funds for public broadcasting, a recommendation enacted in 1967 as the Corporation for Public Broadcasting.

Cable was mentioned only in passing.[24] In one of the Report's supple-mentary papers, an MIT professor, Joseph ("Lick") Licklider, outlined several future scenarios, or "Televistas," for television. One foresaw a multiplicity of television networks aimed at serving the needs of niched audiences. "Here," stated Licklider, "I should like to coin the term 'narrowcasting,' using it to emphasize the rejection or dissolution of the constraints imposed by com-mitment to a monolithic mass-appeal, broadcast approach."[25] The means for delivering these networks would be interconnected CATV systems.[26]

> The cables of CATV will evolve into multipurpose local networks, and the local networks will be linked together to form regional, na-tional, and even international networks. The linking may involve broadcast transmission, with satellite relays playing an important role, and it may involve additional cables, wave guides, microwave channels, and so on.[27]

The author went on to sketch a fully interactive communications system capable of retrieving textual material from libraries on demand, in a narra-tive that came close to describing today's Internet. The resemblance was no coincidence. Licklider was at one time Director of Behavioral Sciences and Information Processing Research at the Defense Department's Advanced Re-search Projects Agency (DARPA), the agency largely responsible for the early development of the Internet, and Licklider is widely credited with being one of the first to envision and articulate what eventually became the Internet. His brief foray into cable Blue Sky conceptualization grew naturally from his computer-centric worldview.

Following a similar conceptual path, in August of 1967, a pair of Rand Corporation economists published a slightly less visionary proposal for what they termed "wired city television."[28] It called for the replacement of the na-tional broadcast infrastructure with a system of local, twenty-channel coax-ial cables system carrying all domestic television. The authors, Harold Bar-nett and Edward Greenberg, argued that such a system would reduce total broadcast costs, improve picture quality, save spectrum space, and, most importantly, substantially increase the number and diversity of program net-works. It would also provide expanded opportunity for educational tele-vision, pay-TV, and specialized advertiser-supported networks, along with political and governmental television. The cable systems, the authors pro-posed, could be constructed by various kinds of companies, including ex-isting CATV firms and electric utilities, but not AT&T for reasons of po-tential monopoly. The report did not dwell on the question of national interconnection of the systems but noted almost off-handedly that such interconnection, perhaps by satellite, would be natural, stating:

> Supplementing the multiplication of local channels and the reduction in cost to use them is the fact that inter-city relay charges—and, indeed, intercontinental relay charges—will decline steeply with development of satellite communications and wave guide. It is likely that a large increase in network programs would develop.[29]

Another influential report came out of New York City. In June 1967, New York Mayor John Lindsay created a special advisory task force on television and telecommunications. It had two assignments. The first was to evaluate the impact of the planned 110-story, twin-towered World Trade Center on city TV reception. The other, more far-reaching task was to develop a plan that took advantage of emerging communications technology to meet the city's expanding needs for modern telecommunications capacity. Lindsay asked Fred Friendly, former president of CBS News, Edward R. Murrow producer, and Columbia University journalism professor, to chair the Task Force.

When it was released in September of 1968, the seventy-five-page report called for an eighteen-channel state-of-the-art interconnected cable system serving all the boroughs of the City.[30] In his letter of transmittal, Chairman Friendly noted that "the promise of cable television remains a glittering one."[31] That promise included cable's ability "to entertain, to educate and inform, to further the participation of the City's citizens in the political process, to foster the City's economic growth, to enhance the protection of life and property in the City, and to enrich the City's social and cultural life."[32] "Communications services to meet the needs of all ethnic communities" were suggested, as well as an almost Orwellian vision of a system that "could be linked to cameras that scan hundreds of public areas for protection against crime, for the detection of fire, for the control of traffic, and for the control of air pollution."[33]

A few months after release of the New York City report, cable made headlines as the subject of a study undertaken by the White House. In 1967, President Lyndon Johnson, sensing a general need for long-term planning in communications policy, commissioned a task force to investigate a myriad of intertwined issues, although the emphasis was to be on communications satellites. The group, formed in August of that year, was headed by Under Secretary of State Eugene Rostow, the policy maker noted above by Streeter. The President's Task Force on Communications Policy completed its work in December of 1968.[34] Formal release of the report was delayed because of a change in administration, but the substance immediately found its way into the press.[35] On December 10, 1968, the *New York Times* ran a front page story describing the study's conclusion that cable television—not satellites—offered the greatest promise for increasing diversity in the nation's TV diet and recommending that the FCC relax its restrictions on the industry.[36] It was only one of several Task Force recommendations, but one that drew much of the attention.

According to the paper, which became known as the "Rostow Report," direct broadcast satellites were technically and economically infeasible, and established terrestrial broadcaster were too constrained by limited channel capacity to provide for the program breadth necessary to serve public needs.[37] Cable television, with its multichannel capacity was the key to the future, declared the Task Force. Satellite distribution of programming to cable systems was specifically seen as one of the most promising avenues for the realization of a diverse national TV fare.

On the slim chance that any elected or appointed officials had missed the point by now, both Rostow and Barnett, among others, brought their ideas to Congress in May 1969, to testify before Representative Torbert Macdonald's (D-Mass.) House Subcommittee on Communications. Barnett extended the vision from "the wired city" to "the wired nation," again underscoring the role that improved communications could play in bringing America's disenfranchised and disaffected citizens back into the national dialogue. Cable and cable-like technologies, it was implied, could bridge the gap between generations, between the classes, and most importantly, at a time when urban riots were sweeping the nation, between the races.

A number of additional studies, several produced in response to an FCC request for comments on the future of cable, were published in the early 1970s.[38] Leland L. Johnson, who served as research director for the Rostow report, subsequently joined the Rand Corporation where he headed a program of study that eventually published more than a dozen papers on cable in the late 1960s and early 1970s. The Electronic Industries Association produced a detailed technical manuscript, "The Future of Broadband Communications," which called, among other things, for the replacement of traditional mail with electronic mail carried via coaxial cable networks.[39]

The Alfred P. Sloan Foundation launched a study in 1970 that culminated in a widely circulated and influential book, *On The Cable: The Television of Abundance*. The Commission was headed by Edward Mason, former Dean of the Harvard Graduate School of Public Administration. Members included the presidents of major universities and public foundations, including the University of Chicago, MIT, the Brookings Institution, and Rand Corp. Its staff members included Monroe Price, Konrad Kalba, and author Ralph Lee Smith. The staff director was Paul Laskin, who had served in the same role for Mayor Lindsay's Task Force. Barnett, Friendly, and numerous others involved in previous or on-going similar studies appeared before the Commission. Following the lead of these Blue Sky visionaries, the Sloan Commission foresaw great things for cable. "If one has any faith at all in the value of communication, the promise of cable television is awesome," it reported. "The power of the existing system is immense; it dwarfs anything that has preceded it."[40] And that promise, predicted the Commission, could be realized soon. Continuing, the Sloan Commission said: "The cable, properly arranged, can read meters, serve as fire and burglar alarms, make market surveys, and even conduct political polls. None of this is fanciful: such services already exist or are being actively planned."[41]

Beyond the academic and public policy statements, the most popular treatment of cable Blue Sky came from writer Ralph Lee Smith, whose quote begins this chapter. Smith turned a lengthy 1970 magazine article in *The Nation* into a book, with both titled, *The Wired Nation*. The title and the idea quickly captured the public imagination.[42] Smith summarized, for the general public, the prophesies and hopes from the growing mountain of reports, as well as passing along the promotional hyperbole of the cable industry itself. He described an exhibit at the 1968 NCTA convention: "It displayed a home communications center, in which the user, through appropriate switching circuits, could enter into two-way exchanges with local stores, could "dial-a-play," and could have at his fingertips the full information contained in vast libraries. This is no dream. The cable could carry it all, and the technology is in existence or soon will be."[43]

Smith called for the creation of a national "information highway." Recalling the U.S. cold-war subsidization in the 1950s of a modern, high-capacity national interstate highway system for cars and trucks, Smith recommended: "In the 1970s, it should make a similar national commitment for an electronic highway system, to facilitate the exchange of information and ideas."[44] Cable television would be at the heart of such a system.

The latter 1960s, in short, saw a dramatic change in the social definition of cable television. Channel capacity had expanded and cable could now deliver, in most markets, a dozen channels. More important than the technology, however, was the new *idea* of cable. While some in the industry had labored to define cable as nothing more than an extension of the customer's antenna, and the FCC had cast the service as little more than an adjunct or extension of the broadcaster's transmission tower, cable was now being promoted as a critical and active participant in the nation's communication network. Some in the industry were even toying with the idea of adjusting the NCTA's name once again to the National Cable Communications Association, to underscore the point.[45]

There was a problem with all the Blue Sky hyperbole , however. Stated simply, it would not work. Public commentators praised the possibilities of two-way digital communication and home services that would include such exotic features as babysitting using cable systems connected to central computers. Frequently, writers would either imply or stipulate, as one columnist described it, a "virtually unlimited" "number of incoming channels."[46] While, in theory, coaxial cable does have the capability of distributing hundreds of channels, it requires electronics capable of exploiting that capacity.

By the late 1960s, the most advanced dual cable technology could handle forty to perhaps fifty channels, a very large number in comparison with the existing over-the-air service, but still nowhere near the exuberant claims of some writers and analysts. More importantly, technological capacity did not translate into direct application. *Television Factbook*'s annual survey of cable in 1970 showed only 86 of the nation's 2,490 system were capable of programming more than twelve channels (see Appendix Table B).

The substantial disconnect between the fanciful daydreaming and the cold reality of technology and economics would be painfully felt by the mid-1970s. In the meantime, however, the rhetoric had its own real-world impact. As sociologist W. I. Thomas pointed out in the 1920s, action is premised in large part on one's "definition of the situation." Parts of this larger vision of a powerful and empowering communications technology would soon seep into federal regulatory thinking and find their way from there to regulatory discourse and finally regulatory action. They would not do so until 1968, however, two years after the Second Report and Order. In the meantime, the Commission was faced with a mounting paperwork problem caused by the report.

Regulating Cable

The FCC, following the Second Report and Order, had looked again to Congress for legislative authority over cable, but came up empty. It was left, as before, to proceed on its own. Unfortunately, as some had predicted, and as even Commissioner Cox had conceded in the Commerce Committee hearings, the consequences of the Second Report and Order were rapidly metastasizing into a bureaucratic nightmare. Although the Report and Order had the general intent of prohibiting importation of signals into the top 100 markets, the Commission, at the same time, said it would consider requests for waivers from the rule, contingent on a showing that such a waiver would be consistent with the healthy development of UHF in a particular market.[47] The industry was quick to take advantage of the loophole, especially when it became apparent that FCC staff was liberal in granting the waivers and, from 1966 through 1968, when the Commission stopped accepting them, operators filed more than 800 such requests. Martin Seiden later analyzed 449 of them.[48] The decision to grant a waiver, concluded Seiden, depended primarily on the status of existing or potential UHF services in the community at issue. Waivers were most commonly granted when no UHF station was involved, or in situations involving only a small geographic area of a top 100 market, typically a large city suburb. Waivers were denied most commonly on the grounds of the possibility of the establishment of a UHF station, rather than the protection of an existing UHF outlet. Seiden noted, in fact, that no waiver request in any top ten market was denied to protect an existing UHF. The Commission was clearly committed to fostering the hope and promise of UHF, despite what Seiden described as the "economically unfeasible UHF television allocation plan of 1952."[49]

The Commission, however, put no new resources into the administration of the 1966 rules. In August of that year, the FCC created a CATV Task Force, but only rearranged existing staff to do so—no new personnel were hired. The Task force faced not only the 800-plus waiver applications filed for the top 100 markets, but also importation requests from operators in smaller markets, requests which, in turn, generated more than 350 objections from broadcasters in those cities. The head of new task force, Sol Schildhause, publicly forsook any hope of dealing with the mounting caseload. He noted that even before the end of 1966 they faced 120 petitions for waiver, 300 requests for special relief, 200 notifications of commencement, and the outlook of literally thousands more reports to review.[50] Without sufficient resources, this huge backlog of cases snowballed. By 1968, the Commission had only settled 168 of them (supporting the cable operators in 147). The rest were left pending in one state or another when the Commission essentially declared it had had enough, and dramatically changed course yet again. The case load was not the only impetus to the change. It was additionally the product of the building Blue Sky chorus and of important decisions out of the judiciary.

Losing in Court: Federal Control and Southwestern (1968)

With conflicting appellate court decisions in *Southwestern* and *Buckeye*, the FCC's authority to impose cable regulations at all was still a question before the courts. The Supreme Court in 1967 agreed to hear the *Southwestern* case out of San Diego. In the meantime, the FCC moved ahead with a hearing on the Midwest complaint. Despite the Commission's full-body press into cable regulation, the FCC hearing examiner rejected all of the complaining broadcast company's arguments and held for the cable systems.[51] "There is no evidence that CATV produced competition to date has had any effect whatsoever on the service offered the public by the San Diego television stations," stated the examiner.[52] Furthermore, it was unlikely that importation of Los Angeles TV signals into the San Diego market, given the Commission's existing program exclusivity rules, would have any impact on the local network-affiliated VHF stations in the future. He conceded that CATV operation could harm possible UHF service, but to date only one poorly run UHF existed in the area and "the loss of UHF service would be virtually unnoticed by the public in San Diego."[53] Southwestern would have only a few months to savor its victory, however.

In June of 1968, the United States Supreme Court issued an historic five to zero decision (Justices Douglas and Marshall abstaining) upholding, for the first time at the highest level of judicial review, the FCC's authority over cable television.[54] Although not ruling specifically on the Second Report and Order, the high court did confirm the FCC's power to issue rules in the San Diego case and thereby effectively authorized the Commission to take control of the new medium.

In pleadings before the Court, cable lawyers, following the FCC's own logic of the 1950s, had argued that cable could be classified as neither a broadcaster nor a common carrier and, therefore, did not fall under the agency's jurisdiction. In fact, in the 1959 Report and Order, the Commission had specifically considered and rejected the contention that they could regulate CATV on the basis of its relationship to (and impact on) broadcasting.[55] Over the years, that view had pivoted 180 degrees, however. The FCC had quite the opposite view now, and the Supreme Court found in favor of the revised position that cable in the 1960s constituted "interstate communication by wire," as per section 152(a) of the Communications Act, and, therefore subjecting it to FCC control. The intent of such control, said the Court, was to serve the advancement of the orderly development of the television system in the United States. As to the extent of FCC authority, the Court was vexingly vague, its only guidance coming in the admonition that any regulation be "reasonably ancillary" to the Commission's fostering of broadcasting. Stated Justice Harlan for the Court: "There is no need here to determine in detail the limits of the Commission's authority to regulate CATV. It is enough to emphasize that the authority which we recognize today under sec. 152(a) is restricted to that reasonably ancillary to the effective performance of the Commission's various responsibilities for the regulation of television broadcasting."[56]

While offering the Commission general regulatory authority over cable, the Court nonetheless limited its additional holdings to the specific facts presented in the San Diego case. It concluded that the FCC was within its authority in issuing the particular restraints in the case, but the Justices pointedly declined to express a view concerning other possible forms or manifestations of FCC regulation of cable.

Industry reaction to the ruling was somewhat understated. Frederick Ford admitted he was not happy with the decision, but thought it was nonetheless inevitable. An NCTA lawyer, attempting to look on the bright side, suggested it would spare the industry from state-level regulation, and Kahn said simply, "We can live very well with it."[57]

With its new flag of Supreme Court-authorized jurisdiction flying high, the FCC met a week after the decision to consider Midwest's original complaint. By a slim four-to-three margin, the Commission found in favor of the broadcaster, overturning its own hearing examiner.[58] In his majority opinion, UHF supporter Lee acknowledged that the complaining VHF station's economic health was probably not at issue, but the Commission nonetheless needed to support its on-going investment in the promise of UHF. The majority concluded that UHF service could prosper in San Diego under the proper conditions (including a planned change of ownership in the existing UHF outlet), but only if such service did not have to compete with several existing Los Angeles UHF independents imported by cable. In dissent, Commissioners Bartley, Loevinger, and Wadsworth characterized the majority's arguments and concerns as "hypothetical and illusory."[59] Nonetheless, as a result, restrictions on the importation of those signals were imposed on the cable systems.

Also at issue was the potential of cable-originated programming to harm existing broadcasters. While declining to rule specifically on that question, the Commission nonetheless denied cable operator claims that it did not have authority under the Communications Act to regulate such programming if necessary, or that such regulation constituted a violation of the operators' First Amendment rights to free speech.

Winning in Court: Copyright and Fortnightly (1968)

June 1968 was one of the industry's most tumultuous months. In the short span of seven days, the Supreme Court handed down decisions on two issues that had nipped at the industry's heals since its inception—federal regulatory jurisdiction and copyright liability. On June 10, the Court in *Southwestern* told cable operators that they would have to learn to live with FCC control. To the extent this dampened industry spirits—no matter how brave the public face—those spirits soared again on June 17 when the Supreme Court, in a decision that surprised and even shocked nearly everyone, found in favor of cable television on the copyright question.

The 1966 District Court decision in *United Artists v. Fortnightly* had been an unpleasant wakeup call for cable. Faced with the Court's decision and with pending legislation in the House of Representatives, both of which promised an end to cable's free programming ride, cable entered into copyright negotiations with Hollywood motion picture executives and the NAB in the summer of 1966. Leading talks for NCTA was TVC's Alfred Stern. All sides were seeking a negotiated settlement on an appropriate level of cable compensation to copyright holders. While negotiations proceeded, a moratorium on copyright lawsuits was honored by the programming side. Cable operators had largely given up on escaping payment entirely and were simply seeking to avoid individual local contracts; they were also bargaining to keep fees within what they considered a reasonable range, about two percent of a system's gross revenue. Some form of copyright fund akin to the ASCAP system was proposed, with hopes that the agreement could be incorporated into Congress's on-going attempt to rewrite the 1909 copyright law. Those discussions, however, soon stalled as each side held to firmly set positions.

The House Judiciary Committee, meanwhile, was growing restless at industry's inability to resolve the problem and, by March 1967, developed legislation—elements of the broader copyright revision effort—that would have brought cable under copyright law. Fortunately for the industry, this effort became entangled in a power struggle between the House Judiciary and Commerce Committees. Commerce Committee Chairman Staggers opposed elements of the cable clause crafted by Judiciary for the new copyright act, claiming it infringed on his committee's jurisdiction over CATV. The cable issue, therefore, became a stumbling block for the entire copyright revision. To move the broader bill along, Judiciary Chairman Emanuel Celler removed the offending CATV-related articles. (Although it would take yet another nine years before Congress actually passed an updated copyright bill.) For cable, it was a reprieve from any action by Congress. Both the industry talks and the efforts of Congress to forge a cable copyright agreement therefore came to naught, but cable was still losing in the courts.

Fortnightly had appealed the district court decision, but would fare no better at the next level. On May 22, 1967, the U.S. Court of Appeals for the Second Circuit unanimously held that cable carriage constituted a "performance" within the meaning of the law.[60] In a clear echo of the district court opinion, the appeals court found that cable was anything but a passive antenna, noting "the expense and effort required to install, operate and maintain the CATV systems' antenna."[61] Once reperformance was established, the precedent of *Buck v. Jewell-LaSalle Reality* was in full play, and therefore, concluded the court, cable had infringed on United Artists' copyright.

Again, Fortnightly appealed. In December 1967, with congressional action and private negotiations at a dead end, the Supreme Court agreed to hear the case, and everyone held their breath. The general expectation was that the Court would uphold the previous decisions; even the cable lawyers arguing the case were apprehensive.

Fortnightly's Sanford Randolph and NCTA legal counsel Strat Smith had been keen to pursue the case, but the NCTA Board of Directors was reluctant. An appeal to the Supreme Court would be expensive and after two unequivocal lower court decisions, the board did not have strong hope for a reversal. Smith was cautiously optimistic, however. The Supreme Court decision to accept the case, following the two lower court decisions, was a good sign for the industry. The Court would not have accepted the case at all if at least some Justices had not harbored doubts about the prior decisions.

United Artists retained the famous attorney Louis Nizer to argue its cause before the Supreme Court. Fortnightly went with the New York law firm of Cleary, Gottlieb, Steen, and Ball. Robert C. Bernard was lead counsel.

In mid-June 1968, Barnard, Smith, or both, would go to the Supreme Court every morning to see if a decision in their case would be handed down that day—the list of decisions was not announced in advance. Both were in court on Monday, June 17, and each caught their breath a bit when they realized the opinion in *Fortnightly* would be read that day.

The Blue Sky rhetoric was just beginning to percolate as the Supreme Court deliberated the copyright future of cable television. If the Justices had been exposed to the early wired city debate, they clearly were not influenced by it. In fact, none of the utopian language could be found in the majority decision. For copyright purposes at least, community antenna television at the Supreme Court was still little more than an extended set of rabbit ears. In a five-to-one decision (Justices Douglas, Harlan, and Marshall abstaining), the Supreme Court surprised lawyers and lobbyists by finding in favor of the cable industry.[62]

As soon as Justice Potter Stewart began reading the opinion, Smith realized they had won. One of the lawyers for the opposing side was sitting directly behind Smith and as Stewart read through the Court's opinion, the United Artists attorney involuntarily gasped an audible "Oh my goodness!" Recalled Smith:

It took some doing to maintain our professional dignities until the Justice finished. Bob Barnard and I just glanced at each other and hung in. It's pretty dangerous to sprint on those marble floors of the Supreme Court building, but to say that we hurried to the telephones would be an understatement. We each called our offices, and I got Bob L'Heureux, one of my law partners. When I said, "Bob, we won, we won!" he was as excited as I was.[63]

Writing for the Court, Justice Stewart revealed that the five-member majority had bought fully into the passive antenna defense forged by Smith in the mid-1950s. CATV systems did not, said Stewart, perform the respondent's copyrighted work within the meaning of the 1909 act. The passive antenna proposition, so thoughtfully erected many years before as a defense against federal taxation, had reached its legal zenith. In fact, it was directly to the tax case of 1956 that the Court looked for foundation guidance on the nature of community antenna television. Citing *Lilly* the Court declared: "Essentially, a CATV system no more than enhances the viewer's capacity to receive the broadcast signals; it provides a well-located antenna with an efficient connection to the viewer's set.[64]

For the Court, the map of television broadcasting could be divided into two hemispheres, that of the broadcaster, or performer, and that of the viewer. Viewers could engage a host of sophisticated devices to receive and watch the signals sent out by broadcasters. Such devices included TV tuners, TV sets, and roof top antennas. In this view, CATV systems were simply an extension of the viewer's existing reception apparatus, said the court. Moreover, viewers do not "perform." A CATV system "no more than enhances the viewer's capacity to receive the broadcaster's signals," said the Court. Cable, in sum, was free of copyright liability.

Dissenting in *Fortnightly*, Justice Fortas concluded that cable ought to be pushed back over to the performer's side of the line and held liable under *Buck v. Jewel* for copyright. At the same time, he added that the case presented a "baffling problem" in the collision of aging law with new technology, offering the court's only nod, and that in passing, to cable as a "new important instrument of mass communications." Finding a balance, however, "calls not for the judgment of Solomon, but for the dexterity of Houdini," said Fortas.[65] Much the same could have been said about the legal entanglements that followed.

The disappointment felt by some in the industry following *Southwestern* was significantly offset by the landmark opinion. The tables had turned radically in cable's favor: Neither the Courts nor Congress had been able to saddle the industry with copyright obligations. The jubilation rippling through the cable community was muted, however. Operators realized that they had to monitor carefully their new-found leverage. Congress was most likely to act in any case and the FCC would keep tight control until the property rights issue was settled. From a political point of view, a negotiated settlement on copyright was essential. Irving Kahn explained, "legislation which is fair to all concerned will be the proper and enduring solution to the copyright problem."[66]

The TelePrompTer suit remained, however. In its action against Tele-PrompTer, CBS had attempted to distinguish its case from *Fortnightly* by arguing that TelePrompTer had used microwave hops to import the programming in question, constituting a more active "performance" of the programming than was the case in *Fortnightly*. It would be 1974 before this suit was resolved; in the interim, it hung over the industry's head like a small dark cloud. Cable operators, in fact, would have precious little time to celebrate their victory in *Fortnightly* because, by the end of 1968, the FCC would act again to effectively drain the protection from the industry.

Blue Skies at the FCC

Through 1966 and 1967, while CATV exemption requests piled up in FCC staff offices, the composition of the Commission continued its never ending churn. And as the chairs revolved once more, so did Commission philosophy. In April 1966, Chairman Henry stepped down and was replaced by Nicholas Johnson. Johnson had been a Maritime Administrator and a former Assistant Professor of Law at the University of California at Berkeley. The young lawyer had the enthusiastic regulatory vigor of Minow and soon began championing efforts to bring the voice of the public more fully into FCC decision-making. This came as something of an unpleasant surprise to President Johnson, who had appointed him. Johnson was aiming for a quieter, less proactive FCC, and with the exception Nicholas Johnson, the President had been generally successful. The activism of the Kennedy administration and its eager appointees, Henry, Minow, and Cox, was being supplanted by a Johnson team much more interested in making the business community an ally. President Johnson himself tended to stay far back, at least publicly, from pronouncements about (or interference in) regulatory matters pertaining to television because, through his wife, he had a significant financial stake in the industry and preferred to avoid susceptibility to accusations of conflict of interest. His political philosophy was given voice, however, by others his administration. Vice President Hubert Humphrey told the 1965 NAB convention: "Government doesn't own you, Government is not your master ... Government is here to help you and serve."[67]

A variety of intertwining and often conflicting forces were at play, therefore, in the FCC's emerging policy posture. On the one hand, the regulatory enthusiasm of the Kennedy Commissioners and the recently-appointed Nick Johnson was tempered with a more restrained view from the White House. On the other hand, the Supreme Court in *Southwestern* had given the FCC authority to regulate cable and, at the same time, had killed efforts via existing copyright law to protect broadcasters' property rights. The Commission had yearned for a favorable copyright decision that would obviate the need for detailed cable regulation, but the *United Artists* decision dashed those hopes. Meanwhile, the wired city frenzy that had been building for two years seemed to be cresting at the Commission itself.

In 1967, Nicholas Johnson publicly outlined his view of cable television; it was a picture washed in sky blue hues. Cable could print newspapers in the home, provide window shopping from the living room, and connect subscribers with distant computers, he suggested. Nationally distributed cable networks, Johnson wrote in the popular *Saturday Review,* could reach a host of specialty audiences not currently served by broadcasters. "Whereas a local broadcaster may not be able to justify programming aimed just at ballet enthusiasts, or the local Negro community, of aficionados of sports cars, a regional or even a national cable network might be developed which could enhance its appeal significantly through such specialized programming."[68]

A year later, fellow Commissioner and cable arch foe Ken Cox joined the chorus in an address to the annual convention of the National Association of Regulatory Commissioners. It had long been Cox's view that the national media system should serve the public and where it did not, the FCC had the responsibility to see to it that it did. He once went so far as to state that the First Amendment was no impediment to Congress applying the Fairness Doctrine to newspapers.[69] Cox, as with many others, now seemed fully consumed by wired nation fever. In Cox's eyes, cable had become a vehicle for electronic delivery to the home of newspapers, books, and magazines; a tool for electronic home banking, shopping, and what would someday become know as "e-mail." "All of this, of course," added Cox, "in addition to the many channels of television and nationwide picturephone service."[70] This in no way is meant to suggest that Cox had become a fan of cable. In fact, the prospect of an information highway controlled by a handful of cable companies concerned him greatly. The solution, he proposed, would be to declare such a network a common carrier, and moreover, to turn control of the nation's cable systems entirely over to the government-regulated telephone companies.[71]

Whether controlled by the telephone or the cable industry, the utopian view of the technology had taken seed, not just in Cox's mind, but now appeared deeply embedded in the thinking of the Commission as a whole. In its annual report for 1968, the Commission, which previously described cable tersely as system for redistributing television signals for a fee, now observed, "What was originally conceived as a mere multiple-channel reception device may develop into a home communications center enabling subscribers to shop from their armchairs ... order facsimile newspapers, or have their meters read—all through a cable connected to their TV sets."[72]

Against this rhetoric, however, was a Commission seemingly constrained by its earlier pronouncements on the harm that cable could do to broadcasting. The agency was also bureaucratically hamstrung by a caseload it could not possibly manage; by June 1968, the Second Report and Order had created a backlog of 500 cases for the CATV Task Force. By the end of year, the FCC appeared, if not confused, at least severely conflicted.

The 1968 Freeze

In an effort to find its way out of the forest, the FCC, in December 1968, announced a Notice of Inquiry in docket 18397, which propounded new interim rules for cable TV.[73] Only two years before, the Commission had stipulated that CATV did "not serve as an outlet for local self-expression"[74] and, in fact, was a pointed danger to the broadcast outlets that did. Things had clearly changed by 1968. The contextual language of the FCC's announcement indicated an agency now mesmerized by the wired nation. The preamble of the Notice quoted from Mayor Lindsay's Task Force report on the "glittering" possibilities of cable communication.[75] It summarized most of the "blue sky" hyperbole, and drew much of its language and tone from the Rostow report, offering an extensive list of potential cable-derived benefits:

> [F]acsimile reproduction of newspapers, magazines, documents ... electronic mail delivery; merchandising; ... access to computers; ... information retrieval; the furtherance of various governmental programs ... special communications systems to reach particular neighborhoods or ethnic groups in a community, and for municipal surveillance of public areas for protection against crime, fire detection, control or air pollution, and traffic ... the provision of a low cost outlet for political candidates, advertising, amateur expression[76]

As part of the Commission's latest about face and standing distinctly out from the Second Report and Order, the FCC now declared cable to be a great hope for enhancing local democracy. Lauding cable's potential to realize the long-standing goal of increasing the number and diversity of local media voices,[77] the Commission proposed rules mandating local origination, and encouraging the provision of additional channels for commercial and community use on a common carrier-basis.[78]

Although seeming to promote, at least from some perspectives, the interests of cable, the Commission acted at the same time to tighten down even further on the industry's actual operational freedom. The Rostow report, which the Commission cited with admiration, had argued for a reduction in FCC controls on cable, but the FCC was moving in a different direction (to the public displeasure of Rostow himself, who commented on it in the May 1969 House hearings). The Notice of Inquiry requested comments, for example, on proposed rules to prohibit cross-ownership of broadcast and cable properties in the same market, limit multiple ownership of cable systems, and, significantly, to restrict the proposed local origination programming to only one channel.

The Notice dealt first, however, with the FCC's paperwork problem; the Commission conveniently swept away the backlogged waiver requests, announcing it would no longer accept applications from operators seeking to import signals into the top 100 markets and would deny those extant.[79] It stated that it was now satisfied that cable could pose a significant threat to UHF television in major markets and proposed regulations that would permit importation only when the cable operator obtained permission from the originating station. Furthermore, it would entertain current requests for importation only from those operators already abiding by this proposed rule. The rules applied to CATV systems operating within thirty-five miles of the main post office in the top 100 markets, although existing CATV operations were grandfathered. In markets below the 100 mark, cable systems could import only enough signals to provide their subscribers with all three networks and one independent, where such service was not available off the air. Any educational station could be imported, but the new rules implemented the anti-leapfrogging provisions discussed in prior proceedings, so systems had to take signals from the closest available distant markets.

It appeared initially that the interim rules meant a cable system in a top 100 market could import a distant station on receiving general permission to do so from that station, and within weeks, Omni-Vision of Pensacola, Florida, presented such a case. It told the Commission it had received permission from two distant stations to bring their signals into Pensacola. The cable company, along with the industry observers, was shocked a few weeks later when the FCC issued a clarification. Simply obtaining the permission of the imported station was insufficient, said the agency. The cable operator was required to obtain permission for every program carried by that station and to obtain it from every party that had a property interest in that program, including the station, the network, the distributor, and the producer.[80] Seiden estimated that the administrative costs alone of attempting to meet such a requirement at $33,000 a year, not counting any actual copyright payments. Of course, the cost was irrelevant because, in practice, the requirement was an impossibility. It was lost on no one that the new rule was designed specifically to overturn through regulation the then-recent Supreme Court decision in *United Artists*. Where the high court had refused to hold cable systems liable for copyright, the FCC had now done so by administrative fiat. Forthnightly's victory turned to dust. The new rules closed any loopholes that might have remained in the Second Report and Order; cable was completely frozen out of the top 100 markets. It would, in fact, become known as the major market "freeze," and as with its namesake from the 1940s and 1950s, it would last four years.

Substantial regulatory irony was seen in the policy and in the contrast between the FCC philosophies of the 1950s and 1960s generally. In the 1950s, the popular criticism of the FCC was that it was too soft on the broadcast industry, too protectionistic. Yet, in its cable policy, the Commission protected and fostered a young but real competitive threat to broadcasting. In contrast, the FCC of the Kennedy years enjoyed a public image of much greater regulatory activism, no friend of the broadcast industry (although its regulatory bark turned out to be much worse than its bite). In its cable policy, however, it could not have worked harder to advance the interests of broadcasters by essentially shutting down the prospects of cable and saddling the industry with impediments to success. The difference between the perception and the reality of regulation in the two eras, at least with respect to cable, was stark.

Although the Commission positioned the 1968 Inquiry as part of larger effort to look at cable, and included lengthy and detailed discussion of visionary wired city possibilities, its real-world stranglehold on cable development was soon palpable. With much on the line for the affected industries and an increasing host of public interest groups, foundations and think tanks all eager to weigh in on the subject, the FCC was greeted with reams of comments and proposals in response to its Notice.

The Politics of Origination

The first set of FCC rules coming out of the 1968 Notice dealt with the origination issue. A year before, as part of its rulemaking in the *Southwestern* case, the FCC, for the first time, had given San Diego cable operators license to experiment with local origination programming, provided they did not sell advertising.[81] In October 1969, the Commission more than extended that license. In keeping with the developing construction of cable as a local communications utility, the FCC ordered all cable systems with 3,500 or more subscribers to create a channel for local origination programming.[82] Part of the rationale was the value of cable as a means of local community expression, especially in service to minority interests. The programming could not be of the automated "time and temperature" variety, and operators were urged to avoid mass appeal fare common to network broadcasting. The Equal Time and Fairness Doctrine requirements of the broadcast industry were also applied. Again, the language used by the Commission was revealing. Cable operators who produced their own material were dubbed "cablecasters," active participants in the nation's telecommunications system. The FCC's cablecasters were allowed to sell advertising on the local channel, reversing the earlier prohibition. But bowing to pressure from broadcasters who opposed any form of cable origination, the FCC limited the flexibility of commercial insertions. Advertising could only come at the beginning or end of programming or during "natural breaks" within the cablecast. The rule limited cable operators to only one origination channel and, in an important side note, the Commission said it wanted to encourage the interconnection of cable systems and the creation of regional and national cable-based television networks.[83] Broadcaster opposition to origination was softened somewhat by a negotiated understanding with the NCTA that cable would not offer entertainment programming on such interconnected networks.

In addition to local origination, the Commission sought to make cable channels available to the public on a quasi-common carrier type basis. It suggested creation of what would become known as access channels for educational, governmental, commercial, and civic groups.[84] The local origination requirement itself was intended, in part, to support the potential for such services by encouraging the development of local production studios.

It was again ironic, given the regulatory history, that the origination requirement attached only to cable systems in the largest communities. Since the 1952 Report and Order, the regulatory philosophy had tilted toward a concern about local expression in America's small towns. But the Commission now conceded that small town CATV operators probably could not afford the necessary production facilities, going so far as to admit the 3,500 subscriber cut-off itself was fairly arbitrary—they labeled it "liberal." The agency promised to seek actual data on an appropriate subscriber number to trigger the requirement. In practice, therefore, the local origination rule touched only a minority of cable systems, about 300 of the 2,200 or more in operation. The affected systems, however, were typically owned by the larger cable companies and their fortunes influenced others throughout the industry. The origination regulations sparked a flurry of activity by operators and potential programmers who sought, enthusiastically or not, to satisfy the requirement. Those activities are discussed in greater detail in Chapter 7. In fact, however, the Commission's origination rules were never enforced and were replaced a few years later by formal public access requirements.

Turn, Turn, Turn

While the cable industry was attempting to absorb the implications of the FCC's latest round of rulemaking, the nation was trying to absorb the impact of a much larger political upheaval. Faced with growing national criticism over involvement in Vietnam, President Lyndon Johnson in the spring of 1968 surprised the country by announcing on national television that he would not seek reelection. A Democratic National Convention in Chicago, rocked by student protests, and aggressive, some said brutal, police retaliation at the command of Chicago Mayor Richard Daly, left the Democratic party in a shambles. Republican Richard M. Nixon beat his Democratic opponent, Vice-President Hubert Humphrey, in the run for the Presidency, and the politics and the culture of the country began a slow but certain swing once again. The political pendulum was moving back toward the right. Most sectors of the country would feel the effects, including cable television.

Nixon had spent years warring with the press. His loss to John F. Kennedy in 1960 was heavily attributed to his poor physical appearance during the nation's first live, televised presidential debate. Through much of his career, the press had been critical, and he considered the national media, especially the three major broadcast networks, his enemies. It was this animosity toward broadcasters that would shade many of the regulatory policies advanced by the new administration and, in turn, pay dividends for the cable industry.

In April 1969, early in the new administration's tenure, the Antitrust Division of the Justice Department, under Attorney General John Mitchell, began pressuring the FCC to encourage cable development. They told the FCC that cable offered great promise for increasing competition and diversity in local television. The Justice Department said cable should be encouraged to produce its own programming and be allowed to sell advertising. It further urged the Commission to relax its rules on cable in the top 100 markets, although it recommended cross-ownership restrictions that included newspapers, to avoid local media monopoly. The Justice Department continued its pressure on the FCC, at least through 1971, arguing, for competition's sake, that cable be allowed to provide pay services and develop commercial advertising. Comments from Presidential assistant (and later convicted Watergate felon) Charles Colson helped confirm White House orchestration of this activity. Colson commented publicly that the administration saw cable television as one hope for breaking up the TV networks, and advocated regulation to foster the cable growth and the concomitant dissolution of broadcast network power.[85]

More directly, the new administration was replacing the Kennedy-Johnson era FCC Commissioners with its own people. In 1969, Commissioners Wadsworth and the conservative Chairman Hyde both stepped down. Nixon nominated Kansas broadcaster Robert Wells to replace Wadsworth. To serve as new Chairman of the FCC, Nixon named Dean Burch, the former head of the Republican National Committee and a one-time campaign manager for arch-conservative Senator Barry Goldwater (R-Arizona). Burch and Wadsworth were sworn in November 1969. The following year, cable's chief regulatory antagonist for more than a decade, Kenneth Cox, retired and was replaced by Thomas Houser, who would serve only seven months on the Commission.

In addition, Nixon created a new Office of Telecommunications Policy in early 1970 to serve as executive branch advisor on communications issues.[86] Presidential adviser Clay T. Whitehead was named head of the new unit. The changes were widely viewed as good news for cable, and bad news for the broadcast industry. According to Krasnow, Longley, and Terry: "Post Watergate research on the Nixon administration shows that Dr. Whitehead's attacks on the networks, public broadcasting and the affiliates were elements of a deliberate assault on the centralized, national media from a White House that had viewed the media as a tormentor."[87]

In November 1969, Vice President Spiro T. Agnew lashed out at the national television networks in a speech before a Republican group in Des Moines, Iowa. The media, he accused, did not represent the views of the American public, but rather the political ideology of "a tiny, enclosed fraternity of privileged men elected by no one and enjoying a monopoly sanctioned and licensed by government."[88] It was a striking overture to what would become a crusade against the established broadcast industry. Following Agnew's speech, the new FCC Chairman, Burch offered his unequivocal support of the Vice President. For their part, cable operators indicated they would have little problem playing the role of an administration whip to sting broadcasters. Burch, through his previous positions in Republican politics, had maintained strong ties to the White House and defended the new administration's active interest in telecommunications policy. Bill Daniels later recalled that the nomination of Burch was a welcome relief for cable operators. Explained Daniels: "Dean Burch was Barry Goldwater's campaign manager, and a lot of us had a pipeline to Barry. I worked on his campaign, Bruce Merrill was close to him, and we were able to get Dean's ear, and Dean started to go to bat for us."[89]

A mixture of personal and philosophical forces was at play, therefore. A Republican administration, by general inclination, would be inclined to a more business-friendly regulatory approach, one that mirrored the Eisenhower era laissez-faire ideology. And the rising rhetoric of blue sky cable would play easily into the hands of an administration seeking to weaken the entrenched broadcasting industry.

The June 1970 Package

In addition to a more cable-friendly demeanor, Burch seemed to bring a new energy and focus to the Commission and sought, among other things, to conclude the long simmering cable dilemma. In March 1970, he stepped in, for example, to the CATV copyright problem. Following the Supreme Court's decision in *Fortnightly*, copyright negotiations had begun anew, albeit this time with cable in the driver's seat. NCTA continued its talks with the NAB and, in early summer of 1969, the NCTA and NAB staff announced a compromise that included a quota system for importing distant signals. The NCTA endorsed the plan, but the NAB Board rejected it following pressure from the large market broadcast association, the Association of Maximum Service Telecasters (AMST), and strong opposition from ABC, the National Association of Theater Owners, and the film industry represented by Nizer.[90] Renewed talks continued through 1969, but no progress was made.

The NAB's recalcitrance angered Senate Judiciary Committee Chairman John McClellan (D-Arkansas), who had been withholding action on a copyright bill pending an industry settlement. In late 1969, the Committee reported out a bill that included a compulsory copyright provision for cable. The proposal stipulated that cable operators would pay quarterly royalty fees ranging from 1 to 5 percent of gross revenue over $160,000. It would give operators in the top fifty markets the right to carry up to seven signals: three network, three independent, and one noncommercial; and give operators in markets below fifty the ability to carry six signals: three network, two independents, and one noncommercial. Broadcasters felt the measure offered too little in the way of compensation, and opposed it.

In March 1970, the FCC, under Burch's leadership, stepped back into the discussion, complaining that McClellan's bill was too detailed and intruded too deeply into FCC territory. Congress listened to the FCC's concerns, but little came of it. In fact, congressional reform of the nation's copyright laws was still years away.

Burch had greater success inside the agency itself. Early in his tenure, he recommended upgrading the Cable Television Task Force to a Cable Television Bureau, coequal with the Broadcast and Common Carrier Bureaus, a plan he executed in early 1970 only over the opposition of Commissioner Cox. Establishment of the new bureau had both substantive and symbolic importance. It was necessary to deal with the increasing number of cable-related issues pressing on the Commission, but was also a recognition of the growing prominence of cable as an industry and a legitimate component of the evolving national communications infrastructure. Working within the existing rules and proceedings, Burch also began to assemble a proposal that he hoped would satisfy the interests of cable, UHF, and educational TV proponents. Some of his job was aided by the changing nature of the affected industries themselves and by the on-going studies of cable's social potential.

In the first instance, expanding broadcast industry investment in cable was making it increasingly difficult for broadcasters to take a monolithic position against CATV development.[91] The cable industry, at the same time, continued to grow in small and medium-sized communities, from about 1 million subscribers in 1963 to 4.5 million in 1970, thereby accumulating a little more political influence in Washington. Cable operators were also more willing to compromise on the issue of copyright liability; while the movement in both the courts and Congress waxed and waned with regard to cable's formal copyright obligations, the political trends were clear and it appeared only a matter of time before cable would have to start paying for its programming.

The cable industry additionally softened on FCC control following a February 1970 Supreme Court decision upholding a Nevada statute that brought cable under state control (see below). (The Connecticut State Supreme Court had also endorsed the state's move to place cable under PUC control). It now appeared that the Supreme Court's decision in *Southwestern*, granting FCC authority, was not going to be a prophylactic against state-level regulation, so the NCTA became increasingly concerned about the possibility of proliferating and splintering regulation across all fifty states, as well as in every municipality, and looked to consolidate more benign regulation under the auspices of the Commission.

Economic studies, meanwhile, suggested that the dire consequences feared by broadcasters as a result of cable growth were not materializing. A series of studies published by the Rand Corp. and presented in FCC hearings during this period may have been influential, especially among UHF advocates such as Robert Lee. The studies suggested that, instead of harming UHF, cable could facilitate the development of UHF broadcasting by giving subscribers a clean UHF picture on a VHF cable channel, putting it on a par with established VHF stations.

Through the first part of 1970, then, observers watched eagerly, even impatiently, while the outlines of Burch's new proposal were leaked to the press, and on June 24, 1970, the FCC unveiled its new cable strategy, a wide-ranging list of rules, initiatives, and proposals that addressed importation restrictions, ownership controls, pay-television and technical issues. The "Blue Sky" hyperbole of two-way interactive communications network delivering specialized programming and serving as a voice to local citizenry, constituted again the gilded frame surrounding the substantive proposals. The Commission invited comments on rules requiring cable systems to have at last twenty or even forty channels and two-way capacity.

The heart of the importation proposal, and the scheme designed to engender support from UHF and non-commercial broadcasting, was something labeled the "Public Dividend Plan." Under the plan, cable would underwrite public television and attempt to bolster the fortunes of UHF TV.[92] Burch, thus, won the favor of UHF supporter, Commissioner Robert E. Lee, and recently appointed Commissioner H. Rex Lee, a strong advocate of non-commercial broadcasting. Nicholas Johnson, also a backer of noncommercial broadcasting, found favor with the proposal, as well. It squeaked by on a four-to-three vote, however, with Commissioner Cox, casting one of his final votes against cable development. Alternatively, cable-friendly Bartley opposed the scheme, thinking cable could be better served under the pending copyright bill.

Many critics saw a plan far too complex for the real world. The Commission would permit the importation of four distant signals into the top 100 markets, but cable operators would have to delete all commercials from those signals and, if requested, replace them with commercials provided by local UHF stations (VHF in some cases). As a condition of importing the signals, the systems additionally would have to pay 5 percent of their gross revenues to the Corporation for Public Broadcasting. A compulsory license clause would have cable operators making copyright payments for programming under a fee schedule set by Congress. Local franchise fees would be capped at 2 percent of gross revenues.

Issued along with the Public Dividend proposal were rules on pay-cable programming. A year earlier, on the same day the Commission had issued its notice of rulemaking in docket 18397, it had also released long-awaited rules on pay-television, authorizing pay broadcast service for the first time, but also controlling it tightly.[93] The FCC fought off legal challenges to the rules in the lower courts and, in early 1970, the Supreme Court declined to review those decisions.[94] This paved the way for the Commission to now extend the rules of broadcast pay-TV to cable.

Although the new rules permitted pay cable for the first time, they also severely limited its scope and economic potential, restricting the most attractive and profitable kinds of programming, movies, sports, and series.[95] The concern of the Commission harkened back to the 1950s, focusing on the economic impact of pay-cable on broadcasters and the worry that pay-cable would "siphon" programming away from them. Such programming, then, would be available only to those who could afford it and deny the material to a majority of the audience. The rules specifically prohibited per-channel or per-program charges for (1) movies more than two years old and less than ten, (2) sporting events broadcast locally in the prior two years, and (3) any serial programming, such as comedies and drama.[96] Sports and feature film programming combined could constitute no more than ninety percent of all programming, and advertising was prohibited.

In addition to the pay-cable rules and the Public Dividend Plan, the Commission also issued new ownership guidelines, contained in the Second Report and Order in Docket 18397.[97] Cable operators had argued that local broadcasters should not be permitted to own cable systems in the same town, and broadcast networks should be banned from the cable business completely. By 1968, broadcasters owned about one third of all the country's cable systems and traditional cable operators saw both broadcasters and the telephone companies as industries eager and willing to swallow up the CATV business whole. Fortunately, for cable operators, the FCC was of a like mind. As a prophylactic against possible monopolistic or oligopolistic control over news and information in a given market, the Commission banned ownership of both a television and cable property in the same market (or more specifically, within the TV station's Grade B contour). Perhaps more importantly, given the interest shown by CBS and to some extent NBC in cable television, the FCC banned broadcast network ownership of cable systems. ABC, which held no cable interests was unaffected by the rule, but CBS and NBC were required to divest their holdings by August 1973 (a deadline that was later extended to August 1975). In addition, the Commission issued a Notice of Proposed Rulemaking on cable ownership caps, suggesting that no company be permitted to own more than fifty systems with 1,000 or more subscribers in the top 100 markets, or alternatively, a total of two million subscribers.[98] In January 1970, the Commission also issued an effective ban on the ownership of cable systems by telephone companies in their service area,[99] shielding the cable industry from the acquisitive tendencies of the telephone industry.

The affected parties lined up in reasonably predictable positions in reaction to the new rules and proposals. Cable leaders had mixed reactions. The larger operators, including Kahn, were enthusiastic. They saw the rules as a liberalization of formerly restrictive FCC controls. Smaller operators were more circumspect. They worried about the financial impact of the proposal.

UHF operators, whom the plan was intended to aid, were perhaps the most colorful in their response. They labeled it an "administrative monstrosity" and "a bad joke," stating that it would cost millions to implement and vowing strong opposition.[100] In addition to the administrative nightmare the plan suggested, commercial broadcasters rebelled against the UHF insertion concept, calling it "insane" and "asinine."[101] They did not think much of the proposals for the public support of noncommercial broadcasting, either.

The volume of response to the Commission's proposal was overwhelming. Hundreds of comments and briefs were filed in the Inquiry and the deadline for submitting material dragged on until February 1971. In March 1971, the Commission held hearings on the Plan. More than 120 companies, groups, and individuals asked to be heard. The demand was so massive that the Commission had to find a larger space to accommodate the proceedings. Oral presentations took five days and, for four days, the FCC held additional panel discussions in a packed auditorium in Washington's National Academy of Sciences building. The proceedings, in fact, were televised to a national audience via a network of 130 public TV stations taking the live feed.

Little new information came from the hearings, as all parties retreated to and reiterated long-held positions. One thing was soon apparent, however. The Public Dividend Plan was rapidly sinking beneath the weight of ridicule. At one point in the hearings, Commissioner Houser remarked that his comments on the Public Dividend Plan were not meant as a defense of the proposal, "I'm only trying to determine how dead it is," he explained.[102] The answer to his question, apparently, was "very," and as the Commission sought to rework the idea, progress toward a new set of rules slowed again. In late April, the House Communications Subcommittee, in a round of routine oversight hearings, asked Burch about the status of the cable regulations and indicated concern about the slow going. Burch assured the subcommittee that the Commission was making progress and hoped to have something by the end of May. Commission staff went to work on revisions, and the basic contours of a new idea, although not the formal plan itself, were presented to Senator Pastore's Communications Subcommittee in June.

While Congress was eager for the Commission to act, it was also eager to make sure the Commission did not do so unilaterally. Both Pastore in the Senate and Congressman Torbert MacDonald (D-Massachusetts), Chairman of the House Communications Subcommittee, had warned Burch that no new rules should be finalized without review by their subcommittees. Burch promised them a comprehensive proposal for their consideration.

Burch was under pressure, not only from Congress, but at least implicitly from the new Office of Telecommunications Policy (OTP). Clay Whitehead was quickly getting air under the wings of the new OTP. Appointed to his post in June 1970, within days of the Commission's release of the doomed Public Dividend Plan, Whitehead was soon deeply enmeshed in cable TV policy. While promising not to interfere in FCC plans or procedures, the cabinet-level proceedings moved in parallel with those of the FCC's and served only to further give a sense of urgency, if not chaos, to the Commission's efforts to finalize the cable rules. Whitehead pulled together a panel of administration executives to review cable policy and construct a set of White House recommendations. The study by the OTP Cabinet Committee on Cable Communications would be several years in the drafting, however.

Meanwhile, and finally, on August 5, 1971, Burch delivered a fifty-five page letter to Congress, addressed to Pastore and the other committee chairs, outlining the FCC's thoughts on the future of cable regulation. The proposal would become the foundation of the Commission's new regulatory framework, but not without some struggle. Whitehead's office was already engaged in talks with the NCTA and NAB, along with copyright holders, in an effort to come to a negotiated settlement on copyright and importation issues. The NCTA had been talking on and off with Hollywood television and film producers and production companies ever since the initial court loss in *Fortnightly*. By June 1971, they reached an agreement. The NAB had not been part of those discussions, however, and sought help from Whitehead to bring the Association into the talks.

The outlines of the new FCC regulatory proposal, however, were more beneficial to cable than the plan on the table in the OTP offices, and the NCTA, reasonably, broke off those conversations. For its part, the NAB, displeased with the Burch plan, indicated it would not support the proposal. Into the political mess stepped Whitehead who put significant pressure on all parties to work out a compromise. He drafted a revised version of the Burch proposal that included only one significant amendment, offering greater program exclusivity protection for broadcasters. He indicated to cable operators that the alternative to compromise was nothing short of an indefinite continuation of the FCC freeze, and both the NAB and NCTA complained that they were being bullied by the administration. Both industry groups, however, were dealing with fragmented and often conflicting internal constituencies. At the NCTA, a rift was growing between the small number of ever-larger MSOs and the numerically superior small town operators. A similar schism was at work in the NAB. Under the conditions, the alternative to the compromise, a continued expensive and protracted policy battle, was very unappealing. Both sides capitulated to Whitehead. As with most good, or at least workable, compromises, the proposal made no one happy, but offered just enough, in addition to the substantial political pressure, that the so-called "Whitehead Compromise" was agreed to by November 1971.

The 1972 Rules

The new cable rules were issued on February 3, 1972. The Cable Television Report and Order in docket 18397 continued, in some ways, the theme of integrating the potential of cable television into a grand scheme of national telecommunications development. It was a 300-plus page monster of often mind-numbing complexity.[103] Where the FCC a few years before had sought to control cable to restrain its growth, the new Commission sought to control cable to exploit its potential, whether the cable operators cared to do so or not.

Under the 1972 rules, new systems were required to have at least twenty channels and two-way capacity. Systems with more than 3,500 subscribers had to offer one access channel, and systems in the top 100 markets had to provide at least three channels for public, government, and educational access as well as a fourth for commercially leased access. One of the many important themes in the Commission's March hearings was that of public access. Public interest groups from around the country, again following the wired city model, argued that local operators ought to set aside channels, on a common carrier basis, for use by civic and minority interests.

The Commission was clearly in harmony with those appeals. It declared that the "fundamental goals of a national communications structure be furthered by cable" goals, which included "the opening of new outlets for local expression, the promotion of diversity in television programming, the advancement of educational and instructional television and increased informational services of local governments."[104]

The public access rules also served to replace the origination requirement of the 1969 order. Those origination rules had been scheduled to take effect January 1, 1971, but were never implemented. Midwest Video Corp., a cable company based in Little Rock, Arkansas, won an injunction in district court blocking the order and a federal appeals court in St. Louis subsequently struck down the requirement.[105] The FCC had been granting waivers from rules, suggesting a softening on the position even before the court decision.

The Commission decided to appeal primarily to maintain its general regulatory authority over cable, but it nonetheless suspended enforcement of the local origination requirement. On appeal, the Supreme Court in 1972 reversed the appellate decision and affirmed the FCC's right to promulgate cable rules.[106] It was an important and telling case from a legal perspective. The Supreme Court supported the FCC's action, but only just barely. In concurrence, Justice Burger suggested that the FCC would have difficulty taking regulations in this area any further. Said Burger: "Candor requires acknowledgment, for me at least, that the commission's position strains the outer limits of even the open-ended and pervasive jurisdiction that has evolved by decisions of the commission and the courts."[107] Later, the public access rules would test those outer limits before the high court.

On the older issues of importation and nonduplication, the FCC's 1972 rules appeared to hold out new opportunities for operators with one hand, while taking them back with another. Under the new regulations, cable systems were granted the right to import distant signals into major markets, but with some restrictions. The number of signals the system could import depended on the size of its market (and the rules applied to any system within thirty-five miles of the given market). Systems in the top fifty markets could carry three network and three independent stations. Systems in markets fifty-one through 100 could import three network signals and two independents. Operators in the top 100 markets could additionally carry two imported "bonus" stations. Systems in markets below 100 could carry three networks signals, but only one independent. Anti-leapfrogging rules severely restricted which signals operators could import. Systems were required to carry signals from all three networks and take them from the nearest network affiliate station. Imported network affiliates had to come from the nearest town.[108] If independent signals were imported from one of the top twenty-five markets, they had to come from the market closest to the system.[109] The importing system had to give preference in some circumstances to independent UHF signals. Existing systems were given "grandfathered" immunity from most of the importation rules.

All significantly viewed local broadcast signals, of course, had to be carried. And, broadcasters were protected by extensive nonduplication and syndicated exclusivity ("syndex") rules, rules not included in the original Burch memo, but added by Whitehead. Whereas the regulation varied somewhat by market size, generally cable operators could not show a program from an imported station if a local broadcaster owned the exclusive rights in that market. Even if no such exclusivity contract existed, the carriage of certain syndicated programs, including off-network series (network reruns) and first-run syndicated programs, was severely restricted in a labyrinth of complicated formulae.

The new regulations required cable operators to file a Certificate of Compliance with the FCC before starting new operations or adding signals. They gave municipalities the right to control rates, but capped franchise fees at 3 percent. Various technical standards also were applied to ensure audio and video signal quality and limit system leakage.

The all-important sports programming rules received their own special scheme for protection.[110] As policy analyst and former OTP official Kenneth Robinson later described it, "Sports broadcasts were protected from siphoning by way of a complex 'high-water mark' rule which allowed cable systems to air only half the number of games of an event that local television stations chose to air, subject to various bizarre qualifications."[111] In short, cable operators were still locked out of any significant sports programming on a pay-basis.

Although not a formal part of the rules themselves, the compromise included a critical promise from all disputing parties to support copyright legislation that would finally obligate cable to pay for the programming it retransmitted. Implicit in that promise was an assumption that Congress, which had been working on copyright reform for years, would actually pass legislation, an assumption that proved to be overly optimistic. Cable's willingness, under duress, to accede to this element, however, helped reduce the opposition of copyright holders to the overall plan.

The 1972 rules were a messy stew of forced compromises on a wide range of detailed issues, all based on uncertain and frequently unfounded assumptions about the future of the technology and the market. Given the complicated and uncertain nature of the regulations that resulted, it is little wonder that reaction to the new regulatory scheme was equally as varied and confused. Interpretations of the 1972 rules ran the gamut from euphoria to doom. Insofar as cable had been subject to near total freeze on development, the rules were characterized by many as a "thaw" and the Commission itself presented the regulations as a liberalization. In this light, the new rules were seen as freeing cable to begin its renaissance, flowering into the communications utility of the next century. The NCTA, putting the best face it could on the rules, admitted they fell "far short of what we consider ideal," but characterized them as a step forward and effectively "an end to the freeze."[112]

Although the rules lightened the burden for some operators, the restrictions remained stringent in the largest markets. Analyst Anne Branscomb wrote at the time that under the new regulations, cable could prosper in only seventeen of the nation's largest 100 cities.[113] Other observers argued that cable was little better off under the new rules than the old,[114] and some thought they were so restrictive as to signal the end of the industry.

Abel Cable Goes to Wall Street

The FCC's seizure of authority and the restrictions of the 1966 Report and Order put the brakes on cable's expansion, slowing but not stopping it. In the industry, businessmen and women worked to adjust, some quickly, some slowly, to the shifting regulatory currents. As cable moved into the later 1960s and early 1970s it was clear to most that the age of Doerfer was over and along with it any hope of a hands-off attitude toward cable TV. A new set of competing regulatory philosophies displaced the older dueling paradigms. In the 1950s, the struggle was between an old-fashioned laissez-faire, free-market view and an activist philosophy that sought government intervention for the protection of the broadcasting system. By the end of the 1960s, the free-market philosophy was long dead. The protectionistic vision remained, but it now challenged a new concept, one in which cable technology was to be fashioned to serve the public interest, even (or in the case of the Nixon administration, especially) when it meant possible damage to the broadcaster. Despite what it might have appeared initially, it was not good news for cable. At least in the 1950s, the industry had one paradigm that left it unfettered by government regulation. Now, cable had to choose between two paradigms, both of which bound it tightly to someone else's agenda.

It was a period of confusion and regulatory flux, and cable's economic fortunes rode out each swell and dip in FCC policy, reacting in addition to trends in the country's broader financial condition. Cable continued to grow, and a reasonable business was still to be made in the industry, but the pace of development slackened. In late 1967, the NCTA reported that, whereas some 350 new cable systems went into operation in 1966, new starts had plummeted to only about sixty-five the following year, reflecting the impact of the Second Report and Order.[115] It also found a drop in the number of new subscribers, from about 900,000 in 1966 to about 650,000 in 1967. The trade press echoed the bleak analysis. *Broadcasting* reported at the beginning of 1968: "Regulation by the FCC, established early in 1966, had has a depressing effect on CATV's growth. No question about that. CATV's booming climb to new records for systems and subscribers, steadily upward for the previous 10 years, subsided to extremely low levels in 1967."[116]

This reaction to the Second Report and Order may have been shaped as much by political posturing as by actual balance sheet figures, however. By mid-1968, *Television Digest* offered healthier figures and reported, "Cable TV grew tremendously last year despite the FCC's near freeze."[117] According to their numbers, system starts reached 230 in 1967, thirty more than the previous year. The 1967 subscriber growth statistics were similar to NCTA's for 1967, about 700,000, but *Television Digest* reported only about 525,000 new subscribers in 1966 compared with NCTA's 900,000.

In retrospect, the 1966 rules did not appear to put a heavy damper on overall cable growth. The wheels of investment and construction had their own inertia. Lenders were wary of the new regulations, but the 1966 Report seemed to leave open a small window for larger market importation and, therefore, hope. Hope also was carried by the copyright talks, which, once settled, were seen by the business community as a means of helping legitimize and stabilize the industry and thereby obviate the necessity of regulating cable in the major markets.

In November 1967, *Barron's* reported Wall Street was still bullish on cable stocks, an optimism based on cable's future rather than on its then-current performance. The article noted that, although cash flow remained solid and price-to-earnings ratios of larger companies ranged from twenty-eight to sixty-five, capitalization and debt service remained heavy.[118] The financial world greeted the Supreme Court's June 1968 decision in *Fortnightly* favorably, with an initial view that the cable was now freed from the serious threat of copyright obligation. Stocks for the larger companies especially advanced sharply, and helped loosen available financing. The exuberance was short-lived, however, because, as many predicted, the decision also set the machinery in motion for regulatory forces to reestablish controls via alternative means. The FCC's decision in December 1968 to tighten cable rules again, and severely, sent industry stocks tumbling. TelePrompTer, which had been trading at around $69, dropped ten points.[119]

By the middle of 1969, cable had fallen into what one analyst called its own "personal recession."[120] Stock prices of major MSOs dropped again sharply in September of that year. Construction was off and hardware suppliers were watching planned and hope-for orders dry up. As those orders fell off, capacity began to outstrip demand, prices were cut in fierce competition and the weaker suppliers were badly hurt. The picture began to turn around the following month (October 1969) when the Commission released its new rules on cable origination and advertising. In the origination rules, Wall Street and some of the larger cable companies saw the possibilities for new programming and, therefore, new revenue. The FCC's actions reining in the telephone companies, establishing the special CARs microwave band for cable use, and authorizing use of microwave links between systems, also helped buoy industry values. Burch's efforts to create the new Cable Bureau and the inability of Congress to move on copyright were also encouraging.

The stock index of cable companies nearly doubled in the last two months of 1969 as industry fortunes rebounded. Driven in part by a planned merger, TelePrompTer and H&B stock soared, TelePrompTer more than doubled from $58.50 a share to $120.50, and H&B leapt from $15 to $25. Jerrold, which reported an employee low of 800 in 1969, was rehiring and had 1,600 workers in early 1970.[121]

The FCC's 1970 regulatory package also buoyed Wall Street's view of cable, with one financial magazine suggesting the Commission's activities portended "the beginning of a new era for the industry," adding that "long term forecasts for the industry range from euphoria to ebullience."[122] The ebullience was softened a bit by a general recession in 1970 and 1971 and, by the end of 1970, cable stocks were still off their all-time highs. In January 1971, *Broadcasting* reported that cable growth had "faltered" and "the early drive for construction, the plans for expansion have demonstrably diminished ... The once fabulous growth rate has slowed."[123] Investment institutions, however, still appeared to be betting on the industry's long-term prospects. The FCC's Public Dividend Plan, details of which began finding their way into print in May, were generally seen as favorable to cable interests. It appeared as if cable were going to be allowed into the larger markets and companies started developing strategies to secure the capital necessary to begin wiring the big cities. Those strategies included a flurry of consolidation and a movement by many companies from private to public corporate status.

The largest firms, TelePrompTer and H&B, were already publicly held, as of course were the broadcasters such as RKO, Cox, CBS, and the telephone companies. As many of the smaller, privately held companies coagulated into larger groups, they too issued stock to leverage financing. A given amount of stock equity could usually generate twice its weight in bank and institutional borrowing. In 1969, there were only nine publicly held cable companies, a figure that nearly doubled by mid-1970,[124] and, by the end of 1972, all of the top ten cable companies had gone public. Sammons was the last to do so in July 1972.

Cable's financial outlook remained stable but guarded after release in 1971 of the 1972 proposal. Wall Street recognized the need for immense capital to build in the cities and the lack of short-term gains, but hoped for a long-term upward trend. Again, however, those hopes were in part a product of the wired city fancies being spun in the public policy arena and included an assumption that the FCC was preparing to unleash cable in the major markets by supporting original programming. *Financial Analysts Journal* in March–April 1970 called cable "an attractive speculative area for investor," an evaluation based in part on confirmed historic performance, steady revenue, and good depreciation, but also on the Blue Sky vision. "Local origination coupled with advertising, leasing channels and the two-way communications highway concept, all add to the financial attractiveness of system ownership," it reported.[125] The chimera of cable's Blue Sky future, in short, drove not only political and regulatory thinking, but substantial investment dollars as well.

What optimism existed then, both among operators and the investment community, hinged on the general assumption that the freeze would be lifted "someday" and when it was, cable would resume its progressively upward march. These hopes were fed by actions in Congress and at the FCC. In Congress, the copyright bill promised cable stability and relief, with generally favorable terms for copyright fees. In the FCC, the changing personnel and the indications of first the Public Dividend Plan offered some hope for the near future.

The optimism seemed confirmed, at least for some, with the release of the 1972 rules. Many in the industry and financial markets declared that the freeze was over and years of slow growth and hard times were at an end. On Wall Street, the cable sector rose substantially, becoming at least briefly the glitter stock of the period. Caveats existed, however, as financial analysts noted that investment dollars would be limited primarily to companies of size—100,000 or more subscribers. Hopes were still contingent on resolution of the copyright problem and development of pay programming. Lending to the larger companies began to flow more freely in the summer of 1972, nonetheless, and some of that money was earmarked for origination facilities.

The Business of Origination

The 1966 NCTA convention and Ford's call for cable origination was an important historic marker, perhaps even watershed, in the industry's view of itself. Through 1967, however, the organization's programming bluster still ran contrary to FCC policy, which discouraged widespread production activity. The Commission's Blue Sky turnabout, its redefinition of cable TV, the 1968 proposal and 1969 enactment of origination requirements, lit a new and hot-burning fire under venture programmers. Entrepreneurs and established firms from a variety of sectors, including equipment manufacturers, syndication companies, film companies, and the cable industry itself, began floating ideas both to satisfy the FCC and to take advantage of the new business opportunity. What had started as political rhetoric soon began firming into proposals for origination programming and equipment.

Cable, of course, had been offering modest, nonbroadcast fare for years, most typically in the form of a cheap camera and weather dial set-up; very few operators actually engaged in local programming. By the mid-1960s, equipment vendors were offering specialty equipment for the automated channels, and some of the devices were fairly elaborate. Television Communications, Inc. (TVC) provided a service from Telemation Corp., an established television equipment vendor, called "Weather Channel 97" which boasted seven time and weather instruments. For the more budget-minded, the company offered a scaled-down "Weather Channel 75" unit, with only five instruments. AMECO had a similar set-up it dubbed the "Weather-Matic" system.

AP and UPI each offered what they characterized as a news service, but these, again, constituted little more than a camera pointed at a teletype machine. In 1965, AP joined with Telemation to offer "News Channel," a service that was soon running on American Cablevision systems around the country. UPI teamed with equipment manufacturer Viking Industries in January 1966 to develop a similar service, marketed as "Video News Service."

With the change in political climate, these modest efforts began. The day after Ford's 1966 convention speech calling for origination, a group of cable businessmen announced creation of a regional network for the distribution of cable programming. Plans called for a microwave system based in the Dallas–Fort Worth, Texas area to feed three channels of material to systems in towns serving up to 200,000 subscribers.[126] To encourage cable origination, for its 1967 Chicago convention the NCTA introduced the new CATV version of Television's Emmy Awards (themselves a copy of the film industry's Oscars). The award was named after cable's venerable Abel Cable. Competition for the first annual "Abel Award" may not have been fierce, but as a marker of industry evolution it was quite telling. The first "Abel" went to Cox Cablevision in Lakewood, Ohio, for its local news coverage. The industry called it "cablecasting."

The 1967 convention also featured the first major presence by TV camera manufacturers, with studio origination equipment exhibited by Sylvania, Packard Bell, and Ampex, among others. Companies showing cablecasting products included International Artists, Inc. of Houston, promising a daily four-hour program block of films and shorts, and International Telemeter Corp., the pay-TV company that had sponsored the Palm Springs experiment in the early 1950s. After leaving cable for more than a decade, Telemeter was trying to get back into the business, going so far as to build its own cable systems. Gulf & Western had acquired Telemeter's parent company, Paramount Pictures, in 1966 and decided that pay cable had a future.

Although origination would eventually come to mean programming developed specifically by the local cable operator, the initial definitions were not nearly so precise and the term encompassed virtually anything that was not a retransmitted broadcast signal, including nationally syndicated films. Some film distributors were accused of selling pirated video-taped prints of popular films to local cable operators, to the great annoyance of the film industry. Other legitimate entrepreneurs saw possibilities beyond feature films. A project called "Cable Television Network" (Cable-Net) was developed in 1967 by a small California MSO out of Palm Springs, Video Communications. Cable-Net was headed by Richard Lubic, who had been national director of cable TV for Time-Life Broadcasters, Inc. Cable-Net planned to provide a 2-hour block of daily programming, featuring films and shorts. Cable operators would pay Cable-Net for the service and retain some advertising time to sell locally.

By early 1969, an NCTA survey reported 329 systems, of 1,048 queried, programming live local origination material.[127] Shows ranged from local sports and news to gardening and arts programs. Some systems ran children's cartoons and feature films. For many of these systems, the origination channel was a loss leader, a marketing tool designed to draw in new subscribers. Nearly 100 of them, however, sold advertising, although none seemed to making more than a few hundred dollars a month on the enterprise; one system in Greensboro, North Carolina, was spending $5,000 a month, hoping advertising dollars would eventually make it worth the effort.

While the 1969 origination order and 1972 rule set therefore left some operators and investors rubbing their hands in expectation of new programming markets to explore and conquer, smaller operators look at the rules with suspicion and even dread. The regulations would mean the expenditure of a great deal of money and the returns were still very speculative.

The Major Markets: "Bring Money"

The FCC's decision to pound cable television into a public policy mold in service to local and national communication needs influenced the economic development of the industry in more than just a general industrial manner. The regulations had an impact on the specific shape and direction of cable deployment across the country.

Cable began its existence in small-town America, but by the mid-1960s was looking to move up. Building in the larger markets promised to be an extraordinarily costly endeavor, however, and the advent of the 1966 rules raised serious questions about near-term profitability there. As a consequence, the industry began more clearly dividing into companies of large and small size, each with different options for development.

As to the larger markets, building a franchise in Denver or Atlanta and certainly New York City was not a job for a mom-and-pop cable shop; only companies with deep pockets could reasonably be expected to take on the task. Operators needed not just the basic financial resources for construction, but personnel with expertise in law, business, engineering, and as the industry would come to know, public relations. All would be necessary to get a foothold in the larger cities and survive there.

Bill Daniels talked to his colleagues about the major markets at the 1968 NCTA convention. He warned them, "when you go big city, be prepared to spend money."[128] He estimated the underground construction necessary for many major market projects at $7,500 a mile. Daniels was a man who would know; in association with RKO, he was seeking the franchise for his hometown of Denver, and estimated the cost of the application process alone at $150,000; actual construction was estimated at $16 million. H&B, which won the franchise for El Paso, Texas, estimated construction costs at $4 million to $5 million. *Barron's* reported cabling costs in New York City at $30 a foot for underground construction.[129] The problems of cabling New York City were very special (and are treated in greater detail in Chapter 7), but the *Barron's* figure comes out to more than $158,000 a mile, high even by Manhattan standards. Other estimates put the New York figure closer to $75,000 to $80,000 a mile, and Kahn himself pegged it at $100,000 a mile.[130] Whether the figure was $160,000 a mile or merely half that, such costs were obviously far beyond the means of the small operator.

Such staggering prices meant that companies venturing into the big cities needed the ability to live in the red for years. Again, that meant substantial corporate resources. It was especially clear that with the freeze in place, any effort to move into the major markets was one based on a longer view of cable's potential. A company had to be willing to absorb the high costs of construction and operation, waiting patiently until the day the industry would be loosed in the marketplace. A major equipment manufacturer described the perspective as early as 1966:

> When the FCC came up with the top-100-market freeze, people began backing away from those markets. Now, the big boys with stamina are moving back in. GE for example, is talking about 10 years into the future. The phone company talks about 20 years. Outfits who can afford it are going to build in those cities, using Grade B signals as permitted by the FCC, prepared to wait out the melting—which is bound to come. Also, competition for franchises in those towns has eased, so those who can afford the gamble face less competition getting permits.[131]

Competition had eased, of course, because there were fewer companies with the size and will necessary to crack those markets. There were several consequences of all of this. Smaller operators were forced to turn their attention to acquiring more customers in established systems and to building in smaller and medium-sized markets not yet served by CATV. Bidding for these franchises intensified in the late 1960s, and it was there that the true expansion in subscriber numbers occurred. Toward the end of the decade, however, the franchising of markets that had either no existing broadcast facilities or a limited number of TV stations was moving toward the saturation point, further forcing increased attention to selling the service in those communities (increasing market penetration).

According to an A. C. Nielsen, cable penetration by November 1969 was heaviest in the small towns and barely existent in the largest. In the rural counties, cable penetration was at about 10 percent. It was higher in small towns, 16.8 percent. In large towns, penetration dropped to 4.8 percent and in major metropolitan areas, 1.6 percent.[132] The tiny penetration of the major metropolitan markets was significant because the vast majority of the nation's viewing homes were in those markets. The fifty largest markets account for about 70 percent of all domestic TV sets and the top 100 markets account for 90 percent. Although cable, therefore, did well in the small towns, it remained a modest business nationally with total U.S. penetration rising from 3 to about 7 percent between 1966 and 1970.

In the larger cities, meanwhile, those cable companies that could, worked to obtain franchises, although not always to begin construction. Despite the freeze, the long-term, often Blue Sky vision helped propel continued interest in larger market CATV. By 1968, franchises had been awarded in Philadelphia; Trenton and Newark, New Jersey; El Paso and San Antonio, Texas; Winston-Salem and Raleigh-Durham, North Carolina; and New York

City. By the beginning of 1971, there were sixteen applicants for franchises in Chicago and a dozen for Washington, D.C.

With revenue potential still in question, however, construction after the 1966 Report and Order slowed and, in many markets in the late 1960s, never got underway. The industry wanted to make sure it had secured the territory when it became profitable to build, but that day was still somewhere off in the future. The franchises granted in Philadelphia in the mid-1960s, for example, still had not been constructed by the end of the decade. Development of New York City continued, but growth there was not contingent on importation of distant signals and so was not in a practical way affected by the freeze.

The need for size, and the inability to attract reasonable financing for any big city project without a certain size, also contributed to the continued creation of ever-larger cable companies. Although the nature of the industry did not lend itself fully to economies of scale, some operators hoped to assemble a sufficiently large audience base to attract regional or even national advertising. Experienced management talent was especially scarce, and larger companies had an advantage in trying to secure that key limited resource. Finally, the intangible psychological factor that size bestowed was often a benefit in negotiations in business and regulation.

The Swelling MSOs

Day by day, merger by merger, what may be the communications medium of tomorrow is passing from the hands of many small operators into the control of a few large corporations ... The meek may inherit the earth, but they are relinquishing control of cable television.[133]

For a number of reasons, the period between 1966 and 1971 saw a rapid acceleration in the development of existing MSOs, along with the entry of several new corporate interests. By the end of this period, more than 1,100 cable companies were serving about 5.5 million subscribers. Following a wave of consolidations in 1971 and 1972, however, twelve companies, each with more than 100,000 customers, controlled roughly half (2.76 million) of those customers, and the fifty largest companies accounted for three fourth of all subscribers. In 1969, the top ten cable companies controlled about 28 percent of all subscribers, but by 1972 that figure had risen to 40 percent. TelePrompTer, the largest company at the time, accounted for 12 percent of all cable customers.

TelePrompTer & H&B

By 1965 H&B was the biggest multiple system operator in the country, with twenty-seven systems and more than 90,000 subscribers,[134] modest figures compared with what was to come. In early 1966 RKO General, which owned about 29 percent of H&B, once again attempted to acquire full control, but the effort floundered, and RKO sold its interest in the company.

Jack Kent Cooke's American Cablevision was the country's third largest MSO in 1966, controlling 77,000 subscribers across nineteen systems. He began selling off systems in early 1967. Some claimed he needed money to finance his growing sports empire; he told the press he wanted to refocus his CATV business in larger markets. Whatever the motivation, he found a willing buyer in H&B, which was looking to expand, and, in 1968, purchased Cooke's remaining cable property. The deal left H&B with sixty-two systems and, along with the acquisition of yet another, albeit smaller, company, General TV, Inc., it claimed some 220,000 subscribers. Cooke himself became the single largest stockholder in H&B as a result of the sale, but did not take an active role in daily company affairs.

The buying was far from over, however. Tremors rippled through the industry in August 1969 when the country's two largest cable companies, H&B and TelePrompTer, announced their intention to enter into an $80 million merger. The companies defended their marriage on the grounds that their combined resources would give them the ability to create and distribute program material not typically seen on the broadcast networks. The FCC and Justice Department found the argument persuasive and approved the merger in the summer of 1970.[135] The deal was consummated in September. H&B became a subsidiary of TelePrompTer, which now controlled more than 100 systems, 400,000 subscribers, and more than 10 percent of the market. Kahn remained president and Chairman of the Board. Cooke, now a major shareholder in the new company, sat on the Board of Directors. Running the cable operations of the merged entity was a young man named William Bresnan, who started in the industry as an engineer, installer, and even pole-climber in Mankato, Minnesota, in 1956. When the small chain of systems Bresnan worked for was purchased by Cooke, Bresnan began climbing the corporate ranks, instead of poles, until he was head of all of Cooke's cable operations.

Although Cooke himself was no longer CEO, events would not allow him to play the silent partner for long. His reemersion into company affairs began in January 1971 when a federal grand jury in New York handed down an indictment charging Irving Kahn with bribery and conspiracy in TelePrompTer's successful 1966 bid for the cable franchise in Johnstown, Pennsylvania. TelePrompTer had been operating in the city since 1961 under a nonexclusive franchise and the system had been one of the company's most profitable one. In 1965, Johnstown passed an ordinance to grant nonexclusive franchises and substantially increase franchise fees. The indictment claimed Kahn paid the Johnstown mayor, Kenneth O. Tompkins, and two city councilmen $5,000 each to secure their support for TelePrompTer's effort to retain and extend its existing franchise. Testifying for the government, the mayor claimed Kahn had bribed them. Kahn admitted paying the officials, but said he was the victim of an extortion scheme. Kahn's troubles were exacerbated in March when a New Jersey grand jury named him as an unindicted coconspirator in a similar franchise-fixing scheme in Trenton, New Jersey.

The Johnstown case had serious financial implications for TelePrompTer. Following the indictment, the American Stock Exchange briefly suspended trading in TelePrompTer stock. In March, Kahn announced plans to step down as President and Chairman of the TelePrompTer, but a dispute soon developed between Kahn and Cooke over Kahn's continuing role in the company and the voting rights to Cooke's stock. Kahn reversed course, reassumed his post as President and Chairman, Cooke resigned from the Board, and a proxy fight was on. Whatever leverage Kahn might have had in the corporate struggle evaporated October 20, 1971, when he was found guilty on three counts of bribery, one count of perjury, and one count of conspiracy.[136] He was sentenced in November to five years in federal prison. Still maintaining his innocence, an unrepentant Kahn nonetheless gave up his executive positions at TelePrompTer and a newly elected Board of Directors settled the proxy fight. Hub Schlafly was named interim president of the firm, but Cooke had effectively taken control, and soon a former Governor of Pennsylvania, Raymond Shafer, was selected to replace Kahn as CEO. Wall Street responded favorably to Kahn's resignation and TelePrompTer stock climbed, ironically increasingly the value of Kahn's personal holdings by an estimated $2 million. Kahn's efforts to appeal his conviction were unsuccessful and he entered federal prison on March 1, 1973, eventually serving just over twenty months before being released on parole.[137] Shortly before heading off to the federal penitentiary in Allenwood, Pennsylvania, he sold his TelePrompTer stock, making millions in the process.

MSO Seedlings

While nothing as colorful or headline grabbing as TelePrompTer's activities hit the rest of the industry, mergers and corporate restructuring nonetheless continued at lower levels. By the mid-1960s, Monroe "Monty" Rifkin had moved from Kahn's TelePrompTer to Daniels and Associates in Denver. There, he headed business and operations duties for a host of Daniels-managed cable systems, taking a minority ownership stake in some. In 1968, a block of systems came to Daniels from an ailing SKL. The equipment supplier had fallen on hard times as a result of its tardiness in converting from tube to solid state technology and was in need of funding. Daniels took those and a number of other systems and created a new cable company. The firm was christened American Television and Communications, or ATC, and was, on its birth, one of the country's largest MSOs. By 1969, it was the third largest cable company in the country with more than 100,000 subscribers and nearly fifty systems. Rifkin was introduced as President and CEO of the new industry giant.

Media conglomerates already in the business were active as well. By the end of 1968, General Tire and Rubber's RKO/Vumore subsidiary had changed its name to Cablecom-General; it remained one of the top five MSOs with more than 100,000 subscribers. The historic Jerrold Corp. continued under its new corporate owner, General Instruments (GI). For Jerrold, the sale had presented an opportunity to garner additional resources that would be necessary for the push into the big cities. Through the late 1960s, the Jerrold division of GI once again worked to expand into cable operations, and it remained a top-ten MSO, with more than 100,000 subscribers, and was an important force in cable affairs; its president, Robert Beisswenger, was elected NCTA chairman in 1968. In 1971, Jerrold, however, decided to abandon the systems field to concentrate on equipment. It began shopping its systems, eventually selling most to Sammon's National TransVideo Corp.

General Electric expanded into cable television, winning franchises in New York state and purchasing systems in Michigan, Mississippi, and West Virginia. Westinghouse Broadcasting and Gulf & Western, among others, bought cable properties or franchises. Blue Sky visions of facsimile-delivered newspapers drove some major publishers into cable, both as a possible extension of their business and as a form of protection from potential competitors. Times Mirror, publisher of the *Los Angeles Times*, began looking at cable in 1966. By 1970, Times Mirror had formed TM Communications as a holding company for its cable operations.

The period saw a number of situations in which small chains congealed into industry-leading corporations. Cable Pioneers Ben Conroy, Jack Crosby (also a principal in Telesystems), and Gene Schneider pooled their systems with other like-minded operators to create General Communications and Entertainment Corp., GenCoE, in 1966. Offering smaller MSOs and individual systems a combination of cash and a percentage of the new parent corporation, the holding company claimed a subscriber base of about 28,000 by late 1966. In 1967, GenCoE merged with Livingston Oil, out of Tulsa, Oklahoma, and the cable operation was renamed LVO. LVO was the eleventh largest MSO in the country in 1972, with some 100,000 subscribers.

Meanwhile, some of the original organizers of GenCoE left LVO in 1968 to begin a new cable enterprise.[138] Crosby and Conroy, joined by businessmen Bill Arnold and Robert Hughes (who had been with a venture capital firm that helped finance GenCoE), formed Communications Properties, Inc. (CPI) and began buying and building systems from Texas to New Jersey. In September 1971, the CPI owners announced a merger with Telesystems, consolidating their systems with others held by Fred Lieberman and Jack Crosby. The newly combined CPI was a top-ten MSO with more than 150,000 subscribers and seventy systems by 1972.

Leon Papernow had headed the community antenna operations division for Jerrold in the late 1950s and early 1960s. When Jerrold sold to H&B, he became Executive Vice President of the new cable firm. In 1965, he founded his own small MSO, Community Cablecasting Corp. A year later he merged his company with another small firm, United Cablevision, to form Cypress Communications. By early 1970, Harriscope Cable Corp., a broadcast company that entered cable TV in 1964, had purchased United Utilities Corp.'s thirty-nine systems for $11 million and soon after Harriscope and Cypress combined in a deal worth $23 million. The new Cypress Corp. had direct or partial control over 165,000 subscribers. This amalgamation would be absorbed by yet a larger conglomerate, Warner Communications, within a few years.

Frank Pellegrin, former President of H-R Television, helped form General Cablevision, Inc. in February of 1966. In April 1971, a group headed by H. Lee Druckman started Century Cable Communications. Druckman began in cable as a western regional manager for Jerrold, and went on to operate Trans-Video Corp., which merged with Cox Broadcasting in 1967. Cox, meanwhile, secured franchises for Atlanta, Georgia, and bought into the San Diego and Bakersfield, California, markets. In fact, Cox's San Diego system was the largest single system in the country, with more than 35,000 subscribers by the end of the decade.

A pair of young men at Harvard Business School in the mid-1960s also were looking at the CATV industry as a possible career path. Amos "Bud" Hostetter had been working in investment banking in 1962 when he became involved in a project to finance a CATV system in Keane, New Hampshire. He met Bill Daniels, who was also involved in the deal, and came to think that CATV might have a future. In 1963, he talked his Harvard classmate, Irving Grousbeck, into looking into the business with him. They spent many nights in their Boston apartment pouring over maps, and drawing radius circles thirty-five to fifty miles around cities, looking, in a manner reminiscent of Shapp's beginnings, for likely CATV markets. They found nothing promising in New England and shifted their gaze westward, finally settling on Ohio. Grousbeck flew out and began visiting small towns, looking for a likely market and settled eventually on Tiffin and Fostoria, Ohio. In the summer and fall of 1967, Hostetter and Grousbeck put together Continental Cablevision. They and their two-person staff, moved to Fostoria and set up shop. Over the next several years, they expanded, building their system and a reputation as a company serious about customer service and local community commitment. Unlike other CATV companies, Continental did not build and sell, but built and held. It stayed private while others went public. Grousbeck left Continental in 1980 to become an instructor at Harvard, later moving to Stanford University. Hostetter moved himself and corporate headquarters back to Boston. In time, Hostetter would become a respected and energetic leader in the industry, active in NCTA affairs, and Continental would become a top-ten MSO.

In 1969, Ralph Roberts' American Cable Systems changed its name to Comcast, combining "communications" and "broadcast." It was still small, even by the modest standards of cable, but it would grow substantially through the next several decades.

In central Pennsylvania, the equipment manufacturing firm of C-COR, was also expanding. C-COR got its start in the late 1940s when a group of engineers from a company called HRB (later HRB-Singer) formed a company to seek a TV license for their area. Unable to get the license, their attention turned to cable and they began wiring small towns such as State College and Bellefonte. They soon branched into equipment design and manufacture and C-COR, headed by James Palmer, became an important supplier for the industry 3; the cable operations side was run by the Center Video division. In the mid-1960s, Center Video, under Robert Tudek and Everett Mundy, began aggressively franchising central Pennsylvania and soon extended into the suburbs of Pittsburgh where the hilly terrain inhibited clear reception of the city's broadcast stations. By 1970, Center Video had acquired nearly seventy new franchises, but soon found that the capital needed to build them was more difficult to obtain than they expected. They decided to either merge the company or sell to a larger firm, one that could provide the necessary construction financing. CBS was interested and the two were in negotiations. Tudek, ironically, was in the plane over Pittsburgh showing CBS executives some of its property the day the FCC issued its ruling requiring broadcast network divestiture of all cable property. "That killed the deal," he observed, with some understatement.[139] Center Video would find another interested party soon, however.

CBS, on the other hand, had problems. The FCC's 1970 rules banning broadcast network ownership of cable systems meant divestiture for it and for NBC (ABC had never become involved in cable). NBC, which only a few years before had purchased Kingston Cablevision's three New York properties, began selling off its five small systems in 1972 and soon was out of the business. CBS maintained a sharp interest in cable, however. By early 1970, it was the nation's fourth largest cable television company, with some 150,000 subscribers, and the H&B-TelePrompTer merger put it into third place. CBS, therefore, crafted a strategy different from NBC's. The FCC, in other rulemaking, had made it necessary for the network to end its domestic program syndication business. The Commission determined that the broadcast networks exerted too much control over prime time programming and, in 1970, enacted a set of rules designed to open the programming marketplace. The Prime Time Access Rule (PTAR) limited networks from providing more than three hours of prime time programming a night, in effect opening a half hour (from 7:30 to 8:00 PM Eastern Time) to non-network providers.[140]

The other rules, the "financial interest" and "syndication" ("fin-syn") regulations, severely limited the amount of programming the networks could produce and control.[141] Broadcasters had been heavily involved in creating their own prime time material, and subsequently offering it in the lucrative syndication market. Program production under the new guidelines would have to be done largely by independent producers.

Legal action delayed formal enforcement of the "fin-syn" rules for many years, but CBS was nonetheless responsive to the orders. It took its syndication operation, combined it with its cable systems, and spun out a new company, initially called Viacom International, Inc. Although most of the stock in the company would still be held by CBS investors, it would nonetheless be a separate corporate entity in compliance with the new FCC regulations. The nascent enterprise struggled a bit getting off the ground and had to overcome FCC review and a court fight waged by dissident stockholders.[142] By 1971, Viacom was an independent new presence in cable operations.

Because of their long-term importance to the industry, the activities of three companies that had not previously been major players are particularly pertinent during this period: Warner Bros., TCI, and Time-Life, Inc.

Warner Bros.

In 1967, the venerable motion picture studio, Warner Bros., which had for some years been in decline, was purchased by Seven Arts Productions, a distributor of films to television. Two years later, Warner Bros.–Seven Arts merged with a hitherto obscure company called Kinney National Services. Kinney owned an assortment of enterprises from car rental agencies and funeral parlors to National Periodical Publications, which published *Mad* magazine and *DC Comics*. The company was headed by Steven J. Ross, an enterprising young man who began his career as an executive trainee in a Riverside, New York, funeral parlor owned by his wife's family. An idea to rent out the company's fleet of limousines in the evening led to an arrangement with a New York parking lot company for limo parking space, which in turn led to a merger of the Kinney parking lot company and Ross's rent-a-limo business. Moving from there through cleaning services and talent agencies, in 1968, Kinney bought National Periodical Publications and set its sights on the entertainment industry. After purchasing Warner in 1969, Ross began restructuring his diverse holdings, selling off nonentertainment properties, and, in 1971, investing in cable television.

In October of 1971, Kinney announced an agreement in principle to buy Television Communications Corp. (TVC) and the cable properties of Continental Telephone. The purchases, consummated in January of 1972, placed Kinney at the forefront of MSOs, with ninety systems and 175,000 subscribers in twenty-five states. TVC president and founder Alfred Stern became CEO of the new company. In February of 1972, Kinney officially adopted the name of its older film studio subsidiary, and the company became Warner Communications, Inc. Ross told the press that the company saw cable as just one more venue for exploiting its film properties. "If they won't go to the movies, let's bring the movies to them," he declared.[143]

Ross moved rapidly to expand his cable reach. In May 1972, TVC agreed to acquire Cypress Communications for $51.5 million. With both Television Communications Corp. and Cypress, Warner became the second largest cable operator in the country behind TelePrompTer.

TCI

When TCI went public in 1970 (initial offering at $30 a share), it was the country's tenth largest MSO. It owned or had an interest in fifty-one systems and 70,000 subscribers in the West. Bob and Betsy Magness controlled 31.5 percent of the company, the Glasmann family 23.8 percent, and George Hatch 10.1 percent.[144] TCI moved to the number six MSO slot in the fall of 1970 when, after the deal with CBS fell through, Center Video sold to Magness. A merger in name only, the arrangement gave TCI more than 130,000 subscribers in seventy-three systems, and a strong foothold in Pennsylvania, especially in the suburbs around Pittsburgh. The company continued its acquisitive ways, buying cable properties in the San Francisco Bay area in mid-1972, and, in 1973, purchasing the cable properties of Foote, Cone, and Belding (FCB), an advertising agency that had been financing cable development for years and had entered the business itself in the late 1960s.

Equally importantly perhaps, was TCI's Western Tele-Communications, Inc., the nation's second largest microwave system behind Bell Telephone. The network stretched over 10,000 miles and provided service to 108 cable systems, twenty-four TV stations, and other facilities in seventeen states. TCI, along with a number of other cable companies, was also seeking entry in the programming market. In April 1971, it bought National Telefilm Associates and the film library of Republic Pictures Corp. The acquisition gave TCI control of 2,000 feature length films, 1,500 TV shows, including hits such as "I Spy" and "Bonanza," and 600 cartoons.

Magness's most important acquisition of the early 1970s would not be a cable system, however, but rather a person, a very talent and driven individual by the name of John Malone. In 1972, that union was months away, however.

Time, Inc. and HBO

Time Life Broadcasting entered the CATV field in 1965, announcing its intent to build at least two systems in Michigan. The noted publisher of some of the nation's leading magazines, Time Life, Inc., bought into broadcast television in the 1950s and was now looking at cable. The company had a large library of programming and sought to augment its distribution system. The company's expressed philosophy was that it was in the information business, not the magazine or television business, and it had no prejudice which vehicle it used to get its product to market. As long as it served the long-term goal of increasing profitability, print, broadcast, and cable were all viable means to the end.

Despite this nondenominational perspective, however, the FCC's new cross-ownership restrictions meant the company had to divest either its broadcasting or its cable properties in colocated situations. Time Life chose to stick with cable. It announced in October of 1970 that it would get rid of its broadcast outlets, selling them in 1971 to McGraw-Hill publishing. As far as the remaining cable properties went, the company had secured systems or franchise rights in cities across the country, including communities in California, Indiana, New Jersey, Texas, Pennsylvania, and elsewhere. Its most historically significant investment, however, was in a company called Sterling Communications.

Sterling Communications was run by Charles "Chuck" Dolan. Born in 1926, Dolan followed the route of other cable pioneers, serving in the Army, graduating from college (John Carroll University in St. Louis), and then returning home to start a career. Home was Cleveland, Ohio, where he created a company to distribute college football highlight films to TV stations around the country. "My wife and I edited the reel each week in our kitchen," he recalled later.[145] But the business failed to take off as he had hoped. "I called up Telenews in New York and said 'Look, I have twenty-some customers. I will trade you the customers for a job.'"[146] They accepted the offer and, at age 25, Dolan moved to New York City. He spent two years as a Telenews producer in New York and, in 1954, became the vice president of the industrial film division of Sterling Manhattan News. Sterling Information Services, owned in part by CATV and TV station operator Elroy McCaw and by William Lear of the Lear Jet Corporation, produced a printed New York visitors' guide. The visitors' guide evolved into a televised version of the same material, distributed to hotels around the city using a coaxial cable network.

By the early 1960s, Sterling became interested in cable television, which seemed to be a logical extension of its existing hotel system. In fall of 1964 Sterling applied for a franchise to wire New York City, and the next year was issued an interim permit for operation in the southern end of Manhattan (the broader story of cable in New York in this period is discussed in the next chapter). Such a massive undertaking required equally massive capital, and Dolan sought investors. One became Manhattan-based, Time Life, Inc., which purchased 20 percent of the venture.

Time Life pumped large infusions of money into the company as Dolan worked to wire the city, and as it did, it acquired an increasingly larger ownership interest in Sterling. In 1967, Dolan asked McCaw and Lear to help guarantee an additional $10 million bank loan needed to continue construction; they balked, and Time Life bought them out, securing control of the company.

As is discussed in the next chapter, however, the Sterling Manhattan franchise was having a difficult time. So much so that Time Life would eventually, and unsuccessfully, attempt to unload it. Other less draconian solutions were first considered, however, in an effort to salvage the financially errant Manhattan expedition. A natural possibility was pay-TV. On a family vacation to Europe in 1971, Dolan sat on his bunk bed aboard the ocean liner, Queen Elizabeth II, and typed out a proposal. It called for the creation of a pay channel consisting of live sports events and old movies. It was initially dubbed the Green Channel and later renamed Home Box Office (HBO).[147]

Time Life was sold on the idea of the Green Channel as another electronic booth for peddling its program library, as well as for helping revive Sterling Manhattan, and it approved the concept in November of 1971. A few months later, Dolan hired a team of young executives for the nascent service. The team included an executive from TelePrompTer, John Barrington, to head up public relations,[148] and a 33-year-old attorney, Gerald Levin, who had experience in contracting for televised films and sporting events.[149]

Gerald Manuel Levin was born May 6, 1939, and raised in suburban Philadelphia. His religious father owned a thriving grocery business and Levin entered the small, well-regard Haverford College as a major in biblical studies. He had a strong interest in religion and might have gone on to become a rabbi, but instead chose a legal career, earning his law degree from at the University of Pennsylvania and joining the New York law firm of Simpson Thacher & Bartlett in 1963. Levin sought a more entrepreneurial career, however, and left the law firm after four years to join Development & Resources Corp., an international consulting and development company. There, among other projects, he supervised the construction of a major irrigation system in southern Iran. Levin developed a reputation as someone who could manage people and money. He possessed a sharp legal mind, a gift for organization, and an excellent memory for detail. He was also known as a movie and sports buff, traits that would serve him well at HBO. He returned from Iran in 1972 looking for a new opportunity, and a friend told him about Time Live and cable television. "I became fascinated with it, " he later recalled, "because the concepts had many of the same principles as the work I'd been doing on massive flood control and electrification. My fascination with that technology is some clue to my subsequent interest. There's very little difference between water, electricity and television."[150]

Dolan hired him initially to help work out the logistics and evaluate the feasibility of the Green Channel concept. "Sterling Manhattan couldn't support all the programming it needed," Levin explained later. "It needed central buying and distribution to a lot of systems to amortize the cost. That, very briefly, was the genesis for the first memorandum."[151] Levin's first title was Director of Finance, Administration, and Transmission, but he soon became Vice President and Director of Programming as well. He was a good negotiator and a strong voice for his company, helping convince Time executives on several occasions of the value of extending the HBO investment.

Home Box Office began service in November 1972 with a microwave feed to a CATV system in Wilkes Barre, Pennsylvania.[152] The inauguration was not without its rough spots, however. HBO had originally signed up John Walson's Allentown system to take the first feed, but the Philadelphia Spectrum Arena owned all the rights to NBA broadcasts within a seventy-five mile radius of Philadelphia, and that encompassed Allentown, Pennsylvania. HBO programming was slated to include Philadelphia 76ers basketball games and so another site was needed to begin service. Walson offered his system in Wilkes-Barre, but even then not everything went as planned. The day of the first transmission saw bad weather, and high winds knocked down the microwave reception dish. A technician was recruited to go to the roof and hold the wounded receiving antenna in place while HBO beamed its first pictures to Wilkes Barre.[153] With little fanfare and few initial subscribers, HBO thereby debuted on November 8, showing a hockey game from Madison Square Garden followed by the film "Sometimes a Great Notion." HBO's first original production, The Pennsylvania Polka Festival, aired from Allentown March 23, 1973.

Avoiding the pay-per-view billing approach as too costly and cumbersome, HBO pursued the strategy of Bartlesville, charging a flat monthly fee for its service, but splitting the revenue with the local operators and thereby providing incentive for local promotion of the service. HBO also helped support the local operator with marketing material and, when necessary, sales personnel. HBO slowly generated consumer interest and, by the end of 1973, it had fourteen microwave-fed affiliated systems. It looked like a good start for Dolan's brainchild, but looks, in business as elsewhere, could be deceiving.

The NCTA Split

Companies such as Time, Cox, CBS, and Westinghouse were corporate giants with national and international wingspan. They cast a long shadow over the hundreds of mom-and-pop cable companies that dotted small-town America and, by the early 1970s, the influx of corporate interest was having a pronounced impact on the character of the cable TV industry. The industry was becoming bifurcated. The cable pioneers were small-time businessmen and women, hometown engineers and entrepreneurs. The new corporate business people were specialized managers, financiers, and lawyers with substantial experience in large-scale business. With them came a fresh perspective on the cable TV business, its future and its relationship with other groups, including the federal government.

The cable industry's 1966 convention was again illustrative. The FCC's Second Report and Order was still glaringly fresh in everyone's minds, along with the lower court ruling in *United Artists* and committee passage of a House bill authorizing FCC control of cable. There was a great deal of hallway grumbling and vows to oppose legislative efforts to halter cable TV. The oppositional stance was widespread, but was especially pronounced among the smaller operators. The larger companies appeared less distraught by the new political reality. One corporate officer looked at the possibility of copyright liability and congressional legislation with an implicit, world-weary shrug: "It's going to add to the complexity and the cost of doing business," he admitted, "but I can live with it."[154]

While the pioneering independents were manning the ramparts to battle governmental interference, the corporate businessmen, many used to operating in the heavily regulated world of broadcasting, were more sanguine. Government regulation was "the cost of doing business." It helped, of course, that the larger corporations could more easily afford to deal with the costs of regulation and tap into the requisite legal expertise. The insertion of major broadcast interests into cable meant a cadre of executives with both high-level business experience and a largely successful history of working within a federally regulated environment. The old guard pioneers may have hobnobbed with senators, but the new players sat at the table with presidents. Leonard Reinsch, head of cable operations for Cox, had deep and strong ties with the Democratic party. In 1960, he was executive director of the Democratic National Convention. He handled broadcast arrangements for Nixon-Kennedy debates, and was an advisor to both Presidents Kennedy and Johnson. He was an established businessman playing at the highest levels, and issues such as copyright concerned him little. "Copyright fees are a normal cost of doing business in many fields," he explained. "As in radio and television today, they would constitute a relatively small element in the total cost of providing a service desired by the public."[155]

The growing corporate presence and increasing industry consolidation had the effect, therefore, of sowing the seeds of a cultural schism between large and small operators. The nature of industry growth was such that the largest number of subscribers were controlled by a relatively few large companies. The smaller portion of subscribers—about 25 percent—was allotted among the hundreds of remaining small operators. As a result, those small operators often felt that their interests were not being adequately addressed in NCTA interactions with the FCC or NAB. NCTA leadership, acutely aware of the problem, attempted an outreach program to the smaller companies in 1972, but it failed to significantly ease the growing rift. These divisions would eventually lead to the small operators breaking off and forming their own organization.

Problems with leadership also nagged at the cable trade organization. In 1969, Frederick Ford stepped down as NCTA president. Despite his service at the FCC and the political ties it provided, Ford did not enjoy unanimous backing among cable people, some of whom felt he was not sufficiently effective on Capitol Hill. Donald V. Taverner was named as his replacement, but proved no more successful in convincing the members of his lobbying skill, and he was forced out in June of 1971 in a low-key, but nonetheless acrimonious, internal political struggle. It took nine months of searching before the Association found a replacement in David Foster, a lawyer and corporate executive with a successful track record in common carrier regulatory politics. Foster was named the new NCTA President in April of 1972.

Social observers and theorists have long noted that successful groups or systems begin to fragment into interdependent, specialized subsystems as they grow. So it was with cable. State cable associations began forming as early as the 1950s. The 1966 NCTA convention boasted a particularly relevant landmark, given the nature of this book—it was a dinner to honor twenty-one cable industry leaders who had been in the business since the early days, a decade (or more) before. The dinner group became the nucleus of a new organization, the Cable Television Pioneers, which sponsored an annual dinner at every convention from that time forward, inducting at each event a new class of cable pioneers. In 1968, the National Cable Television Institute was formed to help promote technical education in the industry, and a year later the Society of Cable Television Engineers (SCTE) was organized. In 1970, the industry formed its first political action committee, headed by Marty Malarkey, and distributed money to the campaign coffers of four senators and two congressmen holding key positions on communications-related committees and subcommittees. In many ways, the industry was expanding, specializing, and maturing.

The Telephone Initiative

By 1967, the telephone industry was busily buying coaxial cable, lashing it to poles and renting the facilities, often at economic gunpoint, to would-be cable operators, or simply running the cable operation themselves. This worked well for those local business people seeking a turnkey entry point into the community antenna business; local newspapers frequently began a cable system in this fashion. The cable industry proper was not happy, however. Luckily, from the point of view of cable operators, the FCC at this point was actively seeking to consolidate its authority over coaxial communications and was taking a close look at cable-telephone company relations. The Commission had indicated in 1966 that it would consider requiring all telephone companies to seek section 214 certification if they wished to run cable "channel service" operations and, in June 1968, the FCC did just that.[156] Section 214 of the Communications Act provided that no common carrier could construct or extend any line without first obtaining a certificate of convenience and necessity from the Commission.

As telephone companies began filing for the section 214 certificates, however, the FCC said it started noticing a pattern: All of the proposed systems were controlled, in whole or in part, by the local telephone company itself.[157] In the midst of an emerging perception of cable as a tool for promoting democratic participation and a companion concern about monopoly control over communications systems, the specter of telephone control of cable was worrisome to the agency. Backed by the Justice Department, the Commission was inclined to oppose local market monopoly of communication service, and declared in January 1970 it would not grant section 214 certificates to telephone operators unless the local cable system was fully independent of the telephone company.[158] The telephone company, in short, could not own the only cable system in town.

In addition to banning local telephone company and cable (teleco-cable) joint ownership, the 1970 ruling sought to retard most of the onerous pole-related practices of the telephone companies. From this point forward, telephone companies that sought section 214 certificates would also have to offer pole space to cable operators on a nondiscriminatory basis and at reasonable charges; tariff restrictions that prohibited service such as local origination of information services, were also forbidden. Local phone companies were free to offer channel service, but only if they also offered pole attachment to those wishing to establish their own system.[159] Nearly all the advantages inherent in the telephone company lease-back arrangements were thereby eliminated.

The actions had their desired effect. Independent telephone companies were forced to divest their local cable businesses. AT&T, having seen the writing on the Commission's walls, announced, even before the January 1970 ruling, in October of 1969, that it would remove its restrictions on CATV and make pole and duct space available to all qualified applicants at standardized prices. By January 1971, General Telephone and Electronics (GTE) decided to abandon the cable television business entirely and sell its twenty plus systems.[160] By 1972, Bell companies had only seventy-eight, section 214-approved lease-back arrangements in operation. Pole attachment agreements among the Bell companies rose from a reported 558 in 1964 to 1,638 in 1971.[161]

The new regulations contained some very large loopholes. In the first instance, the rules restricting predatory pricing and anticompetitive practices applied only where the telephone company sought a section 214 certificate. In other words, the rules only applied if the telephone company sought to provide channel services. Telephone companies that only offered pole space to client cable operators were fully immunized from federal control.[162] In the second instance, even to the extent the rules did apply to portions of the telephone industry, they had no force with respect to other pole-providing utilities, most pointedly the power companies. The battleground, therefore, shifted, but the intensity of the conflict did not abate. Telephone companies and public utilities not looking to enter cable directly were free to negotiate market prices for pole space, and negotiate they did. Even before the FCC issued its ownership order, cable operators were lodging complaints about proposed or enacted rate hikes. Although the Commission would not take direct control in this area, it did put pressure on the colliding parties to work out a compromise. At the request of the Commission, representatives of the Bell operating companies and the NCTA began meetings on pole attachment issues in 1970. The Commission exacted an initial promise from the telephone companies that they would not raise pole rates during the discussions. For the next several years, cable and telephone businessmen would run up hundreds of thousands of dollars in legal fees while attorneys for both sides talked endlessly and fruitlessly over the problem. Talks would break off, resume, and break off. Even the FCC would eventually change its mind on the subject and, in the end, it would fall to Congress to resolve the problem.

The Urban Oil Well

It was easy in the 1950s. If you were a mom-and-pop cable shop, you went to the city council, told them what you had in mind, and without too much fuss, they gave you a Spartan legal document that said you could run wires down the street. The agreement typically indemnified the city from problems caused by the operator; it may have required liability insurance or a bond, and, of course, laid out safety requirements for construction and operation. The franchise period usually ran ten years, but could go for as long as twenty-five. As late as 1964, the FCC reported that 14 percent of the systems it looked at had no franchise agreement at all, in any formal sense, and did not see the need for one.[163]

By the late 1960s, the world of franchising had changed dramatically. The cities were larger and so were the cable companies. As the MSOs approached the more populous markets, the air was filled with Blue Sky expectations of twenty-plus channel systems, local origination, and public access. Money became an issue. Bill Daniels had warned operators to bring lots of it when they went to the big cities. As they quickly found out, not all of it would be spent on coaxial cable and amplifiers—a great deal of it was destined for the city coffers. While some municipal leaders looked at cable and saw the broadband telecommunications infrastructure of the 21st century, many, as often as not, simply saw a river of gold. In one of the more infamous quotes of the period, New York City Mayor John Lindsay proclaimed cable television to be an "urban oil well," capable of gushing millions of franchise fee dollars a year into the ever-financially strapped cities.[164]

Municipalities had levied franchise fees, in various forms, since the 1950s. By the late 1960s, those fees, hovered nationally in the 5 to 6 percent (of gross subscriber receipts) range, although they could go quite a bit higher. The FCC found some fees as a high as 36 percent.[165] Worried that fees could get out of hand, the FCC, as part of its 1972 rulemaking, recommended a range of rates of between 3 and 5 percent, although anything above 3 percent would require Commission approval.[166] Even at this modest level, cable for some municipalities was seen as an easy cash cow. Cable operators for their part did not flinch—much—at the prospect of what was essentially a special municipal business tax. The larger companies, again, considered it the price of doing business, and Irving Kahn once noted that such costs could, and would, simply be passed along to consumers.

Unfortunately, at least in some widely publicized cases, lurked certain "hidden" costs associated with obtaining the franchise, Johnstown being the most famous. Whether it was a bribe or extortion, it was clear that TelePrompTer had illegally paid $15,000 to secure the local license. The broader policy concern was that Johnstown was not an isolated incident. Four city officials in Trenton, New Jersey, were indicted for extortion in connection with franchising activities in that town (and Kahn was named as an unindicted coconspirator). Criminal conduct was also charged in Hamilton Township and in Monmouth and Morris counties, all in New Jersey.[167]

Allegations of improper financial and political ties between cable applicants and city officials were not uncommon in late 1960s and early 1970s. Securing a cable franchise for one's friends and political allies made headlines when exposed. In Tacoma, Washington, a franchise fight led to the recall of the deputy mayor and four city councilmen. According to the local newspaper, the *Tacoma News–Tribune*, one of nine applicants had improper business ties with certain city councilmen. According to the deputy mayor, the *News-Tribune*, which itself was a franchise applicant, had agreed to withhold any criticism of the city council during an election year in exchange for a franchise award. When the council decided to split the franchise between the newspaper and Tacoma Cable, the other company, the newspaper ran a front-page exposé attacking the city council and alleging the improper business ties. The council responded in kind by withdrawing the newspaper's winning bid. Said the deputy mayor, "I am here to tell the *Tribune* and its radio station and its TV station and its financial syndicate friends, that they are not going to be the only cable TV franchise in Tacoma. They have blackjacked their last public official."[168] The final accusation may, however, have been premature. Several months later, the deputy mayor and four councilmen were voted out of office in a recall election supported by the newspaper.

Similar, if somewhat less politically charged, problems arose elsewhere. In Buffalo, New York, a franchise was awarded under circumstances that drew the critical eye of the *Wall Street Journal*, and others.[169] Eight companies applied for the franchise, but, according to the *Journal*, "the democratic political organization that dominates the Buffalo Common Council" voted, without public hearing, to receive, file, and effectively kill, seven of them. Only one application was considered and hastily accepted, that of the local newspaper. TelePrompTer, one of other the applicants, reportedly did not press the issue because it felt the newspaper had an "inside track" in the bidding. Questioned about the hurried proceedings and the lack of public input, the majority leader of the Common Council stated, "We have the votes ... and we have the legislative prerogative to do what we want. If it's illegal, I'm sure we'll be notified."[170]

National headlines of the kind generated by the Buffalo fiasco and Kahn indictment were good news for neither the cable industry nor the nation's municipalities. Even when franchising seemed to go smoothly, with a minimum of political friction, concerns of a different sort were voiced by critics. Especially as the Blue Sky vision of cable began to take form in the late 1960s, some policy analysts worried that small and medium-sized towns could not marshal the expertise necessary to navigate the choppy waters of cable franchising.

Cable, in its new suit of advanced services clothes, was seen as having the potential to bring scores of channels, local origination, and local access to cities across the country. Many of those cities, worried some observers, had insufficient expertise or resources to take advantage of what cable now had to offer. State reports on cable out of New York and New Jersey, along with the influential Sloan Commission findings, all echoed this theme. Cable could now provide up to forty channels, but most franchise agreements did not stipulate a minimal channel capacity. Neither did most franchises require, or even speak to, issues of local origination or local access. The 1971 New Jersey survey revealed that forty-five of the state's sixty-six franchises contained "no provision whatsoever that channel time be made available to the community for public access."[171] Concern was voiced that cities were granting franchises that ran far too long, sometimes twenty-five years or more, and were exercising insufficient oversight over consumer rates.[172]

Concern existed as well about so-called cable "no shows." The practice of franchising a town to secure the license, but awaiting better economic times to actually build the system, was a financial strategy that may had been appealing to the cable operator. It did not, however, sit as well with the city that had issued the franchise. Failure to build a system in a timely manner was unforeseen in many simple franchise agreements, so no clauses were added to compel construction.

The cities, in short, were viewed as either inclined to corruption or as easy marks for slick cable operators. Several solutions were offered. The Mayor's Task Force report from New York City, among others, provided a model contract that set out requirements for channel capacity, local access, rates, franchise lengths, ownership, and so on. To offer expert advice and consultation for cities going through the franchising process, the Ford Foundation and the John and Mary R. Markle Foundation provided $3 million to establish a special cable information center in Washington, D.C. The Cable Television Information Center (CTIC) was created in January 1971. Its mission was to coach cities and towns through the franchising process and to provide a general resource for local officials on cable issues. CTIC acted as a public service consultant for cities across the country, educating them on the possibilities of cable technology and helping draft franchise plans that called for high-capacity systems, local access channels, and production facilities. It did its job so well that it soon became roundly hated by many cable TV operators.

The Hand-Carved Ivory Back-Scratcher

If the cities themselves were insufficiently equipped to deal with the complexities of franchising, some observers suggested looking to higher levels of government, especially the FCC and the states, to relieve cities of the burden. The Commission wanted no part in direct franchise oversight, however. As part of its 1972 rules, it established several franchising requirements for local authorities, including public hearings, construction deadlines, rate regulation, published complaint procedures, and limits on franchise fees. Beyond the newly required FCC "certificate of compliance," the agency clearly had no taste for hands-on control of the thousands of cable systems across the country. As Chairman Burch explained to the Senate in 1971, "we simply do not have the staff and resources to hold comparative hearings in each community to decide who is the best applicant and what portion of a large community he should operate in and so on."[173]

Suggestions, therefore, were that the states take a greater role in the franchising process. FCC regulations did not specifically prohibit state-level oversight and proponents argued that the states, especially through their public utilities commissions and boards, had significantly greater regulatory experience and expertise than did most small towns and cities. It was not a new argument. As noted earlier, cable operators fought fiercely in the 1950s to derail state-level regulatory efforts and were widely successful.

Cable was expanding, however, and in the latter 1960s, its public profile rising. Where state legislatures and public utilities commissions had largely ignored cable in its adolescence, they were now encouraged to examine more closely the often heavily touted "new communications medium." Response from the states varied, as might be expected, based on local political conditions. The industry usually opposed state regulation, preferring even federal oversight to a national patch quilt of inconsistent rules. In cable's seedbed state, Pennsylvania, for example, the industry had early success in promoting laws that expressly forbid nearly all local regulation. Municipalities were prohibited from restricting rates, enacting technical requirements, or requiring access channels. Their authority was limited narrowly to issuing permits for the use of public rights of way.

In an unusual and educational departure from this norm, the cable industry actively sought statewide regulation in Connecticut in the early 1960s, and lived to regret it. In fact, the poster child for failed state regulation—and for cable operators a horror story of bureaucracy, legal fees, and endless hearings—was Connecticut. It was the first state to pass public utility regulation for cable and, by 1970, was mildly infamous for being the only state in the Union that, as a result, had no operating cable system.[174] Part of the problem stemmed from the FCC freeze on cable in the top 100 markets, a restriction that encompassed nearly the entire state. Much of the difficulty lay in a seemingly never-ending hearing process necessitated by state law, and the consequent and equally endless legal challenges that followed. Ironically, cable interests had been strong supporters of state usurpation of franchising authority. Southern New England Bell Operating Company effectively controlled all the poles in Connecticut and refused CATV's access, offering instead prohibitively costly lease-back arrangements. Cable operators hoped state control would help break the telco strangle hold, and Connecticut approved legislation passing authority from the local to the state level in 1963. Connecticut, however, sought to franchise all of its towns and cities within the scope of one lengthy proceeding, and it was clearly not set up to deal with either the size or complexity of such a process. Ultimately, twenty-five cable companies applied to serve seventy of the state's 169 towns. A given company would apply for several different cities, and each city had several cable suitors. The regulations, however, required that cable operators attend every PUC hearing for every franchise bid, and also attend every hearing for every franchise that each competing cable applicant for that town bid. Hearings were only held on Mondays and Fridays, to accommodate court schedules. A Rand researcher looked closely at the history of the Connecticut proceedings, and stated: "The PUC proceedings consumed two years during which 93 sessions of hearings were held, accumulating 9,000 pages of testimony and 10,000 pages of applicant exhibits. Another year passed before the PUC reached its decision about which applicants would be certified."[175]

Then, of course, came the lawsuits. In 1967, at the conclusion of the proceedings, the PUC granted some applications and denied others, but more interestingly, awarded cable operators franchises for towns for which they had not applied. Such towns typically were not viable markets and this was a means for the state to force cable companies to service them. Operators had been warned in advance that this might happen, and the courts subsequently supported the practice. Nonetheless, it would be 1972 before cable service began rolling out in Connecticut, and the events there would haunt many in the industry for years afterward.

In the meantime, through the late 1960s and early 1970s, an increasing number of state legislatures were asked to take up the issue of cable regulation. State cable associations, in concert with the NCTA, worked to beat down such activity, but with mixed results. Laws were enacted throughout the period giving cities, and sometimes counties, the power to grant franchises and set relevant terms and conditions. In most cases, the state did not assert direct authority or set uniform guidelines. A handful, however, did give some or all of the franchising power to their state public utilities boards. Cable operators protested and argued that the federal government had preempted all aspects of cable regulation. The test case for this proposition arose in Nevada in 1968.

Two years before, in 1966, the National Association of Regulatory Utility Commissioners (NARUC) drafted a "Model State Community Antenna Television System Act."[176] The model act offered a legislative template for state utility board regulation of cable television. The model law gave the state utilities commission or board near complete authority over every important aspect of CATV operation, including system size and subscriber rates. The only state to adopt the model, however, was Nevada, in 1967. The law was challenged in court by TV Pix, Inc., a cable company that ran systems in Elko, Carson City, and Stateline, Nevada. On the central issue of whether the FCC had fully preempted state control, the court answered in the negative. The federal district judge held that Nevada could regulate cable in areas not specifically superceded by FCC regulation.[177]

In 1970, the Supreme Court affirmed the district court decision, without offering an opinion on the case, and thereby cemented state and, by implication, local authority over cable in those areas not specifically touched by federal rules. The case was considered a "green light" to state and local regulators who might previously have hesitated to assert control.

Through the late 1960s and early 1970s, dozens of bills were proposed, debated, and challenged. Many died in committee, a fair number passed and were signed into law. Most states, although not all, blithely dropped CATV in the public utilities category, either explicitly through legislative fiat or implicitly by ceding authority to utilities boards and even more commonly to local authorities. On balance, states, with cable's enthusiastic support, typically left the job of direct oversight to local officials. As had the FCC, most state legislatures and PUC boards shuddered at the prospect of attempting to supervise dozens and dozens of individual franchises across the state. A handful of states, however, did sign up for a system of uniform statewide regulation.

States that took direct control of cable through their public utilities boards included Alaska (1970), Connecticut (1963), Delaware (1974), Hawaii (1970), Nevada (1967), New Jersey (1972), Rhode Island (1969),[178] Vermont (1969–70),[179] and, briefly, Illinois (1971), which adopted then dropped state jurisdiction.[180] Massachusetts (1971), Minnesota (1973), and New York (1972) established special state-level regulatory boards for cable television, leaving the actually franchising powers in the hands of the cities and towns.

In addition to authorizing general state or local control, a wide variety of incidental laws were proposed or enacted in this period. The FCC endorsed reasonable local rate regulation by specifying in its 1972 rules that operators had to obtain permission from local authorities before raising rates. A few states, such as Nevada and Rhode Island, specifically asserted control of subscriber rates, although most left it up to local authorities. Theft of service was made a criminal misdemeanor in the states of California (1975), Maine (1973), Indiana (1973), Montana (1974), North Carolina (1973), Vermont (1972), and Washington (1973). Some state agencies were given authority to oversee technical or construction issues, largely for purposes of public safety. Some state laws set conditions for cable sales tax or stipulated terms or conditions of franchises, such as franchise length.

Beyond the applied issues of rate and service regulation, the question of control by state public utilities agencies tugged at a more philosophical question, a question additionally addressed in the *TV Pix* Case. It was, simply, but critically: Could cable TV be defined as a public utility? It was another instance in which the social definition of the technology was called, literally, to the bar. If cable could be defined as a public utility, it fit neatly into an existing mold of regulatory processes in place for decades. It was a legal and conceptual lever with which to squeeze cable into a regulatory system controlled by both state utilities commissioners and local authorities. In most instances where the state had considered the regulation of cable television, it was under the auspices of the public utility commission, and the simple act of finding cable to be a public utility, was sufficient to invoke such control. This, in fact, was part of the state's argument in *TV Pix*. Designation as a public utility, in turn, put cable on the fast track to service and rate regulation. The chief task of public utilities boards, which traditionally oversee power, telephone, and sometimes water businesses, is to assure such things as quality of (and nondiscriminatory access to) service, and reasonable rates.

Public utilities, by classic definition, however, have two essential characteristics. They tend to be, first, necessary services, such a water and power, and, second, natural economic monopolies. The questions of whether cable could be defined as a social necessity or a natural monopoly would be contentious issues before regulators and the judiciary for the next several decades, debated heatedly by scholars, politicians, and public policy groups, and the findings would vary from context to context, as will be discussed.

In *TV Pix*, the Federal District Court concluded that cable used the public streets and probably was a natural monopoly. It, therefore, could be classified as a utility affected with the public interest. The Court concluded: "As the facts appear on the record before us and the controlling precedents, there is no reason to conclude that community antenna service is not monopolistic in character and is not affected with the public interest. State supervision of it as a public utility does not conflict with the Fourteenth Amendment."[181]

It was a controversial finding, but it did fit very well with the wired nation rhetoric of the period. For many of the Blue Sky visionaries, cable as a national information utility was simply too important and too valuable to be left in the unrestrained hands of private industry. It was typical for prophesies about cable's future to be linked to a declaration that the technology and the business should be legally defined, not just as a public utility, but even as a common carrier, along the same lines as the telephone company. Smith's wired nation proposal called for cable status as a common carrier, in which cable operators would be prohibited from exercising any control over content, instead limited to leasing bandwidth on the network in the same fashion telephone operators supplied circuits to callers. The Sloan Commission argued that such an arrangement, in the near term, would suppress investment in (and continued development of) cable, but held out common carrier regulation as a possibility for the future. Within a few years, the OTP would come out with a similar plan calling for a common carrier approach to cable.

Standing starkly out from the social construction of cable as public utility whose necessity to the average citizens was second only to water and power was a more conservative view of the emerging technology. The NCTA tried to be as clear as possible on the matter. "CATV," it stated flatly in 1969 "is not a necessity,"[182] it is just a complement to broadcast TV. Cable is no more a consumer requirement than is television itself, emphasized the trade group, especially when it came to questions of public utility regulation. And in a case out of Ohio, at least one judge agreed. The question before the bench was whether a cable system could be regulated as a public utility when the operator was using a telephone company lease-back arrangement. In *Greater Freemont v. City of Freemont*, local officials had argued that cable possessed all the requisite elements that delineated a public utility, those being significant public need and a position as a natural monopoly.[183] The judge, Don J. Young, in memorable language, disagreed: "The first of these characteristics clearly is not present. The public has about as much real need for the service of a CATV system as it does for a hand-carved ivory back-scratcher. Even if in fact the CATV system is the only one in the market, it is not a monopoly in the economic sense."[184]

While much of the rest of the world was proclaiming cable TV to be the information utility of the next millennium, an indispensable tool for commerce, government and community, Judge Young was looking at the community antenna, with its still-standard fare of game shows, sit coms, and commercials, and suggesting, plainly, that the emperor had no clothes.

So, which would it be? An urban oil well or a hand-carved ivory back-scratcher; a wired nation or a white elephant? The early 1970s opened with one very definite view and closed with another. Over the next several years, cable would continue its roller coaster ride of boom and bust, both in its perceived social role and its real world success. The 1960s had been an interesting adolescence for cable, with memorable successes and failures, but it would be nothing compared with the decade that loomed head.

7
The Cable Fable (1972–1975)

> *All five of our banks were there and the Bank of New York opened the meeting by saying, "we want to be taken out of this credit."... I did something that was a spur of the moment decision. I literally had some keys in my pocket... and I just threw them in the middle of the table and watched them slide all the way down to the end. I got up and started folding my papers and we said, "Guys, that's the key to the door and if you want to go down and take over the company, you've got it."*
>
> —CABLE PIONEER ROBERT HUGHES, COMMUNICATIONS PROPERTIES, INC. (CPI)[1]

It was a story told by more than one cable executive and more than one banker, between 1973 and 1975. The cable company was in arrears; the lenders were demanding their money. The cable operator, saddled with a debt service that dwarfed cash flow, had no money to give them. Keys were dropping on conference tables across the country, as operators, with little real choice, were calling the bankers' bluff. It usually worked, as it did with CPI's Hughes and Fred Lieberman in 1975. Debt was restructured and operations continued, although in a few cases the lenders foreclosed and looked for new management. For the cable industry, the "keys on the table" story was the defining anecdote of the period, and clearly not a promising one.

In Washington, D.C., Marty Malarkey was trying to explain it to the financial community. His consulting firm, Malarkey-Taylor Associates, cofounded with pioneer cable engineer Archer Taylor, was one the industry's largest and most influential. It owed its existence to the growth and prosperity of cable television, and cable television in 1974 was in trouble. Interest rates were skyrocketing, cable stocks were tanking. Banks and institutional lenders were calling Malarkey asking what was going on, what was wrong with cable? They asked the consultant: "What do we do?" Cable operators "are about to drop their keys on our desk and we don't want to run the whole cable system. They're behind on their loan of $50 million, they're not making their payments."[2] Malarkey told them the truth. Some companies were on the verge of collapse. Cable operators, Malarkey said, were "losing their shirts."[3]

It became known as the "cable fable." The *Yale Review of Law and Social Action* was perhaps the first to use the term in its survey of the cable landscape in the spring of 1972.[4] Former TelePrompTer legal counsel Anne Branscomb asked in a 1975 article, "The Cable Fable: Will It Come True?"[5] She recounted the rise and fall of cable fortunes through the first half of the 1970s, beginning with the Blue Sky prophesies, the "glittering" 1960's social construct of cable TV. "Responsible cable operators shook their heads in amazement," she wrote, "while their more apprehensive colleagues pocketed their profits and cable stocks soared ever higher. This was the era of 'science fiction' which Ralph Baruch, president of Viacom, calls most damaging to the industry."[6]

This chapter examines the brief period from 1972 to 1975, first considering the aftermath of the regulatory activities of the late 1960s and early 1970s, the economic challenges subsequently faced by cable, and the reasons for those challenges. The second half of the chapter looks in detail at the technical innovation that pulled cable from its economic difficulties and marked the critical transition of the business from a small-town antenna service to a powerful national and then international news and entertainment utility. That innovation was the marriage of cable and the communications satellite, and the courtship is traced from its origins in the late 1950s to the launch of Satcom I and the inauguration of satellite-distributed cable service in 1975.

In the two or three years before that launch, however, people were taking a second look at cable TV. The 1960s social construction of cable television had come face to face with the local economic and cultural conditions of urban America, and the industry was in a tailspin. The new view was forged in the hot furnace of technical and fiscal reality. A failing Blue Sky vision tainted the fortunes of the industry and put a new gloss on the term Blue Sky itself, which now carried a connotation of unrealistic expectations, social pipe dreams.

The free fall began not long after the mild euphoria that followed adoption of the 1972 rules. The wide-eyed fantasies of urban planners and cautious hopes of Wall Street that cable would now blossom into a deeply embedded national communication matrix of two-way interactive programs and services was running head-long into the thick bog of business and technical reality. The problems were multiple and wide spread, and had much to do with the technical, economic, and political barriers of creating and profitably running cable systems in the nation's largest cities. Warning signs had been building through early and mid-1973 as companies pulled back on franchising and construction activity, but the alarm bells really began ringing on September 7 of that year when the Securities and Exchange Commission (SEC) announced it was suspending trade in the stock of the country's largest MSO, TelePrompTer. The suspension was attributed to an unusual announcement made a week earlier by the company that it was freezing 20 percent of its 1973 construction schedule and reviewing its entire building program. TelePrompTer President Schafer stated cryptically, in the press release, that he wanted to deny any rumors of adverse business developments at the company. The SEC was puzzled and halted trading to look into the circumstances surrounding the company's statement. The circumstances at TelePrompTer, in fact, were not good.

The company had been aggressively pursuing acquisition and system construction, at the expense of developing its subscriber base. The construction campaign meant mounting deferred debt. When the new systems started to come on line they did not generate the anticipated subscriber numbers or cash flow. The estimates had been premised on industry performance in the smaller markets and the underperforming systems were in the unfamiliar urban areas. Although the SEC freeze was initially scheduled to last a week, it was still in effect a month later, amid rumors of accounting mismanagement and even fiscal impropriety that included suggestions the company had improperly capitalized construction costs to increase reported cash flow. The New York Stock Exchange launched its own investigation.

The problems at TelePrompTer were compounded by the fact that it had only begun to recover from the damage done to its reputation by the Kahn-Johnstown incident. With his own heavy investment in TelePrompTer in sudden jeopardy, chief stockholder Jack Kent Cooke flew from his southern California home to New York in October to take personal control of the situation. He immediately installed a new, handpicked team of executive managers, and began a draconian corporate restructuring. There were wide-scale layoffs, some 900 employees were fired, regional offices were closed, and activities in program origination and data processing were suspended. In a report released in November, TelePrompTer's new management said it anticipated 1973 earnings, at best, would be one-sixth of those the previous year. It revealed that at least twenty-four cable systems turned on in 1973 were not making money; it might have to abandon planned construction of twenty-two others and forfeit the franchises.

A recovery plan negotiated with lenders called for TelePrompTer to focus on subscriber acquisition and increase its customer base from 900,000 to 960,000 by year's end, or lose its financing. Cooke managed to meet the subscriber goal through an intensive marketing blitz and, at the same time, satisfy regulators that the company was now in steady hands. The SEC eventually issued the company a slap on the wrist for filing a misleading 401-K and misleading news releases on company performance.[7] TelePrompTer did not contest the SEC allegations, stating it wanted to avoid protracted litigation and get on with the business of rebuilding the company. Trading was resumed and TelePrompTer recovered, albeit slowly. In the meantime, the heavy damage to the industry leader brought attention to the financial condition of all cable television.

The problems of TelePrompTer were extreme, but were shared, to a lesser extent, by many of the other large MSOs. The national focus, therefore, turned away from the potential of cable in the big cities and toward the difficulties. It was dawning on many that the promise of cable had been vastly oversold, and the barriers to success vastly under appreciated. Many of the Blue Sky schemes—the 40-channel, interactive switched systems carrying specialized programming, banking services, and facsimile books—were infeasible from the start, technically or financially. And few, if any, were designed to further the real interests of the industry, which were to generate wealth for owners and shareholders.

There had been warnings. Authors in the above-noted *Yale Review of Law and Social Action*, including Ralph Lee Smith and Kas Kalba, who had served as a staff member on the Sloan Commission, were openly skeptical about cable's economic prospects in the major markets. As early as 1965, economist Martin Seiden wondered aloud what cable operators and investors thought they had to sell in the larger cities. Noting that operators had applied for franchises in markets from Cleveland to Galveston, he mused: "The author is not clear as to what these CATV promoters will offer that makes them think they can gain substantial numbers of subscribers in such areas."[8] By 1975, some operators were wondering the same thing. CPI's Bob Hughes looked at the franchises they had recently won in Louisville, Hartford, and parts of Philadelphia. "As we looked at the programming available. . ., we candidly wondered how we were ever going to sell subscribers in those systems," he said later.[9]

As the FCC unveiled its new, wired city requirements for cable, some operators were quietly pointing out the real-world limitations of cable technology, manpower and money. The kind of switching equipment alone that many of the Blue Sky schemes called for either did not exist or was far too expensive to be considered realistic at the time. The skeptics and "nay-sayers" were swamped by the wired city madness, but in the end proved to be the more worldly. Several foundation stones existed upon which cable was supposed to build the new telecommunications infrastructure. They included the increased availability of imported broadcast signals, the promise of local origination, the development of pay-cable programming, and the successful wiring of the major markets. As it turned out, near-term success was not to be had in any of these areas.

The Cable Recession

Signal Importation

For years, the issue of distant signal importation was the battleground over which cable operators and broadcasters waged bloody war. Cable owners longed for the day they would be free to pull in big-city independents from anywhere in the country and offer hours of alternative programming. Broadcasters shuddered at the thought. The Commission's 1972 rules were seen by some as the first step in cable's slow crawl toward the goal of additional TV fare. It was soon clear, however, that the rules opened a door that led nowhere.

At least two serious flaws existed in the importation plan. In the first place, it was quickly discovered that the FCC's new rules did not substantially expand the possibilities for major market importation. In the second place, and even more critically, even if they had, the additional capacity would have made little difference in the fortunes of either cable or broadcast television.

Analyses showed that the way in which the FCC had crafted the importation rules was such that very few cities actually benefited from the change. If the minimum required service could be achieved using local signals, cable could import only two additional signals into the top markets, and the anti-leapfrogging rules severely limited the choice of those signals. Formally, eighty-nine of the top 100 markets would be able bring in two more independents and eleven could import three signals. The syndicated exclusivity rules further reduced importation possibilities, however, because operators could not bring in signals that duplicated certain local programming. As a result, according to Branscomb's analysis, cable operators could offer attractive additional programming in only seventeen of the nation's top 100 markets.[10]

Even if the FCC had fully thrown open the door to importation, other studies, strengthened by subsequent experience, suggested that additional broadcast TV programming was not the key to wealth in the urban markets. A Rand Corp. study released in June 1972 concluded that, although the exclusivity rules would black out about half of all programming in the largest markets, the impact on cable viewing would be minimal. Because viewers were already relatively well served by existing over-the-air independents, "even full time distant signals will not sell many subscribers."[11] Importation would be a "small but significant plus" in the smaller top-fifty markets, but overall, concluded the author, "the exclusivity provisions severely restrict distant signal carriage in markets where distant signals are not very important anyway."[12]

Television viewers in the nation's larger cities already had access to the three networks, perhaps a public broadcasting station, and one or two independents. In 1972, this pretty much accounted for the total breadth of the nation's TV options. The addition of one or two independents from outside the area might, under the exclusivity rules, mean a few more films. If a viewer was getting a reasonably clear signal with a normal antenna, there was little to justify the monthly cost of cable television. Both cable operators and broadcasters, in short, overestimated the value and impact of importation in the major markets.

Local Origination

If cable operators could not attract new customers by import existing signals from nearby communities, they needed to look to other sources of programming. With its 1969 local origination order, the FCC had presented one possibility. The Commission's intent was to provide programming in the interest of both localism and diversity, going so far as to warn cable operators to avoid mimicking the commercial fare of broadcasters.

As noted in Chapter 6, Wall Street and some of the larger cable operators greeted the origination decree with almost giddy delight. Cable was going to be able to develop its own local programming and, moreover, create a new revenue stream by selling advertising on those channels. The rule applied to only about 300 systems, but they were the largest. The term "origination" was one that, at the time, seemed to cover both local program origination and nationally produced programming distributed to local operators. Each form held out its own set of promises and problems.

Operators, such as Kahn, with a very wide and long view of the industry's future, were enthusiastic advocates of origination and had the resources and will to develop appropriate facilities. On announcing the merger with H&B in 1969, Kahn indicated his next step was to look for a programming partner. TelePrompTer purchased Filmation Associates, a producer of animated features, in June 1969, and was looking to networked distribution of its material. In fact, the FCC approved the merger in part to encourage TelePrompTer's program origination plans.[13]

Other companies sufficiently large also explored the purchase or production of network-quality fare. TCI investigated several avenues for program development. In late 1970, it began feeding Utah Stars basketball games via microwave to systems in Idaho and Montana. As noted in the previous chapter, TCI also purchased its own television syndication company, National Telefilm Associates (NTA), along with the film library of Republic Pictures Corp., folding them into the company's already existing programming subsidiary, TCI Programs, Inc.[14] The library of feature films and off-network TV programs controlled by NTA were planned, in part, to help launch a national programming service, the Cable Television Network (CTN).[15]

The promise of origination also brought forth a swarm of hopeful program suppliers and syndication companies, peddling everything from aging films to cooking shows to bingo game programs. Half of the exhibitors at the 1970 NCTA convention were program vendors. CBS-Viacom, for example, discussed using its cavernous vault of material to develop a dedicated program service for cable operators. National Telesystems Corp., partially owned by TV personality Dick Clark, sold a variety of filmed and live program packages to cable systems, including children's shows, cooking shows, and religious programs.

Megan Mullen, in a detailed study of cable programming development, noted that the film and off-network material made available to cable was not always of the highest artistic quality, "reruns and old movies were inexpensive and already known to be popular with audiences."[16] She quoted one Cox Cable executive, running an origination channel in Lewistown, Pennsylvania, on the economics of the service: "We bought a package of inexpensive movies—believe it or not—Italian-made westerns with English dubbed in. They are pretty bad, but we could afford them and 25 different local sponsors bought them."[17]

Among operators with more immediate goals and lighter wallets, the level of enthusiasm for origination programming was substantially more modest. As the 1971 deadline for compliance with the local origination rule drew near, smaller cable owners became increasingly nervous about the prospects of fulfilling the requirement. The origination rules were going to cost a great deal of money in equipment, personnel, and programming. Mom-and-pop operators were not typically in a financial position to create their own shows, at least not at a production level commensurate with existing network standards. Beyond that, the potential of advertising revenue for the local cable shows seemed problematic at best.

Equipment alone, in even a modest operation, could run into the thousands of dollars. A Palm Desert, California, operator reported putting $100,000 worth of color equipment into a new $40,000 studio, and the major cable corporations were looking at million-dollar equipment bills. H&B American ordered a quarter-million dollars of production equipment from Telemation in 1970, including videotape recorders, cameras, and tapes, in an effort to meet the FCC requirements. Production expenses, including the cost of additional personnel, also had to be considered, and they varied widely along with the quality of the production. One TVC executive reported the combined cost of origination at two of their systems at $8,000 a month.[18]

Even with reasonable financing and equipment, the quality of the production—the look of the programming—was doomed by its very nature to seem amateurish in comparison with broadcast network shows, unless the material was provided by a national supplier. Audiences had long grown accustomed to a certain level of production sophistication, and anything short of that mark was likely to receive little attention. Mullen noted the admirably volunteer, but nonetheless amateurish, nature of origination channels such as one in Honesdale, Pennsylvania, where "the local mortician was cameraman, a high school arts teacher hosted interview shows, a petroleum salesman provided commentary for sports events, and a lawyer interviewed political candidates."[19] Talent shows and high school basketball games drew a small audience and offered a local voice, but they were unlikely revenue producers.

For a variety of reasons, the potential for making any money in local advertising seemed, on closer inspection, to be worrisome. Local advertising meant the need for an advertising salesperson, a personnel cost even if that person worked on commission. Cable advertising incurred production costs when the spots were not supplied by the advertiser. Advertising rates had to be set sufficiently high to offset such costs, but in many markets, the real advertising competitor was the local radio station, which almost always reached more people and sold spots at substantially lower rates, in some cases as little as a few dollars a commercial. Local radio spots, in addition, were exceptionally inexpensive to make. It was unlikely that cable systems, with their relatively higher costs and lower penetration levels, were going to compete effectively against area radio stations or even the local newspaper.

The hope that operators could attract national advertising was simply unrealistic until the advent of system interconnection and the amalgamation of audiences of national scope. The message from country's large advertising agencies, in essence, was "when you have the numbers, we'll talk." Again, a report by Rand Corp., issued in late 1970, cast serious doubts on the financial potential of local cablecasting. The report endorsed local origination, but only as a form of pubic service, and encouraged local operators to act as rental agencies, making equipment and time available to groups to develop their own programming. As to commercial sales, the report concluded "Most cable operators have no experience in selling advertising and no knowledge of how to go about it. In fact, few CATV systems appear capable of preparing a good rate card."[20]

Drawing on the Rand report and laying on his own analysis, Seiden concluded that under the most optimistic assumptions of cost and revenue, only about 230 cable systems, of a total of some 2,700 in 1971, could be expected to break even on advertiser-supported program origination, a number that dropped to fifty-two systems when more realistic base figures were applied.[21] In all but the very largest systems, operating in the most fortunate of circumstances, local origination would be a costly venture for cable operators. In practice, very few systems made any money on local origination or even recouped their production costs. Where origination continued, it was as a loss leader, a promotional, or public relations device to appease local officials and attract new customers.

The lack of federal follow-through on the local origination requirement took some of the wind out of the origination sails. While the 1972 access rules effectively displaced them, they remained officially on the books. The Commission suspended the origination rules[22] following the Eighth Circuit Court's decision in Midwest Video I, and officially abandoned them in 1974[23] only after formal inquiries from the Cable Television Information Center. By then, local origination was described as a "long forgotten" area by the NCTA.[24]

Despite the problems and eventual evaporation of the federal requirement, the inertia of the FCC's local origination rules, coupled with early hope surrounding the service, led to the development of such channels in many of the larger markets. A 1973 NCTA survey showed about 585 systems, about 20 percent of the nation's total, engaged in some form of nonautomated local origination. About 40 percent of the originating systems had fewer than the FCC's required 3,500 subscribers and offered the channels as a public service through some combination of true altruism, public relations savvy, or local franchise requirement. Some 311 systems carried advertising on the origination channel.[25] With few exceptions, however, a 1960s-style, locally produced, local origination program channel existed largely as a public service and public relations tool. It was not the key to large-scale investment and growth.

Pay Programming

When Irving Kahn and RKO entered the cable business in the early 1960s, their primary intent was to use the technology to create a pay-television business. Similar objectives propelled Warner Communications and Time Inc. as they bought into the industry shortly after, with an eye toward exploiting their existing program base. As the failure of the economics of local origination and signal importation in the larger markets began to make themselves clear in the early 1970s, the hopes for pay-television grew.

By 1972, most operators and observers suspected that if cable was going to prosper in the big cities it would have to do so by relying on commercial quality, nonbroadcast programming, especially sports and movies, and offering that programming on some kind of subscription basis. The efforts of HBO described in the previous chapter were only one example. Several other serious attempts to create national pay programming services also evolved, along with specialized technologies that would make pay-cable possible. TelePrompTer began offering pay sports programming—New York Nets and New York Islander games—to its Long Island, New York, customers in October of 1972.[26] Other notable start-ups included Gridtronics, Optical Systems Corp., and Home Theater Network.

TVC unveiled its "Gridtronics" pay service in 1969. Gridtronics offered four channels of films, cartoons, and specialty programming. TVC executive Frank Cooper crisscrossed the country for a year trying to sell Gridtronics to skeptical cable operators. The company goal was half a million subscribers. With that pool, TVC head Alfred Stern figured the company could go to the major studios and convince them to lease films for the pay service, but the studios were not interested. Gridtronics was unable to secure material from Hollywood until TVC's acquisition by Warner in early 1972. With Warner supplying programming from its archives, the Gridtronics initiative began bicycling films to a handful of Warner cable systems in the East and in California in February 1973.

Another pay service, Optical Systems Corp. opened shop in 1972, working with Cox in the latter's San Diego, California, system. Retailing itself as "Channel 100," Optical began as a per-program service, adding a per-channel rate structure in 1974. Where HBO entered into partnership with local systems, supplying programming and splitting the revenue, Optical Systems Corp. leased channel capacity from the local cable operator to run its service, although as did HBO, it also used microwave for program distribution. By mid-1975 it was running in thirty-two markets to 50,000 subscribers in states from California to Pennsylvania.

Richard Lubic, who briefly headed Cable-Net a few years earlier, formed a pay-TV service in 1972 called Home Theatre Network (HTN). Lubic had financial backing from J. Paul Getty and Getty's film-producer son, Ronald. HTN offered two channels of films, sports, and concerts, and Lubic envisioned the potential for a nationally interconnected network for pay-cable (discussed in greater detail below).

By 1972, Telemation had also created Telemation Program Services (TPS), Inc. Its primary business involved acquiring the cable TV rights to films and packaging them for local cable operators. TPS also served the fledgling in-room hotel pay movie market. By the early 1970s, the company had contracts to provide film packages to several of the country's top MSOs, including TelePrompTer, ATC, Cox, TVC, and TCI. The latter, in fact, held an equity interest in TPS.

Other neophyte pay companies included Theatre Vision, Inc., run by former Hollywood producer Dore Schary, which worked closely with Storer, and Cinci Communications, a small Los Angeles area company created by a former Times-Mirror Cable executive.

While many of the larger operators were starting to put money and time into pay program development, the initiatives were still in their very early stages in the spring of 1973. A Hollywood seminar sponsored by the National Academy of Television Arts and Sciences in April of that year brought together most of the important figures in both the business and regulatory spheres. They included NCTA President David Foster, NCTA Chairman and TelePrompTer President Bill Bresnan, Clay Whitehead and FCC Cable Bureau Chief Sol Schildause, Harriscope's Burt Harris, ATC President Monroe Rifkin, and HBO's Gerald Levin, among others. Discussion focused on the half a dozen or so pay experiments that were underway or in the planning stages. Attendance by many of the industry's leading figures testified to the importance placed on the pay services, but it was also apparent that pay programming, at this point, was still very much in the experimental phase, with much of the discussion focusing on tests that companies planned to conduct during the coming summer.

Many of the services built their business plans around a pay-per-program scheme, instead of a monthly charge. Although potentially lucrative, the pay-per-program approach required a more sophisticated technology. Optical Systems used a set-top box that read plastic credit cards and Theatre Vision used cardboard "tickets" in its home terminal. The push for pay generated several technical innovations, most of them involving some form of limited interactive capacity, typically with a home terminal or decoder. One of the earliest was a system designed by Theta-Comm, a joint venture of TelePrompTer and Hughes Corp. As with several similar technologies, its two-way Subscriber Response System (SRS) could be used for services beyond pay programming,[27] such as burglar and fire alarms, polling, and, it was claimed, electronic shopping. The SRS computer-controlled system was demonstrated to rave reviews at the 1972 NCTA convention, and a dozen similar interactive CATV experiments, using a variety of technologies, were planned or underway by the end of that year.

The experiments were hampered by technical difficulties and by the lack of a stable supply of programming, however. Efforts to build cheap and reliable decoder boxes for the pay services illustrated evolutionary diversity in technical development, as a host of techniques were tested, but the technology was slow to come. The interactive boxes were constantly plagued by troubles, and engineers seemed to be forever "trying to work out the bugs." As noted, the availability of attractive feature films was limited by both FCC regulation and by a hesitant film industry.

By May 1974, some form of pay-cable was available in only about forty-five systems around the country serving 67,000 subscribers, often on a limited or experimental basis.[28] HBO itself was struggling to get its business off the ground, reporting about 4,000 subscribers on three Pennsylvania systems in early 1973 and expanding to a modest fourteen systems by year's end. If pay-television was ever going to succeed as a driving force in cable growth, it would have to deal with several large problems, including a steady supply of quality programming, appropriate technology, FCC regulation, and—especially if it were ever going to move into carriage of national sports—the problem of interconnection.

Networking

One of the principal obstacles to making origination or pay programming of any kind work was the lack of national networking. Even if cable operators had been able to obtain a wide diversity of first-run films or off-network shows at a reasonable price, effective distribution stymied growth. Without regional or national interconnection, exploitation of live events such as major league sports was a near impossibility.

The promise and problems of networking were well known to cablecasters, broadcasters, and the FCC. As noted, national interconnection of cable systems had been a pipe dream of cable operators and a source of night sweats for broadcasters since the earliest days of the industry. By 1969, even the FCC acknowledged the potential of regional and nationally interconnected cable systems to develop programming that could compete with the major broadcast networks.[29] An FCC subcommittee in December 1972 went so far as to recommend the Commission require all systems be able to interconnect as a means of encouraging development of national cable networked channels.

Despite the Commission's public service goals, of course, the cable industry's interest in interconnection had little to do with the advancement of social welfare. It had everything to do with the advancement of the economic well-being of the cable system operators. As NCTA's Frederick Ford and others understood, cable's future lay in interconnection. Only by spreading production costs across a national audience could sufficient revenue be generated to make alternative programming possible. The practical methods of national distribution were limited, however. To the extent that any nationwide distribution system existed at the time, it consisted largely in the physical transport of films and videotapes, usually by mail or messenger service, a process known as "bicycling." The only feasible alternative was cable's off-stage nemesis, AT&T. The telephone company owned and operated the nation's only true network of linked telecommunications facilities. Leasing long lines and microwave services from Bell was unappealingly expensive, however. In the mid-1960s, the three broadcast networks were paying AT&T about $50 million a year in networking fees, a figure that doubled by the early 1970s.[30] The broadcasters were also tying together fewer broadcast outlets and using fewer channels than was envisioned in some of the cable proposals.

The other option involved the expansion of cable's neophyte microwave systems. As with most of the industry's evolution, within its political and economic context, the concept of microwave networking was both logical and incremental. The very first substantive cable microwave plan, that of Belknap and Associates in the early 1950s, envisioned eventually expanding the microwave network to serve a dozen communities in three states. By the late 1950s, cable operators were using microwave extensively, with concentrations of activity in the Northwest, Southwest, and Northeast. In these areas and elsewhere, small constellations of CATV systems were served by regional microwave distribution links. It was obvious to many observers that these mini networks could expand, perhaps even to the point where two or more regional systems might merge. It took no great leap of imagination to see a national system slowly accreting in the process. Imagination was cheap, however, and the technology was not.

Despite the potential of regional microwave systems, very high barriers to interconnection existed on a national scale. Building an independent system would require FCC approval for each microwave link, approval that would be opposed at every step by a host of well-known interests, and it would be very expensive. A profound testimony to the necessity and logic of the concept, therefore, was that efforts to create such a system began about this time. One of the most ambitious such plans came from HTN's Richard Lubic in 1972 and 1973. Lubic had reached carriage agreements with a number of major operators, including Sammons, Cablecom-General, and CPI, but his key contribution to the advancement of the networked, pay-program agenda was an alliance with TCI to create the nation's first cross-country cable network facility. TCI, with the largest microwave system outside of AT&T, had been promoting cable networking already and was clearly interested in the idea.[31] Even before joining in Lubic's plan, TCI had trumpeted the potential of cable networking to the rest of the industry, taking out an advertisement in the 1971 *CATV Systems Directory* proclaiming: "Interconnected network cablecasting on a regular basis—It's sooner than you think."[32]

TCI's Western Tele-Communications, Inc. was well suited to the task. It boasted 11,000 miles of broadband transmission capacity. The company was known for its distribution of live sports events in the Rocky Mountain region, especially major league basketball out of Denver and Salt Lake City. To assist with programming, TCI could also draw on its NTA film library. In late 1972, HTN and TCI filed applications for the first stage of the system, seeking FCC permits for a Los Angeles to Denver set of microwave links. The microwave route would run through Arizona, New Mexico, and Texas before turning north to TCI headquarters in Denver. The request for construction permits naturally drew quick opposition from several broadcasters and from the National Association of Theater Owners, but the opponents had little to fear at this juncture; cable's general economic downturn carried the HTN-TCI plans with it. TCI, as detailed below, was hard hit by the cable recession and, as did TelePrompTer, had to abandon a number of initiatives, including this one.

Despite the cable recession, given the obvious industry need, likely some kind of terrestrial networking system would have developed over time, if only in piecemeal fashion, even without the backing of a TCI or a TelePrompTer. How long it might have taken to deploy such a platform is anyone's guess, but most likely three to five years at a minimum. It is fair to suggest, however, that such a system ultimately would have been built, opening the door to the next stage in the industry's evolution. Fortunately for cable, another alternative was in gestation. In 1973, however, that system was still at least two years away and not seriously anticipated by more than a few visionaries. The period 1973 through 1975 constituted tough times for cable and nowhere was it tougher than in the cities.

The Problem of the Cities

In the late 1960s, the big cities were seen as cable's new frontier, even in the face of the FCC's regulatory freeze. The 1972 rules, cast by some as thaw in the Commission's frigid philosophy, only encouraged the dreaming and planning. In July 1972, the Commission issued twenty-four certificates of compliance to operators seeking to build in the major markets. Daniels had been more right than he realized when he told his colleagues to bring money. They would need more than they ever dreamed. Some of the issues associated with obtaining franchises in the medium and large markets were discussed in the previous chapter. Once those franchises were secured, operators faced a new set of challenges in system construction and operation.

While the major markets were the promise and the future for an industry with designs to be more than just a local antenna service, in practice, they were also "minefields" of technical expense, political dogfights, and as often as not, litigation. Even without advertising-supported origination or pay programming, cable might have been able to establish a business in the metropolitan areas by providing a simple service that supplied clear color pictures plus one or two imported stations. This was possible, however, only if the economics of large-market construction and operation were similar to the industry's experience in smaller towns. They were not. The cost of everything mushroomed in the big cities, and the spiraling service demands of local municipalities, in many cases, made the hope of a simple and profitable cable system an impossibility.

The headaches only began with construction costs. In the late 1960s, the cost of laying a mile of cable in a small or medium market ran anywhere from $3,000 to $10,000. In the nation's largest markets, where underground cable was often a municipal requirement, that figure jumped to $50,000 per mile; in New York City, it cost $80,000 per mile. In 1971, Jerrold estimated that $800 million would be needed to fund major market construction between 1972 and 1976. Other estimates pinned the figure at anywhere from $5 billion to $10 billion over a ten-year period for construction and programming.[33] A report issued by Solomon Brothers brokerage firm in early 1972 found major market capital requirements to be 60 to 70 percent higher than smaller markets, and operating expenses to be 30 to 40 percent higher. "Even without adjusting for potentially lower demand," concluded the report, the result "is an unprofitable system."[34]

As described in the previous chapter, operators were also restricted by locally-controlled subscriber rates and mandate services. Until enactment of the 1972 FCC rules, franchise fees were uncontrolled and, as noted, cable looked like an open checking account to some municipalities. As one gloomy operator observed at the 1973 NCTA convention, "A year ago, if you had gotten a franchise for Chicago, your stock would have gone up 10 points. Today it might go down 20."[35]

The troubles in New York City were perhaps the most extreme and, in some cases, special to the Big Apple itself, but Manhattan nonetheless was a glaring illustration of the industry's big city blues.[36] In 1964, six companies had filed applications to build and operate cable systems in New York. They included Sterling, TelePrompTer, RKO General, and, harkening back to the 1940s, a consortium of six apartment master-antenna companies. A local firm called CATV Enterprises, created by attorney and TV producer Theodore Granik, also sought permission to build. By the end of 1965, Sterling, TelePrompTer, and CATV Enterprises had been awarded interim franchises and begun construction.[37] In the subsequent five years, the city took a closer look at cable television, an examination that included preparation and submission of the previously noted New York City Task Force, commissioned by Mayor Lindsay, and issued temporary, short-term franchise extensions. The proceedings culminated in 1970 with New York City awarding the two primary companies twenty-year franchises. TelePrompTer was given the north end of Manhattan, above 86th Street on the East Side and 79th Street on the West Side; Sterling got the southern end of the island. CATV Enterprises was given the smaller, but upscale, Riverdale section of the Bronx (Westinghouse bought 49 percent of the company soon after the franchise was issued). Other companies, including unfranchised Comtel, Inc., developed service areas in New York,[38] but the attention was focused on the two main Manhattan providers.

The difficulties for the cable operators began before the first foot of cable was installed. In addition to establishing limits on subscriber fees and criteria for picture quality, the city of New York set up timetables for construction, but construction was almost absurdly difficult and expensive. Coaxial cable had to be run through the city's underground duct system, which was seventy-five years old. The ducts, built in conjunction with the subway system in the 1890s, were owned by Empire City Subway Company, which in turn was fully owned by the New York Telephone Co., an AT&T subsidiary. AT&T, of course, was not enthusiastic about cable; from the start, the cable companies found it somewhere between difficult and impossible to get timely permission from Empire for access to the underground conduits. For the first two years (1964–1966), Empire City Subway flatly refused to allow any CATV operator into the ducts. Eventually, the city had to intervene and force Empire to issue construction clearances. Even once crews were permitted into the tunnels, installation problems abounded in the city's aging, often ancient, utility vaults. Sometimes the steam pipes that shared tunnel space would get so hot that they would literally melt part of the cable. New York City streets were in a state of constant construction and repair. Cable crews were repeatedly forced to pull out and reroute already installed cable or fix cable sliced in two by city backhoes. A normally simple thing, such as parking for construction and service crews, was impossible on the clogged Manhattan streets and the companies were plagued by parking tickets. TelePrompTer estimated it could install about a half mile of trunk line a week, at a cost of about $125,000 per mile, not counting the parking tickets.

The quest for a solution to the construction problem did result in one important technological advance for the industry. TelePrompTer thought a short-hop microwave system might allow them to skip some of the expensive cable trunk construction by beaming the TV signals over congested streets to mini-hubs located around town. The microwave companies they approached responded with proposals that were too costly or required too much development time, however. Hub Schlafly was explaining all this to an old college classmate at a Notre Dame class reunion when the friend, who happened to work at Hughes Aircraft, told him they were experimenting with a system that might help. As part of its work in early communications satellite technology, Hughes had designed a microwave protocol that could be harnessed to distribute twelve channels of TV programming using one transmitter dish. Schlafly saw the potential for New York and, in April 1965, TelePrompTer called on Hughes.

The result was Amplitude Modulated Link (AML) service, a microwave technology, originally in the 18-GHz band, capable of carrying multiple TV signals over short distances. The FCC approved experimental use and, by 1966, Schlafly was directing tests on the AML system in Manhattan. TelePrompTer and Hughes formed a new company, Theta Corp., to develop the technology. As a result of the partnership, Hughes also invested $20 million in TelePrompTer, acquiring a 17 percent share of the cable MSO and seats on the company's board of directors.

In the spring of 1968, the FCC granted authority to begin using the AML technology on an experimental commercial basis and TelePrompTer was quickly employing it to bypass the expensive underground labyrinth of Manhattan.[39] The FCC formally approved use of the AML service in 1969, moving it to the CARS 12.7- to 12.95-GHz band, and setting aside the 18-GHz range for use by satellites.[40] The AML system, labeled Local Distribution Service (LDS) by the FCC, saved the TelePrompTer a great deal of money. It also proved attractive to cable operators throughout the country who sought the benefit of an inexpensive broadband broadcast relay service.[41]

Despite the eventual success of AML, however, Manhattan seemed to be a place where cable companies, along with tourists, were easily mugged. By the end of 1973, Time had sunk $45 million into its Manhattan operation, TelePrompTer $30 million. Both were burning money at a furious clip, and doing so on problems that went beyond normal construction and operation. The city, for example, required all cable customer service to be done free of charge. Taking advantage of this, some subscribers would call the cable company even when a problem was in their set rather than in the cable, hoping that the technician would fix the problem anyway and save them the cost of a TV repair bill.

Wiring apartment buildings was especially taxing. Permission to enter a building had to be obtained from the building owner and, in the case of absentee landlords, this was often a difficult and time-consuming process. Some landlords simply refused to allow cable into the building because, in rent-controlled New York, it was in the landlord's best interest to make the apartment as unappealing as possible. Only when existing tenants moved out could rents be raised. More typical, and unsurprising, was the landlord who was delighted to bring in cable as long as he or she got a piece of the action—normally a percentage of the monthly gross on the building subscriber fees, plus, of course, free cable service. In fact, cable operators managed to push a law through the New York legislature requiring apartment building owners to allow cable operators to wire the building and setting compensation at a token $1.00, although the law was later overturned by the courts.[42]

Finally, there was the issue of cable theft. Once the wire was in the building, it was no great trick for someone of modest technical skill to illegally tap into the line, which in the case of a clumsy effort would kill or distort the signal to the rest of the building. In addition, new tenants, finding a pre-existing cable outlet still in place from the previous occupant, would often just plug in their set and not bother to pay a bill. Correcting such problems was a chore because service at that time usually had to be turned on or off inside the apartment. Technicians appearing at the apartment of someone not paying their cable bill would often simply have the door slammed in their faces. As cable operators in the big cities soon discovered, they were no longer running wires to their friends and neighbors down the street in Happy Valley, USA.

A significant portion of TelePrompTer's sudden collapse was placed on the doorstep of its Manhattan venture. Manhattan had been equally as un-kind to Time. The company publicly reported losing $10 million on its sys-tem in fiscal 1972–1973 and did everything it could to shed the financial albatross. "We tried desperately to unload Sterling Manhattan," recalled J. Richard Munro, the President of Time Inc. "Sterling Cable was a huge drain on us. It was kind of a disaster."[43]

For a brief period, Time had Warner Cable interested in buying the prop-erty. Steve Ross acknowledged the system's problems, but thought the long-term potential might justify acquisition, then he took a closer look. Explained Munro, "Warner came and took a look and saw what a dog it was [and the deal fell through]." Time sold most of its other cable properties to Rifkin's ATC in 1973, but was stuck with Sterling Manhattan. "There was nothing we could do," said Munro, "except try to dig ourselves out of the hole—to bring in new management and try to make Manhattan work."[44]

In fact, even before Time announced its intent to sell off the Manhattan system, Chuck Dolan had left the company, departing in March 1973. A veteran of the broadcast business, Richard Galkin, replaced Dolan as head of Sterling Manhattan Cable, and Gerald Levin was moved up to head HBO. It would be a few years before HBO got off the ground, and more than one observer inside and outside the company expressed substantial skepticism about the prospects for the nascent enterprise. The same, in fact, could have been said about cable television writ broadly in 1973 and 1974. The crash of cable's fortunes in New York City reverberated back in small echoes from across the country.

The Fallout

In 1972, the National Cable Television Association laid to rest its cartoon mascot of some fifteen years. Abel Cable, they concluded, was an icon for the 1950s, not a symbol for the future. Perhaps they should have kept him around. At least he was smiling. By late 1973, the cable industry was taking on water fast. Some MSOs started bailing, others simply abandoned ship. The problems brought on by the failure of importation, origination, pay, and the melt down in the cities were compounded by a broad national recession through 1974 and 1975.

The Pullback

For about a year after release of the 1972 rules, cable looked like a good risk in the eyes of the financial community. The social construction of the new regulations, for many, was that cable now had an opportunity to exploit the larger markets, the companies only needed the construction capital. They needed a lot of it, as noted above, but some lenders and investors took the gamble. Travelers Insurance Co. and Aetna Life led a coalition of thirteen insurance companies in a $23.5 million loan agreement with ATC. Chase Manhattan Bank, Morgan Guaranty Bank, Allstate Insurance, and John Hancock were also lending money to the industry. Wall Street was taking its share as well. TelePrompTer raised $66 million via common share offerings in 1972; Sammons put up a million shares at $22 a share. Companies dramatically increased their debt and sold stock at prices premised on the expectation of a new surge in building and profitability. As with most balloons, it burst.

By mid-1973 the money market was getting tight again. An unfolding Presidential scandal, Watergate, had shaken the economy as it had shaken the political sphere. The Organization of Petroleum Exporting Companies (OPEC) was raising oil prices and the dollar was sinking on international markets, as was the stock market at home. Inflation and interests rates were on the rise. Obtaining financing at reasonable rates was difficult for even promising industries, and cable was starting to look less promising. Operators at the 1973 NCTA convention heard a representative of the First National Bank of Chicago tell them his firm was raising interest rates the next day, from 7.5 to 7.75 percent, and he expected them to go to 8 percent or more by year's end.[45] He was being optimistic. In late 1972, the prime rate stood at about 6 percent. It doubled by July 1974.

One of the major consequences of all of this for cable was a sudden and explicable disappearance of investor dollars. Savvy financiers, with their attention caught by the TelePrompTer debacle, were soon made additionally nervous by the problems in the cities, general costs of construction, and regulatory noises from the federal government. Not-so-casual proposals from public interest groups and some regulators (discussed below) about the possibility of common carrier regulation and national rate control for cable, only made investors more skittish, as did the growing interest of states in regulating the emerging media.

The stock market took heed. Following the SEC's move against TelePrompTer, Wall Street quivered and the value of most of the industry's firms plunged. In 1973, ATC dropped from year high of $39 a share to a low of $7-1/4; TCI went from $21 to $2-5/8; Cox from $31-3/4 to $7-3/4; Viacom from $20 to $4-5/8, and TelePrompTer itself dropped from about $34 to $3 a share.[46] In fact, TelePrompTer had traded at a high near $120 a share in January 1970. Stocks would begin to recover in 1975, but 1973 and 1974 were the worst stock years in cable's then-short history.

The growing difficulties rippled out into the behavior of the nation's leading companies. The worsening economic conditions, bolstered by a government assault on consolidation, torpedoed a variety of attempts at system acquisition or consolidation. In 1972, President Nixon, some suggest as part of his campaign against the media, had the Justice Department file antitrust suits against all three networks in an effort to force them into compliance with the FCC's 1970 syndication and financial interest rules, the rules governing network ownership of programming. Questions about consolidation of media ownership found their way into cable TV policy. Although the FCC would soon begin showing a general predisposition toward softer overall regulation of cable, it nonetheless began taking a stronger position on ownership issues, in part at the prompting of the administration and the Justice department. In December 1972, an FCC-appointed advisory committee issued a report calling for vigorous action to check what it saw as the swelling tide of cross ownership in telecommunications. The Justice Department became active in arresting the rapid consolidation of the industry. It filed suit in early 1973 to block the proposed merger of Cox and ATC, forcing the companies to abandon their plans.[47] The government argued that, ironically, the merger would have had detrimentally reduced the number of cable companies capable of competing for franchises in the major markets. After being rebuffed by the Justice Department, Cox set its sites on a merger with Livingston Oil's (LVO) cable operation, only to terminate that effort in December of 1973 because of what were termed unfavorable market conditions. Those same market conditions scuttled several other planned cable mergers and acquisitions.

Companies that only a few months before had been captivated by the promise of cable now saw only rising interest rates, construction costs, and debt. Some operators survived off the cash flow from their profitable systems in smaller markets. Some told creditors they simply could not afford to pay them right now, and as did CPI in the anecdote that began this chapter, figuratively or literally threw the keys to the company on the table. Some cable companies simply resolved to get out while they could. Foote, Cone and Belding sold to TCI. Time Inc. which had only a year previously sold its TV stations, announced in May 1973 that it was leaving cable system operation entirely, selling most of its properties, as noted above, to ATC. The massive losses on the Sterling Manhattan project were a principal factor, but so was the lack of any real prospect for a reversal of fortune in the near future. Gulf & Western had organized a cable subsidiary in 1968, Athena Cablevision, and had sixteen systems by 1973. In August of that year, as did several others, it determined that cable was not the way of the future, and reached agreement to sell the property to a still-acquisitive TCI.

Livingston Oil left the cable business, as Gene Schneider took the systems out of the oil company, going public and changing the firm's name from LVO to United Cable Television. United Artists Cablevision, a top-twenty MSO, also decided the time was right to leave. It sold out in late 1972 to Columbia Cable, a company cofounded in 1962 by Robert Rosencrans. Rosencrans began in the TV business in 1953 when he helped form Box Office TV, a closed-circuit theater TV business. The service became Sheraton Closed Circuit Television, owned by Sheraton Corp., which TelePrompTer purchased in 1956. After working under Irving Kahn for several years, Rosencrans gathered nine partners and purchased a cable system in the Columbia Valley of Washington (Pasco-Kennewick), naming the new company after the river. With the merger, UA-Columbia Cablevision, Inc. had about thirty systems and 137,000 subscribers. It would be a top-fifteen MSO through the 1970s; more importantly, however, Rosencrans and his company would go on to make major contributions to the cable industry's development.

Those operators who, by choice or necessity, decided to stick with the business, determined it was time to take a harder and more realistic economic stance, calling, in some cases, for an abandonment of the Blue Sky hyperbole and a return to franchising on the basis of plain vanilla CATV service. Warner Cable Chairman Stern called it a "return to reality."[48] The longer and more hopeful view still argued for acquiring franchises at a high cost and sustaining significant operating losses until sometime in the future when the various obstacles could be overcome. Not even some of the largest companies, however, had the resources or the investor patience to await those prospective happier times. Warner Cable was a case in point. In late 1973, the company essentially declared it had had enough. It had already been designated as the provider of choice by the Birmingham, Alabama, City Council, but the franchise agreement called for many of the Blue Skies attachments typical of the period. The city had requested, and Warner had initially agreed to, such things as a fully interactive system and a minority training program. With the general cable market tumbling out of control, Warner sent a revised franchise agreement to Birmingham. It stipulated a substantial increase in the previously agreed-on subscriber rate, paired with a reduction in the annual fee the company would have to pay the city, from 5 percent to 2 percent of gross receipts. The proposed two-way system was abandoned, as was the minority training program and Warner's obligation to clear its construction plans with the city. The revised franchise, said Warner, was non-negotiable. Birmingham did not negotiate, it moved on to a more willing provider, ATC.

General Electric also took a second look at its franchising activities, withdrawing applications in six cities and declaring it would seek no new sites. It further announced that it would conclude in-progress franchise negotiations with Grand Rapids, Michigan, only if the city made "no unreasonable demands" in the twelfth hour.[49] Storer explained that it did not consider big city franchises economically viable in the near term and had no interest in doing business there. Cities, which a few years earlier had many cable suitors knocking on their doors, now found few and, in some cases, none. Archie Twitchell, City Manager of Boulder, Colorado, sent out letters in 1974 soliciting franchise bids from cable companies. He received this reply from High-Fidelity Cable TV in Great Barrington, Massachusetts: "The blue-sky days of wild franchise bidding are over, Mr. Twitchell. There was a day, perhaps, when cable TV companies driven by the same economic death wish that swept the industry a few years ago, might have vied for the privilege of bankrupting themselves in Boulder. But no more; the industry is growing up.[50]

In existing franchises, rate increases seemed to accompany the industry's growing pains. Many of the larger MSOs raised rates by 30 to 40 percent between late 1973 and early 1975. TelePrompTer, in addition to its draconian corporate cost-cutting, raised rates at many of its systems. Warner hiked average monthly prices from $5.30 per subscriber in 1973 to $7.25 by 1975.[51] Even the NCTA was feeling pinched. Attendance at the 1974 convention was down 10 percent; the Association cut its fiscal 1975 budget by $175,000 and laid off four staff members.

TCI Welcomes John Malone

Deep in the center of the cable slump was Bob Magness. In the late 1960s and in the opening years of the 1970s, Magness had gone on a buying binge. Riding the crest of Blue Sky fever, TCI, as noted earlier, purchased Center Video, FCB, Athena, and cable systems owned by Rust-Craft Broadcasting. It won franchises in medium and major markets around the country and had sites in the San Francisco Bay area, suburban Pittsburgh, and Memphis, Tennessee. It had acquired NTA, was looking to expand its microwave company and, as discussed below, was proposing to launch its own communications satellite. To accomplish all this, Magness had taken on massive debt, $130 million-plus by 1972, and was in need of much more to build the franchises and continue the program of expansion. TCI secured an additional $78 million in high-interest, short-term loans, with a goal of paying off the debt with a $100 million stock offering. Then the cable market crashed. With the TelePrompTer fiasco as a backdrop, the SEC suspended TCI's planned stock offering over concerns about the financial health of its NTA subsidiary. TCI was now precariously close to bankruptcy, and a new company president, brought in by Magness only few months before, was wondering what he had gotten himself into.

Shortly before this, in the fall of 1972, Magness, worried about the complexity of the growing TCI empire and its mounting financial woes, decided he needed help running the company. According to author and *Wall Street Journal* reporter Mark Robichaux: "[A]fter he had spent hours toting up TCI's financial position, Magness finally and fully grasped just how terribly precarious the whole situation had become. He skimmed the numbers, looked up at Betsy, and blurted out, "I'm gonna hire the smartest sonofabitch I can find."[52]

He did, and it was arguably the sharpest business decision Magness ever made. John Charles Custer Malone joined TCI on April Fools Day, 1973, destined to become one of the industry's most respected, feared, sometimes hated, and unquestionably powerful and successful leaders. Malone, the son of a General Electric Co. vice-president, was raised in a quiet upper-middle class manner in Milford, Connecticut. From the start, he was bright and ambitious with a penchant for mathematics and technology. He graduated from Yale University in 1963 with an electrical engineering degree and went to work for Bell Laboratories. Funded by AT&T, he continued into graduate school, earning an MA in industrial management and a PhD in operations research, both at Johns Hopkins University. Although Malone could easily have gone on in engineering, either at AT&T or in higher education, his real interest was in business. Having served his time at Bell and seeing little potential for making a mark in the historically rigid Bell bureaucracy, he moved to the New York consulting firm, McKinsey & Company. General Instruments had recently acquired Jerrold and hired McKinsey & Company to help them figure out how to resuscitate the then-ailing company.

The consulting job called for long hours and long road trips and Malone wanted something that would allow him to see more of his wife, Leslie, and their young children. He found an opportunity at General Instrument, where he was offered the post of group Vice President and President of Jerrold Electronics. Through Jerrold, Malone met and impressed many of the major players in the cable industry, including Magness. According to one story, Malone had the awkward task of explaining to Magness why another cable company was getting a higher discount than TCI, a major customer, on Jerrold equipment. Malone's adroit handling of the assignment impressed Magness. After about three years at Jerrold, Malone was offered executive positions at both TCI and Warner. He chose TCI because he felt he had greater opportunity at the smaller company and because he wanted to get out of New York. In fact, he took a substantial pay cut to move, at age 32, from Jerrold to TCI headquarters in Denver

Malone and Magness were a striking, even disconcerting, pair. Magness was short, buck-toothed, big-eared, and soft spoken. Malone was tall and broad-shouldered, with a squared-jawed, movie star kind of good looks. But both men were very smart, each in their own way. Magness had a homebred instinctive savvy about business generally and the cable business in particular. Malone was renowned for a razor-sharp, analytical mind. Moreover, the two men shared a common goal, to make TCI very large and very wealthy, and, in the process, to make themselves very wealthy as well. At first, however, they labored mightily just to keep the company alive.

Malone became President and CEO of TCI near the start of the cable industry depression. Magness was already renowned for holding on to almost every system he ever purchased. Unlike many operators who played the financial game of the 1960s, buying systems, amortizing the debt, and selling them at their increased values, then repeating the cycle all over again, Magness bought and held. He told people it was easier to pay interest than to pay taxes, but TCI's interest had gotten completely out of hand.

Malone was brought in by Magness, in part, to serve as the public face and the front man for the company, and in part to help generate the necessary financing to help keep the company afloat. Magness and Malone spent the next several years in a dishearteningly unsuccessful search for money. The two spent months almost literally on the road, sharing hotel rooms and knocking on bankers' doors in an effort to secure capital. For the most part, they came away empty handed. TCI came close to going under, according to Malone, who later joked, "I spent half my time going to New York and getting beat up by the banks and the other half of my time going to the cities and reneging on commitments."[53]

The company survived on a continuing series of short-term loans and extensions from creditors. On at least one occasion, Malone, mirroring the frustration of Hughes and other cable operators, threw his own keys down on a conference table, offering bankers the opportunity to take over the company in lieu of settling the growing debt, and then wondering overnight if they would call his bluff. They did not. But Malone cut all but the most necessary spending at TCI, reducing even his own pay.

TCI under Magness had been known as a company that squeezed everything it could from a penny or from a foot of coaxial cable. TCI headquarters was a Spartan, one-story building that looked more like a warehouse than a corporate office, tucked away in a business park in a southern Denver suburb. There were no frills or unnecessary expenses at the TCI home office, and local systems were expected to follow suit. Construction, maintenance, and operation costs were held to the minimal amount necessary to keep the system going. Malone continued and perhaps even extended that philosophy. He kept his attention focused tightly on the bottom line, on cash flow and stock price, eschewing most of the whistles and bells of the Blue Sky rhetoric. TCI's minimalist approach to local service would have consequences down the road for the company, both good and bad. Local officials sometimes lodged complaints about service and, on a few occasions, even tried to replace the company. In doing so, they often ran solidly into another trait for which TCI under Malone would become legendary, hardball business practices. One early incident, for which TCI garnered industry-wide fame, came in 1973 in Vail, Colorado.

The system in Vail was in disrepair and offered substandard picture quality. The Vail City Council voted to terminate TCI's contract. Malone's response revealed much about the troubled times for cable and about his own personal approach to dealing with local officials. His response to the city's action came on Thursday night, November 1, when he pulled the plug on cable TV in Vail. At 6:35 PM, every channel on the system went briefly blank and then up came a message: "If you have any questions about cable service, contact John Dobson and Terrell Minger" the mayor and city manager. Their home telephone numbers were listed. The blackout came at the start of a football weekend, featuring a Denver Broncos–St. Louis Cardinals game, and Vail was a sports town that valued its TV coverage. But there would be no football on the Vail screens through Friday, Saturday, and Sunday, and the telephones rang constantly at the city officials' homes. The city quickly sat down with TCI. Service was restored on Monday and a new franchise negotiated by Tuesday. Malone was unrepentant; a TCI memo circulated later stated, "The ensuing publicity in the trades not only added stature to our image but could possibly serve as an example to other operators with subsequent benefits to our entire industry."[54]

The Vail incident revealed a deeply embedded TCI business philosophy, one that would make the company, generally, and Malone, specifically, a tempting political target in later years. In 1973, the incident was also indicative of the dire straits in which TCI found itself. Some measure of desperation was seen in Malone's effort to save the franchise and keep badly needed revenue coming into the company.

That kind of desperate concern, although manifest in less public ways, was endemic throughout the cable industry between 1973 and 1975. National penetration in 1975 was only 13 percent and it was not clear that it would climb much higher. Cable television was failing in the major markets. TelePrompTer's Bill Bresnan summed it up many years later. "At that time, the industry was dying. The rural areas were wired and we were moving to the urban markets, but we didn't have things to give people that they could get without cable."[55]

The industry needed resuscitation. It needed a product. Many were hoping for some kind of divine intervention. Salvation was on the way, in fact, perhaps not from any divine source, but most certainly from the heavens.

Satellite Salvation: The Quantum Leap

Many benchmarks and many ways exist to segment the evolution of cable television. When pressed, however, the history of the industry and its technology can be distilled into two simple eras—the period before the satellite and the period after it. It is not too far a stretch to describe them as two completely separate industries. In the fall of 1975, Home Box Office began distribution of cable programming by satellite, and the industry entered a new era in its evolutionary development. It was an historic inflection point, a quantum leap in the operation of the technological system, although again, it was the product of decades of incremental advancement in satellite, cable, and television technology; broadly, it was Usher's cumulative synthesis.

The impact of that synthesis on the television industry in the United States can hardly be overstated. The community antenna industry before the satellite left no real mark on U.S. culture or audience viewing habits; it offered a little bit more of the same old thing for a modest number of TV viewers. In the marriage of the two technologies, cable specifically and television broadly began a deep restructuring. The resulting distribution platform would fundamentally alter the nature of television as it had been known since the 1950s, it would help reformulate television content and the nature of television viewing, and it would leave new footprints on the beaches of world politics and culture.[56]

Early Developments

The idea that a television picture could be beamed from outer space first appeared, according to most scholars, in a 1945 magazine article by science fiction writer Arthur C. Clarke.[57] In the *Wireless World*, the article "Extra-Terrestrial Relays," described a means of covering the earth with broadcast signals using three strategically placed, manned space stations in geosynchronous earth orbit.

Realization of the possibilities of satellite communication took a frightening form, at least from the perspective of domestic U.S. interests, when, on October 4, 1957, the Soviet Union launched the world's first artificial satellite, Sputnik I, setting off the Cold War's "space race." The United States' response began with the launch of the thirty-one-pound Explorer I satellite on January 31, 1958. It was followed that December by the Score satellite, which broadcast a taped Christmas message from President Dwight D. Eisenhower. In August 1960, the two-year-old National Aeronautics and Space Administration (NASA) launched Echo, a 100-foot diameter, metallic-coated balloon that acted as a passive reflector of radio signals. Later that year, the military orbited Courier I, which recorded messages beamed to it for playback as it passed over subsequent earth stations. The first truly active communications satellite, in the sense that it carried electronic equipment capable of receiving and retransmitting radio signals in real time, was AT&T's experimental Telstar. Telstar, launched in July 1962, was the first nongovernmental U.S. satellite and the first to demonstrate the power of satellite television by beaming TV signals from Europe to the United States.[58] A similar but more powerful device, RCA's Relay, was launched by NASA later that year.

A critical step in the development of the technology came in 1963 when Syncom-II, built by Hughes Aircraft Corporation, was placed in a geosynchronous orbit.[59] Prior satellites had circled the earth, but an established space communication system required the satellite to remain in a position that was stationary relative to the ground, as originally proposed by Clarke. Geosynchronous, or geostationary, satellites orbit at a speed that matches the earth's rotation (about 6,870 miles per hour) and, thereby, appear to stay in one spot above the equator at an altitude of about 22,300 miles.

Running parallel to the technical achievements were developments in the legal and administrative control of the space communications system. In 1961, responding to the Soviet initiative, President John F. Kennedy issued a call for a national satellite policy. This led to the Satellite Communications Act of 1962[60] and, in turn, the creation of the Communications Satellite Corporation (COMSAT) in 1963. COMSAT, a private, commercial company created by the government, was provided with a state monopoly on the operation of commercial communication satellites. It further was vested with the responsibility of helping develop a similar international body to control global communications satellite activity. That entity, the International Telecommunications Satellite Organization (Intelsat), was formed in 1964. COMSAT was the U.S. representative to Intelsat and its managing partner.

A Focus on DBS

The earliest proposal for combining television and satellite did not involve cable. As with Clarke's scheme, the proposal called instead for a system that beamed TV pictures from orbiting transmitters directly into homes. It would come to be known as direct broadcast satellite (DBS) service. Speculation about the possibilities of DBS began appearing in the trade, scholarly, and popular press as early as the late 1950s, and technical studies were underway by the early 1960s. In 1955, General Electric researchers presented a paper on the use of orbital satellites to transmit transatlantic television pictures.[61] Dr. Richard Taylor, in an exhaustive doctoral thesis on the subject,[62] noted scholarly and industrial discussion of DBS by Stanford researchers in 1961, with subsequent interest by the Atomic Energy Commission and RCA. RCA conducted a series of engineering studies on the feasibility of a DBS system between 1960 and 1962.[63] NASA sponsored several studies in the mid-1960s.[64] Analyses were also conducted by Hughes Research Laboratories, Rand Corp., General Electric, and TRW, Inc.[65]

Commentary on the feasibility of direct-to-home broadcasting also appeared in the trade press. In 1962, *Broadcasting* magazine reported a General Electric Co. engineer's optimistic view of the potential for direct-to-home satellite broadcasts,[66] while an *Advertising Age* article reported an FCC Commissioner's more pessimistic take on the matter.[67] Similar pieces on the problems, promises, and potential of DBS appeared regularly in the trade and popular press throughout the remainder of the decade.

DBS captured the popular imagination. It was a seductive image, one of instant, global television. Social planners, policy analysts, and industry leaders all found themselves captivated by it. In the talking stages of what would become NASA's Applied Technology Satellite (ATS) program, designed in part to use DBS for global education and development, RCA Chairman David Sarnoff spoke glowingly of a system that would "broadcast directly to individual television sets anywhere on earth. They will beam their programs simultaneously over vast areas, and where necessary they will provide the picture signals with a number of sound channels from which the viewer can select one in his own language."[68]

This vision raised serious policy issues, however. The possibility of using TV satellites for international propaganda was the subject of congressional hearings in 1958.[69] In 1959, Arthur C. Clarke sketched a frightening scenario for the popular audience in an article in *Holiday* magazine.[70] In "Faces from the Sky" he imagined a situation in which Russia had developed a powerful DBS system and was using it to capture the hearts and minds of the world's illiterate masses. "In a few years of skillful propaganda, the uncommitted nations would be committed," he warned.[71] Communications scholars, such as Dallas Smythe in 1960, urged caution in the development of international DBS, fearing it could spark off propaganda wars, and he proposed the creation of a new United Nations-sponsored agency to help resolve a host of technical and political issues he saw as inherent in the emerging technology.[72] In contrast, a different set of concerns arose involving the potential for U.S. domination of a DBS system and the possibility of a cultural colonialism that could endanger the indigenous cultures of second and third world countries. A host of competing themes, therefore, echoed through international policy debate over the next several decades, and remain largely unresolved today.

Contrasting sharply with the policy discourse was the absence of any real-world progress in the actual construction and deployment of a DBS system. NASA conducted experiments beginning in 1966, using a series of ATS satellites. It used ATS-6, launched in 1974, for national and international field trials of DBS technology (see Chapter 8 discussion of ACSN). Commercial use, however, was unlikely, given the technology and economics of the day.[73] The power needed to deliver satellite TV signals to small (under three feet) home receivers was on the order of 1 million watts (1 megawatt), which at that time required a communications satellite weighing nearly ten tons. Only the nation's largest booster, the Saturn, was capable of launching such a payload and the cost was estimated at upwards of $1 billion.[74] There was minimal commercial interest in developing direct-to-home satellite television. Such a system would directly threaten the existing terrestrial broadcast TV industry, as well as AT&T, which profited handsomely from the land-based common carriage of broadcast programming. National and international parties, such as the United States and the United Nations, saw potential for DBS in education and cultural development, and hardware manufacturers, such as Hughes, were attracted by any new market for their product. Outside such agencies, however, there was little promotion of the technology.

Television Interconnection

In practice, the earliest commercial applications of satellite communication technology were aimed at telephony, data, and limited television relay capacity. The real business utility of the satellite, for television, was as a relay component in a retransmission system. That utility was demonstrated in June 1965 when COMSAT launched the first commercial communications satellite, the eighty-five-pound Hughes-built, Early Bird, later renamed, Intelsat-1. Early Bird, which was capable of carrying 240 telephone voice circuits or one TV circuit, was responsible for transmission of the first memorable live satellite TV events, including Pope John Paul's visit to New York and a concert by the Beatles.[75]

The potential economic benefits of satellite distribution for the broadcast television networks were clear and immediate. The first request for domestic television use of a communication satellite came on September 21, 1965, from Hughes and ABC. Hughes, looking to stimulate demand for its satellites, approached ABC with an idea for distributing its network programming to national affiliates via satellite. The FCC, however, returned the ABC-Hughes application without prejudice, explaining that the request raised important policy issues that it was not prepared to resolve at the time, including the central question of whether nongovernmental entities should be permitted to own and operate satellites.[76] Instead, it opened an inquiry into the issue of private ownership of domestic communications satellites and their appropriate uses.[77] Initial respondents included COMSAT, AT&T, NBC, and the Ford Foundation.[78] It was an inquiry that would have a significant impact on cable-satellite interconnection, but would take another seven years to resolve.

Cable and Satellites: The Start of an Idea

The earliest uses of the communication satellite, then, were primarily for voice traffic, with some limited trans-Atlantic TV relay service. Discussion about future use focused primarily on the possibilities of DBS and, to a lesser extent, on broadcast network interconnection. Conversation about a union of satellite and cable technology might be heard at the very edges of the debate, but only when listening closely.

Some of the earliest hints of cable-related use placed cable-satellite systems far beyond U.S. shores. A satellite-community antenna distribution system was seen as having potential for rapidly deploying television service in developing countries. Proposals for international satellite television often included the idea of beaming signals to a central "community receiver" in villages in underdeveloped countries. The community receiver might be hooked up to only one TV set, serving the entire village, or it might be the base for community-wide redistribution of programming by either broadcast or wire.[79] The Twentieth Century Fund Task Force on International Satellite Communication in 1969 noted that (second or third world) countries with developed terrestrial broadcast facilities might opt to relay satellite signals from a central receiving antenna "to the home television set by a coaxial cable system or, over the air, by a low-powered repeating station or a regular broadcasting station."[80]

The possibilities of satellite interconnection of domestic cable systems arose primarily with the Blue Sky thinking of the mid- and late 1960s. National cable networking was a recurring theme in speculation about the future of cable, and the idea of using satellites for that purpose arose quickly and logically out of the broader progress in space communications. One of the first to publicly predict a future in which cable would use satellites for program distribution was industry pioneer Leon Papernow (see Chapter 6). Shortly after ABC filed its petition with the FCC in 1965, Papernow penned an article in *Television Magazine* forecasting the delivery of multichannel programming from New York and Los Angeles to cable systems around the country via geosychronous satellite.[81] The idea also took hold in scholarly and regulatory studies.

Many of the Blue Sky reports of 1967 and 1968, such as those authored by Licklider and Barnett, spoke of satellite interconnection of cable systems. In May 1967, Leland Johnson, Rand Corp. scholar and research director for President Johnson's Task Force on Communications Policy (The Rostow Report) offered a detailed proposal for a cable-satellite program distribution system.[82] Leland Johnson said: "In addition to distributing programming from the television networks to their local affiliated stations, it is quite conceivable that (satellite) channels could be leased at low cost to permit groups of CATV systems to pull in programs from a far greater distance than they are able to do today. Technically, the link-up between CATVs and the satellite system would be simple."[83]

The consequences, however, could be far-reaching, he continued: "Combined with a satellite system as described above, greatly expanded opportunities might emerge for a wide variety of programming. We could visualize a producer who tapes a program, leases satellite channels for nationwide distribution, and arranges with a large number of CATV operators to carry the program to the home."[84]

Specialty audiences could be aggregated across the country sufficiently large to justify economically the targeted programming.[85] Johnson noted that such a system would "tend to erode and fragment existing large audiences enjoyed by particular programs today" but the trade-off would be expanded viewer choice.[86]

NASA also took a look at the cable-satellite link. The Summer Study on Space Applications, sponsored by NASA and the National Academy of Sciences, met in the summers of 1967 and 1968 to consider the "Useful Applications of Earth-Oriented Satellites."[87] Panel ten of the summer study reviewed the use of satellites in broadcasting and, among a list of other recommendations, outlined a specific system of satellite networking for existing national CATV systems. The panel envisioned eight national program networks. They included ABC, NBC, and CBS; a public broadcasting network, two additional educational channels, a world-wide United Nations channel, and an eighth channel held in reserve for an unspecified "new service."[88] Satellite receivers would cost about $75,000, by the engineers' estimates. Benefits to the consumer would include a more comprehensive program service and improved picture quality.

The report was drafted in 1968, but not published until 1969. By then, the idea of satellite networking of CATV was spreading rapidly. It was noted in the 1968 New York City Advisory Task Force Report, which sketched briefly but accurately the design for a retransmission system that beamed signals from satellite to earth station with cable TV serving "as the capillaries for local distribution of the signals."[89] Leland Johnson's idea naturally influenced the conclusions of the Rostow report, which advocated the development of satellite cable distribution, and the concept was even endorsed by the FCC in its 1969 origination report.[90]

The Industry Stirs

Although the idea of system interconnection had been kicked around in the industry since the 1950s, the possibility that satellites might prove useful in the task, before 1969, was taken seriously by only a handful of operators such as Papernow. Some even feared satellites initially, concerned that DBS service might spell the end of any need for CATV. For better or worse, by the close of the decade a new champion for the concept of a cable-satellite link had arisen; he was none other than the industry's most colorful and controversial leader, Irving Kahn. Before his difficulties with federal authorities, Kahn was one of the earliest and most vigorous proponents of a satellite network.

While Kahn pointed the direction, TelePrompTer's Schlafly carried on much of the work. Schlafly met regularly with Hughes engineers through the latter 1960s to discuss their joint Manhattan AML project. In one meeting with Hughes satellite chief Harold Rosen, the conversation turned to the issue of transponder capacity on the communications satellites Hughes was building for the government. (AML was adapted from the satellite technology.) The discussion sparked an idea for a practical plan to use satellites to link cable systems.[91] Kahn was immediately attracted to the concept and Hughes began working on a proposal for the construction of earth terminals for cable networking. The report was completed in February 1970.[92]

The relationship with Hughes was natural. The company was a pioneer in the development of satellite technology. Rosen was known as one of the founders of the concept of spin-stabilized synchronous satellites. The company had built the first such "bird," Syncom I, and as noted was joint author of the ABC satellite distribution plan. Following the AML project, Hughes was also now a part owner of TelePrompTer. Hughes was not just a capable technology company, but a savvy marketer; it looked for any opportunity to advance the interests of satellite use and thereby stimulate demand for its product.

The NCTA, meanwhile, was also sharpening its interest in satellites. Spurred by the small, but growing number of proposals, Association President Ford directed his staff to investigate. By mid-1969, the industry was ready to make an announcement. At its annual convention in June 1969, the Association sponsored a General Management and Engineering session on "CATV Via Satellite."[93] Among the presenters were Nathaniel Feldman of Rand (Feldman had been influential in the development of the 1967 paper by his Rand colleagues Barnett and Greenberg), Louis B. Early and Robert Briskman of COMSAT, Frank Norwood of the educational TV association JCET (Joint Council on Educational Telecommunications), and, of course, Kahn. Ford chaired the panel. It was his intent to use the event to launch a bold new proposal for the creation of a national cable television programming service.

In early 1969, just after the release of the Rostow report, Ford directed the NCTA staff to begin working on a plan for a multichannel, satellite-distributed network to serve the programming needs of the industry. Following earlier proposals, it called for a six-channel service that included a channel for Public Broadcasting System (PBS)–type cultural fare, a channel for instructional television, a medical and health channel, a channel for re-rerunning broadcast network (ABC, NBC, CBS) nonfiction programs such as documentaries, and sports, a 24-hour weather channel, and a channel for full-time coverage of Congress. Many of the programming services the industry would develop a decade later were foreshadowed in the plan.

Ford also emphasized that the proposal included only "nonentertainment" programming (excepting the reruns), explaining that the idea had been worked out in consultation with the NAB. Since his 1966 announcement that cable ought to program its own material, Ford had consistently stressed that material should be of a public service nature, fearing the political opposition that a more aggressive position on entertainment programming would initiate. In fact, he was tied to the political agreement with the NAB in which broadcasters would drop opposition to local origination in exchange for a CATV pledge not to develop interconnected entertainment networks.[94] In his opening remarks, Ford also stated that he wished "to draw heavily upon the recommendations of the Carnegie Commission for the development of a national television system that will better serve the public and cultural needs of the people."[95]

COMSAT had been requested by Ford to prepare a small study of the feasibility of such a system and presented its results during the session. As part of the presentation, Early indicated company support for the idea and COMSAT's desire to be the carrier of choice for the new cable networks.[96] Kahn enthusiastically endorsed the proposal, but recommended that the cable industry own and operate the satellite itself, rather than ceding managerial control to COMSAT. "I guess what I am asking," declared Kahn, "is, do we need COMSAT as much as COMSAT needs us?"[97]

Satellite Economics

The attraction of satellites to the industry was not just technical, of course, it was economic. Before satellite communication, any TV programmer, as previously noted, who wanted to interconnect national outlets, including the three broadcast networks, had to use AT&T lines. AT&T charged regulated, but nonetheless monopolistic, prices. Broadcasters, therefore, were among the first to see the benefits of breaking away from the AT&T monopoly, prompting ABC's 1965 request. ABC and the Ford Foundation, in early FCC filings, estimated that satellite distribution would save broadcasters up to 50 percent over AT&T's terrestrial service,[98] and the three networks, at that time, were paying AT&T more than $65 million a year.[99] The savings, especially over time, would more than offset the estimated $25 million it would take to build and launch a satellite.[100]

For cable, the satellite also made greater financial sense than attempting to build a dedicated terrestrial system using microwave, or microwave and leased landlines. In addition to the administrative burden of developing a national microwave system, noted previously, satellites offered a one-time window of deployment. Once the capital investment was in place, the cost of transmission would be equal to all points in the footprint. The cost of adding additional subscribers would only be the cost of the customer reception device. Satellites, in short, enjoyed substantial economies of scale relative to terrestrial distribution. Scale economies also could be realized by increasing the channel capacity of the satellite and distributing fixed costs across multiple channels.

Additional benefits included the scope economies of accelerating system participation. Despite the growth of the MSOs, the cable industry was still composed of hundreds of small and independent operators. Everytime a cable operator joined a networked system, total revenues went up and the costs for all participants dropped. Cost-efficient interconnection, therefore, created a spiraling effect on system participation and revenues. Lower distribution costs would mean more and smaller systems could participate, increasing total and shared revenues.

Satellite distribution, finally, offered improved picture quality over terrestrial networks, which often required several microwave hops and the incremental interference that such continual reamplification introduced. In sum, satellite networking presented a cost-effective method to develop high-quality, specialized programming that might attract subscribers in larger markets. Satellites, in conjunction with AT&T and dedicated microwave networks, also provided another illustration of the process of technological diversity and adoption, because a variety of potential solutions to interconnection were available, but one was clearly superior with the social context.

Open Skies

The conceptual building blocks for a national network of cable systems linked by satellite were well in place by 1969. The feat of actually bringing such a system into existence was another matter. That required action on the part of the FCC to adopt a regulatory structure conducive to cable networking. It also required a resource-rich company from some sector of the cable industry to prime the satellite pump.

The first steps in creating a regulatory environment open to cable-satellite networking began in 1969 and 1970. Following the recommendation of the Rostow report, the FCC was prepared in early 1969 to authorize a COMSAT-proposed pilot program. The experiment called for a limited test of satellite television, controlled by COMSAT. Chairman Hyde took a draft Report and Order of the recommendation to the new Nixon White House. The administration had a different approach in mind, however, and requested a delay in FCC action.[101] The new OTP began its own study, one imbued with an ideological preference for free market solutions, even in issues of space communication. In January 1970, the administration sent the FCC a recommendation for an "Open Skies" policy, one that would permit qualified private interests to own and operate satellite communication systems. Two months later, in March 1970, the FCC issued a Report and Order favorable to a policy of open entry and solicited proposals for domestic satellite service.[102] The Commission said it would take no formal position on the number or ownership of satellite systems until it had reviewed applications, but it did specifically allude to the potential use of satellites by TV broadcasters and cable TV. It was the policy that broadcasters and cable operators had been lobbying for years. The White House and FCC actions ignited a burst of interest by the major players, including the broadcast networks and AT&T, all of whom indicated they would prepare plans for their own satellite systems.[103] One of those drawing up a submission was Irving Kahn.

TelePrompTer

Fueled by the Open Skies proposal, enthusiasm for the idea of satellite interconnection gained cable industry momentum. At an NCTA-sponsored programming convention in May 1970 "cable operators all but cheered at the suggestion that the only answer for the sale of national advertising...must be a national interconnected CATV network."[104] The Association issued a report in 1970 analyzing the growth potential of satellite operations, noting that if TelePrompTer, the nation's largest MSO, interconnected all its systems, it would have a national subscriber base of 450,000 homes, sufficient to begin thinking about new program networks and accompanying advertising revenue.[105]

It was a vision that Kahn very much wanted to make real. Following his appearance at the 1969 NCTA session, Kahn, in October, announced TelePrompTer's intention to create a satellite distribution network. Working with Hughes, the outlines of the system began to take shape over the next several months. Kahn proposed a network of up to ninety ground terminals; Hughes would develop and supply some of the programming, as well as build and operate the satellites.[106] In December 1970, Hughes filed an application proposing two, ten-channel satellites. Specialty TV channels were to include news, public affairs, music, sports, and minority interests. Hughes owned the Hughes Sports Network and TelePrompTer owned Filmation Associates; both would provide material. The service would be priced between $0.25 to $1 a month.[107] Additional capacity on the Hughes satellite would be leased to GT&E for telephone traffic.

Hughes, of course, was not the only applicant. Western Union had been first to file, in July 1970, submitting a proposal for a multipurpose satellite. By March 1971, the filing deadline, several more companies had submitted plans. COMSAT announced a $250 million multipurpose package that would set aside two channels for CATV use (and two for PBS). MCI-Lockheed, a joint venture of Microwave Communications of America, Inc. and Lockheed Aircraft, filed a $169 million proposal that promised to be sufficiently flexible to meet the needs of the CATV industry. RCA Global Communications, Inc. and Fairchild Industries Inc. also filed applications that included provisions for CATV service. An AT&T-COMSAT plan did not allow for cable carriage. The broadcast networks sponsored a well-publicized study that showed they would save millions of dollars in AT&T interconnection fees by either moving to satellite or simply building their own dedicated land-based system. They did neither, preferring instead to await the probable launch of one of the other systems. Finally, and not to be left out, TCI also filed a satellite plan.

Even before TCI, in concert with Lubic's Home Theater Network, began working on plans for a terrestrial national network based on microwave, the company was thinking about satellites. Its filing with the FCC in 1971 called for a $66 million, two-satellite communications system designed, in part, to interconnect cable operators. TCI's satellite plans would crash, along with the cable stock market, in 1973; in fact, it did not get good grades from the FCC even at the outset. An economic analysis of the TCI proposal by the Stanford Research Institute concluded it was not a likely winner in an open skies environment,[108] and an FCC staff report suggested important revisions.[109]

Although ultimately unsuccessful, TCI's involvement nonetheless demonstrated the extent and appeal of the idea, especially among the larger players in the industry. The NCTA, for example, along with other cable interests, prepared lengthy arguments in March 1971 asking the Commission to set aside at least one satellite channel for industry use. NCTA also asked the FCC to allow cable systems to own earth stations and to relax the restrictions on signal importation in the nation's top 100 cities, "so that demand by those CATV systems will be present when it is possible for domestic satellite systems to transmit program material."[110] Time Inc., with its vast programming library, had been in discussions with Hughes about supplying product for Hugh's cable-satellite proposal, and joined in urging the FCC to accommodate potential cable interests, stating: "We believe satellite transmission is critically important both as a means of delivering programs produced by others to Time Life's CATV systems and as a means of distributing programs produced by Time Life to its own and possibly other CATV systems.[111]

Several companies filed applications to own and operate earth stations that could be used in conjunction with a possible cable-satellite program distribution system. One, unsurprisingly, was TelePrompTer, which put in an initial request for five terminals. LVO, the large Oklahoma-based MSO and microwave operator, also requested permission to run five dishes. More surprising was the application for one dish from Twin County Trans-Video, the Allentown, Pennsylvania-based cable system owned by pioneer Bark Lee Yee.[112]

After another year of study, the FCC, in June 1972, finally issued its "Open Skies" order, clearing the way for commercial satellite service.[113] While it declined to approve any specific applications at that time, looking to review them further, it did endorse the staff-recommended concept of private (i.e. noncommon carrier) ownership and operation of satellite earth stations. Ownership of dishes by broadcasters and cable operators was seen as yet another means of facilitating the advancement of satellite-distributed telecommunications.[114]

The first company to receive permission to launch was Western Union in January of 1973.[115] Another five were approved in September 1973. Western Union launched its Hughes-built, twelve-transponder satellite, Westar I, April 13, 1974.

By 1971, the idea of cable–satellite interconnection had a full voice in the Blue Sky chorus. The Sloan Commission that year prophesized that satellite interconnection "will be available to cable television by the end of the decade," and national cable networks would quickly follow.[116] Creation of the physical infrastructure would not be sufficient by itself, however, to support a successful cable network. It would also require programming and a willing cable industry to embrace—that is, buy—the service. TelePrompTer was ready to help supply programming and commit its own systems, but more would be necessary. Despite the dreams of the futurists and the enthusiasm of companies such as TelePrompTer and TCI, successful deployment required acceptance by the hundreds of small systems that made up the bulk of the cable industry, and the smaller operators were not as eager as the larger MSOs.

Their hesitancy stemmed from what was known as "the chicken or the egg problem." To join a satellite system, cable operators would have to spend an estimated $75,000 to $250,000 for a satellite dish. Very few were interested in doing so without some assurance that a steady stream of quality programming would follow. Alternatively, program suppliers were equally as reluctant to spend money on production without an assurance that some critical mass of cable operators would be in place and ready to buy their product. It was an economic "vicious circle."

TelePrompTer's intent was to breach that circle, in part by a real-world demonstration of the satellite promise and, in part, through a plan to organize the industry. While the FCC reviewed satellite applications, TelePrompTer went to work to show the industry that the idea was more than just a pipe dream. Schlafly put out a request for proposals for an earth station sufficiently large to pick up a satellite TV transmission, but small enough to be transportable, and priced under $100,000. As he recalled later, the response from the major manufacturers of such equipment, including General Electric and Raytheon, was deadly silence. Two men from a small Atlanta, Georgia, electronics company approached him, however, saying they thought they could do the job.[117] The company was Scientific-Atlanta, headed by Sidney Topol. Scientific-Atlanta did build the earth station, on time and within Schlafly's budget, although Topol reported taking a loss on the receiver. Topol believed the contract would lead subsequently to larger opportunities, an assumption that turned out to be accurate.[118]

TelePrompTer, meanwhile, obtained the services of another cable satellite advocate, Robert Button. Button worked in government relations at COMSAT for eight years and was a regular exponent of cable-satellite interconnection; he was hired by TelePrompTer in 1972 to help organize their cable-satellite initiative. One of his first tasks was to help plan a live demonstration of the technology, a project they dubbed "Spacecast I." He and Schlafly had the Scientific-Atlanta earth station hauled across country from Atlanta, Georgia, to Anaheim, California, for the NCTA's 1973 convention.

There, on June 18, they helped make television history with the first coast-to-coast satellite transmission of programming designed specifically for cable television. The United States had not yet launched a domestic satellite capable of transmitting the material and TelePrompTer used the Canadian ANIK II. As *Broadcasting* magazine described the historic event: "The first image ever to cross the land via a domestic satellite was that of TelePrompTer's chairman, Raymond P. Shafer, who was at that moment in the studios of WETA-TV Washington (D.C.). A few minutes later, came the image of Carl Albert, the speaker of the House. And after that came a special program put together by TelePrompTer to mark the occasion."[119]

That evening, the satellite link beamed in a highly touted championship boxing match from Madison Square Garden in New York. The feature material was supplied by HBO through an arrangement initiated by TelePrompTer. HBO, led by Gerald Levin, hosted a suite in the Disneyland Hotel to promote the event. The fight between Jimmy Ellis and Ernie Shavers lasted less than one round, however, ending with a quick knockout punch from Shavers. The classic cable tale tells of conventioneers who went off to get their drinks in preparation for the title fight only to come back to find they had missed a piece of history. Levin himself was in a hotel elevator shuttling between suites during the full but brief course of the contest.

Irving Kahn, the man who helped start the project, missed the action as well; by then, he was passing time in a minimum-security federal prison following his Johnstown conviction. TelePrompTer's efforts would continue without him, but only for a few months. By September, the SEC would pull the plug on much of TelePrompTer's activities and the satellite distribution baton would pass to others.

The Cable Satellite Access Entity

The Anaheim satellite demonstration was publicized in the trade press and featured in subsequent articles about TelePrompTer and HBO. Less heralded at the convention was a gathering of the larger cable operators designed to initiate organized industry-wide action on satellites. The NCTA had previously formed a committee on satellites, headed by Schlafly, and Schlafly and Button had been meeting since early 1972 with cable operators, such as John Gwinn, Vice President of Cox Cable Communications, and Bruce Lovett of ATC, to discuss organizing an industry-wide effort to investigate satellite interconnection.[120] Just before the 1973 convention, they sent a letter to cable operators interested in satellite distribution calling on them to convene for a special meeting in Anaheim. The purposes were to organize, solicit financial support, and commission a report on the feasibility of a satellite networking, with a special focus on programming issues. Meeting outside the convention to legally distance themselves from NCTA, more than a dozen companies indicated a willingness to contribute $5,000 each to fund a study.[121] At a subsequent meeting in July, the Cable Satellite Access Entity (CSAE) was formed. The forty-nine-member group was chaired by veteran NCTA board member and former chairman Rex Bradley of Telecable, Inc. CSAE hired the consulting firm of Booze, Allen & Hamilton to conduct the study, which would take another year to complete. In the meantime, Schlafly took his case for satellites, and his earth station, on the road.

TelePrompTer Drops Out

The 1973 demonstration—a technical success—was appropriately lauded in the trade press as a significant step for the industry. Despite the formation of the CSAE group, however, its reception on the floor of the convention center by rank and file operators was tepid. Until operators could be convinced of the business case for the technology, promoters faced an uphill struggle. After the convention, Schlafly literally took the earth station out on the highway, offering demonstrations to individual operators across the country. Although many were impressed with the system, more was needed to make the sale. As TelePrompTer's Button explained in 1973, "The reactions [from cable operators] run the gamut from 'we're too busy getting new subscribers' to 'show us the numbers and when they figure up, we'll go.'"[122] Operators also were skeptical about the supply of programming and concerns were even generated by the novelty of the technology. Some worried that the satellite might fall from the sky. The satellites, of course, were not going to fall from orbit; the same could not be said for TelePrompTer, however. One of the many initiatives axed in Cooke's reorganization after the SEC suspension was the satellite program. By early 1974, the company, at least temporarily, was out of the satellite business.

In September 1973, the FCC approved the satellite applications of five more parties, including American Satellite Corp (80% Fairchild Industries and 20% Western Union), Hughes (National Satellite Service) & GTE; COMSAT; RCA Global Communications with RCA Alaska Communications, and A&T.[123] With TelePrompTer and TCI out of the picture, however, and the entire industry in recession, little on the surface suggested progress on a cable-satellite network. The CSAE, Booze Allen report was finished in August 1974, but was kept as a closely guarded industry document.[124] In fact, it was a lengthy and detailed examination of the technical and financial prospects of satellite distribution. The heart of the report was a finding that such a system was technically feasible, and a market existed for specialty programming. The real problem, according to Booze Allen, lay in the development and financing of the software. It proposed a sample programming plan that included channels aimed at children and women, along with various arts and entertainment programming. In contrast to the view of many in the industry, it downplayed the potential of pay-TV as a primary programming source. It also concluded that an organization such as CSAE was necessary to assure the development of an interconnect system.[125]

The widely held view, it seemed, was that satellite distribution was a good idea, but one for a later time. The industry was in the midst of its 1973–1975 recession; the idea of a satellite system was far too grand to consider at the time and, as with proposals for wiring the big cities, was burning up in the thick atmosphere of the economic downturn. The large MSOs were calling for a return to the basics. The general industry philosophy was to sit tight until conditions improved. With TelePrompTer gone and cable in the doldrums, something was needed to move the satellite project ahead. That something was in gestation at HBO.

Time Life & HBO

Time Inc. was having very little luck with its cable TV investments. The Manhattan system was in deep trouble and the fledgling pay service, HBO, was stumbling. By the end of 1973, HBO was on fourteen systems in New York and Pennsylvania, but its churn rate was exceptionally high.

It was a phenomenon that would come to be common and painful in the cable industry. Churn is what happens when a subscriber signs up for service, especially pay service, frequently with great anticipation and revels briefly in TV exploration and discovery. Eventually, however, the customer settles into a routine, watching only a handful of channels regularly and largely ignoring the rest. In pay-cable, especially in the early days, the variety and quality of the films was mixed. Hollywood turned out fewer than 200 films a year, far too few to fill the program schedule on a daily basis. Perhaps fifty of those films would be heavily promoted by HBO and fewer than twenty would become high-demand blockbusters. HBO, as other pay services that would follow it, relied on older films and "B" movies to fill the time between the more popular films. All the films were run several times during the month, the popular features more than the filler movies, in part because of the scarcity of good material and in part to give customers more than one opportunity to see a film. The programming trick was to determine how much repetition to schedule without having it become redundant and frustrating to the viewer. The built-in problem was that this tipping point was different for each subscriber.

A certain portion of the subscriber base would become quickly bored with what they saw as a high level of repetition, or disappointed by the lack of quality films, or both. As one newspaper columnist observed as late as 1981, "Show me a Home Box Office patron and I'll show you someone who has seen *The Great Santini* 15 times."[126] HBO was not cheap and those who did not want to pay to see *The Great Santini* fifteen times cancelled their service, sometimes after only a few weeks of viewing.

This constant turnover in the subscriber base was called "churn," and churn had hit HBO hard. Despite a promising beginning, HBO was losing subscribers by mid-1973. Obvious solutions included (1) expanding and improving the programming to avoid the heavy rotation of films and (2) expanding the subscriber base. Whereas the latter might not reduce the churn rate, it could at least build total revenue.

New subscribers could be generated, both by aggressive marketing in existing franchise areas and adding new affiliate systems. To achieve the former, HBO began sending its own sales force into existing markets, going door-to-door promoting the service, a strategy that helped pay off in early 1974. Adding new systems, however, presented the challenge of cost-effective distribution. Technical and economic hurdles hampered the extension of the microwave system in New York and Pennsylvania, but they could be overcome. HBO even considered creating several regional microwave hubs as a first step in a nationwide terrestrial network. Expanding the live microwave signal beyond the Northeast, however, was technically challenging and very expensive. An avenue at least worth exploring was satellite distribution.

In fact, the satellite option was not a new concept at Time. As noted, the company had looked into the possibility of working with Hughes in 1971. Dolan included possible HBO distribution via satellite as an eventual option in a January 1972 memo to the Time Inc. board.[127] Levin himself once said that the first time he heard about satellites, it was from Irving Kahn.[128] He explored the potential of the satellite model, but quickly concluded that the Hughes system was not ready for commercial exploitation at that time. "It was too early, it didn't make a lot of business sense," he recalled decades later.[129] They opted for the more practical and familiar approach of microwave distribution. If a regional microwave network proved successful, explained Levin, the system could be extended one city at a time and: "[M]aybe, ultimately use satellite transmission to reach those parts of the country that wouldn't lend themselves to regional networking. There was no domestic satellite activity we could even plan for. It seemed very much a distant thing for us."[130]

Levin's sentiments were echoed at the corporate level. In remarks at the 1973 convention, Levin's boss, Time Life's Barry Zorthian, observed, "The potential for (satellite cable networks) is still limited today. But I admit it's on the come."[131] The company's thinking soon took new shape, however, influenced by both opportunity and necessity. For their own 1973 NCTA convention exhibition booth, HBO initially considered a ticking clock that would show increasing subscribers with every second. "We jettisoned the idea," explained Time Inc. Vice President Richard Munro, "when we realized the subscribership was going down."[132] The problem of churn and the need to expand its market were pressures HBO could not ignore. Satellites were risky, but the alternatives were few.

People, furthermore, were lobbying Levin. Topol, with his own interests in seeing satellites take off, was pressing for someone to take the first step, and Levin was one of his chief targets. He was persuasive; Levine later recalled, "Sid Topol convinced me it could be done."[133] While Scientific-Atlanta was eager to supply the dishes, Levin would still need dedicated satellite capacity to make such a scheme work. Button, according to Schlafly, introduced Levin to the President of RCA Global Communications, Howard Hawkins, to talk about possible use of the planned RCA satellite.[134] As it happened, Levin also had an old friend from Haverford College, Andy Inglis, who was now President of RCA Americom, RCA's domestic satellite operations units. RCA presented an appealing opportunity to try a satellite solution. While transponder time on the Western Union satellite was in heavy demand,[135] RCA's new satellite, Satcom, used a technique called "cross-polarization" that allowed the effectively doubling of their transponder capacity to twenty-four channels, twice that of Western Union's, and they were looking for customers. Building on their long-standing personal connection and mutual business interests, Inglis offered Levin an attractive lease on Satcom time.

From this swirl of economic pressures and interpersonal connections, a plan began to congeal. Levin took the idea to the Time Board of Directors, who gave him a green light to proceed. Time Inc. President James R. Shepley's willingness to back the venture was important in the decision. Levin worked out a six-year, $9.6 million contract with RCA for transponder time that included a substantial discount for the first months of operation.[136] By March 1975, Levin had also retained the services of a newly established satellite communications consulting firm, Transcommunications Corp. That firm, in fact, was the team of Schlafly and Button, recently departed from TelePrompTer. Levin next needed a cable operator to help him solve the "chicken or egg" dilemma. He talked to several, including TelePrompTer. The first operator to decide to take the chance turned out to be Robert Rosencrans, head of UA-Columbia Cablevision, Inc. Levine called him in early 1975 and the two met in New York.[137] Rosencrans said he was attracted to the concept for a number of reasons. UA-Columbia was already an HBO customer, taking the microwave feed at systems in New Jersey and New York, and found the service so promising that he was looking for a means to expand its use. A plan to build a centralized microwave system in Florida, serving a number of his UA-Columbia and unaffiliated systems, was plagued with potential problems, however. Rosencrans explained that it would have been labor intensive, costly, and time-consuming and would have had to rely on videotape shipped to the regional center. Satellite distribution solved those problems and offered a better quality picture than the tape-fed microwave. It also held the important potential for live programming. Following a Friday meeting with Levin, Rosencrans had UA-Columbia Vice President Ken Gunter review the satellite idea. That weekend Gunter reported it was technically sound and the following Monday Rosencrans signed on. He committed seven large systems,[138] giving HBO a sizable initial audience base. Rosencrans also cut a deal with Scientific-Atlanta for a reduced price on satellite dishes, about $75,000 each as opposed to the normal $100,000 price tag.

Levin and Rosencrans announced the agreement on April 10, 1975, at the NCTA's annual convention in New Orleans.[139] Reaction from attending operators was noticeable. *Broadcasting* magazine observed "one could detect both a gleam in the eye and a straightening of the posture of cable operators. An industry chastened by a two-year diet of humble pie began to act like a winner again."[140] The reception may even have been overexuberant, perhaps understandable after several tough years, with some predicting a million pay-TV homes and $100 million in additional annual revenue within two years. It is also worth noting that in the audience at the convention, following his release from twenty months in federal prison, was Irving Kahn, back in the cable business and urging his colleagues once again to be aggressive in their vision and their actions.

Shortly after the initial announcement, ATC signed on to the service. It was somewhat ironic and also testimony to the close knit nature of the industry, at least at the higher levels at that time, that both UA-Columbia's Rosencrans and ATC's Rifkin had spent their early years in the cable business at TelePrompTer, employed by Kahn. Moreover, TelePrompTer, with William Bresnan now at the helm of its cable division, continued to play an important role in the birth of the service. In May 1975, TelePrompTer, still the nation's largest MSO at the time with more than one million subscribers, signed up for the service. The company saw pay-cable as one hope for turning its financial fortunes around. It placed an initial order for fifty Scientific-Atlanta earth stations, bargain priced at $65,000. TelePrompTer's agreement covered eighty-one systems and some 800,000 of its subscribers in twenty-one states. It helped create the critical mass of subscribers the new service would require. HBO helped itself as well; Time Inc. provided millions in low-cost loans to cable operators to help pay for the discounted Scientific-Atlanta earth receivers. Schlafly and Button, through their new consulting firm, also worked as agents for smaller cable operators, bundling dish orders to secure the discounts from Scientific-Atlanta.

HBO and UA-Columbia, finally, had to clear its proposed service with the FCC. As part of the application process, comments were invited from other parties and both Rosencrans and Levine waited nervously for the opposition to file objections. "We were holding our breath," said Rosencrans, through the 30-day window for comments.[141] Both men were shocked when no one filed. Broadcasters and theater operators may have simply thought the idea would never fly; in any event, HBO was clear to begin programming.[142]

The Vero Beach Holiday Inn was packed with 150 officials and guests, all anxiously waiting; a half dozen TV monitors were scattered around the room. There, at 5:25 PM, Bob Rosencrans pulled a switch and the monitors sprang to life. It was September 30, 1975, and the service that Kahn had envisioned and Levin helped make possible had begun. An HBO feed, originating from the company's Manhattan studio, was beamed via microwave to Valley Forge, Pennsylvania, site of RCA's earth station, where it was uplinked to the orbiting Westar satellite and bounced back to earth. The UA-Columbia receiving dish in Florida took the signal and passed it along to the cable system in Fort Pierce-Vero Beach. Some 700 miles to the northwest, the satellite signal was flowing through cable lines in ATC's Jackson, Mississippi, system, while Monty Rifkin presided over the inauguration there.

HBO had to use the Western Union satellite for its initial feed because the RCA satellite was not available until December, but it presented no problem for the opening demonstration. The programming that first day consisted of speeches by FCC Chairman Richard Wiley and Andrew Heiskell, of Time Inc., two films, "Brother of the Wind" and "Alice Doesn't Live Here Anymore," and the featured event, the championship boxing match between Muhammad Ali and Joe Frazier, beamed via satellite from the Philippines to New York, and known as the "Thrilla from Manila." The fight was a particularly shrewd marketing move. Viewer demand for live coverage of the fight was very high, but with the exception of the HBO satellite feed, it was unavailable to viewers in the United States.

This was the beginning of a new age for HBO and for cable television. By the end of 1975 HBO had more than 275,000 subscribers. The largest cable affiliate was Communications Development Corp. in Long Island, New York, one of the systems headed by HBO's founder, Charles Dolan.

HBO adhered to several business and programming practices that proved successful. It stuck with its monthly payment plan, eschewing the complicated and costly per-program scheme proved unsuccessful elsewhere. It partnered with local operators, working together on marketing and promotion of the new service and, critically, splitting revenues with them. HBO normally charged subscribers $8 per month at the time. The cable system received $3.50 of the first $6 and split evenly anything above $6. HBO, furthermore, offered programming beyond just films, including high-profile sporting events, especially boxing, and live concerts. Growth accelerated quickly. By September 1976, it had 765,000 customers. It would be later 1977 before HBO began making money; it reportedly lost more than $7 million through 1974 and 1975, but it was signing up subscribers quickly and, for a more than a year, had the field of satellite-distributed programming all to itself. Even TCI's John Malone once commented that for ten years, they were not selling cable, they were selling HBO. That state of affairs would not last for long, however, as new services began taking to satellite throughout the rest of the decade.

Levin's contribution was significant. HBO, by a combination of daring and necessity, broke the business logjam that, outside of regulatory constraints, held back the development of cable networking. It was a substantial business risk, despite the years of study that promised its feasibility. It required gambling millions on an unproved distribution concept at a time when the industry was in deep financial difficulty. Warner, by way of illustration, had also considered satellite distribution. The company had hired legendary technology expert Peter C. Goldmark (developer of the CBS color TV system and the $33^{1}/_{3}$ long-playing [LP] record) to work on cost-effective distribution plans for their own pay-TV system. Goldmark suggested they look at satellite distribution. As Warner executive, Frank Cooper, later recalled, "Well, that went around the room you know like, sure, let's take on the national debt."[143] It was far too large an economic bite for Warner to consider at the time. Time Inc. had deeper pockets, but it was still a risk. As Levin explained, "I think it was a daring thing for the company (Time Inc.) to do because we were on the hook for that satellite time whether there was going to be one earth station or a thousand."[144]

The launch of satellite-delivered HBO was a watershed in cable's evolution. It was the beginning of cable as a national force in television. Part of the new evolutionary stage was the development of a new and important segment of the industry, the programming companies. Cable was no longer a collection of system operators, it was a two-part business, each half dependent on the other, a symbiotic social context that almost always portends interesting and complicated relationships.

The late 1970s would be more than a renaissance for cable, it would usher in the start of a new era and a new kind of cable industry. The technology would continue to improve along with the general economy. A critical and wide-ranging demolition of FCC regulations would aid in reigniting investor interest in the industry. Old-line cable companies would pass into the hands of new corporate players and a second generation of cable operators would come to the corporate helm. A new breed of cable company—the cable programming network—would be born in a process that would bring with it a fresh cast of influential, and often colorful, personalities. Cable would begin again to seek franchises in the major markets, with both good and bad results, and the social definition of cable would move into another phase. The beginning of this rally, at least symbolically, may have been marked in early 1975, in a speech from cable's old mentor, Irving Kahn.

8
The Phoenix (1975–1980)

[A]s I was saying before I was interrupted...

—IRVING KAHN, FEBRUARY 1975[1]

Irving Kahn looked out on the audience. It was the annual convention of the Texas Cable TV Association, February 27, 1975. He had been free from federal confinement for only a few months and this was his first public address in two years. He began:

> Now, as I was saying before I was interrupted ... my name is Irving Kahn, my business is cable television, and I'm here to assure you that throughout the entire history of the Texas Cable Television Association, you have never had a speaker more pleased to be standing before you than the man you see here today. In case there are any newcomers in the room, let me begin by explaining that if I don't look familiar, it's my fault, not yours.[2]

Kahn told the audience he had had, of late, plenty of time to reflect on the state and future of the industry, and he had concluded that it was time for cable operators to shake off the depression of the past few years and get on with the business of building a new telecommunications system. "No one ever said it was going to be easy," he told the crowd.

Kahn's reappearance came at a propitious moment. Even as he spoke, his old partner Schlafly was working with Levin and Topol to put HBO on the satellite. By the end of the year, the economics of cable television would start looking bright once more. Irving Kahn's prison term, in an odd way, was metaphoric, mirroring in time the industry's two-year slump. His release coincided with the beginning of not just a resurrection of the faltering fortunes of cable but, in fact, an inflection point in its longer evolutionary trajectory. As the mythical Phoenix, dying in a fireball of self-inflicted combustion and arising reborn from its own ashes, cable was in the grips of a powerful transformation. Years of technical and economic evolution had converged in the form of the cable-satellite distribution system and the industry was poised for a quantum leap in its history. Cable, in fact, would go the mythological Phoenix one better, arising not as a single new creature but as two closely related but nonetheless separate bodies. Importantly, this process would now be at least tacitly supported, in distinct contrast with the past, by the forces of law and policy.

This chapter looks at cable in the years immediately following the launch of the satellite distribution system, roughly 1975 to 1980, although it reaches back a few years before Satcom I to consider the genesis of the important regulatory transformations that characterized the later 1970s. An examination of law and policy and the dramatic reversal in regulatory philosophy during this period is the first subject considered. The chapter subsequently examines the rising fortunes of the cable business brought about by both the relaxation of federal control and the harnessing of the satellite. The chapter also introduces the new programming services made possible by these changes, including such notables as WTBS, C-SPAN, ESPN, and Nickelodeon.

"Re-Regulating" Cable

By the end of 1975, HBO finally had given cable operators a product to sell in the big cities. The industry, however, was still constrained by years of mercurial regulatory policy and draining politics. If cable was able to exploit the new satellite technology placed in its hands, those hands would need to be unshackled from the government regulations of the 1960s. As it turned out, both in outer space and on the ground, the times were indeed "a-changing." By 1974, the regulatory philosophies that had worked against cable's interests for more than a decade were dissolving. It has been suggested that the culture of the 1960s did not fade away until the early 1970s. The forced resignation of President Richard Nixon in 1974 may be been one of the last retreating cries in the diminishing echoes of a tumultuous decade. By the opening of the administration of President Gerald Ford, the winds of politics were clearly blowing from a new direction, and the weathervane of regulatory ideology began another slow rotation, pointing once again toward clearer skies for cable.

President Nixon's political assault on broadcasting had worked, more often than not, to the benefit of cable, but a larger change of national culture was at work as well. The years of governmental activism under presidents Kennedy and Johnson, which had abated under Nixon, continued to recede under Ford. In Congress and at the FCC, there was a return to a faith in the self-righting marketplace. Government was increasingly seen, on both sides of the congressional isles, as a natural impediment rather than a stimulant to the efficient operation of the economy and the promotion of the public good. Various agencies of the government debated cable policy, but the new contours of the dispute seemed not to be over whether to deregulate cable but rather how best to do it.

The ink was barely dry on the FCC's 1972 Report and Order when the first signs of the new philosophy emerged. In late 1972, FCC Commissioner Richard E. Wiley, a Nixon appointee on the Commission for less than a year, told an audience of cable operators that he opposed rate-base regulation and municipal ownership of cable. Wiley was considered a moderate conservative. A lawyer, he had spent his early years in private practice, coming to the attention of Nixon through his work in the presidential campaign. Shortly after the Nixon victory, Wiley was appointed General Counsel at the FCC, succeeding Henry Geller, and named to the Commission to replace Wells in 1972. Philosophically, Wiley thought cable operators ought to pay copyright fees, but ultimately felt the marketplace was a better forum than either the FCC or Congress in which to work out telecommunications issues. FCC Chairman Burch, of course, shared this perspective.

A few months after Wiley's 1972 comments, Burch asked cable operators to imagine a world in which new cable networks supplied programming nationally via satellite, paid full copyright fees for the material, and bid against other distributors for the privilege. "The FCC," he said, "can no longer accept existing structures as eternal verities: their truth is subject to constant verification and reverification in the marketplace."[3]

With the departure of the activist Commissioners of the 1960s, it was a philosophical preamble to a multiyear program aimed at dismantling FCC cable rules that had then been in place for only a matter of months. To be sure, there was a practical aspect to the agenda. The existing regulations were complicated and bureaucratically burdensome. The 1972 requirement that cable operators obtain an FCC certificate of compliance meant more employment for communications attorneys, more mail for the post office, and, as in previous rule-making, a backlog of hundreds of applications at a perennially understaffed Commission. Some of the earliest re-regulatory efforts by the agency were designed to reduce the mountain of paperwork it had made for itself. This, however, was something of a sideshow next to the larger policy tectonics again in motion. Between 1974 and 1980, the Commission systematically tore down, brick by regulatory brick, nearly the entire structure of cable policy that it had erected over the prior decade. Coupled with the inauguration of satellite distribution, the removal of federal controls helped clear the path for the emergence of the new industry.

Wiley, especially, would be a major force in the process. He had spent much of his first two years as a Commissioner easing restrictions on radio, in a process he dubbed "re-regulation." Having completed that work in late 1973, he vowed to bring re-regulation to cable. In February of 1974, he was named Chairman of the FCC and began to fulfill that promise. He was joined later in the year by another Nixon appointee, Glen O. Robinson, who indicated early in his tenure that he shared the view that cable had been overregulated.[4] In fact, Wiley enjoyed a remarkable measure of support from his fellow Commissioners. Krasnow, Longley, and Terry, noted that, during most of Wiley's tenure as Chairman, the Commission was composed solely of Nixon appointees, "the first time a president had named all seven members of the FCC since its formation in 1934."[5]

One of the FCC's first moves, in May 1974, was the creation of a cable "Reregulation Task Force." Its public mandate was to review the 1972 rules. Behind the scenes, Wiley's intent was more pointed. According to policy historian Robert Horowitz, the reregulation label was a "calculated misnomer" designed by Wiley to avoid antagonizing proponents of cable regulation in Congress and the broadcast industry.[6] "As Wiley tells it," states Horowitz, "the 'Reregulation Task Force' was in fact an exploration into deregulation."[7] While the Task Force began setting up shop, events elsewhere suggested the FCC would get little interference in its work from either the White House or Congress. In fact, for the first time in many years, it appeared that all the important regulatory players would be pulling their oars in the same direction.

Congress—The Path to Nowhere

The Separations Policy

The social construction of cable television had been a key point of legal and political contestation for years. The industry fought in the early days for its passive antenna definition. The FCC's vision moved from ancillary broadcast service to twenty-first century telecommunications utility. In the states, the issue was whether cable was a social necessity, natural monopoly, and public utility. One of the related, but usually smaller, questions was whether cable should or could be conceived of, and regulated as, a common carrier. In January 1974, the question popped onto the national radar screens. That month, the OTP's Cabinet Committee on Cable Communications released the report it had begun three years earlier.[8] It was one that generated what could euphemistically be termed "vigorous debate." On the one hand, it recommended wide-scale deregulation of cable, proposing that the medium had significant First Amendment rights as a creator of programming content. On the other hand, it recommended a "separations policy" for cable that would deregulate the industry at the cost of separating the content from the conduit. The concept would convert cable systems into *de facto* common carriers on all but a few channels as soon as cable penetration hit 50 percent nationally.

Under the plan, cable operators would be granted control only over channels redistributing local and regional broadcast signals and one or two channels reserved for their discretionary use. Local government and public access channels would be obligatory and the remaining channel capacity would be made available on a leased access basis. The FCC would be prohibited from regulating content, ownership, or rates of cable systems; its only authority would be in setting minimal technical standards, including adequate channel capacity, and in the very limited area of restricting pay-sports programming. Although vague in details, whatever governmental control was left would be exercised at the state and local level.

Despite its radical nature, the OTP's final proposal contained little surprise. The OTP committee began talking about long-term policy for cable in the summer of 1971 and Whitehead had given strong indications about where the committee was headed as early as the fall of that year.[9] The Watergate scandal had taken its toll on the committee membership and on its schedule, however. Meetings were sporadic amidst the rising turmoil of presidential scandal and, in the end, the draft was described as one that, more than anything else, reflected Whitehead's personal philosophy.

Whitehead described the proposal as both a strong affirmation of First Amendment principles and as a commitment to deregulation—a free marketplace in both ideas and commerce. It kept editorial control away from cable operators and, to some extent, from government, at least in comparison to the various fairness doctrine and equal time constraints placed on broadcasters. It flowed seamlessly from a deregulatory market philosophy that stressed private over public control, but prevented what the administration saw as the abuse of editorial monopoly power exercised by the major broadcast networks.

The concept of separating cable distribution functions from editorial or content functions was far from new. In one form or another, it had been knocking around since the Blue Sky rhetoric of the late 1960s. Cable as a common carrier was explicit in the FCC's 1968 Notice of Inquiry and 1969 Origination Order that some channels be set aide on a leased access basis. Ken Cox offered a similar idea, as previously noted, in late 1968, and it was one of the chief recommendations in Smith's popular *Wired Nation* treatise. Converting cable to a public utility in which operators had little or no control over content, in essence a telephone model for cable, soon became a common theme in legal treatises,[10] regulatory recommendations, and legislative proposals. The FCC's hearings in March of 1971 included a variety of suggestions for at least partial regulation of cable as a common carrier, and fell on the sympathetic ears of Commissioners such as Cox, Johnson, and Houser, who indicated, in April 1971, that the concept had substantial merit.[11] While the full Commission found the idea interesting, the 1972 Report recommended against it, at least in the near term, arguing it would inhibit the private investment the industry needed to mature.

The idea was attractive to some for a number of reasons. It was comfortable, flowing easily from decades of experience with telephone regulation. As noted, the social definition of new technologies very often arises from models used to understand and structure existing systems, and many saw similarities between cable and telephone service. The common carrier model answered the concern of those who worried about monopoly control over what could become an important First Amendment forum. If cable were to evolve into a major source for community and national political and social commentary and debate, as the wired nation perspective suggested, then ownership and control of the information resource was a substantive policy concern. Building a thick wall between ownership of the distribution system and control of the content would go a long way toward solving the problem. Common carrier status additionally suggested telephone company-like rate of return regulation and governmental control over the quality of service. It also implied control over subscriber fees.

Industry reaction was surprisingly mild. The NCTA worried a bit about the transfer of control out of federal and into state or local hands, but supported the overall deregulatory theme of the package. Broadcasters, especially ABC, saw it as yet another paving stone on the road to a national cable networking system and a threat to "free" broadcast television. The group most exorcised by the proposal seemed to be the FCC, which saw the plan as a theft of its regulatory prerogatives over cable. The Commission said the plan overly restricted its flexibility in responding to cable activities. Whitehead nonetheless delivered a draft bill to Congress based on the plan.

By August 1974, however, the growing Watergate scandal had finally culminated in the resignation of Richard Nixon and his replacement by Vice President Gerald Ford. Clay Whitehead completed his term of office and left the OTP shortly before the change of administration. His proposal was left on the doorstep of Congress, where another would join it shortly. The Justice Department offered its own cable deregulation bill after declaring the OTP draft too onerous a burden on the industry. Justice implied that the best regulation for cable was probably no regulation at all; if there had to be control, it had best be minimal. Testimony by antitrust division officials went so far as to suggest that any legislative proposal be drafted with sensitivity to cable's financial condition. By 1975, it was clear that cable was having a more difficult than hoped-for time wiring the major cities and the department warned against any bill that would further injure cable's ability to get the financing necessary to complete its urban construction.

The Ford Plan

President Gerald Ford had his own agenda. Early in 1975, Ford initiated a program of regulatory reform that swept across a host of industries, including transportation, retail trade, and communications. He created a special panel, The President's Domestic Council Review Group on Regulatory

Reform (DCRG), to analyze cable policy, along with several other issues, and it spent from August 1975 to March 1976 doing so. Ford had distanced his administration from the Whitehead plan, and part of the assigned task of the DCRG was to craft cable legislation to replace the jettisoned OTP proposals. The deregulatory mandate from President Ford was apparent from the start. In its preamble, the DCRG's cable subgroup stated bluntly that their analysis "rather than building to a conclusion, ... starts from one—that rules on signal use over cable should in good part be eliminated."[12]

Alarmed at the direction taken by the DCRG, broadcasters mounted an extensive lobbying campaign to derail any cable-friendly legislation that might come out of the initiative, and the broadcasters' efforts had some effect. Although generally deregulatory in tenor, many of the conclusions of the DCRG's cable work were simply that more study was needed.

In Congress, attempts at turning either the Justice Department or OTP proposals into law similarly ran aground. In January 1976, the staff of the House Subcommittee on Communications, following a request of the Committee, released the results of a six-month study on cable regulation and policy.[13] The report included a detailed history of FCC action policy, with a special focus on the 1972 rules. Entitled "Cable Television: Promise Versus Regulatory Performance," the report was influential and soon cited frequently in both policy and academic circles. It was nothing short of an indictment of the history of FCC decision-making in the cable area. The FCC "has chosen to interpret its mandate from Congress as requiring primary concern for individual broadcasters rather than the needs of the audience being served," declared the staff report.[14] The staff noted, with no attempt to disguise its irritation, that the FCC had not managed to prepare draft legislation on cable even two years after the Whitehead report and despite repeated requests by the subcommittee. The staff, therefore, offered its own set of proposals. They were very much in keeping with Whitehead's, calling for substantial deregulation of the industry on the one hand, but implementation of the separations policy on the other. Cable, recommended the report, should operate largely as a common carrier. Legislation was recommended "requiring cessation of programming operations by the cable entrepreneurs at a date no later than ten years after [its] enactment" and as a soon as was feasible.[15]

Basement to Penthouse

Despite the strong words from the House, cable legislation remained years away. In April 1976, Congressman Torbert McDonald stepped down as Chairman of the Subcommittee, citing ill health, and was replaced by Congressman Lionel Van Deerlin (D-California). That summer, Van Deerlin announced plans for a major overhaul of the Communications Act of 1934, what he termed a "basement-to-penthouse revamping" of the superstructure of U.S. telecommunications law. It was, in large part, a political move aimed at thwarting AT&T. The FCC had been opening the telephone business to increased competition, threatening AT&T's historic monopoly of the field. In retaliation, the telephone company opened a multimillion dollar lobbying offensive in Congress that culminated in proposed legislation, the Consumer Communications Reform Act of 1976, which would have had the effect of limiting any potential competition. Van Deerlin hoped to derail the AT&T bill by swamping it in a much larger legislative review. He declared that his committee would not consider any piecemeal legislation, such as the so-called "Bell Bill," or, alternatively the OTP or Justice Department proposals on cable, unless an urgent need existed, which of course was not the case. Substantively, the free-marketeer also hoped to bring deregulation and real competition to several telecommunications sectors, including cable TV. A host of skeptical observers suggested the task of rewriting the Communications Act was so large and contentious as to be practically impossible. Van Deerlin plowed ahead, but in the end, the skeptics would be correct.

In April 1977, the Subcommittee staff released a voluminous new set of reports that outlined policy alternatives for cable, broadcasting, common carriage, and spectrum management, along with a revised FCC structure.[16] The "Option Papers" weighed in, as *Broadcasting* critically noted, at six pounds, three ounces, and almost everyone found something to dislike in the tome; criticism from nearly every segment of the telecommunications industry was swift. Specific legislative proposals were drafted and redrafted in the whirlpool of conflicting public and industry interests. A draft bill stemming from the report and unveiled in June 1978 abandoned the separations proposal and offered substantial federal deregulation for cable, but essentially ceded control to state and local authorities while, more critically, also freeing the telephone companies to enter the business. The new set of ideas was no more successful than those before in forging consensus or support.

As with so many previous attempts at legislative reform, Van Deerlin's would come to naught. After several years of hearings, debate, and analysis, Van Deerlin abandoned the rewrite effort in July 1979, noting he could not muster sufficient support for a comprehensive bill even within his own committee. He decided to focus more narrowly on common-carrier reform, but the following year he lost his bid for reelection.

As it had so often done, Congressional action on cable television had seemingly led nowhere. The rewrite, in fact, was simply too large an undertaking for the time. To the extent that sympathy existed for deregulating cable, by this time, the FCC was well on its way to doing so without legislation, as detailed below. Narrower proposals for eliminating federal control across telecommunications also were being introduced and discussed in the Senate. Broadly speaking, Congress was supportive of a regulatory regime more kindly disposed to cable. One important caveat to this philosophy existed, however. Cable still did not pay for its content, and that stuck in the throats of nearly every regulator. Copyright still had to be resolved.

Copyright Redux

If broadcasters felt unduly pressured going into the 1972 Whitehead compromise (and they did), they felt completely hoodwinked after the fact. The NAB was near apoplectic. By almost everyone's account, the copyright problem had to be solved before cable could move on in both the regulatory and financial worlds. And, equally, by nearly everyone's account, the Whitehead compromise had been designed to help bring about a resolution of that issue. Acceptance of the 1972 rule set, especially by broadcasters, was premised in no small measure on the settlement of the copyright issue and the adoption of legislation based on that understanding. Congress, however, despite repeated requests and efforts, had in ten years failed to pass any bill that substantially affected cable television and it appeared that it was not going to start now. Cable, at least initially, appeared to have no intention of going out of its way to force the issue.

A copyright compromise was to have been worked out before the FCC rules took effect on March 31, 1972, but that deadline came and went with no sign of movement. Despite their public denials, the NCTA officials were perhaps fairly accused of dragging their feet on the issue. For as long as a copyright bill was bottled up in the Senate, cable remained free of copyright control. Broadcasters generally, and the AMST in particular, felt they had been swindled by the OTP-forced compromise and without the promised copyright settlement were soon looking for FCC reconsideration of the 1972 rules. Challenges from a number of quarters were turned away, however, and the new regulations took effect on time.

Meanwhile in the courts, TelePrompTer was doing its best to keep the CBS copyright suit in litigation. Kahn had done his math and determined the cost of legal fees was substantially smaller than the cost of a copyright loss.[17] Originally filed in late 1964, the First Federal District Court opinion in *CBS v. Teleprompter* was finally issued in May 1972, upholding cable's right, following *Fortnightly*, to retransmit signals without copyright obligations.[18]

The TelePrompTer decision again only acted to encourage cable's avoidance of the bargaining table and, in the Senate, McClellan was distracted for several months by a tougher-than-expected primary fight in his home state of Arkansas. After narrowly holding onto his seat, he announced that he would proceed on copyright legislation, but that the issues were so complicated and disputatious that he would take no real action until 1973 and even then he declined to promise formal legislation. In early 1973, with the NCTA slyly suggesting they preferred to wait for Senator McClellan to introduced his bill, whenever that might be, the Motion Picture Association of America (MPAA) and NAB gave up any hope of bringing cable to a copyright settlement and abandoned the talks. They went to Congress with their own bill to establish a copyright tribunal, in place of the set fees proposed by McClellan.

The Second Circuit Court opinion in the *CBS v. Teleprompter* suit handed down in March 1973 reanimated McClellan. The decision was a partial victory for both sides, finding that cable was liable for copyright on imported signals, but not on local retransmission or local origination channels, even when using microwave.[19] McClellan soon after formally reintroduced his legislation, which now called for a copyright royalty tribunal, operating within the Library of Congress, to set fees as necessary and collect and allocate royalties. Cable operators would be granted "compulsory licenses" under the bill, which meant they were given a blanket authorization to carry broadcast signals but required in turn to pay for the privilege. The Omnibus Copyright Revision Bill created a sliding scale for copyright payments. The proposal was complicated and, as always, controversial, however. Little chance existed that a reform package of any kind would emerge until at least 1974.

In March 1974, the Supreme Court, in *Teleprompter v. CBS*, announced what would be its final decision on cable's status under the 1909 copyright act. It wavered not an inch from its opinion of six years earlier in *Fortnightly*. Whether or not cable used microwave, declared the Court, it remained a functionally passive extension of the viewer's antenna. There was no reperformance under the law and, therefore, no copyright obligation.[20] The Supreme Court's decision was the push needed to move Congress toward final resolution, although it would take another year and a half for the enabling legislation to work its way through the relevant committees.

In the meantime, many cable operators had begun to change their tune on copyright. The copyright problem had haunted cable from is inception. Because operators did not pay for the signals they retransmitted, they were considered by many to be taking unfair advantage of a loophole in the technology and, following *Fortnightly*, in the law. Broadcasters accused them of theft. OTP Chief Whitehead, among others, labeled them "parasites."[21] While the freedom from liability served their economic interests, it tarnished the industry's public relations position and retarded lobbying efforts in Washington. Even the pro-cable Nixon administration favored copyright responsibility for the industry. Congressmen and the FCC Commissioners told industry representatives bluntly that cable would not be deregulated until the copyright issue was resolved. Commissioner Wiley warned them in 1972: "Without [copyright] compensation, I reiterate my prediction that your expansion will be continually stymied and frustrated along the way."[22] The road to deregulation clearly ran through copyright country.

By the final ruling in *Teleprompter* it was, moreover, fairly clear to all that congressional copyright reform would happen; it was no longer a matter of "if," but rather "when" and in what form. The NCTA did not, and perhaps could not, turn its back to the political wind. It began encouraging its members to support some form of compromise copyright legislation. At the 1975 convention, Association President David Foster and Chairman Bruce Lovett pleaded with members to let them to resolve the issue and put it behind them, which was not to say the Association was going to stand aside and let others shape the bill. The NCTA's strategy was first to delay legislation and second to keep fees in the final package as low as possible. While supporting copyright reform generally, the NCTA, for example, fought to remove a clause creating a copyright tribunal to oversee future rate adjustments. Instead, the organization favored a permanent fee scale set by Congress. It was a point the industry would lose, however.

The NCTA's compromising stance on copyright was having an ancillary effect on internal industry relations, as well. Although most of the larger MSOs (TelePrompTer was an exception) saw the value in settling the aged problem, smaller operators fought energetically against any retreat or compromise on the issue. They went so far to suggest in congressional hearings in 1975 that program suppliers pay-cable operators a royalty fee for adding viewers to their audience. The internal conflict was so great that it caused a split in the Association, discussed further below. In the end, it did not have a significant impact on the resolution of the copyright issue, however.

Following the Supreme Court's decision in *Teleprompter* and after more than two decades of debate and procrastination, Congress finally wrestled copyright to the ground. Legislation moved through the Senate Judiciary Committee with a relatively quick referral to the Senate Commerce Committee. In September 1975, the Senate passed the Omnibus Copyright Revision Bill (S.22). Passage of the House companion bill (HR 2223) soon followed. Liability under the new law, The Copyright Act of 1976, took effect January 1, 1978.

Cable, finally, was held financially responsible for the retransmission of broadcast programming. The compulsory license was adopted for all broadcast signals, local and imported. Beginning July 1978, cable operators would pay a royalty based on system size, market, and use of each imported signal. The newly created Copyright Royalty Tribunal would be responsible for making adjustments in the fee schedule and the Library of Congress Copyright Office would collect the money and distribute it to the various claimants, including broadcasters, film studios and producers, and sports interests. Compensation for pay-cable programming was to be negotiated between individual parties.

The amount of the fee was based on a complex formula that considered system revenues, number of signals carried, market size, and whether the signal was commercial or educational. Small systems would pay a two-tiered flat rate, regardless of the number of signals they imported.[23] Larger systems would pay a fee based on the number and type of signals imported, although even those importing no signals paid a minimum, equivalent to just less than 1 percent of gross revenues.[24]

Of the collected fees, less than 10 percent would eventually go to the commercial broadcasters who originated the signal. Most of the money, about 55%, would be distributed to the producers, writers, and others with contractual or copyright interests in the original programming (represented by the Motion Picture Association of America). The remainder would go to sports interests, including Major League Baseball (MLB), the National Collegiate Athletic Association (NCAA), and the National Basketball Association (NBA), as well as to noncommercial broadcasters. The exact division of money between the participants would become a matter of on-going debate and litigation.[25]

Reaction in the industry was mixed. Some had battled long and hard against copyright exposure and were not ready to give up the fight. Others took a more realistic position. Burt Harris, NCTA Chairman at the time, was sympathetic to both positions. "I was always opposed to copyright for cable because I felt along with many others in the industry that it had already been paid. But then there were those who felt Congress didn't share that view, and I think we realized that some form of copyright payment was inevitable."[26]

It was a new financial burden on the industry, but as with many of the other regulatory reforms of the time, it helped establish cable, economically and legally, as a more stable and, therefore, more viable business. Cable could no longer be accused of being a TV signal "pirate." One of the keys to cable's success in Congress and at the FCC in the latter 1970s was passage of the 1976 legislation.

FCC Re-regulation

Signal Importation

Even before passage of the new Copyright Act, the regulatory fences around cable had begun to collapse. In late 1974, the FCC issued a notice of proposed rule-making following completion of work by a special committee on the relationship between federal, state, and local cable jurisdiction. The Federal/State-Local Advisory Committee (FSLAC) had noted the gradual accretion of cable rules at every level of government, and the FCC declared that it was time to change tact. "It is clear to us," stated the agency, "that cable television is in danger of becoming smothered in regulatory paperwork. For this reason we have already embarked on a program of investigating all our own requirements to see where we could ease this burden."[27] (In mid-1975, the Commission terminated, without taking action, the long-standing examination of the relationship between federal, state, and local control of cable.[28] It did, however, encourage state and local authorities to avoid excessive and duplicative cable controls.)

The actual deregulation of cable began slowly. The Commission exempted systems with fewer than 1,000 subscribers from the syndicated exclusivity[29] and nonduplication rules.[30] In what was largely just a cleaning up of a long-abandoned requirement, it also formally repealed its mandatory local origination rules, but required operators to buy and maintain local origination equipment for community use.[31] In September 1975, the Commission eased cable-broadcast cross-ownership rules, reducing the number of operators affected by the prohibition.[32] In 1976, it removed a 1972 requirement that local franchising authorities regulate cable rates (although cities were still allowed to do so).[33] Most importantly, the Commission began nibbling away at the restrictions on signal importation. At the request of NCTA, the Commission ruled in September 1974 that cable systems could import an unlimited number of distant signals during periods when local "must carry" stations were not on the air, opening the midnight hours to distant signals.[34]

In what would become an important deregulatory move, the Commission dropped its anti-leapfrogging rules in December 1975.[35] Cable operators would now be able to import signals from anywhere in the country. One of the FCC's motives in adopting the restrictions involved a concern that imported "superstations"—successful independent stations out of New York, Chicago, and Los Angeles—would fill up available channel capacity and prevent the carriage of area stations that more closely reflected the local culture and character. In reversing its position, the Commission held such concerns to be overstated. It felt that the high costs of microwave carriage and the low interest of advertisers in regional stations would limit such expansion. (An eccentric yachtsman from Atlanta, Georgia would later prove them mistaken; the FCC had just let Ted Turner into the national cable business.) Additional small holes in the importation dyke were created in 1976 when the Commission opened the way for unlimited use of distant signals from religious and foreign language stations,[36] and for use of network news programs not appearing on stations normally carried by the cable system.[37]

Some of the most important rule changes were those that would directly affect the health and well-being of the still-gestating satellite distribution system, including the rules on pay-TV.

Pay Rules

Broadcasters were distraught by the FCC's 1970 pay-cable rules, arguing that the regulations were far too lenient. A few cablecast operations were beginning to show popular films on a pay basis and the broadcast industry, with good cause, saw this as a dangerous first step in breaking their hold on TV entertainment and sports. For the benefit of policy makers, the NAB cast the problem as nothing short of the creative survival of the national broadcast system, saying, "At the end of the whole CATV [pay] origination road, the Commission will find that it has permitted CATV to become so bloated on the vast financial rewards of entertainment and sports origination, that free television will be uttering its last breath with a bland monotony of cheap, uncreative program fare."[38]

In 1973, the NAB launched a full-scale public relations assault on pay-cable. It formed an anti–pay-TV committee and began extensive lobbying. If pay service was seen by cable as "the key to Fort Knox," broadcasters saw it as "the engine of their destruction." Spending more than $600,000 in the battle, the NAB took out full-page ads in the nation's newspapers warning of the danger of cable and declaring that the public should not be charged for what it already received free. "Keep Free TV Free" was the NAB battle cry. NCTA responded with a $250,000 counter-offensive that eventually included fairness doctrine complaints against several broadcast stations.

Both parties had allies from the film industry. Broadcasters were supported by the National Association of Theater Owners (NATO), which launched its own vigorous attack on pay-cable. The film studios, on the other hand, organized a cheering section for cable. The MPAA eagerly endorsed cable's pay efforts, and MPAA President Jack Valenti became a frequent pro-cable speaker at trade shows and public policy forums. For the studios, which were suffering through their own economic slump, pay-cable was a potential source of additional revenue, in effect a new set of movie screens. The film industry, albeit somewhat cautiously, fed material to the early pay providers. As the first pay experiments began, Hollywood was tentative, offering films that were at least a year old. The time lag between theatre release and pay-cable availability soon shrank, however, with some movies airing on cable within a month or two of their theater showings. The major studios were still not offering a steady diet of their biggest features; many of the films Hollywood provided were losers at the box office and, in some cases, were so bad that they had not even been released into theaters. A few were blockbusters, such as "Summer of '42," however, and these constituted the studios' efforts to test the cable market.

In contrast, theater owners had a predictably darker view of pay-cable. They pressured the studios to restrict the release of films to companies, such as HBO, and the studios had to dance carefully to avoid stepping on too many toes too often, although in the end the profit potential in the secondary markets was too tantalizing to pass up.

While the NCTA and NAB, with their respective seconds, exchanged rhetorical jabs and hooks in the ring of public opinion, the real fight was waged at the FCC. The Commission denied multiple requests to review its 1972 cable rules generally, but it was willing to take another look at pay-cable regulations. It admitted it had erred in extending the broadcast rules to cable without hearings, and offered to review the requirements. The Commission staged hearings in early November 1973, which drew every conceivable aggrieved party. Program producers and distributors from broadcast and cable appeared, along with representatives of the TV networks, all the trade associations, and most of the major motion picture studios, more than 100 participants in all.

The core issue did not differ from what it had been when the FCC considered pay-TV in the 1950s—it was the extent to which pay services would siphon programming away from broadcast audiences and deny that material to what the broadcasters labeled "free TV." They were joined in the 1970s by several public-interest groups. Presaging concerns of the 1990s about information-rich and information-poor, these groups worried that the pay service would reduce the availability of attractive entertainment and sports programs for those who could not afford cable, such as the urban poor, as well as those who could not receive it in unserved rural areas.

Playing the Blue Sky pubic service card, cable operators responded that the revenue generated by pay programming was necessary to support the plethora of social services, such as public access facilities and channels, now required by FCC regulations and increasingly by local municipalities. Cable specifically sought to move the start window on feature films from two years to five, in order to increase the availability of recent, quality movies.

With its normally unhurried pace and with more than the usual amount of intense and prolonged lobbying on all sides, the rule-making dragged on through 1974. Some of the delay was attributed to the strenuous efforts of ABC to kill or at least stall relaxation of the rules. ABC Chairman Leonard Goldenson and President Elton Rule met personally with key members of Congress, who, in turn, pressured the FCC to slow its procedures. Eventually, however, and following a change of Commissioners (Glen Robinson, James Quello, and Abbott Washburn joined the Commission in 1974), a new round of hearings was called in October 1974 and some 100 parties signed up for another two and a half days of oral arguments. By the end of the year, the Commission announced it was easing its regulations, but not by much. When the new rules were released in March 1975, the two-to-ten-year rule was relaxed to a three-to-ten rule, as long as a particular film had not been shown in an operator's market in the previous four years.[39] Series programs not previously shown on broadcast stations could be used, although this limitation was abandoned in November 1975.[40] Big-ticket sporting events, such as the World Series or the Super Bowl, could not be shown on pay-cable if they had been on broadcast television in the previous five years. And carriage of regular sporting events remained tightly restricted in a formula of mind-numbing bureaucratic complexity.

In fact, the FCC's modified rules on pay-cable were modified very little and retained their suffocating effect on the development of pay service. The industry did not waste time requesting reconsideration by the Commission, which it saw as futile. The FCC completed its work on the revised rules on Thursday, March 20, and a coalition of cable operators filed suit in Federal Court in Washington, D.C., the next day. Joining in the action were TelePrompTer, UA-Columbia, ATC, Warner, Viacom, and HBO. The FCC maintained its position on a requested reconsideration, and broadcasters, equally distraught over the FCC's action, filed their own court challenge.

Film studios, including Columbia and United Artists, and even professional baseball filed suits as well. The courts consolidated fifteen different actions into one case, giving HBO the lead title in what became known as *HBO v. FCC*.

In court, HBO argued the regulations went beyond the Commission's jurisdiction and included restraints that violated HBO's First Amendment rights. The D.C. Circuit Court agreed.[41] In a decision handed down in late March 1977, the Court found the antisiphoning rules to be beyond the Commission's authority and unsupported by the evidentiary record. Importantly, and for the first time, the Court also declared that cable television was vested with some measure of constitutional protection. The Court was unconvinced that the pay rules served a substantial government interest sufficient to overcome the inherent First Amendment protection afforded a speaker such as HBO. The decision had important constitutional ramifications, which are explored more fully in Chapter 10, and spoke once more to the evolving legal definition of the medium. The practical implications for HBO and the cable industry were subject to debate. Symbolically, it was a great victory. Cable was now free to negotiate for any film, no matter what its age or nature. Sports events too were now open to competitive bidding by cable interests. But HBO's "bread and butter" was current films, less than two years of age, and original events. Some observers suggested that the decision had little direct impact on HBO's regular schedule and, in fact, concern about cable siphoning had been overstated, for political purposes, from the outset.[42] The FCC appealed the decision, but the Supreme Court declined to review the case and the opinion stood.

Although the D.C. Circuit struck down the cable rules, it upheld the Commission's pay rules as they applied to broadcast television. The Commission, soon after repealed its broadcast pay policy, concluding that it served no public interest to treat broadcast pay programming differently than cable.[43]

Dish Size

Whereas the Court's decision in HBO helped open the spigots of programming, distribution for HBO and other potential satellite providers was still a problem in 1975. The nine-meter (about thirty-feet) dishes mandated by the FCC kept participation in a satellite feed beyond the financial reach of many small operators. In the summer of 1976, CATA petitioned the FCC to permit receive-only earth stations smaller than nine meters. CATA was the new trade organization for small operators (see discussion below) and the rule change was particularly important for the Association. The petition was prompted by an FCC denial earlier in 1976 of a request by a Gaithersburg, Maryland, system to use a smaller dish. CATA head Steve Effros argued that the technology had advanced to the stage where the large dishes no longer were necessary in every situation. Despite opposition to the petition from

AT&T, ABC, and the NAB, the Commission, in December 1976, relaxed its technical rules.[44]

The FCC reduced its size requirement to as small as four and a half meters, as long as minimal technical specifications were met, bringing the price down to as little as $25,000. More operators could afford the dish and the service, and many quickly took advantage of the opportunity. George Gardner, now a successful operator in Carlisle, Pennsylvania, later recalled reaching now to the stars for TV signals using his first dish: "The first satellite reception we got into was probably about 1978 in Carlisle. We bought our first five-meter Scientific-Atlanta dish, found a place to make it work, and got our picture in the paper because of this new technology. I think that one cost us $25,000. I waited until it got down to that price because I just couldn't afford it [until then]."[45]

Access

Cable also enjoyed a victory in the courts over the FCC's system rebuild and access requirements. Cable operators had complained that the 1972 rebuild mandates, which called for up to twenty channels, two-way interactive capacity, and various access channels, placed too great a financial burden on the industry, especially in the tumbling economy. The industry estimated the mandates would cost up to $430 million, and would have to be passed along to the consumer in the form of higher rates. Institutional lenders, such as the John Hancock Insurance Co., joined operators in pleading for delay or elimination of the rules. The lenders told the Commission they saw no near-term hope cable could recoup the investment needed for the upgrades; they were worried about the money they had already lent the industry and were unlikely to provide more under the rebuild rules. The Commission, with its changing perspective on the relationship between government and industry, was sympathetic and, in July 1975, removed the rebuild requirement.[46] The action affected only systems in operation before March 31, 1972; systems built after that date, covering many of the larger markets, were still bound by the technical requirements.

In 1976, the Commission also curtailed its public access rules, allowing operators to offer a single composite access channel where demand failed to warrant four separate ones.[47] In the same Report and Order, it delayed for ten years the requirement that systems achieve twenty-channel capacity by 1977. At the same time, however, the rules were modified such that they applied to all markets (with systems of 3,500 subscribers or more), instead of just the top 100. This had the effect of extending the regulations to systems not previously covered by them, and one of the newly impacted companies was Midwest Video, the MSO that had brought an unsuccessful suit against the FCC in 1971over the local origination rules. Midwest sued again, challenging the Commission's access requirements. This time, they won.

In February 1978, the Federal Appeals Court for the Eighth Circuit struck down the Commission's access and channel capacity rules.[48] The Court found the requirements to be beyond the FCC's jurisdiction and probably a violation of the First and Fifth Amendments, although it expressly declined to decide the issue on constitutional grounds. In Midwest's 1972 local origination case, the Supreme Court had upheld the FCC (*Midwest Video I*), but Justice Warren E. Burger had warned that those rules strained the limits of FCC authority. The access regulations, it seemed, broke through those limits. In April 1979, the Supreme Court, as had the Eighth Circuit Court, concluded that the access requirements turned cable systems into *de facto* common carriers, an action the Commission had no authority to take.[49] The FCC's power over cable sprang from its jurisdiction over broadcasting, insofar as cable was ancillary, following *Southwestern*, to the older medium. But the Communications Act specifically prohibited the application of common carrier-like regulation to broadcasters and, by extension, to cablecasters, according to the court.

Municipalities were free to impose leased, public, governmental, and education access channel requirements on cable as part of the franchise process, and such obligations became common around the country. The Supreme Court's decision, in what became known as *Midwest Video II*, however, brought to an end some of the central legal and industrial manifestations of the original 1960s Blue Sky thinking. It also suggested the social definition of cable in the courts had shed the last vestiges of its passive antenna garb and was looking more like a protected First Amendment speaker (the evolution of cable's First Amendment standing is traced in Chapter 10).

Pole Attachment

Cable in the late 1970s claimed an important victory in its long-term jousting with the telephone industry, as well, although not without some bumps and bruises along the way. The Commission's 1968 and 1970 rules had blocked the telephone companies from a wholesale takeover of the cable service through either direct ownership or lease-back contract. The companies still charged for access to their poles, however, and despite an earlier hiatus in rate increases, pole attachment rates remained a source of cable-telephone company (telco) friction. The FCC had an interest in the on-going talks. Cable, in fact, wanted the Commission to prod the telephone companies, but not necessarily to take control of rates. Industry leaders initially urged the Commission, in essence, to keep out of the negotiations.

The talks began in 1970 and repeatedly stopped and restarted over the next several years without any discernable progress. In 1973, General Telephone Company of California proposed a 100 percent pole attachment rate increase, from $3 to $6, and following cable operator complaints, the Commission indicated it might assert authority of pole attachment rates unless the parties could come together on their own.[50] The FCC instructed its staff to draw up papers asserting Commission authority over pole attachment rates, and then gave the warring parties ninety days to work out a compromise before starting its own rule-making proceedings to implement that authority.[51] The resumed negotiations included utility companies as well as NCTA and the telephone interests.

The FCC subsequently extended to January 31, 1974, its deadline for a private settlement and, for a short time, it looked like an agreement might be reached. In 1974, it appeared that the NCTA and AT&T had engineered a compromise calling for a modest increase in fees, up from the $3.35 figure the negotiating committee cited as the national average. The plan would have resulted in $1 per pole increase over four years in exchange for a roll-back in "rearrangement"[52] costs and other telephone company concessions. The proposal was approved by the NCTA board in January 1974 on a nineteen-to-one vote. The one dissenting vote, however, was cast by the representative from TelePrompTer. TelePrompTer, in the very depths of its financial troubles and strapped for cash, was not of a mind to support a substantial increase in its operating costs. Despite its troubles, it was still the largest MSO in the country and the largest dues-paying member of NCTA. Its cable division was now led by Bill Bresnan, a man not known for soft-spoken diplomacy. The blunt, some might even say pugnacious, Bresnan blasted the NCTA for its "hat in hand" posture in the pole talks, and implied that TelePrompTer might even quit the national cable association over the matter. "If those people [the NCTA] think [a rate increase] is inevitable, then they sure as hell shouldn't be representing us," declared Bresnan.[53] It did not help that Bresnan communicated his displeasure in the form of a press release. Going public with the criticism angered other MSO board members and the feud briefly threatened to split the association. While that rift would eventually mend, TelePrompTer's very public decision to pull out of the talks effectively killed that round of pole attachment negotiations.

The two sides continued to trade proposals over the summer of 1975, again with no progress; by the fall, both parties had given up. Amos "Bud" Hostetter, in charge of negotiations for the NCTA, sent a letter to FCC Chairman Dick Wiley in September saying, in essence, that after more than five years of discussions, a negotiated settlement seemed very unlikely. A week later, AT&T sent Wiley a similar letter. AT&T said it would attempt to settle rate issues on a state and local basis. In response, Wiley offered his own last-ditch proposal, one that called, in part, for a rate freeze in most states while the parties continued to try to work out their differences. In late September 1975, AT&T and the NCTA finally signed on to the Wiley compromise, which included a rate freeze in all states except California and Pennsylvania until January 1, 1979. The relief would not to last long, however. Within a few weeks, the agreement collapsed. The difficulty lay in the scope of the proposal, which covered only AT&T affiliated telephone companies. It did not reach to non-AT&T companies or, more importantly, to electric companies that provided pole space to cable, and these two groups controlled 7 million of the 10 million available poles in the country.[54] Cable operators had battled power companies on a state-by-state basis for several years over threatened and enacted rate hikes. With the electric utility companies declining to sign on, the Wiley compromise disintegrated.

The cable industry looked again to the Commission for help, but this time found none forthcoming. For years, the FCC had pressured the fractious industries to settle their differences and held out the possibility that the government would step in and seize jurisdiction if they could not. Despite the posturing and threats of intervention, the FCC had never formally claimed general authority in pole attachment matters, however. To the extent that it might have been inclined to do so in 1973, by 1976 it had lost the taste. The Commission, in July 1976, ruled it had no authority over power company poles and that it needed to reconsider its position on the question of even telephone company poles.[55] Wiley explained that he pressed for a settlement earlier, in part because he was not sure that sufficient votes on the Commission would support formal intervention. His hunch was correct. In March 1977, the Commission reiterated its position that it had no authority over power company poles—they were not, after all, engaged in any normal form of broadcast or common carriage. Moreover, the Commission stated it now felt it had no jurisdiction over telephone poles, either, reversing the thrust and philosophy of years of previous FCC rhetoric.[56]

The only recourse was to Congress, which responded with surprising rapidity. In February 1978, Congress passed legislation granting the FCC authority to control pole attachment rates in all cases[57] and the Commission soon thereafter enacted a formula (subsequently modified) for determining fair attachment rates.[58] The Commission's action concluded, for the most part, the years of uncertainty and dispute. For a few years, the telephone companies would not be a major concern for cable, but only for a few years.

State Activity

In the 1968 case *TV Pix*, the states were granted the authority to regulate cable where the FCC had not already acted, and a number of them took advantage of the opportunity, as noted in Chapter 6. Eleven states had adopted comprehensive plans for cable regulation and vested authority in state-level boards by 1974. After 1974, that zeal to control cooled. No new state boards were created and, in some cases, states took a cue from federal authorities and eased up on cable regulation. Three states—Massachusetts, Minnesota, and New York—maintained their independent cable regulation agencies, but no other states seemed interested in creating such specialty bodies. In fact, in Minnesota and New York, legislation was introduced to disband the boards and move the regulatory functions into the state PUCs.

States nonetheless addressed a number of more tailored cable-related issues, including service standards, subscriber rates, pole attachment, and taxation. In Pennsylvania, which had long prohibited substantive local control of cable, cities finally won the right, through the courts, to issue franchises and regulate rates.[59] Some states assumed jurisdiction in pole attachment, as per the federal pole attachment legislation of 1978. Cable's status as a possible public utility remained a topic for debate, with most states deciding simply to treat it in a separate category, frequently leaving it ill-defined for legal purposes.

In one important case, the State of New York was prohibited from regulating rates for pay-cable. In 1974, the FCC had reiterated its position that pay-cable prices were not to be controlled by any level of government.[60] Nonetheless, in March 1976, the New York State Commission on Cable Television declared its authority over pay-cable fees. A lawsuit quickly followed. A federal district court, in 1977, enjoined the state's action, finding that the FCC had preempted the field.[61] An appeals court agreed and the Supreme Court declined review. The FCC's authority in this area trumped state and local rules.

Many states had to act, additionally, to protect their municipalities from antitrust lawsuits filed by disgruntled cable operators. The problem began in Boulder, Colorado, in a showdown between city officials and TCI. TCI's local operating company, Community Communications Company (CCC), had been operating for years under a nonexclusive franchise. CCC served only about 20 percent of the city, providing off-air retransmission of signals, most from Denver. In 1979, it began planning for the installation of a new satellite dish, and with the expansion of programming that it promised, sought to wire the rest of Boulder. At about the same time, another company, Boulder Communications Co., requested a franchise to build a competing system in town. Faced with the change in circumstances, the city chose, in essence, to freeze all cable activity while it considered a comprehensive plan for cable in Boulder. It ordered a three-month suspension of CCC construction and invited applications from other companies to franchise the unwired neighborhoods.

TCI's response was direct; it simply ignored the city and began stringing wire. But as quickly as TCI put up new strand, city workers tore it down. At one point, the city had TCI construction crews arrested. The requisite lawsuit quickly followed. *Community Communications, Inc. v. Boulder* was an important case for the industry.[62] TCI sued on antitrust grounds and the city argued that it was immune, as a governmental body, from antitrust action. In January 1982, the Supreme Court held for TCI, finding that municipalities were not immune from antitrust suits unless specifically exempted by the state. Antitrust litigation could, thereby, be used to challenge exclusive franchise procedures and any exclusive franchise granted to a competing cable company. A cable company that felt the city's franchising process had unfairly favored a competitor could, in short, sue under federal antitrust laws. This substantially weakened the position of cities in the franchise bargaining process and, as a side effect, made cable lenders more cautious in light of the increased uncertainty of a granted franchise. In the end, one of the most lasting effects was to send cities scurrying off to their state capitols for legislative authority insulating them from antitrust action. (In 1984, Congress passed legislation largely protecting cities from these kinds of antitrust actions.[63]) TCI and Boulder, meanwhile, finally settled. TCI was allowed to stay in the city with a promise to fully wire the town within two years.

Beyond such isolated incidents, however, no broad-based move was afoot among the states to extend control over cable television. Regulatory activity centered in Washington, D.C., and by the end of the 1970s, the news was almost all good for cable television.

Deregulation

The year 1976 was an important and memorable one for cable. The industry was coming out of an industrial tailspin. HBO was on the satellite and the FCC had helped cut the price of dishes by 75 percent. Congress had saddled cable with copyright obligations, but bestowed political legitimacy in the process. Armed with the new copyright bill, operators renewed their assault on the remaining FCC importation and program exclusivity rules, arguing that that they were no longer needed. The argument found a sympathetic ear in FCC Chairman Wiley, who had characterized the regulations as a "copyright substitute."[64]

The year also saw the election of a new president, Jimmy Carter, replacing Gerald Ford, as the country looked for a respite from the painful years and still fresh memories of Watergate. For telecommunications policy, however, Carter offered little in the way of change. By 1976, the implicit faith in the power of deregulation had become accepted political gospel, even for some Democrats, and Carter was elected in part on a promise to cut back on what was seen as wasteful and intrusive government oversight of private, including commercial, affairs. He campaigned on a promise to reduce government bureaucracy and, as part of keeping that promise, he killed the White House Office of Telecommunications Policy, replacing it 1977 with a smaller agency, the National Telecommunications and Information Administration (NTIA), housed in the Commerce Department.

In the fall of 1977, Carter named Charles Ferris to replace Dick Wiley as Chairman of the FCC. Bill Daniels, who had celebrated the chairmanship of Wiley, was even more enthusiastic about Ferris, declaring, many years later, "Ferris did more for the industry, in my opinion, than any individual in the history of our business."[65] Strong words, but Bill Daniels knew where the interests of the industry lay. Ferris, despite his many years as a staff member in Congress on the democratic side (he had served as Assistant General Counsel to the Democratic Policy Committee and was Chief of State for Senator Mike Mansfield (D-Montana), continued and even extended the deregulatory program of the Commission.

He was aided by other newly appointed commissioners. Margita White was named to the FCC by President Ford in July 1976. She had been his assistant news secretary and Director of White House Office of Communications. Her background also included work for Barry Goldwater and the Republican National Committee. James Fogarty, appointed to the Commission in September of 1976, was a liberal who had worked for John Pastore and served as Staff Counsel on the Senate Commerce Committee. Unlike liberal-minded Commissioners of the 1960s, however, Fogarty sought to release the potential of cable through deregulation. In his confirmation hearing, he told the Committee that one of his objectives was to encourage cable TV growth.[66]

Ferris shared that view, announcing that chief among his limited goals for the Commission was to deregulate where the marketplace would work more effectively.[67] "Remove the Commission from the marketplace," he declared. "If a company has the money to invest in a telecommunications service, let it, and leave it to the public to determine whether the service will survive."[68] Somewhere, John Doerfer had to be smiling.

Ferris lost little time in making good his philosophy. In September 1977, the FCC eliminated most of the federal standards for local franchising, leaving only the 3 to 5 percent cap on franchise fees, and even that was modified to permit revenue from pay-cable and other nontraditional sources to be included in the capped franchise fee base.[69] (The Commission also indicated an interest in eliminating the franchise fee limits, as well.) In December 1977, the Commission eased satellite receiver rules a bit further. Up to that point, cable operators had to seek formal permission for each new signal they downlinked from the satellite. As new programming services began piling on to RCA's Satcom I, applications to receive them began piling up at the FCC. The Commission decided that operators would no longer have to seek formal approval simply to add a new signal if they already had an authorized dish in service.[70] Broadcasters asked the Commission to ban cable carriage of satellite-distributed TV stations, claiming potential damage from the unexpected popularity of superstations; the Commission refused the request.[71] In 1978, the FCC also dropped its certificate of compliance requirement.[72]

The Commission also initiated, in 1978, a special staff investigation into the dominance of television by the three major networks.[73] The two-year study would conclude with a recommendation that the Commission encourage alternative TV distribution systems, including cable television, as a method of bringing greater TV diversity to the public.

Through the latter 1970s, the old cable rules were falling away, but more significant changes were still to come. At each juncture in the rulemaking, broadcasters were given the opportunity to demonstrate that deregulation would result in economic harm to over-the-air TV and injure the public interest. The broadcasters offered studies and testimony, but for the deregulatory minded Commission, the evidence, at each step, was insufficient.

Along with a continuing deregulatory tenor, the 1976 staff report out of the House Subcommittee on Communications helped stimulate FCC review of its cable policy. In late 1976, following passage of the new copyright act, Chairman Ferris helped initiate a study to determine whether the syndicated exclusivity rules still served any important public policy purpose.[74] In June 1977, the Commission opened a second proceeding to review even more broadly the economic relationship between local cable operators and broadcasters and the impact of Commission rules on the financial health of local broadcasting.[75] The studies, the Syndicated Exclusivity Report and the Economic Inquiry Report, were released together in May 1979,[76] and came to similar conclusions: Cable television could not be shown to have a serious deleterious financial impact on local broadcasting. In fact, the Commission's own study, conducted by Rand economist Rolla Edward Park, showed a slightly larger negative impact than did the NAB-sponsored study. The reports suggested importation of signals into the top 100 markets could eventually erode broadcast audiences by anywhere from 6 to 15 percent, but only once the markets were fully wired, a process that would take, at best, many years.[77] The likely conclusion, according to analysts Stanley Besen and Robert Crandall, was that "elimination of the distant signal rules would generate audience losses for local broadcast stations of perhaps 1 or 2 percent per year over a ten-year period in the largest 100 markets."[78] No apparent correlation was found in the analyses between cable growth and broadcast industry revenues, profits, or programming mix. The FCC suggested that any loss in audience caused by cable would be offset, over the long run, by increased population and demand for advertising. Data also indicated that for UHF independents in the top 100 markets, the presence of a cable system actually increased the audience by 5 percent. It was a finding that supported predictions made years earlier by some that cable's ability to move a UHF signal to a VHF channel on the home set would only work to benefit the struggling broadcast service, and it highlighted the irony of years of regulation aimed at restricting cable to benefit UHF. In sum, cable appeared, at most, to constitute a minor threat to broadcast industry and no threat at all to the public well-being.

On day the FCC met to review staff reports, Ferris told the anxiously assembled audience, "Today we meet to gum up the works or bite the bullet. I hope we don't gum the bullet."[79] With the reports in hand, the Commission in April 1979 launched rule-making designed to eliminate the remaining distant signal importation and syndicated exclusivity regulations.[80]

Broadcasters protested, although some observers felt their opposition was half-hearted. Years of difficulty for cable in wiring the larger markets, coupled with the passage of copyright liability may have blunted broadcaster anxiety over the importation rules. The political reality of a Commission clearly bent on deregulation may also have played a role. The NTIA also expressed the opinion that, if the rules were eliminated, cable operators should be required to obtain broadcaster permission (retransmission consent) to carry the signals.

Finally, on July 22, 1980, the Commission ended its investigation and swept away the remaining principal components of the 1960–1972 rule-making.[81] The FCC, on a four-to-three vote, concluded that the old rules protecting broadcasters were unnecessary and, broadly speaking, the marketplace could look after itself. The four-member majority found that elimination of the rules would not significantly harm broadcasters, and the program supply market would simply have to "adjust to a new reality."[82] Broadcasters and the MPAA were sharply critical of the decision, vowing to take their complaints to Congress. One broadcaster, Malrite Broadcasting, challenged the FCC decision in court, but the appeals court upheld the Commission and the Supreme Court declined to review the case.[83] Commissioner James Quello was especially vocal in dissent, arguing there was no justification, especially for the syndicated exclusivity rules, and proposing the agency delay action on those regulations until the new Copyright Royalty Tribunal had a chance to adopt compensation rates. The full Commission declined the proposal, and further denied a motion to require cable operator retransmission consent. Requiring local systems to obtain permission to retransmit local signals, held the majority, was only a surrogate for full copyright liability and Congress had rejected that approach in passing the 1976 Copyright Act.

The Commission maintained its network program nonduplication and sports black-out rules, but otherwise, cable systems were now free to import as many signals as they could carry, from any location, and without need to delete particular programs, except in cases where a specific program would air simultaneously with a broadcast version. Local must-carry obligations also remained. NCTA President Tom Wheeler told reporters he was "extremely pleased" by the vote, and felt it would stimulate program diversity for consumers. Asked if the NCTA would like to see elimination of the remaining cable rules, Wheeler responded diplomatically, "let's not be greedy."[84]

In review, through the latter 1970s, the FCC had eliminated most of its importation rules. Regulations governing pay programming had been abolished with the smack of a judicial gavel, as had any obligations to provide public or leased access channels. Finally, in November 1980, one of the most deregulatory politicians in modern history won the White House. Ronald Reagan was to be President for the next eight years.

Signs of Life: The Cable Business in the Latter 1970s

Irving Kahn entitled his 1975 Dallas speech "Blue Sky through a Green Eye-shade." Green referred to the color of money. This time, it looked like the prophecy might come to pass. The launch of HBO, followed by a host of new cable programmers (discussed below), the easing of federal regulation, an end to "stagflation," and, consequently, more generous attitude among lenders, were the ingredients in a rejuvenating tonic for cable. In late 1974, the industry received a pleasant gift from the Supreme Court, some $4.1 million in court-ordered FCC refunds, which the agency had been collecting to help offset its own operating costs.[85] The high court ruled the regulatory fees to be excessive. This was but a financial appetizer, however. The real story was pay-cable, which at long last seemed to be paying off. HBO, in particular, brought an infusion of badly needed cash into cable. In 1973, only about ten systems in the country were offering a pay service. Largely through the pioneering efforts of HBO and the cable-satellite connection, that figure rose to nearly 1,500 by 1980.[86] The number of pay-cable subscribers swelled apace, from a meager 18,400 to more than 4.3 million by the end of the decade.[87] By 1979, 36 percent of the nation's operators were offering some form of pay-cable and pay was in 6 percent of all U.S. television homes.[88] Revenue from pay-cable grew from about $68 million in 1976 to $785 million in 1980.

The benefits of pay-cable success went beyond the added revenue from existing customers. The pay services helped attract new subscribers. Between 1975 and 1980, the number of cable customers increased from just under 10 million to more than 15 million; penetration rose from 14.3 percent to nearly 20 percent. Total industry revenue grew from about $976 million in 1976 to more than $2.6 billion in 1980.[89] Revenue per subscriber rose, as did cash flow per subscriber.

Even beyond the added revenue, pay services altered the fundamental economics of cable, especially in the large cities. Pay-cable helped increase the revenue-per-subscriber figure, which climbed from about $97 in 1975 to $121 in 1979. With a higher revenue-per-subscriber rate, cable could achieve profitability at lower penetration levels. In smaller, established markets, which generally had historically high levels of penetration—often 70 percent or more—basic cable revenues offset costs, making income from pay-cable pure profit (or at least excess cash flow). In the major markets that cable now hoped to wire, with their higher construction costs and lower penetration rates, the added revenue from pay could mean the difference between a profitable and an unprofitable business. A major market penetration rate of 35 percent may not have been economically viable with only basic cable income, but if 80 percent or 90 percent of that 35 percent took a pay service (or two), the system became sustainable.

With this equation in its hip pocket, cable once again attacked the big cities, now with the backing of the financial community, which was acutely aware of the industry's evolving regulatory and programming context. ATC forged a six-bank consortium, led by Case Manhattan, and secured $24 million in new financing. Bob Rosencrans refinanced UA-Columbia, garnering $9 million, and Cox secured $20 million in long-term financing. The improved outlook for cable additionally meant that companies were able to realign their debt structure, moving from bank-based financing at variable rates pegged to the prime (usually prime plus 2 or 3 percent) to fixed-rate financing from institutional lenders. Breaking, at long last, their own financing dyke, Malone and Magness, in 1977, put together the largest financing package in cable's history (to that point), with a $77.5 million fixed-rate investment from a consortium of pension funds and insurance companies. In 1978, TelePrompTer, still the nation's leading MSO, refinanced $80 million from insurance companies and commercial banks. With the broad restructuring of debt, interest expenses declined substantially. Cable stock, in turn, rebounded. Cable multiples rose and equity markets looked viable once again.

The influx of cash, both in revenue and equity, brought a return to the growth in cable system values. Echoing the early and mid-1960s, a new round of buying and selling (and buying and selling) began. Now, in addition to a return to the attractive cash flow financials of cable, a surging demand for (and limited supply of) available systems further inflated market prices. Systems that sold for $300 to $400 per-subscriber in 1977 were going for $600 to $700 a subscription by the end of the decade. Growing and selling systems became profitable once more; merger and acquisition activity picked up. More than 140 cable deals, totaling $1.5 billion, were consummated between 1977 and 1981.[90] Storer was rebuffed in an attempted forced buyout of Viacom in 1977. A 1980 effort to merge Cox and General Electric, although passing muster with federal regulators, fell apart when the parties could not come to terms. Other large deals went ahead, however. In 1978, Time Inc. reentered system operations, purchasing ATC for $140 million. Time already held 26 percent of the company, as a consequence of the sale of its cable systems to ATC in 1973, but now owned the company outright.

Communications Properties, Inc., the nation's eighth largest MSO at the time, was sold to Times Mirror in 1978 for about $130 million, as some of CPI's principals sought to cash in on cable's new prosperity. When Times Mirror sought to move CPI executives from their Austin, Texas, headquarters to the West Coast, several balked, including Bob Hughes, took their gains from the sale, and started yet another new cable firm. This time they called it Prime Cable, and acquisition commenced immediately with the $9 million purchase of a small cable company called Amvideo.

At least one financial institution saw such promise in the new cable industry that it did more than simply offer capital. American Express Corp. entered into an agreement in December 1979 with Warner Communications to pay $175 million for half of Warner's cable operations. American Express was interested in the possibilities of electronic banking and commerce and may have been partly motivated by Warner's proprietary QUBE technology (see discussion below). Cable also looked like a generally attractive investment at the time. The new company was Warner-Amex Cable. American

Express also agreed to assume $30 million in Warner debt and support a construction line of credit of $250 million.

Another company, Cablevision Systems Development Corp., was beginning to expand. It was the creation of HBO founder Chuck Dolan. Dolan had left Sterling Cable and Time Inc. in 1973 and launched the new company. When he first moved to the New York area, he and his wife settled on Long Island, where they could afford a home. By the mid-1960s, he had won cable franchises for several Long Island communities, and when he left Time, Inc. used them as the base for his new company. Through the 1970s, Cablevision grew from its Long Island base, adding systems in New Jersey, Yonkers, New York, and the Chicago suburb of Oak Park. With a host of powerful New York City TV signals close at hand and available to viewers off-air, Dolan successfully promoted the Long Island systems, and others, with nonbroadcast services such as HBO, Madison Square Garden sports, and the embryonic but growing satellite services discussed below. By the end of 1975, Dolan's Long Island property was the largest single subscriber to the then-new HBO satellite service. He additionally pioneered a marketing strategy that clustered different program packages at different price levels. A customer could get twenty-eight channels of service (not including sports, HBO, action movies, and a weather channel) for $9.50. Adding HBO would increase the price to $16.50, and additional services could be added up to full service at $29.50. He named each level of service after a color (purple, blue, green, yellow, orange, and red), and the full package was called the Rainbow Service. The Rainbow approach came to be known as "tiering of services," and the tiering concept caught on quickly in the industry.

In 1980, one of the industry's first major MSOs ignominiously exited the cable business. RKO General let go of its Cablecomm-General property after twenty years in cable. The buyer was a long-time player in broadcasting, Capital Cities Communications, Inc. Cap Cities was a mid-sized media company that traced its heritage back through media investments by famous 1930s and 1940s radio commentator Lowell Thomas and the Lowell Thomas radio group. By the 1980s, Cap Cities had diversified holdings in radio, television, and newspapers and was now paying about $139 million for the RKO systems. Cablecom-General stated publicly that it was purely a financial decision. Most of its systems were mature, meaning they would soon need upgrading. Many faced refranchising and a raised bar with respect to municipal expectations about channel capacity and services. Instead of a costly round of franchise renewals and rebuilding, the company chose to sell. At the same time, RKO may have been influenced by troubles brewing at corporate headquarters. In June 1980, shortly before announcement of the Cap Cities sale, the FCC had denied renewal of three RKO television station licenses, citing serious violations of FCC regulations and misrepresentation to the Commission. The future of RKO's thirteen other stations was also placed in jeopardy. Before the end of the year, several executives at the parent company, General Tire and Rubber, were caught up in a scandal involving alleged illegal political contributions, and the parent company faced Internal Revenue Service (IRS) fraud charges. Ultimately, the FCC stripped RKO of all its licenses and the properties were sold at distressed sale prices. In essence, RKO was forced out of the telecommunications business, and may have shed its cable properties as part of the process.

From 1975 to 1980 TelePrompTer, with more than 1 million subscribers at the end of the decade, remained the nation's largest MSO. The other top companies included Time-ATC, Cox, Warner, and TCI, each with a half million or more customers. Established firms such as Sammons, Storer, Viacom, and UA-Columbia constituted a second tier of MSO power, with firms such as Times Mirror, Continental, New Channels, and United Cable close behind.

Technology

Along with new products and new money, cable technology continued to evolve. By the mid-1970s, thirty-five-channel, solid-state amplifiers (50 to 300 MHz) were becoming standard in new construction. Alrhough 70 percent of all systems still operated at about a twelve-channel capacity, nearly 12 percent had thirteen to twenty available channels, and another thirteen percent had more than twenty channels of capacity (see Appendix B, Cable Channel Capacity, 1970–2005). The latter typically were the state-of-the-art thirty-five-channel systems being constructed in the major markets. Engineers were beginning to overcome the reliability problems that had plagued the earlier generations of solid-state equipment, and bandwidth continued to expand. In 1979, TRW, Inc. announced the successful development of 400-MHz technology, capable of delivering more than fifty channels.

Higher frequencies also placed greater demands on amplifiers and frequently created more noise in the system. A technique known as "feed-forward amplification" was developed to reduce distortion and, thereby, increase the number of amplifiers that could be cascaded in a trunk line, which in turn meant longer undistorted reach.[91]

Fiber

Irving Kahn had dreamed of pay-cable television and dreamed of satellites. In 1975, recently out of prison but wealthy from the earlier sale of TelePrompTer stock, Kahn was looking around for a new dream. This time it was fiber.

Fiber optic communication uses glass, instead of the copper and aluminum of coaxial cable, to carry information. A single glass fiber is about the thickness of a human hair (50 to 125 μm), although a number of these single threads are typically bundled into a fiber cable. Instead of carrying a signal using electromagnetic radio waves, fiber moves up the spectrum to carry its information on beams of light. Fiber offers a host of advantages over coaxial cable. The higher frequency of the light waves means greater available bandwidth for transmission; fiber optics provides hundreds of times the carrying capacity of a traditional coaxial line. Fiber is also lighter and, because it is made essential from sand (silica), it is, in long-haul applications, cheaper than copper and aluminum-based coaxial wire. Fiber optic signals attenuate less readily than signals in coaxial wire; a fiber optical signal can travel for thirty miles or more before it has to be amplified, thereby reducing both signal interference and the overall cost of electronics. And, because they carry information on beams of light, fiber lines do not leak the electromagnetic radiation that causes interference to other radio devices.

Most contemporary cable systems are based on an architecture called hybrid fiber-coax (HFC), which uses a combination of fiber and coaxial lines to deliver programming. The high-capacity fiber is in the main trunk line; coaxial distribution is used for the "last mile" or the connection into the home. Fiber first made its appearance in communications in the 1970s, but as with the other technologies discussed here, it had a long evolutionary path to the cable industry.

The observation that light can travel through glass is an old one. Experiments with guiding light, first through water and then through glass rods, date back to at least the 1840s.[92] The great international expositions of the late 1800s used an early form of fiber optics to illuminate giant water fountain displays. In 1880, Thomas Wheeler, a young American engineer, filed a patent on a system to light entire buildings using glass pipes fed by an electric arc lamp in the basement.[93] In a classic display of diversity and selection, however, the incandescent light bulb, invented at about the same time, trumped Wheeler's novel proposal.

Some of the early TV pioneers, such as Jenkins, briefly experimented with quartz glass rods as part of their devices, and the ability of a glass pipe to carry an image over a short distance caught the interest of nineteenth century physicians seeking ways to peer into the body. German medical researchers in the 1920s found that thin threads of glass, glass fibers, grouped into bundles, could not only transmit light but were flexible as well. U.S. researchers advanced the work in the 1950s, demonstrating methods to bundle together hundreds of flexible fibers and increase the efficiency of each with special surface coatings. Although the discoveries offered improvements in medical instruments, such as the formerly rigid gastroscopes, there seemed no immediate application to modern telecommunications, at least not until the development of the laser.

Fiber optic communications requires not just the glass thread but a means of modulating light over that thread. Small, relatively inexpensive, semiconductor lasers fill this role today. A common light bulb produces a wide spectrum of light, many of the individual frequencies, or colors, that make up white light. A laser produces a narrow frequency of light and this coherent beam can be modulated to carry information. The first experimental lasers were demonstrated in 1960.[94] In 1962, Massachusetts Institute of Technology (MIT) researchers announced the development of light-emitting semiconductors, setting off experiments in industrial laboratories across the country to create smaller and more powerful solid-state lasers. Bell Laboratories, with a keen interest in using lasers for communications, was one of the interested parties and put its substantial resources behind laser research. In August 1970, Bell announced a headline-grabbing breakthrough in laser technology. The company unveiled a semiconductor laser that could produce a continuous wave, instead of single pulses of light, and could operate at room temperature. By early the early 1970s, companies were producing small, powerful, and fairly reliable lasers capable of industrial applications.

Parallel progress was being made in fiber technology. The chief trouble with glass is that it is normally so full of impurities that light cannot effectively travel very far within it, distortions pile up after a few feet. Work by researchers Charles Kuen Koa and George Hockman at England's Standard Telephones and Cables in the early 1960s suggested that fiber clean enough for practical light-based communications was possible, and, in the early 1970s, Corning Glass Works engineers managed to develop a process for the practical manufacture of "low-loss" fiber.

Using Corning and its own fiber, AT&T assembled a large field test of fiber optic communications in Atlanta in 1975. "On January 13, 1976, they turned on the Atlanta fiber system and spent the next several months waiting for something bad to happen," explained fiber historian Jeff Hetch. "They did not find one problem."[95]

The promise of lasers and fiber communication was fairly well publicized and had become a ritual piece of the Blue Sky mantra in the early 1970s. Its potential capacity and low cost were widely appreciated in the telecommunications world. In February 1976, a long-time communications engineer and businessman, Joseph Stern, made an enthusiastic presentation to the NCTA board on the promise of fiber for the cable industry. The gleam off that promise caught the eye of TelePrompTer. TPT Cable TV President Bill Bresnan explained that they had begun to learn about the advantages of fiber and the plans for several fiber system experiments around the world. Bresnan became a fiber backer, stating fiber "can do everything that coax can do and, what's more, can do it better at less cost."[96] In July 1976, TelePrompTer installed the industry's first run of fiber optic cable. The 800-foot experimental line was used in its Manhattan franchise to carry the HBO feed. The company supplying the fiber cable was a small start-up firm called Fiber Communications, Inc., run by two former Bell Laboratories fiber researchers. Dr. Franklin Dabby and Dr. Ronald Chesler held a number of important patents on fiber technology and, in 1975, began their own small business, producing high-quality fiber cable in a small loft in Orange, New Jersey.

The article in *Broadband Communications Reports* announcing the TelePrompTer fiber experiment was spotted by a regular reader, Irving Kahn. Kahn saw, once again, the future and began with a substantial investment in Fiber Communications. Kahn then engineered a merger between Fiber Communications and the coaxial cable manufacturer, Times Wire & Cable (a division if International Silver Co.). By 1977, Kahn was a director of (and major shareholder in) the new Times Fiber Communications, Inc. A few months later, he entered the laser end of fiber, as Chairman of General Optronics Corp., a company producing light-emitting diodes (LEDs), lasers, and associated hardware for the young fiber communications market. Kahn beat his publicity drum loudly and widely on the promise of fiber, as he had done for satellites, predicting that one day fiber would replace coaxial lines in the nation's cable systems.

Interactivity and QUBE

Engineers and accountants knew what fiber could do for cable over the long haul, but it never captured the imagination of the public the way that the promise of an interactive broadband system did. Two-way communication over the cable system was one of the major notes in the Blue Sky score. It was the technical foundation for cable-based education, shopping, and polling. As with fiber, it was an old idea. In 1960, they called it "participation TV," a technology that, according to the pundits then, "enables the viewer not only to pay for the programs he wants, but to order on the spot the merchandise he see advertised."[97] It was also observed that participation TV would be a terrific market research tool, capturing data on the buying habits of subscribers and cross-tabulating it with consumer demographics, information that could then be sold to advertisers. The vehicle for ushering in the "buy while you watch" home service would be CATV.

Cable operators had been exploring the possibilities of two-way technology since the 1950s.[98] Irving Kahn's 1960 "Key-TV" plan was designed to offer two-way capacity. Spencer-Kennedy Laboratories demonstrated the feasibility of two-way as early as the 1962 NCTA convention. But making interactivity practical and affordable was another question. Cable operators either had to employ the telephone line for its return channel or had to design, build, and install amplifiers that would send a signal upstream from the subscriber's home back to the head-end over the cable system itself. In 1970, operator Richard Leghorn, working with British Rediffusion, tried an interesting experiment at his Cape Cod system.[99] It was a television exchange that allowed subscribers to "dial up" any of eighteen channels (expandable to thirty-six). Coaxial cable was replaced by a dedicated four-wire system from each subscriber home to a television exchange. Looking more in architecture like a telephone switched system than a cable tree-and-branch network, the experiment promised all the standard interactive services, from fire alarms to meter reading. Unfortunately, it was twice the cost of a traditional cable system and would have required ten exchange stations every square mile of plant. The economics clearly were not there yet.

Two-way capacity, as noted, was mandated in the FCC's 1972 rule set. As described in Chapter 7, experimental systems using two-way technology began appearing in the early 1970s. The Hughes-TelePrompTer SRS, computer-controlled, interactive cable demonstration at the 1972 NCTA convention in Chicago was one of the first linking interactivity with computers, allowing direct interrogation of homes. A number of planned trials of more sophisticated two-way services were cancelled in the early 1970s in the depths of the cable recession.

Pay services required control over what the subscriber received; without some form of signal interdiction, all cable channels would be free, and operators did not feel they could rely on the customer honor system to assure payment of the monthly HBO bill. Such control did not necessarily require full interactivity. Pay-cable operators began with passive negative "traps" which blocked specified signals. Addressable taps, which could be turned on and off at the head-end, were introduced in 1975, would eventually evolve into widespread industry systems, increasing control over signals going to the home and decreasing cost. Positive traps, which corrected a jammed or scrambled premium signal, letting it pass to a subscriber's set, were introduced in 1978. The availability of both positive and negative traps helped reduce overall costs for operators. If 75 percent of all subscribers wanted a pay service, negative traps could be used on the nonpremium 25 percent. If the equation was reversed, positive traps could be used for the premium 25 percent (instead of installing negative traps on 75 percent of the customers' lines). Interactive television implied more than such rudimentary technology, however, and the industry was keen to develop its interactive potential.

The company to make the biggest splash with experimental interactivity in the latter 1970s was Warner. They called it QUBE, introducing it in their Columbus, Ohio, system in 1977. Warner used QUBE as a showcase for state-of-the-art cable technology and, more importantly, as a tool for marketing and promotion. Hosted on a thirty-channel cable system, QUBE made its debut in December 1977. It was not, interestingly, the first interactive cable service in Columbus. In 1973, Coaxial Communications, which ran a system in a different part of town, unveiled its own experiment with two-way cable. That somewhat more modest interactive trial, installed by Telecinema, included two channels for films and a third for additional programming, and offered a limited ability to monitor the pay choices of subscribers. It quickly became apparent that plain vanilla pay-cable was a more cost-effective approach to premium service, however, and the Telecinema was closed.

Warner's QUBE service went several steps further. Although technologically primitive by later standards, it allowed subscribers to order pay-movies and special events, look at stock quotes, and register votes and preferences during live, interactive TV programs. At its heart was a central computer that scanned the system every six seconds, recording votes and noting pay-movie orders from the selector boxes in the subscriber's home. The system offered interactive games, local origination programs, educational programs, adult films with parental lockouts, a religious channel, time and weather channel, and a children's channel custom designed for the system, called "Pinwheel." Fire and burglary alarms, and medic alerts also were available. In some ways, QUBE was the Blue Sky vision come to earth. Viewers could shop from home, register with the touch of a QUBE button their approval or disapproval in town meeting programs and vie for cash prizes on interactive game shows. The service costs $10.95 a month and had a subscription base of more than 30,000.

The novelty proved to have weak legs, however. Most subscribers never used the interactive services. The advanced services were a financial loss leader and, by the mid-1980s, QUBE became an interesting but largely failed experiment. Its value for Warner, however, was not necessarily measured by the number of active users, but rather by the number of glowing press reports it received about the service and by the number of cities that it helped the company franchise. QUBE was a powerful marketing device, an example of high-end technology that Warner used to impress officials in cities such as Pittsburgh, Houston, Cincinnati, and Minneapolis.

Public Access

Access channels also were an important part of the franchising process in the late 1970s and early 1980s. Affected operators began setting aside channel space after the FCC's 1972 access requirements went into place and, although the courts eventually struck down the federal rules, they became commonplace in local franchise contracts. Access channels came in four basic varieties—public, educational, governmental (so-called PEG channels), and leased. Local high schools and colleges used the educational channels, and local authorities ran the government outlet, often airing city council or planning board meetings. Leased channels were used commercially on a contractual basis. The channels that usually generated the most interest, and sometime controversy, were the public access channels.

Public access, as noted previously, was a national concern among community interest groups and very much a part of the utopian vision of cable. More than anything else, it was the public access channel that was seen as the spearhead for the cable version of the town square or the electric town hall. Cable operators, of their own volition or as part of a franchise requirement, set up a low-cost access production facilities. The system typically provided portable cameras and video tape recorders for location shooting and, in a few instances, a truck outfitted with production equipment. In larger markets, paid access coordinators would manage the channels, promote the services, and even conduct classes and workshops on video production. The channels did not generate revenue. Operational costs were sometimes passed through explicitly to customers as a percentage of the monthly cable bill, or simply folded into the general cost stream.

By 1980, an estimated 1,300 of the nation's 4,600 cable systems were offering some form of nonautomated access. About 700 had local live programming of some type. About 500 systems offered public access, 500 offered educational access channels, and 340 provided government access.

The presence of access channels gave rise to hundreds of community access groups around the country. The grass-roots organizations programmed thousands of hours of material, creating programs that ranged from local talent shows to public affairs talk and interview programs to outlets for aspiring film makers.[100] In 1977, a number of local organizations coalesced to form the National Federation of Local Cable Programmers (NFLCP). By the end of the decade, it had chapters in forty states, representing citizens groups, cities, libraries, and schools. The NFLCP published a newsletter (the *Community Television Review*), sponsored national conferences and workshops, and became a clearinghouse for information and education on public access.

Unlike origination channels, cable system owners had little control over the content of the programming. Federal law prohibited interference in editorial content and, although the operator was liable for any obscene or indecent material, determining what constituted obscene or indecent material was not always easy. Federal obscenity laws consider, as one criterion, the standards of the local community. The most liberal local standards in the mid- and late 1970s, with little surprise, seemed to be in Manhattan, which was a leader in public access. One of the access categories that received significant attention therefore involved sexually explicit programming.

In 1975, Sterling Manhattan Cable Co. began running a program called "Midnight Blue" on access channel D. It was created by the publisher of *Screw* magazine, Al Goldstein. Nudity and explicit discussion of sexual activity were program staples. *Broadcasting* magazine described some of the program's vignettes: "A professed expert in fellatio describing her specialty in graphic detail, then demonstrating on the cameraman. (The sounds of the act were picked up by the tape deck but the Porta-Pak camera was aimed just wide of the center of the action). Close-ups of the nude breasts of a strikingly endowed young woman, who was rubbing her nipples against a sheet of plate glass."[101]

One of the more noted, if not noteworthy, New York City access programs was "The Ugly George Hour of Truth, Sex and Violence," a program that featured the host, George Urban, approaching women, on camera, on the streets of Manhattan and asking them if they would accompany him back to the studio to take off their clothes for the home viewers. An astounding number of them did. While the New York area access fare captured a few headlines and raised some eyebrows, it generated little legal action. When sexually explicit cable programming began penetrating more conservative bastions of American culture, however, the courts would have to intervene, as discussed in Chapter 10.

Access proponents and coordinators protested the attention paid to the sexually oriented programming, pointing out that 90 percent of the material on the open channels was true to the original intent of the system, which was to offer an avenue of cultural and political expression for the community. Although increasing numbers of viewers received access channels, few of them were watching. Some early promoters hoped access would generate local excitement over the possibilities of community television and the reality of home-based talk shows, information, and entertainment. But the viewership levels did not seem to meet those initial hopes. Access channels held the promise of democratizing television, but it turned out that citizens were more interested in movies, national sports, and network sitcoms. With some important exceptions, very few people regularly watch public access television. Accurate measures are difficult (access channel viewership is not measured in any systematic way), but estimates place viewing of those outlets at well under 1 percent of total TV time. Advocates noted, however, that mass appeal programming was not their intent, and success could and should not be gauged by ratings alone. Simply having a community channel is an important philosophical statement, and citizens reported liking the idea that it exists if needed.

Association News

The entry of larger companies into cable in the 1960s, as previously noted, had begun to create tension between the newcomers and the pioneering entrepreneurs. The widening disparity in power, size, and philosophy of these two groups was exacerbated by the continuing trend toward MSO consolidation between 1968 and 1972. While the top fifty firms controlled most of the nations' subscribers, an estimated 850 mom-and-pop companies, serving a million homes in small towns around America continued to run their community antenna systems,[102] but clearly, they were not part of cable's perceived big city future.

At the NCTA convention in Dallas in July 1973, a small-town operator based in Oklahoma, Kyle Moore, called together some two-dozen kindred spirits to discuss copyright. As noted, smaller operators opposed compromise on the copyright settlement, whereas many larger MSOs and the NCTA itself sought a settlement that accepted some copyright liability. The conversation eventually turned toward a general concern that the national association was no longer effectively representing their interests. Out of that meeting, a small band of about thirty-five NCTA dissidents formed a new organization, one chartered specifically to work for the well-being of the smaller cable operator, those with fewer than about 1,000 subscribers. They called the new organization the Community Antenna Television Association (CATA). Along with copyright, CATA diverged substantially from the NCTA's conciliatory efforts in pole attachment negotiations. Within two years, CATA had purchased office furniture, hired its own legal counsel, and was active in lobbying and testifying in Congress and at the FCC. Moore, at one point, promised that CATA was a temporary movement and would disband when their copyright goals had been reached, however, CATA would continue well beyond the resolution of copyright.

In addition to CATA, a breakaway group calling itself the Ad Hoc Committee of Concerned Cable Operators for a Fair Copyright Law waged a grass roots letter-writing campaign reminiscent of the 1960 fight against S. 2653. Some small operators felt so strongly about copyright that, in addition to forming CATA, they also quit the NCTA in protest. All NCTA members paid dues that constituted the Association's main source of income and, while the loss of money from the smaller companies was not a major blow, it certainly had an effect, especially during that financially difficult period. The rift between large and small operators was not the only point of internal friction. The controlling MSO did not always see eye to eye on policy matters. NCTA President David Foster frequently found himself caught between the differing opinions of the major MSOs and their representative on the Board of Directors. The blowup over pole attachment negotiations noted earlier was only one of the more glaring examples of too many cooks trying to stir the cable pot. Eventually, it became unmanageable for Foster and, in February 1975, he resigned. He left diplomatically and to general praise for his work, but the causes were clear, the internal divisions were making it too difficult to move ahead with a unified vision. After months of searching, the Association named a new president in June 1975. Robert Schmidt was a Washington, D.C., lawyer and a former lobbyist for International Telephone and Telegraph (ITT). Schmidt served as head of the organization until the summer of 1979 when he was replaced by Thomas Wheeler, who had been NCTA's Executive Vice President since 1976 following a seven-year term at the helm of a very different kind of trade group, the Grocery Manufacturers of America.

As cable recuperated through the latter 1970s, other important new industry groups began forming. In 1975, the Cable Television Administration and Marketing Society (CTAM) was organized, holding its first convention in Chicago. In June 1979, Women in Cable Television (WICT) was created to promote opportunities for women in the industry and provide networking opportunities for those already advancing in their cable careers. In the fall of 1980, the Cable Advertising Bureau (CAB) was created to further the industry's need for national and local advertising opportunities.

Programming Pioneers

Between 1976 and 1977, revenue from pay services, mostly HBO, nearly doubled, from about $68 million to $125 million, and nearly doubled again the next year. "We began to realize that we really had something," recalled Bob Rosencrans.[103] What they had, in part, was the start of a totally new industry, the cable programming business. HBO blazed the trail, but that path soon would become a paved thoroughfare and the thoroughfare a high-speed expressway. Cable was evolving into a two-sector business, with cable system operators symbiotically linked to the emerging programming services. It meant the beginning of cable as a national presence, with cable-only programming that would be shared by viewers across the country. It was a new form of national television, one that eventually would build to constitute a true competitive challenge for the traditional over-the-air TV networks and local broadcasters.

The pioneer, HBO, had the skies to itself in 1975, but that would not last. Those already in the business began laying plans to launch their own services, and a host of newcomers from outside cable moved to join the party as well. Over the next several years, the public would come to know networks such as WTBS, CNN, C-SPAN, USA, ESPN, and BET. The important names in cable had been operators, such as Kahn and Daniels, and Magness. A new cable marquee now was being erected and it would feature names such as Brian Lamb, Pat Robertson, Robert Johnson, and perhaps most importantly, Ted Turner.

Ted Turner and WTBS:1976

Ted Turner had just captained the yacht *Courageous* to the 1977 America's Cup, sailing's most prestigious contest. After a spirited, and spirit-laden, celebration at the Newport, Rhodes Island, harbor, Turner was half carried through the admiring throngs to the Newport National Guard Armory where the competing crews and event officials were gathering. The victorious Turner was boisterously joyful, accepting congratulations, beaming with delight, and seriously drunk. He had been downing champagne, beer, and aquavit much of the afternoon and now had a pair of rum bottles close at hand.[104] One of the members of the group that gathered around him thought that the twin rum bottles, sitting on the table in front of Turner, might not communicate the best impression of the new sailing hero and moved the bottles to the floor near Turner's feet. Turner's mood changed quickly. Cursing, "Pike, you dumb fuck. Give me that back," he dove under the table to retrieve his lost property.[105] It was just about this time that the studio lights snapped on and the live, nationally televised news conference began. A now-beaming Turner came up with a rum bottle in each hand. For many in the country, their first close look at Ted Turner came as the drunken sailor rose grinning from beneath the table. "It was a scene," observed one biographer, "to bring the [sponsoring] New York Yacht Club to its well-pressed knees."[106] Turner's good humor returned and when introduced, he wobbled to the podium to proclaim that he loved his valiant competitors, he loved his crew, and in fact he loved everybody in the room. Eventually, his beloved crew hoisted the loquacious Turner to their shoulders and carried him away, much to the relief of the sponsors. The news conference ended, and Turner was last seen that night dashing exuberantly away, through the backyards of Newport.[107] It was only one of many memorable moments in the life and career of Ted Turner.

In the 1960s, Irving Kahn was almost larger than life, and his jail time only served to expand his legend. In the cable industry and elsewhere in the world of finance, sports, politics, and popular culture, no figure ever grew so large or in such a peculiar manner, however, as did Ted Turner. Robert Edward Turner III was born in Cincinnati, Ohio, on November 19, 1938. His father, Robert Edward Tuner, Jr., a dark and formidable influence on Ted's life, was in the billboard advertising business and, after World War II, moved the family to Georgia to begin his own company. The elder Turner was a complex and demanding father. He was a caring but stern and troubled man, with a strong belief in hard work, especially for his children. He also had an unfortunate problem with alcohol. Ted spent his youth being packed from boarding schools to military academies and spent his summers working for his father's company. In his teens, Turner was required to begin paying rent for his room and board at home. The family business prospered, however. Ed Turner made himself a millionaire in the billboard business and Ted, at age eleven, become involved in the costly sport of sailing, enrolling in the junior program at the Savannah Yacht Club. Turner's passion for the sport grew, and he would become a talented sailor.

Ted Turner inherited his father's intensity, drive, and taste for alcohol. He entered Brown University in 1956, majoring, according to most observers, in drinking, sailing, and girls, and excelling in all three. He had a wager with his father that would pay him $5,000 if he could stay sober until he was twenty-one. The young man lost the bet, in fairly short order, and after a couple years of general collegiate debauchery, was thrown out of the Brown for violating rules concerning women in his room (and bed).

On March 5, 1963, when Ted Turner was twenty-four years of age, his father committed suicide following a slump in the family business. The elder Turner had willed the billboard company to his son, but had also sold the company just before ending his life. Company executives counseled Ted to proceed with the sale and take the money. Ted, setting a tone for his career, rebuffed the suggestion. He reacquired the company and, after a few years of financial struggle, returned it to prosperity. Turner probably could have retired at this early point, but he seemed always to need a challenge. In 1968, again against the advice of his business associates, he bought a faltering radio station in Chattanooga, Tennessee, and once again turned it into a moneymaker. He returned to sailing, advancing to prominence in the competitive world of yachting, which he could now afford. He was soon restless and looking for a new challenge. He once said he would turn down an offer to own every McDonald's franchise in the United States because it would be too easy to make money. Very little Turner did seemed to be easy, however.

Turner, early on, showed himself to be as eccentric as he was smart. As one writer put it: "He developed—and seemed to revel in—a reputation as a public lush, a shameless womanizer, and a borderline nutcase."[108] The same writer, almost in the same breath, went on to state, however, "Ted Turner was the most gifted and successful programmer in the history of television."[109] A few might have said that both assessments edged on hyperbole, but only a few. Like some autistic savant, Turner's personal style was often wild and unpredictable, his taste for adult beverages and attractive women, legendary; but his business sense such that not only did he nearly always land on his feet, but those feet usually came down on mountains of money. His business acumen, his willing, even exuberant risk-taking was evidenced at an early age.

In 1969, a local UFH TV station, WJRJ, purchased space on one of Turner Communications Group (TCG)'s billboard, advertising the sale of the station. The station, following the fate of many UHF operations, was dead last in the market, on the verge of financial collapse, and little evidence suggested it could be revived. Turner, naturally, decided to buy it. Friends and financial advisers protested that Ted was betting the firm on an exceptionally risky undertaking. But high-risk ventures were what Turner reveled in. Running the station on a threadbare budget—he would recycle used videotapes instead of buying new ones to make his commercials—Turner began to educate himself in TV programming. He changed the station call letters to WTCG, which formally stood for Turner Communications Group, but in an extensive promotional blitz were converted to the Channel 17 slogan "Watch This Channel Grow." And grow it did, counter programming the six o'clock news on other Atlanta stations with cartoons, aged (and cheap) sitcoms, such as "Gilligan's Island," "Leave it to Beaver," and "Gomer Pyle," along with old movies and even Roller Derby. Audiences looking for an escape from a nightly barrage of Viet Nam, Watergate, and crime news started finding Channel 17. Turner's formula worked.

In July 1970, Turner purchased a second TV station, another failed UHF, WUTV, Channel 36, in Charlotte, North Carolina, buying it with a personal check (because the Turner Board of Directors flatly refused to finance the deal) at an auction on the courthouse steps. He renamed it WRET-TV, after his father, and to help resuscitate the station he conducted a personal on-air beg-a-thon asking for viewer donations and netting more than $20,000. He boasted later that he eventually returned every donation, with 6 percent interest.

Turner believed that sports would draw a strong audience and, in 1973, secured a contract to air Atlanta Braves games, underbidding the local NBC affiliate for the rights. Two years into the contract, with the Braves in the league cellar and ratings for the underperforming team slumping as well, the partnership that owned the team was looking to sell. Turner, fearing the team would either be moved out of Atlanta or keep on loosing, or both, bought the franchise in 1976. Winning local teams attract viewers, and Turner wanted more viewers.

Turner learned early the value and potential of cable television. Andy Goldman was a fellow sailor and friend of Turner's. He was also Marketing Director in Alabama for TelePrompTer. He approached Turner in 1972 about putting WTCG on several southern TelePrompTer systems. Imported stations were not generating much added advertising revenue from extended cable carriage, but Turner was carrying a lot of direct product advertising for such things as the classic Ginzu knifes that could cut through tin cans and stay razor sharp for life. The station took a percentage of every order and each new market meant more Ginzu knife revenue. Coverage of the Atlanta Braves also became popular in southern markets served by WTCG, and Tuner was acquiring a large library of films, which he purchased rather than leased.

Cable carriage became another formula that worked for Turner. As a growing independent, WTCG was picked up off-air by systems in cities near Atlanta, including Macon and Columbus, and was distributed by microwave to cities as far away as Tallahassee. By one count, WTCG and WRET combined were on 150 cable systems serving more than a half-million subscribers by 1976.[110] Although Turner appreciated the value of cable, technology was not his strong suit and he knew little about communications satellites. In late 1974, Sid Topol, whose Scientific-Atlanta was headquartered in Atlanta, introduced Turner to Reese Schonfeld, President of Independent Television News Association (ITNA), a company that offered pooled news coverage to client stations. The Western Union satellite now was in operation and Topol wanted to sell dishes. Schonfeld was interested in distributing his news feeds via satellite and Topol wanted him to help convince Turner to take those feeds. Turner had become wealthy, however, in part, by avoiding news—the unofficial WTCG slogan was, "no news is good news." Turner seemed mildly interested in satellites but not at all in news. Schonfeld left the meeting unimpressed with Turner.[111]

Soon after that Turner received a phone call from his friend Goldman, now working in New York as Vice President of marketing at TelePrompTer. Goldman told Turner about HBO's plans to begin distributing via satellite. According to biographers Goldberg and Goldberg, Turner's initial response was, "What's an HBO?"[112] Turner was soon intrigued, however, by the possibility of distributing WTCG by satellite. After a few more conversations with Goldman, Topol, and through Topol Western Union, he was committed to the concept of satellite distribution and national audiences for WTCG. To operate the satellite uplink he was going to purchase from Topol and to run the distribution service, Turner created a new company, Southern Satellite Systems (SSS), after rejecting equivalent AT&T service as too expensive. He quickly ran headlong into a series of financial and legal hurdles. FCC anti-leapfrogging rules in place at the time substantially limited carriage possibilities for the station. FCC regulations also required Turner to hire a common carriage company to lease the transponders and run the distribution service; broadcasters such as Turner could not also be common carriers for these purposes. With regulatory roadblocks at every turn and an idea not yet proved, even by HBO, capital was scarce.

Turner solved the common carrier problem by selling SSS. The man who purchased it was Ed Taylor, a businessman who knew a great deal about satellites. In the mid-1960s, he headed an AT&T task force to study the competitive potential of domestic communications satellites and, by the mid-1970s, was the youngest corporate vice president at Western Union, where he marketed Westar services. Turner almost leased transponder space on Westar from Taylor. Goldman, however, did not want TelePrompTer systems buying separate earth stations for WTCG-Westar feeds and HBO-Satcom feeds, and convinced Turner to go with RCA. It was an easy but critical decision in the development of the national system. If Turner had used Westar, cable operators would have had to pay thousands of dollars for an additional satellite dish, which they probably would not have done. Going with Satcom I gave Turner instant access to an existing base of satellite-ready cable systems (although operators did bear some cost for additional electronics to receive the superstation). It also helped make RCA's Satcom I the *de facto* bird for subsequent cable programmers seeking to piggyback on the established network.

Goldman clearly knew what he was doing and Turner wanted him to run SSS. When Goldman, and a few others, turned down Turner's offer to head up the operation, he went to Taylor. Turner flew Taylor and his wife to Atlanta where the Western Union executive was wined, dined, and ultimately convinced. The asking price for SSS was $1.

As noted, the FCC's December 1975 decision to drop its anti-leapfrogging rules cleared an important roadblock to Turner's satellite schemes. The cable industry slump and risky nature of the venture kept investors away, however. Even with the growing success of HBO, Turner could only manage $240,000 in capital from a dozen investors in New York. Luckily, RCA was still looking for customers and expressed interest in supporting the plan. Recalls Taylor, "We went $750,000 in the hole building the uplink, and then RCA bought the installation from us and we rented it back from them."[113]

Mid-December 1976 was momentous for Turner. On December 14, the FCC voted to allow cable operators to use smaller and cheaper dishes to receive satellite signals. The next day, the Commission approved SSS's request to begin distributing WTCG nationally via satellite,[114] and two days later, on

December 17, the WTCG signal began flowing through the uplink toward Satcom I. The first feed reached four cable systems,[115] although the numbers grew rapidly. WTCG was the second, after HBO, in what would soon become a cascade of satellite-delivered programming for the cable industry.

Ted Turner pioneered the satellite-delivered "superstation," but others soon followed. As with WTCG, major market independents had been harnessed as regional superstations for years. Despite FCC restrictions on microwave importation through the 1960s and early 1970s, small-town cable operators regularly picked up powerful VHF independents from the nearest large cities. Operators in New York, New Jersey, and Pennsylvania brought in New York City stations, such as WOR, WPIX, and WNEW even in the 1950s. WGN in Chicago was distributed regionally by the early 1960s. Adoption of the 1972 regulations only increased the practice.

In April 1978, the FCC signaled its openness to the expansion of satellite-distributed superstations when it denied a Motion Picture Association of America (MPAA) petition for an investigation into what the studios claimed was the damaging impact of the practice on the film and broadcast industries. The Commission concluded that there was no evidence of the creation of superstations and, even if such evidence were provided, nothing suggested they would be contrary to the public interest.[116] In early October, SSS's Ed Taylor asked the Commission for permission to begin carrying WGN-TV in Chicago, KTVU in Oakland, WPIX in New York, and KTTV in Los Angeles in the same fashion the company retransmitted WTCG to cable systems around the country. Later that month, the FCC approved the applications, at the same time declaring an "open entry" policy for such carriage and approving similar applications for three other firms. Common carrier companies, such as United Video (run by Roy L. Bliss, the son of cable pioneer, Roy Bliss), American Microwave & Communications, and Eastern Microwave, Inc., were soon capturing and distributing big-city independents via satellite.

The stations themselves were compensated by cable operators through the 1976 copyright act. The microwave companies were paid directly for the service.[117] United Video began satellite distribution of WGN in November 1978, charging operators ten cents a subscriber, the same rate SSS received for WTCG. Eastern Microwave put WOR-TV, New York, on the satellite in April 1979. Satellite Communications Systems distributed KTVU-TV, Oakland-San Francisco. WPIX in New York, KTLA in Los Angeles, and WSBK, Boston, became superstations in the same manner. Other broadcasters protested, and Metromedia, which owned KTTV in Los Angeles, complained to the FCC about the unauthorized retransmission of its southern California station, but the Commission was moving in a different regulatory direction at the time, and declined to interfere in the practice.

Turner's operation maintained an important distinction from the other satellite-distributed broadcasters. His vision was to make the channel a truly national media vehicle, selling a transcontinental audience to major national advertisers. Unlike most of the other independents, he owned much of his own programming. In 1979, Turner changed the stations call letters to WTBS, for Turner Broadcasting Systems.[118] By then, the superstation was on 1,710 systems serving 7.2 million subscribers. Satellite Communications Systems' KTVU, San Francisco-Oakland had 140 systems and 700,000 subscribers; WGN in Chicago had 600 systems and 2.3 million subscribers (1.7 million via satellite). WOR in New York had 410 systems and 2.3 million subscribers (400,000 by satellite).[119]

The cable-satellite link, in some ways, was a not-to-distant echo of the earliest CATV systems. It was a technology-business pairing that seemed to be working, and word quickly spread. As with the earliest CATV pioneers, satellite pioneers began assembling financing and casting their signals aloft. HBO was first, Ted Turner second. A third signal was available for cable-casters in 1976, but few operators took advantage of it. The service came from the San Antonio, Texas, studios of Spanish language station, KWEX-TV, and fashioned as the Spanish International Network (SIN). The targeted programming and, more importantly, the fact it used the Westar I satellite, severely limited early use of the SIN signal by cable operators, however. The development of SIN and its cable service offshoot, Galavision, is traced further below. For a cable industry looking for broader appeal material via satellite, after HBO and WTBS, came religion.

CBN: 1977

Marion G. "Pat" Robertson was the son of a U.S. Senator (A. Willis Robertson) and grandson of a minister. His upbringing in Lexington, Virginia, framed by his father's career, was more political than religious.[120] He graduated with honors from Washington & Lee University, joined the Marines, and served in Korea. He earned a law degree from Yale, but failed the bar, and in his mid-20s was a young man in search of a life path. He found one in religion. Influenced by his mother, but catalyzed in his religious zeal during a life-changing dinner conversation with a family friend, Robertson entered seminary school and then took a minister's position in a church in Brooklyn's Bedford-Stuyvesant district, one of the nation's most impoverished inner city neighborhoods. He was there about a year when, in 1959, a message from an old friend changed his direction again. In a letter to Robertson's mother, the friend had mentioned there was a defunct UHF-TV station for sale in Portsmouth, Virginia, concluding, "Would Pat be interested in claiming it for the Lord?"[121] Pat was indeed interested.

With no money and against the advice of his father, Robertson moved his young family to Portsmouth. WTOV (later renamed WYAH-TV), Channel 27, was another victim of the FCC's Sixth Report and Order, a UHF unable to compete, now derelict and for sale. With a combination of fiscal savvy, help from sympathetic local businessmen, a fair amount of luck, and an amazing amount of sheer personal tenacity, Robertson found the financing (about $30,000 in various forms of cash and loans) to get the station on the air. He began broadcasting October 1, 1960, and with steady appeals to the faithful for help and donations, slowly began building what would become the Christian Broadcasting Network. By the early 1970s, WYAH had gone commercial and Robertson also owned WHAE-TV in Atlanta, KXTX-TV in Dallas, and a number of radio stations. He was also buying time on stations across the country to run videotaped, 4-hour blocks of religious programming.

He called his tape-by-mail enterprise the Christian Broadcasting Network (CBN). Powered by contributions from devoted viewers and augmented by an expanding advertising stream, Robertson had a small, but growing, televangelical empire by the mid-1970s. The heart of his programming was the "700 Club." Debuting in 1965, it was a mixture of religious fund-raising, social commentary, interviews, and talk. Robertson supplemented the programming diet with inexpensive off-network shows and old movies.

Robertson knew Ted Turner. CBN bought time on Turner's Charolette UHF and the company competed directly against WTCG in Atlanta. Robertson saw Turner go up on satellite in late 1976 and was clearly attracted by the concept. An important part of the "700 Club" format was commentary on social issues and the bicycled system meant days of delay in responding on-air to current events. "I wanted to be out there the same day or the day after the issues broke," Robertson explained.[122] He purchased an uplink from Scientific-Atlanta and began distributing CBN via satellite on April 27, 1977. Mirroring HBO, which had assisted cable operators financially in setting up receiving stations, Robertson bought sixty Scientific-Atlanta dishes for CBN affiliates, both broadcasters and cable operators, around the country. For cable operators, it was the third programming service available over the still-new satellite system.

In April of 1978, CBN was joined in the cable heavens by another religious programming service, the PTL (People That Love) Television Network. Jim and Tammy Faye Bakker had learned the trade from Robertson himself, working for CBN from 1965 to 1972, and helping fashion the "700 Club" program format.[123] They left after a falling out with Robertson, and joined with Paul Crouch, a friend, former minister, and the general manager of a Christian TV station in Los Angeles. Together, they formed the Trinity Broadcasting Network, based in southern California, in 1973. Again, the Bakkers could not see eye-to-eye with their partners and, in 1974, struck out on their own, founding the PTL service in Charlotte, North Carolina. PTL's first "affiliate" was Turner's WRET, but they successfully leveraged the CBN formula of bicycling taped programming, mainly "The PTL Show," a "700 Club" clone, to other stations around the country. PTL grew quickly; by mid-1976, it was on seventy broadcast stations and twenty cable systems. Within two years, satellite distribution had vastly expanded its reach. Offering around-the-clock religious programming that included comedy variety shows, children's programs, and soap operas, along with the traditional televangelical calls for viewer donations, PTL was seen on more than 200 cable systems by 1980. Robertson would go on to run for President and eventually sell CBN, becoming a millionaire in the process, but Jim and Tammy Bakker were headed for a less auspicious future.

Finally, the multidenominational service, Trinity Broadcasting Network (TBN), also launched to cable in April 1978. Headed by Paul Crouch and based in Los Angeles, the service collected its audience using the bicycle taped distribution system, before leasing a satellite transponder. While never as prominent or controversial as PTL or CBN, it nonetheless grew steadily over the next two decades.

MSG/USA: 1977

Madison Square Garden (MSG) had been feeding many of its sporting events, including New York Nick's basketball and Ranger's hockey, to cable systems in and around New York, via microwave, since 1969. HBO had been running a great deal of the MSG pay-programming, but decided that a full package of almost continual New York area hockey, basketball, and other sports was a little New York-centric for their national satellite service. Bob Rosencrans, the first operator to take a chance on satellite-fed HBO, had a different idea. He met with Joe Cohen, president of the Madison Square Garden Communications Network, about putting their material on satellite, charging ever-decreasing rates for systems further away from Manhattan. "The numbers [charge per subscriber] were small," explained Rosencrans, "but when you multiplied them out and if you could generate enough systems and enough subscribers, it was all incremental revenue to the Garden."[124] Madison Square Garden Sports, a joint venture of UA-Columbia and MSG, launched in April 1977.

The network fit into an expanding interest by Madison Square Garden's parent corporation, too. Gulf + Western began its corporate existence as an oil and gasoline company, but under the control of a colorful, acquisitive and hot-tempered Austrian native, Charles Bluhdorn, expanded into a host of diverse enterprises. It purchased Paramount Studios in 1965 and by the mid-1970s owned 39 percent of Madison Square Garden. In 1977, the same year as the cable network launch, Gulf + Western purchased MSG outright, staking a deep claim in the entertainment industry, as it also picked up the Nicks and the Rangers.

A promising young executive, who had started in franchising with UA-Columbia in 1973, was put in charge of the MSG satellite network. Her name was Kay Koplovitz. In addition to the small, per-subscriber fee the

MSG Network charged cable operators, the service also became the first cable-only satellite programmer to sell advertising. This dual revenue stream model would set the pattern for most basic cable programming networks.

Over the next several years, Rosencrans would use his satellite capacity to add additional services to the MSG feed. None of the earliest satellite services programmed 24 hours a day, not even HBO. Most used their transponders 12 hours a day or less, leaving time available for additional programming. In September 1978, UA-Columbia inaugurated a children's programming service called Calliope, managed by Koplovitz. The 1-hour a day service featured films supplied by the Learning Corporation of America.

In April 1980, Calliope and MSG were merged into a new general-interest, advertiser-supported service, dubbed USA Network. Koplovitz was named President. In addition to MSG sports and Calliope, USA penned an agreement with Major League Baseball to carry games on Thursday nights. With its heritage in "the Garden," USA relied heavily on sports programming in its early years. In addition to "Thursday Night Baseball," USA was the first to sign the National Hockey League and the National Basketball Association to game-of-the week contracts and, in 1981, the first to carry a live college football game. Through the early 1980's it featured sports from bowling to hockey to boxing.

USA carried other important program services in its early years, as well, with Rosencrans providing the seedbed for some of cable's most important fledgling networks, including the Black Entertainment Television (BET) network (discussed in Chapter 9) and a new service called C-SPAN.

C-SPAN: 1979

As early as the 1969 NCTA convention, cable operators had talked about a public affairs channel that would be dedicated to the coverage of Congress, possibly even offering live feeds from the Senate and House of Representatives. In 1977, a reporter for *Cablevision* magazine set out to make the dream real. Brian Lamb, by all accounts, held a strong and real belief in the promise of the U.S. political system, a belief that an informed electorate was central to the promise of democracy, and that television could be harnessed to further those goals. Observers described him as a modest, self-effacing mid-westerner, a bipartisan democrat with a small "d." He worked in his earlier years as a radio reporter for United Press International, served as a volunteer social aide in the Lyndon Johnson White House, was press secretary for Senator Peter Dominick (R-Colorado), and then press spokesman for Clay Whitehead's Office of Telecommunication Policy. His early experiences in the press and in Washington convinced him that the major media were not doing enough to fully cover the broad panorama of American politics and society. "I'd been taught all my life that this was a democracy, and that many voices were better than fewer voices . . . I kept learning by being a part of the system

that there were very few voices."[125] In 1974, he left OTP to become Washington Bureau Chief for *Cablevision* magazine. In 1977, with cable once again
growing and in need of material, Lamb saw an opportunity for a new kind
of public affairs program and began a modest cable service that provided a
weekly stream of fifteen-minute taped interviews with members of Congress.
He signed up fifteen cable operators.

Around the same time, in early March of 1977, House Speaker Tip
O'Neill (D-Massachusetts) initiated a ninety-day test of live TV coverage of
the House floor. Broadcast coverage of Congress had been a hotly debated
topic for years, and Congress was of deeply mixed mind about it. Few
members cared for the edited and filtered coverage provided by the major
networks and some wanted a way to bring their work close to the voters,
improving both informed democracy and their own local visibility. Others
were wary of live coverage, and were especially concerned about who would
control the cameras.

Cameras in Congress were, therefore, a subject of general beltway conversation and, in July 1977, a topic of lunchtime discussion between Lamb
and John Evans of Arlington Cable Partners. Evans mused aloud that he
needed more programming and wished he could tap into the experimental
black and white cameras in the chambers of the House of Representatives.
It was the right comment at the right time to the right man. Lamb and Evans
began their planning. Evans volunteered to provide the cable wiring and
head-end facilities for a service that would bring live congressional debate
to the country. Lamb went to Congress.

As debate over cameras in Congress was nearing a close in October, Lamb
was finishing an interview with a backer of TV coverage, Lionel Van Deerlin.
At the end of the interview, Lamb suggested to Van Deerlin that cable could
provide the distribution system for a national House feed. Van Deerlin liked
the idea so much that he asked Lamb to write a speech for him outlining
the plan. The quickly composed speech was delivered that afternoon, with
Van Deerlin proposing gavel-to-gavel coverage that could be distributed to
American homes via cable TV. The resolution passed; Lamb now had to talk
to the cable operators.

The initial reception from some in the industry was cool. They saw a
service that would cost them money and generate none. Others had a different
perspective, including NCTA President Tom Wheeler, whose job it was to
court the regulators and who saw the political benefits of the idea. Cable had
been struggling with Washington, in one way or another, since its birth. The
development and support of a channel that brought congressmen and women
into every cable TV home and advanced the general cause of political debate
could hardly be a bad thing for the industry. Wheeler urged operators to take
a closer look at the proposal and found receptive ears. UA-Columbia's Bob
Rosencrans already had established himself as a visionary in the industry, and
he was a sophisticated student of cable's broader social and political standing.

As Rosencrans explained it: "I was tired of knocking on congressmen's doors to explain what cable television was . . . So, if nothing else, I thought it would put cable on the map in Washington."[126] He told Lamb, "I'll give you $25 thousand and my name, see what you can do."[127] Lamb took the money and Rosencrans' endorsement, and parlayed them into support from other key cable executives. He raised $425,000 from twenty-two different cable operators, including Warner and TelePrompTer. In addition to dollars, UA-Columbia also helped arrange satellite carriage, allowing C-SPAN to share transponder time with the MSG feed, and Rosencrans became C-SPAN's first Board Chairman. The cable industry further agreed to fund the operation by allocating a monthly one cent per subscriber.

On March 19, 1979, the Cable-Satellite Public Affairs Network, C-SPAN, began live coverage of the House of Representatives. The feed was available on 350 cable systems and in 3.5 million homes. In addition to debate from the floor, C-SPAN went on to provide a full range of public affairs programming, including speeches, congressional hearings and interviews, and even viewer call-in shows. By the mid-1980s, it had grown a national cadre of devoted fans, dubbed C-SPAN junkies, but also established one of cable's truly worthwhile and lasting contributions to U.S. politics, government, and media.

ESPN: 1979

Ted Turner knew that movies and sports brought viewers to the screen. HBO had a jump on the film market so Turner considered an all-sports cable channel. In 1977, he announced plans for a cable sports network, using his Atlanta teams as a programming base. The service never developed, but the idea was powerful. It would clearly draw viewers, but obtaining contracts to carry attractive national sports teams could be difficult and certainly expensive. Many in the cable industry and the public were skeptical then when a New England area sports agent, William Rasmussen, and his son Scott announced plans for an all-sports cable service on July 7, 1978. Rasmussen had been communications director and a play-by-play announcer for the New England Whalers hockey team. He decided it would be a good idea to distribute the Whalers games, along with University of Connecticut sports events, to regional cable systems. Rasmussen discovered that it would cost no more to put the service on satellite than it did to distribute it regionally via microwave. His key success in getting the service off the ground was to convince Getty Oil, which had dabbled in cable previously (backing Home Theater Network) to help finance the venture. The service was initially dubbed ESP-TV, for entertainment and sports programming, and was later changed to ESPN, the Entertainment and Sports Programming Network.

The network began operations in November 1978, with a U-Conn basketball game. A few months later, in February 1979, Getty Oil purchased 85 percent of the enterprise. The Getty money gave ESPN the funding it needed to purchase programming and sustain it through the early years. Rasmussen also brought in seasoned professionals to help run the fledgling operation, hiring Chet Simmons, President of NBC Sports, to be ESPN President, and Scotty Connal, Vice President of Sports Operations at NBC, to be the new ESPN Senior Vice President. In March 1979, ESPN signed its first MSO, United Cable TV, and secured the rights to televise a variety of NCAA athletic events. ESPN began full-time national service in September 1979, with a heavy menu of NCAA sports. College sports looked like an attractive advertising draw to the giant beer maker Anheuser-Busch, which bought $1.4 million of time on the young service that year, the most anyone at that time had every spent on cable advertising. Rasmussen soon came into conflict with Getty ownership over finances and control of the cable channel, and he left ESPN in 1980. An all-sports format turned out to be a hit with subscribers, and the network grew rapidly. Beginning with a roster of events that included everything from slow-pitch softball to hurling to Canadian football, ESPN would grow to become one of cable's most popular services.

SIN & Galavision: October 1979

The Spanish International Network (SIN) began as a broadcast network in the United States in the early 1960s. It was developed by Rene Anselmo and backed by Televisa, the Mexican broadcasting giant that for decades dominated Latin American radio and television. Anselmo, an American-born actor turned businessman, lived in Mexico for 12 years, producing game shows, children's programming, and musical comedies, and developing business ties to Televisa. He returned to the United States in 1960 to help start SIN, taking a 25 percent ownership stake in the new business. In 1961, Anselmo purchased a UHF station (KWEX-TV) in San Antonio, Texas, and soon expanded to outlets in New York, Los Angeles, Miami, and Fresno, California, catering to the large and growing Spanish-speaking population in those cities.

In September 1976, Aneslmo, acting on a tip he had received at a broadcasting convention, placed the KWEX signal on Westar II, making it available to other broadcasters and cable operators across the country. The use of the Western Union satellite limited its appeal, however, and, in October 1979, Anselmo fashioned a new service, Galavision, expressly for cable operators and deployed on Satcom I. Galavision was originally a premium service, but was unable to draw a sufficient pay subscriber base and converted to a basic service. Programming included first-run films, variety specials, sporting events, and a heavy diet of the popular the Spanish TV specialty, tele-novelas. By the end of the decade, it was on seven cable systems serving about 2,000 subscribers.

Nickelodeon: April 1979

When Warner began crafting its highly promoted QUBE system in Columbus, the company wanted a quality children's program, something that would attract young viewers as well as the endorsement of their parents. They hired experts from the Children's Television Workshop, the wellspring of "Sesame Street" and the "Electric Company." Dr. Vivian Horner, who helped create the "Electric Company," was asked to help craft the QUBE schedule. Not surprisingly, what Horner and the committee came up with looked a great deal like "Sesame Street." The Warner program, "Pinwheel House," ran from 7 AM to 7 PM on the QUBE system and featured Sesame Street-like puppets and a commitment to a fun, but educational programming philosophy. Like the QUBE project generally, "Pinwheel" was a loss leader, tied to both a commendable mission to offer quality children's programming to subscribers and to a more practical goal of convincing parents of the value of a monthly subscription, and local franchising authorities to the value of a Warner system. One-time Nickelodeon Vice President Scott Webb later conceded, "In those days, Nickelodeon wasn't about kids, it was about cable bills."[128]

As national programming service began to launch in 1977 and 1978, Warner saw the potential in expanding the "Pinwheel" concept to a national cable audience. The philosophy remained the same. "We are trying to make it be not-television, different from commercial or public television," explained Horner. "The object is not to compete with the commercial networks but to provide an alternative."[129] The service was commercial free, charging operators $0.10 a subscriber per month for the service. When it launched in April of 1979, it was named Nickelodeon.

In 1980, Geraldine Laybourne joined the network as program manager. With a background in television and education and a strong interest in children's programming, she helped redesign the network's programming and lead it to a dominant position in children's television.

More Choices

Cable operators had other satellite services to choose from in the last few years of the 1970s, as established program distribution companies took their products into orbit. Ed Taylor, head of SSS (Ted Turner's carrier), leased another transponder on Satcom I and created the Satellite Programming Network (SPN), offering an eclectic mix of talk shows, variety shows, and films. It began satellite distribution in December 1978. It carried commercials and offered local cable systems their own advertising availabilities ("avails" in the parlance of the industry). Initially used as 3-hour fill-in on Satcom I, it expanded to 13 hours in March 1979 and 21 hours in August 1979.

Modern Satellite Network was run by Modern Talking Pictures, at the time, the nation's largest distributor of free, business-sponsored films. In part educational and in large part corporate promotion—such as cooking shows sponsored by spice and condiment companies. They had been running a tape-based service since mid-1970s and took to the satellite for 5 hours a day in January 1979.

The Appalachian Community Service Network (ACSN) could boast of historical roots nearly as deep as HBO. Its direct ancestor, in fact, was on satellite before the Time pay service. In 1974, a government-sponsored experimental program used NASA's ATS-6 satellite to deliver educational programming and services to fifteen classroom sites, spread from New York to Alabama. The project was controlled by the Appalachian Regional Commission (ARC). The Appalachian Educational Satellite Project (AESP) ran until April 1975 when the satellite was moved to India for further experiments, but was reactivated and expanded when the satellite returned, and was in operation from 1977 to 1979. The success of the AESP project led ARC to create a noncommercial educational programming service for use in schools and community centers across the country and was additionally made available to cable operators for one cent a subscriber per month. ACSN, which was launched in August 1979, provided adult educational programming and a variety of public service materials to commercial and public clients. In November 1980, the channel changed its name to denote more directly its national reach and programming intent. It became The Learning Channel (TLC) and would take a place as a regular on systems across the country over the next several decades.

Finally, the cable industry, along with broadcasters, enjoyed a small boon from The Copyright Act of 1976. The Act took effect in 1978, at which time an estimated 18,000 to 20,000 films that did not come under the new copyright law entered the public domain. They could be had from independent distributors (the Classic Film Museum in Dover-Foxcroft, Maine, was a popular supplier) for as little as $450 a print. It was a deep well of cheap product for the emerging cable pay channels (as well as for broadcasters).

By 1978, the flow of programming product was sufficient to prompt the industry to create a new National Academy of Cable Programming to support and promote the new content. Broadcast television's Academy of Television Arts and Sciences had ruled cable-originated product ineligible for its institutionalized Emmy Awards,[130] so part of the new Cable Academy's mission was to create its own awards ceremony. It instituted the Cable ACE Awards—Awards for Cable Excellence—honoring the first winners at the NCTA's National Show in Chicago in 1978.[131]

Integrating Vertically: Pay Cable Leads the Way

Through the late 1970s HBO dominated pay cable. It had the benefits of being first on the satellite and the deep pockets of Time Inc. Competitors not aligned with larger companies were acquired or withered. In June 1976, HBO purchased the film packaging company, Telemation Program Services, to complement its satellite service. TPS had been moderately successful at packaging films and other pay programming for non–satellite-fed cable systems and for the hotel pay-TV sector. It filled a valuable niche in HBO's marketing scheme.

Optical Systems Corp., despite a brief flirtation with Westar, eschewed the satellite option and paid the penalty in the marketplace. Its Channel 100 service was doing fairly well until the satellite launch of HBO. Relying only on videotape bicycling, however, the service was displaced in system after system by HBO until, at the end of the decade, it served only six affiliates. Another potential competitor, the Home Theater Network, the creation of New England Cablevision, did move to Satcom I in September 1978, but its programming was limited to one film a day. It positioned itself for marketing purposes as the "Extra Channel," a complement to full-service pay channels such as HBO. It also tailored its programming to family-friendly viewing, with films such as "Pete's Dragon" and "Grease." By December 1979, it had 41,000 subscribers on fifty-two systems.

Hollywood Home Theater, formed in 1975, was supported United Artists and 20th Century Fox. Although it served only a handful of affiliates, its customers included the Los Angeles area Z Channel, run by the Hughes-TelePrompTer Theta-Cable, and the Philadelphia area pay service Prism. Prism was co-owned by Ed Snider, principal owner of the Philadelphia Flyers and Spectrum Sports arena, and featured a heavy schedule of Philadelphia sports teams.

None of these services presented a serious challenge to HBO's market dominance. The satellite pioneer turned the financial corner in October 1977, reporting its first monthly profit. By 1979, it had 3 million subscribers, 1,200 system affiliates, and 65 percent of the pay-cable market.

When Kahn, RKO, and later Warner stepped into the cable industry in the 1960s, they shared a vision of owning not just the distribution systems (the local cable systems) but the program content as well, at least a healthy portion of it. In economic terms, they were seeking a vertically integrated company that would give them control over the production, distribution, and exhibition of their product, instead of simply being wholesalers or middlemen in the supply chain. It was an industrial configuration sought and achieved by media industries at earlier points in history, but a structure that had also invited government intervention. The public policy concern was that power over all stages of the supply chain would give a company or set of companies unfair market leverage over competitors, especially when that leverage was coupled with horizontal market oligopoly. That is, if ownership were concentrated horizontally, with only a handful of firms controlling one or more of the stages of production-distribution-exhibition, the vertically integrated firm would have substantial market power to keep out potential entrants.

This was the situation through the 1930s and 1940s in the film industry when the major motion picture firms (RKO, Paramount, 20th Century Fox, Warner Brothers, and Loews) controlled most of the theater screens in the country. In 1948, after a series of legal actions, the government ordered the film companies to divest their exhibition hall holdings. Similarly, FCC regulations for years limited broadcast networks to ownership of a maximum of five VHF and two UHF TV stations, and the FCC's financial interest and syndicated exclusivity rules limited the network's ownership and control of program content. The FCC, of course, had been concerned at previous points in cable's history about cross-ownership issues and about the possibility of broadcast network control of cable. As long as cable was primarily a broadcast retransmission service, without any of its own real content, vertical integration was not an issue, however. That began to change almost as soon as Satcom I had completed its first orbit.

As HBO began to prove itself as a viable business and as the fortunes of the cable industry lifted, Time naturally sought an MSO with whom to partner. Looking to secure the benefits of vertical integration, as well as enjoy the general resurging cash flow of cable system operations, Time, in 1978, as noted above, bought ATC, one of the nation's largest MSOs. Time Inc. and Warner Communications now both had a foot in content production and content distribution. It was the acorn of a new cable industry structure.[132]

Policy concerns were raised during the FCC's review of the ATC purchase. The Commission, pulled by its general deregulatory inertia, however, concluded that cable was still a "young industry" with "ease of entry and substantial growth potential." The issues of media concentration raised by the ATC-Time Inc. merger were not of any great concern for the Commission.[133] The FCC approved the merger in November 1978, just months before the Supreme Court in *Midwest Video II* killed the Commission's leased access requirements.

Showtime and The Movie Channel

Despite Warner's substantial cable holdings and its access to content, it was not the first company to meet HBO in the sky. The HBO monopoly on satellite-delivered programming ended in March 1978 when the CBS spinoff Viacom launched its pay service, Showtime. The decision of other larger media companies to enter the pay-programming field was no doubt influenced both by the success of HBO and by the termination of the FCC's restrictive pay-cable rules. Nonetheless, Viacom Chairman Ralph Baruch lamented their late start against HBO. Company officers below him had determined initially that they did not have the funds—the deep pockets of a Time Inc.—to lease transponder capacity. Showtime, therefore, began its life in July 1976 as a bicycling service, distributing videotapes to Viacom systems and, through an exclusive marketing agreement, to Times Mirror cable systems, primarily on the West Coast. Showtime's early marketing strategy called for supplying pay programs to cable operators unwilling or unable to spend the money to join the satellite network.

Viacom was one of the nation's largest MSOs and Showtime was immediately available to all its subscribers, but its reach was still limited compared with HBO. The lack of a sufficiently large subscriber base coupled with what was proving to be an expensive service meant that, by 1979, Showtime was in trouble. "We were losing a great deal of money," recalled Baruch.[134] The Viacom chief contacted the major broadcast networks to see if they were interested in a partnership. They all turned him down. "I finally contacted Russ Karp (President) of TelePrompTer, and Russ had 300,000 HBO subscribers and we had 300,000 Showtime subscribers," explained Baruch. "I eventually ended up selling half of Showtime to TelePrompTer for $6 million, and I got a doubling of our base, from 300,000 to 600,000 subs."[135] The deal was doubly beneficial to Showtime because TelePrompTer did not simply add the extra service. Critics of vertical integration worried most perhaps about a company favoring its products, in distribution and exhibition, over those of potential competitors and thereby narrowing the field of consumer choice. In a textbook example of this process at work, TelePrompTer, still the largest operator in the country, terminated HBO carriage and switched its subscribers to the new pay partner, Showtime. In one swift move, more than a quarter million customers were taken off HBO's ledger and moved to the plus column for Showtime.

Warner, which went into cable originally to have another outlet for its films, joined the satellite pay business soon after. In 1973, shortly after Warner purchased TVC, the company changed the name of its Gridtronics pay service to The Star Channel. "We decided that the name Gridtronics was unwieldy for advertising purposes," explained TVC's Frank Cooper. "People didn't understand it. We renamed ourselves Star Channel."[136] Warner had found the idea of satellite distribution too expensive in the mid-1970s, but by the end of the decade, economics were improving and the benefits of satellite distribution becoming obvious. In 1979, Warner took its pay service into space, and at the same time changed its name. In December 1979, The Star Channel became the Movie Channel, to convey more accurately its all-movie nature. By 1980, HBO, Showtime, and the Movie Channel constituted the major, vertically integrated pay-cable options, although HBO remained far and away the dominant provider. HBO claimed more than 2.7 million subscribers; Showtime, 825,000; and the Movie Channel, 125,000.

Minis, Maxis, and Exclusivity

The Viacom-TelePrompTer decision to carry only its own pay service was not by any means precedent setting. HBO was the only service carried on ATC systems, not just by technological fiat, but by contractual stipulation.

From nearly the start, HBO required exclusivity with its affiliated cable operators, even non-Time Inc. properties. For a brief period, exclusivity agreements were standard in pay-cable. An NTIA 1980 study showed that as of June 1979, of 101 TelePrompTer systems, ninety-six offered only Showtime; of fifty-four Warner systems carrying pay, forty-seven offered The Movie Channel, only five offered HBO; and all sixty-one ATC systems offering pay carried HBO exclusively.[137] The pattern was not perfect by the time of the survey because, in November 1978, a Wometco cable system in Thibodaux, Louisiana, broke the exclusivity barrier, and began offering both HBO and Showtime, just to see if it would stimulate subscriptions. It did. Moreover, and somewhat to their surprise, they found that people were not just choosing between the pay alternatives but frequently were taking both.

Neither HBO nor Showtime protested very loudly. They were under increasing public and governmental pressure to end the exclusivity practices, and were also rediscovering the historic iron rule of cable television, that more is almost always better. In 1979 and 1980, the vertically integrated MSOs and the independents alike began selling packages of multiple pay services, usually offering a discount for the combination. While HBO alone might be offered for $8.95 a month, and Showtime $9.95 a month, a customer could purchase both for $15.90.

Exerting an additional powerful pull toward packaging, large cities looking to franchise new cable systems came to demand an assortment of pay services from cable suitors. Many cities, seeking family-friendly fare, began asking for pay channels stripped of R-rated films. Because HBO and Showtime offered programming about 12 hours a day, unused transponder time could be put to use carrying companion channels. As a result, new pay services began to appear. With the cities pushing at one end and consumers pulling at the other, the established pay providers began to experiment with new program packages and marketing strategies. HBO and Showtime developed "mini" services, offering reduced content at a reduced price, seeking to attract a market segment unable or unwilling to subscriber to their traditional "maxi" service. Showtime launched a mini pay called "Front Row" in October 1978. It purchased a regional sports service in 1979, "Fanfare" repackaging it as "Showtime Plus." In April 1979, HBO developed, somewhat reluctantly, "Take 2," billed as an all-family oriented mini service. "Take 2" was created at the urging of HBO affiliate operators who, in turn, were responding to local demand.

Part of the marketing strategy was to provide complementary affiliated services. That is, if a customer was going to buy two movie channels, HBO wanted them both to be Time Inc. services. Offering the min-pays was one technique, so was the development of full-fledge maxi-pay sister services. With a critical mass of customers interested in taking two pay channels, HBO abandoned "Take 2" in August 1980 and created in its place, Cinemax, which itself was fashioned from a pay service purchased from Continental

Cablevision called Cineview. Where HBO offered a variety of films, sporting events, and specials, Cinemax would be devoted strictly to film content, primarily, older (and cheaper) films, that the company preferred to describe for marketing purposes as "classic" and "ever-popular" films.

The growing assortment of pay options, along with the increasing number of advertiser-supported services, readily lent themselves to various tiering plans, and subscribers were beginning to be able to choose a wider range of services at varying prices.

The issues associated with exclusivity agreements and vertical integration were not restricted to just program distribution. Access to content, especially for the pay services, quickly became a significant problem. A limited number of major motion pictures were produced each year and cable competition for that product was heating up. The problem was especially acute for the independent pay services, which did not have the resources of Time Inc. or the studio connection of Warner.

Cable Invades Hollywood

For Hollywood, cable was an odd new beast, and the film moguls initially seemed unsure as to how to deal with it. Pay-TV had nibbled at the edges of Hollywood consciousness for years, always a potential threat or a promise, depending on which end of the distribution-exhibition chain one was on. MPAA and the studios initially backed pay-cable efforts, at a significant cost to their relationship with theater owners. For the studios, pay-cable promised a means of tapping market segments that no longer went to the movies in droves. As has been noted, RKO and Warner were early believers in the after-market possibilities of cable television.

The question for the studios was how best to dip into the new revenue stream. After its initial theater run, a new film had traditionally gone next to network television, where the studios charged a negotiated license fee for a limited number of showings. Pay-cable promised to replace network television as the second venue, or window, for films after their run in theaters. Hollywood was especially keen on a financial arrangement that might give them a percentage of every subscriber's monthly payment. The studios, therefore, came to the negotiating table with their palate set for a slice of the pay-cable pie.

The first company they sat down with was HBO, and HBO would have none of it. Time Inc. was better served offering a flat fee for each film or package of films. The company also wanted exclusive rights to the films. No other after-market distributor could use the content for a set period of time after a film's theatrical run and before its release to network television. Both the flat fee and the exclusivity arrangement were unsavory for the studios. HBO paid studios a set price for a film, anywhere from $10,000 to more than six figures, whereas its income was based on subscriber fees, which continued to climb through the late 1970s, and HBO declined to pass any of the growing income along to the studios in the form of either higher license fees or per-subscriber rates. With respect to exclusivity, because HBO also demanded exclusivity with its cable affiliates, it effectively meant that only subscribers served by HBO systems would see the films on cable. From the perspective of the studio, the potential cable audience for the film was thereby dramatically reduced. Instead of a dowry-laden bride, HBO soon became the movie industry's new bête noire.

While the studios certainly did not like it, for a period of several years most felt forced to settle for HBO's terms, for several reasons. HBO, most importantly, controlled the pay-cable market, giving it substantial leverage. In 1978, about 75 percent of all pay-cable subscribers were Time Inc. customers (combined HBO and TPS), with Warner a distant second at about 7.5 percent of the market.[138] Even on a flat-fee basis, the revenue potential was substantial. Some observers also suggested that the studios suffered from a stigma of having missed previous opportunities to exploit emerging technologies, especially the possibilities of broadcast television, and did not want another slipping away.

It was a bitter pill for the studios to swallow, however, made even more distasteful by HBO's approach to contract negotiations, which were exceptionally aggressive. HBO wielded its market dominance freely, even playing the studios off against one another. Thomas Wertheimer, Vice President of MCA/Universal at the time, explained, "We had numerous meetings with HBO representatives who indicated that they only had a need for five studios' product and there were seven studios, or words to that effect. And that if I didn't get on the wagon soon, it would leave without me."[139] HBO developed a reputation as a hard-edged, even abusive negotiator. A clash of cultures between the brusque New York business manner and the laid-back West Coast style no doubt exacerbated the problem, but much of the friction was blamed on the aggressive manner of HBO's programming chief, Michael Fuchs. Fuchs' reportedly abrasive, confrontational personal style (he was once labeled "the most hated man in Hollywood") helped create bad blood between HBO and the studios, and HBO soon gained a sizable portion of business and personal animosity in the Hollywood community.

These problems were compounded as HBO began to invest directly in film production. In the summer of 1976, HBO took a serious look at its future and realized the need for a steady, dependable supply of quality programming at reasonable, hopefully controllable, prices. It was offering as many as a dozen new (to its subscribers) films every month, more than the combined annual output of the major Hollywood studios. Sketching out a closer relationship with its suppliers, especially independent filmmakers and distributors, HBO decided to offer financing for films in exchange for distribution rights. HBO would supply some or all of the necessary up-front money for production; in return, it would receive exclusive rights for the pay-cable distribution of those films. The first such deal it struck was with Columbia Pictures in June 1976. HBO would finance $5 million worth of film production for Columbia and receive pay-cable rights to recently released films.

Cash-strapped independent film producers were drawn to the offer as well as major studios such as Columbia. For mainstream Hollywood, it meant some loss of control over the product that they historically dominated. It limited the potential revenue of a hit film, but offset the financial risk for films that failed at the box office. In 1979, HBO formed a division to finance and coproduce independent films and TV programs, holding exclusive pay-cable rights to the material. HBO's investment in (and control over) increasingly large amount of content only strained relations with the major studios, which saw hundred of millions of dollars escaping Hollywood and flowing to Time Inc. offices in Manhattan.

For the studios, the situation was intolerable. The pay business promised millions to HBO and any other pay-cable provider, and the studios sought their share. Initially, Paramount took a stand and flatly refused to supply films to HBO. In November 1977, the MPAA filed a complaint against HBO with the Justice Department, alleging the company was using its dominant position in the pay-cable business to pressure studios into the exclusive, flat-fee deals. But, the launch of Showtime in 1978 and TPT's switch of 300,000 subscribers to the new service forestalled Justice Department action. After eighteen months, the showdown between Paramount and HBO also ended in the fall of 1979, with an agreement to deal on a nonexclusive basis; Paramount could sell to other pay distributors. In return, HBO landed a fifty-eight picture contract with Paramount, the biggest to that point in the industry's history. With the end of the exclusivity dispute, the other major studios also began supplying more product to pay-cable.

The contracts did not solve what Hollywood saw as the fundamental problem, however, which was control of (and appropriate compensation for) their product. With no sign of help from the government, the studios began series of meetings in 1979 on how to deal with the new programming reality. They discussed a number of options for combining their product and distributing it cooperatively, eliminating the need for HBO and Showtime. The exclusivity arrangements that prevented cable operators from offering more than a one-pay service were showing signs of deterioration, suggesting new pay providers might get a foothold in the nation's cabled homes.

Out of their discussions, in April 1980 Columbia Pictures, Paramount, 20th Century Fox, and MCA/Universal, combined with financial backing from Getty Oil, created their own pay-cable service, dubbed Premier. Cable veteran Burt Harris was tapped as CEO of the new venture. Premier would not only compete with HBO and Showtime, it would be the exclusive first venue for the cable distribution of the companies' movies, which represented more than half of Hollywood's annual film production. After its theatrical run, a film would be available only on Premier for nine more months. A successful film could run in theaters for a year or more, meaning HBO, Showtime, and The Movie Channel could wait nearly two years before it became available and, by then, a large portion of pay subscribers would have already seen it. "We cannot sit idly by watching HBO gobbling up the market with our product. The revenue potential is staggering."[140]

The pay-cable services issued angry accusations of collusion and antitrust, and took their case to Washington. The Justice Department, which had shown little interest in HBO's market power, was much more excited about the Hollywood cabal. In August 1980, Justice filed an antitrust suit against the film companies, alleging price fixing and conspiracy. On December 31, the government won its argument in Federal District Court in New York and, in June 1981, Hollywood pulled the plug on Premier. With the failure of the Premier project, the studios were forced to look at other means of exploiting the pay-cable oil well. Premier's failure in fact set off a chain of events that would lead to a reconfiguration of the pay-cable landscape and the creation of new partnerships and a new studio.

The View from Orbit

In October 1977, HBO reported that it finally was making money and it looked as if the cable-satellite connection might actually pay off. Some 2,500 cable operators had installed earth stations and were down-linking the new channels. Demand for additional programming, as always, was strong and the economic lure of being a satellite provider drew an increasing number of cable gamblers. By 1979, there was a growing roster of satellite-delivered cable TV services available to operators. The shopping list included HBO, Showtime, the Star channel, Hollywood Home Theater/PRISM, Home Theater Network, Cinemerica, Nickelodeon, ESPN, C-SPAN, The Movie Channel, Galavision, Modern Satellite Network, Madison Square Garden, UA-Columbia's Calliope, PTL, CBN, and Trinity. The pay services offered recent-run films and major sporting events. The neophyte basic services included children's channels, sports channels, religious channels, and coverage of the House of Representatives. All-news and weather channels were on their way, and plans for interactive gaming channels (Jerrold's Play Cable) and a service designed for adults fifty years of age and older (Cinemerica) were on the drawing board.

In 1967, the industry introduced its first programming awards, the Abel Award. The winners were primarily local operators producing their own material. By 1979, the industry was creating nationally distributed programming with production values that equaled the broadcast networks, and it adopted a new award and ceremony to celebrate. It was something of a coming of age for cable. It was called the Cable ACE Award, which stood for "Awards for Cablecasting Excellence." The first presentation was held at the

1979 NCTA convention, with honors going to more than a dozen programs, in categories from access programming to pay-cable entertainment.

As with technology, the new programming in many ways was not new at all, but rather an extension of long-standing forms of entertainment, such as films and sport. But the technology offered vastly increased capacity and hence the potential for increased diversity. While the new cable programming was not new in any deep historic sense, there was a lot more of it, and a greater variety. As had been predicted in the mid- and late 1960s by observers such as Johnson, cable's expanded pipeline allowed the medium to target smaller and more homogenous markets with material tailored specifically to their interests. In the evolutionary metaphor, it was the phenomenon of increased diversity and specialization writ across the TV screen.

One of the principal beneficiaries of the expanding programming palette was RCA, which through its Satcom I satellite had become the provider of choice for cable. Satcom I was known as the cable bird, and programmers who attempted an alternative satellite, such as Westar, soon found that few operators were financially able, or willing, to purchase two earth stations.[141] Satcom also had twenty-four transponders compared with Westar's twelve, meaning cable operators could potentially receive twice as many program services using the RCA bird as they could using the Western Union satellite. Programmers, such as SIN and Optical, saw the problem early; SIN moved to find space on Satcom I for its Galavision,[142] Optical simply abandoned the Westar option.

Demand for transponder time quickly surpassed capacity. By 1979, Satcom I was carrying more than twenty different cable channels. RCA announced that it would launch a new satellite, Satcom III, a year ahead of schedule in an effort to meet the demand.[143] The launch was scheduled for December 10, 1979, coinciding roughly with the Western Cable Show, the industry's second-most popular annual event after the NCTA convention. Dismay and shock swept across the convention floor four days after launch, however, with the announcement that Satcom III had disappeared from the tracking screens. RCA's multimillion investment and the hopes of many would-be programmers were, literally, lost in space. System operators and programmers had been counting on the satellite to further expand their menu of saleable goods. The loss of Satcom III was a significant setback for the business. RCA had guaranteed back-up transponders to at least five of the new programmers. Unfortunately, it only had two transponders immediately available. One of the injured parties was Ted Turner, who was counting on Satcom III for distribution of his planned all-news cable network (see Chapter 9). When RCA could not guarantee a quick replacement, Turner took them to court, securing a judicial order that gave him temporary carriage on another satellite. Others had to wait.

It would take two years to build and launch a replacement (Satcom IIIR). RCA leased time on other birds to ease the problem, but significant confusion developed over which satellite would now become the second "cable satellite." Operators were hesitant to buy and install additional dishes until they knew how the situation would be resolved. The result was a brief stagnation in the launch of new programming services.

Despite the pause in growth, cable in the latter half of the 1970s found new life. Channel capacity was expanding and, more critically, operators now had something to sell, something beyond their 1950s broadcast retransmission service, and they were focused on selling that product with vigor. The so-called CATV pirates of the 1950s had tried in the late 1960s and early 1970s to wire the major markets. They found themselves wrecked on the barrier reefs of technical and business reality, but now cable had the satellite. Most of the nation's TV homes and TV dollars were in the big cities, and cable was ready to attempt an approach from the air. It was time to take another run.

9

Cablemania (1980–1984)

> *We don't really have the foggiest idea of how those big
> city bells and whistles systems are going to be paid for.*
>
> —Steve Effros, CATA President, 1981[1]

It was, as Yogi Berra famously put it, déjà vu all over again. Cable franchising activity, which had abated in the early 1970s, mushroomed in the last half of the decade and into the early 1980s. The tumbling of federal controls, the emergence of national cable programming, and the renewed flow of investment dollars meant that cable could try its luck once more in the major markets. Touting forty-, fifty-, and even eighty-channel or more systems carrying HBO, Showtime, ESPN, C-SPAN, and Nickelodeon, cable came back, pounding on the doors of city hall. Only the top MSOs had the resources to take a run at the urban market, but their representatives were there, handing out business cards, and more.

The stakes were high. The major markets, as noted, were where most of the nation's viewers lived. Franchises were long term, ten to fifteen years, and often exclusive. Companies that failed to secure a beachhead in cities such as Dallas, Pittsburgh, Boston, Chicago, and Denver, faced the prospect of being locked out of 70 percent or more of the national subscriber base for years to come. It was a once-in-a-generation opportunity. The competition, therefore, ranged from lively to cutthroat. NCTA President Thomas Wheeler called it a "land rush," *Broadcasting* magazine, a "Gold Rush."[2] In the final days of the 1970s, *Forbes* magazine declared "Cablemania is here again,"[3] and, a few years later, author James Roman took "Cablemania" as the title for a new book on the industry.[4] As the parameters of the assault on the cities became more widely publicized, however, a new label was offered, and stuck. Soon everyone was calling it the "franchise wars."

If any lessons had been learned from the franchising scandals of a decade before, they apparently were forgotten in the renewed frenzy. Pressed by the competition and driven by the fear they would be permanently excluded from most of the nation's television homes, many operators promised far more than they could deliver. As *Broadcasting* magazine later observed: "The attitude of the most aggressive operators was to promise anything to win the franchise and worry about fulfilling the promises later."[5]

Cable's exuberant pursuit of urban America became a massively expensive treasure hunt, damaging the industry's financial health and its public image. Cable faced other challenges in the early 1980s, as well. Equity dollars were costly, and interest rates spiraled so high that operators looked back nostalgically on the 1974 cable depression. This chapter first looks at the franchise wars and the assorted business problems of the early 1980s, as the industry struggled to gain a beachhead in the major markets. It subsequently considers the rise of a new set of potential TV competitors in this period, ironically labeled the "new communications technologies," and the on-going deregulatory movement in Washington, D.C., that fostered them. The second half of the chapter reviews the continuing development of new cable programming services—including the Cable News Network, Black Entertainment Television, MTV: Music Television, and The Weather Channel—and the often-awkward process of creating partnerships between the young programmers and the older distribution side of the business. Finally, the chapter considers the birth of the Cable Communications Policy Act of 1984, an act of Congress that would bring both great benefits and great troubles to the industry.

The Franchise Wars: Wreckage and Response

More than 4,000 cable systems were in operation in the United States by 1980, but public attention focused on fewer than fifty situated in top markets where two thirds of the nation's potential subscriber pool resided. The stalled efforts of a decade before had left most of the big cities still unwired. In 1979, 18.3 percent of the country subscribed to cable, but in the twenty-five largest markets cable passed only 22 percent of the television homes and only about 8 percent took the service.[6] Market size was, as discussed earlier, only part of the attraction. Operators also prized the housing density in the big cities, where more potential customers per mile of coaxial meant increased system efficiency and overall profitability.[7]

While the competition was often intense, it was chiefly limited to the top MSOs. Bill Daniels admonition of the 1960s to operators, to bring lots of money when they came to the major markets, was more true than ever in the early 1980s. Construction costs for pole-mounted systems were running from $7,000 to $10,000 a mile, or more, and as high as $100,000 a mile underground.[8] In 1976, municipal construction was pegged at $500 to $600 per subscriber; by 1982, that figure had jumped to $1,200 a subscriber. Daniels himself won the franchise for his hometown of Denver (a joint venture of Daniels & Associates, AT&C, and a group of twenty-two local investors organized as Mile Hi Cablevision). The estimated cost of construction was $100 million.[9]

Operational costs also swelled in the urban markets. MSOs had to install more telephone lines than they were accustomed to in order to handle the increase in customer inquiries (and complaints). Additional customer service representatives had to be hired to answer those calls. The representatives spent a great deal of time trying to explain the new and more complex programming packages offered in the major markets, packages that, in themselves, were more expensive to assemble. Everything from billing systems to preventative maintenance to promotional campaigns expanded in size, complexity, and price.

The list of MSOs willing and able to spend the money was fairly short. It included firms such as TelePrompTer, TCI, Time-ATC, Warner-Amex, Sammons, Viacom, Cox, Cablevision Systems, and United Cable. There were more than 200 MSOs at the time, but ultimately only a dozen or so would come to control the urban subscriber base and hence the nation's cable marketplace. In this manner, and in absence of legal controls on MSO size, a mix of geography, technology, and economics set the mold for industrial structure in the cable business.

The number of contending companies was pared further by varying corporate philosophies. Some MSOs attacked major franchise opportunities vigorously; a few held back, taking a more cautious approach to growth. Warner and ATC were aggressive, going after multiple markets and hoping to win some reasonable percentage of them. TCI was more circumspect, initially avoiding markets where it felt that, for technical or political reasons, it was unlikely to succeed, and also avoiding whenever possible promises of gold-plated systems.

Gold-plated systems, however, were the order of the day in most negotiations. Sober observers inside and outside the industry characterized it as an orgy of excess and Blue Sky scheming gone wild. Companies offered state-of-the-art (in fact, beyond state-of-the-art), dual cable systems with capacities of up to 105 channels, plus any interactive service that could be conceived. Warner leveraged, to the extent possible, its QUBE technology. Cable companies promised multiple access channels, fully equipped access centers, and a grab bag of revenue sources for the city coffers, all at amazingly low rates.

If cable operators were willing to promise almost anything to get a franchise, some city officials, sharing culpability for the excesses of the period, were just as eager to put together a shopping list for the suitors. The 1960s fantasy of Blue Sky technology and service was tenacious and lingering. It had not dissipated in the minds of urban planners and, often encouraged by cable consultants hired by the cities, became the framework for negotiations. Simple native greed, on both sides, played a powerful role as well. The cities needed funding for playgrounds, libraries, and firehouses. New York Mayor John Lindsay's "urban oil well" had not lost its appeal. Stirred by the utopian rhetoric of some social commentators and the very real needs of their local communities, city fathers and mothers, following the pattern of the late 1960s, demanded the most advanced forms of cable technology and services at city-controlled rates, and this time added perquisites that were often only tangentially related to communications.

The cities, of course, had many legitimate claims and concerns. They had, decades previously, established the reasonable position that safe access to streets and alleys necessitated some form of local oversight and, for the most part, cable operators had obliged. Franchising was a normal part of the business. The question now became, legally, politically, and financially, how far could the franchising claim stretch? To what extent could the cities establish requirements and boundaries for cable service? Could the city, for example, require some form of local universal service?

Most of the nation's metropolitan areas are composed of multiple jurisdictions, sometimes dozens of contiguous urban and suburb municipalities. Los Angeles, for example, is a patch quilt of individual cities and towns, each with its own local cable franchise and franchise process. Operators eagerly trolled the major markets for opportunities in the affluent and middle class suburbs, frequently acquiring, franchising, and building systems there long before the core cities were wired. Within the urban markets proper, well-to-do-neighborhoods were also sought. But, the less affluent sections of town, often the inner cities, were typically avoided, creating what some called a "donut effect," a ring of suburban cable systems surrounding an inner city "cable-free zone." The pattern took the name of "red-lining," a term borrowed from practices in real estate, where lenders avoided placing funds in the less-advantaged parts of town. Cable operators had clear business disincentives to build in such areas. Residents were less able to afford the service in the first instance and more likely to default on payments when they did sign up. There were concerns, legitimate or not, about the greater potential for cable theft and equipment vandalism. While the near-term logic of business militated against construction in the inner cities, enlightened social policy and indeed the long-term best interests of the industry were not well served by the practice. Fights over pay-TV in the 1950s and 1960s had raised issues of television "haves" and "have nots," a separation between those who could afford television and those who could not. This "television rich-television poor" problem now manifested itself in new form, which cities sought to rectify through their franchising practices.

Public policy interests associated with full access to what was seen as an important information resource, therefore, confronted the cable business reality of needing to show a positive cash flow. Officials in Connecticut had required cable to wire cities that otherwise would have been overlooked as unprofitable as a *quid pro quo* for being allowed to build in desirable markets. As cable began, finally and seriously, to approach urban markets, a similar bargain was often required with respect to wiring the inner cities.

The interests of city planners were not always so noble, however. The cable franchising process of the late 1970s and early 1980s went far beyond issues of rate regulation, service area, and channel capacity, and into areas that drew the ire, frequently the ridicule, and increasingly the scrutiny of the press, citizens groups, and federal regulators. The list of demands became a bill of indictment for the franchising process in the popular, legal, and scholarly literature. In Fort Worth, Sammons promised to bring with its franchise a million-dollar community programming facility, three television vans, $175,000 community programming budget, and $50,000 for internships. The Indianapolis franchise went to ATC, which promised four public access studios, two vans, and an $850,000 community programming budget. Cox Cable won the Omaha franchise promising 108 channels (fifty-four video and fifty-four text service) for $5.95 to $10.95 a month. The proposal included a plan for a two-way interactive data exchange system and $7 million worth of local origination facilities.

Cablevision won the Boston franchise in 1981 by promising a fifty-two channel programming package for $2 a month. ATC promised a free, ten-channel community program service to every home in Pittsburgh (and still lost). The poster child for extravagant promises came out of the bidding war in Sacramento, California. There, United Tribune Cable Company won initial approval for a franchise after offering a two-way, 120-channel system that included burglar and fire alarms; ten public access TV studios; and $97 million in community or school-related grants and equipment. United Tribune promised city officials a system that would be able to turn on and off city air conditioners on hot days and, just to cap things off, promised to plant 20,000 trees around the city to aid in municipal beatification. Cooler heads at the parent company headquarters balked at the $200 million price set by the local subsidiary and eventually the Sacramento franchise went to another company (Cablevision of Sacramento), but the United Tribune reforestation plan had drawn headlines. In the end, cable companies ended up paying for libraries, firehouses, and sewer systems.

Cable companies also waged local public relations campaigns, opening storefront offices to promote their systems, speaking at rotary clubs and Better Business Bureau lunches, and initiating letter-writing campaigns. They sought out the most politically well-connected law firms to represent them before the town council, and spent hundreds of thousands of dollars in just the process of applying for the franchise.

One of the more prominent and heavily publicized franchising tactics was the enlistment of local support to obtain the license, a practice that had a faint flavor of the Kahn-Johnstown incident, but now involved money passed over the table. Cities had always privileged local interests in franchise negotiations. Applicants who could demonstrate local ownership, at least in part, had an advantage over those who did not. Operators, therefore, sought out prominent community members, preferably ones with influence on the city council, and made them equity partners in the system, even if it meant loaning them the money to do so. It was dubbed "rent-a-citizen." A group of local, well-connected citizens were typically sold a percentage of the system at astonishingly discounted prices. *Forbes* suggested that TelePrompTer-Comcast investors in Pittsburgh acquired 13 percent of the local company for what amounted to 3 percent of the capital; and ATC sold 20 percent of their Pittsburgh system to local partners for 5 percent of the capital contribution.[10] Going this one better, Warner-Amex simply gave 20 percent of their proposed Pittsburgh company to seventeen local African-American groups. In Milwaukee, local investors were offered stock in a cable contender for one cent a share; if the franchise was awarded, the value would have been $25 a share.[11] Once the franchise was finalized, the cable company typically purchased, at the new and substantially higher market price, the equity back from the local investor, minus whatever the cable company had lent for the purchase of the stock. In some cases, cable companies that entered the franchising process late found that all the local power brokers had been claimed by competing bidders and opted to drop out of the contest for lack of local clout.

Houston may have been the most egregious example of franchising shenanigans. There, the city was divided into five cable districts and separate franchises awarded in each. In four of the five, the awards went to local business people or politicians with close connections to the mayor, including his personal attorney and his campaign manager. The fifth franchise went to an established company, Storer, but only because no one else was interested in operating in that apparently less-desirable part of town. The bidding process was not nationally advertised, and not all of the winners were realistically capable of building a system. Instead, they sold large stakes in their franchises to Warner and Storer a few months after the awards. One of the losing bidders, Affiliated Capital Corp., promptly sued the city and Gulf Coast Cable, one of the winning firms. In February 1981, a federal jury found the city, the mayor, and Gulf Coast guilty of conspiracy and awarded the losing operator $6.3 million in damages.[12]

The results of the franchise wars were predictable. The process became mired in unsavory publicity; winners frequently were unable to fulfill the franchise obligations and returned to the city council to renegotiate. By then, of course, the incumbent system was entrenched, or becoming so, and the municipalities were all but forced to alter the original terms of the agreement.

As in Houston, losers in the franchise wars often filed suit against both the winners and the city. Against the sometimes-outlandish promises of over-eager MSOs, economic reality forced cable, and the cities, to come to their senses; before the decade was too old, many began to take a more realistic look at what it would cost to build and operate the urban systems.

It was a distant but clear echo of the prior decade and another stark example of the clash between the social construction of technology and actual success of technology in gaining a foothold in existing social and economic conditions. Much of the technology demanded and promised simply was not viable and, when it was, it was far too expensive for the real world of 1980s commercial cable television. The cities and operators discovered, for example, that advanced amplifiers promised or stipulated in the franchise agreements would not be ready for months, and the interactive equipment that looked promising on the drawing board was not, in fact, ready for market, and no one really knew when it would be. Some operators found that it was going to take the utilities companies several years just to get the poles ready to accept the new cable. Cable companies frequently had to win municipal approval for even the smallest changes, such as the routing path of a trunk line. Such approval often also required going through several stages of local bureaucracy, including citizens advisory boards, many of which suffered from internal squabbles that delayed decision-making further. At the consumer end, advanced interactive services, when offered, did not sell, and expensive public access centers, when built, sat mostly vacant. In the end, much of the access equipment was simply donated to local schools and universities and the loss written off.

In 1981, CATA's Steve Effros observed that cable technology was an evolutionary not a revolutionary business, and stated plainly that the promises had far outstripped economic reality. "[W]e don't really have the foggiest idea of how those big city bells and whistles systems are going to be paid for," he said.[13]

Like a drunk on the morning after, cable companies began pulling themselves out of a groggy haze of postfranchising frenzy and gazed out at the wreckage left by the wars. Across the country, operators were requesting delays in construction or renegotiating to eliminate many of the fancier services that had been promised earlier. They told the cities they could not, for economic or technical reasons, fulfill the promises made in the heat of competitive courting. Sometimes they ask city officials to discuss the situation, sometimes they just informed the city of their new business plan. Either way, channel capacity was scaled back, institutional networks eliminated, and interactive services dropped. In the most severe cases, the cable operator pulled out entirely.

Some companies, sensing that things were getting out of hand, backed away even before the bidding process was complete. Fairfax County, Virginia, and Boston received only two final franchise applications each; Chicago received only one to three for its five franchise districts. In Baltimore, Cox Cable, considered a leading contender for the franchise, pulled out early. In a suburb of Washington, D.C., the United-Tribune cable company informed officials that it no longer appeared that the system could be made profitable under the stipulated conditions of the franchise and it simply halted construction, predictably triggering a lawsuit by the county. Gus Hauser, formerly of Warner Cable, bought the franchise from United-Tribune in 1986, but not before winning significant concessions on system requirements.

The major "crash and burn" victim of the period may have been Warner Communications. Warner had been among the most aggressive franchise warriors, carrying its QUBE service like a tournament banner through the campaigns and winning contracts in Cincinnati, Cleveland, Milwaukee, Dallas, Pittsburgh, and the suburbs of St. Louis. By 1983, Warner's flag was drooping, however. It was experiencing all the shared problems of the industry in making good its franchise commitments, but had a host of additional burdens specific to the company. QUBE had been expensive. Developmental costs alone were reported at $30 million.[14] Construction and operational costs for the QUBE-equipped systems were high and Warner was losing millions in its cable operations. While QUBE was a technical, promotional, and public relations success, it was a business failure.

Warner's capital expenditures on cable had been supported, in fact, by a flood of corporate cash from another subdivision, but in 1983 that stream dried to dust. Warner had far-flung interests in movies, television programming, records, and book publishing, but its major success in the late 1970s was in the world of video games. In 1976, Warner purchased a promising computer game company called Atari, which had pioneered one of the first home video games, Pong. Atari sold video gaming hardware and software for the home and coin-operated units for arcade market. For several years, Atari, along with similar games, was the must-have consumer product, gushing a fountain of revenue for Warner. In 1980, Atari posted $2 billion in annual profits for Warner. By 1982, the division was responsible for more than half of Warner's total annual income.

The fashion for home video games faded, however. Atari failed to foresee the erosion of the market and, in fact, was wildly overproducing and shipping units just at the time that demand was skidding to a halt. By the end of 1983, the bottom had fallen out of the market and Atari was sitting on an immense inventory of machines that it would never sell. The division lost $533 million that year. Warner stock plummeted from $63 a share in 1982 to $19 by the end of 1983. The corporation began unloading its problems, selling Atari in 1984. At Warner-Amex Cable, operations were scaled back and systems put on the auction block. Through 1983 and 1984, Warner-Amex killed its QUBE service in Cincinnati, Pittsburgh, Houston, and Dallas. In January 1984, it announced it would reduce the scope of its new "builds" in several major markets. Warner-Amex Cable Chairman Drew Lewis flew to Milwaukee to tell city authorities the company could not afford to build the system it had promised to build. He proposed to reduce the planned number of channels by half, eliminate all local origination facilities, cut the promised number of access channels from eighteen to six, kill the QUBE service that had helped them win the bid in the first place, and raise subscriber rates. City officials were livid, but by then unlikely to get a better deal from any other cable provider. The problem was industry-wide. Cox Cable, for example, was quietly renegotiating its contracts with New Orleans, Tucson, Vancouver, and Omaha. Warner-Amex was more public and more aggressive in its retrenchment, but also in greater difficulty. Some Warner franchises were ultimately sold, often at fire sale prices. The Dallas franchise went to Heritage and Pittsburgh went to TCI.

In contrast to Warner and others, TCI escaped the franchise wars with only minimal powder burns. Beginning in the late 1970s, TCI was one of the industry's most acquisitive MSOs, but its initial moves were in traditional, smaller markets or in safe suburban enclaves. TCI kept its distance from the carnage of the big-city franchising wars. Malone and Magness adopted a corporate philosophy that eschewed fancy, hi-tech cable technology. They preferred to buy and operate "plain vanilla" cable systems and were noted for maintaining them in the most cost-effective manner possible. "I like the cable business just the way it is," Malone once said in reference to Blue Skies franchise promises. "It doesn't have to change radically for me to like it. Most of our systems have basic and one, sometimes two, pays, not three or more."[15]

TCI also had a reputation for bare-knuckled tactics in local franchise negotiations, especially in renewal situations. Malone saw a TCI cable system as a multimillion dollar capital investment and any city's effort in franchise renewal or revocation proceedings to take it away as something like an act of governmental theft. But the company's policy of building and running systems on the cheap meant that relations with local customers and officials were frequently strained. One of TCI's more infamous confrontations came in a franchise renewal fight in Jefferson City, the capital of Missouri. Continuing complaints about shoddy service and escalating prices prompted city officials in 1981 to open bidding for new cable providers. TCI, with its system threatened, declared war. It withheld franchise fees and ran newspaper ads claiming that local cable service would end if TCI lost its franchise. It vowed not to sell any of its facilities to new providers, preferring that its cable rot on the poles, and even suggested it would peddle cut-rate satellite dishes against any new cable operator. In an encounter that would haunt the company in the press, in Congress, and in court for years to come, TCI's National Director of Franchising, Paul Alden, allegedly threatened the Jefferson City cable consultant, Elmer Smalling, at the NCTA convention in Los Angeles. In a widely distributed story, Smalling reported that Alden told him, "We know where you live, where your office is, and who you owe money to. We are having your house watched and we are going to use this information to destroy you. You made a big mistake messing with TCI. We are the biggest cable company around. We are going to see that you are ruined professionally."[16] Alden was fired when word of the encounter reached Malone, and TCI attempted to distance itself from the comments and the man, calling Alden a "loose canon" and disavowing any company knowledge of (or support for) his actions. But the harm was done; TCI's already pronounced reputation for aggressive, even ruthless, franchising tactics had only been exacerbated.

Ultimately, TCI retained the franchise, but a losing bidder, Central Telecommunications, Inc. filed suit, charging a TCI with using illegal, anticompetitive tactics to do so. TCI countered, arguing that it did not need a franchise because it had a First Amendment right to wire the city for cable. All of its activities related to holding and developing the franchise, therefore, were protected by the Constitution, and further, the franchise requirement itself was, by extension, a violation of the First Amendment rights of cable operators. (See Chapter 10 for more on the evolution of cable's First Amendment standing.) The courts rejected TCI's argument, upholding in the process the constitutional right of cities to require cable franchises.[17] At the trial level, the court found TCI's activities in Jefferson City to be "nothing short of commercial blackmail" and ordered it to pay $48 million in damages (a sum later reduced to $35.8 million) to Central.

Despite all this, Jefferson City negotiated a new contract with TCI. For Malone, the multimillion dollar settlement was simply the price of doing business, and city officials claimed TCI never took the case seriously. The case illustrated TCI's lean and mean approach to the business. The company also had long-range plans for expansion and that meant it had to look beyond towns such as Jefferson City and toward the major markets. As the smoke cleared across the urban battlefields in the early 1980s, TCI began peeking in to see what it could pick over from the remains, and the pickings were pretty good. Cities were being made to understand that they might have to settle for more traditional cable systems, TCI's specialty. Magness and Malone moved in, buying over-promised franchises at a discount. By 1984, TCI had acquired a system in Buffalo, and secured franchises in Chicago and St. Louis. The case study in the rise and fall of big city franchising in this period, however, may have been Pittsburgh.

Pittsburgh had had difficulties from the start. A citizens' advisory committee established to evaluate the bids had recommended ATC (operating locally as Three Rivers Cablevision) to the city council. But the council, apparently swayed by Warner's substantial equity contribution to local minority groups, voted the contract to Warner. Naturally, ATC filed suit, alleging impropriety in the selection process. Warner survived the challenge and sat down at the planning table with city officials. Pittsburgh prided itself on working with the company to develop what it called a "community communications complex." The head of the city's cable office, a Christian Brother named Richard Emenecker, explained in 1985, "In setting the standards for the design of this system, we expected it not only to set the course of electronic communications in Pittsburgh for the next several decades but also to serve as a benchmark for interactive community-cable systems throughout the country."[18]

It was not to be. By 1984, a struggling Warner-Amex was looking to get out. Warner officials had a meeting with Malone. TCI already had systems in suburban Pittsburgh and the Warner franchise would extend and consolidate its holdings in the market. Malone discovered, to his delight, that the hard-pressed Steve Ross would be happy to sell the system essentially for the cost of the money it had invested. Unfortunately for Pittsburgh, TCI's plain vanilla approach to cable meant a quick end to the high-powered interactive system envisioned several years earlier. As always, Malone was blunt in his analysis: "Two-way has been a franchising tool and a promotional gimmick for 12 years."[19] It would not be a part of the new Pittsburgh cable service. By the time TCI was finished with Pittsburgh, the system had gone from sixty channels to forty-nine; QUBE was history and local cable headquarters had been moved into a tire warehouse, but the cash flow was now there.

To a large extent, Malone, the master tactician and businessman, had gotten it right. Fancy interactive technologies were not ready for market and to the extent that they were attempted as often as not raised little more than a yawn from consumers. The mainstay of the industry continued to be what TCI had banked on for so many years, simple cable TV, offering local broadcast signals, basic networks, and a little HBO. TCI was motivated by straightforward economic calculus, but in terms of the interface between the technology and the market, it came out in just the right place.

By 1984, most of the franchise wars were over and the dust was beginning to settle, although some lawsuits dragged on. TCI was starting to build in Pittsburgh. ATC was in Indianapolis; Continental in St. Paul; Group W in Irving (Dallas), Texas; Warner held on to franchises in Houston and Milwaukee; United Cable in Baltimore; Cablevision Systems in Boston; Prime Cable in Atlanta; United Cable and Colony, among others, were in the divided franchises of Los Angeles; Chicago divided its franchises among Continental, Group W, and TCI.

As the suburbs and central cities were claimed and wiring began, operators also started thinking about interconnects and regional clustering. For three decades, cable operators had staked their claims, sometimes on a competitive basis, wherever they could. As a result, MSO operations were scattered across the country, and any given large market might have two, three, or more companies serving different parts of town. This created problems in efforts to sell local advertising, especially in the larger markets.

In the nation's largest towns, the most lucrative arena for advertising dollars, businesses were accustomed to buying, through the area newspaper, radio, or TV station, access to the entire market. But cable franchises were more commonly split, often among half a dozen or more cable companies, each serving a different suburb or neighborhood. To reach the entire market through cable, advertisers had to negotiate with each company, multiplying the paperwork and time beyond the point that most found worth the effort. In an attempt to overcome this problem, the industry, starting in the early 1980s, began creating cable interconnects, joining together the local systems and opening a single business agency to sell advertising time. Interconnects could be hard-wired, physically joining the different systems by microwave or landline, or they could be simulated, soft interconnects. In the latter, one entity would serve as the representative firm for the local operators, selling advertiser availabilities and coordinating the programming and paperwork. By 1982, cable interconnects of one type or another were operating in nearly a dozen markets around the country, including the San Francisco Bay area, Philadelphia, New York, and New England.

Going one step beyond interconnects was "clustering." It was clear to the cable community that substantial economies of scale could be realized if only one operator served the entire metropolitan market. Head-ends and, along with them, staffing and operations costs, could be consolidated. The process of pulling together, under a single owner, disparate cable firms in a single market was called "clustering." The first tentative steps toward clustering began in markets such as Pittsburgh in the early and mid-1980s, but the process would accelerate into the 1990s.

Although penetration levels in the major markets were still modest, they began to climb as crews started lashing cable to neighborhood poles. Nationally, cable households made up about 22 percent of the TV audience in 1981, and in the top ten markets, about 14 percent.[20] About 42 percent of all U.S. homes had access to cable in 1981; in the top ten markets the "homes-passed" rate stood at about 22 percent. Cities that, for a variety of reasons, were slow in concluding franchise agreements—such as Chicago, Detroit, and Dallas-Fort Worth—had lower subscription levels, under 5 or 6 percent by 1981. Overall, in the counties that held nation's twenty-five largest markets, penetration more than doubled between 1976 and 1981, from 6 percent to 13 percent. These markets accounted for 40 percent of the nation's television homes. In the second tier counties, which held 30 percent of all TV households, cable had increased its penetration from 14 percent in 1976 to about 29 percent by 1981.[21]

Building accelerated in the early 1980s, especially in larger markets where franchise competition abated. From 1976 to the end of 1984, cable penetration had soared from about 15 percent to nearly 44 percent nationally. By 1984, New York City had a penetration rate of more than 33 percent; Los Angeles, 31 percent; Philadelphia, 40 percent; Boston, 37 percent; and Pittsburgh, 58 percent.

The industry also faced another, related challenge, one it had confronted before, obtaining the financing to build the major market cable systems. Dollars flowed freely in the late 1970s, as the satellite was lighting up the skies. System valuations rose from $300 or $400 in 1977 to between $800 and $900 by 1981 and 1982. But capital grew scarce in the early 1980s and, as it had on previous occasion, the industry found it had to work harder for construction dollars and pay a great deal more for them. There were several causes for the changing financing climate. The cost of the build-out, as suggested previously, was going to be staggering, and would cut into overall short-term cash flow and profitability. Industry revenues continued to climb, growing from about $1.8 billion in 1979 to $3.5 billion in 1981, but so did expenses, increasing from about $1.1 billion to $2.3 billion in the same period. Reported net income actually dropped from about $199 million in 1979 to $168 million in 1980 and down to $40 million in 1981.[22] Investors with a near-term window did not find cable attractive.

Financing was becoming increasingly expensive for everyone in the early 1980s. A nationwide recession catapulted the prime rate to more than 20 percent in late 1980 and again in the Summer of 1981. Even those with a longer investment horizon were becoming aware of the mess in the major markets and the difficulty cable was going to have meeting its Blue Sky franchise commitments and making them pay. Until franchises were renegotiated, lenders were hesitant. Finally, and with a longer potential impact, institutional lenders and venture capitalists were starting to shed their view of cable as a monopoly, multichannel television provider. Over the previous several years a roster of new platforms for television program distribution had arisen. These new technologies, with acronyms like DBS, MDS, and LPTV, were seen as ready competitors for the viewers' pay-TV dollar and a threat to the cable business. Collectively, they were labeled "the new communications technology," and one of the greatest admirers of the new technology was now sitting in the Chairman's seat at the Federal Communications Commission.

Technology, Ideology, the Emerging Marketplace

In the late 1960s, "new communications technology" meant cable television, as the medium went through its important process of social redefinition. A similar process was underway in the early 1980s. This time, however, a new set of communication tools was taking cable's place as the current wave in new technology. The term "communications revolution" was being bandied about once again, usually without recognition that it had been much used a decade before to describe Blue Sky cable. Not without a certain irony, however, cable was evolving into a more established communications medium. While less than one-third of the country was connected to cable in 1981, for policy makers and lenders, the industry was losing its new car smell. The new darlings of avant garde technology included a resurrected DBS business, a terrestrial microwave technology called multipoint distribution service (MDS), home video, low power television (LPTV), subscription television (STV), and videotext. Taken broadly, the services and systems were hailed by some as the start of the "information age."

Discussion about and even operation of these new technologies began as early as the mid-1970s. The 1976 staff report of the House Subcommittee on Communications, "Cable Television Promise versus Regulatory Performance," which had praised cable and excoriated the FCC, also pointed to the "new opportunities" held out by evolving systems, such as videodiscs and direct broadcast satellites.[23] By the summer of 1982, the topic of new communications technologies and the challenge they posed to cable had even bubbled up to the front page of the New York Times business section, which noted that the alphabet soup of emerging services had made Wall Street nervous about cable's future—cable stock prices had slipped 19 percent by mid-year—and the industry faced a new competitive reality.[24] The industry itself was pointedly sensitive to the challenge the new communications technologies posed, going so far in 1983 as to launch an $800,000 marketing campaign to promote cable viewing. The NCTA-inspired "Consortium for Cable Information" was backed by the major operators and grew in large measure from the concern about potential new entrants.

In broader terms, the rise of the new communications technologies and the rhetoric that surrounded them was very much an illustration of both the evolutionary diversity of technology and of the social construction of that diversity. Most of these systems were untested in the real world and it would take a few years to determine which would prosper and which would fail. In the meantime, the utopian hyperbole began to build. Like the Blue Sky brocade of cable in the 1960s, the social construction of the new devices and ideas drove business decisions and, more critically, public policy.

In Washington, D.C., close to ten years of deregulation had worked to the benefit of cable television, freeing it from control that hamstrung its operational flexibility. To the extent that the Carter administration and FCC Chairman Ferris had thinned out cable law, the election of Ronald Reagan in 1980 and his subsequent selection of Mark Fowler to head the Commission in 1981, promised a clear cutting of the remaining rules. A communications attorney with a background in radio, Fowler had served as counsel to the Reagan presidential election committees in 1976 and 1980. He was not well known in media circles before his nomination, but his view of the relationship between government and business harmonized with that of the new President. Fowler was one of the most exuberant free market champions to ever sit on the Commission. While Wiley had promised to "re-regulate" cable and Ferris led a campaign to "deregulate" it, Fowler now vowed to "unregulate" cable, as well as broadcasting. He advocated the jettison of the trusteeship model of government regulation in favor of market forces. "I believe in the marketplace approach because I believe in the marketplace," he declared. "What people choose to watch or listen to should be as free as possible from the heavy hand of regulators."[25]

Fowler had been on the job only a few months when he began floating proposals to dismantle decades of broadcast regulation. In September 1981, Fowler's FCC sent to Congress a recommendation to eliminate the Fairness Doctrine and the Equal Time Provisions (section 315) of the Communications Act. Over the next few years, the Reagan-Fowler FCC would repeal comparative license renewals for broadcasters and, in 1985, relax ownership caps, raising the number of broadcast properties a company could own from seven AM, seven FM, and seven TV stations to twelve in each category.[26] License periods were extended from five to seven years. Long-standing policies that encouraged programming for children were tossed away[27] as were rules that encouraged news and public affairs programming.[28] In 1982, the Commission killed its antitrafficking rules, which had required station owners to hold their property for at least three years before reselling. As discussed later, the rule change helped pave the way for the rise of new independent TV stations and a new TV network.

Critical to Fowler's thinking was the power of the open marketplace and the emergence of the new technologies. The grand scheme was to stimulate additional distribution outlets for radio and television, thereby creating a self-righting competitive arena. The new technologies would provide Adam Smith's invisible hand of open market control, serving industry and the public. It was a philosophy deeply reminiscent of Doerfer, and one that did not necessarily cater to the well-being of entrenched broadcasters, or in fact, a now established cable TV industry. The philosophy built on and accelerated a policy perspective that was already in place at the Commission. Ferris had similar if not such extreme views, and in concert with like-minded Commissioners had begun a multipronged program designed to encourage the creation and development of new TV delivery systems. Cable's deregulatory Christmas basket of the late 1970s was only part of a larger effort by the Ferris-led FCC to replace government oversight with market controls (in 1979, the Ferris FCC proposed a wholesale deregulation of commercial radio[29]), an initiative continued and expanded by Fowler. In addition to the DBS, MDS, and STV services noted above, the Commission was also taking a look at freeing the telephone companies to compete in the video marketplace, and an unregulated home video business was building as well.

The 1980s version of the Blue Sky communications revolution was not just the province of political conservatives. Liberals and proponents of a diversified, distributed telecommunications forum also supported development of the new technologies, albeit with overall greater governmental oversight. Both sides, however, would be at least mildly disappointed. Many of the technologies and businesses introduced in this period would wither and die in short order; some, as cable before them, would muscle into a niche market and settle down for the long term; some would rise, fall, and rise again; some would become competitive giants and eventual partners in the information age.

The Changing Competitive Landscape

Low Power Television (LPTV)

In the summer of 1980, the FCC unveiled two new initiatives aimed at broadening the nation's television choice. The first was a plan to license reduced-power VHF TV stations, with sufficient technical restrictions that they could be added to the existing broadcast mix without creating interference with established stations.[30] Nearly 140 of these "drop-in" stations seemed feasible under the proposal. At the same time, the Commission announced plans for a similar, but new service—Low Power Television.[31] LPTV stations were the technical grandchildren of the 1950s broadcast translators discussed in Chapter 4. The FCC decided that translator "stations," both VHF and UHF, would now be permitted to broadcast original material or satellite-based programming, in essence changing the nature of the service. Transmitter power was restricted to less than 1,000 watts UHF, 10 watts VHF, and the range to less than 20 miles. The hope was that the new service would pave the way for the creation of thousands of neighborhood broadcasters. In a magnified echo of 1950s and 1960s "localism," the stations were intended as vehicles for community expression, to be operated by groups serving neighborhood interests too small for the VHF and UHF commercial broadcasters to bother with. LPTV was also seen as a way to extend programming choice in underserved rural areas.

Interest in the idea was so great that the Commission was buried with applications, more than 30,000 piled up at the FCC. In early 1981, the Commission had to issue a temporary freeze on license requests,[32] as they rose ever higher on agency desks, eventually instituting a filing fee it hoped would reduce the number of nonserious parties. Under the Ferris Commission, ownership of the LPTV stations was to be restricted to minority and community voices, but the Fowler administration abandoned this limitation, opening the way for major corporations to attempt to develop national commercial networks based on LPTV distribution. Applicants eventually ranged from true community groups to national organizations seeking to build commercial LPTV networks. The latter groups included Sears and the United Auto Workers. The Commission formally approved the LPTV concept in April

1982,[33] after rejecting various challenges to the plan, but it took several years and the initiation of a special lottery process to work through the thousands of applications.

Subscription Television (STV)

The FCC, as noted earlier, began tossing out its rules on pay broadcast television following *HBO v. FCC*. In 1979, it dropped its prohibition on more than one Subscription Television (STV) station in a market.[34] In 1982, it killed requirements that subscription broadcast services program a certain portion of free material everyday and it opened more cities to pay-TV broadcasting.[35] After decades of political and technological struggle, STV was finally permitted to enter the market in the late 1970s and early 1980s. Stations opened in the New York City area and in southern California in 1977 (KBSC in Corona, California, and WWHT in Newark, New Jersey), with others soon following. By 1981, twenty-four stations were on the air in seventeen cities and STV applications were pending in at least twenty-six markets.[36] STV had a short-term advantage in urban areas that had not yet been wired for cable. Some programmers saw STV as a means of getting pay service to customers before the cable wire had reached the home. Ed Taylor's SSS, among others, entered the business in the early 1980s.

STV was limited in its channel capacity and suffered widespread signal piracy. Charging around $20 for a single pay channel, it was discovered, worked only as long as a multichannel provider was unavailable. As soon as cable, complete with HBO or Showtime or both, came into the neighborhood, the STV set-top boxes were returned or simply thrown out. STV could not compete. Stations in Detroit, St. Louis, and Boston were shuttered by early 1983. One of the major STV companies, Oak Communications (ON-TV), closed its operations in Dallas and Phoenix in April 1983. In September 1982, STV reported some 1.4 million customers. That figure dropped to an estimated 560,000 by the end of 1984. Subscribers and stations slipped under the waves of evolutionary selection over the next several years and, by the late 1980s, STV was a television fossil.

Multipoint Distribution Service (MDS)

Multipoint Distribution Service (MDS), along with its sister service, Instructional Television Fixed Service (ITFS), had been around since the 1960s. They were typically single-channel microwave broadcast conduits, providing occasional special interest programming for colleges, universities, and businesses. MDS and ITFS transmitters broadcast line-of-sight signals to dish-shaped receivers, usually mounted on rooftops, up to thirty miles away.

The FCC first set aside spectrum space for MDS in 1962 for business communications.[37] The bandwidth allocation was expanded to 6 MHz in

1970 to provide for its use in TV broadcasting on a common carrier basis,[38] primarily for closed-circuit business or instructional purposes. ITFS was created to give educational institutions a means of distributing TV content from a central point to schools in a given region. Most schools did not have the technical or economic resources to take advantage of the opportunity, however, and the channels largely went unused.

Although the Commission had specifically doubted the utility of MDS for pay-TV,[39] it quickly became a mainstay of the service and, in 1974, the Commission allocated two MDS channels in the top fifty markets (other markets were provided only one).[40] HBO quickly adopted it to deliver programming to apartments not served by cable in New York and Delaware. The following year, 1975, Microband Corp. arranged to begin a pay service for Cox Cable in twelve large cities. (It is worth noting that the plan also called for interconnection of the markets via Westar.[41]) MDS was also widely used to distribute pay movies to hotels.

Although limited in capacity, MDS was cheaper to deploy than cable, and operators such as Cox and HBO saw it as a means of serving areas that either could not be wired for cable or as the initial incursion into a market slated for future wiring. MDS was operating in more than forty of the nation's largest markets by 1980, primarily offering programming from companies such as HBO, Cox, and Showtime.

In late 1970s, it also became apparent that a number of local microwave channels in the combined MDS and ITFS bands could be bundled together to form an integrated multichannel program delivery system, so commercial interests looking to initiate multichannel programming via broadcast lobbied the FCC to increase the number of frequencies set aside for the services. Consequently, in 1981, the FCC opened rule-making on the possible creation of ten additional MDS channels and a proposal to permit those channels to be combined with existing ITFS frequencies where the channels were going unused.[42] This opened the possibility of creating, in some markets, a twelve-channel broadcast platform. The same year, the Commission lifted restrictions on a related microwave service, private operational-fixed microwave service (OFS), giving those operators a green light to sell video to the public, as well.[43]

Efforts by MDS promoters to capture some or all of the ITFS spectrum met with anticipated opposition, but after much struggle and publicity, the FCC fell on a plan to allow ITFS educators the opportunity to lease excess capacity to MDS businesses. Many cash-strapped school districts were quick to seize the bait. In 1983, the Commission reallocated eight ITFS channels to create "MMDS," or multichannel, multipoint delivery service, setting up two, four-channel systems in designated markets.[44] Within a matter of months, the Commission had received more than 16,000 applications for the new service. The FCC decided to use a lottery system to choose from among the applicants for what would become known as "wireless cable."

Satellite Master Antenna Television (SMATV)

Satellite Master Antenna Television (SMATV) systems are the direct descendents of the apartment house master antenna systems described in Chapters 1 and 2. SMATV's are mini-cable systems, usually fed by satellite, serving apartment buildings, retirement villages, condominium complexes, or similar multidwelling structures. The architecture is simple and cheap, consisting of little more than a dish to pick up the satellite-fed cable channels and a set of wires and amplifiers to carry the signals through the complex. Early 1980s SMATV systems had a bit in common with MMDS technology in that both were spawned, in part, by the failure of cable to expeditiously wire the major markets. SMATVs brought cable programming to apartment houses before full city-wide cable systems were deployed. Even after cable was available, the SMATV service lingered. Cable operators initially put a priority on wiring single-family neighbors, avoiding the more transient apartment dwellers. SMATV was popular among landlords, however, because of its low operational costs, and its ability to generate revenue for the building owner.

SMATV businesses gained relatively easy access to apartment houses by providing the apartment owner with a percentage of subscription fees and, in some cases, entering into a shared ownership agreement with the landlord. But, SMATV was (and remains) almost universally despised by the cable industry. Apartment buildings and similar institutional settings are highly attractive to cable operators because of their high per-mile subscriber density; the industry saw the simple dish systems as a form of "cream skimming." Adding insult to injury, SMATV operators were free from the need to obtain a municipal franchise, pay a franchise fee, or conform to municipal rules on channel capacity or access, and they were outside the oversight of most of the FCC rules that regulated cable operators.

Through the 1980s, cable operators fought SMATV with the same enthusiasm engaged against boosters and translators in the 1950s. They battled on several legal and economic fronts, challenging the SMATV operators' right to operate without a franchise in the first instance and offering expanded services at attractive prices in the second. Cable programmers, aligned with systems operators, began restricting access to their service to SMATV operators. HBO would sell only to a few of the largest SMATV operators and any SMATV system that sought HBO had to first obtain permission from the local cable operator.

Television Receive Only (TVRO)

TVRO stands for "television receive only" and refers to the earth station used to bring in satellite distributed television signals. It was not very long after HBO began beaming movies across country by satellite that TV enthusiasts and radio hobbyists discovered they could, with a little money and a bit of technical know-how, enjoy the programming without the inconvenience of a cable system (or a fee), by taking the feed directly from the satellite.

The first documented TVRO viewer was Dr. H. Taylor Howard, a Stanford University electrical engineering professor. Dr. Howard worked with satellite communications as part of his research, and one day in 1976 a graduate student told him of an odd signal coming off a satellite. The student thought it might be a television picture. Dr. Howard rigged a receiving station in his backyard using a discarded dish he had acquired for future research. It took some time locating the satellite and the proper frequency, but eventually he saw a picture pop up on his reception screen informing him of the proper time and coordinates to pick up the HBO feed off Satcom I. He was soon watching Home Box Office off his own dish and telling colleagues about it. A friend then talked him in to compiling a manual on satellite reception, and the book, *The Howard Terminal Manual*, was soon a hit on the electronics hobbyist circuit.

Word spread quickly and, within a few years, hobbyists around the country were pulling in every manner of satellite-delivered TV signal, from HBO films to back-haul news feeds for the TV networks to Saturday afternoon football games and adult movies. Howard, knowing that HBO was a subscription service, wrote the company asking for a monthly subscription and enclosing a check for $100. HBO returned the check, telling him they only dealt with system operators, not individuals. Thousands of other TVROers were not so conscientious. Following the logic of the cable industry twenty years before, TVRO owners argued that if the signal was broadcast into the public ether, it was free to anyone with the technology to view it. It was not, at the time, strictly legal, however. Ownership and operation of earth stations required an FCC permit. In October 1979, the Ferris deregulation pogrom eliminated this small inconvenience when it dropped its licensing requirement for earth stations,[45] although there was still, again with some irony, the issue of HBO's property rights interests in the programming, a concern that would blossom into legal and public relations headaches down the road.

TVRO technology was called C-Band, after the frequency band in which the television satellites operated. The dishes were not small or cheap. They were still the three-meter (about ten-foot) discs used by the cable industry and other commercial operators. In addition to the dish, owners needed expensive electronic tuners, and the more enthusiastic purchased motorized mounts that allowed the dish to swivel and look at different satellites.

With dish ownership open to anyone with the money and yard space, however, the business community quickly stepped in to market receiving equipment. Dr. Howard was among the first, helping start Chaparral Communications, a TVRO manufacturing and sales company. In 1979, Scientific-Atlanta unveiled its own subsidiary, called Homesat, to sell TVRO equipment and services directly to the public. Scientific-Atlanta advertised a ten-foot dish with two twelve-channel receivers for a mere $20,000. Homesat also offered a bundled programming service featuring the normal menu of movies, sports, children's programming, and so forth, on a sliding scale that ranged from $2 to $180 a month. By early 1980, earth station retailers had opened storefronts across the country and an estimated 5,000 dishes were in home use.

A company out of St. Petersburg, Florida, Starview, sold regional franchises to entrepreneurs who wanted to get into the dish business. Among the people who thought they would give it a try were a trio of friends in Colorado, Charles William (Charlie) Ergen, Jim DeFranco, and Cantey (Candy) McAdam. In 1980, Ergen and Franco, both twenty-seven years of age, along with McAdam, pooled $60,000 in savings, purchased the Starview franchise for Colorado, Utah, and Wyoming, and began selling dishes to farmers and ranchers across the West. Charlie Ergen had been trained as an accountant and financial analyst, earning his BA from the University of Tennessee at Knoxville and his MBA at Wake Forest University. In the late 1970s, he grew bored in his analyst's job at Frito Lay in Dallas, Texas, took his savings, quit the position, and began looking for new opportunities. He found one when DeFranco stumbled on to a display trailer showcasing the Starview satellite dish. Wisely rejecting an initial idea that they name their new company "Sputnik," they settled on EchoSphere (after the Echo I satellite) and opened a storefront in the Denver suburb of Littlejohn, coincidentally just a few miles from TCI's Englewood headquarters. As the TVRO craze spread in the early 1980s, Echosphere prospered. The company severed its relationship with Starview, became independent and expansion minded, entering the wholesale business and establishing its own network of national and then international distributors. In 1982, Ergen and McAdam married. For several years, Echosphere rode the escalator of C-Band popularity. Between 1980 and 1985 dish prices dropped from $10,000 or more to around $3,000, and ownership swelled. More than a million TVRO dishes were in use by the middle of the decade. Unlike most of the small business concerns of the time, however, Ergen and Echosphere would go on to have a significant impact on the cable television industry.

Direct Broadcast Satellite (DBS)

C-Band had an enthusiastic but limited market. The size of the dish militated against its use in most urban, and many suburban, neighborhoods. Even where a homeowner had the yard space for the ten-foot dish, not everyone appreciated its aesthetic qualities, and some zoning ordinances prohibited them. The five-figure cost worked against wider commercial acceptance as did the difficulty in charging any sort of continuing monthly programming fee once the unit was sold.

Direct broadcast satellite (DBS) technology held the potential for changing all of that. As described in Chapter 7, the idea of beaming pictures from space directly into the living rooms across the nation had thrilled and frightened business people, policy makers, and consumers since at least the early 1960s. The promise was compelling, but the technology was not there. In its 1976 inquiry into the feasibility of DBS service, the FCC adopted a "wait and see" position, determining to remain flexible and allow DBS to follow an "evolutionary" path.[46] Several years later, with the success of satellite TV retransmission in late 1970s and the growth of the TVRO market, the idea of high-powered DBS, with dishes as small as three feet across, enjoyed renewed interest. Unlike TVRO, which simply intercepted relatively weak signals with relatively large dishes, DBS was intended for the home market. Technically, that meant higher-powered satellite signals at higher frequencies (the Kurtz-under band [Ku-Band]), allowing for smaller and cheaper home antennas.[47]

The first proposal for such a system came from COMSAT, by way of its Satellite Television Corp. (STC) subsidiary, in 1979. STC outlined a four-channel service featuring films, sports, children's fare, and cultural and public affairs programming. Prompted by the proposal and the general growing interest, the FCC took up the issue and in October 1980 issued a Notice of Inquiry seeking comments on the potential of a commercial DBS system.[48] Something of a DBS fever had gripped some FCC staff members. A staff report released with the FCC's Notice recommended establishment of a new regulatory category for the service. The Commission itself liked several things about DBS, including its potential to reach rural areas not served by cable. The possibility of increasing the number of diverse and specialized TV services, and the potential for stronger program competition, also were appealing.

The Commission's interest was further fueled by the perception that several other distribution platforms for multichannel television were in gestation—including cassettes and videodiscs—and DBS had a narrow window of opportunity in which to become established. The FCC concluded studies on the feasibility of DBS in the fall of 1980 and, in 1981, issued a formal Notice of Proposed Rulemaking to establish DBS service in the United States.[49]

More than a dozen companies moved to take advantage of the opportunity, although the Commission initially accepted only eight applications, including COMSAT's STC proposal. The STC plan called for a two-satellite system (plus a third backup bird) to distribute DBS to what it hoped would be 5 to 6 million subscribers around the country. Home dish-receiver units would be priced at around $200 and marketed through an alliance with Sears (although that partnership soon fell apart). Additional applications came from CBS, RCA, and Western Union. The CBS proposal was historically interesting in that it called for one channel to be used for a high-definition TV signal that would be beamed to CBS affiliates for rebroadcasting in their market. (In March of 1981, CBS, in conjunction with NHK, the Japanese national broadcasting agency, and Sony, offered one of the first demonstrations of HDTV, to policy makers in Washington, D.C., in an effort to promote their DBS proposal.) A second CBS channel would be used for pay programming offered via DBS and through local cable systems. A third channel would be set aside for teleconferencing and business use. Another DBS plan came from Minnesota-based Hubbard Broadcasting, through its subsidiary, United States Satellite Broadcasting (USSB). It called for an advertiser-supported DBS service that would be free to customers (not counting equipment purchases).

In the summer of 1982, the Commission authorized direct broadcast satellite service in the 12.2- to 12.7-GHz range, and adopted interim rules.[50] The FCC granted its first DBS authorization in September 1982 to STC.[51] The start-up cost for DBS was astronomical, $250 million or more to build and launch a satellite. STC additionally planned to spend more than $100 million on affiliated equipment and operations in its first year and another $77 million on programming. The time required to build, launch, and begin operations of a satellite meant that the earliest projection for the inauguration of service was considered to be sometime in 1985 or 1986.

To get a multiyear jump on the market, one partnership proposed initiating service using existing, lower-powered satellites. The idea came from entrepreneurs Richard Blume, Cliff Friedland, and Francisco Galesi. They convinced General Instruments and then Prudential Insurance Co. to back them in a new company, United Satellite Television Corp. (renamed United Satellite Communications, Inc. [USCI], after the Prudential investment). USCI requested FCC permission to begin a direct-to-home service using fixed-satellite frequencies (11.7 to 12.2 GHz), rather than the higher DBS frequencies (12.2 to 12.7 GHz) set aside for that purpose. They proposed leasing an existing Canadian satellite in the Anik series to provide a five-channel, medium-powered service to a four-foot dish.[52] The FCC approved the plan and USCI began the nation's first DBS transmission on November 15, 1983.

Before USCI launched, however, another contender entered the DBS race. He was a millionaire newspaper publisher from Australia with media holdings around the globe. His name was Rupert Murdoch. Murdoch owned a string of newspapers in Australia and Great Britain, including London's *Sun* and *News of the World*. He entered the U.S. publishing market in 1976, buying the *San Antonio (Texas) News* and moving on to New York City where he purchased the *New York Post*, *New York Magazine*, and the *Village Voice*. Murdoch was now looking to break into the U.S. television market.

A start-up company, Inter-America Satellite Television, already was pursuing DBS when it managed to attract the attention and financial backing of Murdoch. In May 1983, he announced formation of his new, domestic DBS venture. Inter-American Satellite Television became Skyband Inc. Murdoch planned to follow the USCI lead, leasing transponders on existing, lower-powered satellites to establish an early toehold in the market. The proposed five-channel service would use IBM's Satellite Business Systems (SBS) III satellite, which had a nationwide footprint, giving Skyband an advantage over USCI, which could not reach all of the country with its Anik-fed signal. Skyband's low-powered system would, however, require six-foot home dishes, a distinct marketing handicap.

Murdoch hoped to quickly acquire programming contracts at affordable prices and beat potential competitors to market. He failed. He could not come to terms with content providers, such as Warner's Showtime, and later complained that the technology was not yet sufficiently mature. Paying a hefty $12.7 million to SBS to pull out of their six-year transponder contract, Murdoch fled the field in November 1983, just before USCI began operations, hinting that he might be back when satellites had more power, more channels, and dish sizes came down.

A week later, USCI announced the start of its national DBS service. USCI began marketing in Indianapolis, subsequently expanding its reach to Baltimore, Washington, D.C., and Philadelphia, all large markets not widely served by cable. The company offered five channels for about $30 a month plus a $300 installation fee. The business quickly grew, but then, almost as quickly, flamed out. It took only seventeen months. In its heady prelaunch publicity, USCI predicted they could garner 1 million subscribers in three years, but by late 1984 the company had only 11,000 customers. While they had neatly avoided the cost of building and launching their own satellite, they nonetheless had expended millions in development, programming, and promotion; the company was on the edge of bankruptcy. Prudential, after failing to negotiate a merger with STC, pulled out entirely. TCI was interested, thinking USCI might be profitable as part of a TCI-backed SMATV distribution platform, and Magness and Malone pumped a half-million dollars into it in a doomed salvage effort. By early 1985, TCI gave up as well. Out of money and unable to pay the mounting transponder fees, USCI pulled the plug on April 1, 1985.

Other applicants meanwhile had been watching and learning. Shortly after the USCI launch, CBS and COMSAT had filed an application with the FCC for a joint twelve-channel service, featuring the high-definition option. Hubbard Broadcasting, in June 1984, signed contracts with RCA for the construction of two high-powered DBS birds. By mid-summer, with the results from the USCI effort becoming clear, a chill swept through the thin air of the DBS industry. Investors, both private and institutional, suddenly disappeared. In the summer of 1984, RCA, CBS, and Western Union all walked away from DBS. STC already had spent $40 million in planning and development, without launching a satellite. Compelled by the USCI and Skyband medium-powered proposals, it was working on a similar plan by May 1983. The company was unable to find funding, however. Partnership talks with a variety of potential investors, including CBS, Paramount, and United Press International (UPI), went nowhere, and when merger talks with a faltering

USCI failed to come to fruition, Comsat, too, decided it was time to flee, closing its DBS shop in December 1984. USSB's Stanley Hubbard, already $10 million into his DBS venture, also was forced to suspend his efforts, citing and criticizing the financial community for not having the courage to back the costly experiment.

The DBS postmortem was not difficult. USCI was charging subscribers $30 to $40 a month for five channels, plus $300 to $500 for the dish (prices varied by installation package). Comparable cable service, when available, offered twenty to forty channels for the same price or less. A four- to six-foot dish was still too large for the tastes of many consumers, and installing them proved more difficult than anticipated. High winds had an unfortunate tendency to rip dishes off of roofs. The programming—two movie channels, a sports channel, a news channel, and a children's or cultural events channel— was insufficient incentive, and it did not provide local broadcast channels or national broadcast networks. Cable offered all the "normal" channels and more, including the popular HBO. DBS did not. At the same time, TVRO was spreading across the countryside. For those homeowners with the space and desire to go satellite, ten-foot dishes with nearly unlimited reception capacity and no monthly fee offered an appealing substitute for DBS. Plus, every TVRO dish installed was one less potential customer for a DBS service. Direct satellite service would need to find a way to decrease its dish size and provide as many channels as cable, at a competitive price, if it were to become economically viable. It would be several years before that became possible.

Videotext

Beginning in the mid-1970s, the idea of using the new technology to deliver interactive or semi-interactive textual material to homes started to come into vogue. The concept was a direct descendent of the 1960s Blue Sky proposals for using cable to deliver instant headlines, sports scores, stock information, home shopping, bill paying, and so forth. While the idea had failed to mature in cable technology, proponents thought that by limiting the services to textual and low-resolution graphic information, bandwidth requirements could be shrunk to the point that the business could be built on the existing telephone and broadcast infrastructure. Several versions of the service were discussed and attempted. The broadcast variety was usually labeled "teletext" and the wire line version, which quickly included telephone and cable, was called "videotext."

Broadcasters planned to harness otherwise unused space in their broad-cast signal—specifically, the vertical blanking interval (VBI). Cable operators could set aside a small amount of channel capacity in existing systems and telephone versions would use a normal switch voice circuit. In most of the plans, a smart terminal and the existing home TV set served as reception units. Broadcast-based teletext was limited in its available content to about

400 pages. Technically, individual pages of text were embedded in a repeating serial cycle, with decoder-equipped subscribers picking off pages, on demand, as they rotated through. Wire-based systems promised greater, even theoretically unlimited, capacity by allowing users to select pages from existing and expandable databases. In many ways, videotext was a forerunner of the 1990s Internet, albeit with a substantially more modest capacity and cruder technology.

Europe took the lead in technical development. The BBC experimented with an early version, called Ceefax, in 1972; by 1983, Great Britain had three different systems in trial (Ceefax, ORACLE, and Prestel). France had is own system, dubbed Antiope, introduced in 1974 and widely deployed by the country's national postal and telephone service which sought to put a unit in every home in the country and thereby eliminate printed telephone books.

AT&T and domestic broadcasters such as CBS were curious about the technology and initiated their own trials in the United States, usually in partnership with an established system such as Antiope. CBS affiliate KSL-TV in Salt Lake City began testing VBI-based teletext in the 1978; Chicago's WFLD-TV, KMOX-TV (CBS) in St. Louis, and KNXT-TV (CBS) in Los Angeles also conducted FCC-approved experiments. The promise and interest were sufficient that the Commission authorized commercial teletext operations in March 1983.[53]

While various electronics industries were excited by the potential of the technology, owners of older media was much more apprehensive. Newspapers saw both the threat and the promise of electronically delivered news—a threat if provided by competing media, a promise if the industry could co-opt the service for itself. The concern was not so much about the possibility of teletext supplanting newspapers as a vehicle for getting the latest headlines, although that was certainly an important part of it. The greater fear was that, as a consequence, competing teletext providers would begin siphoning away advertising, especially local classified advertising. Some of the major newspaper chains moved, therefore, with alacrity into their own videotext trials. Knight-Ridder spent more than $1.5 million installing free videotext terminals in 200 test homes in Coral Gables, Florida, and poured tens of millions into the experimental operations. Times Mirror also pumped millions of dollars and dedicated several years to developing its service, marketed as Gateway, for southern California customers. Time Inc. invested an estimated $30 million in creation of a prototype service, which it never launched. Harte Hanks, Associated Press, and Reuters all tried their hand with some form of the service. The telephone companies, AT&T and GT&E, also plowed capital in videotext, in their own trials and in experiments with other companies.

Cable operators were not left behind. Cox Cable conducted its own tests, while the NCTA set up a study committee and promoted the promise of the technology in terms lifted almost directly from the Blue Sky playbook. The NCTA's 1981 promotional piece, "A Cable Primer," declared "[T]elevision

sets equipped with special decoders can use videotext for informational retrieval, advertising, education, captioning for the hearing impaired, transaction services, electronic mail, home computing, and a range of other services
now under development."[54] There was more sizzle than steak, however. In
nearly every case, newspaper publishers, broadcasters, and cable operators,
all came to the same conclusion: Teletext and videotext, in its 1980s form,
could not be made to pay.

As with DBS, the idea outpaced the technology. Media companies sunk
millions into development, and allocated annual operating budgets that ran
into the hundreds of thousands of dollars. But the customers were not there.
While some companies were still working on teletext, by the end of 1984, the
trade press was already writing its obituary.[55] Again, the problems were easy
to trace. The technology was fairly slow in information retrieval and display,
the graphics were crude by even the standards of the day—chunky block letters in primary colors. Decoder boxes were expensive, $300 to $1,000 each.
The content was limited even in the most advanced systems and, in the era
before the home computer, the combination of the TV set with a keypad
or mini-keyboard just did not work for most consumers. While the FCC
formally approved teletext, it declined to set a technical standard, relying
again on the invisible hand of the marketplace to settle such issues. The
result was unresolved bickering among competing companies and no uniform technical standards for consumer equipment. Potential monthly fees
of up to $30 to $40, for what was inherently a user-unfriendly technology,
doomed the concept. By the late 1980s, domestic teletext and videotext experiments were over and the losses written off.[56] In an obit for videotext,
the managing editor of Times Mirror's Gateway experiment summed up the
problems: "As a medium, it proved to be slow, shallow, intrusive, unreliable and overpriced."[57] It would take a new generation, one fueled by home
computers, to resurrect the idea, and in the meantime, it would pose no
competitive threat to existing media, including cable television.

Home Video and Video Cassette Recorders (VCRs)

While services such as teletext and STV were destined to sputter and fail,
one new product would have a substantial impact on cable, especially on
premium channels such as HBO. Magnetic recording tape was pioneered by
the Germans and brought to the United States by engineer GIs at the close
of World War II. Videotape technology was developed within a few years,
and the first videotape recorders were marketed by Ampex Corp. in the
1950s. The technology was professional grade, size, and price, however, with
bulky $50,000 machines running foot-wide reels of two-inch tape. It took
Japanese engineering to transform videotape into a consumer product. Sony
pioneered the creation of three quarter- and half-inch videotape contained in
a sealed cassette, increasing its convenience and ease of use, and introduced

its Betamax brand to the home market in 1975. Matsushita, parent company of JVC and Pioneer, responded in 1977 with a competing "VHS" (video home system) format. Japanese design and manufacturing brought the price down. The home units were still not cheap, the 1976 Betamax deck retailed for $1,300, but prices continued to drop with accelerating mass production. While some experts believed Betamax to be the superior technology, VHS was less expensive, offered a longer recording time, and Matsushita aggressively pursued movie product for its format. VHS won the marketing battle and soon became the standard for home use.

The first industry to express uneasiness with the home video invasion was not cable, but rather the motion picture business. Hollywood initially perceived the VCR as a threat to its intellectual property and to its established distribution system. Universal Studios and Disney Corp. filed suit against Sony in 1976 to prevent the use of home VCRs on grounds they could record material without studio permission and potentially infringe on copyright entitlements. In January 1984, the Supreme Court ruled, in what became known as the Betamax case, that home taping of video was not a violation of copyright as long as the material was not used for commercial purposes.[58] A variety of other attempts by the MPAA to offset what it saw as a threat to the film revenues also failed. These included efforts to have Congress tax VCRs and blank tapes and use the money to reimburse studios and an effort to install smart circuits that would prevent illegal copying.

While the legal action was proceeding, the studios additionally fought a rearguard action against the growing number of film pirates who were videotaping and illegally reselling current release material on a domestic black market. The studios' response was to under-cut the pirates by licensing the sale their own product on tape, establishing at least some minimal control over their property. Beginning in 1978, the studios began licensing films to video distributors. While they were forcibly dragged into the VCR marketplace in this way, they soon developed a taste for it.

Music Corporation of America (MCA)/Universal worked with Phillips in the development of a videodisc system (see below) for the home distribution of their films. Home users could not record on the discs so the fear of piracy was eliminated. The disc system failed as a business, but the project gave MCA/Universal a corporate structure for home video distribution and the company rechanneled its experience into the VCR market. MCA's "Disco-Vision, Inc.," became "MCA Home Video." Other studios soon followed, creating units to license, sell, and distribute their movies on tape. By 1982, Metro-Goldwyn-Mayer/United Artists (MGM/UA) had formed MGM/UA Home Video, and Columbia and RCA created RCA/Columbia Home Video. CBS joined with 20th Century Fox to develop CBS-Fox Video. HBO, in conjunction Thorn EMI, entered the business in 1984.

Hollywood priced its product fairly high; $80 a tape was typical for the late 1970s and early 1980s. The cost, in turn, helped spark another new industry, the videotape movie rental business. More people were interested in renting a tape for a few dollars than in purchasing one for $80, especially if they were likely to watch it only one or two times. Rental shops, therefore, bought the expensive films and amortized their costs across weeks of evening rentals. Rental stores began sprouting on every street corner and in every strip mall in America. More than 10,000 opened in 1984 alone. Meanwhile, competition began to push down sale prices. Paramount, with a comparatively small library of classic films (it had sold its holdings to MCA in 1948), sought to gain a market advantage by halving the price of its tapes. In 1982, it offered "Star Trek II: The Wrath of Kahn" for $39.95. With a growing availability of inexpensive rentals and declining costs of both players and new tapes, consumers began buying more home VCRs. This, in turn, fed new video sales and rentals in an accelerating upward spiral.

Despite the reduction in price, that they did not share in the rental income directly, and their initial antipathy, the studios soon came to appreciate the revenue from video sales. In 1978, 54 percent of the majors' income derived from domestic box office receipts, the remaining from sources such as overseas distribution and broadcast television rights. Home video and pay-cable accounted for only about 4 percent. By 1983, home video alone accounted for $625 million, or 14 percent of total film industry revenue.[59] By 1986, pay-cable provided 12 percent of all studio revenues and home video 40 percent, easily eclipsing the 28 percent from domestic box office.[60] Total industry revenues were up, and cable and VCRs were largely responsible.

Cable operators were substantially less happy. In the early 1980s, the VCR marketplace was booming. Some 55,000 VCRs were sold in 1976; the figure grew to 805,000 in 1980, more than 1 million in 1982, 4 million in 1983, and over 7 million in 1984.[61] In 1980, only about 1 percent of television homes had a VCR. By 1984, it was more than 10 percent, the next year 20 percent, and by 1987 nearly half the homes in America had a VCR.[62] The VCR enjoyed a diffusion rate that vastly outpaced radio, color television, telephone, and certainly cable TV; only black and white television in the 1950s spread into American homes more quickly. Cable had been around since the early 1950s; VCRs had been on the market for only a decade, but by 1988, more homes had VCRs than had cable television. It was not good news for the cable industry. Pay-cable had, for several years, replaced network television as the second venue after a film's theatrical run. Now, the home video market was replacing pay-cable. Studios, by the end of 1984, were typically releasing their films to the VCR market six months after they finished showing in film houses; films were sent to pay-cable about six months after their home video release. HBO and Showtime, the major pay providers, stood to lose the most from home video rental, and HBO was covering its losses by establishing its own home video subsidiary.

Less successful, at least in its 1980s version, were videodiscs. MCA/Phillips began manufacturing and marketing videodiscs and videodisc players in 1978, and RCA introduced its own videodisc player three years later. The twelve-inch discs spun on a turntable. The MCA/Phillips technology used a laser to read the audio and video information embedded in the disc. The RCA SelectaVision system used a stylus that rode in the grooves of the disc much like a traditional phonograph record. The discs were cheaper to produce and offered superior picture quality relative to videotape, although the players were pricey, $775 for the early MCA/Phillips unit. Unlike tape, consumers could not use them to record programming. Videodisc movies were sold or rented, but the high cost and the inability to record dampened consumer enthusiasm. Sales never took off and the videodisc became a business school case study in product disasters. Lasersdiscs lingered on the market for years (MCA and Phillips sold their interests to Pioneer in 1982), but RCA abandoned its mechanical technology in April 1984 after pouring an estimated $500 million down the SelectaVision drain.

The Birth of the Baby Bells

The telephone system, in its traditional form, was not, of course, a new technology. The industry, however, hoped to harness its existing and evolving technical capacity to enter the emerging telecommunications marketplace. AT&T was also being forced in the late 1970s and early 1980s to respond to mounting legal and political pressure to abandon its decades-old, state-sanctioned monopoly. The emerging political faith in a plurality of competing media systems was having a profound impact on the perception of telecommunication's oldest and most powerful player. The government was seeking to break up the Bell system, and this had implications for cable. Federal intervention in the 1950s, 1960s, and 1970s had stayed the cyclical efforts of the telephone companies to nibble into the cable business, but no one ever doubted that their appetite remained. By 1982, a chain of events set in motion in the 1960s was opening the way for a reconfigured telephone industry to once again pull up a place at the banquet.

In 1963, an energetic, if somewhat disorganized and disheveled, midwestern entrepreneur, John (Jack) Goeken, ask the FCC for permission to construct a microwave link between Chicago and St. Louis.[63] Goeken was a manufacturer's representative for General Electric, selling two-way radios out of Springfield, Illinois. His business plan was to create a communications system to connect truck drivers running between St. Louis and Chicago with their dispatchers. His company, Microwave Communications, Inc. (MCI), intended to provide a cut-rate private line service for customers willing to trade quality and reliability for low prices. The private line would not be tied in to the public switched network dominated by AT&T, but AT&T saw any and every incursion into telephony as a potential threat and brought its massive resources to bear in opposition to the request (General Telephone and Western Union also joined in opposing the proposal). It took six years of legal struggle, but in 1969, the FCC approved the MCI plan.[64] The Commission had spent decades helping protect the AT&T monopoly, but by the late 1960s the regulatory culture had changed. Corporate size was suspect, and few companies were larger than AT&T. Complaints about AT&T service for specialized businesses and the hope that a modicum of controlled competition would benefit the small market MCI intended to serve, also aided the decision.

In 1968, MCI hired a New York-based business consultant to help raise capital and expand the business. William G. McGowan found the venture capital and began filing additional private line applications, hoping to break AT&T's stranglehold on the market.[65] He also eventually took management control of the firm from the less-organized and less-interested Goeken.

The Commission subsequently extended its favorable 1969 MCI decision in a broader policy move designed to encourage completive private line service, especially for a growing market in computer data communications. As part of that rule-making, the Commission ordered AT&T to offer reasonable interconnection to its network for the new companies.[66] Failing to reverse the FCC's ruling in the courts, AT&T sought to frustrate would-be competitors through lengthy and complicated negotiations and requirements associated with interconnection.[67] Aggravated by the delays, MCI filed an antitrust suit against AT&T in March 1974.[68] At about the same time, the Justice Department filed its own suit against the telephone giant, bringing allegations of anticompetitive behavior, including refusal to interconnect other providers, predatory pricing, and other illegal restraints of trade. The Justice Department was especially concerned about AT&T efforts to illegally hamper competition in the long-distant market. It was this suit that became the battleground in the ensuing fight between AT&T and the government.

The Justice Department suit was followed by years of maneuvering in both Congress and the courts, and years of negotiations between the government and AT&T. The Reagan administration, on its ascendancy to power, briefly sought to terminate the action against AT&T, in keeping with its probusiness philosophy. Opposition within its own Justice Department, concern about the political fallout, and an uncooperative federal district judge, Harold Greene, prevented the administration from backing off, however. AT&T did win a victory at the FCC in 1980, when the Commission ruled that basic telephone service should, for regulatory purposes, be separated from enhanced telephone service such as data processing.[69] The former would continue to be subject to service and rate oversight, but the latter would not.

The lengthy and expensive tug-of-war over broader federal control, however, led eventually to a negotiated settlement. In January 1982, AT&T and the Justice Department announced that they had come to terms. By August, Judge Greene, the federal judge overseeing the case, had approved the parameters of the Modified Final Judgment (MFJ).[70] AT&T agreed to divest twenty-two regional telephone subsidiaries. It kept its long lines, or long-distance business, and retained its manufacturing and research companies, Western Electric and Bell Laboratories. The reorganization became effective January 1, 1984. The twenty-two subsidiaries quickly reorganized into seven regional Bell Operating Companies (RBOCs). Each of these "Baby Bells," as they also came to be known, were allowed to provide local service within their operating areas, but were prohibited from offering long-distance service. The remaining AT&T company, meanwhile, was left with the ability to offer long distance, but could not enter the local loop (local telephone) business.

In addition, the phone companies were restricted in their provision of videotex-like information services for at least seven years; they could offer electronic yellow pages, but were effectively prohibited from providing news services and, more importantly, from the perspective of the newspaper industry, classified advertising. The concern, as it had always been, was that the telephone companies could and would use their control over the distribution network and their substantial resources to broach monopolistic control of these businesses. The telephone companies were also prohibited from offering video programming, a victory for cable interests.

Despite the restrictions of the MFJ and existing FCC rules, the new RBOCs retained a strong interest in the cable business, as a well as text-based information services, and began without much delay to prepare the ground for efforts to enter those fields, appealing Judge Greene's line-of-business restrictions and lobbying Congress, formally and informally, to have the legal handcuffs removed.

The cable industry had seen AT&T as a hostile, hegemonic threat since the 1950s, even while it was heavily regulated by federal and state authorities. Now, cable operators feared the giant had been unshackled and began girding for battle. NCTA Chairman Tom Wheeler described the January 1982 tentative agreement as "the greatest deceptive ploy since Br'er Rabbit begged not to be thrown into the briar patch."[71] If telephone companies were ever permitted in, cable operators charged, the RBOCs, as AT&T before them, had the potential to use their monopoly, local loop revenues to cross-subsidize cable service, pushing consumer prices below cost and driving out competition, only to raise prices subsequently to unregulated, monopolistic levels. The NCTA launched a public and political campaign designed to combat what they felt would surely be the RBOCs efforts to penetrate their business and to compete with cable for the provision of enhanced telecommunications services of the Blue Sky variety.

At the same time, the telephone companies were beginning to complain about cable intrusions into phone service. In 1982, MCI's McGowan approached cable operators. He had enlisted Drexel Burnham Lambert to help raise more than $1 billion to finance an assault on AT&T and was seeking cable partners. MCI would supply the long lines and cable the local loop connection to provide local and long-distance service. The plans failed to germinate, but were sufficient to get the attention of the RBOCs. A short-lived experiment by Cox Cable in Omaha to use the cable system's two-way institutional system for telephone service also made brief headlines in 1982. These were among the earliest signs that the historic barriers between cable and the telephone industry were beginning to dissolve, although they would remain fairly sturdy for another decade.

Cable Technology

Against this emerging plethora of television distribution platforms, cable technology continued its evolutionary expansion. Channel capacity increased, techniques for dealing efficiently with pay services improved, and consumer channel choice reached the point where a simple plastic box, just big enough to fit in the palm of the hand, was forced onto center stage in the family rooms of America.

By the early 1980s, cable suppliers were introducing 550-MHz amplifiers capable of carrying more than sixty channels, but that was very much state of the art and considered for only the largest, newest, and most expensive systems. More typical was 400-MHz technology, which could deliver fifty to fifty-four channels, under construction in the larger markets, or in some cases dual cable systems featuring a lashed pair of 300-MHz lines with a total capacity of sixty channels or more (see Appendix B). Warner had planned a dual cable system for Pittsburgh before it fled the city. In 1980, while the franchising or construction process was still underway in many of the nation's urban areas, only eleven of the top fifty markets had more than thirty channels and none had more than forty.[72]

Across the country, the vast majority of cable systems operated at 300 MHz or less and, in many places, did not fill all their available channels. In 1981, only about 35 percent of all cable systems had more than twelve channels. By 1985, however 73 percent of all systems had thirteen channels or more. Nearly 80 percent of all cable subscribers had access to thirteen or more channels and 64 percent of all subscribers were served by systems with thirty or more channels.

While channel capacity expanded in regular increments, some of the larger challenges that came were met as a direct consequence of both the increased channel capacity and the pressing interest in controlling pay services. The proliferation of pay-cable boosted monthly revenue, but posed technical problems. Traps (described in Chapter 8) were inexpensive, but each pay-channel required its own dedicated trap, and two, or at most three, were the practical limit per subscriber line. The introduction of addressable converters in the early 1980s went far to solve the problem. With them, cable operators could remotely activate or deactivate specific channels, such as HBO, for each subscriber.

The converters had their drawbacks, however. They were expensive, and cost-conscious companies, such as TCI, took their time in deciding whether the substantial additional investment was justified. They also were more susceptible than traps, especially pole-mounted traps, to theft of service. Individuals with a modest level of electronics expertise could, and did, defeat the scrambling systems in the converters. Boxes illegally modified to receive premium services became popular black market consumer items, costing the industry millions in lost revenue. Finally, addressable boxes often frustrated those subscribers who had invested in another new and related home TV technology, the cable-ready set.

The cable industry in the 1970s began lobbying the consumer electronics manufacturers to install cable TV tuners in new sets. As noted, the traditional VHF and UHF tuners could not receive all of the available cable channels once operators began moving beyond the VHF 12. The industry could and did build converters, but they were expensive and sometimes cumbersome in the home. Operators hoped to reduce or eliminate the cost and improve consumer convenience by convincing set manufacturers to build in the cable tuners, and the set makers responded. Cable-ready sets started to appear on the market in the late 1970s. By 1981, cable-ready sets accounted for 33 percent of all sets manufactured. Integrate cable tuners were installed in 44 percent of all new sets in 1982, and 66 percent of all new sets by 1985.[73]

Unfortunately, the cable-ready sets could not be made addressable. Operators could not use the built-in technology to block or activate pay channels. If a subscriber wanted HBO, they still needed the cable company's addressable set-top box. Moreover, the tunable converter fed into just one channel on the consumer set—usually channel three or four—and channel changing was done using the converter. The result was a viewer who likely had paid a premium for a cable-ready set and now could use only one of its many channels. The converter's remote control was needed to change channels, but the TV set remote may have also been required for other functions, including volume control. As the VCR craze swept the nation and subscribers tied another piece of equipment into what was becoming known as the home video system, the problems only multiplied. Connecting the various components left many consumers dazed and frustrated. Even those with the skill and patience to assemble their systems were left with a mass of twisting cables and a coffee table littered with remotes. In a number of very interesting and, as will be considered later, conflicting ways, cable was coming to have a special place in the homes and the hearts of the American public.

Before leaving the consumer and the coffee table full of remotes, however, it is worth pausing to consider in greater detail that frequently overlooked but significantly transformative device, the television remote control. There was an obvious drawback, for the consumer, to the increasing number of cable channels. Through thirty years of television, the happy, complacent viewer could and did sit through the evening watching one or perhaps two networks. The need to change channels during the evening was minimal, with only three or four available stations and an established set of viewing preferences. This, in turn, meant little need to get up off the couch, walk to the set, and twist the dial on the tuner. The expanded choice of cable, however, meant a decreasing ability to know what was on each channel at any given time and a curiosity, among some, about what was on every other channel. Into the late 1980s and early 1990s, the practice of "channel surfing," skipping through each channel to sample the programming, would become a staple of TV viewing life, and the source of uncounted jokes, most of them at the expense of men, who seemed to be the major surfers, with women watching in exasperation.

Channel surfing was only possible in any practical way, however, with the advent of the remote control. In fact, the remote, or the "clicker," was so fundamental to the success of the system and arguably so obvious a technology, that one could make the case that its development was inevitable. In real life, there was little chance people, given any choice, would get out of their chairs to turn the TV dial through a dozen or more programming choices, repeatedly through an evening's viewing.

The remote was, and remains, not just a device for program selection, but in many households, an escape hatch from commercials as well. In fact, that was one of its original purposes. Remote devices to control volume and channel made their appearance in radio in the late 1920s, promoted in some cases as means to avoid the advertising messages between musical numbers.[74] These mechanical devices tended to be large, frequently expensive, and always tethered to the radio with a cord that ran under the rug or tight to the baseboard. Similar remotes arrived with the advent of television, marketed with tags such as the "Tun-O-Magic" or "Remote-O-Matic." One TV set manufacturer was particularly keen on developing a consumer-friendly remote control. Eugene McDonald was the founder of Zenith Radio Corp. and he reportedly hated advertising, fearing it would kill the TV industry. He believed that television should and would adopt a pay-programming model of financing. Until it did, however, he planned to offer consumers a method for conveniently skipping the commercials. His first attempt was a traditional device connected by a thin cable to the set. Zenith introduced it in 1950 as "The Lazy Bones."

The intrusive connecting wire and a fairly high price impeded consumer acceptance, so Zenith looked for a wireless solution. In 1955 they brought out the "Flash-Matic." In essence, the Flash-Matic was a fancy flashlight that activated light-sensitive spots embedded in the corners of the television set, each corner controlling a different function, including on-off, channel selection, and perhaps the first mute control. Unfortunately, strong sunlight and inconveniently placed room lamps also tended to stimulate the functions, so it was back to the drawing board. This time, Zenith engineers developed a technology based on ultrasonic sound waves. Four metal rods, each of a different length and acoustic pitch, were set in the unit. The press of a button would cause a smaller hammer to strike one of the rods and the resulting tone—above the level of human hearing—was picked up by a sensor on the set, triggering a change of channel, volume, or simply turning the set on or off. Marketed as the "Space Commander" and introduced in the fall of 1956, the device became the industry standard for remotes and the technology dominated the market through the 1970s. The Space Commander's distinctive click, the striking of the tone bar, also helped establish the remote's nickname, the "clicker."

Advances in microchip technologies and development of smaller, cheaper, light-emitting diodes (LEDs) paved the way for a leap from the mechanical sound-based remote to lighter, less-expensive, and more powerful infrared (IR) light-based remotes. By the late 1970s, the IR remote began to dominate the market. As cable penetration climbed and set-top boxes and cable-ready sets brought increased program abundance, the remote became ubiquitous. In 1981, an estimated 18 percent of all television households had a remote control device. Paralleling, and driven by, growth in cable and in VCR penetration, that figure rose to above 50 percent by 1987 and 84 percent by 1992.[75]

The remote control became a cultural icon of sorts and, by some accounts, a point of contestation and a signifier of power relations within families. Who controlled the remote was a source of academic study and TV talk show humor, as well as a bane to advertisers, perhaps giving some measure of satisfaction to Eugene McDonald.

The Search for Money and Size

Financing

While subscribers struggled to connect VCRs and addressable set-top boxes to their TV sets, the cable industry in the early 1980s went once more in search of money. While the soaring lending rates of 1980 and 1981 dampened some forms of investment, they encouraged others. Looking for necessary funds, the cable industry began exploring alternative, creative forms of financing. Cable television never easily fit the Wall Street mold for publicly held companies, where investors analyzed balance sheets and scrutinized quarterly earnings statements. Cable was a cash flow business, usually heavily leveraged to finance never-ending construction and dependent on accelerated depreciation to write down income while system values matured. The industry survived on massive debt. In 1984, cable reported $9.4 billion in outstanding loans, most of it from major banks, such as Chase Manhattan, and from insurance companies, including Aetna Life and John Hancock Mutual Life. The debt-to-equity ratio for the industry was estimated at 14.7 in 1984, a fairly high number; the comparable figure for the broadcast industry at the time was 2.1.[76]

The equation was not one Wall Street was used to seeing. In the best of times, cable had to sell established investment institutions on the wisdom of lending, and the early 1980s were not the best of times. One alternative involved the limited partnership. Daniels was a pioneer in limited partnerships, using them over many years to buy and grow systems and MSOs. Under the arrangement, partnerships in a cable company, or in a fund to buy cable companies, were sold to private investors, usually for several thousand dollars a unit, but often as low as $1,000. Amounts up to the full price of partnership could be sheltered from federal taxes. Losses from the cable system during its construction phase were passed directly to investors who could use them to write down their individual taxes.

Glenn Jones' Intercable was a leading example of the limited partnership structure. Glenn Jones got his start in the cable business in the early 1960s, working as a young lawyer for operators in the legal matters involving everything from franchising to new system acquisition. After a brief and unsuccessful flirtation with politics (he was defeated in a 1964 bid for congressional office), he returned to law and cable, this time with a thirst for ownership. His first purchase was in Georgetown, Colorado, where he purchased a small, homemade system from the plumber who built it. He used his Volkswagen as collateral for the $400 loan he needed as partial down payment. Jones was typically short on capital as he sought to buy systems and, in the early 1970s, adopted a new approach to financing.

He formed Jones Intercable as a series of limited partnerships created annually in the form of blind investment pools, starting in 1972.[77] Equity generated by the partnerships was used to buy and manage selected cable systems. Start-up losses on the system were passed along to the investors, who used them to write down their own tax liability. The pools, which were especially attractive to high-income earners looking for tax shelters, were promoted aggressively on that basis by Jones and his brother and business partner, Neil. On reaching profitability, the systems were sold to realize the appreciation in value. The partners divided the profits, with 25 percent going to Intercable, which also ran the systems and took a management fee of 5 percent of gross annual revenues. The formula was so successful that by 1985 Jones Intercable was a top twenty MSO with more than half a million subscribers. Tax reform in 1986 reduced the attractiveness of the vehicle, but it remained an industry financing option.

Cable companies continued to issue publicly traded stock; but in addition to a lack of fit with the traditional Wall Street model, some operators did not want to surrender any measure of control over their operations. Even when they maintained majority interest, they were still beholden to large investors and subject to close review by Wall Street analysts. One of several factors behind the growth and success of TCI was its imaginative, if extremely complex, approach to corporate organization and investment, which began as a method of addressing this dilemma. Magness had learned his business craft, in part, at the knee of Glasmann, who built a labyrinth of intercon-nected ventures. Malone learned from Magness and brought his own keen analytical skills to the game. Magness and Malone would buy cable systems under their own names, hold them and then sell them to TCI, for a profit. The company purchased equity interests and entered in joint ventures with other operators and programmers, extending its tentacles across the indus-try. Magness.and Malone once created a holding company called Tiger, Inc., in which to park and protect millions in stock. The company was named after Magness' pet beagle. TCI would later spin out new companies, held as tracking stock by TCI investors, which could generate their own financ-ing without adding to TCI's existing debt. The result of such activities was a corporate structure that many considered mind-numbing in complexity, and seemingly in constant motion. It helped ward off attempts at a hostile takeover by corporate raiders, a popular business tactic in the 1980s, and it helped Malone and Magness maintain tight control over the company.

Malone and Magness also concocted a plan in the 1970s to issue millions of dollars of TCI stock as a means of solidifying their hold on company affairs. TCI created a super-voting category of stock that was purchased in large quantities by Malone and Magness. These Class B shares controlled ten votes per share, compared with the traditional stock at one vote per share. With the ten-to-one leverage, TCI was owned publicly and widely, Malone and Magness controlled only about 11 percent of total stock, but with 60 percent of the Class B shares, they controlled the company and were somewhat insulated from the interference and oversight of large institutional investors. As part of the process, again in a series of complicated financial maneuvers, TCI bought out George Hatch's interest in the company, although its linage to the Glasmann-Hatch family continued through ownership ties with the Kearns-Tribune company.

Finally, the cable industry joined many other business sectors in that period, drinking at the trough of the junk bond traders. The mid-1980s was the era of the junk bond, a high-yield, high-risk financial instrument that provided billions in capitalization for all sectors of a booming economy. An aggressive investment bank, Drexel Burnham Lambert, is credited with developing the controversial instrument and, at the heart of the bank, was a financier named Michael Milken, who would one day do jail time for insider trading. Drexel Burnham became a preferred lender to many in the cable industry in the 1980s, financing much of the expansion on both the operator and programming sides, and Milken became an ally, and even friend, to many in the business.

In late 1982 and early 1983, interest rates began to ease. They rebounded slightly in 1984, then dropped again, slipping to below 10 percent prime and hitting an eight-year low in the spring of 1986. Much of cable's outstanding debt was tied to the prime rate, so the decline meant millions of dollars in debt service savings and a parallel improvement in earnings. Equally as important, money for expansion and acquisition was cheap again. Buyers could pay higher prices for systems and system prices rose. The larger and more aggressive firms could consider leveraged buyouts, which were heavily dependent on borrowed money. This would help drive increased industry consolidation in the second half of the decade, which is not to say expansion and mergers stalled in the early 1980s, for drivers other than the cost of money were prodding cable economics.

Consolidation

As in the past, tight money tended to limit acquisitions and mergers to larger firms, which were better positioned to secure capital at lower rates. The emergence of the cable programming business, however, stimulated a wide desire for consolidation. As already discussed, program suppliers such as HBO sought stable and controllable distribution channels, so they merged with operators such as ATC. Vertical integration was not the only product of the process, however. The desire to guarantee an expanding subscriber base also drove horizontal integration. On the operator side, increased size meant greater leverage in negotiating with unaffiliated cable programmers, and lower overall programming costs. The rise of the programming business meant the development, for the first time, of potentially substantial economies of scale in cable.

The largest business marriage of the early 1980s brought together TelePrompTer and Westinghouse. TelePrompTer, still the nation's largest MSO, had about 1.25 million subscribers, owned Filmation Associates and the Muzak Corp., and held a 50 percent stake in Showtime. Westinghouse Electric was a far-flung conglomerate with holdings in everything from household appliances to jet engines. Its Westinghouse Broadcasting Co. subsidiary owned seven AM, four FM, and six television stations, and its Group W Productions owned cable systems in Florida, Georgia, and New York. In October 1980, the two companies announced their proposed $646 million merger, the largest in the history of electronic media to that time. By some accounts, TelePrompTer Chairman and CEO Jack Kent Cooke was motivated in the sale by a need for cash. He had recently reached a very expensive divorce settlement and at the same time was looking to buy the Washington Redskins football team. For Westinghouse, it was a logical extension of the existing media subdivision.

With completion of the merger, Group W-Westinghouse was a cable industry powerhouse. Company management, interestingly, was quite conservative in its political and cultural philosophy. The company wanted to develop programming and, over the next few years, sought opportunities that focused on wholesome, family-oriented entertainment, or news and public affairs, shedding programming investments that did not comport with corporate values. In 1981, it purchased the family-friendly pay service Home

Theater Network. It sold TelePrompTer's interest in Showtime, and even briefly sought to work with Disney Corp. on the development of the Disney Channel. Group W-Westinghouse had to sell some of its systems, about 12 percent of its subscriber base, to avoid violating FCC regulations against cable-broadcast cross-ownership in individual markets. As a result, it dropped to the number two then number three spot among cable MSOs. ATC, which continued expanding its subscriber base, briefly took over the top spot as the nation's largest cable company in 1981.

Warner, as previously noted, was confronted with a number of business challenges involving the Atari division and the difficulties in the major markets. A weakened Warner had to fight off a hostile takeover bid by Rupert Murdoch. In December 1983, Murdoch purchased 6.7 percent of Warner's stock and then announced his intention to purchase between 25 and 49.9 percent of the wounded company. Although Warner was in difficult financial straits, the company still held the valuable Warner Bros. film studios with its extensive vault of movies and substantial production facilities, as well as distribution through its cable franchises. To fend off the attack, Warner entered into a partnership with Chris-Craft industries, run by financial wheeler-dealer Herbert Siegel. Steve Ross had once done Siegel a favor in a legal battle over Siegel's attempt to obtain the famous Piper Aircraft company. Now Siegel was prepared to repay the debt. Siegel had spent his business career buying and selling companies, starting long before the 1980s fad in such activity. Chris-Craft drew its popular prominence from the handsome boats it produced, but after Siegel obtained the company, the boat business was spun off. Siegel did, however, keep Chris-Craft's six TV stations. In a complicated stock swap, Siegel obtained, on paper, control of Warner and Warner garnered partial ownership of the Chris-Craft TV stations. Federal law prohibits foreign ownership of broadcast properties and the deal effectively killed the takeover attempt by Murdoch, a native of Australia. Murdoch did not leave empty handed, however. Before the takeover attempt, he had purchased Warner stock at depressed prices and, through his hostile bid, had put the company in play, driving up its stock price. When he withdrew, Warner bought back the shares at the higher price, $172 million in total, giving Murdoch a handsome $40 million-plus profit.

Murdoch was not the only aggressor in the field. Through the late 1970s and into the mid-1980s, TCI was easily the nation's most acquisitive MSO. Malone and Magness determined that the key to success was corporate size, and went on a spending spree that had observers gaping, both at the sheer number of purchases and the degree to which those purchases kept raising company indebtedness. Malone later explained:

[W]e became very aggressive in consolidating the cable industry because I saw from the beginning that scale economics was going to determine who was going to survive and who wasn't. We had the advantage of even though we were a public company, we were controlled. Bob Magness, as the principal shareholder, the controlling shareholder, wasn't particularly interested in near term earnings and was willing to really pursue a long-term strategy, which certainly I was, and so we were able to do things that most public companies can't do. We really didn't care about the impact of an acquisition on our earnings, so we would go out and do fairly highly leveraged acquisitions.[78]

TCI, in fact, was caught in something of a vicious circle. It needed the additional subscriber revenue and cash flow to service its massive debt, a debt that increased with nearly every acquisition.

One of TCI's major purchases in the early 1980s was a mid-sized MSO based in central Pennsylvania, Tele-Media Corp. Tele-Media, like many cable companies, had grown through gradual accretion, franchising small and media-sized towns in Pennsylvania and across the country. It had its roots in Centre Video, purchased by TCI in 1970. The Tele-Media acquisition, for $145 million in mid-1983, added 250,000 subscribers to TCI's holdings. TCI also purchased Liberty Communications, adding another 238,000 subscribers. By Malone's own estimation, the company was closing a deal every week through the early 1980s, mushrooming the subscriber base from just over half a million in the late 1970s to more than 2 million by 1984 and more than 3.5 million by 1986. In 1978 it was the nation's fourth largest MSO. By 1983, it was number one.

In addition to fully owned systems, TCI, as noted above, entered into numerous cable partnerships, including one with the Knight-Ridder newspaper chain. The newspaper industry had been involved in cable for years and some of the largest MSOs were the properties of the nation's leading newspaper companies, including Cox and Times Mirror. Newspapers, as noted, had been drawn to cable in the mid-1960s amidst the rhetoric of cable-delivered facsimile news. The head of the American Newspapers Publishers Association urged his members in 1966 to seriously consider buying into the industry, pointing out that many cable operators were pointing cameras at AP and UPI teletype machines and providing for their viewers a "primitive form of electronic newspapers."[79] The industry's enthusiasm for cable cooled with the business downturn of the early 1970s, but enjoyed a small renaissance in the early 1980s. The interest was rekindled, in part, by the revived belief that cable could help deliver information-age services, such as teletext, and by a fear that the telephone companies were looking to poach on information delivery traditionally reserved for newspapers, most especially classified advertising. As a result, several large newspaper groups not previous involved in cable began seeking an entry to the industry.

Knight-Ridder, publisher of more than fifty U.S. dailies, including the *Philadelphia Inquirer* and the *Detroit Free Press*, and Dow Jones & Co., publisher of the *Wall Street Journal,* partnered in an effort to buy UA-Columbia. The United Theater Circuit, which owned 28 percent of UA-Columbia, opposed the sale, however, sparking a heated bidding war. The newspapers' offer climbed to $75 and then $80 a share. A counter offer by United Theaters was initially rebuffed. United Theaters was then joined by Canadian cable MSO Rogers Cablesystems, Inc., and their combined offer of $90 a share was finally accepted. In the summer 1981 Rogers, in effect, purchased the operations side of Bob Rosencrans' company.

Slowed but not stopped, Dow Jones and Knight-Ridder went in search of other entry points into the business. Dow Jones found its opening in Continental Cablevision, investing $50 million for a 10 percent stake in the company. Knight-Ridder, meanwhile, found a partner in TCI and together the companies formed a new cable enterprise, TKR Cable, in late 1981.

Cox, which in the early 1980s owned twenty-three newspapers, more than a dozen magazines, and eighteen radio and TV stations, continued to expand its cable holdings as well. In 1980, it won the large Omaha, Nebraska, franchise. By 1984, it had nearly sixty cable systems, controlled more than 1.3 million subscribers, and was the nation's fourth largest MSO. Newhouse Corp. decided in 1980 to sell five of its broadcast stations, using some of the money to expand its cable holdings. It purchased the latest iteration of Daniel's holdings, Daniels Properties, and acquired Vision Cable Communications for $180 million. With the additions, Newhouse's NewChannels cable subsidiary subscriber base mushroomed from 214,000 to more than 620,000 subscribers and became the nation's eighth largest cable firm.

Even the *New York Times* decided to take the cable plunge, buying systems from Irving Kahn. Fiber optic cable had not been Kahn's only cable enterprise coming out of prison. He also began collecting franchises once again, working, he said later, out of the back of a Hallmark Card shop in New Jersey. By early 1980, he was ready to sell again. In March, the Times purchased Kahn's holdings, which by then served fifty-five franchises in contiguous New Jersey communities just outside of Philadelphia.

From 1980 to1985, the top MSOs included Group W-TelePrompTer, TCI, ATC, Cox, Warner, and Storer, with Times-Mirror, Newhouse, although their relative positions varied. TCI, with its program of aggressive expansion, moved to first place; ATC held steady at number two and Group W-TelePrompTer dropped to third.

Cable and Hollywood

Consolidation and partnership formation also was taking place among cable programmers, especially in the pay-TV category. Following the failure of Premier, the major studios had a backlog of inventory. They had been holding back films in preparation for a scheduled January 1981 launch, but were now eager to get the movies into circulation. Showtime, meanwhile, had been burning through its feature catalog in an effort to cut into HBO's lead during the Premier-induced film drought, and it now needed new material. Showtime moved quickly to secure deals with Universal and Paramount.

The studios, however, were still motivated to seek a more permanent form of downstream integration, and HBO and Showtime needed their product. It was perhaps inevitable, therefore, that the two industries would begin to meet somewhere in the middle, although it was a convoluted journey. Warner, with a foot in both camps, played an opening card. Its financial difficulties in the early 1980s led it, among others, to an effort to sell all or part of The Movie Channel. Warner began talking in 1982 with Paramount and MCA/Universal about possible equity positions in the pay-cable service. The relationships would have brought much-needed cash into financially strapped Warner and secured two more sources for film product, but the talks broke down over a host of small business issues, such as revenue splits and whether to sell the service in noncable markets. Paramount and MCA/Universal then suggested a new plan: Paramount (G&W), MCA/Universal, and Warner would join with Viacom to merge The Movie channel and Showtime into a joint venture similar to the HBO/Cinemax combination.

The deal had to overcome several obstacles. It required, at the outset, purchasing Group W-TelePrompTer's 50 percent share of Showtime. As it turned out, this was not too difficult. The management of Group W, as noted above, had a reputation for conservative values and was bothered by the R-rated films shown on Showtime. Viacom was eager to purchase Group W's interest, but according to Viacom's Ralph Baruch, Group W's Dan Ritchie, head of the broadcast cable operation, was reluctant to sell, despite the concern over content. Baruch, therefore, took an opportunity to talk with Westinghouse CEO Douglas Danforth at a luncheon they were both attending. Baruch said he positioned himself in the service line behind Danforth and began a conversation:

> "I understand we have something in common." "Oh," he says, "What is that?" I said, "Showtime." "Oh," he says, "that little thing." I said, "Well, what may be a little thing for you is a very big thing for us." "Yeah," he says, "I don't like the movies they have on." I said, "What do you mean, the 'R' movies?" He said, "Yes." I said, "As a matter of fact, you know the audience wants that." I had myself well prepared, I figured that would come out. "As a matter of fact, my position is we have to have a lot more 'R' movies, a lot of them!" To make a long story short, several weeks later we were able to buy back the 50 percent of Showtime from Westinghouse for over $60 million.[80]

Other parties were less accommodating. The proposed merger was complicated by negotiations between Viacom and three other companies—Columbia, 20th Century Fox, and ABC—for the purchase of Showtime. That path was cleared when Columbia, purchased by Coca-Cola in 1982, abandoned those talks to forge an alliance with Home Box Office. The remaining combination also had to clear Justice Department antitrust review, however, and in May 1983 the federal attorneys turned thumbs down on the proposal. As a result, Paramount and MCA/Universal dropped out, and the merger went forward with only Viacom and Warner. Justice approved the new TMC/Showtime venture, equally controlled by Viacom and Warner.

The new company was strengthened further in December 1983 when it acquired a third premium movie service, Spotlight. Spotlight was created in August 1980 by a consortium of cable operators seeking to establish their own pay channel. The companies, led by Times Mirror, included some of the industry's largest MSOs—Cox, TCI, Cablevision Systems, and Storer. The combination gave Spotlight tremendous national reach, displacing HBO on many of the affiliated systems. At the product end, however, Time, Viacom, and Warner had corralled most of the attractive recent films through their studio contracts. Unable to secure enough current movies, Spotlight was essentially starved to death. In acquiring Spotlight, Showtime/The Movie Channel, therefore, was really buying Spotlight subscribers, about 784,000 of them. HBO had dominated pay-cable for years. The new combination of The Movie Channel, Showtime, and Spotlight held the potential, for the first time, of providing real competition to HBO. The merger was generally applauded throughout the industry and by consumer groups looking to loosen HBO's grip on the market.

Unable to make a dent with a premium service featuring current Hollywood films, some cable operators tried again, combining to create a pay channel featuring older, classic movies. On October 1, 1984, Cablevision Systems, in partnership with Daniels and Associates and Cox, launched American Movie Classics (AMC). The pay service offered films of the 1930, 1940s, and 1950s, hoping to draw older audiences and film buffs of all ages.

HBO was certainly not happy with the Showtime/TMC merger, but at the same time was forging its own production empire. In 1981, it entered into a deal with Columbia to pay 25 percent of the production costs for Columbia films in exchange for specified exclusive pay-cable distribution rights. Securing film packages from the major studios, while necessary, was still not sufficient to meet HBO's needs. In 1982, Levin said that running a 24-hour schedule, HBO would need about 550 movies a year, far more than they could obtain through just the studio agreements.[81] HBO extended its investments in existing production companies and began to create studios of its own. In 1982, it secured an interest in Filmways (which used the money to buy Orion) in exchange for exclusive rights to certain films. In November 1982, following Columbia's flight from the negotiations for Showtime, HBO combined with them and CBS to form a new film studio, later named

Tri-Star Pictures. Tri-Star began life with guaranteed distribution in all the major venues, theaters, broadcast television, and cable. HBO would get exclusive pay-cable distribution rights to the films, CBS would have them for network TV, Columbia would hold the theater rights, and everyone would share in the profits.

The next year HBO created yet another source for films and income, getting E. F. Hutton to underwrite $100 million for the start of Silver Screen Partners. The new studio would offer limited partnerships to pubic investors, and produce at least a dozen films a year. HBO would have pay-TV rights, 25 percent of the income from network TV sales, five percent of all theatrical profits, and right of first refusal on home video distribution.

In October 1983, it further announced the opening of a new subsidiary, HBO International, in association again with Columbia Pictures, and joined by 20th Century Fox, and Goldcrest Films and Television, Ltd.. These companies, in turn, formed the Television Entertainment Group designed to explore joint opportunities in the United Kingdom. In February 1984, the companies created a new pay-cable service for Great Britain, calling it, interestingly enough, Premiere. Warner Bros. and Showtime/The Movie Channel joined the enterprise in March 1984.

HBO also entered the home video market, partnering with Thorn EMI In November 1984, to acquire and distribute home video in the United States and Canada. By 1985, HBO was one of the largest financiers, producers, and distributors of films in Hollywood.

By then, the pay-cable industry had consolidated into two dominant firms, HBO/Cinemax with more than 15 million subscribers and Showtime/The Movie Channel with more than 8 million customers. Smaller services buzzed about searching for systems and subscribers, including Rainbow's AMC/Bravo and Group W's Home Theater Network, but Time and Viacom/Warner were the major bridges to the Hollywood studios and still the studios' only real conduit to the nation's cable homes. In 1983, Paramount signed an exclusive deal with Showtime/TMC. Columbia meanwhile signed on with HBO.

Finding an aftermarket for film product was not the only concern of the major studios. Hollywood was beginning to make more programming for broadcast television, both first run and syndication (see discussion below), and was thinking about long-term cable outlets for what could become a sizable library of TV dramas and sitcoms. While Paramount and MCA/Universal were not permitted to savor a partnership with Showtime/The Movie Channel, they did have success in another area of the cable business. In the fall of 1981, MSG sold its 50 percent interest in the USA network to Paramount, and Bob Rosencrans' UA-Columbia (which had only a short time earlier sold its cable systems to Rogers) sold its half stake in USA to Time Inc. A few weeks later MCA/Universal bought into USA, which was then equally controlled by MCA/Universal, Paramount, and Time.

Under new ownership, the USA network sought to broaden its programming mix. USA President Kay Koplovitz had worked hard to make a name for USA in sports programming and lamented the shift, but she understood the economics as well. The studios, she explained "frankly saw it as a defensive move, a place where they could play off some of the series that they developed for television and went into syndication."[82] It would be several years before then current TV programming became available to USA. Existing contracts tied the more popular shows, such as "Miami Vice" and "Murder She Wrote," to other distribution venues. For the time being, USA would show "Dragnet" reruns and begin to develop original programming in an effort to stay fresh. But the studios were taking the long view—planning for the future.

Programming Proliferation

The movie moguls were not the only ones probing the possibilities of cable-satellite distribution. The rush to market in the major cities was mirrored by another gold rush of sorts, this one in the world of cable programming. "Cablemania" included a flurry of new ideas for cable program networks in the first years of the 1980s. The factors that drove the urgency in program development in some ways paralleled those that drove franchising. In system construction, the race was to secure a presence in the finite markets of urban America. In programming, would-be cable networks confronted a scarcity of both satellite transponders and cable system channel capacity. Three cable satellites—Satcom IIIR, Comstar D-2, and Westar III—were in operation at the beginning of 1982, hosting a total of about thirty-five channels.[83] More satellites were planned and capacity would increase dramatically over the next several years, but programmers coveted space on satellites for which cable operators already owned dishes. Networks on the primary cable satellite, Satcom IIIR (which replaced Satcom I in late 1981), still had access to more cable operators than programmers on other birds. Even more compelling was the contest to secure channel space on cable systems. Despite the expansion of cable technology and the concomitant increased in channel capacity, thirty- and forty-channel systems were largely available only in the major markets or on the drawing boards of smaller operators. About 75 percent of all systems, representing 55 percent of all cable subscribers had fewer than thirty channels in early 1983, and 20 percent of all customers were still getting by with twelve or fewer channels. Again, channel capacity would expand through the mid-1980s, but it was also worth a great deal in marketing terms to be the first on the local system and first into the home. Successful programmers would have an opportunity to establish brand identities and, with any luck, cable operator and consumer loyalty. Development of that kind of loyalty, or at least habituation, could become a modest barrier to entry, serving to inhibit late-coming competitors.

A variety of business interests therefore sat down to sketch out new cable networks or to buy into existing ones. They included the Hollywood studios and the larger MSOs. Nonaffiliated entrepreneurs also sought to exploit what looked like a promising new business opportunity. Many plans were proposed. Some found financing and launched, some did not. Of those that made it to the operational stage, a portion would flame out within a few months, writing off the losses and, in some cases, merging with other similarly fated networks. A few would find long-term success, but bear substantial operating losses for several years while they established a foothold.

The early 1980s was a period of bruising shakeout for the neophyte cable programmers. It is typical in most emerging industry's that a plethora of entrants will test the waters and hope for the best. Typically, only a few will succeed. Some of the start-ups were well-financed and well-run, some were not. All had to obtain transponder space, cable carriage, and consumer acceptance. As before, programmers aligned with existing MSOs had an advantage in securing space on cable systems. Affiliated services were nearly always privileged over nonaffiliate programmers.

For basic, advertiser-supported services, advertising revenue was a necessity and a problem. The advertising community, entrenched in its own routinized pattern of business practice, gave little heed to the nascent cable programming community, at both the national and local level. Audience figures were assumed to be tiny in comparison to the traditional broadcast networks, although the exact number of people watching was unclear because accurate methods of measuring the new audience were not yet well established. In fact, one of the greatest problems confronting cable was the lack of good audience data, which advertisers nationally and locally demanded. In 1981, the industry established the Cable Advertising Bureau (CAB); its mission in part, to help solve such problems. Advertisers, nonetheless, were loath to put large sums of money into, at-best, soft figures, so advertising dollars that went to cable were often experiments, insurance, or afterthoughts, and always a small part of the overall ad budget. Total cable industry advertising revenue grew from $58 million in 1980 to $595 million by 1984, but this was still a fraction of the broadcast ad revenue, more than $18 billion in 1984.

Despite these problems and challenges, a number of cable networks, including many that would go on to become important, even dominant sources of news and entertainment in the United States and around the world, were established in the early 1980s. Most sought to exploit narrowly tailored target markets, some based on programming content, such as ESPN's sports focus, or on demographics, following the lead of services such as Nickelodeon. In 1983, cable scholars Tom Baldwin and D. Stevens McVoy identified a dozen nonexclusive categories of established and emerging cable programming, including pay, news and public affairs, children, sports, religious, cultural, and ethnic.[84] With the rise of these special interest channels came a new vocabulary for the industry and the culture. In 1967, J.C.R. Licklider, as noted in Chapter 6, had offered a new term to describe what he saw as the future of cable programming. The term was "narrowcasting," intended to provide a contrast to the mass-audience implications of the decade's old term "broadcasting." It is not clear that Licklider's report was the catalyst for widespread adoption of the term; nonetheless, it was a part of industry jargon by the turn of the decade. The nomenclature was also expanded with terms such as "market niche" and "niche networks," again signifying the focus in content or target audience typical of most of the new program services.

The first year of the new decade saw the launch of several of cable's most important and lasting networks. The first, premiering in January 1980, was a programming service intended to serve a long-neglected segment of the American public.

Black Entertainment Television: January 1980

By 1979, cable offered specialty programming for children, the Hispanic community, sports fans, and citizens interested in Congress. Cable companies were struggling to wire the nation's urban markets, with their own diverse ethnic populations. The concept of a cable channel catering to the African-American community seems, at least in retrospect, natural. Long-standing criticisms were that broadcast television had failed to serve adequately the black audience, and now a technical and business opportunity existed to rectify the situation. As before in the history of the industry, many people may have perceived the need and the opportunity, but one in particular had the vision, the determination and the connections to act.

Robert L. Johnson was born in the small town of Hickory, Mississippi, in 1946, the ninth of ten children.[85] His family was not affluent and, as he would later recount, he always worked. He worked his way through college at the University of Illinois, then received a scholarship to Princeton University where he earned an MA in public administration. Johnson and his wife Sheila moved to Washington, D.C., in 1972, where he landed a position as Public Affairs Director at the Corporation for Public Broadcasting. From there he moved to a post as press secretary to Congressman Walter Fauntroy (D-District of Columbia) and then on to the National Cable Television Association, becoming the NCTA's Vice President for Government Relations. Through the NCTA, he knew the heads of the major MSOs and knew their problems. In his own right, he was a bright and determined young entrepreneur and, in 1979, he saw the potential for a programming service for the African-American community, a business opportunity for himself, and a valuable tool for the cable industry. The idea of a channel programmed for the black audience, according to Johnson, had been kicking around for some time, an occasional topic for cable operators seeking to penetrate urban markets and black politicians such as Fauntroy. "There was this idea," explained Johnson, "that cable would create the kind of diversity that the broadcast networks never had. Someone was going to do it. I was already in the cable industry in the late Seventies. Why not me?"[86]

To make it happen he needed capital and he needed carriage. He started near the top, with TCI's John Malone. "Bob Johnson," recalled Malone, "came up to me after an NCTA meeting and said, 'Do you think there would be any hope for a black channel aimed at the black demographic?' And I was very enthusiastic about it because we were trying to build in some markets of heavy black neighborhoods and we didn't have anything to talk to them about. And so we put up what we could afford, a small amount of seed capital."[87]

The seed capital was $500,000, of which $320,000 was in the form of a loan. With the remaining $180,000, TCI purchased 20 percent of the new company. Johnson took out a $15,000 bank loan to establish his own equity stake. He also approached Bob Rosencrans about satellite carriage. Rosencrans was supportive, giving him 2 hours of MSG Sports Network satellite time every Friday night, for free. With Malone and Rosencrans behind him, and assured carriage on TCI and UA-Columbia systems, he also received clearance from most of the other major MSOs, including Warner-Amex, Time/ATC, and TelePrompTer.

It was a fairly easy call for the cable companies. Most of the franchising authorities they were courting held service to local minority interests as an important goal. Parceling out equity in the local system to minority interests was helpful, but offering a programming channel dedicated to those interests was also very persuasive in demonstrating a commitment to serve diverse community needs. Federal authorities, as well, had been critical of the cable industry's track record in minority hiring and service to minority interests.[88] Johnson's proposal would be useful in Congress and in the national court of public opinion, too. In terms of revenue possibilities, cable needed to attract African-American subscribers if it hoped to succeed in the nation's urban centers. Operators hoped the new channel would help drive penetration in those areas. Finally, the deal struck by Johnson called for the service to be free to cable operators. The new channel would not enjoy a dual revenue stream, but for the first several years would have to get by on advertising dollars alone.

It did not hurt that Johnson was politically compatible with the MSO leaders. Malone was candid in his assessment that Johnson would be a safe bet for running the channel. Malone and others were concerned about the potential for a politically radical perspective in both the programming and business operation of such an enterprise, but they saw in Johnson a solid businessman whose long-term interests were consonant in many, if not all, ways with theirs. As Johnson himself explained early on, "It [BET] is a business with a black consciousness, but the accent is on the business."[89]

Bob Johnson began broadcasting Black Entertainment Television (BET) in January 1980, programming from 11 PM to 1 AM on Friday nights and

carried as part of Rosencrans' USA Network. The network struggled financially for several years. A number of major advertisers signed on at the launch,
but Johnson sought to expand and money remained tight. The network left
USA's protective umbrella in August 1982, but the move was costly. It required transferring from the popular cable satellite Satcom I to the new Westar V. Increased transponder time allowed BET to broadcast 6 hours a day,
but fewer operators were equipped to receive it and subscribership dropped
from 10 million to 2 million. An infusion of capital from Taft Broadcasting,
which acquired 20 percent of BET, helped see the company through the next
few years. In 1984, BET jumped satellites once more, moving from Westar
V to the Hughes Galaxy I. The motive, as with the previous hop, was an
expansion of broadcast time, now to a full 24-hours a day. HBO controlled
the Galaxy I transponder and in lieu of cash took an equity position in BET.
To maintain Johnson's majority control of the company, the other investor
MSOs scaled back their stakes, leaving the split: Johnson 52 percent, TCI,
HBO, and Taft, 16 percent each.

CNN: June 1980

Ted Turner and Reese Schonfeld created CNN, the Cable News Network.
"People still ask," Schonfeld wrote later, "'Whose idea was CNN? Yours
or Ted's?' So far as I know, it was Gerry Levin's."[90] Levin had approached
Schonfeld in 1977 about obtaining membership in Schonfeld's satellite news
pool enterprise, ITNA. HBO was considering the possibility of a full-time
cable news channel, tentatively dubbed "News Plus," and would need material from sources such as ITNA. But the ITNA board was composed of
broadcasters who were not drawn to the idea of supplying their content
to a cable competitor.[91] They rejected Levin's request. Levin also had trouble convincing his bosses at Time Inc. of the market potential for a cable
news service. By one account, Levin decided to abandon the project after
mulling it over in the shower one morning. "I get my best ideas in the
shower," he reportedly told a colleague. "I've decided we're not going to do
news."[92]

Full-time news had other precedents. All-news radio stations had been
operating in the major markets for several years. Group W-Westinghouse
took a lead in pioneering the format and its all-news operations were fairly
successful. Cable operators, of course, had had local origination cameras
pointed full-time at news service teletype machines since the 1950s. Both AP
and UPI had been running dedicated text-based cable news feeds since the
mid-1960s. UPI began a satellite-distributed text and graphics service, UPI
Newstime, in July 1978, carried, as it happened, by Turner's own satellite
access provider, SSS, piggybacking on the WTCG signal. News service text
feeds were relatively cheap. A live, 24-hour TV news operation would be
enormously expensive, and no one really knew if the American public would

watch. The Scripps Howard and Post-Newsweek media chains briefly considered all-news cable channels and, like Time Inc., backed away.

Ironically, Ted Turner disliked the idea of an all-news format, or television news in any form. "I hate news. News is evil. It makes people feel bad," he once declared.[93] His WTCG superstation owed much of its success to avoiding and even counter-programming news, offering cartoons and sitcoms as a substitute for public affairs. With the growing success of WTCG and the clear movement by others to begin developing additional cable-programming services, Turner felt pressured to create a new network while the field was still relatively open, and he was casting about for ideas. In brainstorming sessions with his Board of Directors, he explored a number of possibilities, including an all-music format, which was dismissed because no one thought people would want to watch music. At one point, Turner announced publicly he was going to start an all-sports channel, but that plan faded. Eventually, he settled on news. Despite his reputation for extreme and peculiar behavior, he was, again by everyone's evaluation, an ever-astute businessman, and his logic was impeccable. He told Schonfeld: "There are only four things that television does, Reese. It does movies, and HBO has beaten me to that. It does sports, and now ESPN's got that. There's the regular kind of stuff, and the three networks have beaten me to that. All that's left is news! And I've got to get there before anybody else does, or I'm gonna be shut out.[94]

In September 1978, he talked to Schonfeld about creating a cable news network and hiring Schonfeld to run it. Schonfeld had not been particularly impressed with Turner on their first meeting, but times had changed, and Schonfeld had always dreamed of heading up an all-news operation of this kind. He was intrigued by Turner's offer, although no formal agreement was reached at that time. In December 1978, Turner took the idea to the NCTA's Western Cable Show in Anaheim, California, fishing for interest from cable operators. He found none. His pitch to a closed meeting of the NCTA board called on operators to pay $0.15 per subscriber per month and sign commitments for the service on the spot. He was met with skepticism, even some incredulity, and only a few contracts. Turner pushed ahead nonetheless. In April 1979, he called Schonfeld again, this time committed to see the project through. A long-time friend—Braves' general manager Bill Lucas, a man in his forties—had just had a cerebral hemorrhage. Turner was feeling introspective. "None of us is going to live forever," he told Schonfeld, "Let's do this fucking thing."[95] Schonfeld signed on as President, CEO, and sole employee of the new network.

Turner and Schonfeld sparred over formats and philosophies. Turner was thinking in terms of a softer and more sensationalist approach that melded news with entertainment. Schonfeld was committed to a traditional hard news format featuring live coverage and breaking stories. In the end, Turner bought in to Schonfeld's vision. The two talked about talent. Schonfeld thought they could wrest Dan Rather, still waiting in the wings for

Walter Cronkite's retirement, from CBS for $1 million. Turner's reaction was, "who's Dan Rather?" Schonfeld moved on. Working through his Rolodex, he finally decided to approach veteran correspondent Daniel Schorr, who had worked for him at ITNA. Schorr, a highly respected, tough-minded reporter, was, like many, somewhat taken aback by Turner's personal style, but attracted by the possibilities that a 24-hour TV news operation promised.

On May 21, 1979, at the NCTA convention in Las Vegas, Turner, flanked by Schonfeld and Schorr, publicly unveiled plans for the Cable News Network.

There were more than a few skeptics. It would cost, even by Turner's estimate, at least $20 million, and viewer interest was uncertain. Turner, however, had a strong track record and many cable operators were becoming enamored with the idea.

To help finance the venture, Turner sold his UHF station, WRET, in Charlotte, to Group W-Westinghouse for about $20 million. He also secured lines of credit from his banks. He tried to interest Levin and TelePrompTer's Russell Karp in partial ownership; they both declined. Turner, in fact, had bet the company on CNN and, in the first several years, lost tens of millions of dollars on the enterprise; Turner kept it afloat with money from the profitable WTBS. CNN did not turn a profit until 1985.

Turner purchased and renovated a former Atlanta country club, the Progressive Country Club, as the headquarters for the new operation. Situated on twenty-one acres of landscaped grounds, it featured a red brick building with a wide portico and white Doric columns, evoking for many, visions of an ersatz southern plantation from "Gone with the Wind." Schonfeld realized far into the development process that the money was tighter than he anticipated. To save dollars, they decided to hire, at minimum wage, journalism and television majors fresh out of college, instead of seasoned (and unionized) professionals, and to combine job responsibilities for the youngsters, reducing to one position what might have been assigned to two or three people at the networks.

Underfunded and on a frantic schedule to meet Turner's announced June 1, 1980, launch date, Schonfeld and his young staff worked 18-hour-plus days, installing equipment, securing satellite time, creating news formats. Turner worked on obtaining cable affiliates and financing, and squeezing in some sailing. The loss of Satcom III in December 1979 was a blow. It was to be the carrier for the new service, and it took a lawsuit to pry a temporary back-up transponder out of RCA. But, on June 1, a crowd assembled in front of the new CNN headquarters in Atlanta. Turner made a short speech dedicating the new facility. A brass band played the national anthem. At 6:00 PM, inside the CNN studios, a drum roll signaled the start of broadcasting. The cameras cut from the flags outside the building to the adjacent field of satellite dishes and then to a graphic, "CNN, The News Channel." The red light of the camera blinked on and the two news anchors began,

"Good Evening, I'm Dave Walker." "And I'm Lois Hart. Now, here's the news."

Observers at the time gave the young network little chance of success. It was being run on a shoestring budget, staffed with notoriously underpaid and often underexperienced personnel, and it had to compete with broadcast journalism institutions such as NBC and CBS that were delivering news to the American public when Turner's father was a child. CNN struggled through its first several years with on-air gaffes—a cleaning lady once strode across the set and emptied Dan Schorr's wastepaper basket while he was on the air—a continuing lack of quality news flow, and too few viewers. Inside the world of big league TV journalism, it was dismissed as the "Chicken Noodle Network." At its inauguration, CNN had about 1.7 million subscribers, well below the 5 million Turner had promised advertisers. In its first year or two of operation, the network was reportedly losing a $1 million a month.

CNN, however, not only prospered, but by the end of the decade, it was becoming a preeminent force in broadcast journalism. CNN became a first source of breaking news, not just for the public, but often for politicians, policy makers, and even competing news services, who often had one channel in the newsroom always tuned to CNN. Immediate around-the-clock coverage, instant access to breaking or on-going events, became one of its claims to fame. CNN became not just a wildly successful cable channel, but a force in the political world as well. In typical fashion, Turner observed later, "I just wanted to see if we could do it."[96]

BRAVO: December 1980

Many of the obvious specialty markets—movies, news, sports, children, blacks—were, by 1980, tapped. But a few others remained. Two that became popular over the next few years were cultural or arts program networks, aimed at the upper end of the social spectrum, and sexually explicit "adult" programming, aimed at a somewhat different audience. Attracted by the economics of the vertically integrated business model, multiple systems operators Cox, Daniels and Associates, Comcast, and Cablevision created Rainbow Programming Services, taking the name from Dolan's Long Island tiering scheme. Rainbow distributed two niched networks to the combined MSO systems, Bravo and Escapade. Both debuted as pay services in December 1980.

Bravo was positioned as an arts and cultural channel. Its program day included classic foreign films, ballet, and performances by the New York City Opera. Bravo emphasized foreign and domestic arts movies, featuring work such as Kurosawa's *Rashomon* and Zeffirelli's *Romeo and Juliet*. Escapade offered adult material that included nudity and R-rated features (no X-rated films, however). In July 1981, Rainbow split them into separate services and subsequently entered into an arrangement with Playboy magazine, melding Escapade content with material from the nation's business leader in adult entertainment. Although Bravo did not generate a similar business partnership, the idea of a performing arts channel did attract attention from several other important parties.

The Broadcast Networks

In 1980, executives from both ABC and CBS publicly indicated that they might consider owning cable systems, if the FCC ban was lifted.[97] CBS, of course, had entered the industry early on and its Viacom spin-off was one the country's leading MSOs. ABC, in contrast, had expressed nothing but antipathy toward cable since the beginning, and its new position was a greater turnabout. Some characterized the networks' interest simply as a form of, "if you can't beat 'em, join 'em." But there was more to it than just resignation and market opportunism. Federal policy helped nudged the big three toward cable, particularly cable programming.

The networks wanted to participate in the seemingly revived cable TV business. They had substantial programming libraries and resources, and cable offered additional distribution outlets and downstream integration. Talk of cable development hurt their relationships with some local affiliates. Some local broadcasters still saw cable as a threat to the pocketbook. The revenue potential was too great to ignore, however, and many of the larger broadcast groups that made up the army of affiliates were themselves cable owners, which helped diffuse criticism. The main barrier to entry for the networks was the FCC's ban on cable system ownership. There were calls to rescind the ownership ban, including one from the FCC's special staff investigating network control of television programming. Capitalizing on the mood, CBS requested, and in 1981 was granted, a limited waiver from the rule. CBS was allowed to purchase systems, as long as its total reach did not exceed 90,000 subscribers or 0.5 percent of the national subscriber base (whichever was smaller).[98] In March 1982, CBS purchased two small systems in the suburbs between Dallas and Fort Worth, Texas.[99]

In November 1981, the FCC's Office of Plans and Policy recommended abandonment of the network ownership restriction and, the following year, the Commission opened a rule-making proceeding to consider dropping it.[100] The networks, of course, enthusiastically supported elimination of the rule, and cable operators sat quietly but watchfully on the sidelines. Passage of comprehensive federal legislation in 1984 truncated interest, however, and the rules remained in place. CBS sold its small Texas systems to Sammons, and the networks turned their attention to the programming side of the business. Here again, FCC policy encouraged, in unintended ways, the nature and shape of cable programming and broadcast network involvement in the business.

The FCC's 1970 financial interest and syndication (fin-syn) rules (see Chapter 6) were intended to break the network's oligopoly on the creation and ownership of national prime time programming. But the broadcasters fought strenuously in court and in political circles to derail the regulations. In 1972, the Justice Department sued to force network compliance. The networks beat back the government in 1974, wining a court judgment that the Justice Department suit, in essence, had been a politically motivated part of the Nixon administration assault on broadcasting. But the action was reinstated and, in 1977, NBC reached a settlement. Additional legal action brought the other networks in compliance soon after. Cable felt the consequences in several tangential ways.

A substantial portion of television program production shifted from broadcast network studios to the sound stages of the major motion picture companies, which eagerly began filling the programming void created by fin-syn rules. Through the latter 1970s and into the 1980s, the studios dramatically increased their production capacity, output, and program inventory. Some of this product was developed for the networks, becoming available for syndication after its network run. Some of the programming was created for direct distribution in the off-network market (syndication). In both cases, the library of available television content expanded substantially. It would, over time, become a deep well of low-cost programming for fledgling cable TV channels.

In addition, for the broadcast networks seeking to maintain at least a portion of their production business, cable in 1980 and 1981 looked like a possible distribution outlet. Settlement of the fin-syn controversy in the late 1970s, coupled with a rising excitement about the potential for cable, therefore, encouraged all three networks to consider becoming cable programmers as well.

CBS, ABC, and RCA (parent company of NBC) all chose to specialize, following Bravo, in cultural and performing arts programming. Focusing on this genre seemed to offer several financial and political benefits. It was good public relations and played well with policy makers and Congress. It did not directly threaten existing network fare, which relied on shows such as "Dallas" and "Love Boat," which may have helped mollify affiliates. To the extent that it offered competition to any TV service, it was against the weakest player in the industry, public broadcasting, and public broadcasting had become a political target of the new Reagan administration. Reagan's transition team proposed abolishment of PBS the same week the new president was sworn in. While the proposal was purely political—Reagan believed PBS programming was ideologically too liberal—the public justification used by the administration rested in part on the growth of the cable networks. Reagan officials argued that cable programmers such as Bravo, Nickelodeon, and others offered programming redundant to publicly supported broadcasting. The Reagan administration did not put an end to public broadcasting, as it might have liked, but it was successful in reducing federal support. The broadcast networks, on the other hand, were much less successful in their inaugural efforts in cable TV.

The ARTS Channel

ABC was the first to launch. Ironically, through the 1960s and well into the 1970s, ABC had done as much as anyone could to retard the expansion of cable television, but that fight was now clearly lost. ABC entered into partnerships to create two new cable programming services. It joined with Warner-Amex on the Alpha Repertory Television Service, or "ARTS," and with Hearst Corporation to develop a service called "Daytime." Raymond Joslin, head of Hearst Entertainment at the time, said the relationship was a natural.[101] The companies had worked closely for years, the Hearst broadcast properties were ABC affiliates and the heads of the two companies, Leonard Goldenson at ABC and Frank Bennack at Hearst, were good friends. They created Hearst-ABC Video services in early 1981, determining the order of the names on the letter head with a flip of a coin. The new joint venture operated both ARTS and Daytime. The ARTS channel began operation in April 1981. In keeping with its title, it featured programming such as opera, ballet and quality drama, piano recitals, plays, and biographies of artists such as Renoir and Degas.

The Entertainment Channel

ABC announced plans for Alpha about a week before the cable industry Western Show in Anaheim. RCA, parent company of NBC, waited until the cable operators were gathered together in southern California, in December 1980, to unveil its programming project. It was another cultural programming service, developed in partnership with Rockefeller Center. Extensive market research convinced backers that "The Entertainment Channel" (TEC) could succeed, especially in the upscale urban markets that cable was now developing. In addition to the requisite operas and classic concerts, TEC secured a highly publicized contract with the BBC for exclusive use of selected British programming, and its schedule featured a broader and more popularly oriented mix than like services. The new network was headed by former CBS President Arthur Taylor and began service in spring 1982.

CBS Cable

CBS had always been the "Tiffany Network," so named for its well-earned reputation for quality news and public affairs programming—it was the network of Edward R. Morrow—and for the vast wealth that its success in entertainment programming bestowed on it. CBS founder and Chairman, broadcasting legend William Paley, prided himself on that reputation and saw in cable a chance to extend it. Paley championed the idea of a prestigious, cable-delivered arts channel, offering serious film, dance, and theater. CBS Cable was launched in October 1981, and showcased concerts by artists such as Count Bassie, Leonard Bernstein, Aretha Franklin, and Carly Simon. CBS additionally determined to separate itself from PBS and the ARTS channel by producing up to 60 percent of its own content.

None of the networks' cultural efforts would succeed, however. Economics, technology, even personal animosity would all play parts in their collapse.[102] Much of the programming for the services, and for Bravo, came from international sources and by U.S. standards was relatively inexpensive. Nonetheless, all three services spent millions acquiring and creating programming. CBS, especially, spent lavishly to develop original material; ARTS also created some of its own content. Bravo and TEC debuted as pay services, but ARTS and CBS Cable hoped to succeed as advertiser-supported services. (In 1983, TEC also dropped is pay format.)

The audiences for these channels never fully materialized, however. The failure of cable to deploy in the major markets as quickly as some predicted meant that most of the viewer base the cultural services planned to tap was still out of reach, and would likely remain so for several years. The delay in reaching the critical target market, in turn, kept potential advertisers at bay, as did the broader sentiment on Madison Avenue that cable television generally was a poor advertising vehicle. Advertising agencies were additionally skeptical about the drawing power of cultural programming in a country mesmerized by sitcoms and game shows. Even once the cities were wired, they were concerned that performing arts would be unable to attract sufficient viewers (or viewers of an older demographic) to justify advertising dollars.

To the extent that a single arts channel faced substantial hurdles, three or four competing cultural services confronted compounded problems. This was a period in which a host of fledgling services were trying to get off the ground. Cable operators had a glut of programmers from which to choose and, in many markets, limited channel capacity, or cable shelf space. Classic cable systems in smaller towns were not convinced that cultural programming would be a strong draw for their customers. They might allocate one of their increasingly scarce channels to a cultural service, but they were very unlikely to dedicate three or four. Competition for what little advertising there was similarly split across the pool. Finally, and despite its political headaches, PBS remained an established and widely deployed national broadcast network with a loyal fan base and a state-supported (although admitted threatened) financial foundation.

Problems particular to each channel worked against success as well. Following HBO's successful tactic of the mid-1970s, CBS went so far as to underwrite cable operator purchases of the additional dish needed to receive the Westar signal, but this only added several millions dollars to start-up costs. By November 1982, CBS Cable was available to only 16 percent of U.S. cable subscribers. Despite the concerns of skeptics inside and outside the organization, CBS continued to spend substantial, many said outrages, amounts of money in program development, in advancement of founder Paley's vision.

The long-standing animosity between cable operators and broadcasters played at least a partial role in the failure of the services. Years of legal confrontations and fears that the networks now sought to enter and dominate cable programming fed cable operator resistance to carriage of the new services. This sentiment was particularly acute with respect to TEC. TEC's Arthur Taylor, while president of CBS, had spent years antagonizing the cable industry, speaking out publicly about its threat to free television and lobbying vigorously in Washington to stifle cable growth. Taylor once drew trade press attention for berating Ford Administration officials, including Paul MacAvoy, for alleged antibroadcasting bias, in an incident some observers described as an embarrassing temper tantrum.[103] The cable industry, in many ways, was still a small club and its members had good memories. Residual resentment on operators' part left them with little enthusiasm for carrying Taylor's new venture. Shortly after launch, TEC had eighty-four systems and 50,000 subscribers paying $12 a month for a mix of Broadway shows, foreign films, and BBC comedies, and it was bleeding money. Nine months into operation it had lost $34 million and announced it would pull plug in March 1983, although it lingered through mid-year.

In September 1982, William Paley retired as Chairman of CBS. With him went the company's primary backer of CBS Cable, which was costing the company millions. CBS had attempted to find partners for a merger, including with Bravo, but could get no takers. Paley's replacement was CBS president, Thomas Wyman, a seasoned corporate executive with experience at Polaroid and Pillsbury, who Paley had recruited in 1980. Wyman had been skeptical about CBS Cable and, within days of the management change, CBS announced it would shut down the failing service. The struggling cable network ceased operations in December 1982. CBS reportedly lost $30 million on the venture.

ARTS continued to struggle on, expanding its programming to more widely popular material. In December 1983, it picked up TEC's BBC programming options and then merged with the failed service. The newly combined ARTS and TEC relaunched in February 1984 under a new name; The service was now the Arts and Entertainment Channel, or A&E. Production of original material was suspended and the channel shaped its programming to appeal to a broader audience. CBS and Bravo continued to discuss a relationship and, in 1984, CBS purchased 50 percent of Rainbow Service Corp., becoming a partner in Bravo, which also cut back on original production and put forward a heavier schedule of films targeted to a more popular audience.

Lifetime Television: February 1984

ABC's other adventure with Hearst led down a similar path. The partners debuted Daytime in February 1982. Hearst published several magazines oriented toward women, including *Cosmopolitan*, *Good Housekeeping*, and *Harper's Bazaar*, and sought to leverage the editorial material into a cable TV channel. Following the logic of BET and Nickelodeon and targeting major demographics, Daytime was created for women. In June 1982, Viacom, in association with Reiss Video Development Corp., began another specialty service, the Cable Health Network, specializing in programming for health professionals and anyone else interested in health care. Jeffrey Reiss had worked in programming for Viacom, running the Showtime service, and was now striking out on his own, getting started with the support of his former employer. Neither Daytime nor Cable Health Network prospered, however. Neither service ran a 24-hour schedule, sharing transponder time with other programming, and carriage was difficult to obtain. In February 1984, the services were merged to become Lifetime, blending the programming, but eventually emphasizing the female audience demographic. New carriage deals were struck, expanding the launch distribution to more than 16 million subscribers. Operators were drawn not just by the programming but by the fact it was offered free to cable systems, relying solely on advertising sales for its revenue stream.

The Satellite News Channel: June 1982

ABC took one other programming gamble in 1982, a 24-hour news channel. Turner's success with CNN both repelled and attracted the television networks. CNN was doing reasonably well with respect to viewership and appeared as if it might someday even be profitable. For the networks, CNN's success meant the potential loss of viewers and advertisers, but also suggested that a 24-hour news service had promise, and who better to realize that promise than the established television giants? In August 1981, top cable MSO Group W-Westinghouse announced an alliance with ABC for the creation of two new cable news channels.

Group W now owned TelePrompTer and, through it, could guarantee distribution to more than 1.5 million cable homes. It planned to develop its own vertically integrated programming capacity. The leadership of Group W-Westinghouse, as previously noted, was uncomfortable with R-rated movies, prompting the sale of its interest in Showtime. The proceeds from the transaction were partly earmarked for investment in more family-friendly and public affairs-oriented programming. News was a good fit with the corporate philosophy and resources. The proposed channels could tap into the substantial established news-gathering engines of both ABC and Westinghouse Broadcasting, which had pioneered all-news radio. As added incentive for cable operators to carry the new services, Group W announced that, unlike CNN, it would not levy a monthly per subscriber charge; in fact, it would pay cable operators to carry it. The first channel would be formatted in the manner of Group W's all-new radio stations, featuring rotating 20-minute news and headline program blocks. The second, planned for a later launch, would offer in-depth and documentary style programming.

In characteristic high dudgeon, Turner responded by announcing that anyone who went with Satellite News Channel (SNC) was going with a second-rate operation. He accused Westinghouse of pressuring its affiliated systems not to carry CNN. More substantially, within a week of the Group W announcement, he declared he would start a second news channel of his own, CNN-2, later renamed Headline News. Turner, in fact, beat Group W-ABC to the satellite, opening CNN-2 in January 1982. The new Group W-ABC service, SNC, began operations June 21. To counter SNC's economic incentives, Turner sharply reduced the price of CNN and offered CNN-2 for free. In March 1983, Turner took the competition a step further filing an antitrust suit against SNC and Group W charging conspiracy to keep CNN Headline News off Group W-owned systems. Group W countersued, accusing Turner of assorted anticompetitive business practices.

The market thereby found itself with three all-news cable channels, which was at least one more than could be financially sustained. Advertising dollars split three ways in cable news and were then further siphoned off when the three major broadcast networks all launched their own overnight news programs. By the fall of 1983, SNC had lost, by some accounts, more than $60 million. Turner was bleeding as well. Cable operators threatening to switch from CNN to SNC forced him to lower carriage fees and he was looking at annual loses amounting to an estimated $16 million. In the end, it was SNC and Group W that blinked first.

Intermediaries hoping to settle the court battle brought the parties together. The talks, encouraged and aided by NCTA President Thomas Wheeler and by Bill Daniels, moved from the courtroom to the boardroom, where Group W was becoming uncomfortable with its losses. Turner made an offer to purchase his competitor and, in October 1983, bought the ailing SNC for $25 million. He immediately pulled the plug. He not only eliminated SNC, he also gained 6 million new subscribers for his own news networks; as a condition of the sale, Group W-TelePrompTer agreed to carry CNN or CNN-2 on all of its 110 systems. Both parties terminated their legal actions. Turner incurred massive losses in the battle against SNC, but CNN was now the entrenched monopoly provider of cable news and within a month of SNC's demise, CNN tripled its fees to cable operators.

The networks' early efforts in cable programming met, therefore, with mixed results. SNC news and CBS Cable cost the networks millions, but out of the failures would come some long-term successes, such as Lifetime and A&E.

ABC, having failed to capitalize on the strength of its news division with the SNC project, turned to sports, where it also had a strong programming history. A series of events had given ABC a new cable opportunity. Billionaire oilman J. Paul Getty died in 1976, setting off a bitter family feud over the inheritance. The financial, personal, and legal battles lasted for years, concluding in a $4 billion negotiated settlement that broke up the Getty trust. In February 1984, Texaco bought Getty Oil and, with little interest in fields outside petroleum, was open for an offer on the cable programming channel. In early May 1984, ABC wrote a check for $202 million for the remaining 85 percent of ESPN. In October 1984, ABC sold 20 percent of the network to Nabisco.

The Financial News Network: November 1981

One other early 1980s notable was the Financial News Network (FNN). The promise of CNN led venture capitalists to propose a similar, but again slightly skewed service, a program network focused specifically on business and economic news. FNN began life in November 1981, backed largely by Biotech Capital Corporation and Merrill Lynch. With about $22 million in start-up funding, FNN began frugally, with a twenty-person staff, and like others, service was running a substantial early deficit, about half a million a day. It suffered additionally by the loss of President and Chairman Glenn Taylor, who was forced to resign in July 1982 following reports of financial impropriety and failure to pay his income taxes.[104] As with other successful channels, FNN slowly gained subscribers, however, and by mid-1985 reported its first profit.

MTV: August 1981

Before Warner-Amex began suffering seriously from its problems in the cities and from the corporate drag of a failing video game business, it was an active player in the contest to speed cable networks to market. It operated The Movie Channel and Nickelodeon, and was seeking fresh network concepts. A Warner-Amex Vice President, John Lack, had an idea. Lack envisioned a network aimed at the 12- to 35-year-old audience, the late baby boomers and the children of the baby boomers (the "echo"), and featuring video clips of musical artists performing their songs, videos that originally had been used as commercials to sell the music. The idea was to program the videos in the same way that top forty radio stations played records.[105]

Warner-Amex bought the concept and advanced $20 million for development. Warner Bros. Records was one of the larger labels in the business, and the business of popular music in 1980 was in distress. For a host reasons, including earlier industry overspending, aging rock stars, aging rock consumers, and a waning disco craze, the bottom had dropped out of the market in 1979.[106]

An all-music cable channel would help promote, and might help revive, industry fortunes generally and Warner records in particular. One of Lack's first moves was to hire a young executive named Robert Pittman to develop the format. Pittman, then 25, was head of programming for The Movie

Channel, but before joining Warner-Amex had made his name in radio programming. Starting out in radio as a part-time disc jockey at age 15, he had become a star program director for NBC, responsible for the successful formats at WMAQ in Chicago and WNBC in New York, both NBC properties. He made the move to cable in 1979. Pittman knew cable and he knew popular music.

Music Television (MTV) launched on August 1, 1981. Its first video, now a legend in MTV and modern rock history, was pointedly and prophetically, "Video Killed the Radio Star" by the Buggles. Videos from Pat Benatar, Rod Stewart, and Todd Rundgren were part of the fanfare of MTV's first hour. It was popular with its target audience almost from the start, but MTV, as most of the start-ups at the time, lost a great deal of money, an estimated $50 million by some accounts, before it achieved profitability beginning in 1984.[107] In 1984, as part of its effort to deal with its many corporate difficulties, Warner also took MTV and Nickelodeon public, raising $80 million in badly needed cash. Pittman became COO of MTV in 1983 and CEO of MTV Networks in 1985.

While financial success took a few years, MTV quickly accomplished two important things. First, it attracted a new and younger audience to popular music, a generation that had been raised on television. Second, it introduced to this audience a new set of artists and new kinds of music. The latter was as much by necessity as by design. MTV needed music videos, and it found a ready supply in Europe, especially in Great Britain where television was a more common avenue for promotion of records because of the lack of privately owned radio stations. The result was the introduction of novel groups such as Men at Work, Human League, and Stray Cats.

MTV's initial audiences were small, but their buying habits were voracious. Record companies, such as Polygram and MCA, which first balked at supplying MTV with videos, soon saw the wisdom of heavy MTV rotation. As cable penetration grew, so did the power of MTV, and its promotional campaign, "I WANT MY MTV," was soon taken seriously by both the music and the cable industries.

The Weather Channel: May 1982

If cable could provide specialty news and sports channels, at least one company felt the market might be able to support a full-time programming service devoted to the weather. There were more than a few skeptics. The idea for The Weather Channel was developed by John Coleman, at one time the meteorologist for ABC's *Good Morning America*, in concert with Landmark Communications.[108] Landmark was a midsized media company with interests in newspapers, radio, and television. It was also a top twenty cable MSO, seeking, as did other operators, to gain a presence in programming. Under terms of their agreement, Landmark owned 80 percent of the channel, Coleman 20 percent, and Coleman managed the service. The Weather Channel was launched in May 1982. It struggled initially, losing a reported $7 million in its first year of operation. Unfortunately, Landmark and Coleman came to legal blows in 1983 over the start-up costs. Landmark claimed Coleman's management was exceeding the projected and agreed on budget; Coleman denied it. Landmark sued Coleman seeking to end his management and stock agreements. Coleman sued Landmark in an effort to retain control. They settled out of court, with a stipulation that Coleman could purchase the channel for $4 million within a set period of time. He was unable to secure the financing, however, and lost his interest in the venture. He departed in August 1983. The Weather Channel was hemorrhaging money, nonetheless. The costs of maintaining the channel were far beyond anyone's original expectations and it was generating very little advertising revenue. Landmark was close to shuttering the service. The story of its eleventh hour salvation is traced below as an example of the growing symbiotic bonds between the operating and programming side of the industry. With intervention from supportive cable operators, The Weather Channel halted and reversed the negative cash flow problem and began inching toward profitability.

The Weather Channel grew in popularity to establish itself as requisite component of the basic cable lineup. Everyone, it turned out, was interested in the weather. The programmer took raw data from the National Weather Service and packaged it in a standard but attractive format, providing local forecasts, national maps, special features, and live radar, turning more than a few viewers into "Weather Channel junkies." Operators found, over time, that its carriage was a necessity for most systems. It was on systems reaching 6 million subscribers in its first six months, a figure that doubled by October 1984.[109]

Adult Programming and The Playboy Channel: November 1982

Adult-oriented fare had been appearing sporadically on cable for years. As noted previously, some of the first sexually explicit material appeared on public access channels in New York City. Commercial, adult material could be found on a few cable systems in the 1960s. Bark Lee Yee ran soft-core films on a pay-per-view basis on his system in Easton, Pennsylvania. Cablevision pioneered sexually explicit programming on satellite with Escapade. The film production company, Sartori, Inc., initiated a similar service, "Private Screenings," launching about the same time as Escapade, in late 1980. Escapade had operator backing through Cablevision and was available on Satcom I, the cable bird, but Satori used the less-desirable Westar III and lacked the operator ties, and so was handicapped from the start.

As noted, Playboy entered the business by melding with Rainbow's Escapade. The partnership was announced in August 1981. The channel continued to operate as Escapade while Playboy promotional material and then Playboy-branded content were introduced into the program lineup in late 1981 and early 1982. In November 1982, the service made the full evolution to the Playboy Channel, dropping the Escapade logo. Playboy debuted as a pay service and enjoyed some success, reaching 800,000 homes by 1985.

Penthouse magazine partnered with the Telemine Inc. production company to promote the Penthouse Entertainment Television (PET) Network in September 1981. Others included Victorian Video and Atlantis Entertainment Network, EROS, and Quality Cable Network. Depending on the service, a range of content was available, from softer, R-rated adult programming to more explicit, slightly toned-down X-rated films.

The cable industry, as a whole, could safely be described as conflicted over adult pay services. On the one hand, they were, and continue to be, fairly lucrative. Buy rates tended to be higher and more stable than mainstream films, operators typically charged more for adult services and local systems kept a greater percentage of the consumer dollar in comparison to other pay programming. But sexually explicit programming had its obvious drawbacks.

Civil authorities were not always as welcoming to adult programming as were some consumers. Local complaints often followed the announcement that a cable system planned to carry adult material. Some cities tried to restrict such content in their franchise agreements. Most operators were careful not to program hard-core X-rated films, at least in part to avoid running afoul of obscenity laws, but some cities, such as Fort Worth, Texas, additionally banned X-rated programming and required that the cable operator provide, at no charge, security devices to block the adult channels. The First Amendment generally offered protection for soft-core material, or indecent material, and the courts, in a series of decisions (reviewed in Chapter 10) extended that protection to cable operators. But the political and public relations issues associated with adult material continued to plague operators.

Adult programming, like sexually explicit literature, has always been and will continue to be a source of concern for many segments of society and an ongoing public policy issue. Cable operators, however, are part of the social fabric as well and, like the broader population, not of one mind on erotica. In determining whether to carry the Playboy Channel or a similar service, some companies exhibited few qualms. Others were uneasy. Some small operators vowed, as a personal and business philosophy, never to offer adult fare. Even some larger companies were adverse, on ideological grounds, to carrying sexually explicit programming on their systems, as the example of Westinghouse proved.

In ironic contrast to the concerns of more conservative-minded viewers and operators, Playboy and the like programmers suffered economically by not being sufficiently hard core for many viewers. In this regard, the diffusion of videotape would come to cut into Playboy's cable business even more than it impacted the mainstream movie channels. By the mid-1980s, fans of adult material could rent or buy X-rated videos, featuring footage cable operators dared not show. By the latter 1980s, the Playboy Channel would be looking to change its business model.

The Disney Channel: April 1983

At the other end of the cultural spectrum was the Disney Channel. The channel's first president, Disney veteran Jim Jimirro, explained, "We knew there was a tremendously large underserved pay community who didn't want sex or violence in their homes."[110] The year 1983 was a shakeout year for cable programmers, with services such as CBS Cable and SNC closing their doors with alarming regularity, as the glut of programmers hit the market and building stalled in the big cities. But Disney had sizable resources of several kinds. It was one of the preeminent names in family entertainment, a cultural icon in American homes. It had film vaults heavy with decades of popular programming, everything from "Fantasia" to the "Wonderful World of Disney" to the "Mickey Mouse Club." It could afford, and was willing to spend, substantial sums of money on the project. When it launched in April 1983, it was reported to have been the most expensive cable network development to date, with a reported start-up price tag of $100 million over three years.

Disney began as a pay channel. Its only commercials were for Disney-related businesses. Billed as "America's Family Network," it claimed 100,000 subscribers in its first two weeks and, by the end of 1986, was being seen in more than 3 million homes. Programming was geared principally to children, but drew family viewership in many homes. The company had arguably more experience in family entertainment that anyone and drew on that expertise, offering high-quality vintage and original programming and aggressively marketing the service. It was an attractive network for many cable operators looking to appeal to family audiences and useful in combating an image tinged by premium R-rate films and more explicit adult programming.

The channel had been developed in partnership with Group W, which was attracted by its appeal to children and families. It quickly became apparent, however, that the financial partnership, Group W was going to pay half the start-up costs, would not translate into a programming partnership. Disney jealously guarded creative control, so much so that Group W abandoned the joint venture in September 1982, before the actual launch of the service.

The Nashville Network and Country Music Television: 1983

Group W also partnered with The Nashville Network (TNN). Young, but promising, cable channels were already beginning to spawn look-alikes, networks that took a similar concept and shaped it toward a slightly different demographic. MTV was pointedly aimed at the rock and roll youth market, but country and western (C&W) music and culture was big business as well. One of the leaders in the country and western music industry was Nashville's Grand Ole' Opry. The Grand Ole Opry began in the1920s as a radio program on WSM in Nashville. WSM's parent company, NLT Corp., expanded the popular franchise, building TV facilities and a Grand Ole Opry theme park in the 1970s.[111] In January 1982, NLT announced they would team with Group W to create the Nashville Network cable channel. Group W did not have an equity interest, but rather provided satellite transponders, marketing, and promotional support. The advertiser-supported service was a general interest channel with a country and western slant. Country and western music videos were only part of the mix, which also included talk and variety shows. At about the same time, NLT was purchased by the American General insurance conglomerate (NLT itself had large holdings in the insurance business) and sought to spin off the media and entertainment divisions of the company. TNN began operations in March 1983 to 7 million cable viewers. The same year, Gaylord Broadcasting purchased the broadcast properties, theme park, and cable channel.

TNN, meanwhile, was not the only programmer looking to fill the C&W niche. One day before The Nashville Network launch, another Nashville-based network, Country Music Television (CMT), began operation. The start-up was privately held and financed by a small group of entrepreneurs and venture capitalists. Country music videos were so scarce that CMT had to produce some of its own to fill the broadcast day, and in contrast to TNN, CMT made its name early on playing mainly C&W videos. Originating from Video World International, CMT went through several ownership changes in the 1980s before being sold to Opryland USA in 1991.

Other services, of lesser notoriety, took to the heavens in the early 1980s. They included National Jewish Television in May 1981; ACTS Satellite Network Television (operated by the Southern Baptist Convention) in May 1984; and Eternal World Television Network (EWTN) in August 1981.

While not a programming service, one of the more dismal programming-related cable tales of the early 1980s was that of *TV-Cable Week*. In August 1982, Time Inc. announced a plan to start its own printed program guide, a sort of *TV Guide* for cable. It was to list all the available cable fare for each client system on a weekly basis. It was a monumental undertaking given that nearly every cable system in the country offered a different channel lineup, and the publication would have to customize a separate magazine for each of several hundred, if not several thousand, markets it hoped to serve. Time claimed it could pull off this major trick with the aid of sophisticated computing power and the brute force of large amounts of human labor. There were widespread doubts about the feasibility of the project, however, and, in this case, the nay-sayers proved correct. In addition to the technical barriers, the company faced problems of interdepartmental conflict and poor planning. The Magazine Group, which developed the plan, reportedly never talked to the company's own cable people, including Levin, about its potential. And neither did they contact cable operators who, it turned out, had little interest in promoting the publication. *TV-Cable Week* died after five months of operation and $47 million of investment.

Programmer and Operators: A Family Affair

The evolving relationship between cable operators and cable programmers was an awkward thing. If the cable industry, writ large, was still something of a family affair, then the operator-programmer relationship was akin to that of an older and younger brother, alternately protecting and thrashing each other, depending on the circumstances. One of the first difficulties, with little surprise, involved the financial relationship between the two.

Premium operators charged subscribers directly and split the fees with local operators. But, starting with WTCG, cable systems were asked to pay program providers for the service and pass along the cost to subscribers. As non-premium networks began to proliferate, most sought the dual revenue stream of advertising and cable operator compensation. Southern Satellite Services, the WTCG carrier, charged systems $0.10 per subscriber per month, and subsequent programmers followed suit. Nickelodeon, in 1980, was charging operators between $0.08 and $0.10 per subscriber per month, Calliope cost $0.02 per subscriber. CNN initially charged cable systems $0.20 a subscriber, $0.15 if they also took WTBS. In May 1983, Warner-Amex, struggling financially, announced it would start charging $0.10 a subscriber for MTV; it had previously been offered to systems for free. (The measure of Warner's difficulty may have been its move, additionally, to begin selling advertising on Nickelodeon, to great criticism from citizens groups.)

Not every cable operator and MSO welcomed the financial arrangement, and some resisted paying programmer fees, especially as the bills began to mount. TCI's John Malone, was an early skeptic with respect to the fees, but softened his position as he began to see the broader benefits of the programming services:

> I can remember at one NCTA meeting, I was on a panel and it was in the, I think, in the very early '80s and I had changed my mind (on fees). I had really fought tooth and nail with Drew Lewis to prevent MTV from charging us for MTV and the reason was I didn't think they needed the money. They were doing quite well without a subscription fee. But I basically said our company was willing to entertain subscription fees for basic cable channels if the channels fit a unique niche because what we could see was the positive economics of driving subscriber penetration by filling niches and so after that, not only were we willing to talk to people about that, but we were willing to invest our capital in programming ideas that had this dual stream of income.

Even once they bowed to the necessity, and even the long-term benefits, of programmer fees, operators nonetheless sought to keep the costs as low as they could, and contract negotiations between them and the programmers took some interesting turns.

In an effort to gain dwindling channel capacity on the nation's cable systems, some networks, in the early 1980s, began to offer their services for free, hoping to undercut existing fee-based competitors. SNC, noted above, was one of the most prominent examples, offering to pay operators for carriage. ESPN even attempted a short-lived experiment to adopt a broadcast network model of affiliate compensation for the industry. ESPN, in the late 1970s, was charging operators about $0.04 a subscriber (and asking for a five-year commitment for $2.40 per subscriber). But the network's management, as noted in Chapter 8, had been drawn from NBC Sports, and the broadcast networks had a very different relationship with their affiliates, historically paying affiliate compensation to their stations and securing all their revenue through national advertising. ESPN thought cable networks could become fully sustainable through advertising revenue, as well, and follow the same path. ESPN, therefore, declared in mid-1981 that it would begin paying cable operators $0.10 a subscriber a year for carriage. But, as noted previously, the advertising community was not yet ready to spend as lavishly on cable as it did on network broadcasting. And while ESPN was popular, it was also very expensive to program. The sports channel had penned contracts with college football, professional tennis, and the National Basketball Association. By 1982, ESPN was looking at a $41 million loss. Little could be cut in programming, and it was becoming clear that advertising revenue would be insufficient to even cover costs. In January 1983, ESPN President William Grimes announced that they had buried the idea of a fully advertising-supported network. Instead of giving operators $0.10 a subscriber per year, ESPN reverted to the more established formula, and began seeking $0.10 per subscriber per month from operators, a strong increase over its prior fee.

It was not an easy sell, and it did not help that operators were themselves feeling the financial pinch of the 1982 slowdown. Grimes and Finance Vice President Roger Werner went to visit operators. Their first call was to Cablevision's Chuck Dolan, who promptly threw them out of his office. ESPN had to see it through, however, so they told Dolan to pay the fee or stop carrying the network; he refused, prompting the programmer to obtain a restraining order against Cablevision.

TCI's John Malone took a different approach. He not only threatened to take ESPN off his systems, but he opened talks with brewery giant Anheuser-Busch about starting a new sports channel to replace ESPN. Over a year and a half of often-brutal negotiations, ESPN came to settle with every major MSO, including Cablevision and TCI, although their initial asking price of $0.10 per subscriber was cut in half by the time contracts were signed.

A similar chess game was played out in 1984, featuring Malone, Turner, and MTV. MTV had been under attack from some quarters for programming too much sex and violence through its music videos. Notwithstanding, its financial success prompted it to propose an expansion. It was preparing a new service aimed at an older demographic, called VH-1 (Video Hits-1). In August 1984, Turner announced he would begin his own music video channel, in competition with Warner-Amex. It was an eerie echo of his brief Tango with SNC. Using SNC's tactics, Turner told operators he would supply the service for free for five years if they carried it from its inception, and—in reaction to MTV's critics—he would shy away from videos with too much sex and violence. On October 26, 1984, two months before the debut of VH-1, Turner began his Cable Music Channel.

In fact, it was all largely a charade designed to forestall a pending license fee hike by MTV. As Turner explained many years later, operators, especially Malone, had gone to Turner with a request that he begin a competitive MTV-like service. "Malone and a lot of the bigger cable operators felt they were getting screwed by MTV," recalled Turner. "Malone said, 'Ted, do us a favor. Start a music channel and announce that you're not going to charge any fee.'"[112] In outward appearances, the Cable Music Channel was a terrific failure, launching to only 350,000 cable homes and lasting only thirty-six days. By November, Turner sold the service to MTV for $1 million. But it had served its purpose as a bargaining chip for operators in the negotiations with MTV, and the operators were grateful.

Despite the posturing operators and programmers, most, including Malone, realized that subscriber fees were necessary for the survival of the industry. The death of several of the better-financed cable programming start-ups in 1982 and 1983, especially CBS Cable, was sobering for the industry. It demonstrated, for the first time, that one could not simply throw a new network up on the satellite and assume that it would succeed on advertising or a pay basis alone. Operators already knew that they would need the programming offered by new networks if they were to fill their new channel capacity and succeed in the cities. Now, they began to understand that they might have to help pay for it.

The story of The Weather Channel's last-minute rescue from the brink of bankruptcy is a case in point. To help move the channel more quickly toward profitability, Landmark presented cable operators, beginning with Bill Daniels, a simple proposition: either the struggling network would need to begin charging a modest ($0.05 per subscriber) subscriber fee, or it would have to go dark. Advertising revenues would not be sufficient to run the channel, so it began charging operators for the service, asking $0.03 to $0.05 a subscriber per month. The major MSOs accepted the proposal, keeping The Weather Channel alive and a growing number of cable subscribers with one more cable channel.[113]

Copyright also created a mutual problem for operators and superstation programmers such as Turner in 1982. While copyright holders were generally gratified by passage of the 1976 Copyright Act, many felt the fee structure did not adequately compensate them for their programming. Stakeholders, including the NAB, MPAA, ASCAP, and Broadcast Music, Inc. (BMI), lobbied for an increase in rates. Some went further, suggesting the compulsory license itself be abolished and replaced with system-by-system negotiated fees. Efforts by the Copyright Tribunal to disperse compensation funds among the various claimants met repeatedly with court challenges. Parties disputed the allocation formula and the first scheduled disbursement was delayed two years, until 1980, while the validity of the Tribunal's distribution plan worked its way through the courts.[114]

The Tribunal was also given the power to adjust rates and, in January 1980, began an effort to raise them following the FCC's removal of the distant signal importation rules. The panel explained that the rate hike was designed to compensate copyright holders for the increased carriage of distant signals following the removal of the restrictions. Copyright holders and cable operators protested in court.[115]

In 1981, Chairman of the Copyright Tribunal, Clarence James, resigned citing differences with the rest of the panel and urging, as he left, elimination of the compulsory license. Cable operators also sought relief from Congress. In September 1982, the House of Representatives passed a bill (HR 5949) that restructured copyright liability, codified existing FCC must-carry rules, and reimposed syndicated exclusivity restrictions. The bill ultimately died but, as part of the process, the House also stayed the proposed rate increase until the courts determined its legitimacy (or until March 15, 1983, which came first). In court, support went to the Tribunal,[116] and, in October 1982, the CRT set a new 3.75 percent rate (a percentage of the system's basic revenues) on each distant signal imported for large operators. The prior rates ranged from 0.2 percent to just under 1 percent.[117] The increase boosted total copyright bills for some systems by as much as 300 percent. The panel also imposed a special surcharge for systems in the top 200 markets to help offset loss of blackout protection under the old syndicated exclusivity rules.[118] As a result many operators, by some reports hundreds, dropped their distant signals. In most cases, the stations eliminated first were the satellite-distributed WOR-TV and WGN-TV, followed by the regional imports. Tuner's superstation was the last to go, if it went at all.

What may have been bad for the imports, however, turned out to be beneficial to the start-up cable networks and perhaps in the long term to the industry. As cable operators dropped imported regional stations, they replaced them with young satellite-delivered cable channels, such as The Nashville Network, Cable Health Network, MTV and CNN Headline News. "Despite all the yelling, the sky isn't falling," said one cable executive, "we'll live through it. We've lived through worse."[119]

It was a telling and prescient comment. The social construction of technology is by no means always hopeful. Pessimistic, even dystopian, views are nearly as common as utopian ones. As the market slumped and programming start-ups seemed to be folding nearly everyday, it was easy to be glum. By 1984, some in the industry suspected the market could support no more than ten national, basic cable program services and perhaps three or four pay services. "We have seen just about all the programming ideas that will prove viable," predicted Turner Broadcasting President Robert Wussler, "There may be something lurking in the woods, but I doubt it."[120]

In fact, the woods were alive with them. The market would turn around and programming ideas continue to multiply. In 1981, eighteen basic and five pay services were proposed or launched, and, in 1982, four more basic and one more pay service. By the end of 1985, cable operators could choose from more than a dozen premium services, five pay-per-view channels, and some forty basic services. Between 1975 and 1987, the number of satellite-delivered cable networks rose from one to more than seventy.

Few of the basic services were making money, but many were prepared to stick it out. News, sports, movies, programming for women, children, African-Americans, teenagers, and Hispanics was available. In time, the special interest services would include networks featuring 24 hours of cooking, animal shows, home improvement, computer programs, history, health, and travel, to name only a few.

Cable began to climb out of its new round of difficulties in 1983 and 1984. There were a number of reasons, including the resolution of most of the franchising battles, the start of major market construction, the dismal performance of the would-be competitors, and an easing of interest rates. Lenders found cable a more attractive prospect once again and, in no small part, because of the prospect of yet another round of federal deregulation, this time one of historic proportions.

Free at Last: The Cable Communications Policy Act of 1984

The closest Congress had ever come to passing comprehensive cable legislation was the one-vote defeat of S. 2653 in 1960. Subsequent cable bills had been introduced in the 1960s and 1970s, but none penetrated deeply into the legislative process. The Cable Communications Policy Act of 1984 was historic insofar as it was the first time in cable's 34-year history that Congress sanctioned federal control over the medium. It was, at the same time, a very different kind of bill than S. 2653 or any similar proposal designed to authorize the regulation of cable. In fact, the primary thrust of the Act of 1984 was to complete, and in a sense cap, the deregulation of cable that had begun a decade before. More than anything else, the Act freed cable from the last important vestiges of government control and paved the way for much of its explosive growth in the latter part of the decade.

The Act had its roots in the late 1970s. The legislative review of telecommunication policy, the basement-to-penthouse survey initiated by VanDeerlin, never completely vanished, despite VanDeerlin's departure from Congress. Through the late 1970s, the Senate had responded to VanDeerlin's work with legislative proposals of its own and, although none came immediately to fruition, they offered a glimpse of the philosophical direction of the upper house, which in keeping with the times was generally deregulatory.

The 1980 elections, in addition to bringing Reagan and Fowler to power, gave the Senate to the Republicans and placed conservative Arizona Senator Barry Goldwater in the chairmanship of the Senate Communications Subcommittee. Goldwater by then was a senior statesman in the upper house. He did not personally care much for television or its programming, thinking most of what was on it was in poor taste. Asked if he thought the FCC should do anything about it, he held true to his conservative roots. The government should stay out of it, he told broadcasting magazine, "it's not our job" to regulate content.[121] He was relatively new to telecommunications policy, having spent much of his career on foreign relations and broader domestic issues. But his laissez-faire philosophy favored businesses of all stripes, broadcasters as well as cable operators, and Goldwater fit comfortably in the caravan of telecommunications deregulation. Fellow Republican Bob Packwood (R-Oregon) rose to chair of the Senate Commerce Committee, parent of the Telecommunications Subcommittee, and also would become a strong cable ally.

On the House side, Van Deerlin's defeat left a vacancy in the Communications Subcommittee chair, which was filled by Timothy Wirth, a Democratic from Colorado.[122] (The new House Commerce Committee chair was John Dingell [D-Michigan]). Wirth's interest in telecommunications was more than casual. Denver was now the economic epicenter of the cable TV industry, home to some of cable's most powerful players, including Bill Daniels and Associates, TCI, ATC, Cablecom-General, Jones Intercable, and United, and Wirth would look after cable industry interests.

The FCC, over the previous six years, had largely stripped away most of the federal controls cable found burdensome and what was left, principally must-carry, was being challenged in the courts. The chief regulatory threat for cable was coming from the cities, where the industry was locked in an expensive political jousting contest, trying to acquire franchises and build systems without going bankrupt. In the vacuum caused by retreating federal oversight, the cities had moved in to adopt local rules governing everything from system size to programming content, subscriber rates, and ownership. Cable, in response, sought relief from growing local and state control. Led by the NCTA, the industry went to Congress and to friendly members, such as Goldwater, to help enervate or at least blunt municipal authority over cable. The cities responded in kind.

While individual companies negotiated with city officials across the country, the cable industry and the nation's municipalities, therefore, organized broader political campaigns, marshalling their forces for a national face-off. The cities had chaffed at FCC preemption before. In the early 1970s, the head of the National League of Cities (NLC) argued that there should be no limit on the amount a city could charge a cable operator for a local franchise. The FCC's near-complete usurpation of cable regulation, coupled with the moribund state of franchising and construction in the mid-1970s, took cable off the agenda of many city council meetings. But things had changed by 1980. As noted, the Commission had spent the last half of the 1970s dismantling its cable rules, completing the task in July 1980 by eliminating the remaining substantial regulations on signal importation. Receding federal oversight and the land rush of big-city franchising in the late 1970s and early 1980s made cable a focus of municipal debate. The NLC formed a special Task Force on Cable Television and Telecommunications, headed by Seattle Mayor Charles Royer. Royer and the Task Force spearheaded (NLC) activities designed to organize the nation's municipalities, provide information and support in the cable franchising process, and at the same time enter into discussions with the industry about possible congressional legislation.

In May 1980, the NLC opened talks with the NCTA aimed at writing a code of franchising practices that would help create a smoother and more equitable process for all parties. Those talks broke down in the summer of 1980 when the NTCA declared its support for Senate Bill S. 2827, legislation introduced by Goldwater that, among other things, was designed to inhibit local power over cable. The bill would have restricted the cities' ability to regulate rates, to require program origination, to access channels, or to control programming content. It also included controversial proposals to deregulate AT&T. The bill did not move very far in that session of Congress, but everyone expected it to reappear the following term. The NLC, angry at what they perceived as duplicity on the part of cable, not only halted negotiations with cable interests, but also threatened a nationwide moratorium on franchising. A few cities, in response, debated shutting down their local franchise process, but none did.

Moving ahead unilaterally, the NLC by 1981 had drafted its own voluntary "Code of Cable Television Franchising Conduct" designed to help guide cities through the franchising morass and promote a clean and equitable process. The code also served as a kind of municipal cable manifesto, declaring that cities had inherent state power to issue licenses, to control rates, and to require assorted access channels. The NLC intensified its campaign for federal legislation that would formalize and legitimize those asserted powers, but legislative proposals did not always work to the cities' favor.

In May 1981 a growing NCTA moved into new headquarters in Washington, D.C., a spacious seven-story building on Massachusetts Avenue, know locally as Embassy Row. It was a symbol of the success of the industry and its growing need for effective representation in the nation's capital. The NCTA lobbyists did not miss a beat while unpacking their bags. In 1981, the Senate considered legislation, S. 898, which would have taken a significant step toward the deregulation of the telephone industry, allowing AT&T into certain, traditionally unregulated, lines of business. Shortly before the bill was to come to a vote in the full Senate, amendments were introduced by Packwood that would have protected cable operators from local rate regulation and in other ways have reduced municipal power over cable activities. The NLC cried foul, and Senator Ernest Hollings (D-South Carolina) launched a filibuster in protest. Goldwater supported a motion to strip the bill of the offending cable provisions. He argued that the issue of the relationship between cable and the cities deserved its own set of public hearings and a separate review. With the language on cable pulled, the Senate passed S. 898 in October 1981. In December, Congressman Tim Wirth (D-Colorado) introduced a companion measure in the House, HR 5158, the Telecommunications Act of 1981. The bill passed through committee but later died in the House, in large part because of the August 1982 Modified Final Judgment, which contained many of the provisions of the proposed act.

While the House was considering the telephone legislation, Goldwater, in March 1982, made good on his promise to look separately at cable, introducing S. 2172, the Cable Telecommunications Act of 1982. The proposal prohibited foreign ownership of cable (an action directed at Canadian cable companies) unless cable operators were given reciprocal access to foreign markets. It prohibited telephone company ownership of cable systems in their service areas (codifying existing FCC regulations). It required systems with twenty channels or more to set aside 10 percent for public access and 10 percent for leased access; gave cities authority to set rates for basic service and public access but not for pay service, enhanced service, or leased access. The Bill included a signal theft protection clause. It said cities could own and operate cable systems but had to buy them at fair market value. In also provided for automatic renewal of the franchise if the operator had met all the conditions of the agreement. The measure was reported out of Communications Subcommittee, but had received no action from the full Senate as the 1982 legislative session was coming to a close. In an effort to bring the bill to a vote, cable operators—some 300 strong—descended on Capitol Hill in September 1982, lobbying their Senators and arguing their cause. The NLC conducted its own campaign to shape a bill to its liking, albeit it at a somewhat lower political decibel level.

Despite their efforts, S. 2172 died in the lame duck 97th Congress. Goldwater introduced its progeny, S. 66, in January 1983 in the new 98th

Congress. Meanwhile, at the urging of all parties, the NCTA, NLC, and the National Conference of Mayors sat down to negotiations. They met through March 1983 and brought forth an agreement on legislation that Goldwater incorporated into the new bill. The Senate passed S. 66 June 14, 1983, coincidentally during the NCTA convention in Houston. Wirth submitted a companion bill, HR 4103 in the House later that year.

Some of the nation's largest cities balked at supporting the NLC compromise, however, and continued their lobbying in opposition. Under pressure from several big city mayors, the NLC reversed course in late 1983, withdrew its support of the measure, and began working in the House to derail the bill. As a result, the House version stalled in committee, much to the displeasure its author, Wirth. Congressman Dingell (D-Michigan), Chairman of the parent Energy and Commerce Committee, convinced the NCTA and NLC to return to the bargaining table in early 1984.

In the midst of talks, NCTA President Tom Wheeler stepped down from his post to take a job in industry.[123] He immediately was succeeded by association Vice President James Mooney. The impact of the transition on the talks is uncertain, but this time the two sides struck a compromise that would hold. The Telecommunications Subcommittee eased the bill's restraints on municipal control, added a leased access channel requirement, and in the end, approved the legislation. The Energy and Commerce Committee passed the proposal on to the full House, which approved it October 11, 1984, in the final days of its session. A last-minute squabble over details of an equal employment opportunity (EEO) provision threatened once again to derail the bill, but supporters of a section stipulating the percentage of women and minorities that cable operators had to hire backed away. The spirit of the EEO rules was kept, although it was left toothless. The bill finally made it to the floor, in large part thanks to the efforts of Wirth. Cable operators were grateful. "To this day," TCI's Malone recalled later, "I remember thinking all was lost, and then at the end he (Wirth) pulled it out in the congressional session."[124] Differences between the Senate and House versions were resolved in conference committee and public law 98–549, the Cable Communications Policy Act of 1984 was placed in effect.

As with many good compromises, the final bill held something of value for both sides. For the cities, it provided a statutory stamp of approval on the franchising processes, the core authority of the cities to issue licenses and exercise some measure legal oversight. It permitted local authorities to levy franchises fees, but capped them at 5 percent of gross revenues.[125] The Act also permitted cities to require access channels for public, educational, and governmental entities ("PEG" channels).[126] The Act prohibited cable "red-lining," or denying service on the basis of income. It required operators to provide for sale or lease parental lock boxes to safeguard children from adult channels. And the Act "takes the FCC out of our hair," NLC legislative counsel Cynthia Pols told reporters. "The biggest set-back is rate regulation, but what we got there is better than what the FCC has already attempted to give us recently."[127]

For cable, the new law prohibited local authorities from tampering with programming decisions; they would not be permitted to require specific networks, such as Nickelodeon, in exchange for the right to wire the city. Nor would cities be permitted to ban certain program networks, such as the adult services. The 1984 Act provided for nearly automatic franchise renewals; renewal was required unless the cable operator failed to abide by a specific list of basic service and legal requirements. And, in reaction to some of the more egregious abuses of the franchising process, the new law required that any franchise-mandated facilities or services had to be cable-related—tree-planting and firehouse additions did not count. A number of smaller issues also were addressed. Codifying prior FCC rules, telephone companies and television station owners were forbidden to own cable systems in their service area. Theft of cable service became subject to civil and criminal penalties (up to six months in prison and a $1,000 fine for an offense). Subscriber privacy was protected by regulating the collection and use of personal customer information. In an amendment introduced by Congressman Al Gore (D-Tennessee), TVRO owners were permitted to receive unscrambled cable satellite signals.[128] The FCC's EEO broadcast guidelines were extended to cable.

Most importantly, the Act prohibited state and federal regulation of subscriber rates, and authorized local rate regulation only in cases where effective competition did not exist. The Act then left it to the FCC to determine what constituted "effective competition." The standard chosen by the Commission was coverage by the grade B contour of at least three broadcast TV stations, a situation that applied to about 97 percent of all cable systems in the country.[129] The FCC had effectively insulated nearly every cable system in the country from local price controls.[130] Cable had been subject to franchise agreements that specified rates for basic services, pay channels, installation fees, and service charges for years.[131] As of December 29, 1986, when the clause would take effect, cable would be free to charge whatever the market would bear.

The Act left one regulation intact, the obligation to carry all available local broadcast signals. While cable operators, for obvious business reasons, normally carried the prominent local signals—in fact, it would have been foolish not to carry the network affiliates and major independents—the must-carry rules still presented occasional problems. Channel capacity was increasing but still limited, and a growing number of programming services were offering attractive, specialized and general programming options to customers. Carriage of all local broadcast signals meant providing space for even the weakest, infrequently viewed local independents. For older systems with fewer available channels, this acted to consume capacity and sometimes prevent the carriage of a more attractive basic service. Cable operators, therefore, challenged the must-carry law, claiming, in part, that it violated their First Amendment right of editorial control over their medium (discussed further in Chapter 10).

The Act also codified several existing FCC regulations and made a few extant rulemaking proceedings moot. The agency, as a result, closed out several dockets, including those dealing with telephone-cable cross-ownership, regulation of basic rates, and a proposal to abolish the Commission's franchise fee cap. The same day the Congress passed the Act, it also passed legislation heavily insulating cities from antitrust suits of the type made possible by the *Boulder* case.[132]

With the passage of the 1984 Cable Communications Act, most of the important pieces were in place for a new surge of industry activity. Construction was accelerating, new programmers were lining up for a spot on the ever-expanding cable dial. The largest cable networks now were beginning to think in global terms. In 1983, Time formed HBO International, partnering with Columbia Pictures, 20th Century Fox, and Goldcrest Films and Television Ltd., to bring product to the United Kingdom. At the same time, Ted Turner began exporting CNN to Japan and Australia. Encouraged by his successes there, Turner signed a pact in early 1985 with Comsat and British Telecom International to supply a 24-hour news feed via satellite to hotels, U.S. Embassies, and cable systems across Europe. His satellite footprint stretched from Scotland to North Africa, Madrid to Moscow. The news that his service delivered would not always be good. In fact, CNN would, over the next several years, provide live, unedited coverage of heart-wrenching disasters, as well as coverage of events that realigned global politics. In all cases, however, as cable programming matured and spread, the whole world began watching.

10
The Cable Boom (1985–1992)

> *I'm suppose to be so brave and strong, but when I read that CBS and HBO and Columbia Pictures—which is Coca-Cola—were all getting together, and Paramount and Warner and MCA and Showtime and Movie Channel were getting together and, of course, ABC and Group W are getting together—those are some pretty powerful combinations when they're all zeroing in on your area. It's like a big war occurring and none of the players are big enough to play alone.*
>
> —TED TURNER, 1983[1]

Just after 11:30 AM on Tuesday, January 28, 1986, the Space Shuttle "Challenger" began its ascent from the launching pad at Cape Canaveral, Florida. It carried a crew of seven, including the nation's first civilian astronaut, high school teacher Christa McAuliffe. The booster and shuttle rose gracefully from the gantry, soaring spaceward, but at one minute into the flight, the shuttle burst into flame and exploded, as horrified crowds watch from the ground below. The launch was being televised live that morning and viewers across the nation, including McAuliffe's students and colleagues, watched in shocked disbelief, then anguish. As it had in the past, the nation turned to television, first for information and then, perhaps, for some sense of collective reassurance. Less than half the nation had access to immediate, live coverage, however. NBC, CBS, and ABC had been running their normal program schedule that morning. Only one major network carried the launch live, and as word of the disaster spread, millions turned to the Cable News Network.[2] CNN anchors Mary Ann Loughlin and Tom Minter stayed on the air, reporting the story for more than 13 hours.

The explosion was a personal and national tragedy, and a setback for NASA's space program. But it also marked an inflection point in the history of cable television. Recalled Loughlin later, "Suddenly every news editor in the world was tuning in to our broadcast. We became a global news service at that moment."[3]

Cable in 1986 was in about 37 million U.S. homes, 43 percent of all television households. CNN had potential viewers in 33.5 million of those homes, and the network's presence and importance were beginning to expand beyond its subscriber count. CNN was subsequently on the scene of the student uprising in Beijing's Tiananmen Square in June 1989. In November of that year it ran 24-hour coverage of the fall of the Berlin Wall. CNN followed the U.S. invasion of Panama in December 1989, the release of Nelson Mandela from a South Africa prison in February 1990, Boris Yeltsin's coup toppling the Soviet government in August 1991, and the controversial Senate confirmation hearings of Supreme Court nominee Clarence Thomas in October 1991.

CNN's reputation for quick reaction to breaking news and lengthy follow-up coverage grew in harness with its penetration rates. By 1991, cable was in 60 percent of the nation's TV homes. At the turn of the decade, when a major story broke, more than half the country could turn immediately to CNN, and the country made full use of the opportunity in the first month of that year.

The picture is etched forever into the history of international relations, broadcast journalism, and cable television. On the night of January 16, 1991, antiaircraft fire blazed into the darkened skies over Baghdad. U.S. forces had begun bombing the city and the Iraqi military was returning fire. Reporters from the world's largest news organizations were on the scene, many holed up in the Al-Rashid Hotel, with a clear view of Baghdad's night skyline. But, the Iraqi government had been growing suspicious of the world press and, as hostilities began, cut their telephone privileges. Only one news agency had a line out of the country that night—Ted Turner's Cable News Network. The other news agencies established sporadic communication, but for the next 16 hours, only CNN maintained continuous coverage from its correspondents, Bernard Shaw, Peter Arnett, and John Holliman. During those 16 hours, and for days afterward, the world watched and listened to CNN.[4]

Soon, video of the night raid was being aired on all three national networks. Vietnam had been the television war; pictures brought home the stunning brutality of war in a way never before experienced, but those images were usually a day or more old. In Iraq, for the first time, an entire country was watching war break out, transmitted from the enemy capital, in real time. The war itself offered an amazing display of battlefield technology, but the news coverage also marked an evolutionary step in journalism, in cable television, and in the mass media system.

As Operation Desert Storm rolled across American TV screens, those screens were as often as not tuned to CNN. In the first hours of the U.S. assault, more Americans were watching CNN than any other network. It was not just the average family in Spokane that was affixed to the screen. By 1991, the cable service once known as the "Chicken Noodle Network" was a first-choice source for breaking news among the world's political leaders and, as the bombs began falling on Baghdad, Gilbert Lavoie, the press secretary to Canada's Prime Minister Brian Mulroney, called the White House to speak to Marlin Fitzwater, the president's press secretary. Comparing notes, they found they were both were watching CNN. In Washington, D.C., in Moscow, in London, at the Vatican, and in Baghdad itself, presidents and princes relied on CNN to stay current on world events. CNN had begun sending its signal abroad in 1985 and, by the end of the decade, was beamed into more than 100 countries. Great Britain's Foreign Secretary refused to stay in a hotel room that did not have CNN. Then President George Bush once quipped, "I learn more from CNN than I do from the CIA."[5]

By the start of the new decade CNN had become a tool of state in the sensitive game of international diplomacy. The network was used as an instant messaging service to communicate policy positions to political allies and opponents around the world. In the 1992 U.S. presidential race, incumbent George Bush and challenger Bill Clinton appeared frequently on CNN, making heavy use of the network's popular Larry King Live show to present themselves to voters. The King program also became a media springboard for surprise third party candidate Ross Perot.

By taking the point in international news, CNN, over the course of the mid- and late 1980s, had ascended to a position in which the country and the government were in many important ways reliant on the 24-hour service. And with a majority of Americans subscribing to cable, full awareness of that reliance hit home with coverage of the Gulf War. CNN's coverage of Desert Storm, in one sense, was an airburst of social self-awareness, a sudden jolt of realization that the Cable News Network had become a primary source of news for millions.

In January 1992, *Time* magazine named Ted Turner its "1991 Man of the Year," dubbing him "Prince of the Global Village," primarily for his development of the Cable News Network. CNN, stated the magazine, had changed the nature of global communications, adding: "That change in communication has in turn affected journalism, intelligence gathering, economics, diplomacy and even, in the minds of some scholars, the very concept of what it is to be a nation."[6]

The Gulf War coverage spoke not just to CNN's social standing, but perhaps more broadly to the emerging social influence of cable television generally. After all, CNN was the "cable" news network, and what CNN's success brought with it was a concomitant realization that cable, in a number of ways, had woven its way into the fabric of American social and political life. CNN was but one of the more prominent manifestations of this growing use of, and even dependency on, the new television infrastructure. While the whole world watched CNN, large and important segments also watched C-SPAN, ESPN, Nickelodeon, and MTV.

In terms of cultural impact, MTV might have been second only to CNN. Nearly from the start, MTV videos had provided a flashpoint of controversy over sex and, to a lesser extent, violence in music videos. Young people flocked to MTV in the 1980s the way they flocked to rock 'n' roll on AM radio in the late 1950s and to FM and record albums in the 1960s. It was a powerful force in the music and record business; playtime on MTV was a critical factor in the success of a song or an artist. MTV airplay drove record sales as it drove pop culture. MTV and music videos set the style in clothing, hair, fashion, language, and behavior for millions of teens and preteens, first in the United states and then, as it expanded internationally, around the world.

Community antenna television in the 1960s and early 1970s reached only a tiny segment of the American population; most of its customers were in smaller towns and the service was primarily just a means to receive the TV programs the rest of the country got with rooftop antennas. For most Americans, CATV was not a particularly interesting, and certainly not a very important, part of their lives. By the end of the 1980s, however, that had changed. Millions were becoming accustomed to turning to The Weather Channel for the latest storm report, ESPN for a quick check of sports scores, and FNN or the Consumer News and Business Channel (CNBC) for an update on the stock market.

The period was not without its difficulties for the industry. As detailed in the next chapter, the price of cable ballooned and service deteriorated while subscription figures grew. By the start of the next decade, the industry's often-aggressive pricing policy and frequent disregard for its customers would end up costing operators millions. But the problems with rates and service constituted the distracting, albeit important, surface turbulence, artifacts of the deeper penetration of the technology and the business into the national culture and psyche.

The mid- and late 1980s was the period in which cable came to maturation in the United States, establishing itself firmly as a communications medium in its own right and a medium on which citizens would come to rely for the majority of their news, information, and entertainment. It was the period of the cable boom and the subject of this chapter. The boom had several important characteristics, including the massive physical deployment of cable infrastructure, and continued industrial consolidation, as the ever-present pressure for concentration of ownership ground away, and the relationships between operators and programmers grew closer and more complex. It was also characterized by yet another reformulation of the social definition of cable television, a definition that, as always, had implications for business and policy. Because of the accelerated growth of cable in the mid- and late 1980s and the consequent implications for policy, consideration of the period is covered over both this chapter and the next, with the current chapter focusing on the conceptual, business, and industry dimensions of cable growth and Chapter 11 looking at the related issues of politics and regulation.

The Cable Boom

By the mid- 1980s, most of the big city franchise fights had been settled, and the smoke was clearing on the trailing lawsuits. With passage of the 1984 Act, cable was largely free from any meaningful regulation, and the competitive media that some were counting on to keep cable in check had yet to materialize. In medium and large markets across the country, crews were stringing wire. From Boston to Seattle, pole by pole, trench by trench, the golden coaxial thread was rolling into neighborhoods, past homes and apartment buildings. Abel Cable had finally made it to the big city and was greeted, at least initially, with open arms. Houses passed in the major markets and thereby the nation rose at the fastest pace in the industry's history, and so did penetration. In 1978, cable passed 27.2 million homes, about 36 percent of all TV households, and about 14 million subscribed to the service. In 1980, 37 million homes had access to cable television and about 24 percent of all television homes subscribed. By 1985, cable passed nearly 75 percent of all homes and subscribership had more than doubled to 46 percent. Estimates vary on exactly when cable reached and passed the halfway point, signing 50 percent of the nation's television homes to the service, but it probably was sometime in late 1987 or early 1988. By then, penetration reached 41 percent in New York City, 57 percent in Boston, 48 percent in Philadelphia, 34 percent in Chicago, 53 percent in San Francisco-Oakland, and 39 percent in Los Angeles.[7]

Cable, with its expanding menu of specialty programming, also made substantial gains in mid and smaller markets that had not been thought to be financially attractive when cable was only a retransmission service. The number of cable systems nationwide more than doubled, from a reported 4,225 in 1980 to 9,575 by 1990.[8]

With virgin, uncabled territory shrinking to the vanishing point and cash flow accelerating, some operators turned their attention to rebuilds, tearing out aged equipment and upgrading their technology. The need to expand channel capacity to meet the press of increasing numbers of programming services and growing public expectations also drove the reconstruction. Some operators worried that if they did not renovate their 270-MHz systems, an overbuilder would appear at the outskirts of town prepared to lure away customers with expanded service. FCC regulations tightening signal leakage standards, and in many cases local franchise renewal agreements that required increased channel capacity, also spurred reconstruction. In the mid- and late 1980s, industry spending on rebuilds and new construction was running between $1 billion and $1.5 billion a year.[9] Some operators also sought to consolidate their holdings in targeted markets. In early 1985, the FCC approved a system swap between Times Mirror and Storer, setting a precedent for similar trades that helped accelerate clustering in the larger markets.

Expanding Capacity

The foundation for cable's success was, as it had been from the start, the ability to offer more TV content and greater program variety, and this, in turn, was made possible by steady, incremental improvements in technology. New construction in the early and mid-1980s was typically 400 MHz, about fifty to fifty-four channels. In 1984, Times Fibre introduced a full line of 600-MHz equipment (in practice 550 MHz), and two years later a line of 1-GHz technology. By 1987, the standard for new rebuilt systems in the larger markets was 550 MHz, about seventy to eighty channels (although smaller markets usually upgraded to 400-MHz equipment). In 1983, only about 180 cable systems in the country had fifty-four channels or more; that figure rose to nearly 1,000 by the end of the decade (and see Appendix B).

Along with the increase in system capacity came an expansion in the number of available satellites and satellite transponders. By the mid-1980s, satellites operated by Hughes, General Electric/RCA, and MCI were leasing carriage to the growing ranks of cable programmers. In 1980, the industry had about twenty-seven basic and pay cable services. By 1990, operators could choose from sixty-five basic networks, five premium channels, and five pay-per-view services.[10]

Work in fiber technology also advanced significantly. As noted, fiber had been praised in the mid-1970s for its increased carrying capacity and superior picture clarity. Signal propagation in fiber, however, relied on laser or LED devices that used digital or FM analog modulation standards, whereas traditional television signals were transmitted using AM techniques,[11] which were more difficult to achieve for the laser engineers. In 1986 and 1987, a team of Denver-based ATC engineers working on fiber applications discovered that a new laser device from AT&T could successfully transmit up to forty AM TV channels over ten kilometers without additional amplification. ATC worked with Anixter Purzan, cable's largest equipment supplier at the time, to create a version of the distributed feedback laser suitable for general use in cable. Anixter called the new AM LED the Laserlink. The first one was installed in ATC's Orlando, Florida system in 1988. As is usually the case, the cost of the new technology was high—$100,000 per device—but research, development, and manufacturing refinements, plus larger-scale production, soon brought the price down to a level that made economic sense for operators. By the mid-1990s, the AM lasers were selling for less than $5,000 each. Performance, at the same time, increased to the point where a typical optical transmitter could carry eighty analog channels and several hundred compressed digital channels.[12]

Anixter continued to expand, buying smaller equipment manufacturing and distribution companies and, changing its name to ANTEC (from Anixter Technologies), was retailing its new equipment to much of the industry by the end of the 1990s. Telephone companies looking to explore the possibilities of cable TV delivery (see details below) also were experimenting with AM fiber systems. Pennsylvania Bell, U.S. West, and independent United Telephone, among others, were thinking about or implementing trials in 1988, further stimulating interest on the part of cable operators fearful of loosing ground to the telcos.

As AM fiber use grew, the nature of classic cable system architecture began to change as well. Initially, AM fiber was used largely for long-haul feeds. Because of its low attenuation, signals could be sent over longer distances using fewer distortion-inducing amplifiers than was possible with traditional techniques, lowering overall costs. AML links had been employed to send signals to mini-hubs (cable subsystems) but AML was susceptible to rain fade; fiber was not. As fiber was deployed, and experience with it grew through the late 1980s and early 1990s, a new cable architecture began to take shape. Fiber was strung for the primary "backbone" of the cable system, feeding signals from the head-end into major suburbs or neighborhoods around town. Where the system began branching off into smaller and more numerous tributaries, feeding individual streets and homes, coaxial cable took over the job. This combination of fiber and coaxial technology became known as "hybrid fiber coax" (HFC). The largest MSOs, ATC and TCI, praised and promoted fiber and took a lead in buying and installing it. TCI, Time Warner, and Viacom all were building HFC systems by 1992.

The telephone companies also spent billions of dollars in the 1980s and early 1990s replacing their metal with glass. Long-distance telephone companies had installed 95,000 miles of fiber by 1992 and regional telephone companies 188,000 miles.[13]

Expanding Finances

In the mid 1980s, the cable industry faced significant pressure to increase its cash flow. It was highly leveraged, having secured massive financing for expansion. A large portion of the construction in the major markets was finished by the late 1980s. Nonetheless, the debt incurred in that construction remained and cable had to show reasonable return on investment if it was going to service its debt and continue to develop the programming it now required to sell the service. Fortunately, in the eyes of many in the industry and on Wall Street, that opportunity came December 29, 1986, when cable operators were no longer constrained by controls on their subscriber rates.

Passage of the 1984 Act capped in gold the deregulatory phase of industry development. By providing a federal framework for regulation, it gave cable legal, and therefore financial, stability. Within that stability most of the primary controls perceived as onerous by cable and the investment community, including restrictions on programming and subscriber rates, were stripped away. To Wall Street and other investors, cable after the 1984 Act looked very much like an unregulated monopoly, offering a service in high demand with no real restraint on pricing.[14] Cable rates, therefore, began to climb and with them cable revenue. The industry reported nearly $8.8 billion in subscription income in 1985, about $4 billion from basic services and $3.6 from premium. Total advertising revenue that year was a slim $815 million. By 1990, subscription dollars had risen to more than $17.5 billion, with about $10 billion from basic, $4.8 billion from premium, and the rest from sources such as home shopping.[15]

Cable reported national advertising revenue of $634 million in 1985, climbing to more than $2 billion by 1991. A program by the Cable Advertising Bureau to provide national advertising media planners with the tools necessary to buy rationally and efficiently advertising helped at the end of the decade, as did establishment by Group W of the first media rep firm designed specifically for cable advertising. Local advertising revenues experienced particularly strong growth, fueled by expanding cable penetration, the development of improved technologies that facilitated insertion of advertising spots into local programming line-ups, and by the creation of local advertising interconnects that permitted area merchants to buy and run spots on several contiguous, interconnected systems. Overall, local ads sales grew from a modest $14 million in 1985 to more than $100 million by 1990.

In the mid-1980s, interest rates dropped back to more reasonable levels. From a high of more than 19 percent in 1980, the prime rate descended to an eight-year low of 7.5 percent 1986. It rose again to more than 11 percent in 1989, but fell back to less than 10 percent by the start of the new decade. While high interest rates and a weak capital market created sellers, low interest rates and cheap money created buyers. The mid- and late 1980s witnessed a flurry of high-profile acquisitions and mergers across a host of sectors. At the system level, valuations were bifurcated. In 1986, systems were selling for $1,500 per subscriber, up from $300 or so in 1975. By early 1989, large, modern systems in major markets with growth potential, channel capacity, and addressability, were commanding prices at $2,500 a subscriber and higher (as high as $3,000 per subscriber). Classic systems in smaller markets, with lower channel capacity and no addressability, were going for $1,700 to $1,900 per subscriber.[16] Many of the larger transactions were highly leveraged and featured novel and problematic financial instruments. (The impact on the cable industry is discussed in detail below.) The fact that much of cable debt involved floating interests rates tagged to the prime, however, meant that the easing of rates improved overall financials for most of the decade.

Expanding Audience

Cable's material success was not celebrated, of course, by everyone. In the late 1950s, Ed Craney rode through the regulatory landscape like Paul Revere, crying out to all who would listen that CATV was coming and broadcasters should man the barricades. Craney and his western compatriots prophesied the flight of the broadcast audience and advertising revenue as cable swept into town. Craney was not wrong, he was just premature.

It is stating the obvious to note that there are only 24 hours in the day and that expendable income, for most people, is finite. It is from this simple platform, however, that media scholars and economists historically have viewed the changing guard of media displacement. It is a form of the more general phenomenon of technological displacement—the action of new technologies to displace old, for automobiles to take the reins from horse-drawn carriages, and CATV systems to trim the need for roof-top TV aerials. The development of radio put a quick end to the late-1900s wire-based point-to-multipoint telephone systems. In the 1950s, television dramatically reshaped the radio programming, cut into movie attendance, and reduced the attraction of minor league baseball around the country. Sociologists have tracked the changing allocation of domestic leisure time over the course of the century as economics have traced the allocation of consumer dollars. These displacement and reallocation studies have taken one form in media studies in the idea of the "constancy hypothesis." In 1972, media scholar Maxwell McCombs traced consumer expenditures on mass media from 1929 to 1968, examining changing patterns of spending as news media were introduced in the marketplace, arguing that total spending remained relatively constant, with new media gaining ground only at the expense of older media.[17] This "principle of relative constancy" has been debated and modified over the years,[18] but illustrates the broader, generally accepted proposition that, within some wide frame, consumer time and money are limited and distribution patterns change with changing technological and economic circumstance.

What scholars knew from studying the historical evidence, broadcasters knew in their pocketbooks, and they had spent two decades trying to convince regulators. It was not an illogical position to take and FCC rule-making for ten years was premised on its validity. Through the 1960s and 1970s, the dire predictions of audience erosion and declining revenues had failed to materialize, however. With a very few (and those, unconvincing) exceptions, broadcasters could not demonstrate real world economic harm from cable TV development. In the FCC's economic inquiry of the late 1970s, even the NAB's own study showed at best a modest problem. From 1966 through 1978, television industry profits and revenues increased an average of about 4 percent a year in "real" (adjusted for inflation) terms, a substantially greater growth rate than that enjoyed by the economy at large.

As the coaxial line snapped into home after home and people turned with more frequency to MTV, CNN, and ESPN, the big three broadcast networks began feeling the squeeze of technological displacement. Viewers voted with their remote controls; ratings and "share" increased for cable and began a slow but unwavering trek southward for the broadcast networks.[19] The combined prime-time share for ABC, NBC, and CBS fell from above 90 percent in 1977 to 74 percent in 1986 and 62 percent in 1990.[20] The combined share of all basic cable, meanwhile, rose from less than 10 percent to 24 percent in all TV households and up to 35 percent in cable households (see Appendix C).[21]

Cable was not the only cause of the network leakage. The launch of the Fox Network in 1986 (see below), a strike by Hollywood script writers in 1988–1989, and the introduction of a new audience viewing measurement systems by Nielsen (the People meter, which more accurately assessed non-network viewing), also contributed to the slide. Little doubt existed in anyone's mind, however, that the audience fragmentation warned of by Craney so many years before, and now made real by modern cable, was at the heart of the problem.

Audience erosion did not automatically translate into a decline in network advertising revenue. Some national advertising money was beginning to move to cable, from about $487 million in 1984 to just over $2 billion by 1991. But the advertising industry was habituated to network television and was yet to be convinced of cable's reach or power. Moreover, a booming economy in the 1980s and 1990s helped fuel spiraling advertising rates for the networks. Total advertising revenue for the big three networks continued to climb, generally, through the 1980s, from a total of about $5 billion in 1980 to more than $10 billion in 1990.[22] At the same time, those revenues, when adjust for inflation remained fairly flat, and in 1985 actually took a small dip,[23] the first such decline since 1971 when cigarette advertising was banned from broadcasting. The expanding economy helped sustain the networks, but not in the same grand manner as the television oligopoly of prior decades. Network expenses remained high so profitability slackened. In 1986, CBS profits declined 30 percent despite draconian cost cutting and personnel reductions at the Tiffany network. In 1991, NBC publicly claimed that all three networks were losing money.

By the early 1990s, the broadcast networks were becoming increasingly sensitive to advertiser complaints about ballooning rates in the face of declining audiences. The networks defended themselves, explaining that, although they admittedly were no longer the only television game in town, they were still the most heavily viewed of all the available networks, and given their reach, history, and quality programming, deserved the premium they charged in ad rates. The pitch became less convincing with every drop in share point, however.

The Social Inflection Point

From a societal perspective, the sum total of the cable construction of the 1980s, the rising market penetration, the fragmenting and specializing programming, was that, for the first time in the industry's history, cable finally mattered. By the start of the 1990s, cable television had become the dominant source of news, sports, and entertainment, for more than 60 percent of the population. It was now the main artery through which flowed the televised political and cultural lifeblood of the nation. The latter half of the 1980s, in other words, saw another evolutionary stage in cable history as the medium hit the inflection point suggested in Chapter 1 by Usher's cumulative synthesis.

An important part of the evolutionary progression was a concomitant revision of the social definition of cable. The technology and the industry had moved through a number of phases, from retransmission device to Blue Sky communications "uber-utility." The emerging definition of the late 1980s was more nuanced and grounded more solidly in real world experience. It reflected both the growing dependence many citizens seemed to have on cable television and, at the same time, the frustration associated with the service. Cable, in this sense, was coming of age culturally as well as economically.

This maturation was reflected, by way of example, in the development of new auxiliary bodies. In 1985, cable created a center for the collection and study of its own history. The National Cable Television Museum was established at Pennsylvania State University in an effort led by a group of pioneers, including Ben Conroy, George and Yolanda Barco, Joe Gans, and Strat Smith, and aided by University Professor and Administrator Marlowe Froke. (In 1997 the Museum and Center was moved to the University of Denver, long-time hub of the cable industry. Bill Daniels [and others] led fund-raising efforts to construct a building and create ongoing programs in cable history, education, and outreach.) Denver also was selected in 1988 as the new home of a technology research center for the industry. Cable Television Laboratories, Inc. (CableLabs) would be funded by operators and work to advance their technology, later playing a major role in the development of shared technical standards for a host of emerging cable devices.

In many ways, cable was coming to be seen as an independent communications medium with its own history and character. Where cable had been defined in the 1960s as an extension, in one form or another, of the broadcast system, by the late 1980s it was increasingly seen as a communications system related to (but critically distinct from) broadcasting. This raised as many questions as it answered, however. Sketched in the contours of an established and comfortable broadcast model, the characteristics, rights, and responsibilities of cable could be more easily grasped and manipulated. As an autonomous medium, those traits had to be newly defined, especially insofar as they related to cable's legal status. This may have been best illustrated in cable's changing constitutional standing in courts. In fact, the evolution of cable's First Amendment status tracked closely with its broader evolution in social definition, and began making breakthroughs in the mid- 1980s. The key cases arose across a number of areas, including the long-debated issue of must-carry, the question of the city's franchising authority, and government control over cable content, especially the concern about sex on cable TV.

The First Amendment Medium

Cableporn

In 1982, the Mayor of Miami, Maurice Ferre, was visiting New York City when he tuned in to one of Manhattan's public access cable channels. There, much to his horror, he saw "naked bodies." His New York hosts may have been sympathetic, perhaps even amused, but they were certainly not surprised. Naked bodies, and more, had been a fixture of cable access in Manhattan for years. Back in Miami, intent on preventing the beach community from becoming the "Sodom or Gomorrah" of the South, Mayor Ferre proposed a city ordinance to ban such programming from the city's cable system. It passed.

Miami was not the only place in the country where citizens were worried about sex on cable. In Utah, a state heavily influenced by Mormon sensibilities and politics, the legislature in 1981 passed a law banning pornography and indecency on cable. In Roy City, Utah, the elected town leaders passed a similar, local ordinance. This resistance to stronger sexual content was especially vexing to premium services such as HBO. In fact, HBO had difficulties in many communities selling the town elders on a network that would include uncut R-rate films. Time's Richard ("Dick") Munro recalled later:

> I remember Jackson, Mississippi, where the clergy of Jackson boycotted us. We had to go down to meet with the clergy there. We were described as the devil by some. It was hard to overcome that. Four letter words had not been brought into America's living rooms. Of course, you had to pay to bring them in, and we tried to make that point. We tried to make the point over and over again that no one is forcing these R-rated movies into people's homes. They were raising their hand and purchasing it. But early on that was a very, very delicate subject. It was also a delicate subject with our board of directors. We, obviously, were never going to bring X-rated movies in, but we were bringing all those movies that were in your movie theater into your living room.[24]

Bringing X-rated films into living rooms would have created both marketing and legal problems. The Constitution offers wide (but not absolute) protection from government censorship. The First Amendment prohibits Congress and (through the incorporation clause of the Fourteenth Amendment, the states) from enacting laws that abridge the right to freedom of speech or of the press. The Amendment has never been held to be absolute, however. Even the print media are subject to certain restrictions, including reasonable business controls not directly related to editorial content and government censorship when stories might constitute a real and imminent threat to national security. The First Amendment, additionally, does not protect obscene speech (although the legal trick has always been defining obscenity).

More importantly, for present purposes, the Supreme Court has held that different media are privileged to different levels of constitutional protection. So, pure speech and print are typically afforded the greatest measure of First Amendment protection, while broadcasting has been proffered lesser protection. For example, in a case involving a comedy routine by entertainer George Carlin in 1978, the Court held that certain words, Carlin's infamous "Seven Dirty Words," constituted indecent speech.[25] While indecent speech, including the use of such terms as "shit" and "piss," were constitutionally protected in the print media, they were held not to be so protected when distributed through the public airwaves. The question for cable television, then, was whether it was to be legally constructed as a medium more like print, with a stronger measure of constitutional protection, or a medium like broadcasting.

HBO, for both legal and commercial reasons, was concerned about a proliferation of state and local laws that could severely limit the kind of films it could offer customers nationwide. In 1981, therefore, it challenged the Utah law on First Amendment grounds and, in doing so, engaged the broader legal and social issue of the definition of cable TV service. In January 1982, a federal district court overturned the Utah statute.[26] The same court struck down the Roy City ordinance in November 1982.[27] The next year, the Miami ordinance was overturned following a similar suit.[28] In the Miami case, the court held that cable and broadcasting were distinct media with differing characteristics, so much so that different levels of legal protection should be applied. Subscribers, explained the court, had much greater control over cable, through their decision to take the service and through the availability of channel lockout boxes, than did casual broadcast listeners or viewers.

The Theory

The decisions in the Miami and Utah cases were indicative of a changing judicial view of cable. In fact, the story of cable's definitional evolution in the law constitutes something of a microcosm of broader social thinking about the technology. The constitutional evolution of cable TV saw several distinct phases, driven in part by the arguments brought to the courts by cable lawyers. The industry could have made a case for community antenna television as a protected First Amendment speaker from the outset, using the constitution in the early 1960s as a shield to ward off efforts by the FCC to assert general jurisdiction, impose must-carry requirements, or limit signal importation. The "passive antenna" strategy adopted by the industry mitigated against such tactics, however. Cable lawyers could not argue, with logical consistency, that CATV was a passive extension of the home antenna, on of the one hand, and an active First Amendment speaker, on the other. The industry's use of the First Amendment defense, therefore, was held in abeyance, at least by some, until it had won the *Fortnightly* copyright case in 1968, and moved politically and commercially to an industry that sought to develop and distribute its own programming. Cable's judicial definition, like its social definition, was thereby tied to its political posturing, which in turn was driven by its commercial interests.[29]

The First Amendment argument was made by some operators even before *Fortnightly* (with disappointing results), but the industry took up the defense with real vigor beginning in the late 1960s and early 1970s, using it as an argument in a wide range of legal contests. It was one of the legal reflections of the emerging Blue Sky vision of cable TV. Some operators were sincere in their belief that cable was an independent communications service worthy of strong constitutional insulation from government oversight. Others were less ideological, using the First Amendment in concert with any other defense their legal team could conceive, to combat government control of their business operations.

The courts, for their part, were slow to buy into the claim that cable was a First Amendment medium. The judicial conservatism was understandable and, in fact, a part of the institutionalized system of review. Courts make their decisions based, in part, on precedence or "*stari decisis*," the history of case law established in prior, similar cases. It is a judicial form of defining the situation based on previous or existing definitions, a conservatism that arguably helps maintain social stability. For the courts, community antenna television looked very much like broadcast television and, in the earliest cases, nothing that either the cable industry or the FCC brought to the judges worked to disabuse them of that notion. In the copyright cases, cable lawyers had argued that their service was simply a part of the home reception systems. The FCC had argued cable was simply ancillary, in a sense an extension of, the existing broadcasting system. In either case, the courts were channeled into drawing from an existing broadcast model of First Amendment rights. The judiciary had long established that broadcasting did not have the same level of First Amendment protection as the print medium, by virtue of the fact that radio and television used the airwaves, a scarce natural resource essentially held in public trust.[30] Moreover, for the courts in the late 1960s and early 1970s, television was television, whether it came through the air or through a wire. Even though the judiciary came to understand that cable did not, in itself, use the open spectrum, judges nonetheless cited case law in support of regulation that rested on the scarcity argument. Insofar as the FCC was permitted to regulate broadcasting, it could, by extension, also be permitted to regulate cable. As the Fourth Circuit Court of appeals described it in 1968, CATV "is primarily a televisionary [sic] complement," and subject to FCC regulation as such.[31]

At the local level, municipalities, as noted in Chapter 6, additionally argued that cable could be defined as a form of public utility, subject to reasonable local regulation in the same manner as water or power companies. There, the outcome turned first on whether cable could be found to be a necessary public service and whether it was a natural monopoly. As noted, some judges were skeptical that CATV was a necessary social service. Even where cable was held to be a public utility for state and local purposes, however, the constitutional question became whether its standing as an economic monopoly was sufficient to overcome traditionally strong First Amendment protections. In a landmark 1974 case, *Miami Herald Publishing Co. v. Tornillo*, the Supreme Court ruled that even though most newspapers were *de facto* local monopolies—very few cities had more than one daily paper—their monopoly status alone was insufficient to overcome constitutional barriers to government interference in editorial decisions.[32] Although cable might be a natural economic monopoly, therefore, would that be sufficient to open the door to government control?

Not until the mid- and latter 1970s, as cable used the satellite link as a stepping stone to broader deployment and social standing, did a new view of the medium began seeping into judicial thinking. At the federal level, the view of cable as an extension of broadcasting began eroding in 1977 with the appellate court's decision in *Home Box Office v. FCC*.[33] There, the District of Columbia Circuit Court noted that the scarcity rationale was not applicable to the cable TV medium. It further noted that even if cable were found to be an economic monopoly, the Supreme Court's decision in *Tornillo* had killed that as a justification for regulation. Concluded the court, "there is nothing in the record before us to suggest a constitutional distinction between cable television and newspapers on this point."[34] Thus, the court elevated cable television out of its twenty-plus year constitutional status as an auxiliary medium.

The following year, the Eighth Circuit Court also took a strong stand on cable's constitutional standing. The case was *Midwest Video II*, in which the FCC's 1976 public and eased access requirements came under fire. The appeals court analyzed the First Amendment arguments in some detail, finding ultimately that many similarities existed between print and cable media with respect to claimed constitutional rights. The court did not, in the end, decide the case on constitutional grounds; it struck down the FCC rules using a lesser, statutory test. It noted in dicta, however, "Were it necessary to decide the issue, the present record would render the intrusion represented by the present rules constitutionally impermissible."[35]

There was a growing consensus that cable had to be distinguished from broadcasting; it was a medium with its own special character and its legal rights had to be tailored accordingly. Not every jurisdiction saw cable through the same constitutional lens, however. While some courts, within certain contexts, ruled in support of constitutional protection, others concluded that, insofar as cable typically was a monopoly system in its market, there was an economic scarcity that mirrored the technical scarcity in broadcasting and provided the foundation for government control.

As cable moved into the 1980s, it carried the First Amendment shield into a variety of judicial battles. Among the most important were those involving must-carry regulations and local franchising disputes. As the FCC repealed most of the significant remaining cable regulations in 1980s and Congress insulated the industry from onerous local control, one of the few remaining federal rules involved the requirement to carry local broadcast signals. The industry sought to eliminate the must-carry requirement via the courts, using as its principal argument, cable's standing as a protected First Amendment speaker.

Must-Carry

Cable's legal obligation to carry local signals dated back to the 1965 First Report and Order. Operators challenged the rules and, in one case, invoked the First Amendment to argue that must-carry interfered with the operator's editorial discretion to determine system content. But in *Black Hills Video Corp. v. FCC*,[36] the Eighth Circuit Court, in 1968, rejected the early First Amendment plea. Cable was still seen as an extension of broadcasting and, insofar as the government could constitutionally regulate broadcast TV, it could by extension do the same in community antenna TV.

By the mid- 1980s, the public and judicial perception of cable TV had changed radically, as had the industry itself. In the real world of local system operation, the must-carry rules were beginning to pinch as the proliferation of cable programming services began to outstrip channel capacity. Operators, for obvious business reasons, carried most local stations, but carriage of all available local signals, including the weaker independents, ate up channel capacity. Cable programmers began to complain that the channels subscribers most wanted to watch were being threatened with eviction by area broadcasters asserting their legal rights to carriage. One of those threatened programmers was Ted Turner. Frequently at the cutting edge of cable law, as well as business, Turner, in October 1980, petitioned the FCC for elimination of the must-carry rules, arguing that they violated the First Amendment rights of cable operators. The petition languished, however, while legislators and regulators maneuver toward the 1984 Cable Act.

Meanwhile, a small cable operator, Quincy Cable TV, Inc. in Quincy, Washington, decided not to carry three local Spokane broadcast signals, choosing instead to import three stations from Seattle, which it argued were preferred by its subscribers. The FCC, responding to the complaint of one of the Spokane broadcasters, ordered carriage of the local signals and slapped Quincy with a $5,000 fine. In March 1983, Quincy challenged the FCC ruling in federal court. At about the same time, Turner, impatient with the FCC's lack of action on his must-carry petition, filed a second request for repeal of the rules. Again, the FCC failed to act on Turner's request, prompting him in October 1983 to ask the federal court to force the FCC to consider the issue.

By early 1984, the FCC signaled it would review Turner's request and asked the court to remand the case back to the agency. The court did so and at the same time sent the Quincy case back to the FCC as well, following an expansion of the system's channel capacity (a modification in the facts of the case) In April, the Commission ruled against Turner and the following September reaffirmed its ruling in Quincy, rejecting First Amendment arguments in both cases. Turner and Quincy appealed once more to the Washington District of Columbia Circuit Court and the Court consolidated the cases.

Both companies argued that the First Amendment prevented the FCC from controlling the inherently editorial function of determining system programming. Attorney Jack Cole, who had begun his career with Smith & Pepper and watched as the firm helped argue cable's cause before the Supreme Court in *Fortnightly*, now argued cable's First Amendment rights before the District of Columbia Circuit Court, and the court was persuaded.

In July 1985, the District of Columbia Circuit Court ruled the FCC's must-carry rules were an unconstitutional infringement of the cable operator's First Amendment rights.[37] The Court found that the operators did indeed have some measure of First Amendment protection and that the FCC had not met the heavy burden of proof necessary to show that the government's interests were sufficiently important to overcome that shield. The Court did not say must-carry rules were inherently unconstitutional, only that the Commission had failed to make its case that local broadcasting would suffer irreparable damage without must-carry protection and that the rules were narrowly tailored to prevent such damage.

In the area of must-carry, the decision would send broadcasters, cable operators, and the FCC back to the bargaining table, and take them eventually back to the courts.

Franchising

Cable's First Amendment standing also evolved through a line of cases involving the rights of a city to issue a franchise, and how far local control extended into cable operations. The franchise wars of the late 1970s and early 1980s generated several cases in which the cable operator challenged the local authority's power to constrain certain aspects of the business or, in fact, to require a franchise at all.

One of the original purposes of the First Amendment was to assure that individuals would not need a license from the government to speak their minds, either in oratory or by means of magazines or newspapers. Cable argued that franchises were nothing more than unconstitutional government licenses to speak. Even if a franchise were allowable as a form of ensuring the safe wiring of streets and boulevards, the terms should not be permitted to extend into control over services, length of contact, subscriber rates, programming, or access. The cities justified their regulations on several grounds: use of city streets and public rights-of-way, the natural economic monopoly of cable TV, and the city's needs to provide essential public services such as public access. Each position invited its own deeper examination. Did, for example, use of streets justify content control and rate regulation? Was cable, in fact, a natural economic monopoly and even if it was did that outweigh constitutional considerations?

For the most part, the lower courts sided with the cities. As noted previously, in TCI's troubled franchise fight in Jefferson City, Missouri, the courts held that franchises were a reasonable exercise of a city's jurisdiction over use public rights-of-way. First Amendment did not protect cable, or any medium, from normal business rules and regulations. And, in TCI's previously noted legal squabble with Boulder, Colorado, the federal circuit court ruled that reasonable franchise requirements did pass constitutional muster and the print model of First Amendment protection could not simply be thrown over cable.[38] Citing long-standing First Amendment precedent, the court explained, "The nature and degree of protection afforded to First Amendment expressions in any given medium depends on that medium's particular characteristics."[39]

Cable, therefore, was a unique medium in the eyes of at least some judges. The question, however, was what level of constitutional protection would attach to the system? The Supreme Court would be the final arbiter of the question and, through the mid-1980s, had not yet spoken with any clarity on the question. In 1985, the high court finally took up the issue when it heard a case called *Los Angeles v. Preferred Communications* involving a Los Angeles cable franchise.[40] The cable operator in question, Preferred Communications, argued that the franchising process itself and the variety of requirements that typically appended to a franchise agreement were unconstitutional restraints on freedom of expression. In its decision in *Preferred*, the Supreme Court acknowledged that cable did have some level of constitutional protection. The Court failed, however, to set an unambiguous standard for that protection. It could have held that cable, as print, deserved "strict scrutiny" of any attempt by government to interfere in its expression, a standard that offers substantial protection to the speaker. Instead, the Court chose a lesser level of scrutiny that permitted government regulation if the rules did not have an impact on content, furthered an important governmental interest, and were tailored to be no more intrusive than necessary to meet that interest.

For those in the industry and legal community, *Preferred* did not fully satisfy because it did not fully answer the question of cable's First Amendment situation. Following the decision, courts in varying jurisdiction came to inconsistent conclusions about cable's constitutional protection.[41]

The larger lesson, however, was that cable now stood as an independent medium in the eyes of the law and, by the mid-1980s, in the eyes of the public at large. Moreover, it was a medium and an industry that seemed to be multiplying exponentially in size and influence at every turn.

Merger Mania

As cable grew in the mid- and late 1980s, it also continued its ceaseless consolidation of ownership. The small cable companies of the 1960s had matured into the large MSOs of the 1970s and, by the 1980s, many were accelerating their expansionistic ways. Not only were cable companies merging, but the acquisition of programming interests, in full or part, which had begun in the 1970s, was heating up as well. It was often referred to as "merger mania." Cable was not alone in the frenzy; the economy broadly was being driven by a superheated stock market and a wave of new and highly questionable economic-cultural norms, discussed in greater detail below. But, consolidation in media industries carried special social weight and led to a resurgent public policy concern about ownership.

Growth and Public Policy

The ownership and control of the nation's cable television system had been an on-and-off concern of regulators and social commentators since at least the mid-1960s when the FCC first reviewed the issue. Proposals from such agencies as the Sloan Commission and Clay Whitehead's OTP to separate content control from ownership of the hardware—the various forms of structural separation—were intended to remedy the potential for abuse of monopolistic power. As cable grew in size and social influence, anxiety about industry concentration, in both its horizontal and vertical dimensions, grew in step.

In the first instance, the American people have a fundamental cultural distrust of institutionalized size and power, whether it be in business or government. The ancient admonition that power corrupts and absolute power corrupts absolutely is at the core of a host of U.S. legal safeguards, up to and including the three-branch system of checks and balances in the federal government. In business, monopolistic or oligopolistic conditions typically mean poorer service, fewer choices, and higher prices for consumers.

Those concerns are exacerbated in the communications industries, which are counted on to supply the nation with news and information, the lifeblood of the political process itself. The country was founded, in part, on the principle of an open and vigorous marketplace of ideas, where, according to John Milton's classical formulation in the *Areopagitica*, truth and falsity would grapple, for "who ever knew Truth put to the worst, in a free and open encounter?" Concentration of the media generally and the news media in particular therefore raised fears that went to the very heart of democracy.

To a lesser extent, but still critical to the cultural well-being of the society, concerns about the available diversity of art and entertainment, especially given the perceived or hoped-for plethora of video channels, were also an issue. As it was from the very beginning of cable television, the desire on the part of the consumer was for more choice in television content. Concentration of ownership was seen as an impediment to that diversity. When cable television served only 10 percent of the population, ownership issues were not terribly pressing, but by the time cable fed news to half of all the nation's homes, the state of the industry mattered.

Concentration of ownership in the cable industry therefore became the subject of uncounted law reviews, academic studies, books, reports, and magazine articles.[42] Most surveyed industry statistics and trends and held the figures against governmental guidelines. Concentration is typically measured by the percentage of market share held by the top four and top eight firms. It has been calculated in the cable industry in two forms: one using figures based on the number of subscribers served by wholly owned cable systems (consolidated systems) and another measure that also includes subscribers served by systems in which an MSO has partial ownership interest (unconsolidated systems). An additional common measure of concentration is the Hirschman-Herfindahl Index (HHI), used as a guideline by the Justice Department to judge the potential for anticompetitive behavior. Under Justice Department guidelines, a market is generally considered concentrated if the ratio of the top four firms is 35 percent or more or if the ratio for the top eight firms is 50 percent or more.[43] Sectors with HHI indices under 1,000 are considered "unconcentrated" and those with indices between 1,000 and 1,800 are "moderately concentrated."

In their 1988 review of horizontal integration in the industry, researchers Sylvia Chan-Olmsted and Barry Litman found the top eight firms controlled 39.42 percent of the cable market and the top four firms controlled 27.60 percent, with an HHI around 350 points.[44] By 1992, the top four-concentration ratio was about 42.3 percent and the HHI was about 600.[45] All measures were well below that which would trigger Justice Department antitrust review,[46] although the authors warned about the pace and trajectory of industry concentration and recommended continued governmental vigilance.

In the 1960s and 1970s, the Commission responded to such concerns, to the extent it responded at all, with structural solutions, requirements that local cable operators make available educational, governmental, leased and public access channels, and carry all local broadcast signals. It would go no further, however. In the area of horizontal integration, restrictions on the ownership of cable property by broadcasters and telephone companies were codified by the 1984 Cable Act, but proposals to cap the number of systems or subscribers any one company could own never came to fruition. A 1981

FCC staff report, in fact, recommended repealing the Commission's rules restricting cable-broadcast and cable-network cross ownership.[47]

Through the early and mid-1980s, the call for closer scrutiny of cable ownership continued to fall on the deaf ears of an FCC faithful to market-based solutions. The Commission declared formally in 1982 it would not seek limits on the number of cable systems or cable subscribers any one company could control.[48] If concentration of ownership were a potential problem, and the Fowler FCC was not so sure it was, then the solution was increased competition, hence the nurturing of alternative distributions systems such as DBS and MMDS.

The FCC's determination to let business be business was only one manifestation of a broader cultural value at loose in the land. As noted in the opening chapter and demonstrated by the birth of the Blue Sky vision in the 1960s, the social perception of a given technology or business such as cable television is influenced by the broader cultural climate of the time. The Reagan years had ushered in an almost "wild west" attitude of high-risk, high-stakes entrepreneurship in nearly all segments of American commerce and industry. Media—television, film, cable, and telecommunications—were only players in a larger play.

Although it failed to satisfy advocates of closer public scrutiny, this political geology provided fertile soil for increased investment in cable and an unprecedented period of industry expansion. The same forces, therefore, that drove the franchise frenzy in the major markets—a relaxed regulatory climate and increased programming—also drove expansion and amalgamation on a national scale. By the end of the decade, the tsunami of industry consolidation would become a critical factor in cable's downfall in Congress, but regulatory inaction in the early part of the decade only served to foster the activity.

Wall Street Sets the Tone

The 1980s was the era of corporate raiders, leveraged buyouts, green mailers, and white knights. Across the spectrum of American business, companies, progressively expanding through merger and acquisition in the decades following World War II, had become multinational corporate behemoths. Many were composed of a patch quilt of subsidiaries, some related to the core business and some not. The financial game of the day was to assess the market value of the component parts of a corporation. Frequently the sum of the totaled pieces was greater than the value of the mother firm. At that point, the company became a likely takeover target.

The strategy was to purchase the conglomerate, in a hostile takeover if necessary, break the company into its separate pieces, and sell them off. Economically, some companies were worth more dead than alive, at least to the stockholders. Even if a hostile bid was unsuccessful, a corporate raider might be paid a handsome sum, a premium on the stock he or she already had accumulated in the takeover effort, to depart the embattled waters, a phenomenon known as "green mail." Even without green mail, the act of initiating a takeover attempt would, in the parlance of Wall Street, put the company in play, making it open to competing bids and typically driving up the price of the stock. The raider could again cash out at this point, earning a tidy profit. In 1989, the government, as discussed below, would institute rules restricting highly leveraged transactions (HLTs), putting the brakes on much of the activity, but in the meantime, billions of dollars changed hands.

The cable industry played this game with verve. By the mid-1980s, cable TV was no longer the backwoods business that it had been a decade or two before. Many of the companies that started out in cable or entered it in the 1970s had, through internal growth or acquisition, become substantial corporations in their own right, often with interests beyond cable TV. As before, strong personalities edged with varying measures of ego and eccentricity served as wild cards in the process of buying, selling, and trading. It was a frequently wild and unpredictable dance of money, power, and politics. The major actors included Ted Turner, Rupert Murdoch, John Malone, Michael Fuchs, Sumner Redstone, Steve Ross, Dick Munro, and Gerald Levin, among others.

In every case, the sums of money needed in the sport were enormous, hundreds of millions, even billions of dollars. To raise that kind of capital, buyers frequently turned to junk bonds, the high-yield, high-risk, short-term financial instruments pioneered by investment firms such as Drexel Burhnam and its aggressive trader, Michael Milken. The theory was to take on the enormous debt associated with junk bonds in a leveraged buyout, then retire the debt through the dissection and liquidation of the target property. When a company came into play, it was like blood in the waters of Wall Street, and the sharks began circling. Any industry could find itself in these waters, including cable and broadcasting, but in the latter, the leaks in the boat first became noticeable at the three major networks.

The Changing Broadcast Networks

In an historic two-year period, all three networks changed hands, with the impact rippling through U.S. industrial structure and American media culture. At the time, the ownership of each network represented the latest iteration in a corporate lineage that stretched back to the founding of broadcasting itself. An important part of the heritage was a belief that broadcasting, especially network broadcasting, was a public trust and the control of NBC, CBS, and ABC carried with it a social responsibility not associated with other businesses. The corporate culture was not unalloyed. It was, in fact, fed by the immense profits that came with the broadcasting oligopoly. Through the 1970s, the heads of the big three could easily afford to be prosocial and culturally magnanimous. By the mid-1980s, however, the historic dominance of the major networks was, at least in the eyes of many, coming into jeopardy.

The change was in part generational. The men and women who had built broadcasting (and cable), especially the veterans of World War II, were moving into the twilight of their careers, retiring or looking to do so.[49] These pioneers often took seriously the social responsibility frequently attached to radio and television. Now, their sons and daughters were moving in to the corporate boardrooms, a generation schooled more on MBA programs, bottom-line goals, and in too many cases, a sense of personal entitlement. The new generation was much more likely to see the business as just that, a business, not significantly different from selling paper towels or washing machines.

In fairness, the younger executives were confronting a new structural reality in television. The pillars upon which the networks financial security rested were the millions of viewers who, for decades, had tuned in every night to watch "I Love Lucy," "Love Boat," and "Gilligan's Island." In 1976, 92 percent of all television homes watching TV on any given evening were watching one of the networks. As noted above, the coaxial cable was creeping down the street of EveryTown, USA, and those who were not tuning now into cable were settling in for the evening to watch the latest home video rental. Cable, with an assist from VCRs, was recrafting the nation's television diet. The new owners, by necessity, had to respond. In addition to these broader, tectonic movements, each broadcast network confronted its own unique challenges, and in the end, all three would succumb to the changing times.

ABC

The first to feel the pressure was ABC. That network had suffered a number of setbacks in its efforts to react to the changing media environment. Its investment in the Satellite News Channel was written off completely and start-up costs for its other young services were steep. ABC had pumped more than $300 million into its Video Enterprises division and operating losses continued to climb. Network ratings, meanwhile, were off. ABC had slipped back to third place in prime time. It had enjoyed success and profitability for many years with its daytime schedule, but the housewives who made up the bulk of that audience were increasingly being pulled into the work force by a new economic reality of relatively shrinking salaries and rising social expectations. The company's stock price was depressed and, in June 1984, management told its executive staff that it was vulnerable to a takeover.[50] Its breakup value, by the company's own estimation, was twice that of its existing stock price.

Company Chairman Leonard Goldenson was already investigating methods to insulate the company from a financial assault. Goldenson was an "old school" network broadcaster. He had grown up in the industry alongside the founding giants, Bill Paley at CBS and David Sarnoff at NBC. ABC, in fact, had its origins in the nation's first broadcast chain, the National Broadcasting Company. By the 1940s, NBC was operating two of the nation's dominant radio networks, NBC Red and NBC Blue. Government concerns about monopoly of the radio industry led to a forced divestiture of one of the networks, NBC Blue, which in short order and under new owners (Lifesaver candy king Edward J. Noble), became the American Broadcasting Company in 1943. The young network struggled, especially as big-league broadcasting made the transition from radio to television after World War II. In 1951, ABC was up for sale. Paley was interested, but the bidding war was won by the young chief executive of United Paramount Theaters, Leonard Goldenson. Over the next three decades, Goldenson built ABC into a successful and lucrative enterprise. But, in 1984, Goldenson was nearing retirement and had no wish to see his creation seized by others in the sunset years of his career.

Suitors came to Goldenson's door. He was most interested in a potential partnership with IBM, and entered into negotiations with the computer giant. Those talks sputtered, however, as IBM decided not to enter broadcasting. Goldenson hoped for a partner steeped in the culture and ethos of broadcasting, and was open to a proposal brought to him in December 1984 by Thomas S. Murphy, Chairman and CEO of Capital Cities Broadcasting Co. (Cap Cities), which was expansion minded. In addition to its cable properties, acquired from RKO General in 1980, by 1984 it held a full complement of radio stations, ten daily newspapers, three dozen weekly papers, plus a growing list of magazines. The FCC's historic cap on radio and TV station ownership (seven AM, seven FM, seven TV stations) had acted to suppress broadcast mergers and acquisitions, but the Reagan-Fowler deregulation program changed the financial as well as regulatory rulebook. The new ownership limits (12-12-12) were approved in December 1984, and Cap Cities was one of the first try to take advantage of the change, buying additional radio stations and then looking around for TV property.

Goldenson and Murphy both saw a tongue and groove match in the merger proposal. There was little overlap between the companies in local station ownership (FCC rules still prevented ownership of two TV stations in a single market). And four of Cap Cities' seven TV properties were ABC affiliates. Goldenson and Murphy knew each other socially and liked one another. They sparred briefly over price and Murphy sought additional financing. He went to a friend and Cap Cities board member, Berkshire Hathaway Chairman Warren Buffet. Buffet was already legendary for both his investing acumen and his low-key personal style. He offered Murphy $500 million in additional financing. It helped cement what amounted to a $3.5 billion transaction. The deal was made public in March. To the surprise of many, it was structured not as a merger, but as the acquisition of ABC, the national network institution, by the relatively unknown Cap Cities. The *New York Times* subhead read "Little Known Media Company in Surprise Takeover Bid for Broadcasting Giant."[51]

An important part of the arrangement was the promise to keep the network intact. But certain properties had to be shed to comply with remaining FCC regulations. Federal rules still prohibited networks from owning cable systems and, in August 1985, Cap Cities left cable operations, selling its fifty-three systems and 350,000 subscribers to the Washington Post Company for $300 million.

CBS and "Captain Outrages"

The change of ownership at ABC was fairly benign, even friendly, by 1980s standards. CBS would not be so lucky. In July 1983, Ted Turner explained his thinking:

> I'm supposed to be so brave and strong, but when I read that CBS and HBO and Columbia Pictures—which is Coca-Cola—were all getting together, and Paramount and Warner and MCA and Showtime and Movie Channel were getting together and, of course, ABC and Group W are getting together—those are some pretty powerful combinations when they're all zeroing in on your area. It's like a big war occurring and none of the players are big enough to play alone. So I went around to see if anybody wanted to merge with me, but nobody wanted to.[52]

The ABC-Cap Cities merger was still over the horizon, but the broad contours of the terrain were coming clear. Through the early 1980s, the major program suppliers and distributors were teaming at a rapid pace, coalescing into joint ventures of a size and power that gave even Ted Turner, "Captain Outrageous," pause. He determined that he would need to be much larger to compete in the evolving business environment and began looking for a partner in late 1982 and early 1983. He reportedly approached all three networks, plus Time Inc., Gannett, and Metromedia. He was offering to swap his 87 percent in Turner Broadcasting for stock and a seat on the board of the acquiring company. He had a few nibbles but no bites; NBC considered Turner to be a "lunatic."[53]

In January 1985, the arch-conservative republican senator from South Carolina, Jesse Helms, began a crusade against what he saw as the left-leaning news coverage of CBS. He instigated a letter-writing campaign, marshalling his political forces for a highly public, if largely ineffectual, effort to purchase a controlling interest in CBS stock. He told his followers they should aim to become CBS News anchor Dan Rather's boss. Helms demanded a list of CBS stock holders in an effort to purchase a piece of the company and take what everyone was sure would be a very active and heavily publicized role in the company's next shareholders' meeting. Wyman deftly out-maneuvered Helms, delaying release of the list and orchestrating changes in company bylaws that helped protect the firm and defuse the Helms' assault.

But the damage had been done. Helms' action was a catalyst that put the company in play. Other, more serious investors began examining the CBS financial situation. Aggressive arbitrage aficionados, such as the infamous Ivan Boesky (convicted in 1987 of insider trading, he was fined $100 million and sentenced to up to five years in prison), began purchasing CBS stock, positioning for what some believed would be a takeover attempt. As ABC, CBS was vulnerable. It held a rating lead in prime time programming, but just barely. NBC's new comedy lineup, featuring an increasingly popular "Cosby Show," was eating into the CBS margin, and long popular CBS programs, such as "Dallas" and "Falcon Crest" were showing their age. Profit margins at the network's five owned-and-operated (O&O) television stations were lagging those of the other networks. Earlier efforts to expand into other fields, including publishing and toy manufacturing, were yielding disappointing results. The company's stock price was down and its breakup value was estimated well beyond its stock price.

In March 1984, the rumor mills started buzzing about a possible takeover attempt by Ted Turner. CBS dismissed the talk as preposterous. Turner was uncharacteristically quiet. The rumors, in fact, were true. WTBS had become very prosperous, the primary cash-generating vehicle for Turner Broadcasting, but programming was becoming more expensive and difficult to obtain. Programmers were demanding high license fees to lease their product to the superstation. Turner sought attractive, controllable programming and looked first to his old standby, sports. He had wanted to acquire ESPN from Texaco, but the oil company had moved quickly to sell to ABC, without taking bids from Turner or anyone else. The move piqued Turner, who began securing deals for sports events on his own, including a $20 million package of NBA games. He entered a bidding war with ESPN for cable rights to three years worth of games from the young United States Football League. ESPN won, paying $70 million for the contract, but Turner's activities drove up prices for programmers and, in turn, operators, much to their discomfort. Beyond sports, Turner needed entertainment programming and continued efforts to strike a deal with MGM, staging on-again, off-again talks with studio head Kirk Kerkorian.

Through the early 1980s, however, no one seemed interested in a partnership with the unpredictable "Mouth of the South." Turner briefly looked at ABC as a possible takeover target, but realigned his sites when the ABC-Cap Cities merger became public. He now took aim at CBS. On April 18, 1985, Turner summoned reporters to the grand ballroom of New York's Plaza Hotel. He told them of his plan to purchase CBS. The proposal involved a mixed offering of TBS stock, junk bonds, and debentures, totaling an estimated $5.2 billion for 67 percent of the company. No cash was to change hands. In effect, it called on CBS shareholders to finance the buyout of their own company. In exchange for existing CBS stock, they would receive junk bonds and equity shares in the combined, and highly leveraged, TBS-CBS. In true corporate raider fashion, Turner proposed to pay off the debt and added a stock-holder premium by selling off assets, including the company's lucrative O&O stations. Analysts and critics assailed the proposal, especially for its lack of a cash component. *Barron's* reported one analyst's shocked response to the Turner plan as an offer that "frankly borders on the ludicrous."[54] Wall Street odds makers gave the deal little or no chance of succeeding.

Although many observers dismissed the seriousness of Turner's no-cash proposal, CBS was taking no chances. In fact, Paley would later explain that it was not Helms or Turner he was worried about; it was the corporate raiders who stood in line behind them that demanded the company's immediate response. And that response was swift, intelligent, and multifaceted. Some even described it as ferocious in its intensity and speed. One of the weapons was a public relations campaign designed to trivialize, even demonize Turner himself. Wyman repeatedly declared that Turner was morally unfit to run a national broadcast network. CBS, in filings with the FCC and the Securities and Exchange Commission (SEC), charged Turner Broadcasting with all manner of financial improprieties and moral turpitude.

CBS dug additional legal and financial moats around its besieged business. In May, it purchased five radio stations from Taft Broadcasting for $100 million, increasing its overall debt, a traditional defense against corporate raiders. In July, the company proposed spending a billion dollars to buy back much of its own stock, for $150 a share, $32 beyond its current list price, and speculation was that Paley was considering taking the company public. Turner's offer, as its details unfolded, had begun to generate some respect on Wall Street, but it still lacked a critical cash component, and the CBS stock buyback was structured in such as to make a hostile takeover even more difficult. Turner went so far as to file suit opposing the move. How the battle may have resolved itself if left to play out remains a point of speculation. Some observers suggested Turner had a real chance to push the deal through. Others were less sure. The process was interpreted, however, and Turner's attention diverted in July, around the time of the CBS counter-measure, by a phone call from Kirk Kerkorian. He wanted to talk about selling MGM.

Turner flew to Hollywood and began negotiations. Turner was well known for striking quick bargains, and had been frustrated at the pace of the CBS battle. He wanted to settle quickly with Kerkorian and avoid what Kerkorian indicated might otherwise be a bidding war over the studio's assets. MGM was an established and historically important film company. It owned classics such as *Citizen Kane* and *Gone with the Wind*. In 1981, it had purchased the film library of United Artists for $380 million, part of which included features acquired from Warner Brothers in the 1950s. But MGM was a financially troubled, reportedly hundreds of millions of dollars in debt. Kerkorian offered to sell for $1.6 billion, but the price did not include United Artists or the Warner Brothers film holdings. Turner said he would take the price if Kerkorian "threw in" the other old movies. The deal was struck. On August 5, they announced the planned TBS purchase of MGM/United Artists. Turner still needed to raise the money to finance the purchase, however. Junk bond wizard Michael Milken had helped Kerkorian raise $400 million previously for MGM, and was brought in to secure the financing for the TBS acquisition. Drexel Burnham sold $1.5 billion in junk bonds for Turner to buy the studio. The acquisition was completed in March 1986. Turner now owned a huge library of films. He had taken on a massive debt, however; many observers said he overpaid. The studio he purchased was itself in financial freefall, cash flow was meager, and Turner was confronted with servicing a $1.5 billion high-interest loan. To complete the transaction, Turner sold off all of MGM's assets except the film library, including the MGM sound stages, production lot, related real estate, and a film lab. Kerkorian himself bought back the MGM lion logo, the company's TV and home video assets, and United Artists. In the end, Turner was left with some 2,200 classic MGM features, plus the rights to 1,450 Warner and RKO films that had been held by United Artists.

Kerkorian was paid in part with preferred TBS stock, and the agreement with Kerkorian was structured in such a way that if dividends could not be paid in cash, they would be paid in common stock. The notes were due to mature in March 1987, but Turner was so heavily leveraged that there was little or no cash for the payout and he was faced with the prospect of watching Turner common stock start to flow to the preferred shareholders, especially Kerkorian, until he lost control of the company. In late 1986, a worried Ted Turner went looking for more money and, if necessary, new partners. It was estimated that he needed about $300 million to escape the Kerkorian stock clause. He talked briefly with CBS about an equity position in CNN, but a change in management, ironically stimulated by Turner's own attempt to buy the network (see below), and a price too low for Turner's needs, killed the conversation. Increasingly concerned, even mildly desperate, he turned to his allies in the cable business. Malone retained a vivid memory of the early morning telephone call: "I'll never forget the morning the phone rang and it was Ted. It was about 6 in the morning and I was just barely coming awake and he's screaming in the phone, 'You've got to do something!' He says, 'You've got to do something! If you don't do something it's going to be the KNN.' And I said, 'What the hell's the KNN, Ted?' And he says, 'The Kerkorian News Network.'"[55]

Malone and Turner orchestrated a meeting of leading cable operators in New York on January 13, 1987. Turner made his pitch. He needed a bailout, although he would ever after deny the term, and asked for help from the operators. He presented a plan, already stitched together in concert with Malone, to take on the operators as equity partners, raising $550 million in the process.

For many of the MSO executives, including Malone, it was an offer they dare not refuse. They had come to realize how dependent they were on the programming branch of the business generally and Turner's powerhouse WTBS and CNN in particular. They did not care to see the Turner properties fall into cable-unfriendly hands. One of the more significant threats, in the eyes of the operators, was that Rupert Murdoch's News Corp., recently establishing a beachhead in American television, would swoop down and feast on the carcass of TBS. It was a fear vocalized again by Malone, "Turner would have been owned by Rupert Murdoch if the cable industry hadn't rallied and put in some equity capital. We really did not want to see the most valuable programmer converted to some other industry. We made the move defensively to protect the independence of the industry."[56] Turner was wild and unpredictable; he was something of a madman, but from the perspective of the industry, at least he was their madman.

It did not take long for the operators to structure the deal. Fourteen companies initially bought into Turner Communications. Time Inc. was conspicuously absent from the first meeting and it took several months to convince Time to join the consortium. Kerkorian also bought into the pool. In the end, the largest piece, 21 percent, went to TCI; Time bought 17 percent. Both companies were given seats on the board. The rescue squad also included Cablevision Systems, Warner, Continental, Heritage Communications, Jones Intercable, Lenfest Group, Sammons, Storer Cable, TCI-Taft Cablevision, United Artists Communications, United Cable Television, TKR Cable, Viacom, and Times Mirror Cable. In total, the cable industry controlled about 35 percent of Turner Communications.

The Turner Board of Directors was expanded to fifteen members; cable operators would hold seven of seats. Turner would retain equity control of the company, but one objective of the exercise was to manage the unpredictable personality. Turner had mellowed somewhat in 1985, after he began seeing a psychiatrist and taking lithium, which he credited with helping dampen his notoriously wild mood swings. But he was still Ted Turner, and his new partners needed to feel some measure of security. Turner would not be able to spend more than $2 million without authorization from the new board.

When Turner abandoned his takeover attempt at CBS, he sold his stock in the company, at a tidy profit, turned his attention to MGM and did not look back. At Black Rock, however, the shockwaves were still reverberating. The company sold a prized TV station in St. Louis, KMOX-TV, to Viacom to help fund its defensive stock buyback. It also entered into an alliance with Larry Tisch, CEO of Loew's Corp., a restaurant and movie theater company. Tisch had become interested in CBS stock at the onset of the

Turner play, sensing a likely upswing in the stock price. Through early 1986, however, Tisch continued to acquire CBS stock and began speaking out in corporate affairs, flexing his ownership muscle. Tension grew between Tisch and CBS CEO Wyman, who was concerned that Tisch was engaged in a creeping takeover campaign. In an effort to finally insulate CBS from a hostile takeover, by Tisch or anyone else, Wyman sought out potential merger partners or buyers, ultimately recommending to the CBS board that they consider alliances with either Coca-Cola, which at the time owned Columbia Pictures, or with Disney. The board did not hesitate to demonstrate its dissatisfaction with the plan. CBS, it determined, would remain independent, and for good measure, Wyman would have to go as well.

The company patriarch, Bill Paley, by most accounts, hated Wyman. Wyman had orchestrated Paley's ouster from CBS control several years before and had added insult to injury through such petty maneuvers as charging Paley rent for the office space he kept at CBS headquarters. Tisch built bridges to Paley and together they possessed a controlling percentage of CBS stock. In September 1986, Thomas Wyman was ousted from the Chairmanship.[57] Paley came out of retirement to serve as acting Chairman. The company named Tisch, by then its largest stockholder, as acting chief executive.

While Tisch was brought in initially as "White Knight" to help save CBS from Turner, he had not been weaned on the CBS philosophy of public trust, but on the more mainstream corporate culture of steely bottom-line thinking. Coupled with eroding viewership, it meant a tight new hand at the helm. Tisch slashed budgets and fired personnel. The previous November (1985), CBS had closed down its film production operations. In 1986, Tisch cut the CBS corporate staff of about 1,300 by some 30 percent. In 1988, he sold Columbia Records to Sony for $2 billion. He sold the network's interest in Chuck Dolan's Rainbow Services Corp. (owner of Bravo and AMC). He refused to pay more for the rights to NFL football games, letting them slip to Murdoch's expanding Fox network, prompting a number of long-time CBS affiliates to switch to Fox as well.

The most public and publicly debated cutbacks came in the legendary and vaunted CBS news department. In March of 1987 CBS fired 215 news personnel and slashed the division budget by 20 percent. Media observers and critics were aghast; Tisch was pilloried in the press. CBS anchor Dan rather wrote a harshly critical op-ed piece in *The New York Times* entitled "From Murrow to Mediocrity?" all but accusing Tisch of leading CBS into the world of entertainment harlotry.[58] They were noble sentiments, but what Rather and those who shared his perspective were confronting was the new corporate reality in broadcast network television.

NBC
At NBC, that new reality arrived in late 1986 in an ironically historic form. The Radio Corporation of America (RCA) was formed by General Electric

(GE) in 1919 and, by the early 1920s, included as partners Westinghouse and AT&T. NBC, the RCA broadcast subsidiary, was created in 1926, but government concerns about monopoly in the radio industry led GE to shed its interest in 1932.

By the mid-1980s, RCA faced the same financial conundrum as its network brethren; it was a prime target for a hostile takeover. Its stock price floated in the $30 to $50 range, but its breakup value was estimated at $90 a share. GE, under a driven CEO named Jack Welch, had been expanding over the previous several years, acquiring and then frequently shedding, companies at a rapid pace. It had bid for CBS and even tried to buy Cox Broadcasting. Welch, however, did not believe in hostile takeovers, preferring mutually desirable purchases. He worked with corporate matchmakers of the period scouting for promising new acquisitions and, in late 1985, one of them brought him together with Thorton Bradshaw, chairman of RCA. GE and RCA walked quickly through the corporate courting ritual and announced their $6.28 billion merger in December 1986. NBC, in a sense, was back in the hands of its original corporate parent.

As at CBS, head personnel changed and certain corporate assets were shed to generate cash. The respected and successful Hollywood television producer Grant Tinker had been NBC president for several years. In 1987, he retired and was replaced by long-time GE executive Robert Wright. Wright sold off the RCA consumer electronics division as well as the NBC Radio Network (the latter going to Westwood One for $50 million).

Unlike CBS, however, NBC was not intent on simply cutting back to its core broadcasting business. The peacock network was looking to expand its programming reach, embracing the opportunities that cable might offer. Its new initiatives in cable programming are traced later in this chapter.

Finally, in addition to its already lengthy list of challenges, NBC, as well as ABC and CBS, had one more problem to contend with—Rupert Murdoch.

Murdoch and Fox

Murdoch had tried at least twice before to penetrate the U.S. film and television market, first with his stillborn DBS venture and then with his failed assault on Warner in 1983. He was not one to be discouraged by such small setbacks, however. He coveted production and distribution capacity, which was why Warner, with its film studios and cable properties, was so attractive. Failing there, however, he immediately set his sights on another wounded film studio.

Twentieth Century Fox had been one of the dominant Hollywood production companies for decades. It was purchased in 1981 by Colorado oilman Marvin Davis. In his efforts to resuscitate the ailing studio, Davis hired an experienced film company executive, Barry Diller, who had been Chairman and CEO of Paramount Pictures.[59] Diller sought additional investment to help fund development and production of films, but Davis baulked at tapping into his own wealth. Murdoch, still on the hunt for production capacity, supplied the answer. In early 1985, the Australian media tycoon put up $250 million for a half interest in the studio, with a portion of the money going to new operations.

With 20th Century Fox in pocket, Murdoch had his production capacity. He now needed distribution. He found it a few months later, spending $1.6 billion to purchase six of seven large market TV stations that had been placed on the block by Metromedia. Metromedia—formed from the remnants of the failed DuMont network in the 1950s—owned successful, independent VHF stations in New York, Chicago, Los Angeles, Washington, D.C., Houston, and Dallas. The company was controlled by John Kluge and also maintained a successful production and syndication operation. In the early 1980s, the company toyed with the possibilities of using its string of stations and its production capacity as the nucleus of a fourth broadcasting network, but Kluge's attention turned in a new direction. He saw the possibilities of an emerging market in cellular communications and was willing to sell his TV assets to fund the effort. Diller, while at Paramount, also had explored the development of a fourth broadcast network—using the studio's "Star Trek" franchise as its base—and Diller remained interested in the idea. It was at least a part of his thinking when he, along with Michael Milken, who also had worked with Kluge previously, brought together Kluge and Murdoch.

To overcome the FCC's prohibition on non-US citizens holding broadcast licenses, Murdoch took out U.S. citizenship in September 1985. And, by year's end, he purchased the remaining 50 percent of 20th Century Fox from Davis. In October 1986, using the Metromedia stations plus a seventh purchased separately (WXNE-TV, Boston) as his foundation, Murdoch launched a fourth national broadcast network in competition with the big three.

Fox Broadcasting owned stations in New York (WNEW), Los Angeles (KTTV), Chicago (WFLD), and Washington (WTTG). But even this large reach was not sufficient to generate the number of viewers necessary for network-scale program economics. As with the existing networks, Murdoch needed scores of affiliated stations in smaller markets to get his programming to the people. As noted earlier, the problem of TV station scarcity, especially in the UHF band, frustrated creation of additional national networks in the 1950s and 1960s and, in large measure, was responsible for the demise of DuMont. But the technological business and, most importantly, regulatory world of television had changed substantially since those earlier days.

The widespread diffusion of UHF converters mandated by the All-Channel Receiver Act of 1962, plus cable carriage had helped level the playing field for UHF stations and their numbers were rising. UHF proliferation also was propelled by Reagan-era deregulation. The FCC's relaxation of ownership restrictions in 1985 meant broadcasting companies could expand their holdings up to twelve stations, and built new ones when they could not purchase existing ones. Moreover, in November 1982, the FCC rescinded its restrictions on station trading.[60] These so-called "anti-trafficking" rules had required TV station owners to hold their property at least three years before reselling. The intent of the rules was to restrict rapid purchase and resale of stations—trafficking in, or "flipping" stations—which was perceived as working against the goal of localism and broadcast service in the public interest. The Fowler philosophy was that the marketplace was a better judge than the FCC of the propriety of station acquisition and sale. Fowler also felt that stations owners should be rewarded if they could turn a failing property into a profitable one and sell it appropriately.

Deregulation allowed the business community to flip stations, and the market imperative of the 1980s encouraged it to do so. Venture capitalists and leverage buyers saw broadcasting as an attractive cash cow business largely immune from inflation and the increase in demand meant constantly rising station prices. New money, therefore, would buy a station, hold it for two or three years and turn it around for a quick and substantial profit. The cycle prompted an upswing in station construction as well and, because most of the attractive VHF allocations had long since been built, the bulk of new TV outlets were built on previously unwanted UHF licenses.

There were 120 independent stations operating in 1980, a number that grew to almost 300 by 1988. There were seventy-eight UHF stations on the air in 1980, and more than 250 by the end of the decade. One result was that dozens of cities across the nation that previously had only three or four broadcast stations became five- and six-station markets.

Programming became a critical issue for these newcomers. Network affiliations were not possible for most of them, meaning survival based on a diet of cheap syndicated programming, old movies, games shows, and cartoons, with frequently thin profit margins. Murdoch took advantage of the situation, signing on independents, many of them new UHF stations to his fledgling Fox Network. These stations became the lucky ones. Many other independent UHF stations began struggling as the speculation bubble that had driven prices burst. In the last half of the decade many of the UHF independents began folding.

Murdoch had his network, however, a total of ninety-nine stations to start. Anchoring the debut on October 9, 1986, was "The Late Show with Joan Rivers." In April 1987, Fox began offering prime time programming to 108 affiliates. "The Joan Rivers Show" failed, but other Fox offerings succeeded handsomely. Diller designed and ran the network program schedule. He pursued what he saw as an underserved youth market, targeting it with irreverent programs such as "Married with Children" and "The Simpsons."

Even with the affiliated broadcast stations, Murdoch still failed to generate the kind of national coverage enjoyed by the established networks. Seventy or more small and medium markets in the United States were not served by a Fox outlet. By 1990, Fox was making headlines with its nontraditional programming, and audiences in those markets were asking the local cable operator why they could not get Fox. That is when Preston Padden, the Fox head of affiliate relations, received a phone call from TCI. TCI suggested a Fox service for cable operators in the unserved, or "white," areas. Murdoch signed an agreement in September 1990 with TCI that brought the new network an additional million viewers.

The Fox formula was a success. Over the next several years, the network established itself as a serious contender for viewer attention and advertising dollars. For good measure, Murdoch also bought the venerable television institution, *TV Guide*, in 1988, paying $3 billion for the parent company, Triangle Publications.

Meanwhile, the companies that constituted the traditional core of the cable business were themselves merging and congealing into ever-larger units. The important transactions of the period included, on the operator side, Group W's record-breaking swan song and the continuing expansion of TCI. On the programming side, Warner, Viacom, and Time led the field in their own tangled set of eye-popping alliances.

Group W (1985)

In acquiring the cable giant TelePrompTer in 1980, Group W-Westinghouse instantly became one of the industry's largest firms. But, philosophically, Westinghouse was an old guard corporation, and the cable business was for younger, more nimble, and less risk-averse hands. Group W had tried to succeed in cable, but at each turn, the company seemed to stumble: in the failure of the Satellite News Channel, the alliance with Disney, and the sale of The Nashville Network (TNN). Group W also entered the industry during a period that called for massive capital expenditures in the wiring of the major cities.

Group W had poured an estimated $800 million into cable, on top of the $646 million purchase price. As most cable firms, Group W was generating cash flow, but no real income. Unlike most other MSOs, Group W reported to an industrial giant that knew and relied on steady quarterly earned income reports, not the fancy quick step of the cash flow dance. Conservative shareholders were nervous and unhappy. Westinghouse CEO Douglas Danforth saw lower-than-expected profits, potential competitors—especially DBS—gaining ground in the minds of the financial community if not in reality, and he had a strong need to offset recent setbacks elsewhere in the company. He could use an extra couple of billion dollars to rebalance the books and buy back company stock, thereby kicking up share prices and helping prevent any attempt at a hostile takeover. Westinghouse decided to bail out.

In late August 1985, Group W-Westinghouse announced that it was exploring the possible sale of its 150 cable systems (2.1 million subscribers).

By December, it had located a set of buyers. The sale set new records for the industry. Reported initially at $1.7 billion, the price tag later rose to a reported $2.1 billion. The buyers included cable's remaining leaders: ATC, Comcast, Daniels and Associates, Century Communications and, of course, TCI.

TCI

Through the 1970s TelePrompTer dominated the cable TV business. By the mid-1980s, that baton had been passed to TCI. The company was voracious in its appetite for all things cable. It purchased systems and MSOs of every size and snapped up programming services and satellite-related businesses as well. Between 1972 and 1988, TCI completed 482 acquisition or sale deals, an average of one every two weeks.[61] In addition to the Group W purchase in 1985, TCI bought, in whole or in part, firms that included Heritage Communications, Tempo Enterprises, Storer Cable, Televents, Marcus Cable, United Artists Communications, and Cooke Cablevision.

TCI bought Heritage Communications in January 1987. Heritage had been a family affair, established in Des Moines, Iowa, in 1971 as Hawkeye Communications by James M. Hoak. Jim Hoak Jr. was President and his father James Sr. the major stockholder. The name was changed to Heritage Communications in April 1973 as the company grew well past the Iowa state border. By the mid-1980s, Heritage was the nation's tenth largest MSO, with a reputation for integrity and well-run systems. It also had become another 1980s takeover target. The Robert M. Bass Group was an investment organization that had become very active in cable acquisition. In 1986, it began purchasing Heritage stock and was threatening a hostile takeover. Hoak turned to Malone, who bought the company on a handshake, later sending a check for $1.3 billion.

Not all of TCI acquisitions were so straightforward. In 1986 TCI's microwave subsidiary, WTCI, purchased Carl Williams' Televents. In 1988 WTCI merged with Marcus Communications. The new company, WestMarc, then bought Taft Cable Partners in June 1988 for $420 million. In 1989, TCI bought the remaining 25 percent of WestMarc that it did not already own for $206 million. Malone seemed to take special pride in crafting amazingly complex financial arrangements.

In the early 1980s, a pair of Lebanese born businessmen, Robert and Marshal Naify, formed United Artists Communications, Inc. (UACI), eventually expanding it to include twenty-four cable systems, with 750,000 subscribers, and one of the country's largest a chain of film theaters. After several years of courting, TCI in 1986 acquired 51 percent of the company, buying out the brothers' interest, for $420 million in cash and TCI stock.

A year later, UACI announced plans for another merger. The $2.3 billion package would bring together UACI and United Cable (also partially owned by TCI) with Daniels & Associates cable holdings. The three companies were wedded as United Artists Entertainment (UAE) in 1988. With the combined subscriber base, UAE would have been the nation's third largest MSO with more than 2 million subscribers, but those subscribers were tallied at year's end for TCI because the company owned 52 percent of the UAE stock.

The UACI and UAE arrangements illustrated a key TCI business principle. Malone liked to spread the risk and reduce the maintenance on company investments. In addition to its outright purchases, TCI took smaller, noncontrolling interests in dozens of cable businesses. In many ways, TCI became a holding company, with a growing portfolio of diverse cable investments. One of the more important and complicated of this period was the Storer purchase.

In 1985, dissident shareholders at Storer Broadcasting threatened a proxy fight to take control of the company. Their aim was to liquidate the company's assets and reap the cash rewards of the breakup value. To fend off the threat, Storer management called in Kohlberg, Kravis, Roberts & Co. (KKR). KKR was an investment banking firm with a history of cable investment. In 1983, it had purchased Wometco Enterprises—which owned forty-six cable systems with 253,000 subscribers, as well as six television stations—for $842 million. In a leveraged buyout worth about $2 billion, KKR now bought Storer Broadcasting, which included Storer's cable interests.

The investment company quickly sold off the Storer TV stations to Gillette and was looking to unload the Storer Cable company as well. In 1988, a partnership of TCI, Comcast, ATC, and the Robert Bass Group entered into discussions with KKR and an initial agreement was struck to purchase the Storer systems for $2.8 billion. The deal fell apart when the cable partners reportedly attempted to renegotiate a lower price. The Bass group and ATC backed away, and investment bankers at Morgan Stanley swung into action in an effort to salvage the deal. They succeeded. TCI and Comcast agreed to purchase Storer Cable and its 1.48 million subscribers for $1.55 million in cash and $2.1 million in assumed debt. The full package was valued at $2.85 billion, the largest cable TV acquisition to that time. When the ink was dry, Comcast owned 50 percent of Storer Cable. TCI employed its TKR subsidiary, co-owned with Knight-Ridder Newspapers, to run its portion of the Storer properties. Through TKR, TCI controlled 42.5 percent of Storer and Knight-Ridder controlled 7.5 percent.

While of a somewhat smaller scale, another TCI purchase had an interesting historical appeal. Jack Kent Cooke, who had sold his cable empire to Westinghouse in 1981, was attracted again to the business by the middle of the decade. In January 1987, he purchased McCaw Communications, a mid-sized MSO with forty-two systems and 400,000 subscribers, for nearly $800 million. He paid what at the time was considered a near-record price of almost $1,800 a subscriber (although he subsequently bought additional systems at even higher prices). He sought expansion and entered the bidding for larger companies, including Wometco and Rogers Communications, but at nearly every turn, lost to deeper pockets and by the fall of 1988, seemingly unable to grow, decided once again to exit the business. In January 1989, he sold his 700,000-subscriber Cooke Cablevision to a consortium of cable operators for about $1.5 billion. Among the buyers was TCI. Others included Intermedia Partners, TCA Cable, and Falcon Cable TV.

In 1982, TCI had 1.3 million subscribers and was the nation's third largest MSO behind ATC and TelePrompTer. It moved to number one in 1983 and held that position firmly through the decade, reporting 2.2 million subscribers by the end of 1984, 5.1 million by 1988, and 11.3 million by 1990. At the same time, TCI was making similar wide-ranging purchases on the programming side, discussed in detail below.

Malone observed: "we throw a harpoon at everything that goes by," and the fishing in the 1980s was good.[62] In the depths of the cable depression in 1974, a share of TCI stock could be purchased for about a dollar. By 1989, if one included TCI spin-offs, that share was worth $913, an increase of more than 91,000 percent.

Smaller Fish

Other notable mergers and acquisitions of the period included the passage of Wometco Cable television through the hands of several investment groups. In December 1986, KKR sold Wometco to The Robert M. Bass Group for $620 million. A year and a half later, Bass turned the property around, selling it to Cablevision Industries for an estimated $725 million (about $2,300 per subscriber). Cablevision Industries, founded by Alan Gerry and headquartered in Liberty, New York, was one of the acquiring and expanding MSOs. Acquisition of the former Wometco properties from the Bass Group in 1988 took them from about 600,000 subscribers to more than 900,000. Elsewhere, Amos Hostetter's Continental Cablevision purchased American Cablesystems in late 1987 for $481.7 million. Canadian cable company Rogers Communications sold its U.S. properties in 1988. A cable consortium, this one composed of ATC, Paragon Cable, and utility company Houston Industries, paid $1.265 billion for the Rogers' property. Cablevision Systems bought Viacom systems in Long Island and suburban Cleveland and took a 5-percent stake in Showtime/The Movie Channel. Also in 1988, long-time cable executive Ed Allen teamed with Leo Hindrey Jr., who had come out of the publishing business to create a new cable investment firm, Intermedia Partners. It soon would become one of the larger MSOs as well.

While the buying and selling among operators such as Continental and Daniels made industry headlines, such news rarely made it to the popular press. It was a different matter, however, when communications and entertainment giants such as Warner Communications and Time Inc. began courting one another.

Warner and Viacom

Warner's sale of the Atari division in 1984 helped slow the company's financial bleeding, but did not stop it. Warner reported a $586 million loss that year. While its alliance with Chris-Craft was successful in fending off Murdoch, the company was still deep in debt and the new controlling stockholder, Chris-Craft's Herb Siegel, was pressing the company to sell assets to raise much-needed cash. Ross was looking in the opposite direction, coveting a buyout of the American Express interest in Warner-Amex Cable. American Express, which had hoped Warner Cable would be its entrée into Blue Sky financial services, had come to realize the near term improbability of the idea and seemed open for an offer on its stake in Warner-Amex. Siegel and Ross compromised. Siegel gave his approval for the Amex purchase, but only on the condition that Warner sell something to offset the increase in debt. In August 1985, Warner paid $450 million for the American Express interest in the cable company and, within weeks, announced that it was selling most of its key cable programming networks. The buyer was Viacom, with whom Warner had partnered two years earlier in the Showtime/The Movie Channel combination. The sale included MTV, VH1, and Nickelodeon. Viacom also purchased the remaining interest in the Showtime-TMC partnership and Warner's Viewers' Choice pay-per-view service, which had been cultured in the QUBE experiments. The total value of the sale for Warner was $690 million. It was a welcome cash infusion for Warner, and the beginning of a turnaround for the company. Viacom, on the other hand, was on the verge of corporate trauma.

National Amusements Inc. was one of the country's largest theater chains. It was successfully run by Sumner Redstone, who entered the family business when it was just a small collection of drive-in theaters run by his father in the 1950s. The development of cable television, especially pay-cable, and home video convinced Redstone that theaters would not continue to grow as the distribution arm for movies. He began to invest in production, buying stock in the major film studios, including 20th Century Fox, Columbia Pictures, and MGM. In 1985 he began investing, as well, in Viacom.

Following the fashion of the period, a group of Viacom executives, including President and CEO Terrence Elkes, formed, in 1986, an investment group whose aim was to purchase the company and take it private, in part as a shield against potential corporate raiders. With a sizable personal investment in the company, Redstone felt the stock price being tendered as part of the arrangement was far too low, a sweetheart deal for the insiders. He launched a bidding war for Viacom, accompanied by threats of litigation. The war drove up the price of Viacom, from $2.7 billion to more than $3 billion, but in the end, Redstone triumphed, paying $3.4 billion for the company.

In something of an ironic twist, Redstone next cast about for an experienced executive to run the corporation. Those he talked with recommended

Frank Biondi, one-time CEO at Home Box Office. Biondi accepted Redstone's offer and took office as CEO in 1987. He brought with him other HBO personnel, including Winston (Tony) Cox, previously an executive vice president in charge of sales and marketing at HBO. Cox took over as Chairman and CEO of the competing Showtime Networks.

Time Warner Inc.

CEO Steve Ross at Warner, and Time Inc. CEO Dick Munro, along with Time President Nick Nicholas, all watched as the mega-mergers exploded around them. Each company had reasons to be concerned.[63]

Warner, under the frequently irritating but nonetheless effective prodding of Siegel, was regaining its fiscal health. Siegel forced the sale of a host of Warner properties and cut back on corporate expenses. In part through his efforts and in part as a result of some box office hits from Warner Bros. studios, the company began regaining its luster on Wall Street. Being financially sound was not, in Ross's view, enough, however. Ross saw the world, at least in some ways, as did Ted Turner. To compete effectively in the shape-shifting media universe, corporate size and resources were becoming critical. Ross later told those who would listen, "we were too small to be large and too big to be small."[64] It was grow or perish.

Time Inc. determined in the early 1980s that it needed a corporate facelift. The company had been built by the legendary Henry R. Luce on the financial success and public prestige of magazines such as *Time*, *Life*, and later *Sports Illustrated*. By the 1980s, however, one-third of the company's revenue was coming from paper products and forestry and another third from its cable and video units. On Wall Street, Time was seen as stodgy, uncreative, and without clear direction. Merger was one possible path out of the woods so, like Turner several years before, Munro began looking for a partner. He approached most of the major print companies, including The New York Times Co., Dow Jones (publisher of the *Wall Street Journal*), Knight-Ridder, and Gannett. Nothing came of the overtures. Then, in the spring of 1987, Nicholas made a phone call to Steve Ross, just, he suggested later, to get to know one another.

Ross immediately saw a fit between the two companies. Warner had production capacity; Time had distribution resources. Ross floated some joint ventures ideas, although nothing so grand as a full merger. Nicholas and Gerald Levin were intrigued and pursed the possibilities. HBO's Levin, perhaps more than anyone else, saw a Time Warner future filled with movies, produced and delivered to the home under one vertically integrated corporate roof. After weeks of research, consultation, and consideration, Levin penned a memo to Munro in August of 1987 that began, "I am now convinced our primary objective should be merger."[65] It would not be an easy sell, however. From the perspective of corporate culture, Time Warner was not an intuitive fit. Time had a button-down, gray flannel reputation; Warner's was more one of "party hats and horns." Ross was high-spirited, fun-loving, and generous to his friends. He fashioned himself a dreamer more than a manager, spent lavishly on himself, his employees, and his business protégés. For the entertainment of clients, employees, and himself he kept apartments at the Trump Towers in New York and a personal retreat in Acapulco. To whisk everyone around the world in appropriate comfort, a fleet of jets and helicopters was kept on hand. He was a man who exuded calm and elegance and who enjoyed life fully.

By itself, this was sufficient to make the Time board nervous, but there was more. Warner in its corporate form, and Ross personally, had been stung in the 1970s by a criminal probe into the financing and operation of the Westchester Premier Theater in Tarrytown, New York. A host of allegations involving tax fraud, money laundering, and bribery surrounded the development of the theater, allegedly built as a cash cow for organized crime. Warner had purchased $250,000 in stock and another $220,000 was allegedly funneled in through cash bribes. A number of people involved in the theater project were eventually convicted and jailed. Ross was implicated, although never charged in the scandal.

It was the kind of personal and corporate history that made the pinstripe-proper board members at Time turn thoughtfully quiet. Ross was not the kind of person, from top to bottom, who fit the cultural climate of the Henry Luce legacy. There also was the extremely sensitive issue of control—who would command a Time Warner corporation—which neither company cared to cede to the other.

But Levin and Nicholas believed the future of Time was in television and film, not in magazines, and Warner could help take them there. The two convinced Munro and together they eventually swayed the board, although not without caveats. Any deal would have to ensure the continuation of Time's corporate culture; concerns about Ross's personal and business background would have to be assuaged; an iron-bound retirement date for Ross would need to be set and control of the new amalgamation placed in the hands of Time management.

In an effort to appease Ross on this point, it was proposed that he stay on as interim CEO for five years, after which time he would retire and be succeeded by Nicholas. Ross, however, seemed unable to swallow the retirement clause. The talks stumbled. A clumsily worded press release was drafted portraying the merger as an acquisition of Warner by Time. It did not help. The merger talks caved in August 1988 in a reportedly tearful rejection by Ross. All parties walked away assuming it could not be revived (with many on the Time side elated at the thought). All were mistaken.

Word of the discussions leaked to the press, putting Time, once again, in play for corporate raiders, and the company resumed its hunt for partners, talking seriously, albeit briefly, with Cap Cities/ABC. One Time board member, however, continued to believe in the merger with Warner and would not ease his grip on the idea. Mike Dingman had become a multimillionaire by buying and selling companies of every stripe, and he knew the art of the deal. Dingman had dinner with Ross, attempting to rekindle the merger, and found Ross newly receptive. Ross appeared to have a change of heart with respect to the retirement issue, thinking it was perhaps time to throttle back from daily corporate responsibility and spend his hours in broader strategic thinking. He would still need to retain the title of co-CEO, however. Nicholas and Levin went back to work trying to sell the Time board on the proposal. It was easier with Ross's assurance that he would step aside in five years, but once again, as the two side attempted to consummate the contract, it nearly folded as Ross declined to write his retirement formally into the legal language. A new compromise was reached in which Ross would step down as co-CEO in 1994 but remain Chairman of the Board of the company. It was crafted as a legally, nonbinding understanding, but it was enough to satisfy the various factions.

The merger of Time Inc. and Warner Communications was announced in March 1989. It was structured as a noncash transaction, an exchange of stock, valued at $18 billion. The celebrations at corporate headquarters, if there were any, were cut short, however, in June when Paramount's Martin Davis, declaring that the merger put Time Inc. into play, initiated a hostile takeover attempt, offering $10.7 billion for the company.[66] The carpeted hallways of the executive suites at Time rolled and buckled with the shock wave of the announcement. A war room was furnished and Time leadership steeled itself for corporate battle. Time rebuffed the Paramount offer in the most blunt terms imaginable. Munro's letter to Davis included words such as "deceptive," and "rubbish," stating: "Your tender offer raises serious questions about your integrity and motives."[67] Davis responded by raising his bid to $12.2 billion. The press savored each thrust and parry, and the financial struggle morphed into an equally horrific court battle.

Ultimately, Time prevailed. The Paramount bid, however, forced Time and Warner to restructure the terms of their agreement. Time took on $14 billion in debt to finance an outright acquisition of Warner. The massive debt, in turn, restricted its ability to move agilely or aggressively into new competitive markets. Some suggested this handicap was a fallback intention of Davis in the takeover attempt. To help reduce the huge debt, the company, in late 1991, created a new subdivision, Time Warner Entertainment (TWE), selling 12.5 percent of the division to Toshiba Corp. and C. Itoh & Co., for $500 million each. TWE became the holding company for Warner Brothers Film and Records, HBO, and Time Warner Cable.

The new Time Warner Inc. nonetheless became the largest media conglomerate in the world and the consolidation of operating units began. In September 1989, Time Warner merged its cable systems under the direction of ATC Chairman Joe Collins. It was the nation's largest cable provider after TCI, with 5 million subscribers. All the normal economics of scale that attach to corporate size now were at the disposal of the new company. More importantly, with the nation's second largest subscriber base, Time Warner had the critical mass necessary to launch new programming initiatives without having to pass through the monopolistic gates of TCI. The Time Warner merger would be the last major cable acquisition of the period. In October 1989, the U.S. Comptroller of the Currency changed the rules governing highly leveraged transactions (HLTs).[68] The action was intended to restrict the amount banks could lend to debt-ridden corporations, and was sparked primarily by massive failures in the savings and loan industry. The new rules caught cable, along with other industries, in its net, however. Nervousness in the junk bond market splashed over into the broader stock indices. Growing concern over HLTs triggered a major sell off on October 13, 1989; The Dow Jones Index plunged 270 points in a two-day period.

In November, Congress struck a second hammer blow, as Senator John Danforth (R-Missouri) introduced legislation to re-regulate cable. The bill called for controls on rates, audience reach, and ownership and was part of a growing movement to curb perceived industry excesses (detailed more fully in Chapter 11). Wall Street analysts revised their upbeat view of cable and portfolio managers began selling cable stock.

In final weeks of 1989, cable stocks slid an average of 20 percent.[69] Between October 1989 and May 1990, the Kagan and Associates cable stock index dropped an average of 40 percent.[70] Economic contraction gripped the industry. Operators cut back on spending, in turn crimping the construction sector and equipment suppliers. Personnel layoffs were common in the months in early 1990. System valuations sank, and sales and acquisitions slowed dramatically.

The HLT rule change was an important reason, but not the only reason, for ebbing merger activity. The prior frenzy of sales and acquisitions had substantially reduced the number of smaller companies on the market. By the end of the decade, many of the attractive and available systems had been purchased, and a host of well-known names in cable television erased through sale or merger. They included: Storer, Wometco, Daniels, United Cable, Rogers, Taft, Centel, and Cooke.

The Programmers

Mergers and acquisitions on the distribution side reduced the number and increased the size the nation's MSOs. Cable networks, meanwhile, continued to proliferate as channel capacity grew along with consumer demand for new programming. Nearly thirty new services launched between 1985 and 1992. The major start-ups of the period included the Discovery Channel, TNT, and CNBC, which are discussed in detail below, along with one of the biggest splashes in the industry in the late 1980s, the shopping channels. But a host of smaller services were inaugurated as well. C-SPAN II began coverage of the Senate in June 1986. Trans World Airlines (TWA) launched the Travel Channel in February 1987. That same year, Glen Jones Mind Extension University began operations, offering extended education and college credit programming in cooperation with colleges and universities around the country. An interdenominational religious network, VUSN (Vision Interfaith Satellite Network) opened in September 1988. In 1987, the Spanish International Network changed its name to Univision, and the network was purchased the following year by an expansion-minded Hallmark Cards, Inc., backing their commitment with a multimillion dollar programming enhancement plan and the hiring of former ESPN head J. William Grimes as its new president. In 1992, Hallmark sold Univision to a group of investors that included Venvision of Venezuela and Grupo Televisa of Mexico. In July 1991, a niched service focusing on crime and criminal proceedings and feeding off the growing willingness of courtrooms around the country to open their doors to TV cameras began operations. CourtTV, The Courtroom Television Network had heavy support from Cablevision and TCI.

In July 1987, a programming service called Movietime took to the air, focusing on coverage of the entertainment industry, soft news, and features devoted to films and show business personalities. MSO investors included ATC, Continental, Cox, Newhouse, United Cable, Warner, and HBO. The name of the service led many viewers to believe they were getting a new movie channel, however, and in May 1990, the service was relaunched as E! Entertainment Television.

In the mid-1980s, regional sports channels, offering fanatic local fans opportunities to follow their teams for only $10 a month, came into vogue. Prism had been operating in Philadelphia for years and was now joined by regional sports services in New England, New York, and Los Angeles. Dolan's Rainbow Program Holdings purchased several of the larger regional services and combined them in 1986 into a family of networks, sharing costs where possible, under the umbrella of SportsChannel America. Another interlinked family of regional sport networks, Prime Network, was formed by Daniels and Associates, using its existing Prime Ticket network. The choices became so numerous that on-screen program guides, listing the programming on all the other channels, were introduced. United Video inaugurated a set of program guide channels, including the Prevue Guide, EPG (electronic program guide), and EPG, Jr.

Existing services rode the escalator of the cable boom to new heights as well. Between 1986 and 1988 alone, A&E gained more than 14 million subscribers, BET nearly 7 million, CNN 13 million, C-SPAN 13 million, Lifetime 15 million, and the Weather Channel 15 million. Other services made similar gains. By the end of the decade, the leading cable networks, including ESPN, CNN, USA, and MTV, had access to approximately 50 million cable television homes. In 1983, ESPN surpassed WTBS as the nation's most watched cable network, although it would be two more years before the sports channel became profitable.

In the earliest days of television, professional wrestling was a popular programming staple, cheap to produce and a hit with bar patrons across the country. In another recycling of history, professional wrestling arose again in early cable programming. In 1985, USA signed contracts to carry matches of the World Wrestling Federation (WWF). It was a huge success and soon a mainstay of basic and pay-per-view programming (see discussion below). WWF shows became USA's most watched fare. A live, three-hour "WWF Royal Rumble" in January of 1988 was the network's most heavily watched program in history, to that point, with more than 3 million TV homes tuned in and a Nielsen rating of 8.2.

In 1988, with the wire passing more than 50 percent of the nation's homes, cable programming became eligible for Emmy Award consideration and won three statuettes in its first primetime competition. By the end of the decade, more than sixty basic cable programming services and ten premium or PPV channels could think about competing in the Emmy's and on the cable screens of America.

Programming Economics

Creating a successful cable network required, of course, more than just a press conference and a glossy brochure, although some hopefuls had little more. The economics of national cable programming demanded a strict set of base requirements, one of which was a very large bank account; most nascent networks went for years before showing a profit. It also required a business plan that identified a market niche not already staked out by an existing service, a supply of content priced sufficiently low to make the financial equation work and access to a critical mass of eyeballs.

Programmers had to be careful in developing niched services. If they were too narrowly cast, they would not generate sufficient reach to appeal to operators or advertisers. In fact, the most financially successful programmers, despite popular perception, would turn out to be those that mimicked the established general-entertainment broadcast networks. These included USA, TNT, and to a lesser extent, ESPN. At the same time, operators were loath to add what they saw as redundant services, so neophytes had to secure claims in new demographic or concept areas.

On the cost side, programmers had to be mindful of the prices they paid for their raw materials. CBS Cable went bust in part because of the millions it poured into content. CNN owed some of its success to the bare bones economics of its operation. Luckily for many programmers, there was a wealth of cheap material on the market. The 1970 PTAR and financial interest and syndication rules fostered the creation of a deep well of programming, both off-network (network reruns) and first-run syndication shows. The rise of non-network production capacity, both on the part of the film studios and through independent production companies, helped feed the growth of independent broadcast television in the 1970s and 1980s and much of that material later became available for cable operators looking for cheap TV fare.

A critical factor in network success, as noted in the previous chapter, was cable carriage. Operators needed content, but programmers could not survive without adequate distribution. In the late 1980s, critical mass, the audience necessary for a network to become self-sustaining, was pegged at between 10 and 15 million subscribers. When A. C. Nielsen first began measuring cable network viewing, it set its benchmark, the penetration level at which it would begin reporting a network's performance, at 12 million subscribers, about 15 percent of the U.S. TV home population at the time,[71] and only when Nielsen began reporting ratings, could a network have any hope of attracting prominent national advertisers who lived and died by the Nielsen numbers.

Moreover, as the number of programming hopefuls expanded, almost geometrically in the mid- and late 1980s, their numbers began to outstrip average system channel capacity. By early 1991, operators could choose from among more than eighty basic and premium channels, not including the local broadcast stations most carried. Only about 28 percent of all systems could carry more than fifty-three channels, however. About 64 percent of the nation's operators had between thirty and fifty-three available channels (see Appendix B).

Finally, coming full circle, break-even levels of carriage also were a requisite for financing. A programming entrepreneur without assured or probable MSO distribution was much less likely to secure the investment dollars needed to cover start-up costs.

Integrating Vertically

Cable operators, as previously noted, had pursued the benefits of vertical integration since Irving Kahn sold his first pay-per-view fight. Time and Warner, individually, had been pioneers in melding distribution with programming capacity. As cable construction neared national capacity in homes passed in the late 1980s, MSOs looked to feed the beast they had created. The largest companies began forging their own networks and purchasing, in whole or in part, existing ones. The goal, illustrated by the MSO bailout of Ted Turner, was not simply to create additional content, but to control it as well.

Deregulation was helpful in this pursuit. Increased subscriber rates meant increased cash flow and, in turn, greater borrowing power. It all meant more money to invest in content. As Malone explained, "What it [the Cable Act] did was create liquidity, which we used to acquire additional cable operations, rebuild plant and invest in programming. With deregulation you are basically a [program] purchasing agent for your subscribers. If you think it is a good shot to go into and invest in new programming, you can go ahead and do it."[72] And they did.

The larger cable companies became finely honed professionals in the art of developing equity relationships with programmers. There were many benefits. Ownership gave MSOs influence over network design, philosophy, content, and price. Operators could share in revenues from a successful programming service, and they claimed vertical integration gave them a first-line contact with viewers and, therefore, a better feel for consumer programming preferences. Time and Viacom, both top-ten cable distributors, extended and strengthened their programming arms in the late 1980s.

Time Inc.

While the merger with Warner in 1989 constituted Time's biggest programming play of the decade, the company was not idle in the few years leading up to the union. As noted, Time bought into the Turner bailout and, in 1985, took an equity position in BET. Not every initiative paid off, however.

On the pay-TV side, Time failed in its effort in early 1986 to develop a new and purposively targeted movie service. Christened "Festival," Time hoped it would attract the "nevers," people who had never tried a pay service, and perhaps appeal to disconnects, too. It was additionally positioned to appeal to families and featured PG-rated films.

HBO learned, however, that viewers who were not interested in HBO or Cinemax were often not interested in pay-cable at all, and subscribers who wanted child-safe programming were more inclined to watch the known and safe Disney Channel, when it was available. The distinction between Festival and the other pay services was not always made clear to the customer and was frequently promoted only after a subscriber had turned down the more popular channels. Some nonaligned operators also were reluctant to open another channel to an HBO service. Sales ran below expectations and HBO pulled the plug in July 1988.

Festival also suffered from ailments endemic to all pay services in this period. The rising use of home VCRs and VCR rentals drained potential subscribers. Deregulation took a toll as operators increased basic rates, pinching customers who in turn cut back on premium services. Churn also was exacerbated by repeated showings of films and lack of product differentiation among the competing services.

Time continued signing contracts with the Hollywood majors. Between 1985 and 1987, HBO signed licensing agreements with MGM/UA (1985), Columbia (1985), 20th Century Fox (1986), and Warner Bros. Through 1987, Showtime held an exclusive distribution contract with Paramount, but HBO outbid its rival for a new five-year eighty-five picture agreement, paying Paramount an estimated $500 million for the exclusive rights to properties that included the valuable "Star Trek" and "Raiders of the Lost Ark" series.

In addition, Time created its own film studio after withdrawing from the Tri-Star pact. HBO owned 25 percent of Tri-Star Pictures, as did Columbia and CBS. The remaining 25 percent was traded publicly. In 1985, CBS sold its interest in the company to Columbia. Tri-Star was looking to develop film projects for network television and the financial interest and syndication rules prohibited that as long as CBS was a partner. In 1986, HBO followed CBS in selling its shares in the company to Columbia, and formed its own branded HBO Pictures. In July 1987, Time also sold its stake in USA to its partners, Paramount and MCA, which each then controlled 50 percent of the popular cable network.

Viacom

When Viacom purchased the Warner programming properties in 1985, it took control of some of the most important cable channels on the dial and acted promptly to review their performance, making adjustments where it thought necessary.

MTV's original format was based on radio station programming, a constant stream of music videos and commercials, a pure flow approach, as some called it. The format seemed to be wearing thin, however, and Viacom revised the channel, shifting it to a more traditional format of half-hour and hour programs, with specific themes and targeted demographics that could be more readily sold to advertisers.

Nickelodeon had gone through significant changes before the Viacom acquisition. From its inception, Nickelodeon had a reputation as the TV equivalent of green vegetables for kids. Parents liked it for its wholesome programming, but the target audience itself, children, by and large were not attracted. That began to change in the early 1980s under the direction of Geraldine Laybourne, who helped reposition the service to one more appealing to the intended audience and, at the same time, one positioned to generate greater cash flow. In October 1983, Nickelodeon began selling commercials, much to the vocal concern of parents and citizens groups. In 1984, it relaunched as the New Nickelodeon, with a philosophy of being "on the side of the kids themselves." With more humor, more entertainment, and more feedback from their audience, ratings began to climb.

At about the same time, in July 1985, programming was expanded to 24 hours a day, with a large caveat. Adopting a practice of the previous decade, Nickelodeon ran its children's programming only during the day, turning the evening hours over to a new format. It was called "Nick at Nite," which was composed primarily of network television reruns from the early 1960s and 1950s. At 8 PM, programming switched from the children's shows to such TV classics as "The Dick Van Dyke Show," "Green Acres," "Mr. Ed," "Dobie Gillis," and "My Three Sons." It drew new viewers and the low-cost nature of the recycled programming made for appealing financials.

Viacom also shared a 33 percent interest in Lifetime, along with Cap Cities/ABC and Hearst.

Viacom and HBO both pursued the comedy niche in the late 1980s, riding a popular wave of interest in stand-up comedy shows. In May 1989, HBO announced the start of The Comedy Channel. Two days later, Viacom's MTV Networks unveiled its own HA! TV Comedy Network. The Comedy Channel launched in November 1989. HA! began operations April Fools Day 1990. The market could not sustain both, however. Operators remained reluctant to carry what they saw as redundant services and, in December 1990, HBO and Viacom announced a merger of the services. The jointly owned CTV: Comedy Television began operations April Fools Day 1991. A trademark conflict with Canada's CTV Television Network prompted a name change, however, and the channel became Comedy Central in June 1991.

On the pay-TV side, Showtime Networks, as HBO, sought long-term arrangements with major studios to stabilize the flow of feature films. It paid $500 million, for example, for exclusive rights to product from Disney subsidiaries Touchstone and Hollywood Pictures, through 1996.

The film deals constituted a substantial outlay for both companies, however, in a period in which pay-cable audiences were fleeing for the flexibility of home video. Locked into the expensive long-term studio contracts, the premium services were suffering by the end of the decade.

"Independent" Programmers

Programming services not directly created by the major distributors, such as Time and Viacom, had limited options for distribution. Some tried leased access, which local operators were required to provide under the 1984 Cable Act, but leased access was cumbersome. Programmers had to negotiate individual contracts with thousands of systems, and local operators had wide latitude to set terms and conditions. Some critics charged that the operators made unreasonable demands in order to maintain control of their channels and ward off possible competition.[73] The operators responded that it was difficult to set a value for access channels that allowed them to be priced with any rationality. In practice, leased access became a vehicle primarily for home shopping and religious channels (discussed below) that could not otherwise obtain basic carriage.

More typically, programmers sought carriage contracts with the major MSOs, who increasingly ended up as equity partners in the service. Program services that began life as independents, rarely ended up as such.

The largest operators, such as TCI, had powerful leverage to determine which networks were carried, and therefore survived, and which were not. This power became especially important in the latter 1980s when the number of proposed and available programming services began to outstrip the channel capacity of many systems. As shelf space relative to supply shrank, vertically integrated companies were likely to favor networks in which they held an equity interest.

Even when the large operator did not have an ownership tie to a programmer, the sheer size of MSOs, such as TCI or Time Warner, bestowed on them all the standard perks associated with economies of scale. By using their national reach to negotiate for deep discounts, the major MSOs held a competitive advantage over smaller firms.[74] A 1988 NTIA Report revealed, for example, that CNN was charging MSOs with more than 5 million subscribers $0.02 per subscriber per month, but charging companies with fewer than 500,000 customers $0.29 per subscriber per month.[75] Turner, in fact, was one of the sharpest examples of the deeply intertwined interests of some MSOs and programmers.

Turner

By 1988, Turner shared his company with nearly every major MSO, including TCI, Time, Warner, and Continental. He used the cash infusion from the buyout and subsequent refinancing to retire his debt and generate money for new ventures. Turner had long seen himself going head-to-head with the major broadcast networks. Now he had sufficient monetary and programming resources to make the attempt. He was additionally driven by the desire to replace or at least supplement WTBS. In 1987, the FCC was preparing a revision of its syndicated exclusivity rules (see below) that threatened to push WTBS off of thousands of cable systems around the country. For Turner, creation of a "syndex-proof" network seemed prudent. In October 1987, he announced plans for Turner Network Television (TNT). He would use the programming obtained from Kerkorian as the foundation, augmented with material from the sports contracts he had recently signed.[76] TNT debuted in October 1988, following a heavy publicity campaign that featured one Turner's favorite MGM prizes, the film *Gone with the Wind*. Repeated showings of the classic film highlighted the start of TNT, which settled in to a steady schedule of movies, major league sports (including the National Football League), and children's programming.

When he struck the deal with Kerkorian, many industry observers wrote that Turner had dramatically overpaid for the property. Hundreds of the films acquired in the MGM purchase were licensed to broadcasters or other programming services. AMC, for example, had exclusive cable rights for scores of old Warner Bros. and MGM films and Turner eventually paid $50 million to regain those rights. Films under contract to others, however, would not be available to TNT for years. But Turner was looking far down the road. When he began WTBS, he saw the value in owning rather than renting his programming and it was one of the keys to his success. With the MGM deal, he now possessed many of the most important film features of all time, and analysts who a few years before had criticized him were taking a fresh look.

Turner also surprised some critics in 1991 when he purchased Hanna-Barbara productions, the famous cartoon production house responsible for such Saturday morning staples as "Yogi Bear" and "Huckleberry Hound." He paid $320 million for Hanna-Barbara's corporate parent and created his own Cartoon Network. It was one of the most successful cable network launches of all time. The Cartoon Network debuted in October 1992 and capitulated into the top twenty cable network list within a year.

A cable programmer did not need the size of a CNN to attract MSO interest. In some ways, American Movie Classics was more illustrative of the period than was Turner. Following the settlement with Turner over cable rights to classic MGM films, AMC had a $50 million acquisition fund for new content, but it still needed distribution. In March 1986, Rainbow signed an agreement with TCI. It gave AMC the subscribers needed to become financially solvent. In return, TCI took an equity position in the classic movie service. TCI offered AMC as a low-priced addition to pay services, such as HBO, charging only $1 a month, and in 1987, AMC was revamped as a basic service. United Cable subsequently joined the pact and, by 1989, AMC was owned by Rainbow Program Enterprises (Cablevision Systems 50 percent and NBC 50 percent), 50 percent; TCI, 35 percent; and United, 15 percent.

Other significant industry-wide programming investments of the period included the rescue of the Discovery Channel and the brief, but heated shopping channel frenzy of the late 1980s.

Discovery

One of the most important and successful launches of the mid-1980s, and a prime example of MSO investment in programming, was the Discovery Channel. The Discovery Channel opened for business on June 17, 1985. It was the creation of John Hendricks. Hendricks had been a fundraiser at the University of Maryland and a media relations consultant. By 1982, he was running a company that distributed educational audio-video materials to colleges and universities and he saw the possibilities of a cable service featuring nonfiction documentary and educational programming. Through his work in education, he knew that libraries of such material already existed and it was in use in educational institutions and libraries around the country.[77]

Hendrick's had a good concept, building a network on relatively inexpensive and readily available programming. He had start-up dollars from New York Life Insurance Co., venture capitalists Allen & Co., and Westinghouse, but the channel failed to get early traction. Cable carriage and advertiser response were well below that anticipated by the original business plan. Hendricks was near to pulling the plug on Discovery and began looking for additional financing. He ended up on John Malone's doorstep. TCI executives suggested a new business strategy that featured future rebates on advertising revenue to cable operators, but also called for typical network license fees; Discovery had originally been offered free to operators. TCI then talked a number of other major MSOs into investing in Discovery. By June 1986, TCI joined with United Cable, Cox, and Newhouse, pouring a combined $20 million in the service. Ultimately, the MSOs controlled 56 percent of Discovery, but with guaranteed carriage on the nation's largest systems, the new service was well poised for subsequent success. Because of the wholesome nature and eventual popularity of the service, it was an investment Malone took personal satisfaction in recalling as an illustration of operator support of quality programming.

> John Hendricks was a good example. The Discovery Channel idea was his and initially some venture capital money from Allen & Co. and Westinghouse funded it, but they told him that they couldn't put any more money in and so they had withdrawn and he was about to shut down, and John Sie, for us, ran around to see if he could put together an investor group, which he was successful in doing and it was basically us, and United, Gene Schneider, and Newhouse, and Cox and we're still partners in it today. It's been a great business. So there were a long list of programming ideas that blossomed during that period and some of them have been wonderful successes.[78]

With a new business plan and a cash infusion from the major cable operators, Discovery built a programming mix of nature shows, history, adventure, science and technology, and a reasonable measure of original material. It grew rapidly and, by 1992, was on systems that passed more than 57 million homes.

In 1991, Discovery also purchased The Learning Channel, previously owned by FNN parent corporation Infotechnology and by its founding group, Appalachian Community Service Network, although not without some intraindustry friction. Viacom had bid $50 million for TLC, but as the sale moved toward completion, Malone stepped in, announcing that TCI had detected an erosion in the quality of The Learning Channel programming and was considering removing the service from all of TCI's 12 million homes. Without TCI carriage, the value of TLC would have dropped precipitously, a threat sufficient to cause Viacom to withdraw its offer. Malone then suddenly found that TLC was a quality service after all and Discovery, partially owned by TCI, made a successful $32 million bid for the service. Viacom, as might be expected, sued.

Shopping

One of the most lucrative of cable's discoveries in the 1980s was the home shopping concept and the major MSOs, including TCI, were quick to secure stakes in this new breed of consumer entertainment. Home shopping television had its genesis in federal deregulation and in an earlier form of direct-sales television called "infomercials."

For years, the FCC had effectively limited the amount of advertising time on broadcast television to 16 minutes an hour,[79] but in the summer of 1984, the Fowler FCC swept away those restrictions.[80] The Commission additionally eliminated its rules against program-length commercials.[81] Television stations from that time on were permitted to determine for themselves how many commercials they ran each hour, and whether or not to air program-length ads. The direct consequence was the explosion of syndicated infomercials, 30-minute and 60-minute programs designed to sell a particular product of sometimes-questionable quality with often questionable pitches.[82] Carnival barker hosts would pitch the virtues of Ginzu knives, cabbage shredders, and specialty sunglasses. Often cheesy to the point of being humorous, infomercial producers sometimes ran afoul of regulators both for questionable product claims involving weight loss or cures for baldness, and for formats resembling newscasts and interviews that were arguably misleading to audiences.[83]

While some infomercials therefore raised the eyebrows of the Federal Trade Commission and Congress, for broadcasters and cablecasters the economics were exceptionally attractive. Infomercial producers bought program time that otherwise would have gone fallow, usually in the overnight hours from midnight to 6:00 AM or on Sunday mornings. For struggling young cable networks, deep in debt and in need of cash flow, infomercial money was a godsend and became a staple for programmers such as BET and others. Moreover, the infomercials were often amazingly successful. From an estimated $10 million in sales in 1984, the infomercial industry grew to more than $1 billion in sales in 1994. It did not take long for the hour-long programs to evolve into 24-hour-a-day networks.

In 1977, Lowell W. "Bud" Paxon had a problem: too many can openers, 112 to be exact. Paxon along with his partner, Roy M. Speer, managed an AM radio station in Dunedin, Florida, just north of Clearwater. The can openers came from a bankrupt advertiser paying his bill in merchandise instead of cash. Paxon thought they might try to recoup some of the loss by selling the can openers on the air at below retail price. Not knowing what to expect, they were delighted at the enthusiastic response, so much so that they began a daily sales program and turned the front lobby of the station into a merchandise pick-up center.

In 1982, Paxon took the concept to cable, leasing channel capacity on the Vision Cable system in Clearwater. The lesson of radio—that listeners and viewers flocked to the format, calling in the orders and credit card numbers—was repeated on cable. By 1985, with the restrictions on broadcast advertising time gone, Paxon and Speer were ready to go national and launched the Home Shopping Network (HSN) in July.

The Home Shopping Network was a cable phenomenon. It pulled in viewers and dollars like one of its bargain basement vacuum cleaners. In addition to its popularity with a certain segment of the audience, the economics of Home Shopping (the "shoppers" as programs were called) were compelling for the industry.

For the cable industry, the "shoppers" offered a unique and appealing financial scheme. Unlike advertiser-supported or pay channels, the shoppers make money like any retailer, they take a percentage of the revenue generated in the on-air sale of the products. Most of the money goes to the network but local systems and MSOs usually receive about 5 percent of the revenue generated by sales in their area.

The home shopping format was subject to some ribbing, even ridicule, lampooned as the "Zirconium Network[s]," for the prevalence of cheap zirconium jewelry regularly featured. But the shoppers had one important saving grace, like infomercials, they were highly profitable.

The success of the HSN and the relative ease with which its format could be duplicated led to rapid cloning. Among the first to see the possibilities were the MSOs themselves. In May 1986, a consortium of large operators led by TCI helped launch the Cable Value Network (CVN). Irwin Jacobs, a 1980s financier, controlled a Minneapolis-based discount merchandising company called C.O.M.B., Co. He thought he could emulate the success of CVN and he started with a phone call to Malone. TCI and a consortium that included most of the top twenty operators purchased a joint 50 percent interest in the network and signed the all-important carriage contracts, guaranteeing national distribution. At the same time, many of those companies, including TCI, decided not to carry the founding competitor, HSN.

HSN was locked out of many of the nation's cable markets, but it nonetheless reached 15 million subscribers by 1986 and its cash flow was strong. In fiscal 1985–1986, its first full year of operations, it reported $17 million in net income on $160.2 million revenue. It went public May 1986, and the stock immediately headed skyward, climbing from an initial offering of $18 a share to $120 in only a few months. It used its leverage and its potential to go on a broadcast buying spree, acquiring more; about a dozen languishing UHF TV stations in the top twenty markets, in some cases failed STV operations. It soon was offering two 24-hour networks, HSN original, aimed at cable systems and HSN II, featuring more upscale merchandise and created specifically for broadcast outlets as well as interested cable systems.

Within a year, real and proposed shopping networks were popping up everywhere. FNN started its own "Teleshop" in August 1986, and Tempo briefly converted to a shopping format. Before the year was out, the imitators included The American Shopping Channel (Cox Cable), the Video Shopping Mall, Sky Merchant (Jones Cable), Cable Shopper's Network, Shop TV (J.C. Penny), and the Consumer Discount Network. Among the contenders was the Quality Value Network (QVC).

QVC network was created by Joseph Segel, who specialized in entrepreneurial start-ups and had experience in the direct-marketing business. The founder of the Franklin Mint, he was semiretired and living in Florida when he saw an opportunity in the new world of home shopping television. In November 1986, he started QVC, offering small amounts of equity to MSOs in exchange for carriage. Among the early takers was Comcast, one of the few major MSOs that had decided not to take part in the CVN alliance. QVC, in fact, was housed near Comcast headquarters in West Chester, Pennsylvania, a suburb of Philadelphia. By 1989, Comcast owned 14 percent of the network; United and other MSOs, owned 21 percent.

As is frequently the case in the discovery of a successful new product or service, the frenzied round of start-ups was followed, in this case quite quickly, by a shakeout. The weaker businesses failed or were absorbed by the stronger. By 1990, Teleshop had folded (in part a victim of FNN's wider problems), the Fashion Channel was purchased by CVN, the Sky Merchant acquired by HSN, and Tempo, as discussed elsewhere, sold eventually to NBC. The largest merger, however, was that of CVN and QVC.

TCI, which backed CVN, bought 22.7 percent of QVC in 1987. A number of MSOs, including TCI and Time, now were invested in both services. Moreover, with assured carriage, CVN and QVC became the second and third largest cable shopping channels by 1987, behind HSN. In summer 1988, reportedly at the urging of TCI, QVC announced its hopes to buy CVN for estimated $384 million. The deal was consummated in 1989 and QVC absorbed CVN.

At the end of the decade, home shopping was dominated by just two firms, HSN and QVC/CVN, but the consolidation was not yet complete. In early 1992, TCI's Liberty Media (see below) acquired a controlling interest in HSN and bought the company outright in December for $150 million.[84]

Religious

The interlocking tendrils of MSO-programming partnerships extended even into the realm of televised religion. Through the early 1980s, CBN attempted to broaden its programming base, adding family-oriented, off-network programming to distinguish itself from the other religious cable networks that were beginning to proliferate at the time. But its staple of network reruns, especially old westerns such as the "Rifleman," brought it trouble in 1986 in the form of a report by the National Coalition on Television Violence, which announced that CBN rated very high in its index of televised violence. CBN criticized the report, but it was nonetheless a dart in the side of the network's conservative reputation, as was a short-lived attempt at an evening news program, which was attacked for what many perceived as a lack of journalistic objectivity.

Ratings for the network continued to be strong, however, and, in October 1987, Robertson, who had been actively involved in national politics for a number of years, announced that he would become a candidate for the presidency. Whatever the cultural or political legacy of Robertson's run for the GOP nomination, the impact on the cable network was unambiguous and painful. With Robertson on the campaign trail instead of hosting the "700 Club," viewership plummeted and, along with it, contributions, which constituted nearly three-fourths of the network's revenue. The budget was slashed by $34 million, and 645 workers were laid off.[85]

Robertson's bid for the presidency was sputtering at the same time. While generating a great deal of media interest and some early measure of voter enthusiasm, he lost steam as voters learned more of his evangelical background, and scandals surrounding other TV ministries cast their shadow on his activities as well.

In March 1987, Jim Bakker resigned as Chairman of the PTL organization, turning over control of the struggling concern to then-ally Jerry Falwell. Bakker had made national headlines following disclosure of an extramarital affair with a former secretary, Jessica Hahn, and subsequent attempts at a cover up and blackmail. A year later, in February 1988, the popular TV evangelist Jimmy Swaggert fell into public disgrace when his liaisons with a prostitute were disclosed. Swaggert fell from the national scene, but Bakker's troubles continued. An IRS investigation led to charges of tax evasion, a conviction, and jail time for the one-time evangelical minister.

Against the background of televangelistic scandal, Robertson's presidential bid ground to a halt. He gave up his political ambitions and returned to CBN in May 1988. The network almost immediately moved toward an even more secular programming diet, adding shows such as "Gunsmoke," "Burns and Allen," and "Bonanza." The formula worked and CBN began a solid ratings and revenue recovery. That success, however, carried its own consequences. Income grew so quickly that the network's tax exempt status as a nonprofit religious organization came under peril; the IRS began an examination of its operations. In August 1989, Robertson, in an effort to appease the IRS, attempted to distance the now-lucrative TV business from the church by renaming the network, "The Family Channel." The name change alone was not sufficient, however. Robertson decided to more formally separate TV side from the evangelical CBN ministry. He needed capital to do so, however, and hired a cable investment banking firm, Communications Equity Associates, to look for financial partners. The search caught the eye of John Malone. As Malone later explained "Robertson said, 'If you'll make an investment in my channel, I'll be able to restructure, take it out of the church, pay the church for the channel, and retain the format.'"[86]

Malone was always looking for promising investment opportunities, especially in programming, and liked CBN's family-oriented niche and its solid track record with viewers. He invested $45 million, and the new partnership

(TCI controlled 18 percent) formed International Family Entertainment (IFE) in early 1990. IFC, in turn, purchased The Family Channel for $250 million, money it borrowed from CBN.[87] Robertson remained chairman of IFC.

The newly structured "Family Channel" continued to move toward more mainstream programming in the early 1990s, although it retained some of its conservative religious trappings, most pointedly its nightly news-talk program, "The 700 Club."[88]

FNN and CNBC: It's Just Business

Not all the dealings between operators and programmers went smoothly or ended in success. The cable business forged odd alliances, companies that entered into joint operating agreements one year might appear in court against one another the next. It was not personal, it was just business. The creation of one of the important and lasting services of the latter 1980s was a case in point. CNBC was the child of NBC broadcasting, but helping and hindering in the delivery was a roster of cable's biggest players including Malone, Turner, and Dolan.

NBC had tested the cable waters in the early 1980s with The Entertainment Channel. Its failure and subsequent merger with ARTs left NBC with 33 percent of A&E, but the network wanted a channel of its own, exploiting its existing programming capacity, especially in the area of network news. In 1985, unimpressed by the lesson of SNC, NBC began floating the idea of an all-news channel to compete with Ted Turner. NBC was given some encouragement by operators who were stinging from CNN rate increases imposed by Turner following his battle with SNC. But Turner, never one to tarry in responding to a competitive threat, cut new deals with the larger MSOs, including TCI, for deeply discounted rates. He also dropped plans for a TVRO distribution service that had irritated cable operators. Turner wooed the industry back to his side and, by early 1986, NBC saw the wisdom of abandoning its cable news plans. The appetite for a channel remained, however. NBC discussed equity participation with Turner, Movietime and others, without success, then began talking with Malone.

TCI purchased Tempo Enterprises, formerly Ed Taylor's Satellite Syndicated Systems, for $46 million in 1988 following months of negotiation. As discussed in Chapter 11, TCI's primary interest was in Taylor's satellite resources. The company's programming service, Tempo Television, was under-performing and TCI had little interest in it. In fact, *Cablevision* noted that in 1988 only two of thirty-nine basic cable networks had lost subscribers the previous year, Tempo and The Inspirational Network (formerly PTL, the network of disgraced televangelists Jim and Tammy Faye Bakker). TCI decided to spin off the Tempo channel, selling it to NBC for $20 million.

Tempo may not have been pulling in subscribers, but it did have existing contracts with operators and carriage that reached 5 to 6 million subscribers.

To that number, and as part of the sale, TCI agreed to add 8 million basic cable customers. It was a substantial base from which to start and just the kind of window NBC was seeking. In May 1988, NBC announced the acquisition and plans to convert Tempo to a business news service during the day and a sports programming service in the evenings and on the weekends.

NBC Cable had several difficult obstacles to overcome, the first being another confrontation with Ted Turner. Turner still feared NBC as a possible all-news competitor, despite assurances that the network would steer clear of general purpose news. To further mollify Turner and the cable operators who supported him, NBC formally changed the name of the channel to "The Consumer News and Business Channel." Turner was not appeased, claiming CNBC to be a Trojan horse that would enter homes in the guise of business news but over time morph into a service that directly challenged CNN's market share. To guarantee to Turner and his allies that NBC would not pursue such a such a course, the network signed carriage agreements with operators promising they would not go head-to-head with CNN, an arrangement that would come back to haunt cable in Congress later.

CNBC also was given a substantial boost in December 1988 when NBC announced a wide-ranging partnership with Chuck Dolan's Cablevision Systems. The partnership, valued at $300 million, was multifaceted. Most of the headlines went to a joint pay-per-view venture for the 1992 Summer Olympics (discussed in greater detail below), but much more was on the table. NBC purchased a 50 percent stake in Cablevision's Rainbow Programming Enterprises, which held in its stable AMC, Bravo, a new SportsChannel America (launched in 1989) and six regional sports networks.[89] The deal gave SportsChannel access to NBC's sports programming and the resources to compete directly with ESPN. In addition, Cablevision was given $137.5 million and a 50 percent interest in CNBC. CNBC, of course, received carriage on all Cablevision Systems homes.

CNBC's other major obstacle was the Financial News Network. By 1988, FNN was the established leader in cable business news, with more than 30 million subscribers. Despite carriage agreements with several major MSOs, many operators were hesitant to add what they saw as a redundant service, especially when channel capacity was being squeezed. Substituting services was a possibility, but FNN was charging operators less than the proposed CNBC rate. The only good news for NBC was that FNN was in desperate financial difficulty.

From the start, FNN suffered a lack of both capital and cable operator ties. It relied heavily on "infomercials" for income. Problems with reported irregularities in the bookkeeping at Infotechnology, its parent company also plagued the service. By early 1991, Infotechnology was a named defendant in eight civil suits, the target of an SEC probe, and reportedly was $140 million in the red. Infotechnology looked to sell FNN along with its other cable assets, a 51 percent interest in The Learning Channel, as discussed above.

Early bidders included TCI, Cap Cities/ABC, Disney, NBC, Westinghouse and Dow Jones, Inc (publisher of the *Wall Street Journal*). Turner also wanted to put in a bid, but Malone and the Turner Board of Directors talked him out it because of FNN's shaky financial situation.

In February 1991, Infotechnology announced the pending sale of FNN to a joint venture of Dow Jones and Westinghouse. Dow Jones would supply the news content, Westinghouse the business expertise. The high bid was $90 million. Infotechnology, however, needed more money to settle its debts and stave off bankruptcy. NBC made a secret counter-offer, sweetening the price by $15 million. A short time after the announced sale to Westinghouse-Dow Jones, the parent company reversed itself and announced the sale to NBC for $105 million. Westinghouse was livid and a bidding war ensued. The struggle carried over into bankruptcy court where the judge overseeing disposition of the companies held what amounted to an open auction for FNN. In the end, NBC was willing to write the biggest check, $153.3 million. NBC merged the FNN services, subscribers, and staff into CNBC, eliminating its larger rival and moving to more than 43 million subscribers almost overnight. It left the company with such massive debt, however, that Cablevision backed away from the NBC partnership. With liability problems of its own, Cablevision balked at paying half the purchase price for FNN and sold NBC back its interest in CNBC.

Vertical Integration on the Defensive

In the 1960s, regulators not only encouraged but mandated, through local origination requirements, cable operator investment in programming, requirements many in the industry resisted as too expensive. By the mid 1980s, both sides were rethinking their positions. Not all programmers had equity relationships with operators, and regulators as well as smaller MSOs were concerned about the growing vertical integration in the industry.

Regulators and industry critics worried that the local monopoly character of the cable industry, coupled with the kind of national market power exercised by a TCI or Time Warner, could result in contractual agreements between MSOs and programmers that gave price and availability advantages to the large operator not enjoyed by smaller cable companies. As cable economist David Waterman observed, such market power could give a large MSO "effective 'veto power' over individual products that attempt to enter the market."[90] Critics noted that, beyond outright barriers to entry, integrated MSOs also could, and did, give equity networks privileged channel positions and bargain for deep discounts on prices. Finally, there were policy concerns that over time, small, nonaligned firms would be driven out of business, further consolidating the industry overall and reducing national competition.

The industry, of course, denied most of the allegations. In 1989, it sponsored a study that highlighted the MSO investment that had helped foster many of the new networks.[91] The report suggested that no one MSO, even TCI, enjoyed sufficient market power to prevent the successful launch of a new network, and it argued that the larger vertically integrated MSOs offered their subscribers more program choice than nonvertically integrated systems. It concluded that no evidence suggested systematic discrimination against nonaligned programmers by aligned MSOs and cited the success of ESPN, the nation's most watched program service, as an example, along with The Weather Channel, Disney, and USA.

In subsequent years, researchers would have difficulty determining whether the economics of vertical integration worked for the benefit of consumers by expanding program choice and quality or worked against consumer interests by helping restrict competition, or both.[92] While evidence of systematic abuse may have been absent, stinging anecdotes abounded. In 1989, for example, NBC President Robert W. Wright made an uncomfortable appearance before the Senate Commerce Committee.[93] He was asked directly about pressure from cable the operators with equity interests in Turner Broadcasting to shape NBC's cable programming plans. The MSOs demanded that NBC not create a general news service in direct competition with CNN. NBC's agreement to the condition was, in fact, written into CNBC's subsequent affiliation agreements and became legally binding.[94] Malone's power play in the contest for The Learning Channel also was frequently cited as an abuse of the power that TCI's size gave it.

More Family Fights

Despite denials by the vertically integrated companies, MSO-programmer negotiations over price and product could become testy when ownership ties were absent. The key issues involved the rates programmers sought to charge operators and, in the latter 1980s, the relative constriction of channel capacity as the emerging services began to claim the available spots on the dial.

Critical from the operator's perspective was the increasing costs of programming. The operations industry had spent years fostering the development of new services and after rate deregulation was able to pass those costs along to consumers in the form of rate increases. Toward the end of the decade, however, subscribers were beginning to complain. As detailed in Chapter 11, those complaints took form politically in the hallways of Congress and the FCC. Pressured from Washington above and subscribers below, cable operators became less inclined to pass along increased expenses and began pushing back on accelerating program fees.

In New York City, for example, the popular and long-standing Madison Square Garden channel proposed a rate hike and a change of channel position to a lower tier on Cablevision Systems affiliates. Cablevision declined and the dispute led to the operator dropping MSG for several months. Public outcry and threats of legislative action culminated in an eventual settlement, but the dispute illustrated the growing problem.

On occasion, operators were forced to swallow a large increase in fees. In 1987, National Football League Commissioner Pete Rozelle, seeking additional distribution and revenue for football owners, offered a package of Sunday night games to cable. It sparked a bidding war between ESPN, TNT, HBO, Murdoch's new Fox Network, and a consortium of cable MSOs headed by TCI. In March 1987, ESPN entered the winning bid, $153 million for a three-year contract, but parent company, ABC, declared it would not fund the full price. ESPN decided to pass the costs along to cable operators and levied what amounted to a $0.09 to $0.10 surcharge per subscriber. Cable operators met it, at best, with mixed emotions.

Subsequently, in early 1989, ESPN outbid Turner Network Television, SportsChannel America, and USA Network for a 175-game Major League Baseball package. It was a significant programming coup for ESPN, but cost the network $400 million in rights fees. ESPN determined it would not pass along baseball costs, relying on advertising revenue instead, but cable was concerned and Malone called for an industry-wide dialogue between operators and programmers on the issue of rising program costs, especially those associated with the large sports contracts.

One of the most spirited and public disputes between an operator and programmer in this period featured the USA Network and Jones cable. In the fall of 1988, USA announced plans to raise fees by 5 percent in 1989 and another 5 percent the following year. The network had made a number of very expensive programming purchases, including cable rights to the popular "Miami Vice" and "Murder She Wrote" TV programs and had signed a $250 million contract for twenty-four made-for-TV movies. But USA was not allied with a major MSO—it was owned by MCA, Inc. (50 percent) and Paramount Pictures (50 percent)—and operators were not pleased. Viacom and Jones Intercable were especially vocal, suggesting they would consider dropping the service if the hikes were instituted. It did not help that, at the time, MCA and Paramount were complaining about cable monopolies and pressing for industry regulation in Washington, D.C. Jones Intercable dumped USA in October 1988, stating it would be replaced by Tuner's TNT service. Jones claimed program quality, not pricing, was the issue, arguing that USA had been relying too heavily on broad-based network reruns and violent programming. "We don't feel we can reach a 70 percent penetration goal through reruns and Zodiac murder shows," declared Jones.[95] USA sued for breach of contract. In January 1989, a federal court denied the request for reinstatement, but USA pressed on. Eventually, the parties settled and, in 1991, USA was restored to the Jones' systems.

Liberty–Negative Thinking

Disputes of the magnitude reflected in the Jones-USA fight were revealing, but exceptions to typical industry relations. Horizontal and vertical integration was closer to the rule, and in the mid- and late 1980s, the industry was becoming increasingly incestuous. Most of the major MSOs had found a foothold in programming. The all-important Turner services were controlled by a board that featured most of the large operators. Viacom had its own stable of hit networks. The Discovery Channel, BET, American Movie Classics, the new Family Channel, and the major shopping networks all had operator ties. TCI, of course, was a leader in developing these equity relationships and owned a piece of each of the above-listed services.

In the nation's capitol, lawmakers were becoming uneasy. It started with constituents' complaints about rising cable bills after the 1984 Act and spread into issues of ownership and control. All of cable became suspect, but as the nation's largest company and one with a reputation for bare-knuckle business practices, TCI was a special focus of concern. (The root causes and dramatic consequences of this spreading and policy-laden construction of cable television are detailed in the next chapter.) One consequence, however, was loose talk about splitting up cable companies. Since the 1960s, policy makers had debated the common carrier model of cable. Toward the end of the 1980s, some legislators thought targeted divestiture in cases such as TCI was an option worth examining more closely.

One response, thought Malone, was a preemptive strike—symbolically break up TCI before Congress did it for them. In 1991, therefore, TCI piled many of its programming assets (excluding TCI's interest in the Turner networks and Discovery) into a single basket and created a new corporate entity to carry it. The company was Liberty Media Corp.[96] Creating the spin-off had other benefits, as well. Malone and Magness felt the value of their programming interests had been unfairly depressed by the general downturn in cable brought about by the HLT rules change. Splitting off the content investments would permit a more accurate, and an improved, measure of their value. The spin-off also would be helpful in the event TCI found it useful, beyond government pressure, to sell one of the companies. Talk was building of high-definition television (discussed in greater detail in Chapter 12). Malone knew the TCI infrastructure would not support the demands of such new technology without major reconstruction, which would require hundreds of millions of dollars. Dumping the systems side of the business might be a preferred alternative.

According to biographer Mark Robichaux, Malone had one more important side-motive for creating Liberty. Relative to most corporations, TCI lived on a shoestring, and Malone, while doing well, was not making the millions that others in the industry, including his boss Bob Magness, was. He had a modest (by corporate standards) salary, no large stock options, and at the end of the decade, less than one percent of TCI stock. According to Robichaux, Malone's friend Ted Turner teased him about the situation. "Gee, John, I'm getting rich and Bob's getting rich. And the only one that's not getting rich is you."[97]

With his legendary focus and analytical acumen, Malone set out to rectify the problem using Liberty as the vehicle. Existing TCI stockholders were given the opportunity to exchange sixteen shares of TCI for one share of the new Liberty. The press frequently described Liberty as a spin-off company, but in fact the arrangement was much more complicated. It was so complex, in fact, that it drew amazed and bewildered comments from business observers. Unlike a typical spin-off, shareholders were not given interest in the new company to compensate for the reduced value in the parent firm, but rather permitted (if they qualified) to trade equity. Noted Robichaux: "The terms, largely structured by Malone himself, defied understanding by mere mortals. Indeed, the structure of the deal—outlined in a 345-page prospectus—was so Byzantine, it seemed almost designed to obfuscate."[98]

Robichaux and others, in fact, suggested that Malone created a system intended to discourage early investment, allowing him to transfer more of his holdings from TCI to Liberty than would have been otherwise possible. The initial price of Liberty stock was set at a breath-taking $256 a share, high above what analysts could justify as its value. Malone defended the price and initiated a series of labyrinthine maneuvers that left him with a 22 percent ownership stake in the company.[99] As the overall stock market picked up speed in 1991, Malone then decided that the stock price was actually set too high to encourage widespread trading. He split the stock, first twenty-for-one, then four-for-one and finally two-for one. "It was a little schemey," he acknowledged later. "Very tax-efficient. And it worked."[100] All parts of the observation were substantially understated. With each split, trading accelerated and prices rebounded. By 1993, Liberty's value had risen some seventeen to twenty times its 1991 level. It was a $3 billion-plus company and Malone's equity was estimated at somewhere between $600 million and $840 million. Malone was named Chairman of Liberty. A TCI executive and close ally of Malone's, Peter Barton, was named President.

One of Liberty's first substantive business moves, in February 1991, was the creation of a new premium cable network. The new channel was called "Encore!" It was described as a "mini-pay" service and was priced to sell, offered originally at $1 a month. TCI concluded that there was a market niche for a movie service priced below the cost of the dominant Time Warner and Viacom providers. It also was a means of absorbing shelf space on local systems. TCI Senior Vice President John Sie, who like Malone, had once worked for Jerrold, moved to Liberty to head up the service, buying a struggling premium satellite service, Starion, and obtaining distribution contracts for lesser-demand films from the 1960s and 1970s.

As a means of putting Encore! into cable homes as quickly as possible, TCI invoked the marketing practice know as the "negative option." It was not a new idea, just a remarkably inept one. As early as 1975, Jerry Burge of Davis Communications, a Pensacola, Florida, cable company, described the method to his colleagues at the National Cable Convention in New Orleans. It was simple. The cable company gave customers a new service for free for one month, then started charging them unless they specifically asked to have the service discontinued. "Is it legal?" Burge asked his audience rhetorically, "Yes, although you must make it easy for them to say no." He encouraged the conventioneers, "It's not difficult to get into pay-cable [this way]. You'll find it an easy suit to wear."[101] They did. Bob Rosencrans used the negative option to sell the Channel 100 pay service in the mid-1970s.[102] Unfortunately, it created a situation in which customers were billed for a service they did not order, irritating many. Cable was sufficiently small in the 1970s that the unsavory practice caused little notice. When TCI rolled it out on a national basis in 1991, things had changed.

Enraged customers started with calls to the cable company, then rang up state representatives, congressional representatives, state consumer protection agencies, and eventually state attorneys general. The result was a public relations thrashing and lawsuits by attorneys general in ten states. TCI abandoned the practice in short order.

Some might have suggested that the Encore! fiasco was an example of cable industry, or at least TCI, hubris. The 1980s was something like the "Big Bang" for cable, a period of rapid and massive expansion. By nearly every measure—construction, subscribers, programming, and cash flow—cable was experiencing success undreamed of by the early pioneers. Companies such as TCI bought and sold smaller firms on a weekly basis and charged customers, in some cases, whatever they thought the market would bear. Cable, in a sense, was throwing a party for itself. Rapping on the door, however, was a new image, a new social construction of cable. Its avatar would come to be known simply as "The Cable Guy." He was the product of the excesses of the 1980s celebration and he was there to put a hard end to the fun.

11
The Cable Cosa Nostra
(1986–1992)

> *Some operator is going to overplay his hand when it comes to deregulation and end up generating a lot of bad publicity and backlash for the entire industry. I don't know who it will be, but when you give people the freedom they've ask for, someone, somewhere inevitably screws up.*
>
> —BILL DANIELS, DECEMBER 1986[1]

It was December 1986. In a matter of days, rate controls would come off and operators would be free to raise prices at will. Bill Daniels, the reigning godfather of cable television, was issuing a warning— do not abuse the privilege. Many operators paid heed. Many more did not.

Rates began a steady uphill trek. The increases began outpacing inflation and consumers and lawmakers began to wince. Simple avarice was not the sole engine; costs were accelerating, especially for the new cable programming services. But the consumer saw only a swelling bill. Making matters worse, the average customer also discovered cable to be woefully inadequate in its quality of service, so much so that within a few years, it became a standing national joke, epitomized by the icon of "the cable guy."

It was not supposed to have happened this way. Government had placed its regulatory bet on emerging, technology-based competitors, which were supposed to keep cable in check. As the decade wore on, however, it was becoming clear that the bet was not paying off. Through the mid- and late 1980s, the hands-off approach of policy makers and the failure of alternative distribution industries left the multichannel video delivery business largely the sole domain of cable TV.

Economists normally would expect patronage to decrease as rates rose, especially if prices moved to uncomfortably high levels, all things being equal. But while cable rates climbed through the mid- and late

1980s, so did subscriber numbers. All things were not equal. Spiraling view-ership was driven by well-established forces that make subscription levels highly resilient to price increases. The first driver, as always, was demand, which remained robust. The second was expanding reach, as cable passed homes that previously were not served. To some extent, cable had become a victim of its own success. It was not popularly perceived as an expendable luxury service, but rather as a social necessity. Like the Elmira housewife in 1966, the customers of 1986 had "kids and husbands," and "had to have that cable."

The lack of alternative providers made it impossible, moreover, for most viewers to exercise the market option of selecting a different service. Unwill-ing to vote with their dollars and unable to vote with their remotes, customers made their displeasure known in letters and phones calls to the local opera-tors and elected representatives. Lawmakers responded positively, some with gusto. The laissez-faire political philosophy of the Reagan era moderated in the latter 1980s. Regulators were more willing to regulate, and cable came to be portrayed by some as a flinty monopolist charging unreasonable prices for shoddy service. Albert Gore, the Democratic senator from Tennessee, emerged as a leading critic, singling out TCI and John Malone for personal abuse, characterizing Malone at one point as the "Darth Vader" of cable television. Critics painted cable as an evil empire, likening the close-knit world of operators and programmers to the Mafia. Branding the leading companies as "the cable Cosa Nostra" was extreme, but it was not entirely without cause, and, ignoring Bills Daniels' counsel, the industry by the early 1990s would pay a steep price for its lack of self-control. This chapter traces public and political reaction to the cable excesses of the mid- and late 1980s. It looks at the failure of potential competing delivery systems to keep cable prices in check and the congressional response that culminated in restrictive new legislation, the Cable Television Consumer Protection and Competition Act of 1992.

Prices and Service

December 29, 1986, was a landmark day for every cable company in the United States. Champagne corks were popping, figuratively and literally, in back shops and boardrooms from coast to coast. It was the day that the controls came off cable subscriber rates.[2]

Daniels was not the only executive urging caution. Trygve Myhren, ATC President and a man with a reputation for intelligence and integrity, told a cable industry audience that it needed to resist the expected pressure from Wall Street to spike its rates. Increases may look good in the short term, he declared, "but we're in this business for the long term." The industry was not of one mind on the issue, however. Jerry Maglio, Executive Vice President at Daniels & Co. declared: "I hear a lot of people preaching caution, but

I don't agree. Deregulation allows cable to find its equilibrium point, and that's what we'll do."[3]

"Equilibrium points" are substantially different in competitive versus noncompetitive markets, however. Neither the government nor competition provided brakes on pricing. In the first six months of deregulation, the average cost of basic service increased between 10.6 and 14.6 percent, depending on the estimate.[4] Among systems that increased their rates, the average price hike was estimated in one study at 23.8 percent.[5] The average 1983 cable bill, by industry figures, was $8.61 for basic service; it rose to $16.78 by 1990.[6]

The practice of system trading, a staple in the industry since the 1950s, exacerbated the problem. Building or buying a system, writing down the capital costs and selling it at the end of the process for its increased value was a business tradition. Under deregulation, less scrupulous operators would buy a system, raise prices quickly and steeply, then put the system on the block, basing its value on the inflated cash flow from the higher rates. Accelerated system flipping was not standard practice, but, in those instances where it occurred, it further hurt cable's reputation. Critics called it an orgy of price gouging.

The industry defended the rate increases, pointing to rising programming costs, inflation, construction expenditures, and the need for capital to continue expansion. The arguments might have been received more favorably if direct daily customer service had not been so notoriously lacking.

Customer service was fodder for late-night comedians and would-be competitors. The portrait was of an angry cable customer, sitting at home waiting for "the cable guy." Industry routine was to schedule service calls during the morning or the afternoon hours, but those were wide windows and, in far too many cases, the cable guy, as the installer and technicians came to be known, did not show up at all. Or, when he did appear, he would drag his black, coiled cable through the garden, destroying the daisies as he went, then track mud into the house, and perhaps knock over a lamp or two. After several hours of fussing and cussing, he finally would install a line that frequently did not work and subsequently send a very large bill. When the customer tried to call and complain, the phones at the cable office would ring unanswered, or a customer service representative (CSR) would take a message promising a return call, never to be heard from again.

The 1996 film, *The Cable Guy*, starring Jim Carrey, featured a psychotic cable television serviceman that played off the popular image. Some companies had greater commitments to customer service than others and the cable guy stereotype reflected unfairly on many of the better MSOs. But the stereotype did not invent itself. The industry suffered throughout its history with a poor reputation for service. The developmental, construction era of cable was dominated by engineers whose primary aim was to build the system. One pioneer once noted that when his system went down in the 1950s and

1960s and customers began to call in with questions and complaints, the operator made it a practice not to answer the telephones, in part because they were too busy trying to fix the problem and in part because they often did not know what was wrong or how long it would take to repair. As the early "pole climbers" gave way to accountants and business graduates in the 1970s and 1980s, the focus turned away from system construction and toward finances, buying and selling property and negotiating deals for program acquisition. Few operators put resources into making the customer happy.

Even those within the business admitted that it took cable several decades to realize it was a service industry. Always candid and thoughtful, ATC's Myhren told a 1982 Women in Cable convention (their first), "we give crummy service in this industry to our subscribers—let's face it."[7]

The admonitions, however, had limited effect. In addition to their focus on finance and operations, many owners failed to appreciate cable's emerging social role. Critics and policy exponents, since the 1960s, had argued that cable television could (and perhaps should) be treated like regulated public utilities, such as electricity, water, and gas. As discussed in Chapter 6, the courts tended to reject this argument, but it nonetheless persisted in policy debates. Cable's standard reply was that water and power were necessities for modern life. Cable television, on the other hand, was a luxury service. There was no social need for additional channels of television and, therefore, regulation of cable as a public utility was unnecessary. While the operators may have been correct from a purely physical perspective, they were deeply wrong in terms of social psychology. For better or worse, a significant portion of the American public treated television as a social necessity every bit as critical to their daily lives as running water and electric lighting.

The important difference was that over the decades of their maturation, power and water become generally dependable and affordable (largely through regulation). Cable was finding it increasingly difficult to claim either reasonable rates or service. The conflict between perceived social need, on one hand, and price-service performance on the other, more than anything else, was responsible for cable's political problems in the second half of the 1980s.

The Social Construction of Cable Competition

The free market logic of Congress and the FCC flowed from the assumption that cable service would improve and rates stabilize as a result of competition from broadcasters and, more importantly, from what then were seen as developing alternative multichannel delivery systems, including the burgeoning VCR and videotape rental market, MMDS, and DBS. It was the very foundation upon which rested the deregulatory thrust of the 1984 Cable Act.

One of the lessons of cable history, however, is the tendency of utopian thinking to outpace technical and economic reality. For a very brief period in the late 1950s and early 1960s, the capacity of cable technology exceeded local viewing requirements, but the wired city frenzy of the latter 1960s closed that chapter. By then, the social prophets and planners were heralding the dawn of a new age of telecommunications. A few years later, however, most of the wire for the wired city still lay on the warehouse floor. In the early 1980s, the government seized on the idea of the new communications technologies and opened the field to fair play among broadcasters, cable operators, MMDS, DBS, low power TV (LPTV) providers, and the rest. The new technologies were to be the variegated solutions to the problems of video distribution. In the typical pattern of technical development, some solutions found favor in the social and financial conditions of the period and some did not. By the end of the decade, most of the alternative delivery systems had failed. Broadcasters did not offer the variety of content necessary to create widespread cross-elasticity of demand; MMDS and LPTV sank without a bubble, and the direct broadcast satellite industry stutter-stepped through the decade.

Over the decades there had been lengthy debate about whether cable was a natural monopoly or one created by state-sanctioned franchises. Empirical data suggested cable to be a monopoly provider even with state-encouraged competition. By 1991, more than 10,000 cable systems were operating in the country and only 53 of them were subject to direct competition from another cable company.[8] As traced below, the promotion of alternative platforms did not appear to change things much. Only VCRs challenged cable's position. Toward the end of the decade, however, the scales began to fall from the eyes of the politicians and social planners, and the process of financial and regulatory reassessment began.

The VCR Success

Only the VCR could claim any success against cable in the 1980s. The dollar was strong in 1985. The price of VCRs flooding in from Japan was low. Unit sales doubled between 1984 and 1985. By mid-decade, VCR penetration hit 28 percent. It continued to climb for the next five years in one of the fastest diffusion curves in communications' history. By 1990, videotape machines were in 70 percent of all U.S. homes.[9]

Consumer use was dominated by the viewing of prerecorded tapes. A few experiments were conducted in an effort to exploit the time-shifting capacity of VCRs. In early 1984, the ABC network affiliate in Chicago hatched a scheme to broadcast scrambled programming to subscribers' VCRs in the predawn hours. The material would be recorded for later viewing, but the running joke in America was that the viewing public could not even learn to set the clock on the VCR (a problem only really overcome when smart VCRs were programmed to read time codes off cable channels and set themselves). Setting the device to record a 3 AM program was beyond the ability or interest of most consumers, who additionally found that they seldom watched the program even when they did record it. The $15 million ABC experiment, TeleFirst, folded after six months.

The real VCR business was in the purchase and rental of films; as the retail prices of videos dropped in the mid- and late 1980s, discount stores, supermarkets, and even the corner convenience stores began renting and selling them. National chains, such as Tower Video and Video World, were created, seeking the economies of scale that bulk purchasing and national mass marketing could provide. Rentals climbed from a just over 200 million in 1982 to nearly 3 billion in 1988.[10]

Studio revenue from home video surpassed that from pay-cable or broadcasting and, in 1986, for the first time, home video brought in more money for Hollywood ($2.5 to $3.5 billion) than did theatrical release ($1.9 billion).[11] Film studios began releasing their movies to the video market in as little as four to six months after their theatrical runs.

What was good for the film industry was not, however, good for cable. As penetration of VCRs grew, premium cable services such as HBO suffered. In absolute terms, pay revenue continued to climb, in large part because of the tremendous growth in cable penetration, but broadly speaking, VCRs hurt the pay business, and for several reasons. Problems of repetition and churn in cable generally and pay-cable in particular tended to encourage the migration to the video store. The inconvenience of having to go to the store was offset by the greater choice and viewing control of home video. Pay-cable, as with most forms of television, required scheduled viewing; consumers watched the program when it aired. With home video, consumers watched the film on their own schedule and with the ability to stop, start, and rewind at will.

Home video patronage also was encouraged by cable rate increases after deregulation as customers dropped premium service and shifted their dollars to VCR rentals. HBO and Showtime offered operator incentives to keep retail rates for the premium services as low as possible to prevent erosion, but with limited success. In 1983, about 84 percent of all cable subscribers took a pay service. That figure slid to about 78 percent in 1986, recovered a bit then dropped to just over 77 percent in 1990.[12] Premium revenue as a percent of total cable industry revenue hit a high of about 43 percent in 1983 and 1984, and then it began declining through the rest of the decade.

Pay-Per-View

The industry responded to the VCR challenge along two fronts, neither of them terribly successful. Starting around 1987, it began lowering rates for premium service. The average monthly pay rate rose from $9.70 a month in 1983 to $10.25 a month in 1985 and then dropped slightly through the end of the decade. It could afford to do so, in part, because of deregulation. TCI, for example, planned to decrease pay rates but increases basic fees, with the end result being an overall rise in per-subscriber revenue.[13]

The industry also continued its long-term struggle to create an economically viable pay-per-view (PPV) service. Pay-per-view was as old as the Blue Sky dreams of the 1960s and not too far removed from the Bartlesville experiments of the 1950s. The basic idea was that the customer would order and pay for each film or program on an individual basis, rather than on a monthly plan. Films typically were priced around $5, with special events such as wrestling or boxing often many times that price. A customer who ordered only a few films a month would contribute revenue several times the regular HBO fee, making pay-per-view programming the "Holy Grail" of cable television.

Cable operators also saw it as an easy sell. The success of the video rental trade demonstrated that people were willing to pay for individual movies, and, ran the logic, customers would be more amenable to doing so with a click of the remote control or a phone call than enduring the inconvenience of a car trip to the local retailer. Unfortunately, the technology, once again, was slow to materialize. Pay-per-view required a method of sending specific programs to specific homes. Positive and negative traps were the cost-effective way to filter programming and had been successfully used in small systems for high-priced, advanced-notice events, such as boxing championships. (In June of 1982, title fight between Holmes-Cooney was distributed nationally on PPV with most systems using "notch" filters.) Traps had to be installed manually, however, and they were not a technology well suited for impulse buying.

True addressability required converters that allowed operators to identify and automatically send to individual homes requested programming. The technology existed, but it was very expansive and, by the early 1980s, only Warner's loss leader QUBE system was offering pay-per-view in any systematic form.

The upstream communication path that allowed subscribers to order programming was a special problem. The goal was a system that permitted a subscriber to order a film simply by pressing a button on the remote control. Most systems, however, were not engineered at this time to provide consistent dependable upstream, or return path, communication. To order a PPV film, most customers had to use the telephone. In its earliest manifestation, companies had banks of telephone operators taking orders. The development and implementation of automatic number identification (ANI) systems for cable streamlined the process and brought down the cost. The ANI system could identify the caller by the subscriber's phone number and automatically unscramble the pay-per view signal to the home.

True addressable technology was in an infant stage, however, and infant mortality was high. In 1982, Group W's Bill Bresnan reported failure rates in the field of 30 to 60 percent on the early addressable units. "They're getting better, but they have a long way to go."[14] Through the first part of the decade, equipment manufacturers, such as Oak Industries, Jerrold, and Zenith, worked to increase reliability and decrease costs.

Product supply also was an issue. Prize fights were lucrative—the Sugar Ray Leonard–Thomas Hearns rematch in 1989 generated a reported $25 million in pay-per-view revenue—but the staple diet had to be film, and that required studio contracts. Revenue splits had to be negotiated and release windows configured. By 1990, the major studios had overcome their fear of VCRs and become dependent on rental revenue, so much so that they were hesitant to release films to the still small PPV universe.[15] Release windows were established that put PPV forty-five to sixty days behind video, meaning customers could get the film at their local rental outlet for up to two months before it would come to their cable box.

Despite the challenges, cable continued to pursue the pay-per-view dream. Warner's QUBE-based pay-per-view service, Viewer's Choice, offered about twenty-five films a month; $4.50 for a first run film, less for older titles. By the end of 1982, Warner reported about 200,000 customers with pay-per-view capability. Viewer's Choice went to Viacom in its 1985 purchase of various Warner programming assets and, in 1988, it was merged with the Pay-Per-View Network that had been formed by ATC, Continental, Cox, Newhouse, and Telecable. In 1989, its was further joined by Disney and, returning to its roots, Warner Bros. Pay-per-view revenues were split, 50 percent for the film studio, 40 percent for the local cable operators and 10 percent for Viewer's Choice.

Shortly after Viacom purchased Viewer's Choice, it took the service national. At the same time, a competing pay-per-view company announced that it also was open for business. Request TV began operations in November 1985. It had $40 million in backing from most of the major Hollywood studios, including Columbia, Disney, Lorimar, MGM-UA, New World, Paramount, Fox, Universal, and Warner. The studios still sought compensation on a per-viewer basis instead of the flat fee regimen of pay-cable and the one-shot first sale price of the video store, and bet that an industry-supported PPV network might be the answer. Request TV's business plan called for the leasing of satellite transponder time to individual studios. The studios then programmed their own material, splitting the fees 45–55 percent with the local cable provider.

Heading the new company was cable veteran Jeffrey Reiss. A former Showtime Vice President, he knew the industry and had additional backing from his father-in-law, famed TV producer Norman Lear, along with other cable and TV notables. In 1988, Reiss Media Enterprises opened a second channel, Request 2. Request 3, 4, and 5 were added in 1993. In May 1989,

Group W Satellite Communications paid $10 million for 50 percent of the service.

Viewer's Choice and Request TV were the largest, but not the only, pay-per-view services in the 1980s. In July 1985, Playboy announced the start of Playboy's Private Ticket. The marketing plan differed somewhat from the other pay-per-view companies by offering subscribers programming in three-hour blocks instead of on a per-program basis. Unlike many other parts of the cable business, Playboy had difficult times in the mid- and late 1980s. The 1984 Act required operators to supply lock boxes when they offered adult services, and many local communities protested the sexual nature of the programming. As a result, subscriptions to Playboy's monthly premium service dropped in 1986, as operators took Playboy off the system. At the same time, viewers who were interested in more sexually explicit content were siphoned away from Playboy by the growing video rental business and the X-rated material that it afforded. Through 1987 and 1988, Playboy moved aggressively toward the PPV format while maintaining its monthly premium service. In 1989, it converted completely to pay-per-view, rebranding the channel as "Playboy at Night." Other services included Jerrold Electronics' Cable Video Store (CVS), started in 1986, and Graff PPV, which offered traditional pay-per-view film fare and started its own adult entertainment service, Spice, in 1989.

Pay-per-view grew, but slowly, through the latter 1980s. Problems with product supply and technology continued to plague its deployment. By the start of 1987, only 8.5 million of cable's 40 million plus homes had addressable converters and only 3 million of those could be used for pay-per-view (capacity on the others was used for billing and service).[16] Supplemented by older technology, PPV reached 11 million homes in 1988, supported by World Wrestling Federation (WWF) events and prizefights, but by 1990, only about 9.3 million addressable converters were in the field.[17] Buy rates remained a disappointment to operators, stabilizing in the single digits, rather than the double digits they had hoped for, and several planned pay-per-view ventures folded before or shortly after launch.

In 1992, NBC-Cablevision attempted one of the most ambitious pay-per-view experiments to that date, offering expanded coverage of the 1992 Summer Olympics from Barcelona, Spain. The network, which was looking for ways to help offset the $410 million in rights fees it had paid to acquire the games, designed what they marketed as the "TripleCast," three channels of PPV Olympic coverage. Viewers could select from a menu of packages ($30 for a day, $125 for the full 15 days of competition). NBC hoped to gross $100 million in buys, but a large number of cable systems did not have the addressable capacity (or chose not to use it) for three channels of Olympics programming, and consumer interest in the extended coverage did not materialize. NBC and Cablevision lost an estimated $100 million on the venture. It would take another turn of technology in the 1990s to power pay-per-view to greater acceptance and profitability.

LPTV

VCRs gave the consumer a choice and mitigated pay-cable rates. Through the late 1980s and early 1990s, no other TV service could make such a claim, however. By the end of 1986, the FCC still had a backlog of 10,000 LPTV applications, although it was starting to make headway in clearing them out. About 380 LPTV stations were in operation, but most of them (225) were part of Alaska's bush network. By October 1989, 727 LPTV stations were reported on the air, 200 of them commercial.[18]

The concept of locally produced and oriented urban and suburban programming was good in theory, but difficult to implement. Even a low-power station was expensive to run and homegrown programming attracted few viewers. Nonprofit community groups faced the economic and programming realities of daily television, finding both operating income and viewable content difficult to obtain. In some ways, the failure of LPTV echoed the 1950s failure of UHF, and for some of the same reasons. LPTV stations could not harvest an audience of sufficient size to attract the advertising dollars needed for economic viability. As UHF, LPTV received a boost from local cable carriage and, although cable was not required by law to make such carriage available, many did. Eventually, however, many of the original applicants gave up their licenses. Studies in the 1990s showed that only about 8 percent of LPTV licenses were held by minority operators.[19] LPTV stations across the country became outlets for nationally distributed religious and shopping channels with over half of all the stations scheduling satellite-fed programming about two-thirds of the day. It did not position LPTV as a significant alternative to cable-provided, multichannel television.

MMDS

For MMDS, or wireless cable, it was a classic example of too little, too late. Initially encouraged as a technology that could deploy quickly and cheaply, it was seen as a remedy to the problem of lack of cable availability in the larger cities in the early 1980s. FCC licenses for the first MMDS operations were not issued until late 1985 and, even then, the number of available MMDS channels in a given area was limited and usually had to be augmented by leasing ITFS channels from educational institutions. An MMDS operator could assemble twelve to twenty channels of microwave-delivered programming, but might have to compete with an existing cable system offering greater channel choice at a similar price. In addition, cable programmers were reluctant to sell their product to the cable competitors and MMDS owners found themselves shut out of the programming market. The number of MMDS subscribers in 1989 was estimated at anywhere from 325,000[20] to 1 million,[21] but everyone agreed the numbers were falling steadily.

The Telcos

The RBOCs were bound almost hand and foot from entry into competitive, program distribution service. The FCC's 1970 prohibitions on video were codified in the Cable Communications Policy Act of 1984, and reinforced by the prophylactic of the 1982 Modified Final Judgment (MFJ). These did not prevent some companies from trying, however.

While the triple fence of FCC, Cable Act and MFJ orders kept RBOCs from owning and operating their own systems, the section 214 lease-back option of the 1960s and 1970s was still available. (Exceptions to the cross-ownership rule, for "good cause," were possible but difficult to obtain.) One of the first to probe a 1980s version of lease-back was Pacific Bell, which, in 1983, proposed building a state-of-art system for Palo Alto, California. The city rejected the offer, and PacBell turned to an independent operator as its lease-back partner, subsequently receiving FCC approval of the plan.[22] The Chesapeake & Potomac Telephone Company, a Bell Atlantic subsidiary, similarly won FCC approval in 1985 for a lease-back system for Washington, D.C.

Cable operators were particularly unsettled in April 1988 when the FCC's Common Carrier Bureau approved a plan for General Telephone and Electronics (GTE) to build a cable system in Cerritos, California. It was initially presented as a section 214 lease-back; however, GTE proposed to lease 50 percent of the system's capacity to itself, through a subsidiary, for what it described as experiments in advanced technology (fiber) and services (video programming on demand).[23] The NCTA challenged the proposal, charging the lease-back application was a charade, but the FCC, now well motivated to encourage competition in cable services, approved a waiver, with some restrictions, of its cross-ownership rules and construction went ahead.

Palo Alto and Cerritos ultimately were meager and unsuccessful experiments. As will be seen, the telephone companies worked strenuously and successfully in the late 1980s and early 1990s to have the legal restraints removed. Those efforts would not bear fruit for several years, however, and by that time, questions would arise about the industry's true enthusiasm for video service. Through the last years of the 1980s, the telephone companies would provide no more check on cable prices and service than any of the other alternative providers, including the satellite operators.

TVRO and Scrambling

In the mid-1980s, as it had been in the mid-1950s, people wanted television and more was still better. Cable offered attractive new viewing options— MTV, CNN, ESPN, and C-SPAN. But the wired infrastructure was still under construction. Cable reached only about 42 percent of all television homes in 1985, and many areas of the country would never see a coaxial line. In farmlands and rural areas across the nation, the sales of TVRO dishes grew at a healthy clip. Dish ownership rose from an estimated 4,000 in 1980 to more than 1.5 million by the end of 1986.[24]

While initially unimpressed by the small number of home satellite users—as evidence by HBO's response to Howard Taylor's $100 subscription check—cable executives became interested as receivers began popping up in rural backyards around the country. In mid-1985, a consortium of networks, including MTV, ESPN, and Turner, considered packaging their services and selling them through TVRO distributors to areas not yet wired. HBO was looking at a similar plan.

The industry therefore reconsidered as the dishes spread. Drawn by the increased choice and picture clarity of TVRO, urban dwellers began buying dishes and apartment owners installed them to create SMATV systems (typically without compensating the programmers). Studies done at the time suggested that up to one-third of all dishes were being placed in yards where the cable line passed overhead. The potential for consumers to receive cable programming without paying for local cable service became worrisome for the industry and sent programmers to the technology drawing boards.

Pay services especially came to see TVRO as a new form of theft of service, stealing the programming for which most customers paid several dollars a month. In the 1950s, operators argued in courts across the nation that they had a perfect right to intercept and retransmit broadcast signals once those signals were "made pubic" by the act of over-the-air transmission. Now, with no small irony, the argument was thrown back in their faces by dish operators who made the same claim with respect to cable's satellite signals. The issue was resolved legally, at least briefly, by an amendment to the Cable Act of 1984. Introduced by Congressman Albert Gore (D-Tennessee), the clause gave dish owners the right to receive satellite signals such as HBO, WTCG/WTBS, and CNN, as long as use was limited to the home. Apartment houses, taverns, and even fraternities were not protected by the amendment. The clause did not leave cable empty-handed, however. It acknowledged the industry's right to scramble those signals if it wished. The industry, therefore, was keen on deploying scrambling with as much speed as possible.

HBO, along with others in the industry, began thinking about scrambling as early as 1979 when the FCC dropped its licensing requirements for dishes. In February 1982, HBO made public its intention to scramble its feed. It took more than three years and an estimated $15 million to develop the technology and bring it to market. To build the equipment, HBO chose a Massachusetts firm with Defense Department experience in encryption technology. The company, M/A-Com Linkabit, Inc., created a system it called "VideoCipher." In January 1984, HBO installed a prototype in a cable system in Puget Sound/Tacoma, Washington. On July 31, 1985, Time

Inc. turned the switch scrambling its HBO and Cinemax feeds nationally for 12 hours a day. In January 1986, it went to full 24-hour scrambling. Viacom began scrambling Showtime and TMC in May 1986. HBO engineering executive Edward Horowitz later recalled: "During the course of those (early) years, the backyard dish market was growing very, very rapidly. By the time we actually flipped the switch to turn on the scrambling system, there were about 1.5 million backyard dishes installed. We then said, 'All right. Here are people that have had the longest review in the history of HBO. If anyone is going to buy our programming, it's them.'"[25]

HBO was not trying to prevent TVRO use of its signal, but rather to turn TVRO owners into paying customers. They recognized there would be a backlash and took steps to offset it by offering dish owners the opportunity to buy the new VideoCipher set-top box for $395, and to sign up for service at $12.95 a month (with $5 going to the local cable operator as appeasement money). Many took the offer. Many others rebelled. They complained about HBO infringing on their claimed right to receive signals, and accused the cable industry of attempting to stifle competition. Some threatened to sue for restraint of trade. In the most famous protest, a disgruntled Florida dish dealer knowledgeable in electronics was so angered by the HBO move that he jammed the Home Box Office satellite signal. On April 27, 1986, a few months after HBO began scrambling, the movie being played on thousands of subscribers' screens, "The Falcon and the Snowman," blinked off and in its place popped up a message:

Good Evening HBO from Captain Midnight. $12.95 a month? NO way! (Showtime/Movie Channel beware.)

Authorities quickly tracked down "Captain Midnight," who admitted to the prank and paid a $5,000 fine. More vexing to the cable industry was the fast-track development of illegal decoders. The original VideoCipher unit was cloned so quickly that M/A-Com created VideoCipher II. It became the industry's *de facto* standard, but electronics hobbyists cracked its encryption scheme as well and black market boxes multiplied across the nation. By some estimates, up to half of all the decoders in the United States were illegal by 1988, with the "black boxes" openly advertised in electronics magazines and newsletters. Cable worked with federal authorities to stem the flow of pirate devices. Overseas shipments of the decoders or the chip sets at the heart of the descrambling circuitry were seized at port and federal agents raided suspect distributors. Satellite signal theft was never the FBI's highest priority, however, and the industry was left to deal with much of the problem on its own, with support from the FCC and even the legitimate satellite broadcast industry, through its trade group, the Satellite Broadcasting and Communications Association (SBCA). In 1986, the cable supplier General Instruments (GI) purchased M/A-Com's cable division and the VideoCipher technology for $220 million, and GI hired former FBI agents and detective agencies to track down pirate operations and report them to authorities.

Despite the public outcry and the efforts to circumvent the technology, programmers moved steadily to secure their signals. Within a year of HBO's inauguration of scrambling, half a dozen other major cable programmers had followed suit and most of the rest were making plans. The major broadcast networks did the same and backyard dish owners no longer had free television falling from the sky.

Without the incentive of no-cost cable programming, the expensive, bulky backyard dishes became less appealing. Even C-band enthusiasts willing to pay for monthly service felt they were not given access to the full range of cable programming at prices equivalent to those paid by cable subscribers (an issue discussed at greater length below). In the cities, terrestrial cable was finally becoming available, and scores of municipalities unhappy with ten-foot dishes in back yards, front yards, and on rooftops, took to zoning the unsightly antennas out of the city limits.

TVRO receiver sales slowed to a trickle. More than 600,000 dishes were sold in 1985, a number cut by more than half in 1986.[26] By the late 1980s, the boom in TVRO went bust. Dish sales reported at a brisk pace of 45,000 units per month in 1985, dropped to 12,000 a month by mid-1987.[27] Small-scale manufacturers and mom-and-pop dish dealers went out of business and the young industry began to fade. At its height in mid-1980s, TVRO had about 3 million units. By 1989, the FCC estimated dish ownership at about 1.7 million.[28] The remaining users were largely those people in rural areas outside the reach of a coaxial cable and the very hardcore TV enthusiasts. The TVRO business was decimated. It did not go without a struggle, however. TVRO owners and salesmen took their case to Congress, where, as discussed below, they found a sympathetic ear.

DBS Redux

The first round of DBS failures in the early 1980s was very much an example of social construction outrunning technical and fiscal reality. But the technology continued to evolve and business conditions were changing through the decade. The inauguration of scrambling and its impact on TVRO sales helped stimulate fresh interest in DBS. Improvements in transponder power led some to hope they could offer attractive programming packages using smaller, cheaper dishes than those required in either the C-band community or in the previous medium-powered DBS failures. The direct satellite market could include the millions of homes not passed by cable, as well as former C-band customers who, it was hoped, would upgrade to the high-powered service. There also were an estimated half million SMATV customers in the United States in the mid-1980s and those private systems were considered potential DBS allies and users.[29] Some DBS entrepreneurs felt they could even offer additional or competitive service in homes already subscribing to cable. Would-be satellite providers, therefore, saw a business opportunity. Cable operators saw the same thing and moved into DBS as well.

Among the first to try again after the debacle of USCI was, in fact, HBO and RCA Americom. These two companies, which had teamed to create the first cable-satellite link in 1975, jointly proposed a medium-powered Ku-band service designed to provide video feeds to both cable operators and home consumers. Formed in 1985, Crimson Satellite Associates offered free dishes to cable operators in an effort to switch them from C-band to Ku-band technology. The cable industry balked, however. Operators already had too much capital sunk into existing C-band equipment to make a transition economically rational, the smaller dish size was seen as opening an uncomfortable window for additional home piracy and major players, such as Hughes and Viacom, opposed the initiative. Crimson Associates died in its cradle.

Through the early and mid-1980s, other companies filed for DBS authorization, only to withdraw their requests or have them cancelled after failing to raise sufficient funding or make reasonable progress toward launch. A few of those nonetheless filed extensions of their applications, in effect asking the FCC to hold their seat while they attempted to improve the business plan and hunt down financial support.

Staying in the regulatory cue was important because satellite transponders were a finite and contested resource. By international agreement, the United States had a limited number of orbital positions for geosynchronous satellites; of those, only four high-powered orbital slots had downlink footprints that covered the entire continental United States, providing full coverage of the lower forty-eight. These highly prized full-CONUS (continental US) positions were located at 101-degrees W, 110-degrees W, 119-degrees W, and 61.5-degrees W (although other, partial CONUS slots were available as well).[30] The FCC could authorize the use of up to thirty-two channels at any one of the positions.

With renewed interest building, the FCC continued its consideration of DBS and, in August 1989, acted on at least nine different requests.[31] Among the companies affected by the authorizations and extensions were TCI's Tempo Satellite, Inc., Stanley Hubbard's U.S. Satellite Broadcasting (USSB), Charlie Ergen's EchoStar Satellite Corp., and Hughes.

Primestar

TCI flirted with DBS in early 1985, looking into the possibility of investing in the failed USCI venture as a means of bringing cable programming to rural homes. While that effort folded, Malone retained an interest in satellite distribution, in part as a possible adjunct to the terrestrial wire line business and in equal part as a defensive measure against potential satellite-based competitors. If TCI did not claim the scarce FCC-controlled satellite transponders and homestead the DBS ether, Malone knew others would.

As discussed, TCI purchased Ed Taylor's Tempo Enterprises (formerly Southern Satellite Systems, Inc.) in 1988, selling Tempo's cable channel to NBC and bringing the company's satellite assets into TCI's portfolio. SSS, one of the early DBS suitors, had been granted FCC permission in 1984 to pursue development of the service. Along with most of the other early entrants, Taylor pulled back in the mid-1980s, but kept his application alive and, in May1988, the new TCI/Tempo submitted an amended DBS proposal to the FCC.

Tempo told the Commission that the market and DBS technology had evolved to the point where a high-powered Ku-band service was feasible. The company now planned to invest $500 million in the development of a twin-satellite, thirty-two channel service, which it hoped to see operational by 1995. In its August 1989 action, the Commission formally approved Tempo's application, granting it eleven channels at 119-degrees W, but like some cruel bureaucratic joke, the FCC at the same time put the approval on hold.

Tempo's application had been challenged by the National Association for Better Broadcasting and the Telecommunications Research and Action Center (Media Access Project). The groups charged that Tempo's new owner, TCI, was unqualified to hold the license following the $36 million antitrust judgment against the company in the Jefferson City antitrust case. It took nearly two years to resolve the dispute. Even then, the FCC action dragged on for almost another year. Finally, in May 1992, the FCC gave TCI permission to proceed with the Tempo DBS project.

While the Tempo authorization was tied up in the courts, TCI looked for another route to the stars. The company met with GE Americom, builder of the Tempo DBS satellite, and Malone began talking with other cable operators about the possible threat to their business from DBS. By the end of 1989, Malone had rounded up most of the major MSOs, including Time Warner, Cox, Comcast, Newhouse, Continental, and Viacom, and had crafted a plan. GE Americom would supply its medium-powered Fixed Service Satellite, RCA Satcom K-1. The cable partners would provide funding. Taking its name from the bird, the consortium dubbed itself K-Prime Partners. In February 1990, K-Prime announced plans for a ten-channel, "medium-power" (45-watt transponders) DBS service beamed to three-foot (one meter) dishes. The programming would consist of seven superstations and three PPV channels. It would not offer basic cable networks. Operations were slated to begin near the end of 1990. The Ku-band plan was billed as a transitional step leading, at some point, to migration to the full-powered Tempo authorization.

In November, DBS service was inaugurated in about twenty markets, with rollout to the rest of the nation scheduled over the course of the next several months. In December, K-Prime formally changed its name to Primestar Partners, following the brand name adopted for consumer service, Primestar.

Sky Cable

Hughes, purchased by General Motors Corp. in 1985, had been a dominant satellite builder and operator for years. It had the technology, the experience, and the financial wherewithal to take a position at the forefront of the DBS business and it was keenly interested in doing so. It had always looked for help in programming, however, going back to the 1960s when it joined in Irving Kahn's effort to create the original cable-satellite connection. Hughes kept its interest and its FCC applications alive through the 1980s while it searched for a successful DBS formula and for partners. By 1989, it thought it had found both. The FCC's August 1989 DBS grants expanded Hughes' existing, generous DBS authorization, providing it with twenty-seven channels at 101 W. At the same time, it was talking with well-established media companies about a possible joint venture.

Rupert Murdoch backed away from DBS in the United States in 1984, but did not abandon the idea. Instead, he took it abroad. A number of countries outside the United States had been experimenting with DBS since the early 1980s. Japan was a leader in development, and work was carried on as well in Europe. In 1983, the United Kingdom authorized commercial satellite television and, by 1989, the Australian Murdoch had secured a license, a satellite, and the necessary programming. He launched Europe's first major direct broadcast business, Sky Television, in February 1989. Shortly afterward, a competitor, British Satellite Broadcasting, also began beaming signals into TV homes in Great Britain. The British DBS services had one advantage over similar ventures in the United States—the United Kingdom was not heavily wired for cable and multichannel competition was weak. Nonetheless, start-up costs were high and the battle between the two providers was fiercer than either could sustain. Each company reportedly was losing millions of dollars a week. In November 1990, they announced a truce and a merger. The new company, British Sky Broadcasting (BSkyB), was controlled by News Corp.

Murdoch sought a similar outcome in the United States, and a marriage with Hughes was natural. Murdoch wanted entry points in the U.S. television market and DBS was a fast and comparably economic means of doing so. Programming assets were available through 20th Century Fox and, just to make sure they did not run short of material, Murdoch and Hughes signed as additional partners James Dolan's Cablevision Systems and his occasional business ally, NBC. In February 1990, just days after the announcement of the creation of PrimeStar, the Murdoch-Hughes group held their own news conference. Advanced, high-powered (120-watt transponders) Hughes satellites would beam up to 108 compressed channels of video to 12- by 18-inch flat-panel receivers. The dishes and receivers would be marketed through local consumer electronic retailers. The billion-dollar venture was called Sky Cable. The technology would be high definition TV (HDTV) capable by 1993.

The Sky Cable partnership could neither jell nor set, however. Despite the participation of Dolan and the resources of Fox and NBC, Sky Cable had no security that it could obtain the mainstream cable networks required for a profitable operation. Afraid of irritating broadcast affiliates, the NBC and Fox networks declined even to provide their own program schedules. The cable industry, as noted, was proving very reluctant to sell its product to potential competitors and Murdoch clearly wanted to take a bite of cable's lunch. Cable operators were fond of saying they would not sell to competitors, especially Murdoch, the bullets (programming) that would be used to kill them. Trying unsuccessfully to hide in the middle of it all, Dolan spun Sky Cable as a harmless adjunct to cable, providing program distribution where cable did not. There were mixed messages, therefore, out of the Sky Cable partners and skepticism across the industry that adequate financing could be arranged. News Corp itself was deeply in debt following a $3.2 billion purchase of *TV Guide* and it was still spending millions to keep BSkyB afloat. Finally, the project relied on an unproved and suspect technology—digitally compressed television signals (discussed in detail in Chapter 12)—to create the 100-plus channels.

By February 1991, the project was dead. Hughes indicated it would continue to pursue a DBS service on its own.

USSB

It may have been just the opening Stanley Hubbard was looking for. In November 1988, the FCC gave Hubbard the extension he had earlier requested to seek more funding and, in its August 1989 action, the Commission authorized additional channels for USSB. Within two weeks of the grant, Hubbard landed a backer. Nationwide Insurance owned four TV stations and sixteen radio stations through its Nationwide Communications subsidiary. With Nationwide behind him, Hubbard unveiled plans for a high-powered service offering up to 128 channels with expansion capacity to 256. He planned to spend $500 million to $700 million to build and launch the system, and he now needed a technology partner. Hubbard and Hughes quickly found each other.

Both Hughes and Hubbard envisioned a DBS service in direct competition with cable, relying heavily on pay-per-view films supplied by the major studios. With a critical mass of DBS subscribers, they could give Hollywood a direct line to consumers and a more attractive revenue split than that of cable. Hubbard once boasted that a major Hollywood producer told him, "Stanley, you get those satellites launched, and you get just a million homes, and you can put a card table up on Sunset Boulevard, and everybody (all the major studios) will be lined up to sign up, and I'll be at the head of the line."[32]

In June 1991, Hughes and USSB announced their partnership. USSB would buy five Hughes transponders at 101-degrees W for $100 million. Hughes would program the remaining twenty-seven transponders, offering them to other interested parties or using them for its own DBS service.

Advanced and EchoStar

Hubbard was not the only DBS idealist looking for financial and technological support. In Colorado, Charlie Ergen, the C-Band entrepreneur of the early 1980s, was still betting on satellite television, and in Arkansas, Dan Garner, a businessman with an idea and a political connection, also was gazing skyward.

Ergen's first stab at DBS came in 1985 as a junior partner to Gene Schneider's United Cable. United Cable was the creator and lead investor in Antares Satellite Corp. Antares was given interim authorization for service using two, twelve-channel DBS satellites. As with the other DBS hopefuls, United failed to acquire funding and dropped out, but it was an introduction for Ergen. By 1987, EchoSphere had created EchoStar Satellite Corp. and filed for its own DBS license. In the FCC's August 1989 set of grants, EchoStar was given eleven DBS channels at 119-degrees W.[33] It would be 1995 before EchoStar launched a satellite, however.

Arkansas businessman Dan Garner built a successful radio station business, Garner Communications, in the 1970s, and wanted to expand. In 1984, he filed a DBS application with the FCC and, in December, received interim construction permits for a proposed regional DBS system, featuring two satellites and six channels.[34] The FCC gave him, along with several other applicants, until December 1990 to begin operation. Garner had a license, but no money, and no programming contacts.

Garner did know an influential former politician, Wilbur Mills, a former Democratic congressman from Arkansas and one-time Speaker of the House. Mills' political luster faded in 1974 following news reports about his relationship with a stripper named Fanne Foxe. Mills retired from office in 1977 and, in 1988, established the Foundation for Educational Advancement Today in Arkansas.

Garner struck a deal with Mills. Garner's company, Advanced Communications, would dedicate one-eighth of its satellite capacity to the Foundation. In return, Mills would become chairman of Advanced and use his political connections to promote the company, as a means of promoting economic development in his home state. With support from Arkansas, Advanced continued to win extensions of its authorization and even an expansion of its allocation. In 1989, the FCC awarded it twenty-seven channels at 110-degrees W, and half-CONUS channels at 148-degrees W.[35] The company, however, was unable to marshal the financing or strike a deal with a satellite provider. By 1991, the FCC was making it clear that progress had to be shown or it would revoke the authorization,[36] and Mills died in 1992 eroding Garner's political leverage.

Despite all the business and regulatory activity, high-powered DBS remained only a concept at the end of the decade. Consumers who wanted multichannel television could install a thirty-foot C-band dish in their backyard, rent videos, or subscribe to cable television. Once again, the hopes and the fears associated with cable television had been greatly miscalculated and the Blue Sky visions of competing new technologies substantially over estimated.

The Road to 1992

By the late 1980s, politicians, consumers, and would-be competitors faced a new reality: for the time being at least, cable was the only game in town. Few outside of the industry were happy about it. Consumers were unhappy with prices and service. Competitors from the TVRO, DBS, and MMDS sectors complained about program access, charging that ownership, contractual, and even social relations between cable distributors and programmers inhibited their ability to obtain desirable programming. Scholars conceded that program availability and, hence, diversity in TV programming had improved, but argued that much more could be done.[37] Many, pointing to the vertically integrated nature of the industry, noted that, although there were numerous cable networks, control and ownership remained concentrated in a limited number of firms. Broadcasters once again argued that cable held the upper hand in determining which signals it would or would not carry. Station operators and networks wanted must-carry reinstated. They also wanted compensation for carriage of their signals, compensation they did not generally receive under the Copyright Act.

Everyone took their complaints to the government. Individual municipalities could do little under the constraints of the new act, but the National League of Cities launched its own investigation.[38] The National Association of Attorneys General Antitrust Committee formed a five-state task force to look into rate hikes[39] and, in Washington, D.C., both the FCC and Congress began rethinking cable television. The tapestry of the 1984 Cable Act began to unravel and federal action proceeded along multiple fronts.

The free market mood in Washington had not evaporated by the late 1980s, but it had calmed. Following the 1986 mid-term elections, Democrats took control of the Senate (they already controlled the House). Senator Ernest Hollings (D-South Carolina) became Chairman of the Senate Commerce Committee and Senator Daniel Inouye (D-Hawaii) Chairman of the Communications Subcommittee. The emphasis remained on fostering competition, but the hardened political philosophy of *caveat emptor*, which had characterized much of the late 1970s and early 1980s, softened.

Relations between the FCC and Congress had soured under Fowler, who was considered abrasive, ideological, and uncompromising. In April 1987, Dennis Patrick followed Fowler as FCC chairman. Patrick, a former White House personnel officer and NTIA administrator, was named to the Commission in 1983. There was a general hope that FCC-congressional relations would improve under a more flexible Commission leadership, but that proved not to be the case. By some accounts, Patrick managed to annoy congressional communications policy leaders even more than Fowler and was seen as just as tightly bound to deregulation, if not more so, prompting Congress to take a more active role in telecommunications affairs.

In November 1988, Vice President George Bush was elected to follow Ronald Reagan into White House. The Reagan administration had kept its eyes on a guiding ideological star—free enterprise and governmental noninterference. The Bush administration was seen as more pragmatic, less ideological, more willing to consider governmental regulation when it seemed to make sense. The general mood around Washington fell into step, as the political pendulum swung back toward a balance of free market and governmental oversight.

This pragmatism was reflected in Bush appointees to the FCC. Bush's choice to head the agency was Alfred Sikes, at the time the Assistant Secretary of Commerce and Administrator of the National Telecommunications and Information Administration. Sikes moved to the FCC in August 1989 and took with him some very specific ideas about the role of cable in the emerging telecommunications marketplace.

Congress and the FCC, therefore, seemed more willing in the late 1980s to listen to the cable-based complaints rolling like slow thunder toward them. In the end, Bush's Republican preference for market solutions over government command and control measures would remain an ally to the cable industry. By the end of the decade, however, the broader tenor of political philosophy coupled with important changes in the industry itself would create a problem for cable larger than even a presidential veto could overcome.

At the FCC

Must-Carry

Even before the swing in Congress and the sunset of rate control, the FCC was conjuring a harbinger of the near future. Following the 1985 federal court decision in *Quincy*, and for the first time in twenty years, cable was freed from the necessity of carrying local signals. Although nearly all operators continued to provide the dominant local stations, broadcasters howled in anger at both the ruling and the FCC's initial response. The Commission could have attempted to rewrite the must-carry rules, but in keeping with political tenor of the times, chose not to, privileging a market-based solution. The broadcasters looked to Congress for help, and Congress, in turn, pressed the FCC to reconsider its position. In late September 1985, under this congressional duress, the FCC announced it would open an inquiry to hear proposals for revised must-carry rules.

Seeking a negotiated resolution, meanwhile, the cable and broadcast industries began their own talks and, by late February 1986, reached an accord calling for a reinstatement of the must-carry rules while limiting their scope.[40] Building on this compromise, the FCC in November issued new rules. This time, the regulations included a provision that cable systems supply "A-B" switches, giving customers the ability to switch the input on their TV sets between cable service and an off-air antenna. The Commission characterized cable as an independent medium, holding that it no longer was necessary to protect broadcasters, but it nonetheless felt consumers needed a period to adjust to the new A-B switch format and so adopted a continuation of its must-carry rules for five years to shield broadcasters during the transition.[41] At the same time, the Commission announced a sweeping new inquiry into the video marketplace, one that promised to look into cable's compulsory license, the syndicated exclusivity rules, and into the ban on telephone company provision of cable service.

The new A-B switch requirement promised to be an expensive one for operators and, in response to complaints, the FCC, in March 1987, modified the ruling, allowing operators to pass the cost of the switches along to consumers. Turner indicated he was satisfied with the solution, which was officially backed by the NCTA, CATA, and the NAB. But the industry, as often was the case, was not of one mind on the subject, and Century Communications, along with Daniels & Associates and United Cable Television, joined to challenge the new rules on First Amendment grounds. In December 1987, the District of Columbia Circuit Court once again struck down the Commission's regulations.[42] The Court held that the FCC had failed to make the case that consumers needed five years to adjust to A-B switches. Further, according to the Court, the Commission had failed yet again to adequately demonstrate any potential harm to broadcasters. Nothing in the FCC's presentation, concluded the District of Columbia Circuit Court, was sufficiently powerful to overcome the constitutional protection afforded cable communication.

While winning in the courtroom, cable's political position in early 1988 continued to slide. In May, the FCC responded to renewed pressure from Congress with a study of cable's carriage of local signals. Lawmakers favoring must-carry, including Representative John Dingell (D-Michigan), wanted data in preparation for possible legislation. They had decided to seek the evidence of harm to broadcasters that the District of Columbia Circuit Court had failed to find. Broadcasters told lawmakers that cable had abused its power, dropping or denying carriage to local stations, and had several reasons for doing so. Operators, they alleged, would be able to sell local advertising time on the replacement cable networks, or even profit from national ad sales if the operator were vertically integrated with the programmer.[43]

In late August 1988, the FCC released its finding. It had surveyed broadcasters on the issue of must-carry and reported that 31 percent of the responding stations had been dropped or denied carriage following the *Quincy* decision.[44] The survey also revealed stations that had been moved from their on-air dial position and stations that had paid the local cable operator for carriage.

Under increasing political pressure, the NCTA returned to the bargaining table with the NAB. Hopes for a settlement soon collided with a new issue, however, channel positioning. With expanding channel capacity and the legal freedom to carry or not carry local stations, some operators were repositioning lesser-viewed channels on the cable dial. A local station broadcasting on Channel 9, for example, might have been moved up dial to Channel 28 or Channel 34. Channel identity was critically important to local broadcasters and the issue was an emotional one for them. Most of the affected stations were smaller independents, which, through the Association of Independent Television Stations (INTV), led the fight to make "on-channel" positioning a requirement in any new must-carry settlement. A tentative agreement between the NCTA and NAB was scuttled in late1989 when the INTV refused to support the proposed channel positioning provisions, which the organization saw as inadequate. Moreover, the cable industry was rapidly becoming the bete noir of Congress, and broadcasters, feeling the shift in political momentum, became bolder in their demands. The must-carry fight moved to Capitol Hill.

Syndex and Signal Importation

Broadcasters did make progress at the FCC on the issue of syndicated exclusivity. Neither broadcasters nor Hollywood had been happy with the compensation established by the Copyright Royalty Tribunal under the 1976 Copyright Act. Withdrawal of syndicated exclusivity in 1980, which largely eliminated local exclusivity protection, only exacerbated the perceived harm. Efforts never really stopped, therefore, to rectify the problem, and broadcasters, along with copyright interests, were continually lobbying for a revision of the Act. One proposed change involved retransmission consent, an old idea requiring cable operators to obtain formal permission from the local station or the copyright holder before carrying the signal. In 1979, broadcasters and the major studios backed proposed legislation (HR 3333)—a part of the VanDeerlin attempt to rewrite the Communications Act—that would have imposed retransmission consent requirements on cable. The bill died, but broadcasters continued to press their case at the FCC. They complained specifically that cable systems were importing out-of-town stations and superstations that duplicated local programming, programming for which local broadcasters had paid exclusive rights. The duplicate programming, furthermore, drew off audiences and with it advertising dollars. It eroded local efforts of broadcasters to position themselves as the sole providers of certain kinds of programming and some argued it even threatened the syndication market itself.

Superstation operators, such as Tribune Broadcasting, which owned WGN in Chicago and WPIX in New York, argued that reimposition of the rules would damage their efforts to develop first-run syndication programming. Some feared a return of syndication exclusivity (syndex) would spell the end of superstations altogether.

As it promised the previous year, the FCC in February 1987, a month after rate regulation ended, opened rule-making proceedings to consider reinstating syndex. In May 1988, again in response to congressional pressure, the Commission acted on syndex. It explained that cable had grown beyond the expectations of the syndex repeal of the 1980s. Erosion of broadcast audiences to cable was injuring the financial health of local stations and could have deleterious effects on the long-term supply of syndicated programming. The public ultimately would be harmed by the loss of program diversity. The syndicated exclusivity rules were back in place.[45] The effective date of implementation (eventually) was set at January 1, 1990.

Cable operators naturally appealed, but a unanimous District of Columbia Circuit Court upheld the policy in late 1989.[46] The Court found the rules to be well within FCC authority and not in violation of First Amendment principles. Judge Wald ruled the FCC's change of heart on syndex was justified by a change in the industry, specifically the development of cable beyond expectations that served as the rationale for repeal in 1980.

The rules prompted some systems to drop regional imports that duplicated local programming. WTBS and other superstations acted quickly to remove content that might cause duplication problems, limiting their programming to that for which they held exclusive national distribution rights and making themselves "blackout proof."

The Compulsory License

Along with syndex, the FCC took up the question of cable's compulsory license in 1987. The NCTA and the Motion Picture Association of America (MPAA) had held talks through 1985 and 1986 on compulsory license fees. No one was happy with the complicated formula for allocating royalties and, in early 1986, it appeared the two sides had settled on a replacement for the existing structure, with a flat-fee approach akin to that used by cable operators and cable programmers. The deal fell through, however, when the MPAA decided it required an end of the compulsory license itself. The NCTA balked and the film industry declared it would take its case to the FCC and Congress.

In April 1987, the FCC opened its inquiry.[47] The intent was to determine whether the market and the public would be better served by repealing the law and making cable fully responsible for program copyright liability. On October 27, 1988, the FCC, in a divided vote, recommended to Congress that it repeal cable's compulsory license for distant signals. In August 1989, it formally released the conclusions of its study. Consumers would be better served, held the Commission, by elimination of the legal structure for both distant and local stations. Cable, suggested the agency, would be able to negotiate national rights through a number of avenues, and would not have to enter into talks with each individual station.

Rethinking Telephone Competition

When the FCC released its revised must-carry rules in August 1986, it promised to take a new look at its ban on telco provision of cable service, as well. In July 1987, fulfilling its promise, the FCC launched an inquiry.[48] Noting once more that cable no longer was a fledgling industry, but rather a mature and "robustly competitive" one, the FCC wondered aloud if the rules were still necessary. In July 1988, it voted to remove the protective shield from the cable industry, tentatively recommending that Congress do away with the prohibition on telco provision of cable service.[49] A split Commission (2–1) conceded that there dangers to allowing the telephone companies into cable, but that regulatory safeguards could be installed to prevent them from using their great size, wealth, and monopoly position to unfair advantage in the video market. The Commission's position was weakened in early 1989 when Commissioner James Quello changed his mind, citing a concern that allowing in the telephone companies might only lead to substituting one monopoly for another. In the end, however, the FCC could not open the doors to the telephone companies on its own. Repeal of the ban, embedded in the 1984 Cable Act, would require a vote from Congress.

Telephone participation in video service also required a change in the MFJ, and again, cable's lack of pricing restraint worked against it in the court of opinion and the court of law. As cable rates rose, worried politicians and policy analysts looked to promote competition more strongly, and the telephone industry, in contrast with the anemic MMDS business, appeared to have the resources and desire to challenge cable's dominance of the market. One of those supporters, moreover, lived in the White House. The Reagan Justice Department in 1987 released a required report on the status of the MFJ recommending that the RBOCs be freed, in part, to compete with cable, and proposed that Judge Green in his 1987 Triennial Review of the MFJ, ease the line-of-business restrictions. Over the next two years, Judge Green did so, allowing the RBOCs limited entry into the information transmission services (such as voice mail), but he maintained the barrier to content generation and control, including the provision of video services.[50] The RBOCs naturally challenged Judge Green's decision and, in April 1990, the District of Columbia Circuit Court ordered Judge Green to review his ban on information services.[51] The appeals court decision was structured in such a way that Judge Green was required, on review, to repeal the information services ban, issuing the order in July 1991.[52] It was a green light for the

RBOCs to begin offering video programming outside of their service areas (as well as other information services, such as classified ads, yellow pages, electronic banking, and stock quotes, within their service areas).[53]

Congress Looks Again at Cable

Cable was not having any better luck with Congress. The first angry group of consumers to descend on Washington, D.C., was there even before rate controls had been lifted. In the view of TVRO owners, the cable industry had cut off their access to TV choice, and they wanted somebody to do something about it.

TVRO owners had been receiving HBO and other cable services essentially for free for years. The initiation of scrambling in early 1986 was a blow to the home dish industry and greeted as a personal affront by TVRO owners. Even those dish owners willing to pay for scrambled signals complained about cable restrictions. Consumers no longer could simply aim their dishes at a satellite and sit back. To receive legal and unscrambled programming, they had to contract directly with the programmer or go through their local cable operators, who in turn worked with the program network. In either case, TVRO consumers felt they would not (and could not) get competitive rates. Start-up companies that attempted to package programming for the home dish market found that obtaining distribution rights was somewhere between difficult and impossible.

The cable networks, tightly bound by ownership or mutual business interests to cable distributors, were very hesitant to license their product to the TVRO market, which constituted an admittedly small but nonetheless viable distribution competitor. HBO flatly refused to license its programming to third-party vendors, dealing only with cable operators or selling directly to dish owners. In the latter case, HBO also gave rebates to local cable operators, so TVRO customers paid the same amount as cable subscribers, with a portion of their fee going to the local operator, a practice that drew criticism from customers, the dish industry, and legislators.[54] The industry took the same stand with respect to supplying content to most SMATV and MMDS operators.[55] HBO and Disney were notoriously reluctant to sell to those markets and other programmers charged premiums of 15 to 20 percent for programming access. The FCC cited programming services, including TNT, Bravo, AMC, and ESPN, that discriminated against noncable distributors.[56] The programming industry countered, arguing wireless providers were unfairly seeking the deep discounts reserved for mass distribution MSOs.

In March 1986, not long after HBO began its scrambled feed, enraged dish owners flocked to Washington, D.C., for hearings called by Representative Tim Wirth, Chairman of the House Subcommittee on Telecommunications. It was a heated session in which dish owners and suppliers attacked what they considered to be the monopolistic practices of the cable industry. Congressmen and women, especially from districts in rural America, attended closely. As *Broadcasting* magazine reported: "Satellite television has become a way of life in rural America. Said Charles Rose (D-North Carolina): 'You've got to have a pickup truck, a gun rack and a satellite dish on the side of your house.'"[57]

About a dozen scrambling bills were introduced in the House and Senate in this period. Wirth indicated that his interest was not in legislation but rather in putting political pressure on cable to make programming available to dish owners at fair prices. The Senate, however, was more serious about forcing open the programming marketplace. The Senate Commerce Committee held hearings in July 1986, and Senator Gore, along with Senators Wendell Ford (D-Kentucky) and Dale Bumpers (D-Arkansas) introduced Senate Bill S. 2823 to force programmers to make their material available to third-party vendors.

Gore's involvement would become particularly important. He was not the only member of Congress to place cable in the spotlight in the late 1980s, but he was one of the most vocal and most visible. Through the second half of the decade and into the early 1990s, he would hone a reputation as a harsh critic of cable television and even personalize the fight, singling out TCI's John Malone for special attention.

Gore's engagement with cable began early in his political career. His father had been a prominent senator from Tennessee and, after a few years working as a reporter, Gore followed his father's career path, winning an open congressional seat in 1976. Gore had always had an interest in technology. He chaired the House Science and Technology Subcommittee and was an early supporter of the C-SPAN initiative to televise House proceedings. He also had helped his parents install a satellite dish on their farm in Carthage, Tennessee, in a rural area beyond the reach of any local cable service. It brought him into contact with home dish advocates who sought his support for their cause in Washington. Gore even penned a 1985 article in the *Memphis State University Law Review* on the subject.[58]

Gore would never be a friend of cable, but the industry did have supporters in Congress. While TVRO waged what the press described as a "grass roots campaign," the NCTA went to work in its more traditional fashion. It enlisted friendly forces on Capitol Hill, including congressional ally Barry Goldwater (who chaired the Communications Subcommittee), to oppose and ultimately help defeat the programming access bill.[59]

The issue was far from dead, however, and when the Democrats took control of the Senate in 1987, Gore, having won the Senate seat for Tennessee, reintroduced the legislation.[60] Congressman Billy Tauzin (D-Louisiana) offered a companion measure in the House (HR 1885) and both chambers held hearings in July. Gore argued that the cable industry, composed of a tightly intertwined cabal of distribution and programming business interests, was withholding cable programming from potential distribution competitors, including the TVRO industry.

The Bush administration opposed legislation, and the FCC and NTIA released a joint study in March 1987 stating that the marketplace was resolving the problem. In February 1988, the White House Office of Management and Budget also joined in opposition to legislation, but Congress was not persuaded.

While the legislative battled wore on, Wirth continued to work with cable programmers and TRVO vendors to find a compromise. Companies that attempted to obtain distribution contracts, including Amway Corp. and the National Rural Telecommunications Cooperative (NRTC), reported that cable programmers were dragging their heels, asking for exorbitant license fees or flat out refusing to license content. Wirth's pressure on programmers finally bore fruit in July 1988, when the NRTC announced a deal, brokered by Wirth, with the leading cable networks, including HBO, the Disney Channel, Nickelodeon, and TNN. It gave NTRC the right to distribute a package of channels to TVRO owners, the first such third-party deal in the industry.

In the short term, the announcement signaled the beginning of the end of the Gore-Tauzan efforts to pass legislation. Political support evaporated in the presence of what was seen as a preferred nongovernmental solution.

Congress did pass an amendment to the copyright act giving satellite carriers a compulsory license similar to that held by cable operators. The Satellite Home Viewer Act of 1988 gave satellite carriers authorization to provide, subject to royalty payments, superstation and network broadcast signals to viewers in areas outside the reach of a network affiliate signals (unserved or "white" areas).

Framing Legislation

In and of itself, the plight of TVRO owners unable to receive free HBO was not a major issue for Congress. While representatives from rural districts had to take notice, the dish owners numbered about 1.5 million nationally, a sliver of the nation's approximately 90 million TV homes. Television generally, however, was very important to Congress because it was very important to the public, and half the population was now getting TV via cable. Consumers expected service to be reliable and prices reasonable, but service had rarely been reliable and, as prices rose through the late 1980s, so did consumer frustration and anger.

Rate regulation had been effectively dead for less than a year when that anger began spilling into congressional laps and, in March 1988, the Senate Subcommittee on Antitrust opened hearings on a wide range of cable-related topics. Subcommittee Chairman Howard Metzenbaum (D-Ohio) postured the review as a warning shot across the bow of the cable industry. In his opening remarks, he indicated that legislation was not his immediate intent, but rather he wanted to "sound a warning bell and an indication of concern" to the nation's cable operators. The Senator said he had reports from his home state of rate hikes of 50 percent or more in the first six months of deregulation.[61] Noting that many in Congress were beginning to doubt the wisdom of the 1984 provision eliminating rate controls, Metzenbaum told NCTA President James Mooney, "Mr. Mooney, I have to tell you, some of us feel that we were had when we passed that bill."[62]

While rate hikes led the list of congressional concerns, the other issues haunting cable were also aired at the hearing. Broadcasters complained about the lack of must-carry and syndex protection (not yet revived by the FCC). Representatives of the TVRO industry and MMDS operators complained about their difficulties in securing cable programming. Metzenbaum cautioned that if cable did not clean up its own house, Congress would be obliged to do so for them.

Two weeks later, the House Telecommunication Subcommittee, chaired by Congressman Ed Markey (D-Massachusetts), began its own review.[63] It was a softer echo of the Senate proceedings. The themes were repeated, but in a more conciliatory tone. A number of congresspersons expressed the hope that the industry's problems could be solved without recourse to legislation. They also agreed that additional data on the state of affairs in cable television service would be useful, especially on the question of rate increases. Anecdotal evidence from around the country suggested rates hikes of 100 percent or more. The NCTA's own study reported average monthly increases of only 6.7 percent. Markey, for one, wanted better information and indicated he would ask the General Accounting Office (GAO) for a formal survey of cable rates and service. In a follow-up hearing in May 1988, Malone and Turner appeared separately, with Malone striking a conciliatory tone and suggesting the industry was open to modification of the current regulatory structure.

Despite the assurance of some congressional leaders that legislation was not their immediate goal, bills nonetheless began to appear. In March 1988, Congressman Howard Nielson (R-Utah) introduced legislation (HR 4135) allowing telephone companies to own and operate cable services in their own service areas. That same month, Congressman John Bryant (D-Texas) proposed a bill linking the compulsory license to the carriage of all local broadcast signals.

In June 1988, the White House weighed in. That month the National Telecommunications and Information Administration (the former Office of Telecommunications Policy) issued a report on the state of the video marketplace.[64] Staying true to its free market philosophy, the theme of the 127-page report was structural change to increase competition. Reviewing issues that had been circulating in Washington for several years, the report noted growing concern about concentration of ownership in the cable industry, both horizontal and vertical, and cable's increasing power in what was now being termed the "video marketplace." While the NTIA review observed that increasing rates could be the product of increasing programming costs and artificially low rates before deregulation, increased competition would nonetheless be beneficial to the public. To spur that competition, the report recommended a number of policy changes. It endorsed repeal of the prohibition on broadcast network ownership of cable systems. It suggested the FCC review horizontal integration in the industry with an eye toward possible ownership controls. It recommended changes in franchising procedures designed to increase local competition.

The heart of the study, and the source of much of the discussion it generated, however, was a proposal that the telephone companies be permitted to enter the cable television business on a common carrier basis. It was called "video dial tone," and a great deal of credit for the concept was given to then NTIA Administrator Alfred Sikes.

In many aspects, "video dial tone" was old wine in new bottles. It struck many of the same notes as Clay Whitehead's OTP "separation policy" concept of a decade before, and, in fact, cited several aspects of that proposal as a model for policy.[65]

The essence of the proposal was that local telephone companies be permitted to lease broadband capacity to third-party programming providers. Those providers could be new competitors to the local cable company, and include broadcasters, sports organizations, or independent producers. Cable operators, as well, would be permitted to lease lines. Concerns about direct telco control of programming and the long-standing worries about the monopoly power of the telephone industry to cross subsidize service or hedge on pole attachment agreements, kept the NTIA from advocating full telephone company control over local video service. The local telephone company would be permitted to provide ancillary services such as billing and system maintenance.

The NTIA report additionally endorsed reimposition of the program exclusivity rules, and phasing out cable's compulsory license. It noted policy concerns about vertical integration in the industry, but defended operator investment in (and control over) programming networks, arguing they had helped foster new networks and programming choice. The following month, the FCC released its own recommendation, noted previously, that the 1984 prohibition of telephone company ownership of cable be repealed, going further than the NTIA; in September, it delivered its report to Congress on the impact of the repeal of must-carry on local broadcasters.

Congress, therefore, did not lack proposals on how to re-regulate cable television, but had little time to do so as the 1988 legislative session drew toward a close and the politicians returned to their home states to campaign. The general mood was that full re-regulation remained unlikely, but structural changes, such as reimposition of must-carry and allowing the telephone companies into video service, had some congressional support.

As Congress reconvened for the new year, with a new administration in the White House, the press for structural changes began to take legislative form. More than a dozen bills were introduced to either regulate cable, with a focus on rate control, or open the video marketplace to potential competitors. Among those supporting some form of legislation were the broadcast industry, the telephone industry, the National League of Cities, the National Conference of Mayors, the small but vocal satellite and MMDS industries, and a host of consumer groups.

Hearings on cable continued through 1989. Metzenbaum's Antitrust Subcommittee held sessions in April; Inouye's Communications Subcommittee took testimony in June. A bill introduced by Metzenbaum called for restoration of city control over cable rates. Congressman Charles Schumer (D-New York) cited rate increases of 150 percent or more in cities such as Corvallis, Oregon; Denver, Colorado; and Weston, Connecticut.[66] Gore, along with others, sponsored legislation allowing the telephone companies entry and seeking controls on cable group ownership.

The cable industry continued to defend its rate increases, pointing out that the number of cable programming services had increased concurrently: average basic channel service had risen from twenty-seven in 1986 to thirty-two in 1988, and the fees programmers charged operators had accelerated substantially.[67] The industry argued that rates had remained fairly stable, around $0.45, on a per-channel basis, and compared with other forms of news and entertainment, cable was a bargain. NCTA President James Mooney told Congress, in 1990, "For about $15 a month, basic cable service is still the best entertainment value in America."[68] The industry suggested that the post-1986 increases constituted a reasonable attempt to make up ground following years of local price control. One industry-supported study showed that the increase in basic cable service rates between 1972 and 1986 was 89.6 percent, modest in comparison with the 162 percent rate of general inflation.[69]

Arguments based on increased operator costs made little dent in the movement for re-regulation, however, and on August 3, 1989, the GAO released the findings of the study requested in early 1988 by Markey. It was a small bombshell. The report concluded that basic cable rates had risen about 29 percent in the first two years of rate deregulation. Rates for premium service had declined somewhat (a reaction to the growing popularity of home video). Overall, subscriber rates (including pay channels) had risen 14 percent.[70] Critics noted that the rate of increase was roughly four times that of inflation and found in the GAO study evidence that cable was taking advantage of a dominant position in the national TV marketplace. Moreover, a GAO spokesman, in presenting the study to Congress, declared that cable was probably a monopoly in need of regulation.[71]

The report struck like a knife into cable's political defenses, and industry critics were quick to seize the handle and turn. The NAB, in September 1989, sent a letter to Senator Inouye urging substantial re-regulation of cable. The NAB was joined by the Association of Independent Television Stations, led by its articulate and effective president, Preston Padden. (The aggressive broadcaster actions also were a part of negotiating tactics in the broadcaster-cable discussions over must-carry, and a response to a short-lived attempt by Time Warner to program a channel as an independent TV station at its Rochester, New York, system.)

The GAO report raised such a storm that, in October 1989, the House Telecommunications Subcommittee asked the agency for a second, follow-up report. The next month, with anticable sentiment building, Inouye called his Communications Subcommittee into session once more.[72] Testifying at the oversight hearings were NCTA head Mooney, James Robbins, President of Cox Cable Communications; and the star witness, John Malone. It was not a good day for cable.

While some industry observers commented later that the session was less harsh than they expected, it was nonetheless fairly harsh, and memorably so. The day before, Danforth had introduced yet another cable bill, this one more draconian in its regulatory posture than any offered previously. It gave cities authority to regulate rates, capped ownership at 15 percent, and required on-channel must-carry of local signals, among other things. Cosponsor John McCain (R-Arizona) declared: "The public is being ripped off by cable...the little kid on the block has become the big bully."[73] For many congressmen, the biggest bully of them all was TCI's John Malone and, on the day of the hearing, Malone faced his prosecutors.

Inouye had been an ally of cable and a supporter of the 1984 Act, but Time Inc. had recently raised cable rates in his home state of Hawaii—over the strong objections of then ATC Chairman Trygve Myhren. It was more than even the moderate and respected Inouye could tolerate. Inouye stated at the opening, "We can't close our ears to the shouts of our constituents. [Our] mailbags [are] heavy with complaints about rates and shoddy service."[74] Senator Ford said he was frankly skeptical about the findings of the recent GAO report that showed rate increases in the mid-20-percent range. He had reports from Bowling Green of 123 percent rate increases and from Louisville of 204 percent. The fiercest broadside came from Gore who singled out TCI and Malone for special attention. "TCI, for one," said Gore, "is obviously hell-bent toward total domination of the market as it buys up not only more and more cable systems, but more and more programming services, and even more movie studios."[75]

By most accounts, Gore's concern about cable practices had spun out of his earlier involvement in the TVRO problems and out of the actions of a particular cable operator in his home state of Tennessee. A company called Multivision, run by I. Martin Pompadur, a former ABC executive, increased rates by more than 40 percent in what some claimed was a typical system flip.[76] The increase brought a stream of letters and telegrams from unhappy customers into Gore's office, and Gore excoriated Multivision in the press. Multivision later rolled back its fees, but Gore suggested it was too little too late. Multivision was not a national presence, however, and Malone was. Gore blasted cable rates and practices in the November hearings, and once named TCI as the leader in what he termed "the cable cosa nostra." He had called Malone the "Darth Vadar" of the television industry and generally painted a picture of a man obsessed with power and greed and willing to use any means, ethical or unethical, to achieve his ends of dominating the multichannel marketplace. According to Malone biographer Mark Robichaux, Gore, because he made it personal, "became the only man for whom Malone reserved unalloyed animosity."[77]

Gore and Malone spared through the hearings, with the Senator grilling the TCI chief on rates generally and TCI's franchising activities in particular. At one point, Gore asked Malone: "Do you think that your size and power have made your company arrogant and heavy handed?" Malone responded, "I don't beat my wife either. No, I honestly do not believe so."[78] Gore cited the Jefferson City incident as an implicitly common business practice for Malone. Malone did not respond in kind. He said that Jefferson City had been an aberration, adding, "I lost an awful lot of sleep over that one, senator"[79] He defended TCI and the cable industry, noting their investment in cable infrastructure and in programming services.

Malone, Robbins, and Mooney were, on balance, conciliatory in tone, seeking to round off the sharper edges of some legislative proposals. With a strong coalition of Democrats and Republicans supporting regulation, Malone, Mooney, and others put their energies into influencing the shape of the bill. Rate controls were accepted as probably inevitable, but the industry preferred federal oversight to local control. They would not oppose ownership caps, but did oppose divestiture of current holdings. Inouye, for his part, indicated he wanted to work with the industry. Congress adjourned shortly after the November hearings, but committee leaders promised cable bills would be reintroduced the following year.

If cable operators thought they might get a breather over the holiday break, they were gravely disappointed. Seeing cable on the political ropes, the broadcast industry made its most aggressive move to date. In December, an NAB task force met to fashion a new demand. Broadcasters had been pressing hard for must-carry legislation, but now they raised the stakes, arguing that cable operators should not only carry all local broadcast signals, but they should pay for the privilege as well. It was a dramatic turnabout. In some isolated cases, in absence of must-carry obligations, cable operators had exacted payment from broadcasters for carriage. Broadcasters, however, had always felt cable was pirating their signals and the Copyright Act offered insufficient compensation. The NAB now demanded must-carry and must-pay, or as some wags dubbed it, "cash and carry."

The proposal originated at CBS, which brought it to the NAB in mid-1989. CBS was in financial difficulty. It had slipped in the ratings and was bleeding money. Erosion of viewing because of the success of cable was seen as one, although not the only, cause. CBS executives figured cable might provide a badly needed new revenue stream. Cable had enjoyed income from advertising dollars and subscriber fees for decades; broadcasters sought a similar arrangement, asking cable operators to pass some of those fees on to them. The sums to be paid were substantial, ranging from estimates of $2 billion to $8 billion annually, or 20 percent of basic subscriber revenue.[80] Cable executives put on brave face; Amos Hostetter called the idea "wishful thinking."[81] CBS President Laurence Tisch took the case to Inouye himself, who indicated he would put the idea on the table when Congress reconvened.

Coming on the heels of the Danforth bill and a collapse in the must-carry talks, many inside and outside the affected industries characterized the must-pay proposal as a declaration of all-out war. *Broadcasting* magazine editorialized a plea for the two camps to step back from what it termed "Armageddon."[82] But, both sides were marshalling their troops for a fierce legislative battle.

In February 1990, as Congress was preparing to look seriously at cable legislation, the NCTA attempted to undercut one of the long-standing criticisms, a failure to provide quality service to customers. Seeking to reverse the public image and service record, and appease politicians, the NCTA, in February 1990, instituted voluntary customer service guidelines. They called for improved response to telephone calls (calls should be answered within 30 seconds), firm appointments for installation, and quicker execution of service and repairs requests, among other things. The guidelines were greeted with hopeful skepticism but, in fact, a weak record on voluntary compliance would later be seen; by 1993, only 2,000 of the nation's 11,000 systems were certified as meeting the plan. In any event, it did not directly deal with the larger issues, and work in Congress on those issues was getting underway.

Hearings continued and the outlines of legislation began to take shape through early and mid-1990. Elements included the following:

- FCC (not local) oversight of rates
- Some measure of assurance that cable would make its programming available to TVRO, MMSS, DBS, and similar competitor distribution industries
- Required must-carry, with some control given to local broadcasters on channel positioning
- FCC given authority to examine the necessity of ownership caps
- FCC oversight of service quality

At the same time, support for telco entry waned. The specter of telephone company participation in the television business may have been one of the few areas in which cable and broadcast interests met. By mid-1990, they presented a united front, joined by the powerful newspaper industry. It was more than sufficient to hold off the telephone companies—at least for awhile.

The Bush administration opposed most of the key elements of the proposed bills, especially behavioral controls such as rate regulation and program access requirements. The White House, however, did endorse structural changes that would increase market competition, including telco entry. The FCC released its own report on cable in late July 1990.[83] In keeping with administration philosophy, it again warned against unnecessary and burdensome regulation, calling instead for policy to stimulate greater competition. Al Sikes now was Commission Chairman and much of the thinking reflected his video dial tone approach to telecommunications competition. As in the past, the FCC was additionally concerned about the administrative burden that rate regulation would bring to the agency. By its own estimate, one legislative proposal would cost the agency $55 million over the first five years of regulation.[84]

New fuel for the regulatory juggernaut already had come, however, in June with release of the GAO's requested follow-up study of cable rates and service.[85] The headlined finding for the second report showed that basic cable rates had risen between 39 and 43 percent (depending on the definition of basic service) between 1986 and 1989. The GAO report also showed that channel capacity of the average system had increased over the same periods, from thirty-four to forty, and average revenue per channel for cable systems had in fact decreased when adjusted for inflation.[86] The cable industry put on its best brave face, noting that rate increases had slowed and arguing that cable remained a good value. Overall, however, the report was seen as additional evidence of the need for new cable regulation.

That same month, June, the House Telecommunications Subcommittee approved a bill (HR 5267) including most of the provisions noted above. The Commerce Committee passed it on a unanimous voice vote in July. The full House approved its version of cable legislation in September, but in the Senate, Republicans, prodded by the administration, were successful in blocking S. 1880.

Cable operators and the FCC hoped agency action on cable rates might slow the progress on a cable bill and, in August 1990, FCC Chairman Sikes assigned the Office of Plans and Policy and Mass Media Bureau to conduct a review of the broadcasting marketplace to see if FCC rules were still necessary in face of increasing competition. In December, the Commission proposed new rate regulation guidelines.

Under the rules adopted by the FCC in 1985, cable operators with three broadcast stations available in their market faced effective competition and were shielded from rate controls, a situation that existed in more than 90 percent of all cable markets. Under the FCC's revised guidelines, the number of broadcast signals needed to trigger rate relief was increased to six stations in markets where cable penetration was less than 50 percent (rate controls came into effect where cable had more than 50 percent penetration even with six broadcast signals). Operators also could escape rate controls if another multichannel provider operated in the market and reached at least 50 percent of all homes, serving 10 percent of them, or if the local cable system passed a good actor test. "Good actors," provided above average basic service at below average rates. In June 1991, the FCC adopted the new rules.[87] About the same time, the office of Plans and Policy issued a report predicting injury to the health and welfare of broadcast television because of the migration of viewers and advertising dollars to cable television. The FCC also opened wide ranging hearings designed to address the problem. In July, the Commission launched rule-making aimed at reinstating its must-carry regulations and, in October, the NTIA issued another report advocating the removal of cable–telephone company cross-ownership controls.

The agency's activities would be trumped by Congress, however. In early 1991, cable legislation was reintroduced in both houses.[88] By May, the Senate Commerce Committee approved the latest iteration of the bill, S. 12. The full Senate passed the measure January 31, 1992; the House followed in June.

Still opposing the hand of government in industry, President Bush vetoed the legislation in early October. Cable executives at the time may have taken some comfort in the knowledge that the President had vetoed thirty-five bills in his tenure and not one had been overridden. They would not be comfortable long, however, for cable would be the first. It was a measure of the power of the cable issue, the importance and concern with which consumers held cable TV rates and service, that on October 5, 1992, the House overrode the President's veto 308 to 114. The Senate quickly followed suit, 74 to 25, with Gore, now the Democrat nominee for Vice President, taking time out from the campaign to return to the capitol and cast his vote.

The Cable Television Consumer Protection and Competition Act of 1992

Bill Daniels, who warned the industry about the responsibilities and dangers of freedom, had never been more right. The Cable Television Consumer Protection and Competition Act of 1992 would bring controls more harsh than anything that had come before. Congress and the FCC would control, just as a beginning, tiering and pricing.

Under the new law, cable systems were required to offer a basic tier of service that included all local and nonsatellite imported broadcast stations, plus any access channels offered by the operator.

The FCC was pointedly required to bring consumer rates down. Under the conditions of the 1992 Act, cable systems were subject to rate regulation in areas that lacked effective competition. Here, competition was redefined to include only situations in which (1) the system had less than 30 percent penetration, (2) a competitive multichannel provider, such as an MMDS, reached 50 percent of the market and had a 15 percent penetration rate, or (3) the local franchise authority operated its own competitive system. Only a small percent of cable systems in the country met any of these standards.

The Commission, then, was charged with assuring that subscriber rates for the basic service were "reasonable." What constituted "reasonable" was to be determined by the Commission using a list of criteria set out in the legislation and chiefly involving actual cost of service to the operator. Rates for basic service were subject to regulation by the local franchise authority. Rates for additional service tiers could be regulated by the FCC if they proved to be unreasonable. Premium and pay channels were not subject to rate regulation and operators were allowed to require subscription to the basic tier as a prerequisite to purchasing the pay and premium services.[89]

As part of the 1992 Cable Act, Congress authorized the FCC to develop stricter national regulations for customer service, which the Commission, with the NCTA's help, unveiled in 1993. Under the new FCC guidelines, cable operators were required to

- Keep normal business hours
- Maintain a 24-hour telephone hotline for customer questions or complaints
- Answer phones and transfer calls within 30 seconds
- Begin repairs on outages within 24 hours
- Locate service offices within convenient distance of customers
- Execute installations within seven days of an order
- Establish 4-hour appointment windows for home service
- Publicize rate and schedule changes at least 30 days in advance

Local franchising agencies were permitted to require even stricter, but no less strict, customer service standards.

In response to complaints from TVRO and MMDS operators about difficulty in obtaining programming, the legislation required program providers in which operators had "an attributable interest," subsequently defined as a 5 percent equity holding, to make their content available to alternative distribution services on a nondiscriminatory basis.[90]

Must-carry was reinstated. The new law required systems with more than twelve channels to set aside up to one-third of their capacity for local signals and carry those signals on their normal broadcast channel numbers. Broadcasters, moreover, were partly successful in gaining retransmission consent, the right to withhold heir signal or, more importantly and realistically, to negotiate to be paid for it. Local broadcaster could choose to invoke must-carry, forcing the local operator to carry their signal, or, alternatively, they could negotiate for paid carriage. They could not demand carriage and payment both.[91]

Rules governing the ownership of home wiring, restricting indecent programming over access channels, and expanding job categories for purposes of Equal Employment Opportunity reports also were enacted. The law modified the leased access rules to give the FCC greater control over setting rates and terms. Ownership limits on cable were adopted, including restrictions on cable ownership of potential competing media, such as MMDS, but later struck down by the courts (see the discussion in Chapter 12). Finally, the 1992 Act specifically banned the kind of negative option that Liberty Media had attempted in 1991 and had so angered consumers and state attorneys general.

Convergence

The new law was not cable's only problem. The prospect of real competition for multichannel viewers was rising over the far horizon. STV, teletext, and MMDS had failed to make a mark in the 1980s. But new entrants were marshalling their forces for a turn at the market. The largest potential entrants and those most worrisome for the cable industry were the telephone companies. While restricted legally from wide-scale participation in television, the telcos were organizing heavily funded legal initiatives and taking their case to the courts in an effort to have those restrictions removed. DBS pioneers also continued to gnaw away at the financial and technical barriers that kept them from delivering multichannel programming. And, something new began percolating in the telecommunications world of the late 1980s; it involved computers and computer-based communication. As early as 1984, a writer for the industry publication *Channels* magazine wrote, "The convergence of the television screen, the telephone, and the computer cannot be regarded as a merely commercial phenomenon."[92]

"Convergence." A new avatar had arrived.

12

500 Channels
(1992–1996)

> *The person whose face is to be transmitted by the*
> *Telephone Company's new process of television sits in*
> *front of a machine at one end of a telephone line.*
>
> — "SEEING OVER THE TELEPHONE," THE AMERICAN REVIEW OF
> REVIEWS, MAY 1927 [1]

In 1947, engineers working at Bell Laboratories created the first transistor. In the late 1950s, they developed methods to multiply the functions of many of these devices on a single silicon wafer, an integrated circuit, or chip. In 1968, Robert Noyce and Gordon Moore founded Intel Development Corp. to improve and manufacture those chips. In 1975, a small businessman in Albuquerque, New Mexico, Ed Roberts, placed an Intel chip at the heart of the first widely publicized personal computer, the Altair. Inspired by the Altair, two young computer enthusiasts, Steve Wozniak and Steve Jobs, created and promoted the Apple computer. Prodded by the success of Apple and a rapidly expanding personal computer market, IBM, working with a pair of young software designers, Paul Allen and Bill Gates, introduced its own "PC" in 1981. Through the mid- and late 1980s, a growing legion of PC users were exchanging files, messages, and ideas by connecting their computers through telephone lines, frequently tapping into a high-speed data exchange system developed by the government called NSFnet. The computers spoke their own language, the digital vocabulary of zeros and ones. A new technology was evolving, and it would have a profound impact on cable, the broader telecommunications industry, and society.

The "Western Show" was the cable industry's second largest annual event, surpassed only in importance by the NCTA's national convention. Sponsored by the California Cable Television Association and typically held in Anaheim in December, it drew the leading operators and programmers, equipment suppliers, and assorted retainers. John Malone was there in early December 1992, two months after Congress had overridden President Bush's veto of the 1992 Cable Act. Malone had an announcement to make. TCI was placing an order for 1 million new cable boxes with General Instruments. The boxes would harness digital technology to supply the consumer with hundreds of new channels; 500 channels was the number he picked. "This is just the beginning, this first round of products is the first of an evolution," Malone told reporters.[2] The set-top boxes would start going into living rooms in January 1994. "Television will never be the same," he said.[3]

It was not Malone's first public comment on the promise of digital technology to deliver a cornucopia of cable programming. A high-tech company—SkyPix—demonstrated its digital equipment at the NCTA National Show in New Orleans in March 1991 and Malone suggested it could deliver 200 or more channels. But it was an off-hand remark. It was not backed by a formal news conference, an order for a million cable boxes, and a story on the front page of the *New York Times*. "A Cable Vision [or Nightmare]: 500 Channels," read the *Times* headline,[4] and the 500-channel future suddenly was a part of the business and cultural lexicon.

Malone's announcement was widely reported, and sometimes gilded, by the business and popular press; with each retelling, the vision grew in scope and complexity. It was the latest turn in the social construction of cable. Fueled by the emerging power of digital processing and communication, Blue Skies were back, promoted by the same dependable cast—academics, policy analysts, regulators, politicians, and cable leaders themselves—which had driven the rhetoric through each of its earlier manifestations. All were assembled again for the task of redefinition. The new social construction promised hundreds of channels, and more. The social definition was dramatically expanded to encompass not just television, but telephone and data services as well.

In reporting Malone's announcement, the *Times* declared the new technology would likely "accelerate the marriage between computers and television, allowing customers to roam through video libraries hundreds of miles away," and "offer a broad range of interactive shop-at-home services, in which customers order anything from pizza to jewelry by pressing a button on the television remote control."[5]

Malone's 500-channel box would be but one piece of a larger telecommunications infrastructure, one that was churning discussion in every quarter of the television, computer, and telecommunications universe. The system, as sketched in the collective mind of the industry and its observers, used digital technology bound by a global network of satellites, and copper, coaxial, and fiber lines to merge computing, television, data retrieval and exchange, and telephone service. It would be a seamless, unified, information and telecommunications "Uber" system. The new buzz word and social construction was "convergence."

This chapter examines cable television in the grips of convergence, from about 1992 to 1996. It reviews the critical development of digital technology and its application in cable communications, and examines the emergence of real and potential cable competitors, especially the telephone companies and DBS providers. Also examined are the politics of the period, tracing the aftershock of the 1992 Cable Act and the public debate, driven in part by the concept of convergence, which led to a wholesale rewrite of U.S. communications law in 1996.

The Social Construction of Convergence

For decades, cable and broadcasters had provided television, while the telephone companies offered local and long-distance telephone services and computers processed various kinds of information. Convergence, in its simplest terms, was about the integration of these functions and businesses. Convergence had technical, political, industrial, and even spiritual overtones. Technically, as detailed below, it was the use of digital broadband switched technology to bring together the functions previous offered through unique distribution systems. Politically, it was the removal of regulatory barriers to permit companies previously restricted to those walled sectors into a unified and competitive market for television, voice, and data services. Industrially, it was the business opportunity and challenge to exploit the changing technical and political context and expand into new and foreign lines of business, becoming the first to successfully offer consumers the full menu of converged services. Culturally and spiritually, it spoke once again to the ancient human desire for instant, easy access to every manner of entertainment and information. As described by policy analyst Patricia Aufderheide,

> The computer communications revolution brought out an enduring link between technology and spiritual transcendence, resulting in a giddy utopianism. But beside futuristic hyperbole and bombast, this model also provided a vehicle for mobilizing citizen constituencies, launching new corporate projects, and proposing new policy initiatives. It shaped the search for what the Aspen Institute's Charles firestone called "the holy paradigm" for communications policy for a new era.[6]

Civil libertarians became "cyber" libertarians as social and policy analysts once again hailed the promise of an open, flexible world of digital discourse. Mass communication, characterized to this point as point-to-multipoint or one-to-many, was recast: The new model was multipoint-to-multipoint, many-to-many. Anyone could now be a content provider, a source of media programming, as well as a consumer. With web-based servers, users could solicit content and retrieve it on an as-needed basis, directed by one's own schedule, not the schedule of the source. The new paradigm appeared to challenge the existing mass media and telecommunications systems in a host of ways.

Field of Schemes

For those who had been around awhile, convergence had a familiar ring to it. Conceptually, it was 1960s Blue Sky utopianism in updated clothes. It was the Licklider vision taking one more turn around the track. This issue was not whether it was a system that, in theory, people would find useful and desirable. It clearly was, as it always had been. The question was, as it had always been, was the technology really there? More pointedly, was it there at a price that could make it viable in a capital economy? Schemes to create such systems had failed before, and recently, and some skeptics thought they saw the pattern repeating.

For the naysayers, it was fashionable for a period to draw a disparaging parallel between the convergence-laden plans of the telecommunications industry and a popular film of 1989. Actor Kevin Costner starred in a fantasy feature in which his character, an Iowa farmer, carved a baseball diamond out of a corn field in the hopes of attracting great but disgraced baseball legends of the past, especially the famous "Shoeless" Joe Jackson. Titled *Field of Dreams*, a running line and theme of the movie was, "if you build it, they will come." Skeptics of the new social construction of cable heard a similar line in the plans of the architects of the converged information universe, "if we build it, they—in this context, customers—will come." Not everyone was sure and some labeled the endeavor, "field of schemes."

Important differences existed between the convergence vision and schemes of prior eras. The competitive landscape had changed. Venture capital had been circulating for a decade, looking for new distribution platforms, and the kernels of new businesses in DBS and MMDS, despite early failures, were still out there. The telephone industry had undergone a radical restructuring as a consequence of the breakup of AT&T in 1984. Each of the new Baby Bells now eyed broadband, switched systems and convergent business plans with unalloyed lust. The Baby Bells and AT&T were spending millions in what appeared to be an effort to get into television. Much of the money went to lawyers, fighting in courts across the land to pull down legal barriers that blocked their way to the nation's TV sets. A great deal of money and work went into technical and marketing experiments in interactive television distribution. And a fair amount went to purchase or merge with the old enemy—the cable companies themselves.

Convergence, therefore, was not just about the integration of communications technologies and services, it was about the fusion of the related telecommunications industries. As divergent services were seen to collapse into an integrated, unified distribution platform, the formerly disparate companies that offered those services would try to mirror the changes through corporate consolidation. Convergence implied not just a technologically integrated information-entertainment utility, but an economically integrated mega-industry, as well. At the same time, and in sharp contrast, others saw in convergence the hope of accelerated competition among previously separated information and communication providers. This was an especially popular view in Washington, D.C.

The Politics of Convergence

The political temperament of the period proved an incubator for the rhetoric of convergence. In 1992, a Democratic President, William Clinton, took office, the first Democrat to capture the White House in a dozen years. Political culture had changed in the United States since the 1960s and even the new liberals stood behind the power of markets to assure consumer choice and service, although they were more willing to enact policy to force open those markets when they saw them constricting. The newcomers, fashioning themselves in Kennedy-esque trappings, actively promoted an image of youth and vitality. They believed in the future and in the power of technology, and few technologies were more powerful and more immediately appealing to voters than those of television and computers. Clinton was energetic and charismatic, a new face for a new age. His running mate was an equally young and vigorous Albert Gore, former Democratic senator from Tennessee and bitter enemy of many in cable television.

Gore would spearhead the administration's effort to push the United States to the forefront of the information age. He would be instrumental in molding legislation to rewrite the decades-old foundation of U.S. telecommunications law, the Communications Act of 1934. His policy proposals would draw heavily on the rise of the new technological, social, and financial phenomenon—the Internet, the interconnected network of networks that was fast becoming a central component of the larger convergence concept. Gore would propose a new term and a new metaphor—it would be the political manifestation of convergence, eventually embracing the Internet, television, and telephony. It would be the "Information Superhighway," and it would shape once again the fortunes of the cable industry.

Cable Convergence

The social construction of technology frequently spawns its own language, as people struggle to communicate new meaning. In the late 1980s and early 1990s, a new vocabulary was arising out of the technical developments in the Internet and the broader world of data communication. "Cyber" and "virtual" became the preferred prefixes for any given noun, creating virtual reality, cyber communities, virtual intelligence, and cyber cafes.

Much of the cable industry, by desire or necessity, drank deeply from the convergence well, adopting its own new lexicon as the mantra became infectious. In 1967, the National Community Television Association had changed its name to reflect what was then the new definition of the medium, and "CATV" became "cable television." Cable leaders knew that names mattered, representing their face and function to the world. In the early 1990s, the idea, even the definition, of television was expanded to include the integrated services offered by a telecommunications provider. The new codename, therefore, became "telecommunications."[7]

In 1993, the Community Antenna Television Association (CATA) became the Cable Telecommunications Association. By 1995, the industry's dominant trade publication, *Broadcasting & Cable*, proclaimed that the 1994 NCTA convention, heavily spiced with telephone company presence, was history's last "cable" show, and that the industry was "desperately in need of a new name."[8] That same year, Women in Cable and Television (WICT) became Women in Cable & Telecommunications; the Society of Cable Television Engineers (SCTE) became the Society of Cable Telecommunications Engineers, and CTAM became Cable & Telecommunications: A Marketing Society. By 2001, even the National Cable Television Association (NCTA) had changed its letterhead to the National Cable and Telecommunications Association.[9]

Convergence was not just about window dressing, however. Cable spent tremendous sums of money upgrading technology, testing new approaches to delivering the now-promised 500-channel future, and exploring the uncharted waters of data and telephone service. As it had been in the past, the foundation of the reforming telecommunications universe was technological. Driven by business interests looking to harness that new technical capacity for shareholder value, and sculpted by ever-changing regulatory fashion, cable was in the early stage of a new era.

From its inception in the late 1940s until Irving Kahn's dream of a satellite network was realized in 1975, community antenna television had been, in most substantial ways and in the famous words of Justice Stewart, a simple extension of the viewer's home antenna. The inauguration of HBO satellite service marked cable's second life as a global, multichannel video distribution system that changed the nature of television itself and touched the political and cultural life of the nation. The decade of the 1990s would see cable move toward its third major stage of evolution, growing yet again in size and scope, using digital technology to multiply by several factors existing capacity, expanding into broadband computer services and even encroaching into the telephone business.

As always, the change would not come without turmoil, setbacks, and failures. The social construction of technology, even when it accurately gauges the potential of innovation, is, as previously noted, frequently premature and always contested. The soothsayers of convergence, in a retold tale, were years ahead of economic and technical reality, and the Blue Sky vision would crash again before the next round of resurrection. A cycle of boom and bust lay ahead for cable and the broader telecommunications universe. Despite this, the underlying statistics for cable in the early and mid-1990s continued to reflect steady evolutionary progress.

Penetration grew from 61.5 percent to 66.7 percent between 1992 and 1996; revenue climbed from about $20.7 billion to almost $25.7 billion. Channel capacity continued to expand. Most cable viewers had access to well over thirty channels, 14 percent of all subscribers could view up to ninety channels. The public continued to watch ever-increasing amounts of cable programming while ratings for broadcast TV slipped a notch in almost every quarter. In cable households, viewing of broadcast network affiliates fell from a forty-six share in the 1992–1993 season to less than a forty share in the 1996–1997 season, while primetime aggregate viewing of cable networks rose to a little less than seventeen and the basic cable share for the year rose to thirty.[10] The most popular cable channels were still the broader-appeal services such as ESPN, TNN, Nickelodeon, CNN, and MTV. The 1995 murder trial of O. J. Simpson, especially, drove viewership at CNN, the Court Channel, and E!, all of which featured extensive coverage.

As noted, federal controls on highly leveraged transactions (HLTs), significantly reduced the industry's access to money in the late 1980s and early 1990s. Coupled with mild recession from mid-1990 through 1991, the restrictions hurt the industry as system values fell from eleven to twelve times cash flow to around nine times cash flow. Cable turned to other sources of capital, including selling public and private equity (e.g., stocks, bonds, limited partnerships), and once more tapped into institutional lenders, such as insurance companies, pension funds, venture capital firms, and foreign lenders.

Sadly, but inevitably, many of the legends of the industry, men who built and guided cable in the 1950s and 1960s, were passing. In August 1991, the former Chairman of the FCC, Dean Burch, died at the young age of 63. Time Warner Chairman Steve Ross succumbed to cancer in 1992. In 1994, two of cable's founding giants died: Irving Kahn on Saturday, January 22, 1994; and Milton Jerrold Shapp on November 24, 1994. In 1996, TCI founder Bob Magness died of cancer.

The baton was now in the hands of the next generation, held firmly by executives such as Malone, Turner, Levin, Dolan, and the Roberts. Under their guidance and despite the substantial sting of the 1992 Cable Act, the industry would adapt to the changing regulatory environment and the competitive challenges, both imagined and real, that lay ahead. Fundamental to riding the new wave was the ability to deal with the quantum technological leap from analog to digital.

Being Digital

Malone's 500-channel promise in one respect was off-handed, an admitted rough guess of what might be possible, but the basic idea was not cut from whole cloth. The 500-channel prophecy had a firm basis in technical innovation. It is, in fact, hard to overstate the significance of the shift from analog to digital technology. In terms of the evolutionary metaphor outlined at the beginning of this book, no change better or more fully embodied the concept of the quantum leap. Nor did any technology better illustrate, at the same time, the evolutionary nature of technical advancement and Usher's cumulative synthesis that must precede such a leap. The shift represented the confluence of decades of work in multiple worlds of engineering and mathematics, and the consequences of the transition, for media, for business, for society, were overwhelming.

The term "analog" derives from the idea that the electromagnetic carrier wave that transmits audio and video signals resembles in some way, or is analogous to, the original sound wave that it carries. Phonograph recordings offer perhaps the best example. Sound waves from voice or music travel through the air compressing pressure-sensitive elements in a microphone. The microphone translates the sound waves into variations in electrical current. Those electrical variations then are embossed in corresponding bumps cut into the grooves in a record, which can reverse the process and reproduce the sound at playback. The electrical signals and the grooves are said to be analogous to the sound waves themselves. Similarly, sound and sight can be converted to electronic or electromagnetic waves and the waves continuously modulated (AM or FM) and demodulated to carry a signal via radio and television.

Digital communication employs a different signal processing strategy. Digital processing involves the conversion of all information into the zeros and ones of machine language, the on-off switches of microcircuits. In this a binary world, information such as a TV picture is sent out in streams of digitized numbers that are translated into viewable images at the end user.

Digital communications holds a number of advantages over analog. In traditional cablecasting, analog signals weaken and degrade as they push through the miles of coaxial cable. The amplifiers boost the signal, but also introduce addition electronic noise, which can lead to snowy and unstable pictures on the viewer's screen. Digital transmission, by comparison, maintains its original quality throughout the process and, when working properly, is as strong and clear when it reaches the home set as when it leaves the studio.

Another advantage of the digital signal is its malleability. Once the signal is digitized, it can be manipulated in innumerable ways. Video, audio, and data can be mixed, paving the way for interactive consumer new services. It also means that in the long term, the technical differences between the computer and the TV set vanish.

HDTV

Digital television found its way into cable and Malone's 500-channel promise primarily through the worldwide effort to create high definition television. In February 1981, the Japanese national television network, NHK, demonstrated a new kind of TV transmission system in the United States. Instead of the normal 525-line NTSC resolution standard, the NHK system offered a 1,125 line, high-resolution wide screen analog system. The high-definition picture was dramatically sharper in image quality than had been previously available, but at a technical cost. It required six standard 6-MHz TV channels to deliver the signal. Despite its hunger for bandwidth, the NHK high-definition system set off an international race to develop and implement practical high-definition TV service.[11]

Through the economic boom years of the mid- and late 1980s, there was a growing national awareness that business and industry were increasingly operating in a global rather than domestic arena. There was accompanying sensitivity in U.S. business and political circles that American industry could fall behind in the global marketplace. A number of countries, including Japan and several in western Europe, offered substantial support to their domestic industries, economic props not typically available in the United States. One of the flashpoints was high-definition television (HDTV). Some in business and politics feared that U.S. firms would be left behind in the creation of this potentially multibillion dollar consumer electronics market.

The Japanese system, Muse, had been in development since the mid-1960s and, by the early 1980s, U.S. industry had begun to respond. In 1982, a consortium that included the NAB, NCTA, Institute of Electrical Engineers (IEEE), and Electronic Industries Association (EIA) formed the Advanced Television Systems Committee (ATSC) to encourage development of HDTV in the United States. It also was given the mission of proposing universal standards to the International Telecommunication Union (ITU). Their initial proposal became ensnared in international politics, as global powers maneuvered for position in the standard-setting contest. By 1987, several European countries also had banded together to develop their own high-definition technologies and standards.

The FCC formally joined the debate in July 1987 when it issued a Notice of Inquiry to consider the technical and policy questions surrounding creation of a high-definition system, although the Commission used the term "advanced television" to describe its interest.[12] By November of that year, the Commission had created a twenty-five-member task force composed of representatives from cable, broadcasting, consumer electronics, and government to study advanced TV and craft policy recommendations. Former FCC commissioner Richard Wiley was asked to chair the new Advisory Committee on Advanced Television Service (ACATS).

Between 1987 and 1990, more than a dozen companies submitted advanced and HDTV proposals to the committee. Firms began working on like systems, and the number of competing formats thinned to six by the end of 1990. The engineering challenge was made more difficult for the competitors in September 1988 when the FCC declared that any new system would have to be compatible with existing NTSC receivers and fit within the established 6-MHz transmission band so as not to make obsolete every TV set in the United States.[13] At stake, however, were billions of dollars in licensing fees and equipment sales for the firm that could create the anointed standard.

To this point, all the proposals relied on analog technology. General Instruments had a different idea. As noted in the previous chapter, GI purchased MA/-COM in 1986 and acquired with it the company's VideoCipher II scrambling technology. But GI knew that high-definition technology was a potential business problem for them. The earliest high-definition proposals (before the FCC's 1988 order) called for bandwidth beyond the traditional 6-MHz channel. Conversion to any such system would require much larger and more expensive consumer dishes and reception equipment. GI set a team of engineers on the problem of squeezing an HDTV signal into a traditional 6-MHz channel. One possibility was to digitize the TV signal.

VideoCipher II already employed digital processing to scramble the audio portion of the TV signal, but making a practical operating system capable of HDTV was a significant challenge. In addition to converting the signal from an analog to a digital format, the engineers needed to find a way to compress the wide-band high-definition signal into the standard 6-MHz pipe. Absent compression, a digitized NTSC television would consume more than five times the bandwidth of an analog signal.

By 1989, the engineers at GI had crafted a workable compression algorithm. It was based on the notion of transmitting video information in a picture only when that information changed. In analog transmission, every frame of every picture is sent, even if the picture elements from frame to frame, second to second, do not alter. And, a great deal of picture information in a given TV scene does not change quickly. When two people are talking in front of a wall or building, for example, the picture elements of the wall or building remain constant for a given period of time. Digital compression was able to take advantage of this, in effect, by telling the reception device to maintain the existing picture element until further notice, sending new data only as needed.

On June 1, 1990, GI unveiled its work. Dubbed "Digi-Cipher," it was the first all-digital HDTV system for terrestrial television. With all the caveats that this volume must attach to the label, it was fairly described as revolutionary by most parties. At the moment of its introduction, all the previous HDTV proposals were rendered obsolete. In the laboratories of the leading firms, the analog high-definition plans, the product of years of work, were dumped in the shredder (except for the Japanese). All hands turned to the digital solution. By March 1992, four all-digital proposals had been submitted to the Advanced TV Committee, including those of GI, Zenith, a GI-MIT joint venture, and one from a consortium that included Phillips, Thompson, and RCA. NHK also still had an analog plan in the hunt.

In July 1992, the FCC outlined the parameters for conversion to the new advanced TV standards, including a proposed timetable, and a decision to keep TV broadcasting in the existing VHF-UFH spectrum. In September 1992, it issued a revised timetable calling for full conversion to HDTV by 2008.[14] The committee reviewing the proposals came to the conclusion in February 1993 that the competing plans were roughly equivalent in performance and promise and suggested, among other things, a grand alliance that would meld their ideas into single proposal. The Grand Alliance was formed in May 1993 and a unified plan hammered out and submitted by December 1994. After more than a year of testing and deliberation, the FCC in December 1996 approved the package of Grand Alliance standards.[15]

It was with digital compression in mind that Malone was able to make his 1992, 500-channel prediction. While compression could squeeze a wide-band HDTV signal into a standard 6-MHz space, it could also squeeze a standard TV signal such that several could be sent through the same 6-MHz channel. Digital compression dramatically expanded one of cable's most precious commodities—channel space—by multiplying many times the carrying capacity of existing bandwidth.[16]

The form for expressing compression became the number of digitized non-HDTV channels that could be fed through an existing NTSC space. A five-to-one compression ratio meant compressing five digitally processed signals into one analog channel. A ten-to-one compression ration meant that 500 channels could be transmitted using the cable bandwidth previously used for fifty. In the early 1990s, state-of-the-arts amplifiers were running at 750-MHz and 1-GHz amplifiers were becoming available. A 750-MHz system could carry more than 100 channels and therefore, in theory, 400 digitally compressed channels at a four-to-one ratio. By the late 1990s, compression ratios of twenty-to-one or more were feasible,[17] although cable systems would use much of the expanded bandwidth for pay-per-view (PPV) films and the increased capacity would not be starkly apparent to the consumer.

The Internet

High-definition television presented cable with its first digital challenge and opportunity. Those challenges and opportunities, however, were arguably small in comparison with another manifestation of the digital transition—the growth of computer networks, specifically the development through the 1980s and early 1990s of what was first known as APRAnet, then NSFnet, and then, the Internet.

In the 1960s, the link between cable and the computer was based on a technical and social construction of centralization. Computers were large, astoundingly expensive and required a small army of skilled technical for their upkeep. Only large businesses and institutions, such as universities and government agencies, could reasonably afford them. Individuals who wanted to use a computer went to it, physically bringing decks of punched computer cards to the computer sites, or for the hard-core researchers, distantly accessing the computer through a dumb terminal, a device much like a teletype. As the interest in (and need for) access increased and computer terminals became more widespread, computer engineers turned to the concept of time-sharing, allocating precious and expensive computer time to the many users accessing the system at any given time.

But Moore's law, along with Intel's microchip, brought down the size and cost the technology to the point that, by the mid-1970s, knowledgeable computer fans could start building their own computers at home. In 1975, a New Mexico electronics dealer, Ed Roberts, announced the Altair, the first commercially available home computer. The Altair had no keyboard and no monitor. It had no input or output ports of any kind, could not be attached to any other device, and could perform no known practical function. The computer fans loved it. By laboriously flipping a row of switches on the front of the machine, they could program it to blink its row of small lights in some specified pattern. It was a start.

To make the machine more practical, Roberts needed to be able to program it in basic computer language. Two students at MIT learned of Robert's interest in a basic program for the Altair and said they could write it. Bill Gates and Paul Allen successfully wrote the primitive software, keeping the property rights to the code, and went on to found their own company, Microsoft, in their hometown of Seattle. Meanwhile, in the San Francisco Bay area, two members of the Home Brew Computer Club were looking at the Altair and developing their own designs. Steve Wozniack designed and built his own elegant machine and another club member, Steve Jobs, said he could help market the device. In 1976, they talked venture capitalist and retired Intel executive Mike Markkula Jr. into investing $250,000 dollars in Apple Corp.

Apple computers were a small hit. Software was introduced that allowed businesspeople to construct dynamic spreadsheets, and word processing packages also became popular. In 1981, IBM entered the growing market, enlisting Gates and Allen to produce the software. Within a few years, IBM's PC or a PC clone was on hundreds of thousand of business desks and in dens around the country. Almost from the start, PC users tried to communicate with one another and access remote databases. They used phone lines and a device called a modulator-demodulator (or modem), which converted the binary computer language to acoustic tones that could sent over the twisted pair lines.

Government and university researchers had been communicating via computer for a decade using a system developed as part of a national defense initiative. Created on orders from President Dwight D. Eisenhower in the late 1950s, the Defense Agency Research Project (DARPA) was asked to develop advanced technologies for national security purposes. One of the tasks was to fashion a communications network linking national defense sites and capable of surviving a nuclear strike. Engineers designed a plan for a distributed, redundant, computer-based network employing a digital packet switching form of communications. Using this process, all information— voice or data—was digitized and broken into binary packets. If necessary, these electronic packets could disassemble, each finding its way through a wired network, and then be reassembled at the final location. If one or more cross-country telephone lines were knocked out—whether by tornado or Soviet ICBM—the packets would be rerouted through existing viable lines. Unlike traditional telephone traffic, no single dedicated, and interruptible, circuit would be created between the caller and the receiver. By the mid-1960s, DARPA scientists looking for means to communicate and share data called on the packet-switching scheme to create ARPAnet, a means of interconnecting widely scattered computers. The first nodes were installed in major research universities, such as MIT and UCLA, in 1969. As computer use and demand for interconnection, especially in university settings, grew through the 1970s, the ARPAnet usage expanded beyond its original defense brief.

In 1985, the National Science Foundation called for a revised and expanded network, opening NSFnet the next year. With the growing popularity of personal computers, traffic on the network accelerated and software designers began developing programs to more easily move through the networks to locate and access information.

In 1989, a scientist at CERN, the European Particle Physics Laboratory in Geneva Switzerland, Tim Berners-Lee created a new approach to sharing information on what was becoming known as the Internet. It was a method of organizing and presenting information using what he called "hypertext." By simply clicking on a highlighted word or phrase, the user would be instantly taken to a new database. He also created a program called a "browser" to navigate the Internet and access hypertext materials. He called his interconnected hypertext network, the "World Wide Web" (WWW).

Berners-Lee was a scientist and a purist and saw little need for pictures or fancy graphics, but not all users were so focused. In 1992, a young programmer working at the National Center for Supercomputer Applications at the University of Illinois organized several colleagues and created a new browser, one that could quickly present quality pictures and graphics, making the WWW more accessible, more powerful, and more fun for the average user. Marc Andreeson released his MOSAIC browser in January 1993. In 1994, having graduated, he formed a new company backed by venture capital to market the browser under a new name. It was now called "Netscape."[18]

By the mid-1990s, the World Wide Web was a powerful cultural and economic force. Millions of people were going online everyday. Businesses sprinted to create websites; electronic or e-commerce became first a catch phrase, then an industry. The question for cable was how to become a part and take advantage of this new world order?

Digital technology and the rhetoric of convergence was a double-edge sword for cable. It held the potential for expanding classic cable service and opening the door to new businesses, including data and even telephone service. At the same time, digital technology and the convergence metaphor put the same power into the hands of cable's rivals, which included some of the world's largest corporations, The unregulated glory days of the 1980s were over. By 1992, DirecTV—backed by General Motors—was planning true DBS. The behemoth telephone companies were in court and in the laboratories working to get into the video business, and the FCC was sharpening its pencils and sitting down to the task of implementing Congress's punitive 1992 Cable Act. Before cable could think about spending hundreds of millions on new technology, it would have to deal with the impact of the 1992 legislation.

Implementing the 1992 Cable Act

Cable Rates

Congress, in the 1992 Act, wanted cable rates brought into line, but left it to the FCC to determine the level of that line and how to set it. It was an unenviable task made more difficult by tight congressional deadlines. The Commission began with a survey of 748 cable systems to determine rates, dimensions of service, and local competition as of October 1992. Of the 748 inquiries, it received 687 usable responses, representing 1,107 separate community franchises. From this group, the Commission found 145 that faced some form of "effective competition," as defined by the Cable Act,[19] and used them to set the benchmark levels for rates.[20] The FCC determined that consumer rates in such situations were 9.4 percent lower than those in noncompetitive situations. The Commission would use this as the benchmark for its regulation of basic rates.

The rules primarily sought to control rates for basic tier service. Rates for enhanced or extended tiers would be regulated only following a complaint by subscriber or municipality and a determination by the Commission that those rates were "unreasonable." Rates for services offered on a per-channel (premium services such as HBO) or per-program (PPV programs) were not subject to regulation. Following the mandate of the 1992 Act, the Commission also issued orders limiting rental costs on customer equipment such as cable boxes and remote controls.

On April 1, 1993, the Commission ordered a freeze on all cable rate increases and then began a systematic rate rollback, mandating decreases of up to 10 percent.[21] Specifically, cable rates that exceeded 10 percent of the benchmark were ordered reduced, although no cuts could exceed 10 percent of the subscriber's average bill. Systems with prices less than 10 percent higher than the benchmark were required to lower them to the benchmark. The rate caps took effect September 1, 1993.[22] The complaining and maneuvering began even before the start date. To begin, the formula used by the FCC to determine rollbacks or levels for any particular franchise was a wonder of bureaucratic obfuscation, and was derided by cable operators and municipal franchise officials alike.[23] The actual cutbacks produced an industry-wide grimace. Operators called them unreasonable, even draconian. Worse still, as operators began adjusting their prices in response to the new law, consumers and regulators watched in shock and bewilderment as rates, instead of coming down, began to climb. An FCC survey of rates between April 5, 1993, and September 1, 1993, showed an average decrease in regulated rates of 6 percent, most of which came from reductions in equipment rentals.[24] But, for the top fourteen MSO's, 31 percent of subscribers experienced rate increases following adoption of the new rules.

Operators blamed the new law for the rate hikes; regulators blamed operators. Both sides could make a case. The regulations required uniform rates in a given franchise area. Some operators had been offering discounted service to some groups, such as senior citizens, and free or low-cost lifeline tiers to others. Operators, responding to the uniform rate requirement, began charging for these services. The law strictly limited charges for multiple sets, a traditionally lucrative added revenue stream for operators. Arguing that they needed to recoup their costs for the lost income (and could under the law), operators sought to raise rates for the basic service. The FCC charged that the industry searched for any and all cracks in the regulatory façade through which to slip additional consumer charges. Because general regulation applied only to the lowest tier offered by the company, some operators moved popular channels, such as CNN and MTV, to an unregulated expanded tier, or even offered services on an *a la carte* per-channel basis, raising prices for those channels in the process. Some companies, before issuance of the rollback order, had retired, moving channels from basic to extended or premium tiers and recalculating prices in expectation of the FCC cap.

An internal TCI memo leaked to the public did not help the industry's cause. Penned by TCI Chief Operating Officer Barry Marshall, the August 1993 note advised TCI system managers not to be overly concerned about the rate controls because the company could (and would) recoup some of its lost revenue by charging for all itemized customer service changes, including upgrades, downgrades, even VCR hookups. The memo concluded with particularly unfortunate language, declaring "We have to have discipline . . . . We cannot be dissuaded from the charges simply because customers object. It will take a while, but they'll get used to it . . . the best news of all is, we can blame it on re-regulation and the government now. Let's take advantage of it."[25]

Subscribers, consumer groups, and members of Congress were outraged, pointing to such tactics as proof of the industry's malevolence. TCI issued an immediate apology. Malone wrote a somewhat defensive, three-page letter to acting FCC Chairman James Quello detailing the various ways in which TCI was complying with the Act, and noting pointedly how much the company was spending to do so. But the damage was done. News headlines spoke not about rate reductions but rather rate hikes. Consumers, congressmen and women, and senators, individually and in groups, penned blistering letters to the FCC. The freeze on rate increases had been set to end November 15, 1993, after which operators would have been permitted to pass along increased operational costs associated with inflation, programming, and technology.[26] Ten days before the expiration date, the Commission extended the rate freeze. The next step was uncertain. The FCC declared it needed time to study the problem. Before the end of the month, the issue would fall into the hands of a new Chairman, Reed E. Hundt.

An antitrust lawyer by trade, Hundt had attended high school with (and was a good friend of) Al Gore. He advised Gore during his 1984 Senate campaign, 1988 Presidential campaign, and 1990 Senate campaign. Hundt also was a Yale Law School classmate of Bill and Hillary Clinton and an economic adviser in the 1992 Clinton-Gore Presidential run. As the new administration began organizing its choices for the myriad of governmental positions it would have to fill, Hundt asked Roy Neel, Gore's designated Chief of Staff, to lunch. Over soup, he told Neel, "Roy, I would like to be chairman of the Federal Communications Commission."[27] A few weeks later Gore said he would do what he could to fulfill Hundt's request.

On November 29, 1993, Vice President Gore swore in Hundt as Chairman of the FCC. On visiting his office for the first time, the new commissioner was struck by two things. The first was the dated, shabby decorating and the thick haze of dust that had accumulated in the eleven months that the office had been unoccupied.[28] The second was the stack of letters from "several hundred" congressmen and senators smoldering on his desk. The source of the heat was the perceived failure of the FCC's rollback of cable rates.

Despite his interest in the FCC post, Hundt was not experienced in the cable or broadcast arena. He could read numerical tables, however, especially those charting the rise in cable prices despite the 1992 Act, and he understood the need to respond to the yelping for action from Congress. He hired a University of California Berkeley economist, Michael Katz, to analyze the problem and develop a new pricing structure. As Hundt wrote in his 2000 memoir, "He [Katz] invented a multifactor polynomial that set prices equitably in every one of the fantastic variations of cable packages and prices that populated the market. Dazzled by the impenetrable brilliance of our economist, and oblivious to the risks of our formula's complexity, Blair and I sold this solution to Commissioners Quello and [Andrew] Barrett."[29]

The sales job took a little negotiation. Hundt's initial recommendation, based on Katz' new regression equation, called for an additional 18 percent rate cut beyond the 10 percent benchmark of the original standard. Commissioner Barrett opposed any additional rollbacks, other Commissioners sought a milder revision. They compromised, and, in February 1994, the FCC unveiled its revised pricing formula.[30] The FCC said operators not facing effective competition would have to lower their basic tier prices by an additional 7 percent, bringing the total 1992 Act rollback to 17 percent for affected systems.

The reaction was immediate and hostile. The new rules may have been the most maligned cable-related orders ever issued by the FCC. Critics assailed them on two counts, mind-numbing complexity and *post hoc* rationalization. Forms and instructions for operators ran to more than 1,000 pages and just one part of a system's obligation for determining its new rates called for use of the benchmark formula:

$$LAR = .204 + .07(MSO) + 8.14(RSS) - 1.45(RTC)$$
$$+ .253(PNB) + .103(PAO) + .172(PRM)$$
$$+ .057(PT2) + .353(PTC) + .069(LIN).$$

System operators had to figure several "natural logarithms" and a number of complicated weighted sums as part of the process.[31] In addition, observers noted that the new benchmark had been a political compromise driven by congressional ire and that the fancy math had been manipulated simply to reach a predetermined figure. As policy analysts Robert Crandall and Harold Furchtgott-Roth sarcastically observed:

> Not only had the commission appeared to manipulate its empirical analysis to accommodate political pressure, but it was forced to publish a regression it claimed was "too complicated" for cable system owners to understand. These operators, presumably lacking access to calculators or personal computers, would probably have to ask the highly skilled FCC staff to estimate the benchmarking for them. In fact, when finally released, the regression results were embarrassing.[32]

Embarrassing because of their complexity and, in part, because under the new formula, cable systems owned by the large MSOs would be able to charge more for equivalent service than smaller, independently owned systems. While he defended the rate adjustment on the grounds that consumers would benefit in the long run, Hundt quickly learned the cost of angering the industry and its friends.

More important, perhaps, was the accusation that the rollbacks of both the first and second rounds had damaged cash flow needed to support investment in the information highway. This was an issue of great sensitivity to Hundt and his assay of the consequences of the rate order would soon prompt him to reevaluate the Commission's philosophy on telecommunications regulation. The industry mounted the requisite legal challenges to the rate controls, but the courts upheld the Commission's actions.[33]

The rate rollbacks of 1992 and 1994 had a number of consequences—almost all of them damaging from the perspective of the industry. Stable or lower rates were good for consumers, but ripped the bottom out of the revenue pot for many operators. Losses were estimated in the hundreds of millions of dollars, by some estimates reaching a billion dollars or more. The drop in revenue meant, in turn, decreased funding for construction, upgrades, and investment in new program networks. TCI said the cuts would reduce their annual cash flow by $300 million and take $500 million from its 1994 capital budget of $1 billion. Time Warner reduced its construction and upgrade commitments by $100 million.

Lenders were reluctant to invest money in an industry with tightly controlled prices, and equity pools began to evaporate. Programming costs continued to rise, however, and with them, debt. System prices stalled. One study showed that prices slowed even before rate retrenchment in expectation of the new regulations. Average per-subscriber prices peaked in 1989 at $1,884, then dropped to $1,619 in 1991 and $1,522 in 1993, although they began to climb again in 1994, to $1,748.[34]

On Wall Street, cable stocks felt the full force of the dropping economic dominoes. Between November 1993 and May 1994, TCI stock plunged 34 percent; Cablevision Systems, 35 percent; and Falcon Cable, 41 percent. The Kagan Cable MSO average fell from more than 1,500 to 975.

Despite these setbacks and the strong and continuing protests of the industry, historically, all the major statistics of industry development continued their steady climb. Between 1991 and 1995, national penetration increased from 60.6 percent to 65.7 percent. The average monthly rate for basic or extended basic services rose, despite rate regulation, from $18.10 to $23.07. And industry revenues from basic service climbed from about $11.4 billion to $16.8 billion. Only the unregulated rates of pay services appeared to drop consistently over this period, moving from an average per-subscriber figure of $10.27 a month in 1991 to $8.54 a month in 1995.[35]

Rate regulation hurt the industry, but the depth of the wound was difficult to assess, and it would heal. More lasting were the consequences of the Act as they affected cable's relationships with its long-time rivals in the broadcasting and the telephone industries.

Cable Programming

Few things were more vexing to cable than the must-carry retransmission consent rights won by the broadcasters in the 1992 Act. Judicial challenges were assumed and prompt. Operators had some justification in hoping the new must-carry provisions would fare as well as the old attempts, which had been overturned by the courts. They were disappointed. In a series of challenges brought by Turner Broadcasting, the courts held for the government, accepting the argument that local broadcasters not carried on cable were placed in economic harm's way, and if they went dark as a result of not being on the cable, noncabled homes in the area would be deprived of their service.[36]

Local broadcasters, therefore, could demand carriage (must-carry) or opt for retransmission consent and negotiate a price with the local operators, or in the case of national programmers, with the large MSOs. The resultant agreements could be renegotiated every three years. The smaller stations, noncommercial or religious broadcasters, and small-audience ethnic or shopping stations, chose must-carry, forcing local operators to add channels they would not have otherwise carried, including signals from nearby cities that were available over air and duplicated existing local channels. Only about 10 percent of the nation's broadcasters chose this option, however.[37]

The TV stations considered essential by audiences and cable alike, including the affiliates of the major broadcast networks, sat down to the bargaining table to discuss retransmission consent fees.[38] Broadcasters had lusted for decades over cable's dual revenue stream. But MSO chiefs, such as Malone, had made their fortune in eyeball-to-eyeball business confrontations such as these and many were born gamblers. As the two sides glared at each other over the table, the MSOs steadied their gaze and declared that no money would pass over, or under, that table. Cable stated coldly and publicly that it would drop any station that demanded retransmission payment and if people really wanted to watch local broadcast TV, they could buy an antenna. Amos Hostetter, head of Continental Cablevision, went so far as to order 1 million A/B switches, which he was prepared to install in the homes of his customers so they could receive the local broadcast channels.

In drawing this line in the sand, the industry, already well out of favor with regulators and the public, was risking additional pummeling in the media and in Washington. Many MSOs knew (or should have known) that they would be blamed if millions of cable subscribers could no longer receive "Monday Night Football." In the end, it was less of a blink and more of a wink that resolved the stand-off. Broadcasters would receive no cash, but would obtain something that was arguably more valuable in the long run— channel space on the nation's largest cable carriers.

The major broadcast groups, much like the major film studios, controlled decades of old programming and had substantial production capacity. What they required to exploit those resources were additional distribution outlets. The 1992 Act gave them the lever to pry those outlets out of cable and inaugurate a new era of broadcaster-owned cable programming services.

It began with Fox. Murdoch's young Fox Network was the smallest and weakest of the major broadcasters and carried the least weight in negotiations with cable. Murdoch and Malone had bargained before, with TCI agreeing

600 / Chapter 12

in 1990 to carry the new Fox Broadcasting Network. They met again and, in May 1993, TCI agreed to carry a new Fox cable channel, the fX network. TCI would pay carriage fees of $0.25 per subscriber and, to help appease his broadcaster affiliates, Murdoch would turn $0.05 of that monthly amount over to them, as well as giving the affiliates a 25 percent ownership stake in the new service. fX launched in June 1994.

The agreement became the template for similar deals. ABC was the next to come to terms, followed by NBC and several smaller broadcast groups. As a consequence, ABC launched ESPN2 in October 1993. NBC began a news-talk service called "America's Talking," a CNBC spin-off. Scripps-Howard, a major newspaper and broadcast station group, started the Home and Garden Television Network. Another newspaper-broadcaster, the Providence Journal Co., began the Food Network in November 1993.

Only CBS refused to bargain with the industry, holding out for direct payment, but the operators would not move and CBS was forced to take the must-carry option. Independent broadcasters who could not work out immediate agreements for carriage frequently settled for local cross-promotions, ad sales alliances, or coproduction partnerships.

With limited system capacity, some operators, therefore, were forced to drop existing services to make room for the newly mandated broadcast signals. One of the victims was C-SPAN II. The service did not produce revenue, making it especially vulnerable to cuts, and severing the service was pointedly designed to send a political message to Congress. Programming services in the planning stages, especially those lacking MSO affiliation, found that national shelf space had dried up and were forced to abandon or delay their launch. In 1994, these included S The Shopping Network and TV Macy's, Classic Sports Network, the Golf Channel, BET on Jazz, and the History Channel (although some services, including GEMS, The Outdoor Life Channel and TRIO, did begin operations in 1993 and 1994).

The 1992 Cable Act, and the consequent 1994 rate deregulation in particular, had one more significant impact on the cable industry. It involved the relationship between cable and its long-time adversary, the telephone business.

Telco TV

In the 1920s, AT&T was one of the world leaders in the development of television technology, holding the famous 1927 demonstration discussed in Chapter 1 and noted in the quote at the beginning of this chapter. Regulation, for the most part, had kept AT&T and then the RBOCs out of the TV business since AT&T abandoned its spinning disc technology in the 1930s. The telephone industry, nonetheless, had been snapping at cable's heels for decades, held in check only by federal fear of the monopoly power of the telephone company, a fear that still lingered after the Bell breakup. A number of proposals were floated during the debate over the 1992 Act to allow the telephone companies into cable, but concerns about the size, power, and business history of the Bell system had prevented them from gaining traction. The telephone business dwarfed cable, by almost every measure. Total revenue for the cable industry in the mid-1990s was about $26 billion compared with about $200 billion for the telephone industry. But the steady and dependable telephone business was considered to be "mature," a code for unlikely future growth. The staple of the business, plain old telephone service, or POTS, was estimated to be growing at only 4 or 5 percent annually—steady but not spectacular. Newer services, such as call forwarding and call waiting, were seen as boosting the bottom line, but only modestly, and changes in technology and competition were driving down long-distance prices and revenues.

To expand, the phone companies were attempting to exploit the explosion in devices that increased total lines, including FAX machines and computer modems, as well as the booming cellular phone business. Revenues from expanding data and Internet traffic were seen as particularly promising. Separated by law, the RBOCs were nonetheless also desperate to find a way into long-distance service, and AT&T harbored a similar desire for local loop service. Local exchange carriers, such as the RBOCs, made up to a third of their income, nationally around $25 billion a year, charging AT&T and other long-distance carriers access fee to connect to their local networks. Entry of AT&T into local service, therefore, meant not just a new competitor for the local business but an overall revenue loss for the incumbent local exchange carriers (ILECs) and a cost reduction for AT&T.

The provision of multichannel video, therefore, was only part of a larger business strategy. Some observers suggested that the RBOCs' interest in cable service was, in fact, just a smokescreen designed to conceal their real interest in the long-distant market. Politicians and regulators were more concerned about competition in the video marketplace than about long-distance telephone rates, so pledging participation in video was a method of pushing for deregulatory legislation (that would include repeal of local and long-distance restrictions). Such a read may have been too cynical, however. Video was worth pursuing in part because of the raw cash flow it could generate and, more importantly, because of the promise of integrating the bundled service prophesized by convergence. The ultimate goal was to be the full-service monopoly provider, bundling voice, data, and video over one wire and sending customers one bill at the end of the month.

Telco Technology

Technically, the telephone companies had several distribution tools at their disposal. Adopting existing cable equipment and overbuilding extant systems in targeted markets was prevented only by legal and regulatory prohibitions.

Multichannel distribution services, the microwave-based MDS systems that came up short in the 1980s, also offered possibilities. The digital compression technology of the early 1990s held the possibility of multiplying by many fold the MDS channels and creating a platform for wireless competition. The telephone industry also was working on methods to send video over its existing narrowband, twisted pair lines.

Cable's broadband coaxial transport capacity had always given it an advantage over broadcasters and the telephone carriers. The twisted pair wires of traditional telephone service were insufficient for even high-quality audio communication, much less any form of video.

Digital technology pioneered by Bell Laboratories, Stanford University, and British telecommunications helped rectify that that problem, however. An asymmetrical digital subscriber line (ADSL) was capable of increasing the carrying capacity of a typical 51-Kbps telephone line to 6 Mbps even 8 Mbps by the mid-1990s.[39] That was sufficient to transmit four compressed video channels and still retain room for a traditional telephone call. It had its downsides. ADSL was expensive—estimated at $4,000 per line in its earliest applications, and limited in the distance it could transmit a program. But it was a beginning point. With broadband or broadband-like capability, the telephone companies could exploit the one great technical advantage they had always held over cable—the switched system.

The simple switch, first an electromechanical device, and later a small, sophisticated computer, directly connects any subscriber with any other subscriber or with any data base, information server, web site, or voice mail box. The classic telephone "star" system connects every home to a central neighborhood switch and then interconnects these neighborhood hubs—the local loop—to the larger public switch telephone network (PSTN) that spans the globe. Cable's tree-and-branch system, even in its more advanced HFC form—lacked this switching capacity in the early 1990s—programming normally flowed from the cable head-end to each subscriber's home. Interactive capacity, to the extent it existed, was limited to a simple return path from the subscriber back to the head-end.

One of the logical consequences of the convergence discourse of the late 1980s and early 1990s was the observation that the strengths of both systems could (and should) be joined, combining cablelike bandwidth with the switched flexibility of the telephone platform. On the drawing board, this led to the dissolution of the differences between the two—cable and telephone could become like systems, each including voice, video, and data exchange. The trick, as it had always been, was to make it practical and make it financially viable. Each industry moved from its place of strength, cable companies seeking means to create or mimic switched and interactive applications and telephone companies to expand the size of their narrowband path to the home. Technical changes, however, would be pointless for the telephone industry without changes in the legal environment.

Telcos in the Courts

When the NTIA released its Video Dial Tone report in 1988, Alfred Sikes was head of the agency.[40] A year later, Sikes was promoted from his administrative post to the Chairmanship of the FCC and he took his video dial tone (VDT) idea with him. In 1992, the Sikes' Commission voted to implement a VDT plan. This retrofitted telco separations policy authorized telephone companies to build multichannel distribution systems and lease channel capacity to third-party programmers, although it restricted the telephone companies themselves to less than a 5 percent interest in the program services.[41] At the same time, the FCC again recommended to Congress that the 1984 Act be amended to permit telephone companies easier, albeit controlled, access to the video marketplace.[42]

Some telephone companies attempted VDT systems, but most did not want to be limited to this option alone. As previously noted, the restrictions placed on the Baby Bells by the 1982 modified final judgment were rescinded in 1991 so the only remaining roadblock to full participation in the cable business was the 1984 Cable Act.

In assaulting this final barrier, the RBOCs hoisted a powerful legal weapon, one used at least somewhat successfully by the cable industry itself on occasion. It was the First Amendment, and an argument that the various legal restrictions on telephone company provision of video services constituted unconstitutional restraints on the RBOCs protected the right to freedom of expression.

It was a legal strategy that fit well with existing political philosophy, as well. Despite the Clinton win, the view that government should, as much as possible, stay out of the path of business was popular in both parties. The idea that purveyors of electronic information and communications services had important First Amendment rights was gaining ideological ground. Historically in law, strong First Amendment protection was reserved for speech (in the classic form of the lone orator) and print, including newspapers and magazines. As previously noted, broadcast and cable television were seen as having a smaller portion of First Amendment protection and, insofar as the telephone company was a common carrier, it had no real constitutional standing in this regard at all; it was not, in fact, a speaker.

Cable had made inroads into this view, however, and the courts had concluded that it deserved some measure of First Amendment protection. Broadcasters successfully rode the First Amendment argument to the death of the fairness doctrine and the elimination of a host of other long-standing government controls. The telephone companies, taking a page from those books and marrying it with the new view of a convergent, unified telecommunication system, carried its case for business freedom back to the courts and Congress.

In 1993, Bell Atlantic successfully challenged, on First Amendment grounds, the cross-ownership ban in the 1984 Act.[43] Bell Atlantic's victory was only the first in a series of judicial decisions pursued by its sister companies that struck down the Cable Act prohibition nationally.[44] The telephone companies now were free to begin exploring the provision of cable service in a variety of forms.

Ma Bell Makes Her Move

Telephone company experiments with cable television began, as previously noted, as early as the mid-1980s with updated versions of the old lease-back arrangement. By the mid-1990s, the telephone industry was funding scores of experiments, establishing test beds, and forging joint ventures. They were looking at technology issues, marketing problems, and customer relations questions. They were using traditional coaxial cable, ADSL, and MMDS.

In 1991, U.S. West and AT&T announced video-on-demand trials. The next year, U.S. West said it planned to upgrade many of its major markets, including Denver, Minneapolis, Phoenix, and Seattle to deliver cable services. It started in Omaha with an advanced fiber-to-the-curb (FTTC) system using analog and digital distribution and featuring seventy-seven basic cable channels plus video-on-demand programming.

Some of the first efforts invoked the VDT option. Bell Atlantic, in 1992, said it would attempt, in conjunction with Sammons Cable, a VDT network in northern New Jersey. In 1993, the company also began an ambitious ADSL test in the Washington, D.C., suburb, Alexandria, Virginia. Marketing the service as "Stargazer," it offered near-instant access to a server-based library of more than 550 programs including 200 movies. Prices ranged from as little as $0.49 for an older television feature to $4 for a current full-length film.

Southern New England Telephone Company (SNET) opened a VDT test in Hartford, Connecticut, in 1994. On the West Coast, Pacific Bell in 1993 proposed a $1.6 billion, fiber and coaxial rebuild of its entire California network, promising multichannel video as well as voice and data.

Ameritech, one of the most active and successful in bringing competition to cable, did it the old fashioned way, with traditional 750-MHz HFC systems and an over-builder philosophy. It had intended to use VDT, but the cumbersome paperwork and lengthy approval process, coupled with the legal ability to pursue more traditional routes, led the company to withdraw its VDT requests in 1995 and begin an aggressive series of franchising moves throughout the Midwest. Starting with its first franchises in Plymouth, Michigan, (a Detroit suburb) in 1995, it went on to win construction rights in cities in Ohio, Wisconsin, and Illinois.

In the same manner, Bell South pursued traditional cable franchises, winning approval to wire towns in South Carolina, Alabama, and Florida. U.S. West and Bell South also invested in systems offering voice and cable services in Great Britain, looking to test operations oversees before larger deployment in the United states.

For both the cable and the telephone industries, spending billions to retrofit or fully rebuild its physical plants and enter into a new and unknown business with uncertain outcomes, was daunting. It would certainly be easier, and perhaps even cheaper, to realize the benefits of convergence by a more direct route. In February 1993, therefore, SBC Communications (formerly Southwestern Bell) became the first RBOC to purchase an existing cable company, buying two systems in the Washington, D.C., area from Hauser Communications for $650 million.

That May, U.S. West purchased 25.5 percent of Time Warner Entertainment Co (the Time Warner cable subsidiary) for $2.5 billion. U.S. West also moved into Atlanta, Georgia, in 1994, buying systems from Wometco and Georgia Cable TV for a total $1.2 billion, and rumors were floating that it had its eye on even larger cable investments. Nynex invested $1.2 billion in Viacom and, in late 1993, SBC said it was planning a $4.9 billion merger with Cox Cable, a top ten MSO. In December 1993, Bell South purchased 22.5 percent of Prime Cable, along with an option to buy the entire company.

Each new announcement of a telephone company purchase sent small shock waves through the cable industry, but nothing could compare with the tsunami that rolled through the cable world in October 1993 when John Malone and a man named Ray Smith called a news conference.

TCI Meets Bell Atlantic

The telephone industry and cable television had been nearly sworn enemies for decades. The mergers and telco purchases of the early 1990s were buffeting the old culture of the cable pioneers. Long-time cable operator Bob Hughes intentionally placed himself in the line of fire. Hughes and his business partners decided, in 1992, just after the passage of the 1992 Act, to set up a consulting business, Prime One, to advise companies looking to get into cable. "Some of our cable brethren started calling me on the phone," recalled Hughes. "Jim Robbins, Brian Roberts, and others saying, 'Hughes, you've become a turncoat. You're consulting for these big phone companies.'"[45] He was, and he took a contract to run the Hauser Washington, D.C., systems for SBC.

Despite these advances by the phone companies into cable operations, little could have prepared the industry for the October announcement that cable's largest and arguably most powerful firm, TCI, had agreed to merge with Bell Atlantic, the biggest of the RBOCs. It appeared as if the historically crossed swords of the two industries now would be raised to cover a wedding procession.

As John Malone told reporters and biographers later, he was tired. He felt bruised by the political battles leading to 1992 Act. He did not like being called Darth Vadar, although TCI employees tried to make a joke of it. He had received death threats at home, hired security guards, and his wife was understandably distraught. He thought about taking the Liberty programming wing of the company and selling out the distribution side — the cable system empire he and Bob Magness had built. Biographer Mark Robichaux wrote that on a business flight on TCI's corporate jet in July 1993, Malone took out a yellow note pad and began jotting down his goals and objectives. They included retreating from the public eye, reducing stress, and having more fun (as well as securing financial stability for himself and his family). Among the possible means were retiring from TCI but retaining Liberty and staying on as its Chairman; most of his personal fortune was tied up in Liberty, not TCI, stock.

The talk of convergence, digital television, and an interactive HDTV future also were on his mind. He saw the Baby Bells moving on numerous fronts to contest the cable market. To respond effectively, MSOs, including TCI, would be forced to spend millions in technology upgrades—money that the always heavily leveraged TCI did not have. Malone realized that a partnership with one of the telcos might have to be in TCI's near future. (Some suggested that Malone's 500-channel promise was only a ploy to pump up the price of TCI in anticipation of a planned sale.) Malone would take no action until he had consulted with his friend and mentor Bob Magness, but soon thereafter he began considering his options, including extending a conversation already begun with another telecommunications executive.[46]

From all appearances, Ray Smith was a true believer in convergence. He was the chief executive of Bell Atlantic. In the early 1990s, he saw a united telecommunications system delivering voice, data, and video to all the customers in the Bell Atlantic service region and beyond. He was a strong supporter of deregulation and sent his legal troops into successful battle against the various laws that kept his company out of the affiliated markets. He was also shrewd and experienced businessman. The 1982 divestiture had broken up the Bell monopoly, but none of the RBOCs or AT&T believed their sister companies would be content to sit placidly in their assigned territories and collect dividends from POTS. Each company was inspecting the new telecommunications landscape and pondering means of expansion. AT&T was a worry for all the RBOCs. Its interest in penetrating the local loop was open and eminently reasonable. Legal restrictions—which it would also seek to reverse—kept it out of direct landline completion, but in 1992, AT&T announced it would purchase a third of McCaw Cellular for $3.8 billion, upping the stakes in 1994 when it acquired all of McCaw, by then the nation's largest wireless carrier, for $12 billion. With McCaw, AT&T would have a system in place that could give customers the ability to replace their landlines with AT&T wireless phones, connecting directly to AT&T long distance and potentially by passing the local exchange carriers (LECs).

The RBOCs, such as Bell Atlantic, stood not to lose just the revenue from the cellular call itself but also from the access fees it charged to connect local calls to the AT&T long distance lines—revenue that accounted for up to one-third of an RBOC's total bottom line. Smith also saw SBC buying cable systems in Washington, D.C., and U.S. West investing heavily in Time Warner, each creating cable beachheads in Bell Atlantic territory. Once other Baby Bells had secured positions in broadband systems, the fear was that the advancing digital cable could then be turned to provide local phone service, if the law permitted.

Bell Atlantic, of course, could avail itself of the same business opportunities. The question was whether it would wait for the sharks to arrive, or join the pack. In May 1993, Smith placed a call to Malone. The two knew each from previous meetings and said later they enjoyed each other's company. Negotiations continued over the next several months and culminated in a meeting between Malone and Smith in a suite at the Waldorf Hotel in New York City where the final details were resolved. On October 13, they held a news conference at the Macklowe Hotel in Times Square to announce the deal. Bell Atlantic would purchase TCI for $33 billion, $23 billion in stock and $10 billion in assumed TCI debt.[47]

At the time, it was the largest corporate merger in U.S. history. Bell Atlantic had 18 million phone lines, TCI controlled 10 million cable homes. Annual revenue for the new company was estimated at more than $16 billion. By the end of the decade, the combined companies and soon-to-be-combined technologies would be delivering the convergent video, voice, and data services to millions of customers around the country. Bell Atlantic would commit billions to the necessary upgrades and new technology.

Malone had attempted, during the negotiations, to separate himself and Liberty from the deal. His hope, as per his plans, was to sell the TCI distribution arm to Bell Atlantic and, with Liberty under his wing, leave that side of the business behind. Smith, however, would not take TCI without Liberty or without John Malone. Smith knew the substantial value of each. In preparing for the sale, TCI had reabsorbed Liberty in a $3 billion stock transaction in August 1994. The new Bell Atlantic would include the programming properties and be structured such that Smith remained CEO while Malone could take a less public and quieter position as Vice Chairman.

To the extent that telephone company interest had been building for several years, it now accelerated. Convergence seemed real and companies scrambled not to be left behind. In Washington, D.C., regulators and members of Congress expressed deep concern about the anticompetitive implications of the merger and the Senate scheduled hearings.[48]

There is a process in the cable business, as elsewhere, called due diligence. In the sale of a business property, due diligence includes review and inspection of the property or properties involved. Bell Atlantic sent out representatives to conduct their due diligence and to inspect the TCI systems that would soon serve as the backbone of Bell Atlantic's thrust into the information age. They reported back with disappointing news. Although the vision of a broadband interactive future was enticing, it would not be realized by the physical plant that TCI had in place. The company's long-standing philosophy of providing the minimal necessary system to service basic needs had caught up with it. In far too many cases, TCI's wires were aging and in disrepair. The plant overall could not support the voice and data applications Bell Atlantic had hoped for. It would take, by Bell Atlantic's estimate, $15 billion to $20 billion dollars to bring the TCI infrastructure up to date. Even before the bad news began filtering back to Bell Atlantic headquarters, industry analysts and stockholders had begun to doubt the wisdom of the purchase. TCI, with its billions in debt, had been sold for a premium $2,350 per subscriber. Many thought Smith had overpaid, and Bell Atlantic's stock, which enjoyed a brief surge at the announcement of the acquisition, began to slump. By February 1994, it was down $20 from its October high.

On February 21, 1994, Bell Atlantic informed TCI it was adjusting the purchase price to about $30 billion. Malone acquiesced. Then, on February 22, Smith sat in his office and watched C-SPAN coverage of the FCC's second round of rate reductions for cable. It was the deathblow to the merger. Smith and Malone met the next day, agreed that the FCC's additional restrictions would cut too deeply into TCI's cash flow to make the deal viable for Bell Atlantic. On February 23, they announced the collapse of the agreement.

Failure of the TCI-Bell Atlantic merger attempt was nearly as traumatic for the industry as was its original announcement. The alliance was portrayed and perceived as both the substantive and symbolic realization of the promise of convergence, the creation of the twenty-first century integrated voice-data-video infostructure. Two months after the implosion, SBC and Cox announced they would kill their planned $4.9 billion merger, as well.

Once again, the vision of what was technically possible had outpaced financial and political reality. The parties took a step back and revised their plans. With the rate rollbacks in place, telco enthusiasm for mergers and cable purchases cooled. The RBOCs turned their attention to alternative distribution strategies, several of which already were in development. VDT options, which gave the telephone companies only partial control over the system, lost favor. Instead, the Baby Bells looked to put more resources into creating futuristic broadband switched networks, employing digital MMDS options, or simply overbuilding with traditional technology, with strategies and preferences varying from company to company. The news was not good for cable, and did not seem about to improve anytime soon.

DBS—The Death Star

In 1992, the only operating DBS system in the United States was owned by the cable industry. The Hughes-Murdoch Death Star, as its Star Wars namesake, had exploded. Hughes and Hubbard had announced their plans to launch a combined system. Hubbard would control his USSB service and Hughes would create its own DBS company, DirecTV. It would be another two years before they started selling service, however. Advanced and EchoStar needed money to turn their dreams into active transponders and no one seemed interested in lending them the millions necessary to make it happen. There was a reason.

From the day the first coaxial cable dropped into the first home, the commodity in question was television choice—variety—and the seller offering the greatest number of channels at the lowest price, won. By 1991, 64 percent of all cable systems had thirty to fifty-three channels; 28 percent of all systems, accounting for most of the nation's subscriber base, had more than fifty-three channels. A Ku-band direct broadcast satellite held thirty-two transponders, or thirty-two standard 6-MHz TV channels. Moreover, the FCC had declined to authorize the full thirty-two frequencies to any one licensee; Tempo was approved for only eleven; Hughes had twenty-seven. Because no DBS provider could offer the same number of analog channels or the range of variety as even a middle-grade cable system, competing head-to-head with cable was not a viable business plan. Would-be DBS services told lenders that the market, therefore, was the countryside where the cable did not run—the old C-band stomping grounds, plus the people at the margins who declined to purchase cable service or, alternatively, would buy any and all TV programming available. Primestar itself offered little more than a barebones schedule with none of the mainstream cable networks and, in any event, it was largely a pawn in the chess game to block competitors from acquiring transponder space. As late as 1992, Malone was rightly skeptical of the chances for a true DBS business in the United States.

With compression, the rules changed on a digital dime. It not only gave Malone the ability to hyperbolize about multiplying a fifty-channel cable system into a 500-channel one, but it also gave direct broadcast satellite business the power to daydream about transforming a thirty-two-channel Ku-band bird into a 320-channel one.

Daydreaming was about all anyone could do in 1992 because, as with most Blue Sky technologies, the idea and the rhetoric were much further advanced than what any engineer could actually create on the work bench. But also, as with past technology, real progress was being made.

DirecTV

Hughes mounted its first DirecTV satellite on a French Ariane rocket and placed it into orbit on December 17, 1993. It had twenty-seven authorized frequencies at 101-degree W; USSB had another five. Using a fully digitized signal, Hughes could generate a four-to-one compression ratio.

The task of organizing the new service fell to Hughes executive and engineer Eddy Hartenstein, who had been given the project when Sky Cable fell apart. Hartenstein, a former mission planner for the Viking and Voyager space probes, understood complicated problems, but one of the complications for DirecTV had little to do with technology. The refusal of the cable industry to sell product to competitors, or to do what they could to raise barriers to sale, was one of the driving forces behind enactment of the 1992 Cable Act. More pointedly, the Primestar group, according to the government, entered into an agreement in violation of antitrust laws to control access to programming, sharing information and fixing prices to ensure that Primestar always received the best deal in programming negotiations.

A Justice Department antitrust investigation concluded in a consent degree signed with Primestar in June 1993[49] in which the company agreed to abide by the program access provisions of the 1992 Cable Act. With the 1992 Act in its back pocket, Hughes and Hubbard could obtain the stable of programming needed to attract customers to its multichannel service.

In June 1994, Hughes began selling service. It was not a consumer dream come true, but it was pretty good, and more than a little competitive with existing cable service. Upfront costs were high. The equipment—the satellite dish and special set-top box—cost $699 to $899, not including installation. Monthly rates ranged from $6 to $30, depending on service level. The signal was subject to some interference by rain, but the dishes were small and, to some extent, inconspicuous, at least in comparison to the C-band sails of a decade before. "Pizza-sized" eighteen-inch dishes, was the often-heard description. For the price, and when it was working well, subscribers received a crystal-clear digital signal, 175 channels of basic, premium, and PPV channels.

DirecTV successfully targeted consumers in noncabled rural America, promoting the Disney and Turner networks. To the surprise of some, it also did well in the urban markets by providing a level of service quality and choice that existing cable systems were hard pressed to match. Movies and sports had always been major television draws and DirecTV looked to satisfy the consumer thirst for both by offering an extensive film menu and specialized sports packages. Its PPV schedule included fifty-five channels at $3.99 a showing, moving closer to video store convenience than any cable system.

Despite the high upfront costs, many customers loved it for the signal quality and program variety. Some (mostly noncable affiliated) programmers loved it as an additional distribution outlet and revenue stream. Regulators loved it because, for the first time, it offered a tangible competitor to cable, one that would, it was hoped, finally help keep consumer prices down and quality high. The cable industry hated it because it was surprisingly effective at taking away customers and, more importantly, taking away the high-end, high-paying subscribers that cable valued most.

In March 1994, several months before the rollout of DirecTV service, Primestar began its own digital feed, upgrading to seventy-five channels, but marketing, by intention, was meek. Cable operators were afraid of cannibalizing their own subscribers. By the end of 1994, DirecTV/USSB had 400,000 customers; Primestar reported about 250,000.

Primestar and Company

Even with its own digitally compressed signal and improved programming package, Primestar was handicapped by its inferior, medium-powered technology and bulky twenty-seven to forty-eight inch home dishes.

If the cable operators were going to stay even with DirecTV/USSB in the battle for the DBS dollar, they would need to move to a high-powered service. TCI, through Tempo, still controlled the necessary transponder allocations (at 119-degrees W), but Tempo had only been allocated eleven frequencies, too few to mount a real challenge to the Hughes-Hubbard capacity (thirty-two). TCI, therefore, went looking for someone with transponders for sale.

Dan Garner's Advanced Communications had advanced very little in its effort to build a DBS service, and the FCC had warned it that its license would be in jeopardy if it continued to fail to meet reasonable goals for securing financing, satellite construction, and launch support.

EchoStar's Charlie Ergen, meanwhile, had been talking with Garner since 1989 about a possible merger. Ergen controlled twenty-one frequencies at 119-degrees W; Garner had twenty-seven frequencies at 110-degrees W. They signed a letter of intent in 1992 to combine their assets, but two years of haggling over strategy and control left them no closer to an operational system and Garner began casting about for new partners. The Primestar board represented most of the major cable operators in the country, each with a different business philosophy, and not all of them were enthusiastic about DBS. Acting on his own, therefore, Malone reached out to Garner and, on September 29, 1994, Tempo announced it had purchased the Advanced DBS license for $50 million.[50]

Garner had been living on FCC extensions, but the Commission's willingness to grant additional time to applicants was diminishing. In his successful 1991 extension request, Garner had been warned by the FCC that, in the future, it would be more strict in applying its requirements for progress.[51] A month before announcement of the TCI agreement and with no progress to show toward a functioning system, Garner again went to the FCC to ask for more time.[52] This time, the Commission said "no." Finding no evidence of satellite construction or preparation for launch and operation, the FCC, in April 1995, took back the DBS frequencies allotted to Advanced and subsequently sold to Tempo.[53]

The FCC then determined that it would be in the best interest of the public and the nation's bank balance to auction off the frequencies to the highest bidder. In this, the Commission was following a recent mandate of Congress to auction off spectrum space.

The FCC-approved bidders included EchoStar, TCI (through TCI Technology Ventures, Inc.), and a new player, MCI Telecommunications Corp. As had the Baby Bells, the long-distance telephone company had been lured to the siren call of the multichannel video business by the rhetoric and promise of convergence. Moreover, MCI chairman Bert Roberts had been talking with Rupert Murdoch. Shut out of DBS in two previous attempts, News Corp. was trying once more. Brought together by financier Michael Milkin, Roberts and Murdoch planned yet another competitive multipurpose interactive distribution system, with Murdoch supplying the content and MCI the technical infrastructure. In May 1995, MCI paid $2 billion for 13.5 percent of News Corp.

In the first round bids, Ergen offered $125 million (the minimum allowed), MCI, $175.2 million, and TCI $201 million. TCI dropped out after eleven rounds. Ergen had determined prior to the auction that he would go as high as $650 million. He did. But MCI and Murdoch, better financed and perhaps hungrier, beat it. MCI's winning bid for twenty-eight channels at the 110-degree W slot was $682.5 million. Ergen did win a parallel auction for twenty-four channels at the 148-degree, partial CONUS slot for $52.3 million.

Frustrated with its inability to win a high-powered DBS seat, TCI went north of the border. It arranged a deal with Telesat Canada in 1996. TCI would sell Telesat the two satellites it already had built, for $600 million. TCI then would buy transponder space on those satellites from Telesat to create a DBS service for the United States. The proposal failed to win FCC approval, however, and TCI once again was left with only Primestar.

EchoStar

On December 28, 1995, Ergen launched EchoStar I from a Chinese Long March rocket. On March 4, 1996, he began his campaign to undercut the competition. Positioning his service as a low-cost version of DirecTV and Primestar, he opened a marketing campaign with a youthful contemporary feel, branded as "Dish TV." Ergen's plan was to procure customers using deep discounts, and a fierce price war broke out between DirecTV and DishTV. The cost for dishes and installation began dropping, first to $200, then $100, then $50. It got to the point that Preston Padden, Murdoch's satellite chief, famously joked, "Take this free dish and we'll buy a house to bolt it onto."[54] Within a year, EchoStar had signed 350,000 customers.

By the end of 1995, the cable industry, for the first time in its history, was confronted with something that looked like real competition. DirecTV was signing new subscribers every day. EchoStar had launched a discount DishTV service, and Murdoch and MCI were planning an additional, well-financed

DBS service of their own. Despite the collapse of the TCI-Bell Atlantic deal, the telephone companies were still pursuing the broadband switch system promised by convergence and the voice, video, and data package that came with it. Cable had to respond.

The Cable Response

Malone's 500-channel news conference was more than a simple, if startling, proclamation about the future of television. It was a move in the business and political chess game that cable generally and TCI in particular was playing in the early 1990s. Some years before, in the mid-1980s, cable looked to investors like an unregulated monopoly; it did not look like that anymore. Wall Street was getting nervous and Malone was looking to assure financial analysts that cable still held the lead in cutting edge technology and services, if only to prop up the value of the stock while he looked for a buyer.

The industry responded in a variety of ways, varying from company to company, each with its own corporate personality and economic constraints. A battered John Malone sought to escape entirely by selling to the phone company. Cox attempted a partnership with SBC, as Dolan explored a possible DBS alliance with Hughes and Murdoch. With the collapse of each of those deals, however, the companies were pushed to adopt a more confrontational stance, one already taken by a number of cable firms, which by choice or necessity, had determined to fight rather than switch sides.

Digital technology gave the telephone companies and the direct satellite providers the tools they needed to consider systems capable, at least in concept, of competing with cable television. But the cable industry had access to the same tools. Digital compression allowed it to multiply its channel offerings. This meant not only new niched programming networks, but more importantly, additional bandwidth for current films, one of the products most prized by consumers. With sufficient channel capacity, popular films could even be shown on multiple channels, each with a different start time. Even more advanced systems called for digitized films available in full video-on-demand (VOD) form.

With either traditional or digital switching techniques, cable also could consider using its local network to offer telephone service, challenging the telephone companies in a way not previously possible. The first line of defense was to continue to increase channel capacity by improving amplifier power and driving fiber deeper into the system, bringing it as close to the consumers' homes as economically feasible. Traditional HFC configurations sent several fiber lines directly from the head-end into local neighborhoods, terminating at neighborhood "hubs" that serve up to 2,000 homes each. More advanced fiber-to-the-node or fiber-to-the-curb designs proposed in the late 1980s and early 1990s moved fiber more deeply into the system, with each node serving 500 to 2,000 subscribers, or in advanced systems as few as 125 homes. As the number of coaxial amplifiers servicing each home was reduced, the signal quality was proportionally enhanced. As the number of homes at each node decreased, the amount of bandwidth available to each customer increased because overall system bandwidth is divided among fewer people. HFC systems using 1-GHz amplifiers began to appear in the early 1990s, although 750 MHz was a more typical ceiling and provided sufficient bandwidth for most purposes, including up to 110 analog channels, or eighty to 100 analog and several hundred compressed digital signals.[55] By 1995, the cable industry had strung more than 81,000 miles of fiber, and was laying more every day. The cost was estimated in the tens of billions of dollars.

A number of proposals for fiber-to-the-curb systems were floated by both the cable and telephone industry in the early 1990s, although in practice, traditional HFC systems dominated through the decade. It was cheaper to employ the multiplexing power of digital compression. At least one famous experiment was conducted, however, in New York City.

Time Warner in Queens

James (Jim) Chiddix had a reputation as one of the industry's best and brightest. He was Time Warner's chief engineer, known for his ingenuity and integrity. Chiddix learned his cable craft in Hawaii, where he developed his own homemade approaches to advertising insertion, PPV, and fiber interconnection. The Honolulu company for which he worked, Oceanic, was purchased by ATC in 1980 and, by 1986, Chiddix had moved back to the mainland where he went to work with an ATC research team in Denver. There, he helped develop the AM fiber technology and cable's modern HFC system architecture.

John Malone was a skilled businessman, tactician, and salesman. His 500-channel promise served a variety of purposes, including planting the seed of doubt into an investment community considering supporting potential cable competitors. It would be many years before the technology was sufficiently advanced to bring those 500 channels to market, however, and many in the industry knew it. Steve Ross at Time Warner wanted to see what they could do with the best state-of-the-art analog technology in 1991 and directed Chiddix to build a prototype. The company selected Queens, New York, as their test bed. They would use 1-GHz amplifiers, recently developed by C-Cor Electronics, and an architecture that drove fiber more deeply into the system—a fiber-to-the-node configuration. In March 1991, they announced that they would upgrade the seventy-seven channel system to 150-channels and offer forty PPV channels, along with interactive video games and similar experimental services.

Time Warner was signing up customers by the end of 1991 and showing off the facility to curious cable executives from across the country. In

1992, Viacom began its own experimental 1-GHz system in Castro Valley, California, and TCI began testing advanced services in four cities around the country. It was a very expensive venture, but some companies saw it as the system of the future, others simply saw it as one of the few means of survival.

Multiplexing

By the turn of the decade, the possibility of additional bandwidth, either through improved amplifiers and architecture or through digital compression, started to become real in the minds of programmers, especially those associated with operators, such as Time Warner, which were developing advanced systems. So, while Jim Chiddix was drafting plans for 150-channels in Queens, the HBO chiefs at Time Warner put their minds to securing and exploiting the new shelf space.

In May 1991, Time Warner announced plans to program three channels each of HBO and Cinemax. It was called multiplexing. There were a number of objectives. The programmers hoped, in the first instance, to reduce churn by providing greater film selection. Secondly, the outlines of the 1992 Cable Act were coming into focus and it appeared that rates for premium channels would be not regulated, suggesting expansion of that part of the business. Finally, entrenched programmers sought to secure additional channels in part for their own programming but in equal measure as a defensive strategy to prevent them from being claimed by potential rivals.

The earliest multiplexed services were planning well ahead because cable capacity on a national basis had not yet begun its digital expansion—and would not do so for several years. The launch of DirecTV, with its 150-channel-plus carrying capacity and national reach, ironically opened the doors for multiplexed channels by providing the necessary national bandwidth.

Viacom launched FLIX, featuring films from the 1970s and 1980s and promoting them as "Cool Classics," in August 1992. In May 1994, Viacom's Showtime Networks unveiled plans for a new set of themed movie channels, including Showtime Family Television, Showtime Action Television, Showtime Comedy Television, and Showtime Film Festival. Two months later, TCI's Encore responded creating a fleet of niched film services, including Encore Love Stories, Westerns, Mystery, Action, True Stories, and WAM!, America's Kids Network.

The Full Service Network

The "Holy Grail" for cable programmers was video-on-demand. There had been efforts at VOD in the past, usually based on clumsy and unreliable banks of videotape machines controlled by computers or even in some experiments fed by employees zooming from storage cabinet to tape machine on roller skates.

Expanded bandwidth by itself could not provide true VOD, which required a high level of digital interactivity. But bandwidth could move closer to something that looked like video-on-demand by reducing the time a subscriber spent waiting for a PPV movie to begin. Additional channel capacity allowed programmers to stagger the start of any given film and thereby cut wait times down to less than an hour. It was called near-video-on-demand (NVOD).

For cable, true VOD, which involved connecting an individual subscriber to a particular movie, starting at a time specified for only that subscriber, required a dedicated pathway between the subscriber and the movie, and a switch of some type to create that pathway. As noted, telephone architectures had such switches; tradition cable tree-and-branch architectures did not. It was, luckily, another problem that could be rectified by the use of digital technology.

Computer information was increasingly being stored on high-capacity disc drives, often stacked together to afford massive storage space. These devices could hold libraries of books, data, even video information, and the technology was advanced to the point that multiple users could access the data simultaneously, with the electronics serving the desired material to each client on an individual basis. In theory, a company could place digitized movies on these computer "servers" and, using packet-switching techniques, allow an individual subscriber to link with an individual film. The flexibility of the system was sufficient, moreover, to create VCR-like controls, allowing the subscriber to stop, start, pause, and rewind the film on commend. It was the architecture for full video-on-demand. With dedicated upstream bandwidth, digital VOD could be implemented using cable existing HFC topology.

The economics were uncertain, however. Packet switching was cheaper than building circuit-switched systems from scratch. But such a platform would require advanced set-top boxes that could translate the digital video into analog signals that the users' set could understand—the same challenge TCI had given GI in placing its order for digital boxes. The cost of such boxes, again, was uncertain.

Someone would have to be the first to create a prototype system, someone with a lot of money, someone willing to take a risk. Gerald Levin had been one of those willing to take the risk, in 1975, buying satellite time for HBO. He decided to roll the dice again.

In January 1993, Time Warner announced it would construct what it called a "Full Service Network" (FSN) in Orlando, Florida, home of Disney World, Epcot, and site of the company's second largest cable system. Technically, it would be an HFC architecture with analog and digital capacity. Interactive services, including VOD, would be provided by a packet-switching protocol called ATM (asynchronous transfer mode). In addition to the traditional cable networks and new VOD services, Time Warner's Full Service Network would offer all the interactive services the Blue Sky imagination could conjure—games between customers, interactive news, sports, weather and information services, electronic mail, and most importantly, shopping. Subscribers would be able to purchase anything from automobiles to postages stamps at the FSN "Video Mall."

At the same time, the FSN computers could track viewing and purchasing activity. The information would be useful to advertisers looking to target customers interested in particular products and, of course, the advertisers would pay Time Warner for these valuable sales leads. Using the leads provided by the cable company, automobile dealers or sewing machine manufacturers then would bombard the targeted customer with additional advertising and promotional material. At the time of the announcement, Time Warner said the FSN would be up and operating in the first quarter of 1994. It was December of that year before the first customer was signed up and the unveiling took place.

Levin appeared to be a true believer in the service. Chiddix saw it as a useful test of equipment and methods that could serve the company down the road. What it was not, as was soon clear to everyone, was financially viable. Customers were slow to engage all the whistles and bells offered by the system. Fewer than expected customers took advantage of the video shopping opportunities.

A certain number of early adopters explored the potential of the FSN, but not enough to justify the capital costs of the system. One of the chief problems was the expense of the set-top box. The varied reports on the cost of the FSN box ranged from $3,000 up to $10,000 each. The FSN was an interesting, but very expensive experiment.

Cablephone

At first glance it looked a bit like turn-about as fair play. If the telephone industry could get into the cable business, then why could not the cable industry offer telephone service? Legal restraints had kept the telephone companies at bay for decades, but both industries had a wire going to the home. Cable did not possess the requisite switch necessary for traditional local loop service, but there were alternative forms of entry, and exploration of telephone service was yet one more response of the industry to the challenges of the early 1990s.

The move was not driven by any sense of irony, however. Cable executives, such as Malone and Roberts, were not quixotic. They knew that local telephone service generated about $60 billion a year in revenue, dwarfing total cable industry of about $27 billion in 1993. If cable could take even 10 percent of the business, it would dramatically increase its annual revenue stream.

Some of the first experiments involved a marriage of the local cable system with a wireless form of telephony called personal communications service, or PCS. The engineering involved spacing numerous wireless transceivers on poles or similarly accessible sites throughout the cable system. Customers would use wireless telephone units to communicate with the transceivers, which in turn, would pipe their calls through the cable lines and on to the local exchange carrier switch. Traditional cellular telephone service began in the United States in 1984 and, by the close of 1995, some 33 million people were using cells. Cable saw their participation in PCS, which offered smaller, lighter handheld units than the typical cellular handset in use at the time as a means of taking part in the growth of the wireless sector.

Cox Communications was one of the more aggressive cable firms to pursue local telephone service. It made a small splash in February 1992 when it debuted its prototype PCS technology in San Diego with Cox Enterprises CEO James Kennedy placing a PCS call to FCC chairman Al Sikes in Washington, D.C.

PCS service was not the only path to telco service pursued by Cox and others. For decades the business world had complained about service provided by the monopoly AT&T and then its offshoot Baby Bells. In the late 1980s, a small company called Teleport (Teleport Communications group), capitalizing on the RBOC weakness, offered business a dependable, high-capacity, fiber link to the long-distance network, at discount rates. Teleport was one of a number of businesses known as competitive access providers, or CAPs. As demand for its service expanded beyond its line capacity, Teleport went to the local cable provider, leasing coaxial capacity to augment its own plant. In 1992, the same year of the Cox PCS demonstration, the company acquired an interest in Teleport from then-owner Merrill Lynch. TCI also bought in.

At the local level, Time Warner announced plans in 1994 to offer residential service in its cable markets, starting with a trial in Rochester, New York. TCI also joined with Sprint, Cox, and Comcast to form Sprint Telecommunications Ventures, a Sprint-cable alliance aimed at harnessing PCS technology with cable systems at the local end of service and Sprint as long-distance provider. In March 1995, Sprint Telecommunications Ventures spent $2.1 billion to secure twenty-nine PCS licenses in cities ranging from New York, Detroit, Dallas-Fort Worth, to San Francisco.

Take the Money and Run

Time Warner was one of the world's largest and most diversified media companies. It could afford to finance the forward-looking gambles of executives such as Ross and Levin. Cox was also in a position to experiment with untried technologies and unknown markets. Most cable operators were not so privileged. Even TCI was feeling the discomfort of changing market conditions in

1994. The FCC's rate reductions had narrowed the arteries of cash flow and dampened the interest of the lending community. The RBOCs seemed to sit at the very border of the cable business, probing for soft spots, and Congress began considering legislation in 1993 that could remove all legal barriers to telco provision of cable services. Malone, after the collapse the Bell Atlantic merger, had withdrawn from the public eye. Rumors were he had fallen ill or, at the very least, had lost his taste for the cable business (see Chapter 13). For some operators, the future seemed problematic.

Bill Daniels' 1968 warning about the cost of building one-way coaxial systems in the nation's biggest cities echoed down the years and grew louder with each reverberation. Driven by the social construction of telecommunications convergence, operators by 1994 were running spreadsheets on the cost of interactive, digital HFC systems with server-based VOD, data communication, and residential telephone functionality. The capital costs were immense, the probable success uncertain. Many of the large and mid-sized MSOs had been built by pioneers now nearing retirement age. The companies they created were now worth hundreds of millions, even billions, of dollars. As Malone in his contemplation of a sale to Bell Atlantic, some began thinking that this might be a good time to take the money and run.

The confluence of regulatory, technical, and social forces was nudging old-line cable firms and larger companies whose investment in cable was at the corporate margins, to sell out instead of trying to keep pace with the upgrades required to stay competitive. Cablevision Industries was a case in point. Alan Gerry founded the company in 1956. Always privately held, expanding gradually one small market at a time, Gerry built Cablevision into an MSO of 1.3 million subscribers. By 1995, it was the nation's sixth largest operator. Gerry, however, had become disillusioned by the political atmosphere surrounding and constraining the industry. Time Warner was buying, and it offered Gerry $1 billion in Time Warner stock and the assumption of the company's $2 billion in debt. Gerry cashed out and Cablevision Industries was added to the Time Warner stable. Time Warner similarly acquired mid-sized operators Summit Communications and Houston Industries.

Newhouse, like Cox, was a diversified communications company founded largely on newspaper publishing. Unlike Cox, however, Newhouse did not see cable as central to its future. In 1995, it entered into a partnership with Time Warner, turning over operation of all its systems to the larger MSO. Other newspaper companies sought shelter as well. Times Mirror merged its cable division into Cox in June 1994 in a deal valued at $2.3 billion, moving Cox to a position as the nation's third largest MSO. Continental had courted Times Mirror, but having lost the bidding war to Cox, turned to another target. The Providence Journal Co., another newspaper firm, sold its Colony Communications cable division to Continental for $1.4 billion. Knight-Ridder also left the industry, selling its joint interest in TKR Cable to its partner, TCI.

Comcast was one of the active purchasers. It paid $1.6 billion for the cable interests of the E. W. Scripps Company. Comcast also bought the cable systems held by Maclean Hunter in the United State for $1.27 billion.

Charles Sammons had purchased his first cable systems from Bill Daniels in the late 1950s and, by 1995, the company controlled the fourteenth largest MSO in the country, but cable was not its core enterprise. Sammons was involved in insurance, steel, and manufacturing and, in March 1995, it began dealing out its systems, selling most of them to Marcus Cable for $1 billion. Cable companies exiting the business in 1994 and 1995 also included: Crown Media, Columbia International, Gaylord, United Video Cable, and Multimedia.

MSOs determined to stay in the business and expand included Time Warner, TCI, Cox, Adelphia, and Comcast. TCI purchased the mid-sized operator TeleCable Corp. in 1994 for $1.4 billion in TCI stock, garnering an additional 740,000 subscribers. Adelphia paid $63 million for 75 percent of Telemedia Corp., the State College, Pennsylvania, company founded by cable pioneers Robert Tudek and Everett Mundy. The purchase gave Adelphia close to 1.7 million cable customers and made it the nation's seventh largest MSO.

Influential in the acquisition strategy of these firms was the goal of increased clustering. The purchases were aimed, in part, at consolidating regional systems and thereby enhancing the economies of scale and the general cost savings that regional interconnection provided. Clustering was seen as particularly important in creating the broadband urban systems necessary to provide converged data and telephone services. In 1995, Time Warner, illustrating the point, sold fifteen systems in areas that did not provide target clusters, building a large consolidated operation in the New York and Los Angeles areas. TCI swapped systems with Post Newsweek Cable, Multimedia Inc., Cox Communications, and Continental. Continental strengthened its presence in the Northeast while TCI consolidated systems around St. Louis and the U.S. Northwest.

Among the most noteworthy transactions of the period were the tumultuous departure of Viacom from the operations side of cable and the headline-making acquisition of Turner Broadcasting by Time Warner.

QVC, Paramount, and Viacom

The Viacom and Turner sales touched programming more than distribution, with the former engaging a legal brawl between Barry Diller and Sumner Redstone, with important roles for John Malone and Brian Roberts.

Diller was the programming executive largely credited with formulating the youth-oriented format that propelled Fox Broadcasting to its early success. Talented and ambitious, Diller had been number two man at Paramount in the late 1970s and early 1980s, but decided to leave Fox to run his own shop. He was a powerful figure in the film and television industry and the gossip mills churned while he decided where to land. Those mills ground to a befuddled stop in December 1992 when he announced he would not be taking over a major entertainment company, but rather had decided to run a cable shopping network, QVC. He explained that he was attracted by the massive cash flow and its potential for expansion. Investing $25 million of his own money, he took over as CEO. Moving quickly, he hatched plans in the summer of 1993 to purchase QVC's major rival, HSN. Antitrust concerns and a Justice Department investigation derailed the strategy, however. Both TCI and Comcast held major interests in the shopping channel.

In September 1993, with backing from Malone, Diller set his sites somewhat higher and tendered a bid for Paramount Pictures, only days after Viacom had announced its own plans for an $8.2 billion merger with the film and media giant. The bidding war escalated and turned nasty. Redstone filed a federal antitrust suit seeking to block QVC's takeover attempt. In the suit, he accused Malone specifically of attempting to monopolize the cable industry. The vitriolic nature of the broadside made the political rhetoric leading to the 1992 Act look tame by comparison. "In the American cable industry, one man has seized monopoly power. Using bullyboy tactics and strong-arming competitors, suppliers and customers, that man has inflicted antitrust injury onvirtually every American consumer of cable services and technologies. That man is John C. Malone," Redstone stated in the suit.[56]

Looking for additional funds to power the purchase of Paramount, Viacom sealed deals with the video rental chain Blockbuster Video and the New York area RBOC Nynex. Blockbuster invested $600 million in Viacom and Nynex put up $1.2 billion. Bidding that began at about $70 a share rose to $85 a share by the end of 1993. Malone had promised $500 million in backing for Diller, but as TCI's own negotiations with Bell Atlantic gathered steam, Malone's interest in the Paramount purchase faded. Ultimately, Redstone prevailed and Viacom purchased Paramount in February 1994 for $10 billion.

Diller moved on to yet another target, announcing in June 1994 an agreement to merge with CBS. Under terms of the $7.2 billion deal, CBS would purchase QVC and Diller would replace Larry Tisch as chairman. Comcast, however, opposed the plan, and to roadblock the sale made its own $2.2 billion offer for outright ownership of QVC, scaring away CBS which feared a costly bidding war. TCI decided to join Comcast in the QVC purchase and, in August 1994, the two MSOs struck a deal. Comcast now owned 57 percent and Liberty 43 percent of the shopping channel. Diller was invited to stay on as head of QVC, but determined to seek different, if not greener pastures, he left the company in February 1995 (having made a $91 million profit on his original $25 million QVC investment) and took over as head of TCI's other shopping channel, HSN.

Viacom was left with massive debt from the Paramount purchase and, in November 1994, announced it was selling its cable systems, and 1.1 million subscribers to a consortium of operators led by TCI and Leo Hindrey's Intermedia Partners for $2.25 billion. In addition to the need for cash, Viacom explained that it had estimated it would take $15 billion in system upgrades and expansion to compete in the prophesized convergent distribution business and was now unable to fund such an effort. (It took more than a year to complete the acquisition, however, due to political complications. TCI and Intermedia were part of group led by Mitgo Corp., which was owned by African-American businessman Frank Washington. Washington's involvement was designed to take advantage of federal tax law [the minority tax certificate program] that provided tax breaks for sales to minority owners. Congress, concerned the tax break was not being used as intended, however, repealed the law.)

As a consequence of the sale, Redstone dropped his antitrust suit against Malone; TCI and Intermedia agreed to carry Viacom's MTV and Showtime networks and Malone and Redstone declared their good friendship. The Viacom systems were divided such that TCI and Intermedia obtained a large cluster in the San Francisco Bay Area and controlled every major market in Washington, Oregon, and Northern California except for Sacramento.

The Broadcast Networks

By 1990, the broadcast networks were feeling the pinch. As noted above, cable was fragmenting the audience and broadcast ratings were eroding. Programming costs, at the same time, continued to escalate. The networks responded along a variety of fronts. While they were no longer the only television game in town, they remained the largest. No single cable channel could offer the national coverage of NBC, CBS, or ABC. The combined primetime ratings of the top twenty basic cable networks in 1995 was only 22.4, less than half that of the combined ratings for the three broadcast networks.[57] The USA network, consistently one of the most watched cable channels, drew a rating of only about 2.3. Entrenched industry practice worked in the networks' favor as well. Advertising planners who determined where to place ads for national clients found calculating total reach across various fragmented cable programming bases difficult and foreign. Buying network time was the known, comfortable, conservative, and easy route. Demand for network ad space, driven by an expanding economy and especially in the mid-1990s the dot-com bubble, therefore grew. The networks raised ad rates about 15 percent annually through the mid-1990s, far outstripping the pace of inflation.

At the same time, the industry worked to arrest spiraling programming costs by successfully seeking repeal of the long-standing financial interest and syndication (fin-syn) rules, which had severely restricted their production and content ownership since 1970.

With the success of cable television in the 1980s and the rise of Fox, the networks argued that they no longer could create a bottleneck in program diversity nor exercise monopolistic control in the programming marketplace. It was an argument met with sympathetic ears. In May 1991, the FCC relaxed the rules,[58] but they remained too stringent for the broadcasters, who successfully appealed the modified regulations. In November 1992, a federal appeals court struck down the restrictions,[59] and a year later, a U.S. district court judge lifted the fin-syn consent decrees that the networks had entered into with the Justice Department years before.[60] Prodded by the courts, FCC controls were eased over the next several years to increase network ownership of the prime time schedule at ABC, NBC, and CBS and ultimately dissolve the all restrictions on nonprime time programming.[61]

Relaxation of the fin-syn rules stimulated Hollywood's appetite for control of national broadcast distribution. In 1993, Warner Bros. and a partnership between Paramount and Chris-Craft Broadcasting both announced the start of their own broadcast networks. Warner Bros., in partnership with Tribune Broadcasting, launched the WB network in 1994, and Paramount and Chris-Craft started the United Paramount Network (UPN) in 1995.

Walt Disney Company announced it too was getting into the TV network business, albeit by a different route. Michael Eisner was the head of the Disney Corporation. A programming executive at ABC through the late 1960s and early 1970s, he joined Disney as CEO in 1984. He envisioned a global reach for Disney, leveraging its trusted entertainment brand across every possible media platform. One of his goals was to own a television network and he talked in briefly and unsuccessfully with NBC about a possible purchase in 1985. (The deal was contingent on RCA spinning off the broadcast property, but at that time, RCA was unwilling to part with the generous cash flow created by the network. The talks withered, although shortly thereafter, NBC was sold to General Electric.) In 1996, Disney announced it was acquiring CapCities/ABC for $19.2 billion.

Ownership of CBS passed hands in the same period. In November 1995, Larry Tisch orchestrated the sale of CBS to Westinghouse for $5.4 billion, personally making $2 billion on the deal. Westinghouse had a deep history in broadcasting; it had been one of the original investment partners in NBC in the 1930s. Its activities in the 1990s extended into manufacturing, government services, and even nuclear power generation. With the purchase of CBS, the company decided to focus exclusively on the communications sector, however. In December 1997, it officially changed its corporate name to CBS Inc., and shed its nonmedia businesses. To extend its reach in broadcasting, it purchased radio groups including Infinity Broadcasting in 1996 for $4 billion and American Radio systems in 1997 for $2.6 billion. In cable, it bought Gaylord Entertainment in 1997 for estimated $1.5 billion, obtaining Country Music Television (CMT) and TNN, the two major country music cable channels.

Time Warner and Turner

When Ted Turner was on the road seeking a friend with $300 million to bail him out of the MGM debt in 1985, one of the places he visited was Time Inc. Time management was conservative and suspicious of the over-excitable Turner, who appeared to have bitten off more than he could ever chew with the MGM deal. Time Inc.'s Nick Nicholas offered Turner $225 million up front with a possible $75 million later, but Turner needed all of it and moved on. Nicholas later described his decision as the worst he had ever made in business and longed for a second chance to purchase the ultimately successful Turner company. In 1996, Time got that opportunity.

Turner once again saw competitors coalescing and expanding around him. Viacom had purchased Paramount, Disney had purchased ABC/Cap cities, acquiring the lucrative ESPN franchise and creating a production and distribution giant. Turner's lust for a TV network had not eased. He sought an established national distribution platform to complement his expanding programming presence. His first target was NBC, but Time Warner sat on the Turner Board of Directors and Levin vetoed the buyout proposal. An angry Turner let his displeasure at the blocking action be known in terms startling even for the famous "mouth of the south."[62] He next tried, once again, to purchase CBS. The network was in play. Westinghouse had made a $5 billion offer to buy the company. Turner hired disgraced financier and old friend Michael Milken to raise $10 billion for a competing bid. Malone, according to most reports, encouraged Turner. Time Warner was not any more comfortable with a CBS buyout than it was with an NBC bid, however, and in any event, Gerald Levin had another idea.

In February 1992, Nicholas was pushed out of Time Warner after an internal corporate power struggle. Following the Time Warner merger, Levin and Nicholas shared power under Ross, but Ross was diagnosed with prostrate cancer and was in rapidly failing health. Levin reportedly engineered a palace coup, securing the support of Ross and key members of the Time Warner board to force out Nicholas, sending him news of the action while he was on a skiing vacation with his family in Vail, Colorado. Levin became president of the company. On December 20, 1992, Steve Ross died of prostate cancer and Levin took the helm of Time Warner.

As CEO of Time Warner, Levin sought to improve company finances by upgrading its technology, building toward an integrated, broadband model of cable communications. He spent $4 billion on capital improvements to the company's cable infrastructure, and ordered the Full Service Network in Orlando. He was not universally liked in his own company, with a reputation as fiercely intelligent, but also detached, even isolationist and mystical. By many accounts, Time Warner itself was a troubled company composed of nearly autonomous and constantly squabbling fiefdoms.

Its motion picture division, Warner Bros., was established and reliable; the record division, Warner Music, highly lucrative on the strength of records in the "gangsta rap" genre (drawing political and social criticism). HBO, run by Fuchs, generated steady cash flow, as did the cable division. The print side, with the well-respected *Time* magazine and *Sports Illustrated* franchises were stable. But it was immensely leveraged—$16 billion in debt. The 1993 sale of 25.5 percent of TWE to US West was designed, in part, to bring in much needed dollars to reduce debt and fund improvements in the cable systems. The massive obligation dragged on the stock price and the internecine warfare between divisions helped make the company a potential target for a hostile takeover. Time Warner did not appear to have a clear direction and many questioned Levin's leadership.

Levin's answer, which critics characterized as an attempt to buy time and distract from the company's growing problems, was to fly to Bozeman, Montana, in August 1995 and meet with Turner and his wife, Jane Fonda, at their sprawling ranch. Levin offered to buy all of Turner Broadcasting for $7 billion in Time Warner stock. Turner would personally get 11.3 percent of the merged company, putting his own net worth at about $2.5 billion. It was an offer Turner decided he could not refuse. On September 22, 1995, Levin and Turner announced the deal.

Turner was made Vice-Chairman of the corporation and given a seat on the board. He was largely taken out of the daily affairs of company management and, by his own description, was a corporate officer without portfolio. He quickly became dissatisfied with the post, but, at the same time, he was enormously wealthy, so much so that in September 1997 he pledged to donate $1 billion over ten years to the United Nations.

To seal the Turner acquisition, Levin had to get by Malone. TCI still controlled 21 percent of Turner Broadcasting and held veto power over any large-scale transaction. Levin, in effect, bought him out, paying a premium over what Time Warner offered other shareholders for TCI shares, as well as penning long-term carriage agreements guaranteeing TCI stable rates for the Turner networks. The sweetheart deal angered Turner and other TWE shareholders, prompting an avalanche of lawsuits from U.S. West, Continental Cablevision, Comcast, and others. The government also feared the concentration of power created by an alliance between Time Warner and TCI. Only after bruising court battles[63] and a promise to the government that Malone would be a passive investor, was the purchase approved and completed in October 1996.

More Consolidation

The result of this surge in MSO sales was a further constriction in the ownership structure of the industry. Large MSOs had begun buying other large MSOs; more of the country's cable systems were being held by fewer companies. *Cablevision* magazine in 1996 reported that, for the first time, the number of MSOs had shrunk to below 200 because of the acquisition binge.[64] The contraction was vividly illustrated in the trend in industry concentration ratios. Reviewing almost twenty years of data, analyst Sylvia Chan-Olmsted concluded that cable was more horizontally concentrated than it had ever been.[65] Ratios for the top four (Top 4 CR) and top eight (Top 8 CR) firms nearly doubled between 1977 and 1995, from 24.5 percent to 49.4 percent (Top 4 CR) and 36.4 percent to 63.6 percent (Top 8 CR). The HHI had tripled.[66] By 1996, the top five MSOs controlled 66 percent of all subscribers; the top twenty companies served 85 percent of all cable customers. TCI alone held a 20 percent market share in 1995.

Congress had attempted to rein in cable consolidation, but those efforts were drowning in litigation. As part of the 1992 Act, Congress instructed the FCC to consider ownership limits on cable. In response, the Commission, in 1993, proposed limiting subscriber reach for any one cable company to 25 percent of the national cable audience (the same cap in effect at the time for broadcast television). After further and sometimes heated debate, the Commission settled on a 30 percent cap.[67] The FCC also looked at vertical integration, establishing a 40 percent limit on the number of wholly or partially owned program channels a given operator could carry. Finally, the FCC proposed prohibiting cable operators from purchasing SMATV or MMDS properties.

Over the next several years, the FCC ran the anticipated legal roadblocks with difficulty. In 1993, a federal district court struck down the 30 percent ownership cap.[68] In May 2000, the District of Columbia Circuit Court reversed that decision, affirming the constitutionality of the ownership clause of the 1992 Act and the FCC's authority to promulgate the rules.[69] In February 2001, the U.S. Supreme Court declined to hear an appeal. A week later, however, the District of Columbia Circuit Court, while reaffirming the Commission's authority to promulgate ownership rules, nonetheless struck down the rules they had established.[70] The sum of the legal machinations meant that cable operated for years without any substantial restriction on horizontal integration.[71]

Political Convergence

In 1984, lawmakers had severed cable's regulatory bonds, trusting to the forces of the marketplace and emerging communications technologies to ensure quality service, stimulate the creation of new programming, and hold consumer prices at competitive levels. It had not worked out. A decade later, they decided to try again.

Those who remembered may have appreciated the failure of the previous attempt, a consequence of a failure of the technologies themselves. But that was years ago. By the early 1990s, it appeared as if high-powered DBS might be viable, the telephone companies were actively probing the flanks of cable, and, most importantly, convergence and the information highway were perceived as pouring the foundation for a new explosion in integrated communications service.

At the head of the political army marshaled to promote the convergence agenda was newly elected Vice President Albert Gore. Gore saw himself as one of Washington's most knowledgeable and experienced proponents of communications technology, and few would dispute his experience in the field.

At the outset of the new administration, Gore determined that his vice-presidential specialties would be the environment and communications policy. He became a born-again champion of new communications technology, trumpeting its potential to usher the United States and then the world into the information age and at the same time crush the old technology oligopolies that had dominated the telephone and cable industries. According to Reed Hundt: "We wanted to permit the cable companies to compete in the telephone business, and the telephone companies to sell video, encouraging these industries to rebuild the networks to sell multimedia content against each other."[72]

Within a few months of taking office, the Clinton-Gore administration organized an Information Infrastructure Task Force (IITF) chaired by Secretary of Commerce Ronald Brown. In September 1993, the IITF released its report, the *National Information Infrastructure: Agenda for Action*.[73] Capitalizing on the emerging computer network and echoing the ideas articulated by Licklider and others in the 1960s, it conceptualized a broadband, digital, fully integrated, open and universally accessible communications network— a national information infrastructure (NII). Benefits would include the improvements in health care, education, and commerce. The system would be supported by the government largely through the National Science Foundation (NSF), but over a few years move toward near-total privatization, with only those regulatory safeguards necessary to ensure fair and uniform access and pricing. Gore almost immediately began his NII publicity campaign. In speeches at the National Press Club in Washington in December 1993 and at UCLA in January of 1994, he outlined the administration's plans, advancing an old metaphor in the process.[74]

In 1970 (as noted in Chapter 6), Ralph Lee Smith, drawing on the model of the nation's commitment to the construction of the interstate highway system in the 1950s, called for the creation of a national "information highway." Cable television, declared Smith, would serve as the backbone of an "electronic highway system, to facilitate the exchange of information and ideas."[75]

One of the key political figures responsible for the development of the national highway project after World War II was the powerful U.S. Senator from Tennessee, Albert Gore Sr. Albert Gore Jr., now Vice President of the

United States, saw himself standing on his father's shoulders as he advocated the creation of a twenty-first century public highway system, not of physical transport but of communication. In his National Press Club address, Gore invoked the ideas laid down decades before by both his father and Smith.

One helpful way is to think of the National Information Infrastructure as a network of highways much like the Interstates begun in the 1950s. These are highways carrying information rather than people or goods. And I'm not talking about just one eight-lane turnpike. I mean a collection of Interstates and feeder roads made up of different materials in the same ways that roads can be concrete or macadam—or gravel. Some highways will be made of fiber optics. Others will be built of coaxial or wireless. But—a key point—they must be and will be two-way roads.[76]

In the latter sentiment, Gore was underscoring his commitment to extending the information superhighway to all levels and classes in society, with emphasis on education and universal service. In a March 1994 speech before the World Telecommunication Development Conference, Gore took the vision international, calling for creation of a Global Information Infrastructure (GII). From his bully pulpit, he helped organize efforts at the NTIA, in Congress, and at the FCC. FCC Chairman and Gore ally Reed Hundt later explained, "My duty, as I saw it, would be to fulfill Al's vision for the information highway."[77] To do so, however, Hundt would need to modify FCC philosophy and work with a cable TV industry that had developed a robust hostility toward the Commission, generally, and himself, in particular.

Going Forward

Six months after the Bell Atlantic-TCI implosion, John Malone sat down for an interview with *Wired* magazine, a glossy new periodical devoted to coverage of information age news. He appeared on the cover dressed in the garb of an "Infobaun" warrior, playing off the image of actor Mel Gibson in the film "Road Warrior," complete with leather jacket and sawed-off shotgun. Malone was asked by the interviewer, who would be building the information superhighway? "Us. We're the guys building it," replied Malone. "We've got 35 percent of the country done right now... And by the end of [19]96, we'll be completely done in terms of fiber and coaxed deployment—terrestrial network that is the superhighway" adding later in the interview that he would make a commitment to Vice President Gore to complete the job by 1996. "All we need is a little help you know, shoot Hundt. Don't let him do anymore damage."[78]

Not everyone saw the humor in Malone's comment and Hundt supporters in Washington denounced what they characterized as a threat on the life of a government official. Malone later apologized to Hundt; his sentiment had been expressed, by his own admission, poorly. But Hundt knew that it was only the tip of an iceberg of industry animosity calved by the Commission's 1994 rate rollback.

Hundt characterized reaction to the FCC's second round of rate cuts as "the firestorm" that "nearly destroyed my chairmanship,"[79] and it, along with conversations with Vice President Gore, caused him to revise his thinking. In Hundt's new view, the consumer had benefited only minimally from the rate reductions. "Most Americans, indeed, never knew that their cable bill had been lowered by, say, a dollar a month," and "much of the savings went to buy an additional premium channel."[80] Meanwhile, investment in competitive services had been, in his new analysis, substantially harmed. The revised FCC strategy, more harmonious with White House policy, would be to encourage investment that stimulated competition.

In November 1994, the FCC declared that its previous rules may not have left sufficient incentive for operators to develop new services and expand capacity. It was adopting what it called "going forward" rules that would allow cable operators to create tiers of programming for new services at generally unregulated rates.[81] Operators were not permitted to move existing services to the new product tiers (NPTs) or change rates for existing basic or extended basic services if they added NPTs. Operators were not required to obtain regulatory approval when adding an NPT, but did have to notify the FCC of the addition.

In addition to permitting unregulated NPTs, the Commission also revised its rule for adding channels to existing service tiers. Operators were permitted to charge $0.20 per month for each extra channel added to regulated tiers, up to six new basic channels over a three-year period.

On the strength of the going forward regulations, a number of planned cable networks that had been delayed by the impact of rate regulation and retransmission consent agreements began operations. In November 1994, Fox unveiled its own classic movie channel, Fox Movie Channel, drawing from the vaults of 20th Century Fox. The next month, Sony Pictures Entertainment launched its Game Show Network, featuring decades of syndicated material. Arnold Palmer helped organize MSO participation (including Comcast, MediaOne and Cablevision Systems) to create the Golf Channel in January 1995. C/Net: the Computer Network opened in April 1995. Cable veteran Glenn Jones spearheaded the creation of a new country music station, Great American Country, in December 1995.

The Commission negotiated one additional compromise with the industry. The FCC's rules required the Commission to respond to consumer complaints about service and rates, and it took little time for the FCC's mailboxes to overflow with letters. To resolve the avalanche of cases generated by the complaints, the Commission adopted a "social contract" approach, entering into agreements with individual cable companies designed to resolve pending cases. Under the social contract agreements, companies were required to upgrades systems, reduce rates, and provide additional public benefits, as negotiated in each case. Continental Cablevision signed a social contract with the FCC in August 1995, agreeing to invest $1.35 billion in system improvements and offer refunds to subscribers in specified rate complaint cases. Time Warner did the same, pledging $4 billion in system upgrades, $4 million in customer refunds, and free service to public schools. Comcast, Cox, TCI, Cablevision Systems Corp., and others entered into similar agreements with the Commission, clearing off many of the pending complaints against the industry.

The Commission's late 1994 actions were the regulatory manifestation of the larger White House policy designed to repaint the landscape of telecommunications law, open markets, and speed deployment of Gore's information highway. FCC actions, however, were a mere sideshow compared with the main act, which was being played out in Congress.

The Telecommunications Act of 1996—If at First You Don't Succeed

The White House wanted to remove barriers to entry in historically walled business sectors, specifically telephone and cable services. The goal was increased competition, investment, and the creation of the broadband, twenty-first century info structure.

In many ways, the administration's position looked a great deal like the laissez-faire, free market philosophy of the Reagan-Fowler years. Clinton and Gore, however, sought to preserve safeguards for consumer welfare and especially sought to create regulations that extended the benefits of an online world to the socially disadvantaged. One of the main thrusts of Gore's initiative was to make sure schools, libraries, and similar public institutions had access to the "Infobaun," even if the industries themselves had to pay for it, and they would. The administration also wanted provisions that would prevent one industry segment, such as the RBOCs, from entering, for example, the long-distance sector, without the reciprocal ability of the long-distance companies to enter the local loop business. It was an attempt to maintain a balance of industrial power while encouraging competition, and included efforts to limit concentration of ownership.

The emergent view held that competition between communications providers might better serve the needs of the society than heavy governmental oversight. It was not by any means a universally shared position. Concerns about the monopoly power of global business interests, such as AT&T, fed a long and vigorous debate over the appropriate relationship between government and the communications industries. In the end, however, the majority political sentiment leaned toward relaxation of control.

The RBOCs eagerly, albeit less altruistically, embraced the general concept. Most observers understood that their interest in video, while real, was not their primary motivator. Competing against cable, claimed some, was a public relations Trojan horse designed to tap political energy directed against cable and mask their true objective, which was entry into the long-distance market.

Long distance was a multibillion dollar market. In the mid-1990s, AT&T alone had annual revenues of more than $50 billion, $20 billion more than the entire cable industry. And, telephone service was the core expertise of the RBOCs; their technical and business structures were established and their experience and expertise nearly a century old. Cable television, on the other hand, was foreign country, with programming, business, consumer, and technical characteristics almost completely unknown to the old world Bell companies. The RBOCs sought access to the long-distance business, and interexchange carriers, most especially AT&T, wanted the freedom to provide local loop service.

Cable had few options in the face of this ideological sandstorm. Deregulation certainly promised strong benefits for cable, chief among them the possible elimination of the hated price controls, but it also meant the possibility of competing head-to-head with the telephone industry. That notion was unsettling, but industry leaders knew the political train was leaving the station and they could be on it or under it. The NCTA chose a new president in 1993, replacing the controversial James Mooney. Decker Anstrom had served previously as Executive Vice President of the Association and was liked and respected. He proposed that the industry "embrace competition" and attempt, as it had in the past, to do what it could to shape legislation he knew they could not prevent. Cable, suggested Anstrom, would support deregulation, leaving the telephone industry to enter the cable business with the *quid pro quo* that cable operators would be free to offer telephone and data services in a regulatory neutral environment. He presented his proposal to the NCTA Board of Directors in the late summer of 1993, later recalling, "They took a deep breath, looked around the room, and said 'Okay.'"[82]

Bills to deregulate the telecommunications industries began hitting the House and Senate floors in 1993.[83] Republicans, especially those closely aligned with the telephone industry, offered their own ideas about how to reshape the world of electronic communication. The Democrats, following the lead of the White House, pursued a more nuanced path, but there were more similarities than differences in the overall tenor of the bills, which, in one form or another, called for removal of decades-old business restrictions and requirements.

In November 1993, Congressmen Jack Brooks (R-Texas) and John Dingell (D-Michigan), both closely aligned with the local Bell operators, introduced HR 3626, designed to open the long-distance market to the RBOCs with minimal restrictions. A competing bill, HR 3636, was introduced by

Congressmen Ed Markey (D-Massachusetts) and Jack Fields (R-Texas). Representing a more balanced administration position, it would have permitted RBOC entry into long distance, but with much more severe restrictions. It also proposed to allow telephone company provision of cable service, but only on a common carrier basis. Additional articles protecting universal service and requiring reduced pricing for service to schools and libraries also were part of the package. Differences between the bills were negotiated and they merged as HR 3626.

Senate bill, S. 1822, was introduced by Senator Ernst Hollings in February 1994.[84] More in keeping with Clinton administration policy goals and harmonious with the Markey-Fields' measure, it included provisions to protect competition and the consumer interest. Before RBOCs would be permitted to enter the long-distance business, real competition in local services had to be demonstrable. Moreover, that competition had to be based on the existence of competing firms using some of their own equipment—known as facilities-based competition—as opposed to alternative providers leasing and reselling existing ILEC facilities. The bill also required the provision of broadband services to schools (K-12), hospitals, libraries, museums, and health care facilities, and it contained cross-ownership restrictions limiting joint cable-telco control of services in a given geographic area.

The restraints on RBOC latitude in the long-distance market brought opposition from the Baby Bells and spawned an alterative, RBOC-friendly bill, S. 2111, from Senator John Breaux (D-Louisiana) The RBOCs also were a long-time ally of Senate minority leader Robert Dole (R-Kansas), who helped organize opposition to the Democratic measure. Moreover, as the country moved toward the November 1994 elections and it appeared that Republicans might make gains in Congress, Dole initiated an across-the-board political barricade of all Senate legislation forcing Hollings to withdraw S. 1822.

November was not a good month for Democrats. When the 104th Congress convened in 1995, the Senate and House both were in the hands of the Republicans for the first time in fifty years. Flush with unfamiliar power, the new majority set out to advance its own legislative agenda, including work on telecommunications reform.

Senator Larry Pressler (R-South Dakota) introduced the Telecommunications Competition and Deregulation Act of 1995, S. 652. It called for a complete rewrite of the country's foundational Communications Act of 1934. On telephone service, the bill would preempt state and local constraints on new providers, such as cable, from offering local telephone service and require the LECs to provide interconnects on reasonable terms. It also called for elimination of most cable rate regulation passed by Congress only three years before in the 1992 Act. It also abolished local cable-broadcast cross-ownership rules. It would permit telephone companies to own cable systems in their own service area. On June 15, 1995, the Senate voted 81 to 18 to pass the telecom reform bill.

The process was messier in the House, now led by fiery and controversial majority leader Newt Gingrich (R-Georgia). Democratic-backed bills from the previous Congress were introduced, but quickly killed. HR 1555, "Communications Act of 1995," was introduced in May and amended in August. The changes, inserted by Gingrich and the House majority leadership, substantially loosened restrictions on the RBOCs, permitting them easier and faster entry into new markets and giving them greater flexibility in charging for access to their local loop.

For cable, the House bill called for elimination of rate regulation immediately in smaller systems (communities under 10,000) and after eighteen months for all other systems. Rates for some basic tiers of service would remain regulated. A clause that permitted telephone companies to offer cable service only on a common carrier basis was dropped. The bill won House approval on August 4, and Democrat efforts to modify the legislation moved to the Conference Committee.

Clinton threatened to veto the final bill unless it was adjusted to address concerns about rate regulation and concentration of ownership. The administration also sought to preserve the clause requiring connection to classrooms and libraries included in the Senate bill but not in the House version. The combined measures constituted a large and complex regulatory framework that covered subjects from the implementation of digital broadcast television, to indecent content on the Internet to cable rates and telco competition. Most of the industries affected had something to gain from the legislation and few wanted to see it smothered in the cradle. Compromises were reached, strengthening consumer and competitor protection for RBOC entry into long distance, slowing down cable rate deregulation, and maintaining the Senate's universal service language. On February 1, 1996, the House and Senate passed the Conference Committee bill and, on February 8, President Clinton signed it into law in a ceremony held in the ornate reading room of the Library of Congress.

The Telecommunications Act of 1996 was a dramatic, lengthy, and complex reconfiguration of U.S. telecommunications policy. It touched cable, broadcasting, and telephony in areas ranging from ownership to business operations to programming, but its spine was constructed to improve and extended integrated service by fostering competition.

Before the Act, ILECs were permitted to provide local service within their territories, but could not provide long distance. Long-distance companies, such as AT&T, Sprint, and MCI, could not offer local service. The new law removed those barriers to entry. The RBOCs were freed to offer long-distance service outside their regions. They also could offer long distance inside their service regions using a separate affiliate, but only after opening their own facilities to interconnection with competing providers and satisfying an FCC checklist of criteria. To encourage competition in the local loop, the legislation required ILECs to permit connection to their network facilities and additionally mandated that they sell service to new entrants at wholesale prices, giving new companies an opportunity to resell the service at retail.

The Act gave cable the legal authority to begin offering telephone service, while providing a reciprocal authorization to the telephone companies to offer video service. The ban on cable-telco cross-ownership was largely maintained in most service areas, but lifted in rural areas and small towns.[85] The FCC's Video Dial Tone provisions, which prohibited a telephone company from programming the local system, were replaced by a concept called Open Video Systems (OVS). Under OVS, a telephone company could create a local cable distribution platform and run its own programming on up to one-third of the channel capacity, leasing the remainder to an outside entity. OVS additionally insulated the company from traditional cable franchising requirements. Both industries, along with others, were freed to offer data communications and Internet access to the business and home markets.

Attached to these broad guidelines was a myriad of conditions and qualifications designed to shepherd the industries onto a level competitive playing field, although many of the regulatory mechanisms were hotly contested in the courts. Critically for the industry, the new law relaxed rate regulation. Small systems—those with no ties to major MSOs and with fewer than 50,000 subscribers—were freed from rate control immediately.[86] Rate regulation for any tier other than basic on all systems would end on March 31, 1999. Systems that faced competition could request removal of controls before that date. Effective competition in the 1996 Act was defined as service offered by any comparable, unaffiliated (non-MSO), multichannel provider, except DBS.[87] Once again, regulators believed competition would keep rates in check.

Congress still worried about the ability of cable operators, through their ownership interests in programming companies, to restrict access to popular programming to potential competitors. Therefore, one of the few provisions of the 1992 Act left intact by the new law was that of preventing programmers from discriminating against new entrants by restricting access to their product.

The Act substantially deregulated broadcasting, as well. License periods were extended to eight years for both radio and TV. Ownership restrictions were eased and the ban on TV network ownership of cable systems was removed. The Act also provided additional spectrum space to allow broadcasters to begin development of digital broadcasting services, especially HDTV.

The bill contained provisions governing indecency in the new media. Title V of the Telecommunications Act, known more commonly as the Communications Decency Act (CDA), strengthened restrictions on indecent programming on cable and via the Internet. The CDA required any multichannel provider to scramble the audio and video signals of all sexually explicit,

adult-oriented programming to prevent reception of the signals by nonsub-
scribers. The CDA also established a schedule of fines and jail time—up to
two years—for transmitting "indecent" or "patently offensive' material to
minors over the Internet. A separate clause engineered in the Conference
Committee mandated establishment of a national TV rating system that included a requirement that all TV sets contain a "V-Chip" that would give
parents the power to block offensive programming from the home. The system used special codes embedded in programming and a chip built into the
TV set. The "V" stood for violence, although parents and regulators sought
to protect young audiences from inappropriate language and sexual themes
as well.

The Telecommunications Act of 1996 looked as if it would usher in a
new and much more competitive age for cable television. The question was
whether cable would be ready.

Dr. No's Basement

In the wake of Malone's 500-channel promise and spurred by the prospects
of invigorated competition, especially from DBS, TCI set about building a
state-of-the-art digital programming distribution system in Littleton, Colorado, not far from TCI headquarters. The $100 million National Digital
Television Center was capable of converting analog to digital programming
from nearly any source and was designed as the base of operations for TCI's
proposed Headend in the Sky (HITS) service. HITS would beam via satellite
digitally compressed cable channels to systems around the country, helping
to provide even small operators with the digital resources necessary to remain competitive in a convergent video universe. Industry wags, noting the
Center's high-tech control rooms and walls of video monitors, dubbed the
facility "Dr. No's Basement." At the signing of the Telecommunications Act
of 1996, the facility was yet to be powered up, however. Digital signals were
of little use to operators without digital set-tops boxes to decode them, and
more than three years after Malone's proclamation of a digital age for cable,
the boxes were yet to appear. The new legislation appeared to open the doors
for well-financed rivals and, without the boxes, cable was entering the battle
only partially armed. Malone began making phone calls to the other cable
CEOs. He wanted to talk about organizing a business expedition to seek
help. He was looking to the West and the home of the silicon revolution.

13

"What's Gonna Be Next?" (1997–2005)

> *You know, I have around $10 billion in cash.*
> *I could do that.*
>
> —BILL GATES, 1997[1]

ill Gates, the richest man in the world, was having dinner with friends. The cable industry had about $38 billion in combined subscriber and advertising revenue in 1997. Gates himself was worth about $40 billion. Microsoft was awash in cash and Gates had $10 billion he did not quite know what to do with. One of the options was to put some of it into cable television and that was exactly what Brian Roberts, the young CEO of Comcast, had just suggested between bites. It was April 1997. Assembled at the table were Gates, Roberts, John Malone, Cox Communications head Jim Robbins, Joe Collins of Time Warner Cable, and CableLabs chief Richard Green. It was a dinner that would take on legendary status in the cable industry.[2]

The cable moguls were in Seattle on a pilgrimage and marketing tour. In the spring of 1997, the industry, once again, was in trouble. It appeared that the mammoth telcos were gearing up to compete for video subscribers. DBS was nibbling away at cable's customer base. Malone's famously promoted digital boxes had become embarrassing hyperbole and, in the summer of 1996, Intel CEO Andy Grove had pronounced cable's national infrastructure too old and feeble to handle the demands of high-speed Internet traffic. Wall Street had lost confidence in the industry and cable stocks were stumbling. Cable no longer was seen as instrumental in the construction and operation of the digital future.

The leaders of the cable industry gathered at the table that night knew they needed to forge business and technology alliances with the hardware and software firms that were leading the social and economic phenomenon coming to be known as the dot-com boom. In early 1997, Malone helped organize a tour of the West to talk with the leaders of those industries. The stops included California's Silicon Valley and the Microsoft campus in Redmond, Washington. One of the goals was to convince the Internet and computer community that cable's technology was the best broadband route into the American home. They had gone with charts and graphs and numbers to illustrate the hundreds of millions of dollars they were spending for Internet-friendly upgrades.

At dinner following the business meeting, Gates began asking questions. It started out as light banter. Gates said he saw merit in the industry's pitch. Roberts suggested, half-jokingly, "Gee Bill, if you believe in us, why don't you buy 10 percent of everyone in the room tonight?" Everyone smiled, then Gates asked, more seriously, "How much would that cost?" After a little mental calculation, Roberts said, "I don't know, let's say $5 billion."

"You know," said Gates, "I have around $10 billion in cash. I could do that."

The response brought more light laughter; the cable leaders did not seem to be taking Gates seriously, and the conversation turned to Gates' vacation plans. After a few minutes of discussing his upcoming trip to the Amazon, Gates returned to the cable question.

"Would there be any regulatory problems?" he asked. They talked a bit more about it, but still it was considered casual dinner conversation and later Malone kidded Roberts, "Brian, it sure was sad to see you on bended knee asking Bill Gates to bail out the cable industry."[3]

Returning to Comcast corporate offices in Philadelphia the next day, however, Roberts found a note waiting for him. Microsoft's Vice President for Corporate Development, Greg Maffei, had received an e-mail from Gates asking him to follow up on the cable investment idea. A month later, Roberts flew back to Washington to talk with Microsoft. In June 1997, Comcast and Microsoft had an announcement—Microsoft was buying 11 percent of Comcast for $1 billion.

It was a turning point for the cable industry. Technical and financial observers had lost faith in cable, but Bill Gates was a Wall Street demigod, and investment dollars followed his lead. Gates, of course, had his own agenda. He had determined that the Internet was becoming a, or perhaps the, dominant form of global electronic communication and to move increasingly large volumes of materials, especially video, over the digital platform, high-speed broadband functionality would be essential. He saw television and the Internet merging in classic convergence style and, shortly before the cable visit, had spent $425 million to purchase rights to WebTV, a technology that would allow subscribers to surf the Internet using their television rather than their computer. WebTV and Bill Gates were going to need cable. It was fortunate for the industry. Over the next week, Comcast's stock rose

17 percent, and the rest of the industry followed in tow. Standard and Poor's index of cable stocks climbed 13 percent in a matter of days. For better and worse, cable was now a dot-com player.

This chapter considers cable television development from about 1996 to 2005. It examines the impact on the industry of the dot-com stock bubble of the late 1990s and the consequences of the bubble's burst in 2001. It looks particularly at the relationship between cable, the telephone industry, and the DBS companies as each attempted to create integrated voice, video, and data businesses, sometimes in direct competition with one another and sometimes through mergers and acquisitions of historic proportions. Considered more broadly and once again are both the evolutionary nature of change in cable communications and the influence exerted on that evolution by society's conceptual notion of what the medium was and, more importantly, what it could become.

The Dot-Com Boom

By the mid-1990s, the information superhighway was the Internet and increasingly that portion of it known as the "World Wide Web," and in almost every dimension, the web was exploding. Use was expanding exponentially. It was called the "dot-com boom." In 1996, about 35 percent of the nation's homes had a computer and as many as 20 percent of all households had a modem; some 13 percent used the Internet and its World Wide Web.[3] It was estimated that people were signing up for online services at a rate of 150,000 per month.[4] Thousands of web sites were being built daily. In 1996, there were about 627,000 registered web sites; by 1998, more than 3.5 million sites were registered.[5] A new industry sprouted just to handle the business of creating company web sites. By the end of the decade, an estimated 42 percent of American adults used the web on a regular basis and data traffic was doubling every ninety days.[6] The Internet was generating $300 billion in annual revenues in 1998, more than the $270 billion in the telecommunications sector.[7]

A staggering amount of venture capital fueled what became a dot-com frenzy. Speculation in technology stocks drove the market to heights few had dreamed of the previous decade. The Dow Jones Industrial average rose from around 4,000 in 1995 to more than 10,000 by the end of the decade. The NASDAQ, home of many of the new high-tech enterprises, including the digital industry giants Microsoft and Intel, moved up on an almost daily basis, climbing from 800 in early 1995 to around 4,000 by the end of 1999.

Electronic commerce, or "e-commerce," became a critical part of the business equation, as well as the title for books and college courses. To support the transition to an online global economy, telecommunications companies, both old and new, were pouring billions of dollars into construction of national and international high-speed networks. Fiber was going into the ground at a staggering pace and companies such as Sprint and Global Crossing were heavily promoting their glass lines for voice and data.

The social construction of technology was in overdrive. It channeled investment and colored political debate. Entrepreneurs with little more than a saleable idea and a good PowerPoint presentation could build a web site, lure venture capital, and become dot-com millionaires, sometime billionaires, overnight. Bill Gates, one of the most savvy and successful technology businessmen of all time, had now told wary investors, cable should be a part of their thinking.

In the last part of the 1990s, cable industry revenue would continue to climb, albeit at a slower pace. The success of DBS in luring away subscribers, especially while cable systems waited for the digital boxes, would cut into cable growth. Some programming success at the broadcast networks also would retard, but ultimately not stop, the slow seepage of viewership from broadcast to cable. Cable consolidation would continue with fewer but larger mergers as the pool of available MSOs shrank. Better news would come toward the end of the decade as the desperately anticipated digital boxes finally began to move off the assembly lines and into homes. Channel capacity followed, then more programming choice and, critically, expanded pay-per-view (PPV) choices.

In the repeating cycle of political and financial Blue Sky speculation, however, cable, Gates and the rest of the world, within a few years, would confront the clash between the myth of the technology and its real world viability. The dawn of the new millennium would bring a crash and burn story on a scale not previously seen in telecommunications. In the first three years of the new decade, the bubble would burst and billions of dollars in equity would simply evaporate, much of it belonging to the cable and telecommunications industries. Every market index would drop precipitously, and web sites, dot-com start-ups, and even major telecommunications corporations would go dark. Business scandals endemic to the frenzy of the age also would touch cable, and one of the industry's oldest families would find itself before a sentencing judge.

Even before the bottom dropped out, the convergence vision would begin to erode as companies moved from hyperbole to application. The telephone industry would discover that the business of providing entertainment to millions of Americans was significantly more taxing than they thought, especially when attempting it with almost no experience in the business. The cable companies would find penetration into the telephone business similarly daunting. More critically, the digital boxes, on which the new era of program choice and interactive services largely depended, would become a case study in "vaporware" and, when they finally did ship, would be years behind the promised schedule. Ever iconic, TCI in many ways would mirror the industry-wide turmoil. For John Malone, the period between the collapse of the Bell Atlantic deal and 1998 would be one of his most challenging.

TCI

Rumors ricocheted around the industry in 1994. Some suggested John Malone was perilously ill; others were just mystified by his disappearance from the cable scene. He was not coming to the office, not going to the conventions. Day-to-day TCI operations were left in the hands of COO Brendan Clouston. Malone's biographer, Mark Robichaux, explained that the TCI chief fell into a deep depression after the failure of the Bell Atlantic merger and retreated—physically, emotionally, and psychologically—to his 200-acre estate near Boothbay Harbor, Maine, and was spending a lot of introspective time on his custom-made yacht, Liberty.

Investors wanted information on TCI's plans after the death of the merger, but Malone was not around to provide it. The company said it was building the information highway, and Clouston spent hundreds of millions of dollars upgrading TCI technology, purchasing state-of-the-art billing systems, and hiring marketing consultants to prepare for the rollout of integrated, digital service. But the January 1994 deadline to implement the 500-channel promise had come and gone, and one of the company's most significant problems, one shared by the entire industry, was the delay in the digital box technology that was supposed to make all the magic happen.

The promise and the hope, again, had outpaced technology. The boxes, which required the sophisticated ability to handle compressed digital video, voice, and data, proved to be more difficult to engineer and produce than the business and policy community had anticipated. A delay in the settlement of industry standards for digital compression and changing functionality requirements from TCI also hindered development. Once the hardware and software designers began to clear the hurdles, the production costs were found to be far higher than what made sense for a consumer market. At a 1996 trade show, a frustrated Malone chided General Instruments with a joke about a woman who had been married three times but was still a virgin. The first two husbands perished before the marriage could be consummated and now she was wed to a General Instruments' salesman who "just constantly tells me how good it's going to be when I get it,"[8] Malone told his audience. And while they waited impatiently for the boxes, TCI's $100 million National Digital Television Center in Englewood sat dark and unused, a silent testament to the failure of the digital dream.

Clouston, nonetheless, promised investors that TCI was positioning itself to be a leader in the post-Telecommunications Act world of convergence, and he was spending thunderously to get there, building TCI's debt to a record $15 billion. The TCI facade began to tremble, however, on October 25, 1996, when the company announced its third quarter earnings. Instead of the increases in subscribership and cash flow that Clouston had promised would emanate from the convergence plans, and from a widely criticized 13 percent rate hike the previous June, TCI had lost 70,000 basic and 308,000 pay customers, many to DBS. Cash flow was up, but only by a disappointing 3 percent. Investors charged for the lifeboats on the news, selling TCI shares as they went. The stock was the most heavily traded on the NASDAQ that day, plummeting to a fifty-two week low of $11.56, down 40 percent from its 1995 high.

In 1973, Jack Kent Cooke flew from his California home to rescue a badly damaged TelePrompTer Corp. Twenty-three years later, Malone, alerted to the pending financial news, flew from Maine to Denver for similar reasons and with similar urgency. In Kent Cooke-like style, the cost-cutting began even before the public figures were announced. By the end of October, Malone had suspended all equipment purchases for the remainder of the year. In December, he announced plans to lay off 2,500 workers, about 6.5 percent of the workforce. The four corporate jets were sold and Malone said company executives, including himself, would be taking hefty pay cuts. He sold Carolco Pictures to News Corp. for $50 million.

The company also announced plans in December to drop a half dozen cable networks it claimed were under-performing and to add only networks that would pay for carriage. Among the victims were Viacom's MTV and VH1. It was a cultural and a business mistake. The music services, especially MTV, were considered cable staples, with powerful supporters in the music and entertainment industries, and Viacom's Sumner Redstone was as tough a corporate street fighter as any cable pioneer. Viacom orchestrated anti-TCI rallies in various cities, including a large one in TCI's backyard, Denver, featuring pop music stars such as Jewel and John Cougar Mellencamp. It was a public relations disaster for TCI. Redstone made it personal, telling the press that Malone had promised him carriage of the affected networks and that he did not believe Malone would renege on such a promise. A week after his comments were published, the networks were reinstated.

All of this was made more difficult and much sadder when, on November 15, 1996, company founder and industry legend Bob Magness died at the age of 72 of lymphoma. Complicating an already difficult period, Magness had failed to leave a formal, detailed will (to the surprise of many) and, over the course of the next several months, a family struggle ensued over the estate, estimated at $1 billion. The conflict pitted Magness's two sons, Kim and Gary, against his second wife, Sharon Magness.[9] So, in addition to the emotion turmoil—Malone and Magness had always been close—Malone was confronted with the possibility that the heirs might need to sell TCI stock to pay the estate taxes. Magness and Malone had always controlled the super-voting shares and such a sale could lead to Malone losing control of TCI.

In typical Malone fashion, the financial engineer worked out a complicated scheme in which he paid between $125 and $200 million (reports varied) for the right to buy family shares if they were ever sold, with a proviso giving Malone authority to vote those shares. The heirs could pay their taxes and Malone retained control of the company. Lawsuits between the widow and the sons were settled out of court, for an undisclosed sum, in late 1997.

In the meantime, Malone continued to work to turn the company around. He liked and had supported Brendan Clouston, but felt a different management approach was now needed. In February 1997, he went to Intermedia Partners' Leo Hindrey Jr., a cable veteran and Malone ally. Hindrey's style was something different for TCI, the well-known hardball player of the industry. Calming, friendly, and conciliatory, Hindrey soothed wounds, rehired some employees who had been let go, and reorganized the company, stressing local control and customer service.

In 1997, the company reorganized a number of its joint ventures, selling and transferring systems in deals that involved Cablevision Systems, Time Warner, MediaOne, and Bresnan Communications, among others. It was able to pay down $5 billion in debt and gain equity in a number of other MSOs as a result, but the loss of subscribers dropped the MSO into the number two position for the first time years.

Cable Retrenchment

In April 1997, Time Warner announced plans to suspend operations at the much-vaunted Full Service Network in Orlando, closing the system officially in September 1997. Time Warner characterized the shutdown as the end of a successful and informative experiment, and much was learned on the technology side. Financially, however, the lessons spoke to huge consumer apathy toward interactive cable and untenable operational costs. The early digital compression techniques were not yet ready for prime time and the cost of the cutting edge equipment was far beyond what made economic sense. As with television in the 1930s, digital TV remained "just around the corner."

For cable, failed promises, failed mergers, and failed technology were confronting a new world of cable-unfriendly regulation and potentially vibrant competition. At the close of 1996, while the broader stock market seemed stuck at full throttle, cable stocks were simply stuck. The Kagan cable stock index dropped from 1,506 to 937 between November 1993 and the end of 1996.

It was, however, the nadir. Gates' billion-dollar check to Comcast was a catalyst for renewed investor interest, and cable began to rebound. Gates's interest was supported by changes in technology and policy in the last few years of the decade. After years of frustration and failure, GI shipped its first digital boxes in October 1996. That month, TCI announced the beginning of digital service to customers in Hartford, Connecticut. Branded as ALL TV, the service featured 150 channels, including forty PPV. At the same time, it began operations at its National Digital Television Center, promoting its new "Head End in the Sky." Cable modems also were coming to market, and the rate deregulation of the 1996 Act promised an end to the onerous 1992 caps. With Gates' backing, cable could claim to play a major role as the premier conduit for information superhighway traffic, and even Gates' partner, Microsoft cofounder Paul Allen, would start buying cable systems.

Critical to the turnaround was the not-so-slow realization that, after all the fire and smoke, the telephone companies were not going to be confronting cable on the video battlefield. In fact, after a few exploratory passes, it appeared the telcos had decided their money and time were better spent elsewhere.

Telco Retrenchment

Early 1996 looked promising for Telecommunication Act backers. Bell Atlantic launched one of the telephone industry's most heavily publicized video trials, a commercial broadband, switched FTTC network offering voice, data, and video to the residents of Toms River, New Jersey. It began as a VDT trial in direct competition with an existing Adelphia cable system. Bell Atlantic contracted with programmer FurtureVision to provide seventy-seven channels of cable programming. Other content providers were offered space on the system as well. Bell Atlantic later purchased FutureVision, running the programming itself and operating as an OVS system under the 1996 Act. Adelphia responded with a furious rebuilding and marketing campaign. It upgraded to a 750-MHz system, cut prices, and began a massive promotion. The head-to-head battle was followed with keen interest by most of the industry.

Bell Atlantic also was testing wireless cable using digitally compressed MMDS technology. MMDS had been a TV stepchild for decades, but compression offered the possibility of conjuring a 200-channel digital system out of a twenty-channel analog one practically overnight and at a fraction of the cost of a coaxial system.

In 1995, Bell Atlantic and Nynex invested an estimated $100 million in MMDS provider CAI Wireless and announced plans for wireless cable operations in Boston and Hampton Roads, Virginia, beginning in 1997. Pacific Telesis purchased the Cross Country Wireless Cable system in Riverside, California, for $175 million with the stated goal of developing a multichannel MMDS platform to serve much of Los Angeles and Orange counties. And Bell South spent $12.1 million in early 1996 to purchase a bankrupt MMDS system in New Orleans that it intended to use as a test bed for wireless cable services. By 1996, 200 MMDS systems were reported in operation, serving 1 million subscribers and gathering in about $272 million in annual revenue.[10]

But, the technology and, perhaps, the interest were not there. MMDS, the tests demonstrated, still had rain fade problems. Even with compression, it could not compete with cable for channel capacity, and there was no return path, short of a standard telephone line, which would be needed for interactive capability. Before the close of 1996, Bell Atlantic and Nynex suspended their business agreement with CAI wireless, dropping plans for a 1997 launch of wireless cable in Boston and Hampton Roads. Pacific Telesis also scraped its $175 million deal to buy wireless systems in San Francisco, Seattle, San Diego, Tampa, and elsewhere, although Bell South continued to pursue MMDS, acquiring rights in Atlanta, Miami, and New Orleans.

It was more than just a technological issue, however. Within a year of passage of the 1996 Act, the telephone industry's entire business agenda appeared to be shifting. The idea behind deregulation was that telephone companies, such as U.S. West would begin offering cable television service, and cable companies, such as Amos Hostetter's Continental Cablevision, would be begin offering telephone service and both would offer Internet access. In theory, it had been the law that had prevented such competition from blossoming in the first place. Reed Hundt, Al Gore, and hundreds of congressmen and senators had forgotten one simple thing, however. Very few businessmen really wanted to compete. In fact, any alternative that led to prosperity was preferred. Competition meant lower prices, which meant lower dividends and frequently led to declining stock value. What business really sought, in most cases, was a monopoly position in a high-demand sector. For decades, cable and telephone operators had bathed in the milk and honey of nearly pure monopoly in services no American could or would easily do without. They had very little experience with true competition and less interest in learning. Four weeks after the signing of the 1996 Telecommunications Act, U.S. West and Continental therefore did the most reasonable thing imaginable—they announced their merger.

U.S. West, succeeding where Bell Atlantic had failed, purchased Continental for $10.8 billion. Continental served 4.2 million cable subscribers and had $1.8 billion in revenue. U.S. West provided telephone service to more than 14 million customers in fourteen states. It had a reported $11.7 billion in revenue in 1995. With its existing stake in Time Warner Entertainment, it now had access to more than 14 million cable homes as well.

Hostetter was one of the industry's pioneers, and like Alan Gerry, Sammons, and a number of others, he had determined that the cost and regulatory challenges of building a twenty-first century interactive broadband system were best assumed by companies of larger scale. He was taking the opportunity to cash in.[11] He ascribed the decision to the changing dynamics of business and technology and to the new law. "This is going to be game of large players that will occur at an accelerated rate," he explained. "Clearly, the passage of the Communications [sic] Act of 1996 has created a whole new world."[12] "Companies of size will be the ones that survive. It's going to be the waltz of the elephants, and you want to be sure you don't get stepped on."[13]

The assumption on the part of most observers in cable and in Washington was that the telephone companies would use their resources to pioneer convergence and create structural competition. In championing telecommunications deregulation in the early 1990s, the Baby Bells, by word and deed, had committed to that vision. In the months following passage of the law, however, whatever telco interest had truly existed in cable television seemed to evaporate as a mirage does upon approach.

Soon after the purchase of Continental, U.S. West shut down its seventy-seven-channel VOD experiment in Omaha as too demanding and expensive.[14] By 1998, it was getting cool even on the Continental enterprise. Apparently finding cable an awkward fit with its existing organization, it reorganized all its cable properties into a company called MediaOne in June, and then spun it off.

In 1997, SBC Communications purchased Pacific Telesis and began cutting back on all its video experiments. It pulled out of its MMDS operations and shut down its 8,000-subscriber full-service network trial in San Jose, California. It killed a PacTel cable project in San Diego and sold its Virginia and Maryland cable systems back to Prime Cable for $637 million.

The midwestern RBOC Ameritech had been one of the most aggressive in overbuilding cable systems in its territory. By 1999, it was under new management, however, purchased, as noted below, by SBC. Shortly after the acquisition, the not-so-Baby Bell announced it was leaving the cable TV business, shuttering or selling systems and franchises in 114 communities in Illinois, Ohio, and Michigan.

In their public pursuit of the consumer's cable dollar, the telcos also had created new programming companies, promoted as cooperatives tasked with obtaining contracts with the cable network community and developing new programming capacity. In 1994, Ameritech, Bell South, Southern New England Telecommunications (SNET), and GTE announced they would team with Disney to create "Americast." In 1995 Bell Atlantic, Nynex, and Pacific Telesis founded a program acquisition clone called "Tele-TV," hiring former CBS Broadcasting President Howard Stringer to head the firm. With a combined war chest of $800 million, the companies soon were signing contracts with program producers and most of the major cable programmers. By late 1996, however, the waters no longer were churning. The funding slowed, then stopped, and both companies went dark.

Capping the retreat in late 1998, Bell Atlantic called a halt to its ambitious Toms River FTTC system. When the wires fell silent, Bell Atlantic had spent more than $70 million to sign up about 2,600 of the potential 110,000 customers in the franchise area. The company did not even stick around to collect the $1,500 set-top boxes it had placed in its customers' homes. Abandoned Bell Atlantic subscribers gave the boxes to the Adelphia installers as they switched back to the cable provider. Adelphia stored the set-tops in a backroom, telling the press that the telephone company could have them back, "all they have to do is call." Or maybe, they mused, "we can donate them to the cable museum."[15]

The MMDS industry became collateral damage in the telco course change. In 1998, two of the major wireless companies, CAI Wireless and Heartland Wireless, filed for bankruptcy.

As noted, some critics had suggested that the telephone industry's professed interest in video had been little more than a smoke screen to mask its true ambition. The RBOCs, they had argued, wanted into long distance and AT&T sought local loop revenue. The video pullbacks lent support to their arguments, but the companies also discovered that they lacked the experience, expertise, and technology to compete in video. They knew the telephone business, but they did not know the cable business. That fact, plus the relative size of the two markets, led again to logical business conclusions.

Long distance was the primary target for the Baby Bells, but under the Telecommunications Act, they could enter the market only after competition had been established in their local service area. The FCC established severe guidelines—a checklist of items—necessary to demonstrate local competition, and the RBOCs were hard pressed to meet all the demands. They additionally accused the dominant interexchange carriers, AT&T, MCI, and Sprint, of stalling in contract negotiations and technical compliance, specifically to keep the LECs out of the long distance market.

The RBOCs determined that increasing their own corporate size and reach would significantly improve their opportunities in both the local and long-distance sectors and they wasted no time in doing so following passage of the 1996 Act. In April 1996, SBC announced its acquisition of Pacific Telesis, the RBOC serving California and Nevada. Two weeks later, Bell Atlantic, having recovered from its attempt to acquire TCI, announced its merger with Nynex. Under terms of the $25.6 billion deal, Nynex would surrender its name, while corporate headquarters were moved to New York City. The new Bell Atlantic would be the primary local loop provider for most of the northeastern seaboard. The original seven Baby Bells were summarily reduced to five, and the consolidation was not complete. In February 1988, SBC acquired the large regional company SNET and, a few months later, Ameritech, for $79 billion, leaving four remaining RBOCs: Bell Atlantic, SBC, U.S. West, and Bell South. Sweeping up some of the smaller pieces, Bell Atlantic purchased the largest non-Bell ILEC, GTE, in 1998 for $53 billion and then in 2000 renamed the corporation Verizon.

Consolidation also was taking place in long distance. With the success of MCI and younger start ups, such as Qwest and WorldCom, the business had become increasing competitive. Aggressive construction of fiber capacity in the long-distance sector and equally aggressive pricing and promotion drove down prices and hence revenues through the latter 1990s. The eroding long-distance business and a convergence-driven business climate pushed all the long-distance carriers to seek partners. MCI had made its fortune challenging AT&T starting in the 1960s, but by the mid-1990s had become nearly as bureaucratically bloated and complacent as its old nemesis.[16] In 1998, WorldCom, headed by Barnard J. Ebbers, purchased controlling interest in MCI for $30 billion. In 2000, Qwest purchased U.S. West for $48.5 billion.

By 1998, it was clear that the regional Bell systems were inwardly focused, with the various companies forming alliances to compete first in voice service—local and long distance—and secondarily in data. Video might come later, but no one was talking about when "later" might be.

Cablephone

Cable's inroads into telephone service progressed no better than the RBOCs efforts at video. TCI conducted trials with "friendly" users in Illinois and Connecticut. Jones Intercable provided service to a few apartment buildings in Alexandria, Virginia. Cox launched its commercial PCS service in San Diego in late December 1996. Its partner, Sprint, provided long-distance service. The MSOs maintained their interest in CAPs service. Teleport Communications Group was owned jointly by TCI (49.9 percent) and Cox (50.1 percent). Adelphia had a CAPs division called Hyperion.

Under the 1996 Act, cable, as well as other competitive local access providers (CLECs), could offer "facilities-based service" in which the company owned its own lines and switches, or they could offer "service-based" plain old telephone service (POTS) on a resale basis. The latter involved buying carriage and switching capacity from the incumbent LEC, at a discounted price, and reselling the service to the local customer. Congress had hoped that under the new Act, competitive local access facilities would be constructed and true competition brought to the local loop, but they were aware such facilities-based construction would take massive investment. The access and resale provisions were designed to accelerate competition. Cable confronted substantial challenges in both the facilities and the resale options.

Cable knew the cost of building a circuit-switched facility was massive, although Cox made the effort. Offering primary service meant providing expensive powered lines and 911 service. The resale option, alternatively, meant the local cable operator had to negotiate with the incumbent LEC, a difficult and complex process under the best of circumstances, and ILECs had little reason to move quickly in resolving the difficult technical, legal, and financial issues when the end result was a new competitor using their physical plant.

Moreover, just as the telephone companies had little experience in the cable business, cable looking to compete in telephony had to confront the complexities of an unfamiliar technology, business, and regulatory terrain.

Billing systems, customer service, government relations, and legal issues were all foreign, and the costs of providing cable telephone service were estimated at $750 to $1,000 a line.[17] Cable also had its image problem working against it. The U.S. telephone system had built a reputation for reliability that almost no other technical system in the world could match. Phone calls went through almost every time and the service was always there. Few customers were willing to rely on the local cable guy for primary telephone service.

By 1996, there were only two wire-based, local cable phone networks, one run by Cablevision Systems Corp. on New York's Long Island and one by Time Warner in Rochester, New York.[18]

In 1996 and 1997, confronted by the difficult reality of running a local telephone system, most MSOs pulled back on their phone plans. Some looked at offering "second line" service that would not require self-powering or 911 capability. In 1998, TCI, Comcast and Cox removed themselves from their Sprint PCS partnership, selling their interests in the company for Sprint stock. In 1999, Comcast sold its cell phone operation (800,000 customers) to SBC for $400 million and $1.3 billion in assumed debt.

Cable did have another option, however. The rise of digital technology and the architecture of the Internet provided the industry with an enticing alternative to circuit switching. The Internet used packet switching to move its text and data around the globe. Speeding information across town by breaking data into individual packets, sending them through cyberspace, and reassembling them at the receiving end worked well for text and still pictures, but doing the trick for voice, and later video, was a more daunting task. Engineers and Blue Skyers were confident they could make it work, and even make a business out of it, however. The technology became know as "Voice over Internet Protocol," or VoIP. It was tremendously appealing to cable because it did not require a local switch. Cable could use its tree-and-branch platform, connecting directly to the public switched network (PSN) and bypassing the local loop provider. VoIP was vastly cheaper than building the infrastructure necessary for a traditional circuit-switched service and less messy legally than resale. And, due to a quirk in the law that the FCC did not want to fix, it was for all intents and purposes, an unregulated service.

In 1997, CableLabs began working on standards and protocols for cable-based VoIP software and hardware in its PacketCable program. In 1998, TCI announced that it would forgo any efforts at resale-based telephony to explore VoIP options. The next year, several cable companies began testing VoIP service. Watching with interest was the grandfather of all U.S. telephony, AT&T.

AT&T and TCI

AT&T needed attention. Income from long distance, which constituted 75 percent of the company's revenue stream, was in steep decline as a consequence of increased competition and improved technology. In the ten years after the breakup of AT&T in 1984, residential long-distance rates fell by more than 50 percent, a drop that promised to accelerate in the face of deregulation and the surging RBOCs.

Telephone companies for decades had provided a dependable flow of dividends for a fabled army of widows, orphans, and stock market coupon clippers. AT&T, however, did not draw the interest of youth looking to invest in the information age. With eroding revenues and a lack-luster Wall Street persona, the company sought a new identity and new leadership. In 1997, it hired a new CEO. Michael Armstrong had grown up in the electronics industry, serving thirty years with IBM and moving in 1992 to Hughes where, as chief executive, he had oversight of DirecTV. At AT&T, Armstrong went quickly to work, cutting budgets and staff and selling auxiliary units.[19] In early 1996, AT&T had purchased a 2.5 percent interest in DirecTV with an option to buy more. In December 1997, Armstrong sold the interest back to Hughes.[20]

In addition to the belt tightening, Armstrong drew up plans to invade the two markets he and others thought the most promising—local loop service and Internet access. Like cable, AT&T rejected competitive resale access as too cumbersome and uncertain, and facilities-based competition as prohibitively expensive (estimated by some at up to $100 billion). AT&T chose the popular "buy" option.

The easiest and most obvious route to an integrated long-distance, local loop company was to merge with an existing RBOC. Even before Armstrong joined the company, AT&T, led by CEO Robert Allen, began talking with SBC. But rumors of the discussions raced across the industry and to the ears of Washington, D.C., regulators. FCC Chairman Reed Hundt's immediate reaction was "[T]his isn't what we wanted ... AT&T is supposed to compete in the local phone business ... It has to take on the Bells, not merge with them."[21] Gathering support from procompetition members of Congress, Hundt made it known publicly that such a combination would not be welcomed in Washington and the talks collapsed.

Armstrong sought smaller targets, buying McCaw Cellular with the idea of creating wireless local service that could connect to AT&T's long lines. AT&T also looked at the purchase of existing competitive access providers. One of the CAPS that interested him was Teleport, so Armstrong sat down to talk with Malone and, in January 1998, AT&T purchased Teleport from its cable owners for $10.3 billion.

Armstrong was keen to get a beachhead in the Internet access business and talked with AOL about a relationship. He also discussed a possible purchase of @Home. Malone was not interested in selling @Home, but offered another proposal.

Through 1997 and 1998, with the nod of approval from Bill Gates and a slow trickle of digital boxes coming to market, TCI and the cable industry were on the comeback trail. Cable stock was rising. From hovering at around $16 dollars at the end of 1996, TCI shares were moving toward $40 by the summer of 1998. Despite the recovery, Malone still sought to escape the distribution business and become a programmer. The question of whether TCI's infrastructure could really deliver on the convergence vision lingered, and it was unclear how long the window of opportunity opened by the stock surge would last. Malone had reason to move quickly to attempt once more to sell TCI. He had failed with Bell Atlantic, but AT&T now offered a new hope.

Armstrong's interest in a TCI merger was pulled by the convergence vision. He saw cable's last mile of coaxial cable to the home as the local loop AT&T needed to offer integrated local and long-distance telephone service, plus video and Internet access. He would use the promising but untested VoIP, in lieu of traditional switched local phone service, to compete with the RBOCs, which themselves were moving rapidly for an assault on his long-distance business. Malone and Armstrong, therefore, came to terms quickly. On June 24, 1998, one of the nation's dominant cable companies announced its merger with the country's historic telephone giant. AT&T was purchasing TCI for $48.3 billion in stock, cash, and assumed debt, paying a premium $50 a share (and $55 a share for the super-voting class B stock held by Malone and the Magness heirs).

The purchase left Malone with 1.5 percent of all AT&T stock, making him the company's single largest shareholder. It also left him with control of the programming wing of the company while allowing him to dump the distribution side of the business, his original aim in the Bell Atlantic deal. To help guard against a recurrence of the Bell Atlantic collapse, the agreement included a clause that required AT&T to pay TCI $2 billion if the merger was not consummated.

The TCI acquisition did not complete AT&T's cable play, however. In March 1999, just as the AT&T-TCI deal was closing, Comcast announced a bid to buy MediaOne for $58 billion. Within a few weeks, the offer was accepted. MediaOne had been looking to expand its reach and, failing to find a large cable company interested in selling, it decided itself to become a seller rather than a buyer.

Comcast had become increasingly acquisitive in the 1990s (see below) and, with the MediaOne purchase, was poised to become the nation's largest cable provider. It was a position Armstrong had thought they had staked out for themselves, and AT&T considered a counter offer. Malone counseled against it, fearing the massive debt load that the combined TCI and MediaOne purchases would put on the balance sheets. Hostetter and other MediaOne investors lobbied for AT&T entry, hoping for a bidding war that would drive up their share prices.

Armstrong, supported by Hindrey and other AT&T executives, decided they would not be the number two cable company in the nation and tendered a $62.5 billion bid for MediaOne. Comcast chose to step away from the deal, but took with it a $1.5 billion breakup fee plus the rights to buy certain cable properties controlled by AT&T, including systems in the Philadelphia area,[22] which furthered Comcast's long-sought goal of consolidating the hold on its home market.

AT&T did have to clear regulatory hurdles, including local and federal challenges on the issue of opening its local system to competitive ISP providers (detailed below). It also had to shed some systems to comply with the FCC's ownership caps. The MediaOne purchase meant AT&T now had a 22.5 percent equity interest in Time Warner Entertainment and the Time Warner cable properties. Combining TCI, MediaOne, and TWE, gave AT&T entrée into nearly 30 million cable homes, exceeding the FCC's total ownership limits of 30 percent. As noted, the ownership rules were in litigation and alternatively upheld and struck down by the courts, but the MediaOne sale was nonetheless stalled until AT&T sold systems.

The acquisition of TCI by AT&T may not have been perfect historical irony, but it was close. John Malone had sold TCI to the company he began his career with and which he left because of its historic rigidity and lack of opportunity. More broadly, the company that helped invent television, then backed away from it in the 1930s, the company most hated and feared by the cable industry for decades, was now the nation's largest cable TV provider.

DBS Consolidation

The latter 1990s also was a period of consolidation in the DBS industry. With the auction of the Advanced Communication orbital slots to Murdoch and MCI, Congress and the FCC had hoped for additional competition in the combined cable-DBS market. But, as with the telephone industry, the forces of economics, colored by leadership style, were pushing in a different direction.

Charlie Ergen could be focused and fierce in management and in competition. Price wars between EchoStar and DirecTV erupted almost the minute EchoStar set up shop. Not only did prices drop precipitously, but in August 1998, EchoStar began offering retailers a $100 bounty on every DirecTV customer they could convert to the Dish Network. DirecTV responded by doubling the reward to $200 for every Dish customer switched to DirecTV.

Meanwhile, Murdoch was plotting his own satellite strategy. He had programming and production capacity along with the 110-degree west satellite slot, but he sought the security of additional DBS channel capacity, and the only way to get it was by aligning with one of the existing licensees. Murdoch approached Primestar about a possible partnership, but Time Warner was not interested. That left EchoStar.

Ergen had satellite capacity and an aggressive marketing style but was so low on cash that most analysts gave him slim odds of survival without a partner, and he promised he would find one. Michael Milkin helped bring together Murdoch and Ergen to form a new DBS company. They shook hands and signed the paper in February 1997. The mating of Ergen's cash-strapped DishTV start-up with the powerful programming resources and business experience of Murdoch seemed natural. Together, they would control enough frequencies for 500 digitally compressed channels.

On February 24, 1997, Murdoch announced the news at an investors' meeting in Los Angeles. News Corp. would buy 50 percent of EchoStar for $1 billion in cash. The new company would be called "Sky." After Murdoch's proclamation of a new competitive start for satellite and cable TV, his satellite chief, Preston Padden took the stage. Padden had spent most of his career as a professional opponent of cable. As President of the Association of Independent Television Stations in the 1980s, he earned a reputation as an aggressive and effective lobbyist, battling cable on numerous regulatory fronts. Now, he was positioning himself, it seemed, as Murdoch's head of cable competition. He promised that Sky would be the much-needed antidote to the cable industry's shoddy, overpriced service. He told the audience cable would be "calling Dr. Kevorkian," in reference to the controversial physician who famously assisted in suicide attempts.

Murdoch and Padden also proposed to overcome one of the primary consumer complaints about DBS, the lack of local broadcast signals. Limited transponder capacity and copyright concerns prevented the DBS operators from offering the local broadcast stations most customers were used to seeing on cable or with their rooftop antennas. Customers who wanted the benefits of DBS and local stations were forced combine traditional off-air antennas with their rooftop dish and set up awkward A-B switches. But advances in compression meant Sky would technically be able to offer "local in local" programming.

Padden's comments and the apparent full-scale assault were not met kindly by the leaders of the cable industry. While cable executives did not always see eye to eye, it was still a small club, and getting smaller. The programming access provisions of the 1992 Act meant Murdoch could obtain programming for the new Death Star, but the Fox programming channels also needed carriage on the nation's cable systems to survive and that access soon became more difficult to obtain. As discussed below, Murdoch was attempting to start a cable news channel in competition with CNN, but without cooperation and clearance from the nation's major MSOs, the success of such a start-up was problematic. Malone went so far as to suggest TCI and News Corp. end their partnerships in the fX programming service and the regional sports channels. At the same time, Murdoch was feeling some nervousness from his own investors about the wisdom of a very expensive and somewhat late entry into a market dominated by one of the world's largest corporations, General Motors (GM).

As relations between Murdoch and the cable industry grew more strained, Murdoch, according to reports, reconsidered his position. He once again approached the Primestar partners about a possible relationship. TCI supported the proposal, but Time Warner (motivated in part by personal animosity between Murdoch and Levin and Turner) was reluctant. Murdoch went to Malone for advice. Malone counseled him to abandon his merger with Ergen and his ambitions to own satellite distribution, focusing instead on growing his programming assets. As part of the move, News Corp. could partner instead with Primestar. Not only would such a switch end the feud, but it would help assure Fox of carriage of its broadcast and cable programming assets.

An effort by Murdoch to acquire the Family Channel from Malone's Liberty also had been snagged in the DBS dispute. And if added incentive to make the switch was necessary, it was supplied by the increasing friction between Padden and Ergen, who were forced to work together, but agreed on very little and came to such a management impasse that Padden quit the project.

Murdoch ultimately decided to follow Malone's advice. He dumped Ergen and switched to Primestar. Malone helped sell the Primestar board, including Time Warner, on the plan, which was finalized in June 1997. Murdoch would become a nonvoting partner in Primestar in exchange for the News Corp.-MCI satellite assets. The same day, News Corp. also announced the purchase of the Family Channel, which would be added to the growing Fox family of cable networks.

Ergen immediately sued for breach of contract. But he was not the only concerned party. Regulators did not like the deal, either. In 1998, the Justice Department filed an antitrust suit seeking to block the combination on the grounds it reduced overall competition in the multichannel TV industry. In the government's view, DBS was supposed to compete with cable, not be owned by it. The antitrust action effectively killed the merger and, with it, Primestar. With only a medium-powered satellite, lukewarm marketing, and increasing competition from Hughes and Ergen, the cable industry gave up. In January 1999, Primestar announced the sale of its DBS assets, including its 2.3 million subscribers and eleven high-powered Tempo frequencies at 119-degrees west, to DirecTV for $1.36 billion.[23] Murdoch and MCI took the same path, selling the 110-degree slot and the associated business assets to Ergen for $1.25 billion in stock, ending Ergen's $5 billion lawsuit against News Corp. at the same time. When it was done, only two DBS providers remained in orbit.

Providing a boost to DBS marketing efforts, Congress in November 1999 enacted legislation amending the Satellite Home Viewer Act of 1988 to require DBS operators to carry all local signals in markets they chose to serve ("local in local") by January 2002, and to permit them to import distant network signals in certain situations. DirecTV and EchoStar signed retransmission consent agreements with major broadcast networks and began rolling out local in local service over the next year. Customers with local DBS reception no longer had to switch manually between their satellite dish and a roof-top antenna to get local channels, although they frequently had to install a second dish to receive them.

Cable retained some technical advantages in bandwidth and interactivity, but DBS was making substantial gains. EchoStar ended the decade with 3 million subscribers; DirecTV had 7.1 million and was the nation's third largest multichannel video provider behind AT&T Broadband and Time Warner. In 1999 cable's share of the multichannel marketplace dipped 3 percent (to 82 percent) as DBS ascended.

Digital Efforts

Critical to the plans of both the cable and the telephone industries was entry into data service, the Internet, and e-commerce—the heart of the dot-com boom. Cable efforts to exploit its broadband capacity, therefore, were not just an attempt to realize the revenue potential of the triple play, but to convince the financial community and others that it was a player in the new digital world.

Cable and the Internet

To enter the Internet access business, cable needed a data equivalent of the set-top box, a modem that could be adapted to the high-speed architecture of the coaxial tree and branch system. The man credited as the "Father of the Cable Modem," Rouzbeh Yassini, first developed the idea in 1986 while working for a data networking company. In 1990, he founded LANcity and offered his first product. Unfortunately, the initial sticker price was $15,000, but, by 1995, he had engineered a modem suitable for the home for less than $500. Selling the company in 1996, Yassini went on to head up CableLab's cable modem interoperability testing and certification operations. The cable technical consortium took the lead in creating shared technical standards for cable modems, assuring interoperability and making it possible even to sell cable modems directly to the public through electronics retail outlets.[24]

Businesses and institutions could purchase high-capacity "T1" lines from the telephone company, but the expense of the service prevented its expansion into the rapidly growing consumer market. This was the market cable hoped to exploit. Cable modems offered home users amazing speed. A moderately complex World Wide Web page took 32 seconds to download via a 28.8-Kbps dial-up phone line, giving rise to the impatient phrase, "World Wide Wait." The same page required only 2.7 seconds to download using a cable modem.

The device had other advantages. Using a cable modem meant that customers did not have to tie up their home telephone line each time they went online. The cable connection also was always on and available. The disadvantage for customers and operators was that the system was a type of shared

Ethernet, so the more users online in a given neighborhood, the slower the service became. Early introduction also was hampered by outdated cable plant. Intel's Andy Grove had been at least partially correct in his assessment of the condition of the nation's existing cable lines. Analog signals are very forgiving of leaky wires and aging amplifiers—digital signals are not. Cable systems had to undergo significant upgrading and tightening to deliver clean digital signals, both for data and for digital video. Much of cable's return path capacity was especially poor. Cable modems could deliver downstream data at up to 40 Mbps, but in many early applications had to rely on a telco twisted pair for upstream interactivity.

Despite the challenges and driven by economic necessity, the industry inaugurated cable modem service at mid decade. Continental Cable began offering Internet access, using proprietary technology, in March 1994, with speeds up to 10 Mbps for $125 a month. Viacom began testing approaches in California in 1994 and most of the major MSOs were running small-scale trials by 1996. The tests proved sufficiently successful and the popular enthusiasm for the Internet so great, that by the end of 1996, cable companies had placed orders for hundreds of thousands of modems. Motorola alone shipped more than a half-million to cable companies in 1996. By 1997, operators were charging anywhere from $15 a month for a basic service to as much as $40 a month for expanded services and Internet connection. By the end of 1998, the industry had a modest 500,000 cable modem subscribers nationally. The figure rose to 1.6 million by the end of 1999 and reached nearly 4 million by the end of 2000.[25]

Providing the hardware, the cable interconnection service, was only part of the Internet business, however. Customers typically used commercial on-line providers as their gateway to the Internet, electronic mail services, and web access. The leading Internet service providers (ISPs) were dial-up services, America Online (AOL) and CompuServe. At mid decade, more than 20 million Americans were paying $10 to $20 or more a month for news, sports, entertainment, information, and games, plus access to the more open universe of the Internet.[26]

The major MSOs created their own Internet portals. Customers, when logging on, would see first the page of this ISP and, from that home page, be offered extra services, and advertising, for a nominal monthly fee. In 1996, a consortium of operators led by TCI and including Cox, Comcast, Cablevision Systems, Intermedia Partners, Marcus, Rogers, and Shaw Communications, launched @Home. In January 1999, @Home announced the purchase of the young web portal and search engine, Excite, for $6.7 billion in stock. Time Warner set up its own service, marketed as RoadRunner. In 1998, Time Warner and MediaOne partnered in RoadRunner and Microsoft invested $425 million in the service.

Other companies created services and technologies that used the viewer's TV set, instead of a home computer, to enter the web. As noted above,

Microsoft offered WebTV. General Instruments marketed SURFview, a $99 set-top box that provided Internet access, and Worldgate produced Internet Over TV.

Along with most businesses in the United States, every program network, MSO, and affiliated association put up its own web site. Some were more elaborate than others. Disney, for example, joined with dot-com start-up Infoseek to create a web portal called the "GO Network." It offered access to all Disney-ABC web content, plus news, shopping, Internet search, and e-mail services. The QVC and HSN shopping channels were naturals for online sales. Both established an "interactive division" in 1994, taking their electronic retailing expertise online. Some companies launched new technology divisions and initiatives. Discovery was a leader, developing online and compact disc (CD)-based programming.

Telco and DBS Data

The telephone companies also looked for methods to develop high-speed data services, recalling two technologies unsuccessfully employed earlier for video. The primary tool was ADSL. With the family of DSL technologies, the telephone companies could dramatically increase the bandwidth to the home. While it still failed to match cable modem speeds, it marshaled data rates of more than 1.5 Mbps and, by 1998, was being marketed as high-speed internet service.

And while the RBOCs had dumped MMDS for wireless cable use, the long-distance carriers snapped them up on the cheap to use for distribution of wireless Internet traffic. MCI-WorldCom spent $500 million for CAI Wireless, Wireless One, and Prime One, providing MMDS coverage to about half the nation's major markets. Sprint spent $1 billion for various MMDS assets.

Even the DBS operators tried their hand at Internet service. In 1999, DirecTV teamed with AOL in a $1.5 billion plan to offer high-speed access via satellite (DirecPC), using a telephone line for its return path. In 2000, EchoStar spent $50 million to launch a partnership with an Israeli-based satellite company, Gilat Satellite Networks. The Gilat-to-Home Internet service planned to employ two-way satellite communication using the same dish for video and data traffic.

Open Access

Convergence, and the possibility that cable could sell high-speed Internet access, was not without legal entanglements. ISP service was a growing, multibillion dollar business. ISP competitors, such as AOL, were concerned that cable would exercise its monopoly broadband hold on consumers and block alternative providers from their networks. They formed an OpenNET coalition, demanding open access, and AOL spent a great deal of time and money in 1998 and 1999 lobbying for legislation that would require cable operators to open their systems to competitive ISPs.

For their part, most cable companies took the position that they should be permitted to control the coaxial conduit they had built and paid for. Providing Internet service was no different, they argued, than providing cable service, and they should be permitted to determine what their systems offered, charging appropriately. They worried, additionally, that open access would mean competitive broadband providers would be able to offer a host of additional digital services using the local cable plant, including competitive VoIP and even streaming video.

The open access debate grew teeth in 1998 after the announcement of the AT&T-TCI merger. In Portland, Oregon, municipal authorities, as a condition of transferring the local franchise from TCI to AT&T, required the company to guarantee open access to alternative ISPs. AT&T challenged the move in federal court. A federal district judge in June 1999 held in favor of Portland, but the FCC stepped in, successfully petitioning the Ninth Circuit Court to overturn the decision on the grounds that it was a federal question subject to FCC jurisdiction, not a local one.[27]

Congress, with its antenna always up for charges of monopolistic practices on the part of the cable industry, held hearings on the issue and even introduced legislation.[28] It was clear that public and political pressure was building against AT&T. The company had operated as common carrier for a century and was less resistant to the basic concept, and Armstrong wanted the merger to succeed. In December 1999, AT&T, therefore, abandoned its opposition and agreed to open access by providing one of the nation's largest ISPs, Mindspring, entry to its systems. It helped clear the way for final federal approval of the acquisitions.

Digital Boxes

While AT&T began to puzzle over how to exploit its new national broadband resource and the RBOCs rethought their convergence strategies, the evolution of the technology—slower than most hoped or planned for—nonetheless began to make its presence felt. First at a trickle, then a stream, digital boxes were leaving plants, shipping to warehouses, then to trucks, and then into homes. The first shipment of GI boxes in the fall of 1996 counted about 40,000 units, but production costs remained high. To bring down costs by ramping up mass production, a consortium of major MSOs entered into a $4.5 billion agreement with GI in December 1997 to purchase 15 million set-top boxes over five years. They would be powered by Intel Pentium II chips and capable of handling HDTV, Internet access, and VoIP. Software would be supplied by Microsoft and, to retard any monopolistic cable ambitions on the part of Bill Gates, Sun Microsystems. It was the critical economic-technical equation necessary to propel cable into the digital age.

TCI's October digital opening in Hartford was followed by other MSO deployments. In fall 1997, Comcast began digital service in Orange County, California, and in Philadelphia. By 1998, it was offering twenty-four premium and forty digital PPV channels. MediaOne was providing up to seventy-two digital channels in some markets.

By the end of 1998, NCTA estimated the number of digital subscribers at 1.5 million, a figure that reached about 5 million by the end of 1999 and 9.7 million by the close of 2000.[29] Average channel capacity rose as well. By 2000, more than 60 percent of all subscribers had fifty-four or more channels, compared with less than 10 percent of all subscribers in 1985. Almost 5 percent of all subscribers had access to ninety or more channels (and see Appendix B).[30] As capacity expanded, programmers arose to provide the content.

Programming

Programming activities, which had stalled in the wake of the 1992 Act rate regulations, began recovering in the late 1990s, first as a result of the November 1994 going forward rules, then through the broader deregulation of the 1996 Act, and finally with the rollout of digital.

Cable followed DBS in using much of its new capacity for pay-per-view films. Pay-per-view had fallen short of operator expectations in the mid-1980s. Big-name boxing matches, the wrestling extravaganzas, and a few concert specials could be counted on to draw viewers and revenues. But the daily dose of PPV film revenues cable had hoped for was hampered before the digital conversion by limited bandwidth and clumsy ordering technology. A decision in 1994 by Hollywood to increase the release window between home video and PPV from 60 to 90 days only exacerbated the problem. The PPV field was dramatically reduced in 1998 when Request TV closed shop after failing to secure a merger with its major rival, Viewer's Choice. The next year, Playboy purchased the Spice adult channels for $100 million. In January 2000, Viewer's Choice, still controlled by the major MSOs, changed its name to InDemand.

DBS and digital compression helped improve the PPV picture. By 2000, DirecTV was running fifty PPV channels, capacity used not just to offer a greater selection of titles, but to run current, popular films at frequent intervals, providing start times every half hour or so and bringing the system closer to a video-on-demand model. Combined DBS and digital cable penetration meant that PPV service was available in 55 percent of all multichannel households by the end of the decade.

The premium services used the growing channel capacity to extend their multiplexing strategies. In June 1998, Time Warner's Cinemax launched its MultiMax package, including MoreMax, ActionMax, and ThrillerMax. HBO reconfigured existing networks and added others creating HBO Plus, HBO Signature, HBO Family, HBO Zone, and HBO Comedy. Viacom splintered Showtime, opening Showtime Extreme in 1998, Showtime Beyond in 1999, and Showtime Family Zone, Showtime Next, and Showtime Women in 2001.

For advertiser-supported programming, the new capacity did not suggest a host of new companies entering the business. The mature networks had established relations with distribution companies, and frequently, as has been seen, were owned fully or in part by the major MSOs. They had production facilities and personnel experienced in all the details of the enterprise, from hiring and script development to writing contracts to bookkeeping and finance. Digital compression meant that even satellite transponder time was already included in the existing overhead, making the creation of a new service a relatively cheap add-on. New entrants, alternatively, would have to spend millions to establish the kind of infrastructure already enjoyed by the incumbent programmers.

The potential for competition at the same time was an important driver of programming proliferation. As noted, for established programmers, new product development was a defensive measure designed to block, to the extent possible, others from gaining access to the added shelf space. So, entrenched companies began casting about for niched services that were somehow related to their core programming agenda, sufficiently inexpensive to be financially viable, and not as yet exploited by competing firms.

Leveraging his dominant position in cable news, Turner, for example, launched his own financial news service, CNNfn, in December 1995, and teaming with Time Warner's *Sports Illustrated* created CNN/SI in December 1996. He also started an international news channel, CNNI, in January 1995, and a Spanish-language network, CNN En Espanol, in March 1997. BET began its own music channel, BET on Jazz, in January 1996, and its own premium movie service, BET Movies, in February 1997.

Discovery began life with a potpourri of nonfiction programming, ranging from history to science to nature shows, and was among the most ambitious in exploiting the new digital opportunities. Each of its major subject categories became a platform for a new 24-hour network. In October 1996, it launched an all-animals network, Animal Planet, a social history service, Discovery Civilization Channel, a home improvement and hobbies channel, Discovery Home & Leisure, the Discovery Kids Channel, and Discovery Science Channel. In July 1998, it added an additional digital layer to its offerings with Discovery Health Channel and Discovery Wings, and then a Spanish language version of Discovery, Discovery En Espanol, in October 1998. A&E followed suit, spinning off niched versions of its popular nonfiction programming, including The History Channel in January 1995, and in November 1998, History Channel International and The Biography Channel.

In November 1996, Disney's ESPN added a 24-hour sports news service, ESPNEWS. Classic Sports Network thought it could make a business showing reruns of sporting events from the past. It launched in May 1995 and was acquired by ESPN (which outbid a joint Fox-Liberty media offer) in October 1997, becoming ESPN Classic. ESPN EXTRA, a PPV service, and ESPN NOW, a barker channel, began operations in September 1999. As was the case for other notable programmers, such as Discovery, ESPN leveraged its brand into businesses beyond cable TV, creating ESPN Radio, ESPN magazine, and even a chain of ESPN Zone sports bars. The Disney-Hearst Lifetime channel started the Lifetime Movie network in July 1998, and drawing on its decades of cartoon material, TOON Disney launched in April 1998.

Viacom's MTV-VH1 franchise spun off its popular, and very cheap, Nick at Nite evening programming block into its own channel in April 1996. In 1997, the company announced development of what it called a "digital suite," unveiling MTV S (targeting the U.S. Hispanic market), MTV X (heavy metal), MTV 2 (all music), VH1 Country, VH1 Classic Rock, and VH1 Soul, all of which launched in August 1998.

Chuck Dolan's Bravo networks developed the Independent Film Channel in September 1994 to provide an outlet for the work of independent filmmakers. In February 1996, it was joined by a similar service, the Sundance Channel. Robert Redford, long-time promoter of independent filmmaking and internationally known for his Sundance Film Festival, worked with Viacom's Showtime Networks and Universal Studios to create the service.

Fox

News Corp. was busy on several fronts. In 1996, as part of TCI's corporate restructuring, Malone joined TCI/Liberty's fifteen regional Prime Sports networks with the sports network assets of News Corp.[31] To assure carriage of the new Fox Sports Network, Fox offered operators an unheard-of $10 per subscriber incentive at launch. In 1999, News Corp. became full owner of the network, purchasing Liberty Media's 50 percent stake for $2.8 billion in stock. In addition, following TCI's sale of its interest in the Family Channel, News Corp. relaunched the service in August 1998 as the Fox Family Channel.

In a much more public and contentious move, Murdoch also went after a news channel, doing what others thought unthinkable—creating a competitor to the entrenched CNN.[32] According to reports, Murdoch had long lusted after CNN, but it had been taken out of his reach by the Time Warner acquisition, so in early 1996 he revealed that Fox, with backing from TCI, would develop its own cable news network. Ted Turner, anticipating the announcement, made trade press headlines, declaring, "I'm looking forward to squishing Rupert like a bug."[33]

Murdoch had a reputation for conservative politics and a tabloid treatment of news. A popular feature of one of his British news-gossip publications was a daily picture of a topless model. Murdoch hired Roger Ailes to run the new network. Ailes was a former Republican political consultant who had worked as a media adviser for Presidents Nixon and Reagan as well as then Vice President George H. W. Bush. But he also had made a successful career in TV programming and, most recently, had served as the head of NBC's CNBC and America's Talking cable networks. His cable experience, pugnacious personality, and conservative political ideology made him a perfect fit for the news operation that Murdoch had created, at least in part to counter what he felt was a liberal political bias at CNN.

Fox News launched on October 7, 1996, gaining carriage, again, by buying it, paying MSOs up to $13 per subscriber for space on the system. The amount was staggering. Said one industry executive, "It scares the [heck] out of people that Murdoch's willing to spend hundreds of millions to establish a beachhead."[34] The first to take Murdoch up on the offer was TCI. In effect, News Corp. would pay TCI more than $100 million, and offer an option for a 20 percent equity stake in Fox News, to carry the channel. Cap Cities/ABC, which had announced plans for a similar all-news network earlier, dropped out, reportedly deciding that it did not want to enter the bidding war for channel space.

The carriage bounty was not sufficient, however, to overcome Time Warner's resistance. Time Warner controlled 1.1 million of subscribers in the nation's largest market, New York City, and Levin had earlier agreed to give Murdoch channel space in New York, considering it as a possible replacement for NBC's ailing American's Talking (AT) network. NBC had carriage contracts in place, however, and was leveraging them for the AT successor, MSNBC. Levin told Murdoch about a month before the launch of Fox News that he did not have sufficient channel capacity for both Fox and MSNBC, and had chosen the latter. Murdoch did not take it well. The loss of a million New York City subscribers and exposure in the nation's largest media market would be a damaging blow to Fox News' chances. Murdoch also saw the move as a personal betrayal by Levin.

Murdoch claimed that Time Warner was only trying to protect CNN from competition, and filed a $2 billion breach-of-contract suit in October 1996. The dispute quickly turned ugly, with Turner, now Vice-Chairman of Time Warner, calling Murdoch a "scumbag" and comparing him to Adolph Hitler. Murdoch responded in kind, using News Corp.'s *New York Post* to wage war in its own way—one memorable headline read, "Is Ted Turner Nuts? You Decide."[35] The business spat became a political soap opera when New York City Mayor Rudolph Giuliani joined in, accusing Time Warner of an attempt to restrain trade and announced he would use the city's public access channel if necessary to carry Fox News. Time Warner sued the mayor, alleging an improper and illegal use of the access channels, and won a preliminary injunction prohibiting such use.[36]

The mud fight, however, was just one skirmish in the larger conflict between Murdoch and the cable industry over News Corp.'s DBS plans. As noted previously, cable industry executives, especially those at Time Warner, were pinching off Murdoch's distribution access following announcement of the News Corp.-EchoStar DBS project. Once that alliance dissolved and Murdoch signed to work with Primestar, access to the nation's cable customers came more freely. In July 1997, Fox News was added to the New York City system on a channel provided by the city, and the Time Warner family of channels was assured access for its programming to Murdoch's global satellite television system.

Filling the Digital Tier

In 1996, broadcasters were given an opportunity to negotiate once again for channel capacity, following the first, mandated three-year agreement cycle under the must-carry retransmission consent provisions of the 1992 Cable Act. Most of the discussions went smoothly. Some did not. Having missed its opportunity in the first round of talks in 1993, CBS, under new Westinghouse management, won carriage for a service dubbed "CBS Eye on People," with documentary and feature footage from the CBS archives. "Eye on People," however, generated little viewer enthusiasm and, in 1998, CBS sold the network to Discovery, which rebranded it as "Discovery People."

General Electric's NBC, looking to forge a connection with Bill Gates, joined with Microsoft to create its own 24-hour news channel, MSNBC. The service began in July 1996, replacing NBC's disappointing "America's Talking" talk-news network. Promoted as an example of converged programming services, Microsoft developed a companion online MSNBC news site.

News Corp.'s request for digital carriage of fX, Fox Family, and Fox World Sports was hung up, as noted, in the DBS struggle. In one showdown, Cox Cable briefly pulled the local Fox broadcast stations off its systems in Washington, D.C., Cleveland, Dallas, Houston, and Austin.

In the 1999 round, a similar fight broke out between Disney/ABC and Time Warner, after Disney demanded carriage of its SoapNet (soap opera) channel and repositioning of the Disney Channel from a pay to a basic tier service. Failing to come to terms, Time Warner made front-page headlines by pulling the plug on the ABC flagship station in New York City. The loss of one of the major networks in New York, home of all three networks, was exacerbated by the sudden absence from home screens of ABC's wildly popular "Who Wants to Be a Millionaire." The FCC had to step in and order that ABC be reinstated; eventually the two sides resolved their differences.

A variety of other networks—most negotiating equity interests with the major MSOs—took the opportunities of the mid- and late 1990s to inaugurate specialty services, usually to fill the expanding digital tier. They included: Trio (September 1994), Ovation (April 1996), Speedvision (January 1996), The Military Channel (July 1998), BBC America (March 1998), Pax TV (Bud Paxson, August 1998), TEN—The Erotic Network (September 1998), and Style (October 1998). Reflecting the high tech infatuation of the period, publisher Ziff Davis began a network in May 1998 revolving around all things cyber. ZDTV was subsequently purchased, appropriately, by Paul Allen's Vulcan Ventures, Inc., which changed the name to Tech TV. In February 2000, former Nickelodeon head Geraldine Laybourne led a group that included investors Oprah Winfrey and Paul Allen in the development of Oxygen, the first major challenge to Lifetime's dominance of niched women's programming.

In one of the more interesting transformations of the late 1990s, CBS renovated The Nashville Network, dropping its original country music format and repositioning it as TNN: The National Network, a USA-like mass appeal channel. Failing to attract the desired ratings, CBS then moved it to a format featuring a harder line of action and soft sex content aimed at young men, proclaiming itself, in contrast with the women's networks, as the first network for men. To further push the new brand, CBS dropped the TNN logo in 2003, changing the name of the channel to Spike TV, and setting off a short court challenge from film producer-director Spike Lee, who contended viewers might mistakenly think he was, in some way, backing the channel.

Digital Politics

Compared with the first half of the decade, the latter 1990s were relatively quiet in the area of federal regulation, as Washington awaited the outcome of the 1996 Act. The industry confronted issues spawned by the new law in several important areas, however, including must-carry, content regulation, and rate control.

Digital Must-Carry

It was sufficiently irritating to cable operators that broadcasters had won must-carry rights in the 1992 Cable Act. In propagating rules governing the transition to digital television, however, the FCC had provided broadcasters with a new frequency to be used in addition to their existing analog channel as part of the conversion process. Once stations had made the transition to digital broadcasting, they would, in theory, surrender their old analog channels back to the government.

Under the original federal transition plan, broadcasters were required to begin broadcasting in digital by May 2002, but permitted to retain their analog channel until December 31, 2006.[37] The first digital towers were fired up in November 1998, with about twenty-two stations going on the air in the nation's largest markets, but the pace of executing the conversion was painfully slow and astronomically expensive. The broadcast industry had to convert nearly every piece of equipment it owned, from cameras to editing facilities to transmitters and towers. The task was enormous and retarded by both the expense and a lack of trained personnel to construct the towers. Broadcasters sought postponement of the transition deadline. The timetable for returning analog frequencies was set back, initially until 85 percent of the stations' viewers had digital receivers, either through satellite, cable, or broadcast, and then, as noted below, until 2009. Even as digital towers began operations, however, few were watching. The number of TV sets capable of receiving a digital signal was estimated at no more than a few hundred thousand in 2000, and with unit prices beginning in the $2,000 to $5,000 range, rapid diffusion seemed unlikely.

The implications for cable involved broadcaster requests for carriage of multiple signals, first the dual analog and digital signals carried during the transition and, looking to the future, all the possible multiplexed signals that the digital technology would provide.[38] In 2000, broadcasters pressed the FCC to require cable carriage of their digital channels. The cable industry's response was unsurprising, arguing that a carte blanche carriage guarantee for multiplexed digital broadcasting would force the displacement of existing cable networks. The FCC began an inquiry in 1998, eventually holding in 2001 that cable would be obligated to carry only one digital channel and not the full array of multiplexed signals.[39] Several years later, in 2004, the slower-than-hoped-for rollout of broadcast digital service spawned an FCC staff proposal to force operators to carry additional digital channels; however, in February 2005, the Commission reaffirmed its decision to require carriage of only one.[40]

CDA and the V-Chip

The Communications Decency Act (CDA) drew almost instantaneous court challenges. On review in June 1996, a special three-judge panel in federal district court in Pennsylvania struck down the section of the law dealing with indecency on the Internet, explaining that less restrictive means were available to achieve the goal of protecting minors. That decision was appealed directly to the Supreme Court, which upheld it in June 1997.[41] In 1998, a federal court struck down the provisions pertaining to cable television, also concluding that alternative means, such as channel-blocking devices existed to protect children from the harm. The Supreme Court affirmed the decision in 2000.[42]

The industry response was markedly different with respect to the V-Chip requirements of the 1996 Act. The major networks and the NAB initially objected on First Amendment grounds to the federal government's assumed authority to craft a ratings standard if the industry failed to do so voluntarily. The cable industry, however, quickly lent its support to the system and, when Fox broke broadcast ranks to do the same, other broadcasters soon followed, agreeing to work on a voluntary system. Under the system, programs were given age-based ratings akin to those used in the motion picture industry, later adding additional information about levels of violence or sexuality. Parents could program their V-Chip to filter inappropriate programming, but by the early 2000s, there were questions about whether the filters were being widely used.

Rates

Cable rates continued to climb after the 1996 Act, outpacing inflation even before the sunset of controls in 1999. Rates rose 53.1 percent between 1993 and 2003, compared with a 25.5 percent increase in the Consumer Price Index.[43] Consumer groups complained and some members of Congress took action, but not much. Senator John McCain (R-Arizona) and Congressman Billy Tauzin (R-Louisiana) were among the leaders criticizing the increases. McCain held hearings in 1998 and Tauzin, with Congressman Edward Markey (D-Massachusetts), introduced legislation designed to maintain some prices controls.[44] Cable operators responded, blaming the increases on rising programming costs, adding that the industry had poured $75 billion into system upgrades between 1996 and 2002, and pointing out that inflation-adjusted prices had dropped when figured on a per-channel basis. Despite the criticism, the Republican-controlled Congress was not interested in reversing its deregulatory philosophy, and was distracted by larger domestic and international issues (including a Presidential sex scandal and subsequent impeachment proceedings.) The rate control scare was minor and short-lived and the March 31, 1999, expiration of rate regulation generated little notice in public or governmental circles.

More Merger Mania

Like Hostetter and Gerry, many of the older cable operators decided to take advantage of the sellers market and cash out. At the same time, those companies that chose to stay were forced by the nature of the market to expand. In 1999, for example, veteran operator Cox bought TCA Cable for $3.26 billion and purchased $2.8 billion worth of systems from AT&T (see below), then began swapping and clustering in concert with other MSOs. Family-owned Adelphia made a modest purchase in Telemedia Corp. for $85 million, then went on a buying binge at the end of the decade. It purchased FrontierVision Partners, Century Communications Corp., and Harron Communications Corp. In 1999 alone, it spent more than $11 billion on acquisitions, doubling its subscriber base to more than 5 million by the end of the decade and becoming the nation's sixth largest MSO.

Overall, in 1999, half of the top ten cable MSOs were bought out.[45] In the process of consolidation, company names that had been dominant in the industry for years—TCI, Continental, Cablevision Industries, and Sammons—were wiped off the chart. By 2000, only seven MSOs had more than 1 million customers each. The leaders included AT&T Broadband, Time Warner, Comcast, Cox, Adelphia, Cablevision Systems, and a newcomer, Charter Communications. AT&T, Time Warner, and Comcast alone accounted for about 50 percent of all cable subscribers. Concentration ratios increased for the top four firms from between 41 to 49 percent in 1995 to 60.1 percent in 2000. The ratio for the top eight firms increased from around 60 percent to 84.8 percent in the same period.[46]

Consolidation had characterized industry development throughout most of cable's history, but through the 1970s and 1980s, the number of cable companies continued to grow as entrepreneurs sought out new, uncabled markets. By 2000, however, no room remained for such expansion, more than 95 percent of the nation had been cabled and consolidation was not offset by the formation of new firms. As a result, the 1990s saw a drop in the total number of MSOs and even more overpowering industry dominance by the top six or seven. Of the leaders, two deserve special consideration for their 1990s growth spurts, AT&T, discussed above, and Charter Communications.

Charter

Paul Allen was one of the richest men in the world. Along with Bill Gates, he founded Microsoft Corp. and, in the 1990s, was expanding his interests beyond just software. Gates had signaled his interest in cable buying with a billion dollar investment in Comcast. Allen also saw cable's broadband pipeline as the distribution channel of choice for the digital age. He knew the handicaps of other technologies in the provision of high-speed Internet access and the bandwidth gluttons, such as video streaming. He chose a more direct route to participation in the cable market for his venture capital firm, Vulcan Ventures, and in 1998 began buying cable systems.

Paying premium prices (up to fifteen times cash flow when the going rate was closer to eleven), Allen went on a buying spree, purchasing 94 percent of Marcus Cable Partners for $2.8 billion in April 1998 and then Charter Communications for $4.5 billion in July, maintaining the Charter name. In February 1999, Allen bought about two-thirds of Greater Media, Inc.'s cable properties. In May 1999, he acquired Fanch Communications for more than $2 billion and Falcon Cable for $2 billion. He also purchased cable properties from InterMedia Partners, Rifkin and Associates, Bresnan Communications, and Avalon Cable, spending a total of $12 billion. Charter also spent $1.65 billion for a percentage of cable competitor RCN Corp. in 1999 (see discussion of RCN below). By 2001, Charter was the fourth largest cable company in the country with more than 6 million subscribers.

Comcast

Comcast had not exactly been flying under the radar. By 1999, it was the country's fourth largest cable operator. But it never generated the publicity of a TCI or a Time Warner. Ralph Roberts was a solid, mainstream businessman, not in the same mold, for better or worse, as a John Malone, Ted Turner, or Gerald Levin. Comcast plodded a steady path of investment and expansion, relatively free from colorful trade press headlines. It had participated in major systems acquisitions in the 1980s, buying a large piece of the Westinghouse cable division (the old TelePrompTer properties) in 1986 and expanding further by buying 50 percent of Storer Communications in 1988.

In February 1990, Ralph Roberts' son, Brian, was named president of the company. The 30-year-old Roberts had learned the business at his father's knee. Like the children of other cable pioneers, he had spent a portion of his youth working for the company during the summers, helping to string wire and sell service. His father took him to business meetings, where he sat listening quietly while Ralph Roberts negotiated with bankers, programmers, and other system operators. Brian Roberts graduated Wharton Business School, and his father encouraged him to find a job outside of Comcast, but the son wanted to be a part of the family company. His father agreed and Brian continued his managerial apprenticeship, working as an assistant general manager at a Comcast system in Flint, Michigan, then as a general manager in Trenton, New Jersey. He learned the business from the amplifiers up and, according to most reports, won the respect of local and corporate employees along the way.[47]

Comcast's philosophy differed from that of TCI and many other cable operators who sought complex interlocking partnerships and were willing to cede control in exchange for equity. Malone had built a career buying non-controlling parts of cable distribution and programming companies, seeing TCI and Liberty in part as cable holding companies. Ralph Roberts was known to privilege close managerial control of his assets and he avoided any alliance or business relationship that did not give him substantial say in decision-making. Like TCI, however, Comcast had a super-voting class of stock such that Ralph and Brian owned less than 3 percent of the company, but held 80 percent of the voting shares.

Comcast's steady but measured growth through the 1970s and 1980s seemed to accelerate after Brian Roberts became President. As noted in Chapter 12, the company purchased MacLean Hunter's U.S. cable systems and E. W. Scripps cable in the first part of the decade, and successfully jousted with Barry Diller, who Brian Roberts had helped hire to run QVC, for control of the shopping channel.

Comcast's increasing aggressiveness was starkly illustrated in 1997 when Brian Roberts teamed with Bill Gates to make a secret unsolicited bid for control of TCI just after the death of Bob Magness. Through an intermediary, they offered to buy what would have been controlling shares of stock bequeathed to the Magness heirs. The effort failed and when Malone learned of it later, he was furious at Roberts. Roberts apologized, then Comcast went searching for more cable systems.

In 1998, Comcast bought the Bell Canada interest in Jones Intercable, following a legal dispute between Jones and Bell. It then exercised an option to buy the super-voting shares held only by Jones himself, effectively taking control of the company. In February 1999, it purchased cable systems serving 80,000 subscribers in its home territory of Philadelphia from Greater Media, Inc. (Charter purchased the remainder of Greater Media.) The next month, it made its most ambitious move to date, in its offer to buy MediaOne. AT&T won the brief contest to control MediaOne, but Comcast would walk away with a $1.5 billion consolation prize as well as important cable assets that helped solidify its coverage of the Philadelphia area. Cable was not the company's only local interest, however. By 1999, it owned 66 percent of Comcast-Spectator, the sports holding company that controlled the Philadelphia 76ers basketball team, the Philadelphia Flyers and Philadelphia Phantoms hockey teams, the CoreStates Spectrum, and CoreStates Center sports and entertainment arenas.

Between 1996 and 1999, Comcast added 2.4 million subscribers to its base, and remained on the lookout for new opportunities. It would not be long before the Roberts' patience was rewarded.

Consolidating at Home and Abroad

For years, regional clustering had given cable an opportunity to increase efficiency and utilize economies of scale. By the mid-1990s, clustering offered operators efficiencies not just in areas such as personal, marketing, and advertising sales, but also in emerging technologies such as fiber ring topologies, digital servers, and ever-smarter routers. MSO system swapping designed to help create clusters, therefore, continued at a healthy clip in the latter 1990s. By 1998, clustered systems accounted for more than one-half of all cable subscribers in the nation.

Time Warner held systems in New Jersey, Pennsylvania, and Illinois that were coveted by TCI. TCI had systems in Florida, New York, and Maine desired by Time Warner. Complicated systems exchanges allowed Time Warner to consolidate its clusters in New York City, Tampa, Florida, and Los Angeles. TCI (and subsequently AT&T) began to dominate service in northern California, Oregon, and the state of Washington. Cablevision was consolidating in the Long Island, New York, area and Comcast was doing the same in and around Philadelphia.

As the clusters grew, the number of individual cable systems, for the first time in the industry's history, began to decline. A city that at one time had several urban and suburban cable companies saw those properties amalgamated through purchase and trade into one unified metropolitan provider. As a result, the total number of cable systems in the United States fell between 1994 and 1999 from an historic high of about 11,200 to about 10,500.[48]

The nature of corporate consolidation also experienced a qualitative change. Horizontal integration had long been a concern for regulators, as MSOs expanded every decade. With the advent of satellite-delivered programming, vertical integration—ownership ties between networks and distributors—entered as an additional antitrust concern. By 1999, some of the largest media companies in the world, which had ownership interests in programming and distribution across several media platforms, had become typical. Companies such as Disney, News Corp., and Time Warner, owned cable television networks, Hollywood production companies, television networks, and publishing houses.

One goal of these multidimensional corporations was to fully exploit the revenue potential of their key asset—intellectual property. Intellectual property included all television programs, scripts, books, films, even story concepts. Owning distribution systems in all media forms, including TV, print, film, and online, meant reaping profit from a given piece of property in all those venues, as well as maintaining stylistic and marketing control. Companies began using the term "repurposing" to describe the reuse of content in differing venues. The classic example was Time Warner's ownership of the Batman franchise. The company continued to sell Batman in its original comic book form, but also peddled it in films, television, and even records, not to mention the ubiquitous ancillary merchandise such as children's lunchboxes, t-shirts, bed sheets, and key chains. Disney was especially successful in extending content such as "The Lion King" across a host of merchandising outlets.

In addition to leveraging existing property, conglomerates hoped that by bringing together content creation and multiform distribution, new ideas would emerge from the mix. The new catchphrase, therefore, became "synergy," and companies sought the synergistic rewards of merger and integration.

In part, with such ends in mind, Viacom sought once more to expand. To help pare down its debt, Viacom sold its half stake in USA Networks to Seagram Co., (owner of Universal Studios) for $1.7 billion in 1997. Then, in September 1999, Viacom announced it was acquiring CBS for a $37.7 billion. The small syndication operation that the Tiffany network had spun off in 1970 as a consequence of the FCC's fin-syn rules had grown and now was purchasing its birth parent. In 2000, Viacom also bought

Chris-Craft Industries' 50 percent interest in UPN. The purchase of the CBS and UPN owned-and-operator stations placed Viacom out of compliance with FCC ownership regulations, but the company received a waiver and, in May 2000, the Commission approved the historic merger. By the turn of the century, Viacom controlled CBS, UPN, Paramount Studios, cable channels MTV and VH-1, Nickelodeon, and BET, Blockbuster video stores, and Simon & Schuster book publishing.

In a related chain of sales, Barry Diller's Home Shopping Network purchased the USA Networks (USA and SciFi Channel) and other Universal Studio assets from Seagram for $4.1 billion. Murdoch, at about the same time, took advantage of the success of Fox to spin off the broadcasting and cable assets into the publicly offered Fox Entertainment Group. The initial public offering (IPO) was the third largest in corporate history, raising $2.8 billion. While Murdoch was selling, Malone was buying. Malone invested $2.1 billion News Corp. in 1999 and, with about 6 percent of the company, became its second largest shareholder after the Murdoch family.

Each of these corporations, moreover, had global reach. CNN, Discovery, CBNC, and many of the Fox channels, were distributed by satellite and cable across Europe, South America, and parts of Asia. In October 1993, for example, MTV inaugurated a Spanish-language service to Latin America. In November 1996, Murdoch opened Fox Sports Latin America, a Spanish-language premium service for the same region, and in November 1997, News Corp. introduced Fox Sports World.

Decade's Close

By 1999, the cable industry was dominated by AT&T, Comcast, Time Warner, Cablevision Systems, Charter, and Adelphia on the distribution side, and companies like Fox, Time Warner, Disney, and Malone's Liberty on the programming side. Consolidation even touched equipment manufacturing, as Motorola purchased GI for $14 billion in 2000.

With the number of smaller cable companies dwindling, CATA closed its doors on June 30, 1999. Those small firms that did remain had formed the Small Cable Business Association (SCBA), creating a business cooperative to help negotiate lower package prices with national programmers. SBCA later changed its name to the American Cable Association to avoid confusion with SBCA, the satellite industry association.

Nationally, homes passed stood at about 97 million at decade's close, penetration was at about 68 percent. DBS was making inroads, but digital boxes were on the move. The industry reported 4.9 million digital subscribers in 1999 and 9.7 million in 2000. Some 1.6 million subscribers took cable modem service in 1999, 3.9 million in 2000.

Cable invested more than $14 billion in system upgrades in 2000 and another $8.8 billion in programming. The movement into mainstream programming was rewarded by success in the Emmy competition. In 1996, cable networks won one-third of all nominations. HBO alone garnered sixty-six nominations and won twelve Emmys. With its success there, the National Academy of Cable Programming abandoned its cable-only ACE Awards program in 1998.

Elimination of rate control was helpful to cable's standing in the financial community. Revenue from basic and pay subscribers rose to more than $40 billion in 2000. Advertising brought in an additional $13.7 billion. Investors and lenders were more comfortable placing their money in the industry; assisted by dot-com speculation, cable stocks gained, although not to levels commensurate with those of the Internet technology companies. In 1998, the stock value of the top ten publicly traded MSOs soared by 80 percent, with some groups increasing by an average of 30 percent in 1999. Per-subscriber system valuations, which stood at $1,753 in 1992, rose to $2,877 by 1998 and nearly $3,900 in 1999.

Almost fifty years previously, Bill Daniels had seen the financial promise of cable television. Daniels & Associates became the leading brokerage firm in the industry and an incubator for MSOs that, in time, would grow to share dominance in the business. It was near the height of the dot-com boom and cable's latest round of successes that Bill Daniels passed away on March 7, 2000. More than 2,000 mourners, including most of cable's preeminent figures, attended his memorial services. It was a sad day for the industry, and not its last that spring.

The New Millennium

The Dot-Com Bust

The dot-com boom had given young engineers and entrepreneurs the opportunity to retire as multimillionaires before they reached the age of 30. The superheated stock market, driven by the Blue Sky technology frenzy, rose nearly every week for several years. Investors thought that all they had to do was pour in money and sit back. But financial bubbles are susceptible to bursting. The carnage for the technology sector began in the spring of 2000.

Venture capital-backed tech start-ups were burning through millions of dollars a week attempting to gain a foothold in the market. Inexperience and sometimes lavish over-indulgence were widespread; revenues were not. Price-to-earnings ratios mushroomed to the point that moderate investors and fund managers began getting nervous, while bearish observers issued ever-louder warnings about betting only on potential. The market began to slip just before the close of 1999. It rallied through the first half of 2000 and, in August, began a jagged descent. Worried venture capitalists began cutting back on the river of money they were pumping into the tech sector. As the capital thinned and with no real revenue, companies began to slash expenses, then lay off workers, then close shop. The slide became a plummet.

As the technology stocks dove lower, they dragged the rest of the market with them. The NASDAQ peaked at a record 5,048 in early 2000, and then began to stumble, rallying briefly at mid year, before dropping almost without a pause through early 2001. By fall 2001, it had fallen to below 1,500. The Dow Jones and S&P followed. In early 2001, the Dow dipped below 10,000, and dropped to below 7,600 by September 2002. The fortunes so easily made by many were lost. More painfully, average investors who had placed their life savings or retirement in the market found some of it, and sometimes all of it, now gone.

The dot-com flameout singed nearly every corner of the economy. No sector was hit harder, however, than telecommunications, especially those companies that had invested millions in building what they thought would be the hardware backbone of the information age. Firms such as MCI-WorldCom, Adelphia, and Global Crossing had plowed hundreds of millions of dollars worth of fiber into the ground in anticipation of near-term heavy use. But the traffic did not materialize and the fiber sat dark. The debt that companies had incurred to install the fiber remained, however, and with the fall of the market, investors looked more closely at the highly leveraged telecom industry. Equipment and hardware manufacturers were badly bruised as construction orders dried up.

The problems were exacerbated by corporate fraud and scandal. The go-go years of the dot-com boom were fertile ground for a more casual approach to ethics and oversight. Everyone seemed to be making money so, in many cases, no one looked too closely to see how it was being done. In some instances, financial slight of hand, rather than sound management, was responsible for cheery quarterly reports, especially as the market began to unravel. Some executives were more interested in padding their personal portfolios than in the long-term health of the company. Some companies sought to mask the problems from investors by cooking the books, overstating earnings, and disguising losses through often complex and creative, but nonetheless illegal, accounting schemes. The Securities and Exchange Commission (SEC) and state attorneys general began probes and the daily headlines became a billboard of corporate malfeasance and scandal.

The Texas-based energy firm, Enron, after manipulating the ledger to hide its own mounting losses, much of it from a failed foray into broadband telecommunications, declared bankruptcy in December 2001. Enron executive Jeffrey Skilling and company founder Kenneth Lay later would be convicted on multiple counts of fraud and conspiracy. Closer to home, MCI-WorldCom declared chapter 11 in July 2002. Company head Bernie Ebbers went to trial on charges of accounting fraud.[49] AT&T faced the dot-com bust honestly; there were no charges of fraud or financial skullduggery, but the fiscal trauma was real and would have severe consequences for the company.

AT&T Departs

No issues of impropriety existed at AT&T, just problems of poor luck and, critics would charge, poor management. For the first time since 1972, Malone was no longer in control of his cable systems. He advised Michael Armstrong to spin the TCI cable systems off into a tracking stock. The systems needed massive capital for information highway-level upgrades, spending that would cut into AT&T's historically reliable dividends, an important perceived gauge of the company's fiscal health. Structurally separating the systems would allow for cable's notoriously dividend-free economics to be kept off AT&T's books. But Armstrong saw Wall Street, at least initially, as rewarding the company for its cable acquisition and decided to leave well enough alone.[50] Malone chose not to press the issue.

Armstrong proceeded with his plan to use the combined TCI and MediaOne cable assets, now AT&T Broadband, as a springboard for unified, branded data and VoIP phone service. He hoped, additionally, to convince other MSOs to partner with AT&T in data and telephony and thereby create a national local loop system using cable's last mile plant. He sent Leo Hindrey to talk operators into the plan. They did not respond favorably.

AT&T wanted agreements with the major cable firms to use their plants to deliver AT&T-branded voice and service. But, for reasons of both history and marketing, the cable operators were not buying. In November 1998, Cablevision told the press it was uncertain about the deal. In February 1999, Adelphia's John Rigas told reporters, "Ten years down the road, do you want the brand name for that telephony service to be established as an AT&T product or as an Adelphia product?"[51] Publicly, the operators said they would study the proposal. Privately, everyone knew it was dead on arrival.

The relationship between AT&T and Hindrey was not working out well, either. He had stayed on as head of the new AT&T Broadband & Internet Services, but the telephone culture was a very different one than he was used to, and his operating style did not gel with that of his new employers. In October 1999, Hindrey emptied his desk and left AT&T, subsequently taking over as head of an Internet company, GlobalCenter, Inc., a subsidiary of Telcom firm, Global Crossing, Ltd.

AT&T also was having difficulties with Comcast and Cox in their combined Excite@Home broadband venture. The partners had differing ideas about the use of the high-speed service. AT&T saw it as a base for developing local VoIP telephony and wanted to spend millions in upgrades. The cable operators were skeptical about the steep price tag. The system itself was having technical difficulties. Frequent outages and poor service generated more consumer complaints than revenue. Excite@Home, however, was a dot-com company and its stock rose with its market kin through 1999. Tired of haggling with the cable operators over business strategy, AT&T began negotiating for full control. The company threatened to suspend funding for improvements if it could not buy out the partners, and ultimately paid a premium of $48 a share, $30 above its trading price at the time (about $2.9 billion), for the service. If the market had continued to climb, AT&T would have made a good deal. The agreement was announced in late March 2000, just two weeks after the NASDAQ had reached its all-time high. The dot-com crash took out the once-promising ISP. Before the end of 2001, the company would declare bankruptcy; shares that once traded for about $100 could be had in the end for one cent.

About the same time, AT&T was discovering that its MediaOne purchase would cost much more than originally projected. At a record-setting $4,900 per subscriber, critics said AT&T had paid far too much. The $62 billion price was to have been offset by the sale of the company's international assets and its 25.5 percent stake in TWE. The international properties brought in less than anticipated, however, and AT&T could not agree on terms with Time Warner for the TWE position. AT&T had to take on substantially greater debt than anticipated to fund the purchase and much of that debt was in the form of short-term loans.

On May 2, AT&T issued a press release stating revenues for that year would probably be lower than originally expected, largely as a consequence of declining long-distance earnings. In the midst of a downward turning stock market, the news sank AT&T share price by 14 percent. AT&T then began discussing the possibility of spinning off the consumer long-distance division, one source of the fiscal stress. The company already had decided to spin off its wireless operation. The internal debate logically expanded to a discussion of the possibility of a full breakup of the company, much to Armstrong's discomfort given his original strategy of using AT&T's full telco and wireless resources to deliver bundled, integrated voice-data and video services to customers (one wire, one company). The debt load, however, had to be dealt with.

In the fall, the company's board reluctantly concluded that a breakup was the only viable means of saving AT&T. On October 25, 2000, it went public. AT&T would divide into four new companies: AT&T Broadband (the cable and internet divisions), AT&T Business Services, AT&T Consumer Services (consumer long distance), and AT&T Wireless. (The Liberty media assets that came with TCI's prior sale to AT&T also were being spun off into a tracking stock under John Malone's control). The company's position was that the softening long-distance business was pulling down the other more promising portions of the corporation, including the cable assets, and the breakup would, among other things, unleash the true value of those businesses.

Even that value was in question, however. Armstrong bought into cable in part out of necessity and in part out of an apparently sincere belief that he could make the "one wire, one company" equation work. Once again, the social construction of the technology led by several years its true capability. The VoIP technology especially was not yet ready for commercial deployment. AT&T might have weathered that problem, holding its cable assets while the technology matured. With the MediaOne purchase, the company, however, had taken on more debt than it could sustain in a declining market.

Along with the announcement of the division, AT&T said it was reviewing its dividend plans, a signal that its historically safe and stable track record of issuing checks to stockholders was threatened by its precarious financial condition. On the news, AT&T stock dropped heavily once again and now was down 54 percent on the year, vaporizing an estimated $105 billion in company value. Some declared it to be the death knell of a great American business icon.

By issuing a separate tracking stock for Broadband, AT&T also was effectively putting the company up for sale, and buyers were waiting. AT&T had not intended to solicit offers for the cable division, but Comcast, having lost MediaOne to AT&T, was ready to try again. Ralph and Brian Roberts approached the company, but AT&T ignored the overture, hoping Comcast would go away. The Roberts had no intention of doing so, however. Comcast had been serious when it attempted to buy a piece of TCI after Magness' death, serious in its bid for MediaOne and serious now. "The Roberts don't take 'no' for an answer," Reed Hundt told the *Wall Street Journal*, "They repeatedly don't take 'no' for an answer."[52]

In early 2001, Comcast began preparing a hostile takeover of the Broadband division of AT&T. Several meetings with AT&T executives during which Comcast expressed its interest had gone nowhere. On July 8, Comcast tendered an unsolicited bid of $58 billion, including $13 billion in assumed debt, for the cable assets. AT&T Broadband now was in play. AOL Time Warner and Cox entered offers of their own, to some extent at AT&T's prompting, in an effort to bid up the price. The dance of the elephants—AT&T, Comcast, Time Warner, and Cox—continued through the end of the year, but Comcast would not be denied this time. In December, AT&T accepted the Roberts' offer of $47 billion in cash and assumption of $25 billion in AT&T debt, buying the cable systems that had cost AT&T $100 million.[53]

Comcast added more than 11 million homes to its subscriber base and was now the nation's largest cable company, with 22 million customers in forty-one states, 5 million digital cable subscribers, and 2.2 million cable modem customers. AT&T, after only a few short months, was out of the cable business.

John Malone left with his prize, the multibillion dollar program holding company, Liberty Media. With interests in dozens of cable networks and media firms, including News Corp., Malone looked also to new horizons and began investing in cable properties overseas, finding new regions of the world in which to pioneer cable television. The first years of the new millennium also posed challenges for most of the other leading cable firms, including Time Warner, Cablevision, Comcast, and especially for Adelphia.

AOL Time Warner

No cable company better symbolized the expansion and contraction of the dot-com bubble and the social construction of digital technology than did Time Warner. Gerald Levin, ever the faithful devotee of a technological solution, had bet time and time again on the promise of new technology, and no technology was more promising at the end of the 1990s than the Internet. Time Warner explored various ways to enter the game, creating its RoadRunner cable modem service and launching web sites for most of the Time Warner brands. As early as 1994, the company created a site called "Pathfinder" as a gateway to all Time Warner digital content. The company failed to gain financial traction with any of its online initiatives, however. Wall Street did not perceive Time Warner as a dot-com company. Cable had superior broadband capacity, but cable was not the Internet. Without a public perception as a company poised to exploit the potential of the digital era and without a clear plan for becoming one, Time Warner's stock languished. Levin and the board felt the corporation was undervalued, but also knew that to move its stock price (holding steady in the $60s in December 1999), it would need to change its image and that meant making a larger play in the Internet world.[54]

In 1999, AOL was the largest Internet service provider in the country. Some 20 million AOL subscribers used low-speed telephone dial-up modems, but its relative ease of use and growing familiarity to customers made AOL an online favorite. Its portal offered a host of information and telecommunications functions, including e-mail service and an emerging form of online communication increasingly popular among young people, Instant Messaging. Advertisers were discovering AOL and it was enjoying the cable-like benefits of both subscriber and advertising revenue.

The company was established in 1983 as Control Video, selling modems that could download video games and seeking to exploit the enthusiasm for home video game consuls, such as Atari. The company's fortunes waned along with the first wave of electronic home games in the mid-1980s. It changed its name to Quantum Computer Services (QCS) and its business plan to one focusing on software that could help home computer users go online. In 1985, it started a monthly pay service offering interactive services. QCS struggled through the late 1980s, but stayed afloat with venture capital and a small subscriber base. In 1991, it changed its name to America Online and in 1992 it went public. By then, it had about 150,000 subscribers and was poised at the edge of the Internet explosion. Steve Case started as a part-time marketing consultant for the company, largely through the efforts of his brother, Daniel Case III, an investment banker who helped provide funding for the start-up. Case rose through the ranks in the later 1980s and became CEO in 1991.

The Internet began taking off in the early 1990s and AOL was prepared. It marketed aggressively and, by February 1995, it had 2 million subscribers and a market capitalization of nearly $1.5 billion. In November 1995, with 4 million subscribers, Steve Case was named Chairman, replacing long-time company head, James Kimsey. AOL's subscriber count continued to accelerate, hitting 5 million in February 1996 and ten million in 1997. Its value did the same, passing $15 billion in April 1998 and $63 billion by the end of that year. It used its size to buy related and competing businesses, including its chief rival, Compuserve, in 1998, and Netscape in 1999.

As with the major players in the cable industry, AOL was motivated to grow in part out of fear of competition, although in this case the concern was focused on only one company, Microsoft. Case also was sharp enough to realize the market was hyperextended and at some point, perhaps soon, AOL stock would stabilize or even erode. Now was the time to make a move. Growth and diversification, therefore, were on the minds of AOL leadership in 1999.

A variety of potential partners were considered, including the RBOCs and several media conglomerates, but Time Warner stood out both for its variegated content assets and, importantly, for its broadband cable lines. The Internet and AOL had expanded largely on the strength of dial-up modems, but to handle the next wave of Internet traffic, especially streaming video, the system would need broadband capacity. AOL saw in Time Warner what AT&T saw in TCI, high-speed access to millions of U.S. homes.

Case and Levin began talking informally in September 1999, and moving quickly, Case proposed a merger in mid-October. Negotiations over price, corporate identity, and leadership continued through December, reaching what appeared to be an impasse at least once. On January 10, 2000, however, they made their announcement. AOL was buying Time Warner for $183 billion—it was the largest corporate merger in history. AOL would control 55 percent and Time Warner 45 percent of the new company. Steve case would be Chairman, Levin, CEO.

The two sides had discussed the contractual safety mechanism that automatically adjusted the purchase price or, in some cases, killed the deal, if the stock of the either company wavered beyond a specific limit. AOL refused the "collar" and Levin, confident the Internet was the way of the future, acquiesced. "I'll take responsibility for it," Levin said later, "because I believed it was the right thing at the time."[55]

At the time AOL had a market cap of about $160 billion, about twice that of Time Warner. It had only a fraction of the revenues of the established media conglomerate, however. Skeptics pointed out the dangers of betting on a possible market fad and noted the potential clash of corporate cultures. AOL was based in Dulles, Virginia, just outside Washington, D.C., but vibrated with the Silicon Valley style of frenetic casual. It had grown spectacularly in a few short years, making paper millionaires out of many of its employees. The company leadership was young, brash, some even suggested arrogant and obsessed with both making and spending as much money as it could as quickly as it could. Melding to Time Warner's more established suit, tie, and Wall Street style would be a substantial challenge, and was reminiscent of the difficulty in bringing together the Time and Warner cultures a decade before.

The merger was announced in January, a period during which the market was beginning its implosion, and the grand plan began to unravel even before it could be implemented.

The merger was based on the new social construction of synergy. Convergence of technologies and business would lead, it was claimed in the late 1990s, to synergistic relations between the different arms of the previously disparate industries. In this case, Time Warner's massive catalog of content, from its records division to *Time* magazine to CNN and the Cartoon Channel, now would be leveraged through the Internet portal of AOL. Content would be "repurposed," or resold through different distribution channels, print, cable, and Internet, multiplying revenue streams with little added production or distribution costs. The synergy never seemed to gain traction, however, in part because of conflict within the company.

Friction between AOL and Time Warner personnel started immediately and grew more intense as business conditions deteriorated. On paper, AOL was the dominant firm and Case was in charge. The brusque, cool Case and the perceived hubris of the AOL faction irritated those at Time Warner. AOL-ers said they felt like second-class citizens, even though they constituted the future of the company. Relations grew tense. The notoriously independent fiefdoms of Time Warner meanwhile squabbled over how to deliver content over the Internet and, more critically, how to allocate costs and revenues.

AOL had a growing revenue problem, as well. Although subscriber counts had mushroomed through the late 1990s, much of its revenue growth came from advertising, as Madison Avenue poured millions into the cyberspace phenomenon. As the bubble began to deflate, however, advertisers were among the first to flee, cutting back dramatically on their Internet expenditures. In addition, start-up dot-coms with advertising and promotion contracts with AOL were falling into bankruptcy on a monthly basis, wiping millions in anticipated income off the AOL books. The overall growth promised by AOL, therefore, failed to materialize.

The merger faced serious regulatory challenges as well. Companies as small as regional ISPs and as large as Microsoft, Disney, and AT&T were quick and clear in expressing to the government their concern that the new company would discriminate against competitive online businesses. Would

AOL Time Warner permit alternative ISPs on their system? If so, under what financial terms? And critically, who would have "first screen" rights? In other words, whose screen would a customer see first when going on line? The screen was like the front page of a newspaper and would be the door through which all AOL Time Warner customers had to pass to get further into the web. It was a prime location for both internal AOL promotion and, more importantly, advertising. As competing companies and public interest groups began filing briefs with the regulatory agencies examining the merger, FTC Chairman Robert Pitoksky, observed, "I don't believe in my seven years [at the FTC] there has ever been a transaction in which more third parties filed more white papers, challenging the validity and the antitrust laws...There was practically a queue around the block of lawyers and CEOs."[56]

Case had been a leader in the open access fight, challenging the AT&T purchase of TCI and lobbying for federal legislation that would require cable operators to open their facilities to all Internet providers. The irony of Case's new position escaped no one, including Case and Levin. In February 2000, Case and Levin went before the Senate Judiciary Committee and took the open access pledge.

Gaining regulatory approval for the merger nonetheless took most of the year, as the companies carefully picked their way through the antitrust minefield. Finally, in December 2000, the companies agreed to FTC terms on open access, promising to offer equitable access to all providers and even signing a carriage contract with its chief ISP rival, Earthlink. The FCC gave its approval in January 2001.

While the merger worked its way through Washington, the stock market was sliding. In July 2001, AOL Time Warner officials said the company might not hit its financial targets unless advertising improved in the second half of the year. Its stock dropped nearly 10 percent. In August, the company fired 1,700 employees to cut costs and, by the end of the year, Ted Turner had gone public with his anger. AOL Time Warner stock had dropped from about $55 a share in January 2001 to $35 by the end of the year, costing investors an estimated $100 billion. Turner had personally seen $7 billion in worth evaporate. He complained that selling Turner Broadcasting to Time Warner was the worst mistake he had ever made.

As AOL Time Warner stock sunk lower and lower, Turner and other investors looked to assign blame. Turner focused on Levin; Case, seeking to turn any attention away from his own culpability, encouraged the idea. Levin, by his own admission, was tired. The stress was affecting his health; and his detachment from others, his practice of making decisions in isolation from the board and even other executives, irritated many.[57] He did not consult widely in making Time Warner's unsuccessful bid for the AT&T cable assets in 2001 and the move created a rift between Levin and Case. Instead of fighting on, Levin announced his retirement. One of the industry's enduring and most influential leaders was stepping down. On December 5, 2001, he wrote to employees: "I felt that once my work was completed and I was satisfied with the company's direction and progress, I'd invoke that [six month's notice] provision and turn my full energies to the moral and social issues I feel so passionate about. That time has arrived." Adding later, "I'm not just a suit. I want the poetry back in my life."[58]

Richard Parsons was named to succeed Levin as CEO. Parsons took over in May 2002. It did not help the stock price, and company market value dropped to about $90 billion by June. In July, news reports began circulating of potential improper financial transactions at AOL just before the merger, as the company made an effort to maintain its stock value. The SEC began an investigation[59] and COO Robert Pittman, one of AOL's corporate leaders,[60] resigned under pressure. News of the accounting irregularities prompted another drop in share price and, by August, AOL Time Warner, which had been worth about $240 billion in January 2001, was valued at about $40 billion. Investors had lost, by some estimates, up to $200 billion in eighteen months and they began filing class action suits. Anger and attention now turned toward Case, portrayed by some as a digital shyster who had sold Time Warner an empty bag of dreams. Under pressure from powerful investors, including Turner, Case announced his resignation as Chairman of AOL Time Warner on January 12, 2003.

A few weeks later, one of cable's most overpowering figures decided it was his time to leave, too. In an interview for CBS's "60 Minutes II" in February of 2003, Ted Turner told Mike Wallace that he had been reluctant to endorse the merger with AOL, but felt he had little choice to do anything other than support the decision, and claimed he had personally lost between $7 and $8 billion dollars in the subsequent stock decline. Shortly after taping the interview, but before it aired, Turner announced he was retiring, leaving AOL Time Warner. He said he wanted to pursue his philanthropic interests. One of cable's most memorable, transforming giants was departing the industry at age 64. He told Wallace that when the time came, he wanted only one simple sentence on his tombstone, "I have nothing more to say."

AOL Time Warner, which traded for more than $50 dollars a share in 2000, sunk to a low of $9 before beginning a small rebound in the first part of 2005, rising to nearly $17 by June. In August 2005, the company reached a tentative agreement to pay $3 billion to settle lawsuits filed by unhappy investors. Seeking to distance itself from the AOL debacle, the company dropped the AOL logo, reverting to its Time Warner moniker. But the troubles had not ceased. By the end of 2005, billionaire investor Carl Icahn, whose fund owned about 3 percent of the Time Warner stock, launched an assault on the firm, buying more shares and calling for Parsons' resignation and the break up of the firm in the effort to "unlock the company's value." Case resigned from the board in 2005 and joined Icahn in lobbying for the breakup of the company. The investment community did not seem impressed by Icahn's efforts, but he continued his work to dismantle the corporation into early 2006.

Adelphia

John Rigas opened his first cable system in Coudersport, Pennsylvania, in 1953. By June 1973, he had systems in Pennsylvania and New York and brought them together to create a new company, Adelphia. Over the next three decades, he and his family built Adelphia Communications into the nation's sixth largest MSO, with more than $21 billion in annual revenues and 5 million subscribers. It owned Adelphia Business Solutions (ABS) and the Buffalo Sabres hockey team. Its tiny home base, Coudersport, was from top to bottom a company town, owing its comfortable existence to the Rigas family business. The three Rigas sons held key positions in the company. Tim was Adelphia's Chief Financial Officer. Executive Vice President James ran the data and telecommunications division. Michael served as COO.

As with other companies, Adelphia had extended its business into telephony and, through its ABS subsidiary, established a state-wide fiber telecom presence, spending millions in its construction. As with similar, larger fiber-based telecommunications companies, however, the failure of the dot-com promise sent ABS into a market tailspin. Adelphia's $20 billion debt was enormous even by the lofty standards of the cable industry, and the ABS decline was difficult to offset.

Clouds had gathered over the world of cable and telecommunication in 2002, and that spring it began raining bad news in Coudersport. In late March, on a conference call with stock analysts discussing Adelphia's latest earnings report, one analyst asked about a puzzling footnote in the company's press release. The footnote involved loans to the Rigas family guaranteed by Adelphia. Tim Rigas' answers seemed only to confuse the issue and bring more questions about the company's financial condition. Adelphia's stock dropped 18 percent after the conference call. On April 1, Adelphia failed to file its annual report with the SEC, requesting more time and triggering another downturn in its stock. The next day, stockholders began filing lawsuits and by mid-April the SEC had opened an investigation into the company's accounting practices.

Adelphia reported that it had guaranteed $3.1 billion in bank loans to the Rigas family. Much of the money was used to buy Adelphia stock. That stock, which had traded at more than $80 in March 1999, dropped to below $13. Under increasing pressure, Rigas and his sons resigned from the company in May and the NASDAQ halted trading of its shares. Temporary management was named to direct the company and grand juries were convened in Pennsylvania and New York to consider criminal charges. Through May and June, the company failed to make required bond and stock dividend payments totaling more than $140 million and the new management released a list of questionable company expenditures, including $13 million to build a golf course on family property. On June 25, Adelphia filed for chapter 11 bankruptcy protection.

On July 24, 2002, authorities arrested John Rigas, handcuffing and escorting him into a police car in front of his daughter's apartment in New York City. Sons Tim and Michael were arrested at the same time. Reporters had been notified in advance of the arrests and were waiting with cameras. The pictures of the 78-year-old, now-disgraced cable executive being led to jail adorned the nation's newspaper and trade press.

In court, federal authorities charged the Rigases with securities fraud, bank fraud, wire fraud, and conspiracy. They accused the family of looting Adelphia like "a personal piggy bank" and misreporting company finances to keep its stock price buoyant. In July 2004, John Rigas and his son Tim were found guilty on the fraud and conspiracy charges. The elder Rigas was later sentenced to 15 years in prison, his son, 20 years. Michael was acquitted on some charges and the jury deadlocked on others.[61] (In March 2006, Michael was sentenced to 10 months of home confinement on reduced charges.)

Throughout the ordeal, Rigas claimed his innocence, declaring, "we did nothing illegal. My conscience is clear on that."[62] In 2005, in a settlement with the government, the family agreed to fund a $715 million compensation pool for investors to recoup some of the billions they lost in the collapse of the company. The assets of the bankrupt firm were ultimately sold off to pay creditors. Company headquarters were moved to Denver during the transition and the $30 million Adelphia building in Coudersport was left shuttered and vacant. In 2005, most of Adelphia's systems were sold to Time Warner and Comcast for $17.6 billion in cash and stock, further adding to the consolidation of the industry's structure.

Cablevision Systems

Cablevision stocks were pummeled in the market in 2002. It had declined to carry a new Yankees Entertainment and Sports network (YES) on its New York system, causing thousands of loyal Yankee fans to defect, many signing with DirecTV. The company lost a reported 45,000 customers. In response, Cablevision cut costs, laying off 7 percent of its work force, and, in part to raise much-needed cash, sold its successful Bravo cable network to NBC for $1.25 billion.

Cable pioneer and company founder Charles Dolan saw what he thought was a new opportunity in high-definition television and, in 2002, announced development of a DBS system exclusively for the HDTV market. The high-definition only satellite company was called "Voom." Launched in October 2003, it offered twenty-one HDTV channels. But national deployment of HDTV-capable sets was meager at the time and some cable operators already were offering competitive HDTV service. To protect Cablevision from potential losses associated with Voom start-up, Dolan spun off the Voom company along with Cablevision's Rainbow Media programming division. Voom never stopped losing money, however, and uncomfortable investors began encouraging an exit strategy.

Cablevision, as Comcast and Adelphia, was family business. Charles' son James was named Cablevision CEO in 1997. Unlike the father-son combination at Comcast, Charles and James could not agree on business strategy. In January 2005, the younger Dolan led a group of board members in a vote to kill the Voom project and sell its assets to EchoStar. The 78-year-old founder responded by exercising his political and equity interest and engineering the ouster of three board members who voted against him. He then brought in an old ally, John Malone, to join the Cablevision board and lend support to his plans (along with three other new board allies). It was a doomed effort, however. Voom was a financial catastrophe that Dolan had been supporting out of his own pocket. In April 2005, he reportedly reconciled with his son and voted with the rest of the board to shut down the service, which at that point had lost an estimated $661 million.

Comcast

Comcast moved briskly in the first years of the new decade to incorporate the AT&T systems into its network, taking advantage of its increased size and repairing the financially tattered ends left by the acquisition. It fired 5,000 employees and, in July 2003, it sold its 57.5 percent interest in QVC to Malone's Liberty for almost $8 billion to help offset the cost of the purchase. It upgraded equipment (although resuscitation of the AT&T systems proved less expensive than feared) and used its new reach to renegotiate improved carriage fees with programmers such as ESPN and Viacom. The company also began heavy promotion of its VOD and Internet services.

In February 2004, it announced a hostile takeover bid for Disney Corp. Beyond just the historic family-friendly theme parks and cartoon empire, Disney held program-rich resources that included ABC and ESPN. It was seen as a good source for much-needed video-on-demand product. Disney head Michael Eisner had alienated a number of Disney board members, including members of the Disney family, and Brian Roberts calculated that he might be able to exploit the rupture. Despite the internal friction, however, Eisner and the Disney board summarily rejected the $66 billion offer, and Brian Roberts took a beating in the press for reaching too far and too high, while Comcast stock dropped 20 percent. By April, Comcast had abandoned its pursuit of Disney, moving on to new acquisitions, specifically looking to purchase part of troubled Adelphia.

Despite the failure with Disney, by 2005 Comcast was the nation's largest cable TV company, expanding rapidly into high-speed Internet service and VoIP telephony. It continued to expand its own programming, buying, for example, Tech TV in 2004 and folding it into its gaming channel G4. Sadly, company cofounder Daniel Aaron died in February 2003, after a long battle with Parkinson's disease. He was 77.

News Corp.

Murdoch's News Corp. continued to look for investment opportunities, especially those that linked developing technology with content. One of its targets was a company called Gemstar. Gemstar was quickly becoming a dominant force in the provision of interactive program guide (IPG) service. To navigate the broadening terrain of digital programming, cable companies and independent companies, such as Gemstar, developed various types of program guides. The most simple, a scrolling guide placed on one of the channels, listing programming for the next several hours, was a standard service on most systems. But digital boxes permitted interactivity and presented the opportunity to create systems that permitted viewers to navigate the guide, selecting specific channels, looking more closely at content, and clicking directly to that channel or setting reminders that could automatically prompt later viewing. The guides also sorted content by category, from sports to movies to children's shows.

Gemstar was a leader in IPG field. It was led by Henry Yuen, a lawyer, mathematician, inventor, and businessman who moved to the United States from Hong Kong when he was 17 and developed a system for automatically videotaping TV programs at home (VCR+) and then proprietary technology for interactive program guides. He cofounded Gemstar in 1989. Yuen was quick to file patent infringement suits against any and all companies that contested the field in the mid-1990s and, with the rollout of digital cable in the late 1990s, Gemstar seemed poised to dominate the sector.

Gemstar's major competition was *TV Guide*, which was owned by News Corp. and offered its own on-screen listing service. In 1999, Murdoch brokered a deal with Yuen in which Gemstar purchased *TV Guide* for $9.2 billion in stock. News Corp. thereby owned about 41 percent of the merged company. Malone also took part in the deal. He had purchased a heavy stake in *TV Guide* and, with the Gemstar acquisition, owned 21 percent of the combined company. In late 2000, Malone exchanged that stock for 12 percent of News Corp., giving him a total of 18 percent of the company and making him the largest shareholder after the Murdoch family holdings.

Shortly after the Gemstar-*TV Guide* merger, however, an internal audit discovered accounting problems, and the company later released revised revenue statements. Gemstar's stock collapsed on news of the revenue overstatements and News Corp. negotiated a separation agreement with Yuen. Eventually, Yuen became the target of an SEC probe and a criminal investigation.

Murdoch had better luck in his twenty-year quest to own satellite distribution in the United States. In September 2000, GM confirmed rumors it was exploring the possible sale of its Hughes Electronics division, which included DirecTV. It gave Murdoch another chance, and he was nothing if not tenacious. Also eager to write a check, however, was EchoStar's Charlie Ergen, and the bidding war commenced. To raise a war chest, Murdoch sold the Fox Family Channel to Disney for $5.2 billion. (Disney relaunched it as ABC Family). While Murdoch negotiated with GM, however, Ergen made an unsolicited bid of $30.4 billion in stock and $1.9 billion in assumed debt.

Murdoch chose to drop out of the contest, and GM reached agreement with Ergen, selling DirecTV for $25.8 billion in 2001. The DirecTV-EchoStar merger raised antitrust concerns similar to those of the failed Murdoch-Primestar combination, however. It would instantly reduce the country's DBS duopoly to a monopoly, and antitrust flags were flying almost from the start, with Murdoch doing his best to fan the winds of regulatory discontent. Predictably, both the FCC and the Justice Department rejected the proposed merger and the companies terminated their agreement in December 2002.

Waiting, of course, near the now-unused altar was Murdoch. Aiding in his new bid was financial support from Malone's Liberty Media, which had invested an additional $500 million in News Corp., raising Liberty's ownership interest in the company to near 20 percent. Murdoch purchased 34 percent of Hughes Electronics for a bargain basement price of $6.6 billion. This time, antitrust was not an issue, at least not horizontal integration. Some lawmakers were concerned about the vertical integration issues raised by the Fox-DirecTV production-distribution pairing, but the acquisition promised to maintain two national DBS firms and strengthen an existing alternative to cable. Many regulators were pleased with the outcome, which passed federal scrutiny and was finalized in December 2003.[63] Cable, of course, was not as happy. Vertically integrated MSOs had worked for years to restrict access to their programming resources by competing distributors, such as MMDS and DBS companies, prompting the programming access safeguards of the 1992 Act. The tables now were turned. Fox controlled some of the nation's most popular networks, including fX, Fox News, and Fox Sports. Cable operators feared Murdoch would privilege his new DBS platform over the traditional terrestrial wires of cable, and were not assured by his pledge to abide by the FCC's program access regulations.[64]

Viacom & NBC

In 2005, Sumner Redstone split Viacom into two companies. CBS and Viacom's broadcasting and publishing interests were spun off into a new corporate entity, CBS Corp. Viacom's motion picture division, Paramount Studios, was combined with its lucrative cable programming assets to form the new Viacom, which also purchased Dreamworks studios for $1.6 billion in 2006. Redstone retained controlling interest in both companies.

General Electric (NBC), meanwhile, purchased 80 percent of Vivendi Universal Entertainment, forming NBC Universal in 2004. It also purchased USA Network and, within a few years, had a cable programming stable that included CNBC, MSNBC, Bravo, and the SciFi Channel as well. It additionally held equity interests in A&E, the History Channel, the Biography Channel, National Geographic International, and the Sundance Channel.

The RBOCs

For the old-world RBOCs, the telephone business was not what it used to be. Long-distance revenues had been falling for a decade as technology and competition drove down costs, then cash flow. As the new millennium opened, the local loop business was suffering as well. It was a classic case of technological displacement. Cell phones were becoming ubiquitous. Business had been the first to take advantage of the mobility that cell phones offered the legendary industry "road warrior." But closely following were millions of young people, typically first adopters of new communications technology. Instant contact with friends and family, plus anywhere-anytime, personalized, downloadable ring tones that could be keyed to identify specific callers, and built-in directories that meant an end to memorizing or even writing down telephone numbers, were quickly absorbed by the wired generation. The cell phone became so handy and powerful that by mid decade, customers were abandoning their landlines, converting completely to wireless. Between 2002 and 2005, landline usage was shrinking about 4 percent a year.[65]

There also was a healthy migration to VoIP. The VoIP start-up Vonage, propelled by a powerful marketing campaign, low prices, and reasonable and improving voice quality, was steadily nibbling away at the telcos' customer base. An FCC ruling in 2004 holding VoIP service to be outside the scope of traditional state public utility-like regulation only added to its allure.[66]

Cable also was responsible for some of the lost landline customers. In 2003, nearly all of the 2.8 million subscribers to cable telephone service were using a circuit-switched system.[67] But cable operators were becoming aggressive in rolling out VoIP service. Time Warner announced the start of VoIP service in 2003 and, by 2005, was offering it in twenty-seven states. Comcast began a major push into VoIP service in early 2005. In April of that year, AOL and MSN network said they would launch their own branded VoIP service. Cox had been successful with traditional circuit switch technology in telephony but, in 2003, also said it would begin working with VoIP in some markets. Because of the power and flexibility of the digital processing, most of the cable VoIP services offered expected consumer features such as voice mail, caller ID, call forwarding, and 911 emergency service. A typical package gave residential customers unlimited local and long-distance service for a flat $49.94 a month. Operators, such as Cox, also were aggressively peddling service to the business community, offering bundled voice and high-speed data packages. In November 2005, Comcast, Time Warner, Cox, and Advance/Newhouse joined with Sprint-Nextel to offer Sprint Cellular service as part of their package as well. It was an alliance designed to set the stage for wireless delivery of news and entertainment content to cell phone users.

The telephone industry, therefore, was compelled to seek new business opportunities, and it was not without its own resources and strategies. In 1984, AT&T was broken into the long lines division and the seven RBOCs. Following the 1996 Act, the centripetal force of economics began pulling the Baby Bells back together. By 2005, there were four: SBC, Verizon, Qwest, and Bell South. In the interim, the cellular business had consolidated. Verizon controlled its own branded service (in partnership with Vodaphone). SBC and Bell South owned Cingular and, in 2004, they purchased AT&T's cellular business, AT&T Wireless, for $41 billion. (Later that year, Nextel announced the purchase of Sprint for $36 billion.) The loss of AT&T's cable properties and the subsequent sale of its cellular company left the former giant a struggling business relic. It was ripe for purchase and, on January 31, 2005, SBC announced that it was buying what was left of AT&T for $16 billion. The deal that Reed Hundt had helped kill several years before was back.

The acquisition significantly strengthened SBC's national and international telecommunications presence. Despite its difficulties over the previous years, AT&T retained an overseas presence coveted by SBC. Without end-to-end overseas capacity, SBC had found it difficult to court corporate clients. Following the acquisition, SBC served 50 million local customers in thirteen states throughout the West, Pacific West, and Midwest; it offered long distance through AT&T's network, was the nation's leading provider of long-distance and data communication to corporate America, and with Bell South provided nationwide cellular service through Cingular. It was the largest telephone company in the nation, surpassing Verizon. It was beginning to resemble the AT&T of old, and two more business decisions would reinforce the similarity. In November 2005, appreciating the value of the famous AT&T brand, the SBC board voted to officially change its name to AT&T. Then, in March 2006, the new SBC/AT&T again reached out to touch someone, telling the press it would buy Bell South, coming close to reassembling most of the former U.S. telephone monopoly providers. (The FCC approved the $86 billion merger in December 2006.)

Spurred by SBC's purchase of AT&T and seeking to secure its own long-distance capacity, Verizon announced in February 2005 that it would buy MCI for $6.75 billion in cash and stock. MCI had emerged from chapter 11 in 2004 a seriously weakened company, but it retained its physical plant and the merger helped Verizon complete the local, long-distance, and cellular package it needed to remain competitive.

Eliminating competition by absorbing competitors in the long-distance and switched-local loop sectors was insufficient by itself, however. With those businesses in decline, the RBOCs needed to expand in broadband data and video. The telephone companies flirted with cable television throughout the 1990s, advancing, teasing, then backing away. The idea of convergence was sound. It was efficient, in the abstract, to offer related telecommunications services via one wire and one provider. But technological change is always local and contingent, and the technical and business contingencies of the mid- and late 1990s were disincentives to enthusiastic telephone company pursuit of television customers. The telephone companies sought first, therefore, to stabilize their core voice business, then, as the technology matured, to restart their broadband strategies. In data services, that meant posting DSL technology against the technically superior cable modem service, competing primarily on price. It was, however, just a holding action. Full, fiber-based broadband capacity to or near the home seemed to be the end-game strategy. Fiber still promised the delivery of the convergence vision and, by the early 2000s, the remaining and much larger telephone companies, especially Verizon and SBC/AT&T, appeared poised to try again.

In June 2004, SBC announced plans for development of a high-speed infrastructure, running fiber into neighborhoods and using existing copper lines for the last 3,000 feet to subscriber homes. In October 2004, Verizon said it would inaugurate a more ambitious plan to spin fiber directly to 16 million homes, nearly half the company's subscriber base. The cost of the endeavor was estimated at $18 billion, an investment that unnerved stockholders and drove down share prices. But, Verizon Chairman and CEO Ivan Seidenberg was committed to capital expansion for future services, noting the Verizon platform would provide more bandwidth than that of SBC/AT&T. Verizon also began offering fiber-based voice, data, and video services at a test site in Keller, Texas, in 2005.

Both Verizon and SBC/AT&T were again negotiating with content providers, seeking carriage contracts with the major cable programmers and alliances with Hollywood studios. In this, the telephone companies were at a disadvantage, lacking the decades-old business and ownership ties, and discounted rates enjoyed by companies such as Time Warner and Comcast. Working for Verizon and SBC/AT&T, however, was the desire of non-aligned programmers to stimulate competition between distributors, driving up prices for their product. Programmers also had a thirst for the new distribution outlets provided by cell phone users, screens that content providers were hoping to colonize on both a subscription and advertiser-supported basis.[68] SBC/AT&T also partnered with EchoStar to market Dish Network programming and delivery it through SBC/AT&T lines. Verizon and Qwest formed their own marketing partnerships with DirecTV. The RBOCs' path had wavered over the previous ten years, but the direction continued to point toward competition with cable television.

Industry Summary

By the end of 2005, many of the pioneers who had built the cable industry—engineers and entrepreneurs like L. E. Parsons, John Walson, Bob Magness, Milt Shapp, and Irving Kahn—were departed. It may have impressed, possibly even pleased, them to know that the enterprise they helped start now brought hundreds of channels of television into the homes of most Americans. In June 2002, prime time cable viewing drew a 54.0 national share (and see Appendix C). It was notable as the first time in cable's history that more than half of the country was watching cable programming in the evening.

The challenges and the promise of the industry remained. The total number of subscribers leveled off, even dropping by some counts, as DBS continued to pluck away customers. In the FCC's annual review of competition in the multichannel video marketplace, the Commission reported 92.3 million households subscribing to multichannel video providers as of June 2004.[69] Cable's share of that audience had settled at about 66 million customers, while DBS continued to expand, growing from 20.4 million subscribers in 2003 to 23.2 million in 2004. Cable, therefore, controlled about 71.6 percent of the multichannel video program distributor (MVPD) audience in 2004, down from 73.6 in 2003, while the DBS share grew from 22.7 percent to 25.1 percent in the same period.[70] At the end of 2005, cable had a reported 65.4 million subscribers, DBS about 27.1 million or 29 percent of all multichannel video households. The top three multichannel television providers that year were Comcast, DirecTV, and EchoStar.

Merger mania had subsided, with the number of available corporations diminished by previous acquisitions.[71] In 2002, Advance/Newhouse reentered the cable business, dissolving its partnership with Time Warner and taking back control of cable systems serving 2.1 million subscribers. Interestingly, an ambitious overbuilder, RCN Corp.,[72] struggled out of bankruptcy in 2004 and held on to more than 400,000 customers to retain its position as a top-twenty MSO. Founded in 1997 by entrepreneur David McCourt, the RCN mission was to offer competitive bundled service in some of the nation's largest markets, including Boston, New York, Chicago, Los Angeles, and Washington, D.C. Fueled by dot-com optimism, it attracted venture capital, including millions from Paul Allen, but fell with others in the crash. Unable to sustain a debt that reached more than $1.6 billion, it filed for chapter 11 protection in May 2004. Stock once valued at $72 a share sank to below $0.10. By the end of 2004, it had reorganized and was moving ahead with discounted cable-telephone-Internet service that had lured enough customers away from entrenched providers, such as Comcast, to keep it afloat. The list of top MSOs remained stable, however, with Comcast leading, in order, Time Warner, Charter, Cox, and Cablevision Systems.

As noted, the new AT&T and Verizon were once again laying plans to enter the programming arena. Cable was responding to the DBS and telco industries with billions of dollars in digital upgrades, spending about $9.6 billion in 2005 for infrastructure improvement. The NCTA reported 10.9 million digital subscribers in early 2001, climbing to 22.2 million at the end of 2003 and 28.5 million, about 27 percent of all cable households, in 2005.

Digital deployment led to expanded channel capacity. In early 2004, according to FCC estimates, the average cable operator was offering seventy-three analog and 150 digital channels.[73] The total number of channels offered on the average system increased from about 210 in January 2003 to 223 in January 2004. More than 85 percent of all system had bandwidth of 750 MHz or greater.[74] Capacity and demand meant more networks. Companies continued to buy and sell existing channels and start new ones. The *New York Times* bought 50 percent of Discovery Civilization in 2002, later rebranding it the Discovery Times Channel. In 2003, Viacom purchased Time Warner's interest in Comedy Central for more than $1 billion. The NCTA reported about 106 national video programming services in 1994, 283 by 1999, 339 by the end of 2003, and 531 in 2005.[75] Cable viewing rose along with increased program choice and perhaps a generational swing away from conditioned broadcast network viewing. Total daily share for basic cable programming reached 48 in 2005, whereas the combined daily share for ABC, NBC and CBS fell to 24.4.

Despite vigorous DBS competition, cable rates continued to outpace inflation rising to more than $36 a month for extended basic in 2003. And subscription to the premium services, which peaked at about 36 million in 2001, began to decline, dropping to 35.3 million in 2002 and 34.8 million in 2003. The industry continued to nudge out its cable telephone services. It reported 3.5 million customers in 2004 and 5.6 million in 2005, modest but growing numbers. The cable modem business was more robust. There were a reported 21 million cable modem users in 2004 and 25.4 million in 2005.[76]

Cable's financials remained solid, with a reported $69.5 billion in revenue in 2005, as programming choice and services expanded. In 2004, companies such as Comcast began talking about issuing dividends to shareholders, a dramatic change from cable's history of reinvestment and capital upgrades, a sign that investors were starting to press for cash return on investment, but also that the industry was maturing financially.

Content and Culture

Over the course of its history, cable forged a new framework for business practice and industrial structure in U.S. television, and subsequently for television globally. The recast TV structure, in turn, brought changes in programming and content, many of which have been detailed in the last several chapters. Specifically, the introduction of cable television, as scholars and popular writers began noting even in the 1960s, worked to fragment the audience, nudging the classic TV model away from one based on a mass media, three network system. By 2000, the NCTA listed more than 300 premium and advertiser-supported cable networks available for local distribution.

Researchers discovered in the 1980s that viewers typically settled into a routine of viewing a dozen or fewer channels, what was labeled the individual's "channel repertoire."[77] But that repertoire was distinct to each person, and very much not the 1960s experience of a nationally shared TV menu served up by ABC, CBS, and NBC.

Programmers sought market niches to dominate, starting in the early 1980s. Those niches needed to be sufficiently small to be substantively distinguished from existing alternative programmers, but also sufficiently large to sustain the financial requirements of national production. The programmatic homesteading of hard news by CNN, children's TV by Nickelodeon, and sports by ESPN were among the dominant examples.

In an effort to separate themselves from the competition and extend exiting audience bases, programmers also pushed the boundaries of taste and culture that historically had circumscribed the limits of acceptability in U.S. television, and before that radio. Topics, language, and visual material that once had been unacceptable to regulators and to many television consumers began to seep into the programming mix. What viewers began watching on cable in the 1980s, in substance and style, was often radically different from the programming of television's first two decades, arguably a consequence of changing cultural values, but also encouraged by the new economics of cable television.

The premium services led by HBO in the mid-1970s began the exploration by airing unedited R-rated films (much to the displeasure of some, as we have seen). By the end of the 1990s, sexually explicit, soft-core pornography was available nearly every night on some of the leading premium channels, and PPV services, such as Playboy, offered more graphic fare.

Advertiser-supported channels seeking to lure younger viewers pushed the cultural envelope in the areas of sex and language beginning in the 1980s. Interestingly, it was a broadcast company, Fox, in its effort to appeal to a younger demographic, that pioneered new styles and formats in the 1980s, led by irreverent shows such as "Married with Children" and "The Simpsons."

In 1993, MTV upped the ante with a cartoon program featuring a pair of maladjusted, scatological teens called Beavis and Butt-Head. Comedy Central also used a cartoon to explore the outer reaches of social propriety, launching a controversial cartoon program for adults in August 1997 called "South Park." Sometimes brilliantly written social satire was anchored in the lives of a group of suburban Colorado children with a penchant for obscene, and again scatological, language and sensibilities. Comedy Central offered a late-night program in 1999 designed to attract the difficult-to-reach young male audience. "The Man Show" featured among its primary topics of interest, beer drinking and scantily clad women, running its closing credits over videotape of "women on trampolines."

Audience fragmentation encouraged programmers to expand their cultural horizons in other ways. Despite the self-promotion of Murdoch's Fox News channel as "fair and balanced," most observers suggested it took a much more politically charged and unapologetically conservative slant on news and analysis, ruffling the feathers of many mainstream reporters and editors raised on the mother's milk of journalistic neutrality and objectivity. One could argue, however, that Fox News was simply exploiting, and quite successfully, an ideologically defined market segment. The preindustrial press of the U.S. colonial period was characterized by political party affiliation. The notion of an objective, mass appeal approach to journalism was the product, in part, of later industrialization. As cable also helped alter the mass market audience of network broadcasting, the centrifugal forces it unleashed arguably could be seen as leading to additional news channels defined on the basis of ideology, in something of a return to the party press of the colonial era.

As with any business, TV programming is financially successful when revenues exceed costs, but niche programmers faced the issue of how narrowly they could slice the audience pie before production and distribution expenses exceeded the revenue supported by the thinning numbers. One of the variables at work in the equation was, of course, production cost and, as competition squeezed audience size and profit margins, programmers looked to reduce the amount they spent on programming. Cheap material in the form of syndicated game shows, network reruns, and educational nonfiction programming had been a long-standing industry solution to this problem, but in the 1990s a new approach gained popularity. It came to be known as reality programming. Reality programs placed volunteer subjects (and sometimes involuntary subjects) into contrived and sometimes challenging staged circumstance, then simply recorded and edited the results. The new format required minimal writing, employed no professional actors and, consequentially, was very inexpensive when compared with traditional fiction programming. One of the pioneers of the format was MTV, which brought a group of young men and women together into a single household, taping their interactions over the course of several months. MTV's "Real World" became a ratings hit and, in classic television style, gave birth to programmatic siblings. MTV's "Road Rules" took "Real World" into recreation vehicles (RVs). The format became a national sensation when the major networks adopted it and CBS started "Survivor" in the summer of 2000. The cultural and political implications of the rise of cable television, especially in the impact of CNN and even MTV, can only be hinted at here. Details of those changes could and should take another book to explore.

Regulatory Fallout

While provoking something of a popular stir in its first few years, by the early 2000s, the general tenor of the more problematic forms of programming seemed to have come to be taken for granted by many Americans, and what would have generated national controversy decades before was given little notice. Much of the change could be accounted for by the mysteries of changing cultural values, but much was also driven by the demands to seek new and lucrative specialty markets, the launch of several gay-themed channels being but one example.

The 1996 Act and the Clinton administration hoped to deal with the rise of many of the more controversial programs, the increase in sex, violence, and problematic language, technologically, using the V-Chip to empower parents to control the viewing of their children. In the latter 1990s and early 2000s, the V-chip was discussed less and less and many suspected parents either did not know how to use it, found it too troublesome to bother with or, occasionally, did not know it existed. At the same time, the levels of sex and, to a lesser extent, violence, continued to rise in cable and then on network broadcasting. A politically conservative administration and Congress grew more sensitive to the concerns, especially those vocally critical of the world of TV programming, and in March 2005 FCC chairman Michael Powell, who had taken a classic libertarian approach to content issues, was replaced by a more stridently conservative, Kevin Martin.

What was explained as a "wardrobe malfunction" at Super Bowl XXXVIII on February 1, 2004, devolved into a regulatory problem for cable. When singer Justin Timberlake grabbed a piece of Janet Jackson's costume during the Super Bowl halftime show, the "wardrobe malfunction" led to the brief unveiling of her right breast, to the subsequently expressed shock and dismay of millions. Within days, conservative citizens groups and members of Congress were donning their battle fatigues in the war against televised indecency. Broadcast and cable networks beefed up their standards and practices departments and instituted five-second delays on live programming. Congress considered legislation in 2004. Some legislators proposed broadcast-like indecency rules for cable. Some proposed a la carte program availability to protect families from unwanted, questionable content. To deflect the more drastic regulatory proposals, cable and broadcasters quickly initiated multimillion dollar public relations campaigns to remind customers of the blocking V-chip technology and to encourage its use.

Content controls raised First Amendment issues that some legislators were reluctant to broach, but tiering continued to be an issue, stimulated by the indecency furor and by ever-present concerns about cable rate hikes. Pressure in Congress in 2004 to legislate a la carte pricing, led by Senate Commerce Committee Chairman John McCain, died for lack of support, and the FCC under the leadership of Chairman Powell concluded that a la carte programming would increase consumer costs. But his replacement, Chairman Martin, championed a new push for unbundled programming, reversing the Commission's position on the issue.

It led to FCC pressure on cable, first to ensure against indecency and then to create "family friendly" programming tiers. Some of the nation's largest MSOs, including Comcast and Time Warner, as well as EchoStar, complied in late 2005, promising family packages of thirty-five to forty channels priced about $30 a month.

In February 2006, the FCC recommended the industry adopt a la carte program offers, issuing a report that concluded consumers would save up to 13 percent on their cable bills if they could buy individual channels. Cable distributors and the programming community expressed concern about the proposals, noting that tiering provided a form of cross subsidy that supported younger and more finely niched channels.

The Evolving Revolution

The motive force behind the widespread adoption of technology is its ability to extend human power and control—the power of reach, the power of transport, the power to see, hear, know, and communicate instantly across great distances. The achievement of those deeply situated human ends is contingent on the successful mating of the technology with extant social conditions, which include cultural, political, and, critically, economic factors. The introduction of digital technology offered one of those historic inflection points in which the cumulative synthesis of technology presented powerful and flexible new tools for the control of communication and information resources. Digital cable and the Internet were the preeminent manifestations of the digital platform, but several important ancillary devices and functions grew from the transition, presenting cable with both new opportunities and new challenges. These included long-sought Blue Sky services such as HDTV and true VOD, and unexpected devices, such as digital video recorders (DVRs). The greatest long-term challenge to cable, however, was likely to be broadband Internet service, which brought people ever closer to the ultimate goal of nearly unlimited program choice with substantial viewing time control.

HDTV

By 2004, broadcasters were offering standard digital and high-definition signals to 99 percent of all U.S. television households. Although the signals were available, the pace of the digital transition was slowed by the lack of digital TV sets in consumer homes. Broadcasters complained that while they were beaming digital signals to most of the country, very few people were watching them. By the end of 2003, the FCC reported that between 7 and

8.7 million TV homes had digital TV (DTV) sets of about 108 million TV homes.[78]

Consumer adoption of the new technology was slowed in part by the public's lack of awareness of the change and by the high cost of DTV sets. To help rectify these problems, the FCC began a public relations campaign in 2004 designed to inform people about the transition and convince them of its wisdom. It also adopted regulations designed to ease and accelerate the conversion. Manufacturers, in an effort to hold down costs, were selling digital sets without the tuners necessary to receive off-air signals. They reasoned that the additional $250 estimated for the off-air tuners was unnecessary for most consumers who would receive their digital signals through a cable or DBS set-top box and have no need for the device. The FCC, in August 2002, however, ordered installation of digital tuners in all sets over 13 inches by 2007, extending the order in November 2005 to cover sets of all sizes.[79] Recognizing that the pace of change was much slower than anticipated, but nonetheless seeking to set a firm end date, Congress in February 2006, extended the deadline for completion of the transition to February 17, 2009, the date on which broadcasters would have to turn their analog frequencies back over to the government.[80]

In 1999, Time Warner became one of the first cable operators to begin providing high-definition programming, with the test of an HBO high-definition channel for subscribers in Tampa, Florida. Other operators began rolling out limited HDTV distribution soon after and, in 2002, the top ten MSOs pledged to begin offering a package of HDTV channels within a year. The plan hit choppy waters when programmers such as Discovery and ESPN informed the MSOs that the higher costs of producing high-definition programs would be passed along to operators in the form of additional programming fees. By 2005, however, all the major broadcast networks were offering most of their evening programming in a high-definition format, as were many of the major cable programmers, including HBO, Discovery, ESPN, and Showtime. High-definition cable service was available to 96 percent of all TV homes by the end of 2005.

VOD

Advances in digital compression, storage, and switching and the steady diffusion of digital boxes into U.S. homes gave operators the increasing ability to satisfy consumer desire for the "anytime" movie choice provided by video-on-demand. True, full-powered, server-based VOD had been the Holy Grail of cable for decades. It began its national rollout in 2001 and within two years all the major MSOs were offering the service. It gave subscribers access to dozens of current-run films, along with programming that included news, sports, and general entertainment. Some companies offered content on a pay-per-item basis; others offered it as a subscription VOD (SVOD) option that provided access to the full library of material for a monthly fee.

VOD came with VCR-like functionality that allowed customers to pause, rewind, and fast forward films with the touch of the remote. By 2003, more than 12 million cable subscribers had access to VOD, another 8.2 million to SVOD capability.[81]

DVRs

At the heart of the VOD system is the digital server, a high-capacity storage device, the electronic grandchild of the basic computer hard drive. Commercial servers could hold hours of digitized video and host hundreds of subscribers, giving each separate, controllable access to the recorded content. One part of the evolutionary process of technical advancement involves the manner in which new technical possibilities make themselves apparent only after, but as soon as, ground has been broken by preceding developments. With each step up the hill, new vistas appear over the expanding horizon. As the cost and physical size of servers shrank, additional applications became possible and it was no great conceptual leap to consider taking the servers from ISP and cable back shops and placing them into consumers' homes. Home-based servers could perform much like VCRs, but with added digital power and flexibility. The new devices were called digital video recorders (DVRs) and they debuted as consumer devices at the 1999 Consumer Electronics Show in Las Vegas. Also known as personal video recorders (PVRs), two companies offered the product along with a subscription service, TiVo and ReplayTV (although ReplayTV ultimately failed in the marketplace).

To promote the DVR as more than just a digital VCR, the companies linked the devices with subscription services allowing customers to preselect programs for later playback. One of the important selling points was the DVR's ability to give a viewer enhanced program control even when watching a live event. Using the DVR, a viewer could pause the action to attend to other business, then return to the program at his or her leisure. It also allowed the viewer to skip the commercials, and to rewind and replay the program at will.

TiVo partnered with DirecTV to market the utility to potential DBS customers, incorporating the DVR into the DirecTV set-top box. Cable quickly caught on and operators began adding DVR functionality into their set-top boxes, as well. By early 2006, an estimated 7 percent of U.S homes had DVRs.[82]

DVRs broke the shackles of what became known as "appointment viewing," watching a program in traditional style at the regularly schedule time. Viewers no longer had to make that appointment but, again, could automatically record it and watch it at their convenience. Many DVR users vowed they would never return to appointment viewing. For a TV and advertising industry locked hand and foot to a dependable, nonflexible programming schedule, however, this new consumer power was somewhere between worrisome and terrifying. It meant that programmers no longer had surety about when, or even if, a given program was being viewed. Nielsen Media Research responded in late 2005 by creating a service that reported ratings in their traditional form, calculating live viewing, but also adding viewership for programs in the 24 hours after they originally aired and for seven days after airing.

DVRs also were forcing advertisers to change their decades-old media strategies. DVR users could, and did, skip the commercials in their recorded programs. One response was to increase product placement activity, integrating branded soaps, cereals, cars, and colas into the content of the programs themselves. Madison Avenue also was reconsidering the general future of the classic 30-second TV spot and exploring new methods of reaching the American consumer.

Internet Service

VOD and DVRs helped satisfy the consumer's appetite for improved command over the television environment, but they were limited and modest enhancements compared with the potential power of program distribution over a broadband Internet. The national diffusion of broadband access was a double-edged sword for cable. As a leader in high-speed service, cable modems offered an important revenue stream and helped position the industry as one of the principal portals for consumer communications and information services of the future. Program producers could use that platform, however, as a new mode of content delivery, augmenting and, potentially bypassing, the classic broadcast and cable distribution systems.

The importance of the related issues grew along with cable modem penetration. In late 2001, the number of cable modem customers stood at about 7.2 million. The figure grew to 16.1 million in late 2003, 18.5 million in June 2004,[83] and 25.4 at the end of 2005.[84] In 2004, cable Internet service accounted for about $9 billion or 15 percent of all cable industry revenue.[85] By late 2005, more than 73 percent of all U.S. adults used the Internet, according to the Pew Internet and American Life Project. About one-third used low-speed dial-up connections, two-thirds used some form of broadband link. The public was making increasing use of the Internet for all manner of communications, information, entertainment, and business, driving a migration from dial-up to high-speed connections. The increasingly sophisticated and, therefore, bandwidth-hungry content itself additionally encouraged the switch to broadband.

Cable continued to compete with the telcos in a contest that pitted superior cable modem speed with the lower price of telco DSL service. The pressure for greater speed also gave cable, and others, an entrée for offering bandwidth as a commodity, setting escalating price points for tiered levels of speed.

With increased technical capacity and a growing Internet audience base, content owners began experimenting with online program access and distribution. In the first years of the new millennium, peer-to-peer, file-sharing software gave rise to widespread online swapping and distribution of popular music. The MP3 format made it easy for millions of young people to copy and then share music files using utilities such as Napster. That most of the copying violated copyright law and was grossly illegal did little to impede the popularity of the activity. And while the music industry fought a rear guard action to halt the ubiquitous piracy, the increasing power of the software and hardware provided users the ability to begin treating visual content, specifically movies, in the same manner. By the early 2000s, films and TV programs were a small part of Internet traffic, but a growing one.

In an effort to both control their intellectual property and to seek a means to take commercial advantage of the situation, film and TV production firms began, tentatively, making their products available online. In early 2006, John Malone's Liberty Media introduced Vongo, a Starz Entertainment Group web site offering hundreds of concerts, films, and sporting events for $10 a month. NBC Universal moved its struggling Trio channel off cable and onto the web in early 2006, while the company's Bravo division began creating additional internet-only programming sites.

In March 2006, AOL introduced what it promoted as the first broadband TV network on the web, AOL Television, or In2TV (for Internet 2). It offered a large helping of vintage Warner Bros. TV content such as "Growing Pains" and "F Troop." Other commercial sites offered cartoons, short films in the public domain, and thousands of home videos. Viacom offered online access to popular Nickelodeon programming, such as "SpongeBob SquarePants," as well as to clips from CBS news. Google, one of the industry leaders, opened video.google, featuring TV programming that included "I Love Lucy" shows, sports, and films.

An important additional level of complexity and opportunity was provided by the evolution of compact, mobile storage devices such as Apple's iPod and by national broadband wireless capability. Cell phones initially used only for voice communication were given, at first, narrow band data functionality that permitted customers to check their e-mail and conduct limited Internet searches. The spread of high-speed wireless networks, wireless hot spots, and Internet cafes, opened possibilities to move beyond just downloading sports scores to downloading sports clips and even entire games, along with news and entertainment programming.

By the end of 2005, many of the major studios were offering clips of programming and even entire shows as video downloads available to anyone with the right software and a mobile laptop computer, an iPod or a cell phone.

In fall 2005, Disney began making individual ABC programs, such as the popular "Lost" and "Desperate Housewives" available for downloading to iPods. In December 2005, Sprint announced it would begin streaming movies to its cell phone customers for $6.95 a month. In early 2006, News Corp. formed a new division dedicated to producing entertainment content exclusively for distribution to cell phones, allowing customers to purchase content directly from it instead of going through the telecom provider. CBS was pursuing similar projects to create a monthly subscription service to send news and entertainment to cell phone customers.

Finally, what some considered the democratizing nature of the Internet was making its presence felt not just at the consumption end, but at the creation end as well. Before the Internet, studio-quality film and television production was remarkably expensive and distribution tightly controlled. Digital production and distribution technology was rapidly tearing down those barriers to entry. The increased power and decreased cost of digital video meant the price and complexity of individuals creating their own TV program was falling like a stone. Home-brewed video programs of every description were being posted on the web; teenagers were trading video files as their grandparents had traded baseball cards. Commercial sites began accepting customer product as an adjunct to their own fare. In announcing its video.google site, Google declared that it would allow users to upload their own programs as well.

Leading the way was a generation raised on computers and cell phones, rather than cable television, and as comfortable with and dependent on IMing, text messaging, Google, and MySpace as previous generations had been with the landline telephone and AM radio. It was described as a primordial soup of culture, technology, and economics. What kind of creature or creatures would arise from the brew was very much an open question, but savvy businesspeople continued to speculate. In 2005, Murdoch purchased the wildly popular youth-oriented web service MySpace for $580 million.

"What's Gonna Be Next?"

The rush by cable, telephone, and entertainment companies to lay plans for the delivery of Internet and video content to mobile and wireless devices vividly illustrated, once more, the underlying mechanics of technological evolution and the major themes of this book. Progress in the creation and deployment of these services was built on the incremental improvements in speed, capacity, and flexibility of the technical platform. Ushers' cumulative synthesis could be seen in the conversion from analog to digital environments. The continuing efforts to expand high-speed, broadband capacity across the nation and the globe also suggested a likely inflection point in the qualitative nature of the system.

At each step, the process has been channeled and shaped by local social and economic conditions. It also has been powerfully influenced by the human imagining of what it could become, the ideation of the business people financing and managing it, and of those in political power with the ability to promote or impede certain kinds of development through policy and regulation. As the pilots of cable television and the affiliated communications industries charted their course into the early twenty-first century, the cultural constellations of the social construction of technology continued to offer up the guiding stars by which they steered.

Whether or not a new generation of customers would flock to view news, sports, and movies on the two-inch screens of their cell phones and iPods remained, in 2006, an open question. Whether the always-emerging communications system would follow the course plotted by cable operators, movie studio chiefs, and politicians remained, as always, to be seen. A review of the history of cable television suggests that surprises, not all of them pleasant, lurk behind the swell of every new product launch, and one must approach prediction of future development very, very cautiously.

At the end of 2005, Comcast was the nation's largest cable operator. It had 21.7 million subscribers, twice that of its closest rival, Time Warner. It provided broadband Internet service to 9 million homes and phone service to 1.5 million. Its patriarch, Ralph Roberts, was one of the few industry leaders left from the pioneer days of the 1960s. He had built the company slowly and steadily, handing it on to his son Brian, who accelerated its expansion. The company had taken one of its largest strides in late 1985, purchasing a part of the Westinghouse cable assets and doubling its subscriber size in one step.

Two days after the deal closed, the senior Roberts gathered his executives in a conference room at corporate headquarters in Philadelphia. He had a question for them, a question concerned specifically with Comcast's near-term growth plans, but ripe with broader meaning for the cable industry, an industry that had marched, over the previous fifty-plus years, across the United States and then the world, changing the face of television, touching international politics, news, entertainment, and culture, turning rough-shod small-town businessmen and entrepreneurs into billionaires and helping lay the foundation for the broadband digital communications infrastructure of the next century. Resonating technically, culturally, and politically far beyond the Philadelphia boardroom, Roberts' simple question to his cable colleagues was, "what's gonna be next?"[86]

Appendix A: Selected Cable Statistics (1948–2005)

YEAR	SYSTEMS	SUBSCRIBERS	BASIC CABLE/TV HH (PERCENTAGE)	SUBSCRIPTION REVENUE ($ MILLION)	PAY CABLE REVENUE ($ MILLION)	AV. MONTHLY BASIC RATES
1948	1–3					
1949	3–5					
1950	10					
1951	30–50					
1952	70	14,000	0.1			
1953	150	30,000	0.2			
1954	300	65,000	0.3			
1955	400	150,000	0.5	15		$5.00
1956	450	300,000	0.9	n.a.		n.a.
1957	500	350,000	0.9	n.a.		n.a.
1958	525	450,000	1.1	n.a.		n.a.
1959	560	550,000	1.3	n.a.		n.a.
1960	640	650,000	1.4	45		5.00
1961	700	725,000	1.5	n.a.		n.a.
1962	800	850,000	1.7	n.a.		n.a.
1963	1,000	950,000	1.9	n.a.		n.a.
1964	1,200	1,085,000	2.1	n.a.		n.a.
1965	1,325	1,275,000	2.4	90		5.00
1966	1,570	1,575,000	2.9	n.a.		n.a.
1967	1,770	2,100,000	3.8	n.a.		n.a.

(Continued)

YEAR	SYSTEMS	SUBSCRIBERS	BASIC CABLE/TV HH (PERCENTAGE)	SUBSCRIPTION REVENUE ($ MILLION)	PAY CABLE REVENUE ($ MILLION)	AV. MONTHLY BASIC RATES
1968	2,000	2,800,000	4.4	n.a.		n.a.
1969	2,260	3,600,000	6.1	n.a.		n.a.
1970	2,490	4,500,000	7.6	337		$5.50
1971	2,639	5,300,000	8.8	n.a.		n.a.
1972	2,841	6,000,000	9.6	n.a.		n.a.
1973	2,991	7,300,000	11.1	n.a.		n.a.
1974	3,158	8,700,000	13.0	n.a.		n.a.
1975	3,506	9,800,000	14.3	764	29	6.50
1976	3,651	10,800,000	15.5	887	68	6.75
1977	3,800	11,900,000	17.3	1,025	125	7.00
1978	3,875	12,500,000	17.1	1,167	239	7.25
1979	4,150	13,600,000	18.3	1,355	435	7.50
1980	4,225	15,200,000	19.9	1,649	785	7.85
1981	4,375	17,830,000	22.3	2,100	1,336	8.14
1982	4,825	24,900,000	29.8	2,579	2,081	8.46
1983	5,600	28,320,000	34.0	3,096	2,787	8.76
1984	6,200	32,930,000	39.3	3,627	3,410	9.20
1985	6,600	36,340,000	42.8	4,353	3,787	10.25
1986	7,500	39,160,000	45.6	5,156	3,876	11.09
1987	7,771	41,690,000	47.7	6,553	4,084	13.27
1988	8,413	43,790,000	49.4	7,664	4,398	14.45
1989	9,050	52,565,000	57.1	8,792	4,814	15.97
1990	9,575	54,871,000	59.0	10,174	4,882	16.78
1991	10,704	55,786,000	60.6	11,418	4,968	18.10
1992	11,035	57,212,000	61.5	12,433	5,108	19.08
1993	11,108	58,834,000	62.5	13,528	4,810	19.39
1994	11,214	60,495,000	63.4	15,170	4,394	21.62
1995	11,218	62,956,000	65.7	16,860	4,607	23.07
1996	11,119	64,654,000	66.7	18,395	4,757	24.41
1997	10,950	65,929,000	67.3	20,405	4,823	26.48
1998	10,845	67,011,000	67.4	21,830	4,857	27.81
1999	10,466	68,538,000	68.0	23,146	4,930	28.92
2000	10,400	69,297,000	67.8	24,142	4,861	30.08
2001	10,300	72,958,000	69.2	26,324	5,201	31.58
2002	9,900	73,525,000	68.9	27,690	5,226	34.52
2003	9,339	73,365,000	67.7	28,962	5,192	36.59
2004	8,875	65,250,000	67.0	n.a.	n.a.	n.a.
2005	7,926	65,400,000	66.3	n.a.	n.a.	n.a.

Number of systems between 1948 and 1951 are author's estimates. Number of systems and subscribers after 1952 are drawn from *Television & Cable Factbook*, Services Volume, Washington, D.C.: Warren Publishing, annual. Basic revenue, pay revenue, and average monthly basic rates from 1955 to 1975, compiled by Paul Kagan Associates, Inc., as reported in *1983 Cable TV Financial Databook*, Daniels & Associates, Inc., 46. Basic revenue, pay revenue and average monthly basic rates from 1976 to 2003 from NCTA, *Cable Television Developments*, annual.

Appendix B: Cable Channel Capacity (1970–2005)

CHANNELS	1970	1975	1979	1985	1990	1995	2000	2005
125 +							13	126
91–124							90	147
54–90				396	808	1,558	2,190	2,087
30–53			358	2,455	4,846	6,376	5,716	3,441
20–29		382	758	1,881	1,370	1,104	735	330
13–19	86	354	103	319	283	353	218	94
6–12	1,720	2,415	2,793	1,203	1,081	558	247	120
5 only	459	195	129	53	25	10	7	5
Sub-5	61	32	22	10	7	4	5	0
No report	164	27	17	527	1,192	1,133	1,022	1,576
Total	2,490	3,405	4,180	6,844	9,612	11,126	10,243	7,926

Television & Cable Factbook, Washington, D.C., Warren Publishing, annual. Data as of October 1.

Appendix C: Viewing Shares for Basic Cable Networks, Broadcast Network Affiliates and Independents (1983–2003)*

YEAR	BASIC CABLE NETWORKS	BROADCAST NETWORK AFFILIATES	IMPENDENT BROADCAST STATIONS
1983–84	9	69	19
1984–85	11	66	18
1985–86	11	66	18
1986–87	13	64	20
1987–88	15	61	20
1988–89	17	58	20
1989–90	21	55	20
1990–91	24	53	21
1991–92	24	54	20
1992–93	25	53	21
1993–94	26	52	21
1994–95	30	47	22
1995–96	33	46	21
1996–97	36	43	20
1997–98	40	49	12
1998–99	41	46	11
1999–2000	46	44	12
2000–01	46	42	11
2001–02	53	39	11
2002–03	57	38	12

*Share of total television households. Beginning in 1997–98, Fox Broadcasting affiliates, previously counted as independents, were counted as network affiliates.

From A. C. Nielsen Company and the Cable Advertising Bureau, *Cable TV Facts*, as reported in, NCTA, *Cable Television Developments*, Washington, D.C.: National Cable Television Association, annual.

Notes

PREFACE

1. See, for example, Ron Garay, *Cable Television: A Reference Guide to Information* (New York: Greenwood Press, 1988); Christopher Sterling, James Bracken, and Susan Hill, eds., *Mass Communications Research Resources: An Annotated Guide* (Mahwah, NJ: L. Erlbaum Associates, 1998); Don LeDuc, "A Selective Bibliography on the Evolution of CATV," *Journal of Broadcasting*, 15:2 (1971): 195–235; and see Felix Chen, *Cable Television: A Comprehensive Bibliography* (New York: Plenum Publishing, 1978).

2. Mary Alice Mayer Phillips, *CATV: A History of Community Antenna Television* (Evanston, IL: Northwestern University Press, 1972).

3. Kenneth Easton, *Thirty Years in Cable TV: Reminiscences of a Pioneer* (Mississauga, Ontario: Pioneer Publications, 1980).

4. Kenneth Easton, *Building an Industry: A History of Cable Television and Its Development in Canada* (Lawrencetown Beach, Nova Scotia, Canada: Pottersfield Press, 2000).

5. Archer Taylor, *History Between Their Ears* (Denver, CO: The Cable Center, 2000).

6. Mark Robichaux, *Cable Cowboy: John Malone and the Rise of the Modern Cable Business* (Hoboken, NJ: John Wiley & Sons, Inc., 2002).

7. Joseph DiStefano, *COMCASTed: How Ralph and Brian Roberts Took Over America's TV, One Deal at a Time* (Philadelphia, PA: Camino Books, Inc., 2005).

8. Porter Bibb, *Ted Turner: It Ain't as Easy as It Looks* (Boulder: Johnson Books, 1997).

9. James Roman, *Cablemania: The Cable Television Sourcebook* (Englewood Cliffs, NJ: Prentice Hall, 1983); Thomas Baldwin and D. Stevens McVoy, *Cable Communications* (Englewood Cliffs, NJ: Prentice Hall, 1983); Patrick Parsons and Robert Frieden, *The Cable and Satellite Television Industries* (Needham Heights, MA: Allyn & Bacon, 1998).

10. Martin Seiden, *Cable Television U.S.A.: An Analysis of Government Policy* (New York: Praeger Publishers, 1972); Don Le Duc, *Cable Television and the FCC: A Crisis in Media Control* (Philadelphia: Temple University Press, 1973).

11. Ralph Lee Smith, *The Wired Nation* (New York: Harper and Row, 1972); Ralph Negrine, ed., *Cable Television and the Future of Broadcasting* (New York: St. Martin's Press, 1985).

12. Steve Keating, *Cutthroat: High Stakes and Killer Moves on the Electronic Frontier* (Boulder: Johnson Books, 1999).

13. Baldwin and McVoy, *Cable Communications*; Christopher Sterling and Michael Kittross, *Stay Tuned A Concise History of American Broadcasting*, 2nd ed. (Belmont, CA: Wadsworth Publishing, 1990).

14. Thomas Southwick, *Distant Signals* (Overland Park, KS: Primedia Intertec, 1998).

15. This book draws on many of the oral histories collected by the Center. Those oral histories are available online at: http://www.cablecenter.org/education/library/oralHistories.cfm.

CHAPTER 1

1. Edward Bellamy, *Looking Backward* (Signet, Books, 1960), 86–87.

2. See, for example, George Basalla, *The Evolution of Technology* (New York: Cambridge University Press, 1988); and John Ziman, ed., *Technological Innovation as an Evolutionary Process* (New York: Cambridge University Press, 2000).

3. Abbott Payson Usher, *A History of Mechanical Inventions* (Cambridge, MA: Harvard University Press, 1929), 68–69.

4. See, for example, Richard Hubbell, *4000 Years of Television: The Story of Seeing at a Distance* (New York: G.P. Putnam's Sons, 1942); and David Fisher and Marshall Fisher, *Tube: The Invention of Television* (Washington, D.C.: Counterpoint, 1996).

5. E. Stratford Smith, "The Emergence of CATV: A Look at the Evolution of a Revolution," *Proceedings of the IEEE 58* (July 1970): 967–982.

6. Alex Roland, "Theories and Models of Technological Change: Semantics and Substance," *Science, Technology & Human Values*, 17:1 (Winter 1992): 79–100, 84.

7. Thomas F. Carter, *The Invention of Printing in China and its Spread Westward*, 2nd ed. (New York: Ronald Press, 1955).

8. Elizabeth L. Eisenstein, *The Printing Press as an Agent of Change* (Cambridge: Cambridge University Press, 1979).

9. Thomas P. Hughes, "The Evolution of Large Technological Systems," in *The Social Construction of Technological Systems*, eds. Wiebe Bijker and Thomas Hughes and Trevor Pinch (Cambridge: MIT Press, 1987).

10. Brian Winston, *Media Technology and Society* (London: Routledge, 1998).

11. Erwin Krasnow, Lawrence Longley, and Herbert Terry, *The Politics of Broadcast Regulation*, 3rd ed. (New York: St. Martins Press, 1982).

12. See generally, Bijker, Hughes and Pinch, *The Social Construction of Technological Systems*; and Janet Fulk, "Social Construction of Communication Technology," *Academy of Management Journal*, 36:5 (1993): 921–950; and more broadly, Peter L. Berger and Thomas Luckmann, *The Social Construction of Reality: A Treatise in the Sociology of Knowledge* (Garden City, NY: Doubleday, 1966).

13. Edward R. Murrow, address to the Radio Television News Directors Association (Chicago, IL: October 15, 1958).

14. The manner in which radio evolved from "wireless" telegraphy and telephony into "broadcasting" is traced by authors such as, Erik Barnouw, *A Tower in Babel: A History of Broadcasting in the United States to 1933* (New York: Oxford University Press, 1966), *The Golden Web: A History of Broadcasting in the United States, 1933–1953* (New York: Oxford University Press, 1968), *The Image Empire: A History of Broadcasting in the United States from 1953* (New York: Oxford University Press, 1970); Hugh G.J. Aitken. *Syntony and Spark: The Origins of Radio* (New York: Wiley/Interscience, 1976), and *The Continuous Wave: Technology and American Radio, 1900–1932* (Princeton, NJ: Princeton University Press, 1985); and Christopher H. Sterling and John Michael Kittross, *Stay Tuned: A History of American Broadcasting*, 3rd ed. (Mahwah, NJ: Lawrence Erlbaum, 2002).

15. Susan Douglas, *Inventing American Broadcasting, 1899–1922* (Baltimore, MD: The Johns Hopkins University Press, 1987), xvi–xvii.

16. Robert Hilliard and Michael Keith, *The Broadcast Century and Beyond*, 3rd ed. (Boston: Focal Press, 2001), 3.

17. Alvin Harlow, *Old Wires and New Waves: The History of the Telegraph, Telephone, and Wireless* (New York: D. Appleton-Century Co., 1936), 37.

18. Ibid., 38.

19. Ibid., 39.

20. See generally, Gerald Brock, *The Telecommunications Industry* (Cambridge, MA: Harvard University Press, 1981), 55–88.

21. See, Milton Mueller, *Universal Service: Competition, Interconnection and Monopoly in the Making of the American Telephone System* (Cambridge, MA: MIT Press, 1996); and, Winston, *Media Technology and Society*.

22. M. D. Fagen, ed., *A History of Engineering and Science in the Bell System: The Early Years 1875–1925*, vol. 1. (New York: Bell Laboratories, 1975), 22–23. The memo was written sometime in late 1877 or early 1878 while Bell was on a honeymoon trip in Great Britain.

23. Elliot N. Sivowitch, "A Technological Survey of Broadcasting's 'Pre-History,' 1876–1920," *Journal of Broadcasting* 15 (Winter 1970–71): 1–20, 2.

24. Colin Cherry, *The Age of Access* (London: Croom Helm, 1985), 62.

25. *The Daily Graphic*, New York, March 15, 1877, 1.

26. Bellamy, 86–87.

27. Sivowitch, 2.

28. Hilliard and Keith, 4.

29. Carolyn Marvin, *When Old Technologies Were New* (New York: Oxford University Press, 1988), 209–31.

30. Woods, 201.

31. Marvin, 210.

32. Woods, 201.

33. Sivowitch, 3.

34. Marvin, 229–30.

35. "Asks Right to Put Music on Phone Lines," *New York Times*, July 29, 1931, 24.

36. "40 Cafes Get Music by Wire," *New York Times*, February 18, 1936, 27.

37. "Wired Radio System Operating in Ohio," *New York Times*, August 15, 1935, 11.

38. Grayson Kirk, "Supplying Broadcasts Like Gas or Electricity," *Radio Broadcast*, May 1923, 35.

39. Kirk, 37.

40. Robert Martin, "Giant Receiver gives radio to Whole City," *Popular Science*, September 1932, 8–19.

41. *Mutual Broadcasting, Inc., et al. v. Muzak Corporation*, 30 N.Y.S. 2d 419 (Sup. Ct. 1941).

42. Susan Opt, "The Development of Rural Wired Radio Systems in Upstate South Carolina," *Journal of Radio Studies* 1(1992 1–2): 71–81.

43. Asa Briggs, *The History of Broadcasting in the United Kingdom*, vol. 2 (London: Oxford University Press, 1961–1979), 330; see also, R.H. Coase, *British Broadcasting: A Study in Monopoly* (London: Longmans Green, 1950), Chapter 4.

44. Kenneth Easton, *Thirty Years in Cable TV: Reminiscences of a Pioneer* (Mississauga, Ontario: Pioneer Publications, 1980), 18.

45. James Welke, "Wired Wireless: British Relay Exchanges in Their Formative Years," *Journal of Broadcasting* 23:2 (Spring 1979): 67–178.

46. Winston, 305–306.

47. Briggs, 331.

48. Quoted in Briggs, 331–332.

49. Ralph Negrine, "Cable television in Great Britain," in *Cable Television and the Future of Broadcasting* (New York: St. Martin's Press, 1985), 103–106.

50. Easton, 15.

51. Winston, 305.

52. John Rutherford, "Equipping the Waldorf-Astoria with Radio," *Radio News*, December 1930, 517, 559, 571–572.

53. *Buck v. Jewell-LaSalle Realty Co.*, 283 U.S. 191 (1931).

54. John Borst, "One Solution of The City Antenna Problem," *Radio News*, December 1931, 473–474, 572.

55. Borst, 473.

56. "Television Antennas for Apartments," *Electronics*, May 1947, 100.

57. Borst, 572.

58. Fisher and Fisher, *Tube*, 11.

59. Willoughby Smith, *Journal of the Society of Telegraph Engineers* 5 (1873): 183–184.

60. "Punch Almanac for 1879," *Punch* 75 (December 9, 1878): 11.

61. Albert Abramson, *The History of Television, 1880 to 1941* (London: McFarland & Company, Inc., 1987), 23.

62. Abramson, 23.

63. Andrew Inglis, *Behind the Tube: A History of Broadcasting Technology and Business* (Boston: Focal Press, 1990), 158.

64. Marcuse Martin, "Wireless Transmission of Photographs," *Wireless World*, June 1915, 62–165.

65. Contemporary analog television sets in the United States have a standard 525 line format, but not all of those lines carry picture information. Digital high definition TV sets contain up to 1,080 lines.

66. Shelford Bidwell, "Telegraphic Photography and Electric Vision," *Nature*, June 4, 1908, 105–106.

67. Abramson, 11.

68. Ibid., 13.

69. Abramson, 81.

70. Ibid., 128.

71. Fagen, 784–785.

72. Ibid., 793.

73. "Far Off Speakers Seen as Well as Heard Here in a Test of Television," *New York Times*, April 8, 1927, 1.

74. Fagen, 795.

75. Abramson, 199–200.

76. Ibid., 127.

77. Fagen, vol. 7, 122.

78. "Television: Sight Broadcast Takes a Step Closer to Reality," *Newsweek*, August 3, 1935, 24.

79. Staff Report of FCC Investigation of the Telephone Industry. Pursuant to Public Resolution No. 8, 74th Congress. 1938, p. 239.

80. "Movies Transmitted by Wire in First Coaxial Cable Show," *Science Newsletter*, November 20, 1937, 326–327.

CHAPTER 2

1. Milt Shapp, Oral History, The Cable Center, Denver, Co. (Note, all the oral histories cited in this work are available online at: http://www.cablecenter.org/education/library/oralHistories.cfm).

2. The anecdote from Shen-Heights is drawn from personal interviews with the Shen-Heights operator, Marty Brophy, son of Frank Brophy, who founded the system, and with Thomas McCauly, an older Shen-Heights technician, September 29, 1993, Shenandoah, PA.

3. "Visual Broadcasting Still an Experiment," *Radio News*, February 1931, 761.

4. FCC Order No. 19, 4 FCC 30 (1937).

5. William Boddy, "Launching Television: RCA, the FCC and the battle for frequency allocations, 1940–1947," *Historical Journal of Film, Radio and Television* 9: 1(1989): 45–57, 54.

6. "Networks for Television," *Radio News*, Nov. 1947, 39–41, 120–126.

7. "Television: Sight Broadcasting Takes a Step Nearer Reality," *Newsweek*, August 3, 1935, 24.

8. "AT&T., Farnsworth, Television," *Business Week*, Aug. 14, 1937, 18.

9. "1939, Television Year," *Business Week*, October 29, 1938, 31.

10. "Television Doldrums," *Business Week*, June 8, 1940, 46.

11. "Television Payoff," *Business Week*, February 1, 1941, 28.

12. "1948–Year of Television," *Radio & Television News*, December 1948, 35–38.

13. *Television Digest* reported the average cost of TV set: $575 in 1947; $375 in 1948; projected $275 in 1949. *Television Digest*, March 20, 1948, 2.

14. Christopher Sterling and John Michael Kittross, *Stay Tuned: A Concise History of American Broadcasting* (Belmont, CA: Wadsworth Publishing Co., 1978), 290.

15. *Television Digest*, March 20, 1948, 2; April 10, 1948; and August 21, 1949, 1–2.

16. Christopher Sterling and Michael Kittross, *Stay Tuned A Concise History of American Broadcasting*, 2nd ed. (Belmont, CA: Wadsworth Publishing, 1990), 209.

17. *Television Digest*, vol. 2, no. 1, January 5, 1946.

18. Sterling and Kittross, 290.

19. Fred Hamilton, "Spot Radio News," *Radio News*, May 1948, 18.

20. "The March of Television," *Newsweek*, October 25, 1948, 66.

21. "Young Monster," *Time*, January 3, 1949, 31.

22. "Videotown," *Time*, July 18, 1949, 74.

23. "Housing Officials Discuss Television," *New York Times*, September 10, 1948, 4–6:6.

24. William Boddy, *Fifties Television: The Industry and its Critics* (Urbana, IL: University of Illinois Press, 1990), 22–24.

25. Michelle Hilmes, *Hollywood and Broadcasting: From Radio to Cable* (Urbana, IL: University of Illinois Press, 1990), 122–125.

26. "Stocks: The Television Boom," *Newsweek*, May 1, 1950, 62.

27. "1948–Year of Television Progress," *Radio & Television News*, December 1948, 35, 38.

28. "Young Monster," *Time*, January 3, 1949, 31.

29. Numbers of stations on-air, especially for 1948–1949 vary widely according to source and should be treated with caution.

30. Andrew Inglis, *Behind the Tube: A History of Broadcasting Technology and Business* (Boston: Focal Press, 1990), 193.

31. Boddy, *Fifties Television*, 51.

32. As of November 1, 1948, "1948–The Year of Television Progress," 36.

33. "TV Today," *Sales Management*, January 15, 1949, 44.

34. Boddy, *Fifties Television*, 51.

35. "TV Today," *Sales Management*.

36. See, for example, "When will we get our TV?" *TV Digest*, August 14, 1948, 1; and J. A. Stanley, "No Television in Your City?" *Radio and Television News*, December 1951, 40–41.

37. "Television Antenna Booster," *Popular Mechanics*, January 1949, 232; "Boosters for Television Sets: A Preliminary Report," *Consumers' Research Bulletin*, February 1949, 17–18.

38. "Television Antennas for Apartments," *Electronics*, May 1947, 96.

39. Ibid.

40. "Housing Officials Discuss Television," *New York Times*, September 10, 1948, 4–6:6.

41. "Apartment Owners Ban Television Aerials," *Science Digest*, May 1947, 55.

42. "Television Antennas for Apartments."

43. "Apartment Owners Ban Television Aerials."

44. Ira Kamen, "Television Master Antennas," *Radio and Television*, April 1949, 31–34, 136–140.

45. See, "Video for Hotel Rooms," *Electronics*, February 1948; "Central Television Distribution System Engineering," *Tele-Tech*, January 1948, 46; "Multiplex TV Antenna Systems for Stores," *Radio and Television News*, April 1950, 45; "Multi-TV Receiver Operation From Single Antenna," *Radio and Television News*, May, 1950, 79.

46. "Video for Hotel Rooms."

47. Kenneth Easton, *Thirty Years in Cable TV: Reminiscences of a Pioneer* (Mississauga, Ontario: Pioneer Publications, 1980), 22.

48. Ibid.

49. Ibid., 25.

50. "Apt Owners Ban Television Aerials."

51. See, for example, "Video for Hotel Rooms"; "Master Television Aerials to be Installed for tenants in 3 Multi-Family Projects," *New York Times*, November 30, 1949, VIII, 5; Ira Kamen, "Television Master Antennas"; Robert Donaldson, "Multiplex TV Antenna Systems for Stores," *Radio and Television News*, April 1950, 45, 110–113.

52. "Master TV Antenna System for Apartments," *Popular Mechanics*, November 1950, 230–231.

53. "Big Video Opening Ceremony Received by local Company," *Franklin News Herald*, January 12, 1949, 11.

54. Easton, 48.

55. *Radio & Television News*, June 1951, 126.

56. Personal interview with George Yuditsky, September 29, 1993, Mahanoy City, PA.

57. Patrick Parsons, "Two Tales of A City: John Walson, Sr., Mahanoy City, and the 'Founding' of Cable TV," *Journal of Broadcasting and Electronic Media* 40: 3 (Summer 1996): 354–365.

58. Personal telephone interview with Bob Tarlton, March 27, 1991.

59. Taken from interviews with Jim Davidson, and from "First Television Here Saturday," *Tuckerman Record*, November 18, 1948, 1.

60. *Television Factbook* (Washington, D.C.: Warren Publishing, 1964–1965), 155-C.

61. Rufus Crater, "TV Booster Outlets," *Broadcasting-Telecasting*, April 18, 1949, 25.

62. No relation to the author.

63. There are several different accounts of the origin of Parsons' efforts; this is the one reported in the press in mid 1949, one of the most contemporaneous accounts. Virgil Smith, "Astoria Brings Seattle's Video to Home City," *Portland Oregonian*, July 28, 1949, 1; and L. E. Parsons, Oral History, The Cable Center, Denver, CO.

64. Parsons, Oral History.

65. Parsons, Oral History.

66. Parsons, Oral History.

67. "UFH experimental TV station to feed Astoria (Ore.)," *Television Digest*, December 3, 1949, 4.

68. KAST letter on file at The Cable Center, Denver, CO.

69. Letter from Parsons to the FCC, August 14, 1949, on file at The Cable Center, Denver, CO.

70. In early 1950 the FCC denied his request. *Television Digest*, February 25, 1950.

71. "TV 'Satellite,'" *Broadcasting-Telecasting*, August 1, 1949, 45.

72. "Soon to apply..." *Television Digest*, August 13, 1949, 6.

73. "Small Town Television," *Popular Mechanics*, April 1950, 114–117, 268, 270.

74. Letter on file at The Cable Center, Denver, CO.

75. Reported start date, August 23, 1950. See, *Television Factbook*, no. 18, January 1954.

76. Victoria Gits, "Parsons' Dream," *CableVision*, July 22/26, 1985, 60–63, 63.

77. Letter from G. L. Davenport of Cox Corp. to Mary Alice Mayer in her book: Astoria businessmen, 13 stockholders, reduced to seven on November 1, 1954 and 50% of the stock sold to a group of CATV investors from Seattle and Aberdeen, Wash. This group eventually sold the system, under the name Clatsop Television Company.

78. Personal telephone interview with Abe Harter, October 15, 1993.

79. "Community TV Antenna System," *Electronics*, September 1950, 182.

80. Personal interview, Abe Harter.

81. Robert Tarlton, Oral History, The Cable Center, Denver, CO.

82. Ed Gildea, "When Lansford Got TV," *Valley Gazette*, Lansford, PA, March 1988; Tarlton, Oral History.

83. E.D. Lucas, "How TV Came to Panther Valley," *Radio and Television News*, March 1951, 31–34, 107–111, 107.

84. E. D. Lucas, "How TV Came to Panther Valley."

85. J. Gould, "TV Aerial on Hill Aids Valley Town," *New York Times*, December 22, 1950, 30 LT.

86. Micahel Saada, "Stretching Television: New Utilities Deliver TV to Towns Outside usual Reception Range," *Wall Street Journal*, January 8, 1951, 1.

87. Patrick Parsons, "Two Tales of A City."

88. "Facts About Television in Mahanoy City," *Record American*, Mahanoy City, PA, September 29, 1950, 5.

89. "Luther Holt adds new channel to TV reception here," *Record American*, Mahanoy City, PA, July 5, 1951, 5.

90. Saada.

91. Bureau of the Census, U.S. Dept. of Commerce, *Census of Housing for 1950*, vol. 1, part 5, Table 20 (Washington, D.C.: U.S. Government Printing Office, 1953).

92. "Mountain-to-Mohammed Eases Freeze," *Television Digest*, January 13, 1951, 2–3.

93. Sixth Report and Order, 41 FCC 148 (1952).

94. Communications Act of 1934, sec. 307b (48 Stat. 1084).

95. This is true for stations clustered in either the High or Low bands. The frequency separation between the two bands means that Channels 6 and 7 will not create adjacency problems.

96. Christopher Sterling, Sidney Head and Lemuel Schofield, *Broadcasting in America*. 7th ed. (Boston: Houghton Mifflin Co., 1994), 58–59.

97. See, Stanley Besen and Paul Hanley, "Market Size, VHF Allocation and Viability of TV Stations" *Journal of Industrial Economics* 24 (September 1975): 41.

98. House of Representatives, Committee on Energy and Commerce, Subcommittee on Telecommunications, Consumer Protection and Finance. *Telecommunications in Transition: The Status of Competition in the Telecommunications Industry.* Staff Report. 97th Cong., 1st Sess., November 3, 1981, Committee Print 97-V, 247.

99. R.E. Park. "New Television Networks." Santa Monica, CA: Rand Corp., December 1973, R-1408-MF.

100. *Electronics*, December 1952, 106.

101. Earl Abrams, "Is Community TV Here To Stay?" *Broadcasting-Telecasting*, February 9, 1953, 82.

102. Milton Shapp, Oral History.

103. Abrams.

CHAPTER 3

1. Note that the time periods indicated in the chapter titles from this point forward are general time frames and the text itself may include relevant material from a few years before or after the stated dates, as is appropriate to the subject matter under examination.

2. "Are Community Antennas a 'Sleeper'?" *Television Digest*, May 26, 1951, 4–5.

3. Archer S. Taylor, *History Between Their Ears* (Denver, CO: The Cable Center, 2000), 15–16.

4. "Community Antennas Bring TV Programs," *Science Newsletter*, June 2, 1951, 345.

5. *Who's Who in Cable Communications* (El Cajon, CA: Communications Marketing, Inc., 1979), 154.

6. Archer Taylor, "My View: Milton Jerrold Shapp 1912–1994," *CED: Communication Engineering & Design*, February 1995, 78.

7. Don Kirk, Oral History, The Cable Center, Denver, CO.

8. Ibid.

9. Milton Shapp, Oral History, The Cable Center, Denver, CO.

10. Mary Alice Mayer Phillips, *CATV: A History of Community Antenna Television* (Evanston: Northwestern University Press, 1972), 35. The sample shown in Chicago was actually a prototype and the only one in existence at time, according to Kirk.

11. Mayer Phillips, 35.

12. Shapp, 4.

13. See, Taylor, *History Between Their Ears*, 9.

14. "System and Method for Sending Pictures Over Telephone Wires," Patent No. 2,222,606, 1940.

15. For a history of these manufacturing firms, and others, see, Taylor, *History Between Their Ears.*

16. Micahel Saada, "Stretching Television: New Utilities Deliver TV to Towns Outside usual Reception Range," *Wall Street Journal*, January 8, 1951, 1, 5.

17. Committee on Interstate and Foreign Commerce, Communications Subcommittee, *VHF Booster and Community Antenna Legislation*, 86th Cong., 1st sess., pt. 1, June 30, 1959, Milton Shapp testimony, 689.

18. Shapp, Oral History.

19. Milt Shapp, unfinished draft biography, chapter 12, 5. On file at The Cable Center, Denver, CO.

20. Shapp, Oral History.

21. George Gardner, Oral History, The Cable Center, Denver, CO.

22. Roy Bliss. Oal History. The Cable Center, Denver, CO.

23. Personal Interview with Bark Lee Yee, May 22, 1994, New Orleans, LA.

24. *Teleservice Company of Wyoming Valley v. Commissioner of Internal Revenue*, 254 F.2d 105 (1958), 106.

25. Senate Committee on Interstate and Foreign Commerce, Communications Subcommittee, *VHF Booster and Community Antenna Legislation*. Part 1. 86th Cong., 1st Sess., 1959. Testimony of A. J. Malin, 536.

26. *Teleservice Company of Wyoming Valley*, 106; and Walter Buschbaum, "Community TV," *Radio & Television News*, May, 1954, 35, 37.

27. Senate Committee on Interstate and Foreign Commerce, *VHF Booster and Community Antenna Legislation*, 535–536.

28. "Wired TV," *Broadcasting-Telecasting*, February 11, 1952, 66.

29. Taylor, 83.

30. Committee on Interstate and Foreign Commerce, *VHF Booster and Community Antenna Legislation*, 686.

31. Gardner, Oral History.

32. Personal interview with Robert Tarlton, March 27, 1991.

33. These were "pressure taps," so named because one could clamp the tap to the main line and push a "stinger" into the cable. The stinger would contact the center conductor while shorter metal teeth in the tap would bite into the main line's outer conductor. Jerrold's deign required cutting into the mainline to prepare it, while Entron's "FasTee" model did not.

34. Robert Cooper Jr., *CATV System Management & Operation* (Thurmont, Maryland: TAB Books, 1966), 116.

35. Gardner, Oral History.

36. "Community TV Antenna System," *Electronics*, September 1950, 182.

37. Tarlton, Oral History.

38. "Community TV Antennas: A Nationwide Survey," *Colorado Municipalities*, June 1954, 122.

39. "Township Rejects Proposal for TV Antenna at Heights," *Shenandoah Evening Herald*, Shenandoah, PA, November 25, 1950, 1.

40. Marty Malarkey, Oral History, The Cable Center, Denver, CO.

41. Bruce Merrill, Oral History, The Cable Center, Denver, CO.

42. Personal Interview with an anonymous cable operator.

43. L. E. Parsons, Oral History, The Cable Center, Denver, CO.

44. Committee on Interstate and Foreign Commerce, *VHF Booster and Community Antenna Legislation*, 686.

45. See, Robert Cooper Jr., *CATV System Management and Operation* (Thurmont, MD: Tab Books, 1966), 85.

46. *United States v. Western Electric Co. and American Telephone and Telegraph*, Civil Action 17–49 (D.N.J. 1956).

47. Shapp "VHF Booster and Community Antenna Legislation," Hearings before the Communications Subcommittee of the Committee on Interstate and Foreign Commerce, 86th Cong., 1st sess., pt. 1, June 30, 1959, at 692.

48. Shapp, "VHF Booster and Community Antenna Legislation," Hearings, p. 686–692.

49. *Antenna Systems Corp. v. New York Telephone Co.*, 17 RR 2145 (September 29, 1958).

50. Malarkey, Oral History.

51. Tarlton, Oral History.

52. Edward Allen, Oral History, The Cable Center, Denver, CO.

53. Bark Lee Yee, Oral history, The Cable Center, Denver, CO.

54. Ray Schneider, Oral History, The Cable Center, Denver, CO.

55. "MD Community TV Seeks Advertising," *Broadcasting -Telecasting*, August 22, 1955; and "Problem for Advertisers," *Broadcasting-Telecasting*, May 12, 1958, 39.

56. National Community Television Association (NCTA), Transcript, Annual Convention, 1952.

57. Ibid.

58. Gardner, Oral History.

59. Joseph Gans, Oral History, The Cable Center, Denver, CO.

60. Ibid.

61. George Mannes, "The Birth of Cable TV," *American Heritage*, Fall 1996, 47.

62. The 1930s patent belonged to Dr. W. S. Percival, chief scientist of the British electronics firm EMI.

63. Community Engineering, founded by engineer Dr. Walter Brown, in fact, began as an attempt to obtain a TV license for the State College area. Like an earlier effort by Ed Parsons, Brown's bid for a license was turned down, and in 1951, along with several others, he started a CATV system in the area. An equipment company was formed soon after to provide necessary equipment for the system.

64. Gardner, Oral History.

65. Tarlton, personal interview.

66. Notes from the organizational meeting taken by George Bright, September 18, 1951. Minutes on file at The Cable Center, Denver, CO.

67. Minutes of the National Community Television Council, September 26, 1951. The Cable Center, Denver, CO.

68. Minutes of the National Community Television Association, January 16, 1952. The Cable Center, Denver, CO.

69. *Edwards v. Cuba Railroad Co.*, 268 US 628 (1925).

70. 27 T.C. 722 (1957).

71. *Teleservice Company of Wyoming Valley v. Commissioner of Internal Revenue*, 254 F.2d 105 (1958).

72. Ibid., 112.

73. Ibid.

74. *Lilly v. U.S.*, 238 F.2d 584 (1956), 588.

75. Ibid., 587.

76. *Pahoulis v. U.S.*, 242 F.2d. 345 (1957), 347.

77. Ibid., 346.

78. Ibid., 347.

79. "Community TV Users To Get Excise Refund," *Broadcasting-Telecasting*, August 26, 1957, 52.

80. *Re: Bennett*, 89 PUR (ns) 149 (1951); and see, "Master Antenna," *Broadcasting-Telecasting*, August 6, 1951, 78.

81. *Opinion of the Attorney General*, 12 RR 2094, October 18, 1955.

82. *Opinion of the Attorney General*, 10 RR 2058, April 19, 1954.

83. *Opinion of the Attorney General*, 14 RR 2059, November 2, 1954.

84. *Television Transmission Inc. v. Pubic Utilities Commission*, 301 P2d 862 (1956).

85. *Cokeville Radio & Electronics Co.*, 6 P.U.R. 3d 129 (Wyo. Pub. Serv. Comm. 1954),11 RR 2041 (1954).

86. See, for example, *Rawlins Community Television Co.*, 12 P.U.R. 3d 208 (Wyo. Pub. Serv. Comm 1956).

87. *Community Television Systems of Wyoming*, 17 RR 2131 (Wyoming, D.Ct. 1958).

88. Richard Gershon, "Pay Cable Television: A Regulatory History," *Communications and the Law* (June 1990): 3–26.

89. Notice of Proposed Rulemaking, FCC Docket no. 11279, FCC 55–165; and see, "Would Toll Kill Free TV?" *Broadcasting-Telecasting*, June 13, 1955, 27.

90. Third Report in Docket 11279, 26 FCC 265 (1959). The FCC authorized experimental service in 1957 (First Report and Order, 23 FCC 532 (1957)) but rescinded the order a year later (Second Report and Order) at the request of Congress, which was conducting its own inquiry.

91. See, Edwin James, "Pay-As-You-See," *Broadcasting-Telecasting*, May 3, 1954, 67–68, 72.

92. The owner of the Sun-Air Drive In argued that Paramount's control over production, distribution and exhibition of the films constituted a violation of a then-recent government antitrust decree that forced film companies to divest their exhibition (theater) interests. See, Michelle Hilmes, *Hollywood and Broadcasting: From Radio to Cable* (Urbana, IL: University of Illinois Press, 1990), 127.

93. Milt Shapp, "The Case for Community Television," *Broadcasting-Telecasting*, February 9, 1953, 83.

94. See, David Ostroff, "A History of STV, Inc. and the 1964 California Vote Against Pay Television," *Journal of Broadcasting* 27: 4 (Fall 1983): 371–386.

95. "TV Comes over the Mountains," *Fortune,* October 1954, 155.

96. See, for example, Shapp, "The Case for Community Television."

97. Senate Committee on Interstate and Foreign Commerce, *VHF Booster and Community Antenna Legislation*, 683–684.

98. *U.S. v. Jerrold Electronics Corp.*, 187 F. Supp. 545 (E.D. Pa. 1960); see also, *Jerrold Electronics v. Wescoast Broadcasting*, 341 F.2d 653 (9th Cir. 1965).

99. "Toll TV Near in Oklahoma," *Broadcasting-Telecasting*, February 18, 1957, 28–29.

100. "Broadcaster and Family Missing," *Broadcasting*, August 19, 1960, 76.

101. "Crash Victims," *Broadcasting*, September 5, 1960, 87.

102. Bruce Merrill, Oral History, The Cable Center, Denver, CO.

103. Bill Daniels, Oral History, The Cable Center, Denver, CO.

104. Daniels, Oral History.

105. The operator was Joe Saricks in Bradford, PA.

106. Robert Magness, Oral History. The Cable Center, Denver, CO.

CHAPTER 4

1. Senate Interstate and Foreign Commerce, Committee Hearings on S. Res. 224 and S. 376. *The Television Inquiry*, pt. 6, 85th Cong., 2nd sess., 1958, 3507

2. E. Stratford Smith, Oral History, The Cable Center, Denver, CO.

3. "FCC Interoffice Memorandum," March 25, 1952, as cited in Senate Interstate and Foreign Commerce, *The Television Inquiry*, pt. 6, 3490.

4. "Antenna Systems: FCC Awaits Staff Report," *Broadcasting-Telecasting*, July 16, 1951, 59.

5. "City Antennas," *Broadcasting-Telecasting*, December 31, 1951, 63.

6. "FCC Interoffice Memorandum," Senate Interstate and Foreign Commerce Committee, *The Television Inquiry*, pt. 6, 3490–3500.

7. Ibid., 3498.

8. Roger Noll, Merton Peck and John McGowan, *Economic Aspects of Television Regulation* (Washington: Brookings Institution, 1973), 125–26.

9. Memorandum Opinion and Order in Re: J. E. Belknap and Associates, 18 FCC 642 (1954).

10. "Community TV, Legality Issue Before FCC," *Broadcasting-Telecasting*, October 15, 1951, 72.

11. The exact date of the filing is unclear. The 1952 FCC memo cites a September 24 date, subsequent FCC documents cite the July 1 date.

12. "Community TV: FCC to Hear Belknap Case," *Broadcasting-Telecasting*, July 28, 1952, 88.

13. For a review of the proceedings in this case, see generally, Senate Interstate and Foreign Commerce Committee, *Television Inquiry*, pt. 8, 4120–4133.

14. See, Senate Interstate and Foreign Commerce Committee, *Television Inquiry*, pt. 6, 4133.

15. Don Le Duc, *Cable Television and the FCC: A Crisis in Media Control* (Philadelphia: Temple University Press, 1973), 65.

16. Ibid.

17. Senate Interstate and Foreign Commerce Committee, *Television Inquiry*, pt. 8, 4572; and, Christopher Sterling and Michael Kittross, *Stay Tuned A Concise History of American Broadcasting*, 2nd ed. (Belmont, CA: Wadsworth Publishing, 1990), 357.

18. Noll, Peck and McGowan, 112.

19. "Mountain-High power TV urged for Montana by Craney," *Broadcasting-Telecasting*, April 6, 1953, 58.

20. "Antenna Systems: FCC Awaits Staff Report," *Broadcasting-Telecasting*, July 16, 1951, 59–62, 62.

21. Earl Abrams, "Is Community TV Here to Stay?" *Broadcasting-Telecasting*, February 9, 1953, 82, 83.

22. "TV Outposts," *Broadcasting-Telecasting*, January 5, 1953, 48.

23. *C.J. Community Services v. FCC*, 246 F.2d 660, 661 (1957).

24. E. Stratford Smith, personal interview, March 3, 1997.

25. "A Small Town + TV Repeater Idea: FCC Headache," *Broadcasting-Telecasting*, November 22, 1954, 82.

26. Ibid.

27. "TV Satellite, Booster Competition Spotlighted at NCTA N.Y. Sessions," *Broadcasting-Telecasting*, June 13, 1955, 69.

28. "Wash. State Community Group Answers Broadcast Bureau," *Broadcasting-Telecasting*, November 28, 1955, 77.

29. C.J. Community Services, 20 FCC 860 (1956).

30. "FCC Opens Gates on TV Translators," *Broadcasting-Telecasting*, May 28, 1956, 75.

31. See, LeDuc, 87.

32. *C.J. Community Services, Inc. V. FCC*, 246 F.2d 660 (1957).

33. Ibid., 664.

34. See, FCC Report and Order, 18 RR 1505 (1958).

35. Marty Malarkey, Oral History, The Cable Center, Denver. Co.

36. Ibid.

37. "Community TV Assn. Fights KOA-TV Move," *Broadcasting-Telecasting*, December 20, 1954, 58.

38. "Another Community TV Flaunts Station," *Broadcasting-Telecasting*, June 20, 1955, 80.

39. Ibid.

40. Ibid.

41. "Beacom Asks Regulation of Community TV Systems," *Broadcasting-Telecasting*, August 30, 1954, 68.

42. "Rules and Regulations Incidental to Restricted radiation Devices," subpart D, pt. K5, Docket 9288, 13 RR 1546a, July 11, 1956.

43. Lawrence Lichty, "The Impact of FRC and FCC Commissioners' Backgrounds on the Regulation of Broadcasting," *Journal of Broadcasting* 6 (Spring 1962): 97–110: See also, William Wenmouth, Jr. "Impact of Commissioner Background on FCC Decisions: 1962–1975," *Journal of Broadcasting* 20: 2(Spring 1976): 239–259.

44. Erwin Krasnow, Lawrence Longley, and Herbert Terry, *The Politics of Broadcast Regulation*, 3rd ed. (New York: St. Martin's Press, 1982), 41–48.

45. Eric Barnouw, *Tube of Plenty: The Evolution of American Television* (New York: Oxford University Press, 1990), 152.

46. Ibid.

47. James Baughman, *Television's Guardians: The FCC and the Politics of Programming, 1958–1967* (Knoxville, The University of Tennessee Press, 1985).

48. Lucas Powe Jr., *American Broadcasting and the First Amendment* (Berkeley, CA: University of California Press, 1987), 77.

49. Baughman, 13.

50. Ibid.

51. Krasnow, Longley, and Terry, 100.

52. William Boddy, *Fifties Television: The Industry and its Critics* (Urbana, IL University of Illinois Press, 1990), 215.

53. Ibid.

54. Boddy, 123.

55. John Doerfer, "Community Antenna Television Systems," *Journal of the Federal Communications Bar Association* 14 (1955): 4–14, 6.

56. Ibid., 8.

57. Ibid., 13.

58. Ibid., 14.

59. *FCC v. Sanders*, 309 US 470 (1940).

60. Ibid., 474.

61. Ibid.

62. "Freedom of Competition for NCTA – Doerfer," *Broadcasting-Telecasting*, June 13, 1955, 69–70.

63. Baughman, 38.

64. "Interoffice Memo," in Senate Interstate and Foreign Commerce Committee, *Television Inquiry*, pt. 6, 3493–3499.

65. Ibid., 4133.

66. This would be E. Stratford Smith, attorney in the Common Carrier bureau; Stanley Neustadt, in the Broadcast Bureau, and Bernie Strassburg, Chief of the Telephone Division of the Common Carrier Bureau.

67. Senate Interstate and Foreign Commerce Committee, *TV Inquiry*, pt. 6, 4135.

68. *Frontier Broadcasting v. Collier*, 24 FCC 251 (1958).

69. "Interoffice Memorandum," Senate Interstate and Foreign Commerce Committee, *Television Inquiry*, pt. 6, 4142.

70. Ibid., 4146.

71. In re: Intermountain Microwave, 24 FCC 54 (1958).

72. Ibid.

73. Senate Interstate and Foreign Commerce Committee, *Television Inquiry*, pt. 6, 4147.

74. Commissioner Robert Bartley abstained from the vote; Commissioner Robert E. Lee did not participate.

75. *Frontier Broadcasting v. Collier*, 254.

76. Ibid., 256.

77. "FCC Disclaims CATV Control," *Broadcasting*, April 7, 1958, 66.

78. "Collision on TV Delivery Routes," *Broadcasting*, May, 12, 1958, 33–40.

79. Notice of Inquiry in Docket 12443, FCC 58–493 (May 22, 1958).

80. Smith, Oral history.

81. Senate Interstate and Foreign Commerce Committee, *Television Inquiry*, pt. 6, 3507.

82. Ibid., 3555.

83. "FCC Ought to Rope, Brand CATV, Western TV Operators Tell Hill," *Broadcasting*, June 2, 1958, 56.

84. Senate Interstate and Foreign Commerce Committee, *Television Inquiry*, pt. 6, 3563.

85. "CATV Backfire Lit at Capitol," *Broadcasting*, June 16, 76.

86. Senate Interstate and Foreign Commerce Committee, *Television Inquiry*, pt. 6, 3847.

87. Kenneth Cox, Special Counsel on Television Inquiry. "The Television Inquiry: The Problem of Television Service for Smaller Communities." Staff Report prepared for the Senate Committee on Interstate and Foreign Commerce, 85th Cong., 2nd sess., December 26, 1958, Committee Print.

88. Ibid., 18.

89. *Carroll Broadcasting Co. v. FCC*, 258 F.2d. 440 (1958).

90. Ibid., 443.

91. Cox, 20.

92. Ibid., 33.

93. Ibid., 46–48.

94. Inquiry into the Impact of Community Antenna Systems, TV Translators, TV "Satellite" Stations, and TV "Repeaters" on the Orderly Development of Television Broadcasting, Report and Order, 26 FCC 403, 18 RR 1573 (1959).

95. Ibid., 1576.

96. Ibid., 1595.

97. Ibid., 1606.

98. Ibid., 1607.

99. Ibid., 1595.

100. *Congressional Record*, 69 Cong. Rec. 2880, February 3, 1927.

101. Ibid.

102. Inquiry into the Impact of Community Antenna Systems, TV Translators, TV "Satellite" Stations, and TV "Repeaters," 18 RR, 1601.

103. Ibid., 1610.

104. Ibid., 1607.

105. Senate Committee on Interstate and Foreign Commerce, Communications Subcommittee, *VHF Booster and Community Antenna Legislation*, 86th Cong., 1st sess., pt. 1, June 30, 1959, 341.

106. Ibid., 399.

107. Ibid., 627.

108. "CATV Probe Ends With Fireworks," *Broadcasting*, July 20, 1959, 72.

109. Senate Committee on Interstate and Foreign Commerce, *VHF Booster and Community Antenna Legislation*, 724.

110. Ibid., 730.

111. Ibid., 349.

112. Ibid., 347.

113. Ibid., 574.

114. Ibid., 573.

115. Senate Committee on Interstate and Foreign Commerce, Communications Subcommittee, *VHF Booster and Community Antenna Legislation*, pt. 2, 86th Cong. 1st sess. October 27, 28, 29, 30 and December 15 and 16, 1959, 869.

116. Ibid., 1092.

117. Ibid., 1095.

118. Ibid., 1100.

119. "Regulation by FCC approved by CATV group," *Broadcasting*, February 15, 1960, 68.

120. Minutes of the NCTA Board of Directors, February 7, 1960.

121. National Community Television Association, *NCTA Bulletin*, May 9, 1960, 85.

122. Letter on file at The Cable Center, Denver, CO.

123. "CATV Bills Set for Senate, " *Broadcasting*, May 16, 1960, 85, 86.

124. "CATV Bill Fails By Single Vote," *Broadcasting*, May 23, 1960, 84.

125. "Senators in Clash Over a TV System," *New York Times*, May 18, 1960, 33.

126. U.S. Senate, S. 2653, *Congressional Record*, May 17, 1960, 10417.

127. Letter on file at The Cable Center, Denver, CO.

CHAPTER 5

1. Chris Welles, "The Tangled Tower of CATV," *Life*, November 18, 1966, 54–7, 57.

2. A. J. Malin, Address: "NCTA: The 9th Annual Exhibit and Symposium," June 21–24, 1960, transcript, A1–A11.

3. FCC Second Report and Order, 2 FCC 2d at 738 (1966).

4. Martin Seiden, *An Economic Analysis of Community Antenna Television Systems and the Television Broadcasting Industry. Report to the Federal Communications Commission.* (Washington, D.C.: Government Printing Office, 1965) 52.

5. Archer Taylor, *History between Their Ears* (Denver: The Cable Center, 2000) 133.

6. Bruce Merrill, Oral History, The Cable Center, Denver, CO.

7. Taken in 1966 after release of the Second Report and Order and published in early 1967.

8. Seiden.

9. Christopher Sterling and John Kittross, *Stay Tuned: A History of American Broadcasting*, 3rd ed. (Mahwah, NJ: Lawrence Erlbaum Associates, 2000) 864.

10. Shaffer, Helen. "Community Antenna TV," *Editorial Research Reports*, December 16, 1964, 923–40.

11. "CATV: It's Coming in Loud and Clear," *Los Angeles Times*, November 29, 1963, 11–2.

12. J. T. Murray, "The Golden Antenna of CATV," *Dun's Review*, May 1965, 44–6.

13. Seiden, 82.

14. Ibid., 54.

15. Murray.

16. "Television's Lusty Offspring," *Newsweek*, February 1, 1965, 62–4.

17. "CATV's Pass 1,400—Still Going," *Broadcasting*, November 30, 1964, 52.

18. "Television Tempest," *Wall Street Journal*, December 15, 1964, 1.

19. Ibid.

20. Ibid.

21. "Strong Reception," *Barrons*, September 24, 1962, 5.

22. DeLeon, and Mineral Wells, TX; Anaconda, Butte, Dillon, Bozeman, Helena and Miles City, MN.

23. Shapp also made an unsuccessful bid for the 1976 Democratic Presidential nomination.

24. "Beacon of Stability and Innovation," Extra, Extra! *Cable Hall of Fame 2000 Induction Broadsheet*, May 7, 2000, 18.

25. "The Official Facts & Figures of CATV," *Television Digest*, December 19, 1966, 1.

26. "Community Antenna Enters the Big TV Picture," *Fortune*, August 1965, 242.

27. "Ranking the Cable Operators," *Television Digest*, October 25, 1965, 2.

28. Patrick R. Parsons, "Horizontal Integration in the Cable Television industry: History and Context," *Journal of Media Economics*, 2003, 23–40.

29. Acquisition by Television Broadcast Licensees of CATV Systems, Notice of Inquiry, 29 F.R. 5416 (April 16, 1964).

30. "Hassle For TV Grant Shapes Up Between Applicant, CATV Group," *Broadcasting*, May 19, 1958, 81.

31. Acquisition by Television Broadcast Licensees of CATV Systems, Notice of Inquiry, 29 F.R. 5416.

32. Acquisition by Television Broadcast Licensees of CATV Systems, First Report, 1 FCC 2d 387 (July 27, 1965).

33. "CATV's Pass 1,400—Still Going," 52.

34. *Television Factbook*. 1969–1970. Washington, D.C.: Warren Publishing, 79-a.

35. "AT&T Ready to Serve Wired Pay-TV," *Broadcasting-Telecasting*, June 3, 1957, 27, 28.

36. Bill Daniels, Oral History, The Cable Center, Denver, CO.

37. Irving Kahn, Oral History, The Cable Center, Denver, CO.

38. Ibid.

39. Ben Conroy, Oral History, The Cable Center, Denver, CO. It is also worth noting that Kahn, in New York to oversee production of the Patterson-Johansson fight, appeared before the NCTA delegates via a TelePrompTer Theater TV link.

40. "Strong Reception," 19.

41. Shaffer, 929.

42. "CATV—Growth Industry," *Financial World*, July 29, 1964; Gould, Jack. "Network of Receivers." *New York Times*, December 27, 1964, sec. II, 15.

43. "Collision on TV Delivery Routes," *Broadcasting*, May 12, 1958, 33.

44. "Television's Lusty Offspring," 63.

45. Bruce Merrill, Oral History, The Cable Center, Denver, CO.

46. "Television's Lusty Offspring."

47. Albert Kroeger, "CATV Revisited," *Television Magazine*, September 1964, 9–37.

48. "Television's Lusty Offspring."

49. John Johnsrud, "Antenna Systems in TV are Gaining," *New York Times*, October 22, 1961, 15.

50. Seiden, 10.

51. Ibid., 16.

52. Ibid.

53. "Strong Reception," 5.

54. Welles, "The Tangled Tower of CATV," 57.

55. Andrew Simons, "A Look At CATV," *The Journal of Commercial Bank Lending*, April 1968, 42–7; and Daniel Hake, "Financing Community Antenna Television," *The Journal of Commercial Bank Lending*, March 1970, 43–55.

56. "Regulation Slows CATV Growth," *Broadcasting Yearbook*, 1968, 16–21. Reported figures are only rough averages, however. Seiden notes that one company estimated the cost to build a system in Torrington. Connecticut, at $208,115, whereas another company bid the same town for $536,000. A similar difference was found for bids in Winstead, Connecticut, with a low estimate at $115,739 and a high estimate at $320,000. Seiden, 24.

57. "Cable TV Leaps into the Big time," *Business Week*, November 22, 1969, 100–8.

58. David Loehwing, "How much Interference," *Barron's*, November 13, 1967, 25. Cable pioneer Sanford Randolph offers the same figure in his Oral History.

59. "Cable TV Leaps Into the Big Time."

60. Seiden, 36.

61. "Collision. . . " *Broadcasting*, May 12, 1958, 33.

62. Seiden, 30–1.

63. Ibid., 27.

64. Daniels, Oral History.

65. "TelePrompTer Obtains $4 million Unsecured Loan," *TV & Communications*, August 1964, 8.

66. Murray, 118.

67. Kroeger, 24.

68. Robert J. Cooper Jr., *CATV System Management & Operation* (Thurmont, MD: TAB Books, 1966), 121.

69. Anthony Lamport, "The Community Antenna Television Industry," *Financial Analysts Journal*: January-February 1965, 53–5.

70. George Gardner, Oral History, The Cable Center, Denver, CO.

71. "Strong Reception," 5.

72. Kroeger, 24.

73. "Mesa Microwave Sues to Force FCC Action," *Broadcasting*, October 20, 1958, 68.

74. "What the Shouting is About," *Broadcasting*, June 29, 1959, 40.

75. *Television Factbook*, 1965, 204–5c.

76. "KXLJ-TV Loses Petition Fight," *Broadcasting*, December 22, 1958, 45.

77. They sold the company to Magness around 1965.

78. *Mesa Microwave, Inc. v. FCC*, 262 F.2d 723 (D.C. Cir. 1958).

79. Carter Mountain Transmission Corp., 24 F.R. 5402, June 29, 1959.

80. Mesa Microwave, 18 RR 678 (June 29, 1959), and Carter Mountain, Docket 12931, 24 F.R. 2402 (June 29, 1959).

81. John Palmer, James Smith, and Edwin Wade. "Community Antenna Television: Survey of a Regulatory Problem," *Georgetown Law Journal* 52 (1963): 136–76.

82. In re: Amendment of Part 21. FCC 59–762, July 24, 1959. See 47 CFR 21.709 (Supp. 1963).

83. "Weigh CATV Problems," *Broadcasting*, October 26, 1959, 102.

84. Opinion and Order, Docket No.12931, FCC 60–564, May 20, 1960.

85. Initial Decision of Hearing Examiner, 32 FCC 468 (1961).

86. Carter Mountain Transmission Corp., 32 FCC 459 (1962); *aff'd, Carter Mountain Transmission v. FCC*, 321 F 2d 359 (D.C. Cir. 1963), *cert. denied*, 375 U.S. 951(1963).

87. Report and Order, 26 FCC at 431–432 (1959).

88. *Carter Mountain*, 32 FCC at 465.

89. Ibid.

90. Ralph Lee Smith, *The Wired Nation*. (New York: Harper & Row, 1972) 45.

91. Erwin Krasnow, Lawrence Longley, and Herbert Terry. *The politics of broadcast regulation*, 3rd ed. (New York: St. Martin's Press, 1982).

92. James Baughman, *Television Guardians: The FCC and the Politics of Programming, 1958–1967* (Knoxville, TN: University of Tennessee Press, 1985), 166.

93. Ibid., 167.

94. Minnow, Newton. Address before the National Association of Broadcasters, May 9, 1961. The full speech is reprinted in Newton Minow. 1964. *Equal time: The private broadcasters and the public interest*. New York: Antheneum, 52.

95. "FCC in New Tack on CATV Problem," *Broadcasting*, January 16, 1961, 70.

96. "FCC's 1952 TV Order 'Catastrophic'—Lee," *Broadcasting*, February 19, 1962, 9.

97. Krasnow, Longley, and Terry, 178–81.

98. "Catv attracts interest of broadcasters," *Broadcasting*, June 25, 1962, 66, 68.

99. "Joint meeting on CATV?" *Broadcasting*, October 22, 1962, 79.

100. Business Radio Service. Docket 14895. Notice of Proposed Rulemaking, FCC 62–1285, 27 F.R. 12586 (1962).

101. In re: Television Montana and Capital City Broadcasting, 20 RR 1533 (1960).

102. *Carter Mountain Transmission Corp. v. FCC*, 321 F.2d 359 (D.C. Cir. 1963).

103. Notice of Rulemaking, FCC Dockets 14895 and 15233, 28 F.R. 13789 (December 13, 1963).

104. "FCC Eases Up On CATV Rulemakings," *Broadcasting*, December 16, 1963, 71–2.

105. "CATV Growth Paced by Problems," *Broadcasting*, July 26, 1965, 31–4.

106. First Report and Order, 1 FCC 2d 897 (October 18, 1965). In 1966, the FCC granted the first Community Antenna Relay Service (CARS) license to Santa Maria Valley Cable TV.

107. Franklin Fisher and Victor Ferrall Jr. "Community Antenna Television Systems and Local Television Station Audience," *The Quarterly Journal of Economics*, May 1966, 227–51; and "The Economic Impact of CATV," *Broadcasting*, October 26, 1964, 74–5.

108. Quoted in Shaffer, 931.

109. Ibid., 932.

110. Seiden, 3.

111. Ibid.

112. Second Report and Order, 2 FCC 2d at 744 n. 30.

113. "Broadcasters Disagree on Regulation of CATV and Pay-TV," *TV & Communications*, June 1964, 7.

114. "Television Tempest," *Wall Street Journal*, December 15, 1964, 24.

115. Wenmouth Williams, Jr., "Impact of Commissioner Background on FCC Decisions: 1962–1975," *Journal of Broadcasting* 1976 (20): 244–56; and Lawrence Lichty, "The Impact of FRC and FCC Commissioners Backgrounds on the Regulation of Broadcasting," *Journal of Broadcasting* 1962 (6): 97–110.

116. Baughman, 124–25.

117. Repeated requests for a judgeship were turned down by Johnson.

118. Gerald Flannery, ed. *Commissioners of the FCC, 1927–1994* (University Press of America, Lanham, MD: 1995) 135.

119. Shaffer, 924.

120. First Report and Order in Dockets 14895 and 15233, 38 FCC 683 (1965, April 23).

121. Ibid., 706–7.

122. Ibid., 749.

123. Ibid., 758.

124. Notice of Inquiry and Proposed Rulemaking, 1 FCC 2d 453 (1965).

125. Ibid., 464.

126. Ibid., 500.

127. LeDuc, 145.

128. "FCC's Regulation Plans Basted at Denver," *Broadcasting*, July 26, 1965, 32–34.

129. Shaffer, 935.

130. Second Report and Order in Dockets 14895, 15233 and 15971, 2 FCC 2d 725 (1966). The order itself is dated in March 8, but the Commission announced its substance in a press release February 15.

131. Ibid., 727.

132. "How the FCC Takes Control," *Broadcasting*, February 21, 1966, 30–31.

133. Second Report and Order, 2 FCC 2d, 775.

134. Ibid., 783.

135. Ibid., 733.

136. Ibid., 808–819.

137. Ibid., 809.

138. "Wild Escalation in CATV Fight," *Broadcasting*, February 21, 1966, 25–28.

139. Ibid., 28.

140. Staggers had replaced Rep. Harris, the strong critic of the Commission's cable policy, who was preparing for appointment as a federal judge.

141. HR 13286, 89th Cong, 2nd sess. (1966).

142. "The Pros and Cons of CATV," *Broadcasting*, March 28, 1966, 84.

143. "Another jolt for CATV operators," *Broadcasting*, June 13, 1966, 27–30.

144. "Plans to Wire Nation Charged," *Broadcasting*, February 21, 1966, 34.

145. Midwest had been very active before this in seeking FCC restrictions on cable. It had filed lengthy comments in the FCC's rulemaking leading up to the Second Report and Order and, in fact, had filed an unsuccessful court challenge to the Second Report and Order within 24 hours of its public announcement in February 1966. Midwest argued the FCC's action did not go far enough to control the harm allegedly caused by cable importation.

146. In Re. *Midwest TV*, 4 FCC 2d 612 (1966).

147. *Southwestern Cable Company v. United States*, 378 F.2d 118 (9th Cir.), *cert. granted*, 389 U.S. 911 (1967).

148. *Buckeye Cablevision, Inc. v. FCC*, 387 F.2d 220 (D.C. Cir. 1967).

149. Ibid., 225.

150. Ibid., 225, n. 19.

151. *Z-Bar Net v. Helena Television*, 125 USPQ 595, 20 P&F RR 2010, Dist. Ct. Mont. 1960).

152. *Intermountain Broadcasting & Television Corp. v. Idaho Microwave, Inc.*, 196 F. Supp. 315 (1961).

153. *Cablevision, Inc. v. KUTV, Inc.*, 211 F. Supp. 47 (D.C. Ida. 1962); 335 F.2d 348 (9th Cir. 1964), *cert. denied*, 379 US 989 (1965).

154. *International News Service v. Associated Press*, 248 US 215 (1918).

155. *Cablevision, Inc. v. KUTV, Inc.*, 211 F. Supp. 47 (D.C. Ida. 1962).

156. *Cablevision, Inc. v. KUTV, Inc.*, 335 F.2d 348 (9th Cir. 1964).

157. "Community TV Assn. Fights KOA-TV Move," *Broadcasting-Telecasting*, December 20, 1954, 17.

158. Stanley Besen and Robert Crandall, "The Deregulation of Cable Television," *Law and Contemporary Problems* 1981 (44): 77: 91, n. 71.

159. House Committee on the Judiciary, *Supplementary Report of the Registrar of Copyrights on the General Revision of the U.S. Copyright Law*. 89th Cong., 1st sess., 1965, Comm. Print, 40.

160. "Is CATV's Free Ride Over?" *Broadcasting*, May 30, 1966, 31.

161. *Buck v. Jewell-LaSalle Realty Co.*, 283 U.S. 191 (1931).

162. *Mutual Broadcasting, Inc. et al. v. Muzak Corporation*, 30 N.Y.S. 2d 419 at 420 (Sup. Ct. 1941).

163. *United Artists Television, Inc. v. Fortnightly Corp.*, 255 F. Supp. 177 (S.D.N.Y. 1966)

164. Ibid., 195.

165. Martin Seiden, *Cable television U.S.A.: An Analysis of Government Policy* (New York: Praeger, 1972), 36.

166. *Ceracche Television Corp. v. Public Service Commission* (49 Misc.2d 554, 267 N.Y.S. 2d 969 (Sup. Ct. Albany County, 1960) ruling pole space rental not subject to PSC regulation and effectively overturning *Antenna Systems Corp. v. New York Tel. Co.* (25 Pub. Util. Rep 3d (PUR) 316, 1958); *International Cable TV Corp. v. All Metal Fabricators, Inc.*, 66 Pub. Util. Rep 3d (PUR) 446, 463, 1966 (California); *Consolidated Cable Service Inc. v. Leary* (382 S.W.2d 78 (Ky. 1964) (Kentucky); *Twin Cities Cable Co. v. Southeastern Tel. Co.*, 200 So. 2d 857 (Fla. Dist. Ct.App. 1967) (Florida).

167. Quoted in Hart, Thomas. "The Evolution of Telco-Constructed Broadband Services for CATV Operators," *Catholic University Law Review* 1985 (34):697–735.

168. "Struggle for Control of CATV hardware," *Broadcasting*, July 26, 1965, 66–7.

169. LeDuc, 160.

170. Hart.

171. Seiden, *Cable Television U.S.A.*, 36. The FCC activity during this period may have helped arrest any substantial increase in poles rates, which remained steady at about $3.00 (but ranged from about $1.50 to $4.00) per pole between 1964 and 1971.

172. Hart, 702.

173. "Southern Bell to Refuse Pole Rentals for CATV," *TV & Communications*, November 1964, 11.

174. Hart, 703.

175. "GTE Will Get Out of CATV," *Broadcasting*, January 18, 1971, 35.

176. *Television Factbook*, 1969–1970, 79-a.

177. "Struggle for Control of CATV Hardware," *Broadcasting*, July 26, 1965, 86.

178. Common Carrier Tariffs for CATV Systems, 4 FCC 2d 257 (1966); in 1968 it extended the requirement to all telephone companies (General Tel. Co. of Calif., 13 FCC 2d 448).

179. Ed Allen, Oral History, The Cable Center, Denver, CO.

180. Stanley Searle, "How TAME is TAME?" *TV & Communications*, April 1964, 30.

181. Richard Phalon, "Cable-Television Systems Seek to Tap New Market by Selling Service in Big Cities," *New York Times*, February 7, 1965, 1.

182. "Television's Lusty Offspring."

CHAPTER 6

1. Ralph Lee Smith, "The Wired Nation," *The Nation*, May 18, 1970, 582–606.

2. "FCC's Regulation Plans Blasted at Denver," *Broadcasting*, July 26, 1965, 33.

3. "NCTA Prepares Its Battle Plan," *Broadcasting*, June 27, 1966, 36.

4. Ibid.

5. "CATV'S New Look—Local Origination," *Television Digest*, July 4, 1966, 2.

6. The Association voted approval of the name change shortly after its 1967 convention.

7. Brian Winston, *Media Technology and Society* (London: Routledge, 1998), 206–16.

8. Bruce Merrill, Oral History, The Cable Center, Denver, CO.

9. Archer Taylor, *History Between Their Ears*. (Denver, CO: The Cable Center, 2000), 119–20.

10. Taylor, 65.

11. Stanley Gardner, Oral History, The Cable Center Denver, CO.

12. Taylor, 67.

13. FM radio is also carried in a piece of the low band, 88–108 MHz.

14. "CATV for Rural Areas?" *Broadcasting*, December 14, 1964, 47.

15. Morris Gelman, "Will Wire Take Over?" *Television Magazine*, December 1965, 28–31.

16. "CATV—A Booming New Industry," *Electronics World*, May 1966, 6.

17. "Big Changes Ahead In TV," *U.S. News and World Report*, April 4, 1966, 92–3.

18. Chris Welles, "The Tangled Tower of CATV," *Life*, November 18, 1966, 55.

19. Thomas Streeter, "The Cable Fable Revisited: Discourse, Policy, and the Making of Cable Television." *Critical Studies in Mass Communication* 4 (1987): 174–200.

20. Ibid., 178.

21. Ibid.

22. "NCTA Prepares Its Battle Plan."

23. The Carnegie Commission on Educational Television. 1967. *Public television: A program for action*. New York: Bantam Books.

24. Ibid., 63–4.

25. J.C.R. Licklider, "Televistas: Looking Ahead Through Side Windows." In: *The Carnegie Commission on Educational Television*, 212.

26. Ibid, 201–25.

27. Ibid., 213.

28. Harold J. Barnett and Edward A. Greenberg, "A Proposal for Wired City Television." Rand Doc. P-3668, 1967; and *Washington University Law Quarterly*. Winter 1968: 1–25.

29. Barnett and Greenberg, 20.

30. New York City, The Mayor's Advisory Task Force on CATV and Telecommunications. *A Report on Cable Television and Cable Telecommunications in New York City*, September 14, 1968.

31. Ibid., Letter of Transmittal, September 17, 1968, 5.

32. Ibid., 73.

33. Ibid., 12.

34. President's Task Force on Communications Policy. *Final Report* (Washington, D.C.: Government Printing Office, December 7, 1968).

35. Johnson himself never acknowledged completion of the report and his successor, President Nixon, released it in May 1969, only after a pointed request from Representative Torbert McDonald (D-Mass.) Chairman of the House Commerce Subcommittee on Communications.

36. John W. Finney, "Panel Would Lift Curbs on Cable TV," *New York Times*, December 10, 1968, 1.

37. President's Task Force on Communications Policy. Chapter 7.

38. The Sloan Commission on Cable Communications. *On the cable: The television of abundance.* (New York: McGraw-Hill 1971); Walter Baer, *Interactive Television: Prospects for Two-Way Services on Cable.* Memorandum R-888-MF. (Santa Monica, CA: Rand Corp., 1972).

39. "Special Issue on Cable Television." *Proceedings of the IEEE*, July 1970.

40. Sloan Commission, 167.

41. Sloan Commission, 33.

42. Smith, *The Nation*; and Ralph Lee Smith. *The Wired Nation.* (New York: Harper & Row, 1972), 45.

43. Smith, *The Nation*, 584.

44. Ibid., 602.

45. Steven Rifkin, "The Changing Signals of Cable TV" *The Georgetown Law Review*, 60 (1972): 1475–1511, n. 25.

46. C. H. Simonds, "Wanted: Black Jackie; Electronic Pipeline." *National Review*, September 10, 1968, 917–18.

47. Second Report and Order, 2 FCC 2d, 782.

48. Martin Seiden, *Cable Television U.S.A.: An Analysis of Government Policy* (New York: Praeger, 1972) Note, that of the 449 he examined, 105 were granted, 91 denied, and 253 still pending when new rules were proposed in 1968.

49. Seiden, 105.

50. "Mass Action Set for CATV." *Broadcasting*, October 31, 1966, 52.

51. In re: Midwest Television, Inc., Initial Decision of Hearing Examiner Chester F. Naumowicz, Jr., 11 RR 2d 273 (October 3, 1967).

52. In re: Midwest Television, Inc., 11 RR 2d, 303.

53. Ibid.

54. *U.S. v. Southwestern Cable Co.*, 392 U.S. 157 (1968).

55. 1959 Report and Order, 18 RR, 1599–1601 (1959).

56. *U.S. v. Southwestern Cable Co.*, 392 U.S., 178 (1968).

57. "New Sweep of FCC Powers." *Broadcasting*, June 17, 1968, 23.

58. In re: Midwest Television, 13 RR 2d 698 (1968).

59. 13 RR 2d, 734.

60. *United Artists Television, Inc. v. Fortnightly Corp.* 377 F.2d 872 (2d Cir., 1967).

61. 377 F.2d 872, 878 (1967).

62. *Fortnightly Corp. v. United Artists Television*, 392 U.S. 390 (1968).

63. Stratford E. Smith, Oral History, The Cable Center, Denver, CO.

64. *Fortnightly*, 399.

65. Ibid., 402.

66. "Is CATV's Future in FCC's Hands?" *Broadcasting*, June 24, 1968, 19–22.

67. James Baughman, *Television guardians: The FCC and the politics of programming, 1958–1967*. (Knoxville, TN: University of Tennessee Press, 1985), 143.

68. Nicholas Johnson, "CATV: Peril or Promise." *Saturday Review* November 11, 1967, 87–98.

69. "At Center Stage for Last Act." *Broadcasting*, August 24, 1970, 19.

70. "CATV and Interconnection of Tele[phone] Devices—Where Do we Go From Here? Address by FCC Commissioner Kenneth Cox before the 80th Annual Convention of the National Association of Regulatory Utility Commissioners, November 13, 1968, 9–10.

71. Jules Tewlow, "CATV Revisited." *ANPA R.I. Bulletin*, September 25, 1968, 382.

72. FCC. *Annual Report, 1968* (Washington, D.C.: Government Printing Office, 1968).

73. Notice of Inquiry and Notice of Proposed Rulemaking in Docket 18397, 15 FCC 2d 417 (1968).

74. Second Report and Order, 2 FCC 2d, 775.

75. Notice of Inquiry, 15 FCC 2d 417, 419.

76. Ibid., 420.

77. Ibid., 421

78. Ibid., 427.

79. Ibid.

80. Matter of Amendment of pt. 74, subpart K, of the Commission's Rules and Regs. Relative to Community Antenna Systems, Docket 18397, FCC 69-45, January 17, 1969.

81. Midwest Video, 13 FCC 2d 478, 510 (1968).

82. First Report and Order in Docket 18397, 20 FCC 2d 201 (1969).

83. Ibid., 203.

84. Ibid., 202–5.

85. "Charles Colson predicts dearth of TV networks." *Broadcasting*, February 5, 1973, 11.

86. Creation of the OTP was proposed by President Nixon in February 1970 and came into existence on April 20, 1970.

87. Erwin Krasnow, Lawrence Longley, and Herbert Terry. 1982. *The Politics of Broadcast Regulation*, 3rd ed. (New York: St. Martin's Press, 1982), 71–2.

88. "Broadcast Journalism Under Siege." *Broadcasting*, November 17, 1969, 27–8.

89. Bill Daniels, Oral History, The Cable Center, Denver, CO.

90. "Wrecking Crews Hit Cable Compromise." *Broadcasting*, June 23, 1969, 44–6.

91. Seiden, 7.

92. Second Further Notice of Proposed Rulemaking in Docket 18397-A, 24 FCC 2d 580 (July 1, 1970).

93. Subscription Television, Fourth Report and Order, 15 FCC 2d 466 (1968).

94. *National Association of Theater Owners v. FCC*, 420 F.2d 194 (D.C. Cir. 1969), *cert. denied*, 397 U.S. 922 (1970).

95. CATV: Memorandum Opinion and Order, 23 FCC 2d 825 (1970).

96. Ibid.

97. Second Report and Order, 23 FCC 2d 816 (1970).

98. Community Antenna Television Systems, Diversification of Control, Notice of Proposed Rulemaking, 35 F.R. 11,042 (1970).

99. In re: Applications of Telephone Companies for Section 214 Certificates, Final Report and Order, 21 FCC 2d 307 (January 28, 1970).

100. "Cable Package Draws Mixed Reviews." *Broadcasting*, June 29, 1970, 26–8.

101. "Cable Package Draws Mixed Reviews," 26.

102. "At FCC: A Spectacular on Cable TV." *Broadcasting*, March 15, 1971, 50–1.

103. Cable Television Report and Order, 36 FCC 2d 141 (1972).

104. Ibid., 190.

105. *Midwest Video Corp. v. United States*, 441 F.2d 1322 (8th Cir. 1971). In an interesting historical footnote, Midwest, in the early 1960s, purchased the Poplar Bluff, MO, system involved in the earlier Belknap importation case.

106. *United States v. Midwest Video Corp*, 406 U.S. 649 (1972).

107. Ibid.

108. Or the nearest town in the state with a qualified signal.

109. This applied to the system's first two importation choices. If the operator had a third importation choice it had to be a UHF station within 200 miles, if one was available. 37 F.R. 3265 (1972).

110. Showing of Sports Events, Report and Order, 34 FCC 2d 271 (1972).

111. Kenneth Robinson, "Introduction and General Background." In Paul W. MacAvoy, ed. *Deregulation of Cable Television: Ford Administration Papers on Regulatory Reform* (Washington, D.C.: American Enterprise Institute for Public Policy Research, 1977), 9.

112. "Thaw: New Rules to End the Cable TV Freeze," NCTA Report, Washington, D.C., February 23, 1972; transmittal letter.

113. Anne Branscomb, "The Cable Fable: Will It Come True?" *Journal of Communication*, Winter 1975, 44–56.

114. Smith. *The Wired Nation*, 62–3.

115. "Regulation Slows CATV Growth," *Broadcasting Yearbook*, 1968, 16.

116. Ibid.

117. "CATV's Annual Growth Pattern," *Television Digest*, May 13, 1968, 3.

118. David Loehwing, "How Much Interference? Nobody Likes Cable Television but Wall Street and Main Street," *Barron's*, November 13, 1967, 3, 19–25.

119. "Heard on the Street: Effect on CATV Stocks of FCC Proposed Inquiry," *Wall Street Journal*, December 16, 1968, 33.

120. Paul Kagan, "Exciting Picture: Prospects Have Brightened for Makers of Cable TV," *Barron's*, March 9, 1970, 11, 16, 18.

121. "Kahn plans a satellite for CATV," *Broadcasting*, February 9, 1970, 40.

122. "CATV—A Pioneering Industry," *The Magazine of Wall Street*, September 12, 1970, 22–4.

123. "CATV Operators Chafing at Bit," *Broadcasting*, December 28, 1970, January 4, 1971, 61.

124. "Happiness is CATV's Trip to Chicago," *Broadcasting*, June 1, 1970, 26–8.

125. Herbert Goodfriend and Frank Pratt, "Community Antenna Television," *Financial Analyst's Journal*, March-April 1970, 48–57.

126. One of the investors was John P. Cole, a prominent cable attorney in Washington, D.C. "Regional Program Center Set for CATV in the Southwest," *Broadcasting*, July 4, 1966, 32.

127. "CATV's With Ads Now Number 98," *Broadcasting*, September 22, 1969, 46.

128. "Caution Signals on CATV's Open Road," *Broadcasting*, July 8, 1968, 32–7.

129. Loehwing, 23.

130. "Teleprompter Gains Speed with FCC 18GHz Grant," *Television Digest*, May 20, 1968, 2.

131. "CATV Manufacturers Gauge Industry Outlook," *TV Digest*, July 11, 1966, 3.

132. "Pinpointing 3.7-million CATV Homes," *Broadcasting*, June 1, 970, 57, 58.

133. "Will the Mighty Inherit the CATV Earth?" *Broadcasting*, March 20, 1972, 21–2.

134. "CATV Industry Cites New 'Firsts'," *Sponsor*, March 15, 1965, 16.

135. Teleprompter Transmission of Kansas, Inc., 25 FCC 2d 469 (1970).

136. *U.S. v. Kahn*, 340 F. Supp. 485 (S.D.N.Y. 1971), *aff'd*, 472 F.2d 272 (2d Cir. 1973). TelePrompTer itself was fined $10,000 dollars.

137. Kahn surrendered to federal authorities in New York on March 1, 1973. He served eight months at Allenwood Prison and was then transferred to the federal "model-prisoners" program, serving the remainder of his time managing the marina and working at the cable system at Elgin Air Force base in Florida. He was released on November 12, 1974.

138. Ben Conroy, Oral History, The Cable Center, Denver, CO.

139. Taylor, 181.

140. Network Television Broadcasting, Report and Order, 23 FCC 2d 382 (1970).

141. Competition and Responsibility in Network Television Broadcasting, Report and Order, 23 FCC 2d 382 (1970).

142. *Iacopi v. FCC*, 451 F 2d 141 (1971); Network Television Broadcasting, 35 FCC 2d 411 (1972).

143. "A New Name for Kinney," *Broadcasting*, February 14, 1972, 49.

144. "Cable Promoters Start Thinking Big," *Broadcasting*, January 26, 1970, 54–7.

145. Arthur Unger, "Chuck Dolan: The Reluctant Gatekeeper," *Television Quarterly* 24 (1990): 15–30.

146. Ibid.

147. George Mair, *Inside HBO* (New York: Dodd, Mead & Co., 1988), 4.

148. "Sterling Expands," *Broadcasting*, December 18, 1972, 48.

149. Thomas Whiteside, "Onward and Upward in the Arts," *New Yorker*, May 20, 1985, 61.

150. Richard M. Clurman, *To the End of Time: The Seduction and Conquest of a Media Empire* (New York: Simon & Schuster, 1992), 149.

151. "A young Philadelphian puts a big circle around September 30 on HBO's calendar," *Broadcasting*, September 8, 1975, 73.

152. Mair; and Richard Gershon and Michael Wirth. 1993. "Home Box Office," In: Robert Picard, ed. *The Cable Networks Handbook* (Riverside, CA: Carpelan Publishing, 1993), 114–22. Details of events and HBO history.

153. Mair, 8.

154. "NCTA Prepares Its Battle Plan," 40.

155. "How a Broadcaster Spreads its Net," *Business Week*, December 3, 1966, 57–8.

156. General Tel. Co, 13 FCC 2d 448 (1968), *aff'd sub nom*, *General Telephone Co. v. FCC*, 413 F.2d 390 (D.C. Cir. 1969), *cert. denied*, 396 U.S. 888 (1969).

157. Application of Telephone Companies for Section 214 Certificates, 21 FCC 2d 307, 308 (1970).

158. 21 FCC 2d 307, reconsidered in part, 22 FCC 2d 746 (1970), *aff'd sub nom*, *General Telephone Co. S.W. v. U.S.*, 449 F.2d 846 (5th Cir. 1971).

159. Final Report and Order, 21 FCC 2d 307 at 326–327 (1970).

160. "GTE Will Get Out of CATV," *Broadcasting*, January 18, 1971, 35; although not all systems actually sold until 1972.

161. Seiden, 36–40.

162. Thomas Hart, "The Evolution of Telco-Constructed Broadband Services for CATV Operators," *Catholic University Law Review*, 34 (1985): 697, 711.

163. Seiden, 44.

164. "Burch Says FCC Doesn't Want to Regulate Cable by Itself," *Broadcasting*, February 12, 1973, 68.

165. Cable Television Report and Order, 36 FCC 2d 141, 209.

166. 37 F.R. 3281 (1972).

167. "4 Indicted in Trenton Cable TV Award," *New York Times*, March 25, 1971, 78; Stanley Penn, "Controlling the Cable: Scandals and Allegations Intensify the Bid to Divest Towns of Authority over CATVs," *Wall Street Journal*, April 20, 1971, 38.

168. "Deputy of Tacoma Blasts 'Media Monopoly'," *Broadcasting*, January 26, 1970, 66.

169. "A Controversial CATV Ploy in Buffalo, N.Y. Could Hasten a Move to Federal Regulation," *Wall Street Journal*, February 2, 1971, 32, and February 3, 1971, 23.

170. Ibid., February 2, 1971, 32.

171. *Crossed wires: Cable television in New Jersey*. 1971. Princeton, NJ: A Report by the Center for Analysis of Pubic issues, 62.

172. Stephen Barnett, "State, Federal and Local Regulation of Cable Television," *Notre Dame Lawyer*, April 1972 (47): 685–814.

173. Barnett, 709; and see Cable Television Report and Order 36 FCC 2d 141, 207.

174. Although two small cable companies had been operating in New Haven in the early 1960s.

175. Michael Mitchell, *State Regulation of Cable Television* (Santa Monica, CA: Rand Corp., 1971).

176. NARUC Bulletin, No. 13-1966.

177. *TV Pix, Inc. v. Taylor*, 304 F. Supp 459 (D. Nev. 1968), *aff'd mem*, 396 U.S. 556 (1970).

178. In 1968, the Rhode Island State Supreme Court ruled cable was not local in nature and, therefore, not subject to local licensing. The Court held that the state bore full responsibility for issues affecting streets and highways, including use by CATV wires. *Nugent v. City of East Providence*, 238 A 2d 758 (1968).

179. Although New Jersey and Delaware gave municipalities the authority to grant franchises.

180. On September 9, 1971, the Illinois Commerce Commission assumed jurisdiction over cable (3 CCH Util. L. Rep. Sec. 21,604.11 (1972) and the following January (1972) issued a notice seeking comment on a comprehensive set of rules for cable. Commission jurisdiction was eventually struck down in the state courts (*Illinois-Indiana Cable Television Association v. Illinois Commerce Commission*, 302 N.E. 2d 334, 1973).

181. 304 F. Supp at 467. At one point in his decision, the judge characterized CATV as a "parasite on the national network of television broadcasters," at 467.

182. National Cable Television Association, "The CATV Industry & Regulation," Washington, D.C.: NCTA, December 15, 1969.

183. *Greater Freemont v. City of Freemont*, 302 F. Supp. 652 (N.D. Ohio, 1968).

184. Ibid., 665.

CHAPTER 7

1. Robert Hughes, Oral History, The Cable Center, Denver, CO.

2. Martin Malarkey, Oral History, The Cable Center, Denver, CO.

3. Malarkey.

4. "The Cable Fable," *Yale Review of Law and Social Action*, 2 (1972): 3. The thrust of the review, a response to the Sloan Commission Report, was that commercial development of cable alone would not lead to a realization of the promise of cable as a social good. It argued for a greater public voice in cable policy and a firmer governmental hand in shaping the new medium for public purposes.

5. Anne Branscomb, "The Cable Fable: Will it Come True," *Journal of Communication*, Winter 1975, 44–56.

6. Branscomb, 46–7.

7. TelePrompTer, after lengthy negotiations with the SEC, accepted a consent decree in July 1974, and agreed not to contest the Commission's allegations, stating it simply wished to "avoid protracted litigation."

8. Martin Seiden, *An Economic Analysis of Community Antenna Television Systems and the Television Broadcasting Industry. Report to the Federal Communications Commission.* (Washington, D.C.: Government Printing Office, 1965), 84.

9. "Reflecting Upon a Historic Decade," *Cablevision*, July 22/29, 1985, 82–8.

10. Branscomb, 48.

11. Rolla Edward Park, "What's Left of Distant Signals after Exclusivity Blackouts?" Rand Corp. Paper P-4848, June 1972, 6.

12. Park, 8.

13. The vote came despite an FCC rulemaking proceeding at the time to restrict the growth of cable MSOs, sparking a dissent by Commission Johnson. Teleprompter Transmission of Kansas, Inc., 25 FCC 2d 469 (1970).

14. "TCI Pays the Price for Programs," *Broadcasting*, April 26, 1971, 49.

15. Peter David Shapiro, "Networking in Cable Television: Analysis of Present Practices and Future Alternatives." PhD diss., Stanford University, 1973, 124.

16. Megan Gwynne Mullen, "The Revolution Now in Sight": A History of American Cable Television Programming. Ph.D. diss., University of Texas at Austin, August 1996, 126.

17. Mullen, 127.

18. "CATV Headed for Ad-Supported Network," *Broadcasting*, May 4, 1970, 23–4.

19. Mullen, 129.

20. Nathaniel Feldman, "Cable Television: Opportunities and Problems in Local Program Origination." Rand Report R-570-FF, October 1970; and "Dim Future Forecast for CATV Programming," *Broadcasting*, October 5, 1970, 55.

21. Martin Seiden, *Cable Television U.S.A.: An Analysis of Government Policy* (New York: Praeger, 1972), 34–5.

22. Suspension of Community Antenna Television Mandatory Origination Rule, 36 F.R. 10876 (May 27, 1971).

23. Report and Order on Cable Television Service, 49 FCC 2d 1090 (1974).

24. "Facts on Origination," *Broadcasting*, July 9, 1973, 34.

25. "585 Systems Turning Out Local Fare," *Broadcasting*, September 10, 1973, 58.

26. The games of the Nets basketball team and Islanders hockey teamed were actually being carried free to TelePrompTer's Manhattan customers, but a contractual arrangement with the Nassau Coliseum where the teams played their home games, in effect, forced the pay arrangement on Long island.

27. SRS was installed in TelePrompTer's El Segundo system by 1973.

28. Benjamin Compaine, ed, *Who Owns the Media* (New York: Harmony Books, 1979), 310.

29. First Report and Order, 20 FCC 2d 201 at 207–208 (1969).

30. Seiden, *Cable television U.S.A.*, 134.

31. Shapiro, 41.

32. Cited in Shapiro, 129.

33. "Cable Begins to Look As Good As Gold," *Broadcasting*, July 31, 1971, 50.

34. "Private Hour with FCC: Cable Owners Plead for Less Control," *Broadcasting*, October 8, 1973, 21.

35. "But Big Cities Don't Always Mean Big Money for Cable," *Broadcasting*, June 25, 1973, 34.

36. "The Process of Cable Television Franchising: A New York City Case Study," Communications Media Center, New York Law School, 1980; an excellent review of the New York experience.

37. TelePrompTer began operations in Manhattan in June 1966.

38. Comtel provided service primarily to midtown area hotels using lines leased from New York Telephone Co., but was forced to relinquish its cable operation in 1970.

39. Opinion and Order, 12 FCC 2d 936 (1968).

40. Report and Order, 17 RR 2d 1650 (Dec. 23, 1969).

41. AML does have its disadvantages, including susceptibility to interference by heavy rain.

42. A building owner challenged the law and, in 1982, the Supreme Court struck it down as an unconstitutional taking of property. The Court said such a statute might be valid if it provided reasonable compensation to the landlord. *Loretto v. Teleprompter Manhattan CATV Corp.*, 102 S.Ct. 3164 (June 28, 1982).

43. Richard Munro, Oral History, The Cable Center, Denver, CO.

44. Thomas Whiteside, "Onward and Upward with the Arts," *New Yorker*, May 20, 1985, 63.

45. "But Big Cities Don't Always Mean Big Money for Cable," *Broadcasting*, June 25, 1973, 34.

46. "No Upward Trend Seen in Spurt by Cable Issues," *Broadcasting*, January 28, 1974, 47–8.

47. The deal initially had been opposed by minority groups. The media companies managed to reach agreement with the challengers. But the Justice Department pressed it suit, worried by that time about consolidation in the industry.

48. "In Cable Franchising, the Cards are Being Played Closer to the Vest," *Broadcasting*, February 4, 1974, 48.

49. "Grand Rapids Nervous about GE Pulling Back on Cable Operations," *Broadcasting*, July 16, 1973, 32.

50. "Retrospection," *Cable Television Business*, February 1, 1984, 128.

51. "Cable's Success in Raising Rate Base Detailed for NCTA," *Broadcasting*, April 21, 1975, 28.

52. Mark Robichaux, *Cable Cowboy: John Malone and the Rise of the Modern Cable Business* (New York: John Wiley & Sons, Inc., 2002), 17.

53. John Malone, Oral History, The Cable Center, Denver, CO.

54. Stephen Keating, *Cutthroat* (Boulder, CO: Johnson Books, 1999), 65.

55. "Shape of Things to Come," *Cable Television Business*, February 1, 1984, 40.

56. An earlier version of the discussion on the development of the cable–satellite network appears in: Patrick Parsons, "The Evolution of the Cable–Satellite Distribution System," *Journal of Broadcasting & Electronic Media*, 47 (March 2003): 1–17.

57. Arthur C. Clarke, "Extra-Terrestrial Relays: Can Rocket Stations Give World-Wide Radio Coverage?" *Wireless World*, October 1945, 305–308. Although accounts of communication by radio and satellite had appeared in earlier science fiction writing, such as: Hugo Gernsback, *Ralph 124C41+*. (Boston: Stratford Co., 1925).

58. The first trans-Atlantic satellite signal was beamed from AT&T's remote Andover, Maine, ground station via Telstar to receiving sites in Pleumeur-Bodou, France, and Goonhilly, England. The first picture sent was of the American flag, waving in front of the Andover facility.

59. Syncom-I failed in orbit in 1963, and, in fact, Syncom-II wobbled a bit though its orbit. RCA's Relay II and Hughes' Syncom-III inaugurated true geosynchronous service.

60. Public Law 87–624, 87th Congress, H.R. 11040, 31 August 1962 (76 Stat. 419).

61. "Satellites As TV Relay Points Pondered," *Broadcasting-Telecasting*, November 21, 1955, 96.

62. Richard D. Taylor, "Satellite Direct Broadcasting: The Prospect for Development." EdD thesis. Columbia University, 1977.

63. Donald S. Bond, "A System for Direct Television Broadcasting Using Earth Satellite Repeaters. Five papers by Members of the Technical Staff of the Radio Corporation of America." Paper presented at the 17th Annual Meeting and Space Flight Exposition of the American Rocket Society, Los Angeles, CA, November 13–18, 1962.

64. Jerome F. Prochaska, "Key Factors for Domestic Policy in the Establishment of Broadcast-satellites for United States Television." DBA thesis. George Washington University, 1974, 17–24.

65. Prochaska; and Taylor, 48–53.

66. "GE Engineer Envisions Space-to-Home TV," *Broadcasting*, November 12, 1962, 66.

67. T.A.M. Craven, "Global TV Unreal: Pursue Practical Uses," *Advertising Age*, June 25, 1962, 72.

68. "Putting Space to Work to Educate the World," *Business Week*, December 25, 1965, 17.

69. House Select Committee on Astronautics and Space Exploration. *Summary of Hearings*, April 15–May 12, 1958.

70. Arthur C. Clarke, "Faces From the Sky," *Holiday*, September 1959, 48–9.

71. Ibid., 49.

72. Dallas Smythe, "Space Broadcasting: Threat or Promise?" *Journal of Broadcasting* 1960 (Summer):191–8.

73. Craven; and "Douses Satellite-to-Home: J. V. Charyk," *Broadcasting*, April 4, 1966, 100.

74. President's Task Force on Communications Policy. Final Report, December 7, 1968 (The Rostow Report).

75. Patrick Parsons and Robert Frieden, *The Cable and Satellite Television Industries* (Boston: Allyn & Bacon, 1998), 52.

76. Federal Communications Commission. Non-Acceptance of Application by American Broadcasting Cos., 2 FCC 2d, 671 (1965).

77. FCC, Notice of Inquiry, Establishment of Domestic Non-common Carrier Communications Satellites by Non-Governmental Entities. Docket No.16495, 31 Fed. Reg. 3507 (March 2, 1966).

78. The Ford Foundation proposed a novel plan to use satellites for distribution of both commercial network and public broadcast programming and use the revenue from the service to help support the noncommercial broadcasting system. Ford Foundation. Comments and Legal Brief and Comments Before the Federal Communications Commission in the matter of the Establishment of Domestic, Non-Common Carrier Communications Satellite Facilities by Non-Governmental Entities. Docket 16495. Pub. Date, August 1, 1966.

79. Wilbur Schramm et al. "A Pilot Test of An Educational Television Satellite," Ford Foundation, 1966; and Harold Rosen, "A Satellite System for Educational Television," *Astronautics and Aeronautics*, April 1968, 58–63.

80. Twentieth Century Fund Task Force on International Satellite Communications, *Communicating by Satellite*. New York, 1969, 67–8.

81. Leon Papernow, "One Man's Opinion," *Television Magazine*, December 1965, 30–1.

82. Leland Johnson, "The Impact of Communications Satellites on the Television Industry." Paper presented at the 13th Annual Meeting of the American Astronautical

Society, Dallas, Texas, May 1–3, 1967. Rand Corp. paper. P-3572; and Leland Johnson, "New Technology: Its Effect on Use and Management of the Radio Spectrum." *Washington University Law Quarterly*, 1967 (521)

83. Johnson, 6.

84. Ibid., 8.

85. Ibid., 7–8.

86. Ibid.

87. National Academy of Sciences, *Useful Applications of Earth-Oriented Satellites: Broadcasting*, Report of Panel 10, Summer Study on Space Applications. Washington, D.C., 1969.

88. Ibid., 23.

89. New York City, The Mayor's Advisory Task Force on CATV and Telecommunications, *A Report on Cable Television and Cable Telecommunications in New York City*, September 14, 1968, 9.

90. First Report and Order, 20 FCC 2d 201, 207–208 (1969).

91. Hubert Schlafly, Personal interview, October 20, 1998.

92. "Study Report for TelePrompTer: CATV Satellite Earth Terminal." Hughes Aircraft Corp., February 1970.

93. National Cable Television Association, 18th Annual NCTA Convention, Official Transcript, *Rx: Cable, Prescription for the Future*, June 22–25, 1969, San Francisco Hilton, 668–745.

94. In his 1970 *Nation* article, Smith blasted the NCTA-NAB agreement as nothing less than "collusion in restraint of trade," Smith, Ralph Lee. "The Wired Nation," *The Nation*," May 1970, 589.

95. NCTA, 18th Annual Convention, transcript, 669.

96. Ibid., 693.

97. Ibid., 731.

98. Roger Noll, Merton Peck, and John McGowen. 1973. *Economic aspects of television regulation*. Washington, D.C.: The Brookings Institution, 246–50; and "TVN also sees satellite use by program syndicators, *Broadcasting*, January 27, 1975, 26.

99. Seiden, *Cable Television U.S.A*, 134; and "Everybody to Cut the Pie in Space," *Broadcasting*, February 2, 1970, 23–4.

100. Nathaniel Feldman, "Interconnecting Cable Television Systems by Satellite: An Introduction to the Issues," Opening remarks. 112th Technical Conference of the Society of Motion Picture and Television Engineers, Los Angeles, CA, October 26, 1972, 16. (Rand paper P-5035.)

101. Second Report and Order, 35 FCC 2d 844, 864. (Concurring opinion of Commissioner Nicholas Johnson).

102. First Report and Order, In the Matter of Establishment of Domestic Communications Satellite Facilities by Non-Governmental Entities, 22 FCC 2d 86. Docket 16495 (1970).

103. "Everybody to Cut the Pie in the Sky?" *Broadcasting*, February 2, 1970, 23.

104. "CATV headed for ad-supported network?" *Broadcasting*, May 4, 1970, 23.

105. John Lady, "Analysis of the Potential Growth of the CATV Industry," NCTA Report, 1970, 4.

106. "Kahn Plans a Satellite for CATV," *Broadcasting*, February 9, 1970, 38.

107. "Hughes Files for CATV Satellite System," *Broadcasting*, December 28, 1970, 9.

108. D. S. Allen, J. L. Bossert, and L. I. Krause. "Economic viability of the proposed United States communications satellite systems, Final Report." Menlo Park, CA, Stanford Research Institute, 1971.

109. Proposed Second Report and Order, 34 FCC 2d 9 at 40 (1972).

110. "Up in the Air over Satellites," *Broadcasting*, April 5, 1971, 71.

111. Cited in, Shapiro, 149.

112. Proposed Second Report and Order, 34 FCC 2d 9 at 22 (1972).

113. Second Report and Order on Domestic Communications Satellite Facilities, 35 FCC 2d 844 (1972).

114. Ibid., 855–6.

115. FCC Common Carrier Action, Public Notice No. 1000 (Jan. 4, 1973).

116. Sloan Commission, 42.

117. Hub Schlafly, Personal interview, October 20, 1998.

118. Sidney Topol, Oral History, The Cable Center, Denver, CO, June 20, 1991.

119. "Domsat Show is High Note of NCTA's All-stops-out Convention," *Broadcasting*, June 25, 1973, 25.

120. "Cable Men get Serious about Satellites," *Broadcasting*, May 28, 1973, 49.

121. "Domsat Show is High Note of NCTA's All-stops-out Convention," *Broadcasting*, June 25, 1973, 25.

122. "How Teleprompter Figures to Weave a Cable Network," *Broadcasting*, March 19, 1973, 114.

123. FCC Common Carrier Action, Public Notice No. 1270 (Sept. 13, 1973).

124. "Nothing to Say," *Broadcasting*, August 5, 1974, 5.

125. Booze, Allen & Hamilton, "Final Report: Satellite Interconnection Feasibility Study, Cable Satellite Access Entity," Norfolk, VA, August 1, 1974.

126. George Mair, *Inside HBO*. (New York: Dodd, Mead, 1988), 43.

127. "HBO: The First Twenty Years," Time-Warner, Inc., 1992.

128. Gerald Levin, Comments at the Second Annual Cable Television Hall of Fame induction ceremonies. Chicago, IL, June 13, 1999.

129. Levin. "25th Anniversary of Satellite-Cable Programming," C-SPAN, November 28, 2000.

130. "HBO: Point Man for an Industry Makes It into the Clear," *Broadcasting*, October 17, 1977, 51.

131. "Domsat Show is High Note of NCTA's All-stops-out Convention," *Broadcasting*, June 25, 1973, 25.

132. "HBO: Point Man for an Industry Makes It into the Clear," 51.

133. Levin. Comments at the Second Annual Cable Television Hall of Fame, induction ceremonies.

134. Hub Schlafly, Personal correspondence, May 21, 1999.

135. Mair, 24.

136. "Bird is in Hand for Pay Cable, " *Broadcasting*, October 6, 1975, 26.

137. Robert Rosencrans, Telephone interview, October 20, 1998.

138. Fort Pierce and Vero Beach, FL; Fort Smith, AR; Laredo, TX; Yuma, AZ; El Centro, CA; and Pasco and Kennewick, WA.

139. "Time Inc. Unit to Use satellites to Deliver Programs to UA-Columbia Cable Systems," *Wall Street Journal*, April 11, 1975, 21; "Mr. Levin's Giant Step for Pay TV," *Broadcasting*, April 21, 1975, 16.

140. Ibid.

141. Robert Rosencrans, Personal interview, May 20, 1999.

142. In Re: Florida Cablevision, 54 FCC 2d 881 (1975).

143. Frank Cooper, Oral History, The Cable Center, Denver, CO.

144. "HBO: Point man," 51.

CHAPTER 8

1. Irving Kahn, "Blue Sky Through a Green Filter." Speech before the Annual Convention of the Texas Cable TV Association, Fairmont Hotel, Dallas, TX, February 27, 1975.

2. Ibid.

3. "Burch to NCTA: May the Best Medium Win," *Broadcasting*, June 25, 1973, 26, 27.

4. In May 1974, Nixon nominated Abbott Washburn and Glen O. Robinson to FCC, and named Robert E. Lee for fourth term.

5. Erwin Krasnow, Lawrence Longley, and Herbert Terry. *The Politics of Broadcast Regulation*, 3rd ed. (New York: St. Martins Press, 1982), 45.

6. Robert Britt Horwitz, *The Irony of Deregulation* (New York: Oxford University Press, 1989), 256.

7. Ibid.

8. Cabinet Committee of Cable Communication. "Cable: Report to the President." (Washington, D.C.: Government Printing Office, 1974) (The Whitehead Report).

9. "Whitehead on Access: Cable as Common Carrier," *Broadcasting*, September 20, 1971, 43.

10. Charles O. Verrill, "CATV's Emerging Role: Cablecaster or Common Carrier?" *Law and Contemporary Problems*, 34:3 (1969): 586–609.

11. "Cable Regulation According to Houser," *Broadcasting*, April 26, 1971, 22.

12. Paul W. MacAvoy, ed. *Deregulation of cable television: Ford administration papers on regulatory reform.* (Washington, D.C.: American Enterprise Institute for Public Policy Research, 1977), preface.

13. Staff of the Subcommittee on Communications of the House Committee on Interstate and Foreign Commerce. *Cable Television: Promise Versus Regulatory Performance*, 94th Cong., 2d sess., 1976, Subcomm. Print.

14. Ibid., 3

15. Ibid., 7.

16. House Committee on Interstate and Foreign Commerce, *Options Papers*, 94th Cong., 1st sess., 1977, Comm Print.

17. Irving Kahn, Oral History, The Cable Center, Denver, CO.

18. *CBS, Inc. v. Teleprompter*, 355 F. Supp. 618 (S.D.N.Y. 1972).

19. 476 F.2d 338 (2d Cir. 1973).

20. *Teleprompter, Inc. v. CBS*, 415 U.S. 394 (1974).

21. "Whitehead Touts Cable As Great White House Hope," *Broadcasting*, February 19, 1973, 44.

22. "Cable Finds a Friend in FCC's Wiley," *Broadcasting*, November 6, 1972, 46, 50.

23. Systems with gross receipts under $292,000 pay 0.5% of the first $146,000 and 1.0% of revenue beyond that.

24. 0.893%

25. *NAB. V. Copyright Royalty Tribunal*, 675 F.2d 367 (D.C. Cir. 1982).

26. "20 Years of Cable Television," *Cable Television Business*, February 1, 1984, 60.

27. Cable Television—Additional Rules, 49 FCC 2d 1199 (1974) at 1207.

28. The proceeding arose following the FCC's 1972 Cable Report and Order. Questions about the role of state and local agencies in the control of cable led to the formation of an FCC Federal-State/Local Advisory Committee (FSLAC). The Committee was divided on the issue of how to avoid excessive regulatory duplication, and subsequent FCC hearings ended without any action being taken. See, Report and Order Terminating Proceeding in the Area of Duplicative and Excessive Over-Regulation of Cable TV, 54 FCC 2d 855 (1975).

29. Report and Order in Docket 20482, 55 FCC 2d 529 (1975).

30. First Report and Order, 52 FCC 2d 519 (1975).

31. Report and Order, 49 FCC 2d 1090 (1974).

32. Cable Television Systems, Second Report and Order, 55 FCC 2d 540 (1975).

33. Subscriber Rates—CATV, Report and Order, 60 FCC 2d 672 (1976).

34. Rules and Regulations Relevant to Carriage of Late-Night Television Programming. Report and Order, 48 FCC 2d 699 (1974), and Memorandum Opinion and Order, 54 FCC 2d 1182 (1975).

35. Report and Order in Docket 20487, 57 FCC 2d 625 (1976).

36. Rules and Regulations Relative to Specialty Stations. 58 FCC 2d 442 (1976).

37. Re: Rules and Regulations Relative to the Carriage of Network News Programs on Cable Television Systems, 57 FCC 2d 68 (1976).

38. "Old and New Fires Stoked to Life in Pay-Cable Comments," *Broadcasting*, November 6, 1972, 50–51.

39. Cablecasting and Subscription TV Rules, First Report and Order, 52 FCC 2d 1 (1975).

40. 35 R.R. 2d 767 (1975).

41. *Home Box Office v. FCC*, 567 F.2d 9 (D.C. Cir. 1977), *cert. denied*, 434 U.S. 829 (1977).

42. Peter J. Fadde and James C. Hsiung, "Pay TV: A Historical Review of the Siphoning Issue," *Communications and the Law*, 9 (1987): 15–26.

43. Report and Order, 41 RR 2d 1491 (1977), 42 RR 2d 1207 (1978).

44. In re: American Broadcasting Co., 62 FCC 2d 901(1977).

45. Stanley Gardner, Oral History, The Cable Center, Denver, CO.

46. Docket 20363, 54 FCC 2d 207 (1975). (The rules had mandated the upgrades by March 31, 1977).

47. Report and Order 59 FCC 2d 294 (1976).

48. *Midwest Video Corp. v. FCC*, 571 F.2d 1025 (1978).

49. *FCC v. Midwest Video Corp.*, 440 U.S. 689 (1979) ("Midwest Video II").

50. Memorandum Opinion and Order, 40 FCC 2d 1138 (May 22, 1973).

51. Pole Attachment Proceedings, 38 Fed. Reg. 21957 (August 3, 1973).

52. Telephone companies often have to rearrange existing lines on their poles to make room for the addition of cable TV lines. Costs for this rearrangement process were often a sticking point in pole attachment rate negotiations.

53. "Teleprompter's Bombshell Sends Shockwaves Through Cable," *Broadcasting*, February 5, 1974, 43, 44.

54. Thomas Hart, "The Evolution of Telco-Constructed Broadband Services for CATV Operators," *Catholic University Law Review* 1985, (34):697–735.

55. California Water & Tel. Co., 37 RR 2d 1166 (July 28, 1976).

56. Memorandum Opinion and Order—California Water and Telephone Co., 64 FCC 2d 753 (1977).

57. The Communications Act Amendment of 1978. Pub. L. No. 95–234, sect. 6, Stat. 33, 35 (1978). 47 U.S.C. sect. 224.

58. Second Report and Order, Cable Television Pole Attachments, 77 FCC 2d 187 (1979); constitutionality upheld in *FCC v. Florida Power Co.*, 480 US 245 (1987). The formula was maximum rate = space occupied by CATV total usable space × (op. expenses + capital costs of poles).

59. *Borough of Scottsdale v. National Cable Television Corporation*, 368 A.2d. 1323 (1977).

60. Clarification of the Cable Television Rules, 46 FCC 2d 175, at 200 (April 15, 1974).

61. *Brookhaven Cable TV, Inc., et. al v. Kelly*, 428 F. Supp. 1216 (N.D.N.Y. 1977), *aff'd*, 573 F.2d 765 (2d Cir. 1978), *cert. denied*, 441 U.S. 904 (1979).

62. *Community Communications, Inc. v. City of Boulder*, 485 F. Supp. 1035 (D. Colo. 1980), *rev'd*, 630 F.2d 704 (10th Cir, 1980), *rev'd*, 455 U.S. 40 (1982); 496 F. Supp 823 (D. Colo 1980), *rev'd*, 660 F.2d 1370 (10th Cir. 1981), cert denied, 456 U.S. 1001 (1982).

63. Local Government Antitrust Act, Pub. L. No. 98–544 (1984).

64. "Armed with New Copyright Bill, NCTA Will Seek Fresh Look at FCC's Rules on Signal Carriage," Broadcasting, October 11, 1976, 48.

65. "Reflecting Upon a Historic Decade," *Cablevision*, July 22/29, 1985, 82–88.

66. Gerald Flannery, ed. *Commissioners of the FCC* (Lanham, MD: University Press of America, 1995), 184.

67. Ibid., 178–179.

68. "The Laissez Faire Legacy of Charlie Ferris," *Broadcasting*, January 19, 1981, 37–42.

69. Report and Order in Docket 21002, 66 FCC 2d 380 (1977).

70. In re: Micro-Cable Communications d/b/a Florida Cablevision, 67 FCC 2d 339 (1977).

71. Memorandum Opinion and Order, 68 FCC 2d 57 (1978).

72. Report and Order, 69 FCC 2d 697 (1978).

73. Commercial Television Network Practice, 69FCC 2d 1524 (1978).

74. Notice of Inquiry in Docket 20988, 61 FCC 2d 746 (1976).

75. Notice of Inquiry in Docket 21284, 65 FCC 2d 9 (1977).

76. Report in Docket 20988, 71 FCC 2d 951 (1979); and Report in Docket 21284, 71 FCC 2d 632 (1979).

77. Stanley Besem and Robert Crandall. "The Deregulation of Cable Television," *Law and Contemporary Problems* 44: 77 (1981): 103–105.

78. Ibid., 104–105.

79. "Fame, Fortune and Franchising," *Cable Television Business*, February 1, 1984, 78.

80. Notice of Proposed Rulemaking in Dockets 20988 and 21284, 71 FCC 2d 1004 (1979).

81. Cable Television Syndicated Program Exclusivity Rules and Inquiry into the Economic Relationship Between Broadcasting and Cable Television, 79 FCC 2d 663 (1980).

82. "FCC Now All But Out of Cable Business," *Broadcasting*, July 28, 1980, 25.

83. *Malrite TV. FCC*, 652 F.2d 1140 (2d Cir. 1981), *cert. denied*, 454 US 1143 (1982).

84. "FCC Now All But Out of Cable Business," 26, 27.

85. The Commission had long collected money from broadcasters and cable operators to help offset its operating costs, but raised its annual levy in 1970 under pressure from Congress and the White House to fund 100 percent of its budget from industry fees. In 1974, the Court overturned the 100 percent fee schedule and order refunds for cable operators.

86. Technology and Economics, Inc. "The Emergence of Pay Cable Television, Volume II: Final Report." Prepared for the National Telecommunications and Information Administration, July 1980, 15.

87. "The Emergence of Pay Cable Television, Volume II: Final Report," 16.

88. Ibid., 20.

89. NCTA. *Cable Television Developments 1990* (Washington, D.C.: The National Cable Television Association, 1990)

90. George Mair, *Inside HBO* (New York: Dodd, Mead, 1988), 55.

91. Taylor credits early development of the technology to Vic Tarbutton, an engineer with Anaconda Wire and Cable, later Century III. George Ray and Bill O'Neil, who left SKL to form their own company, Amplifier Design Service (ADS), advanced the design and O'Neil was one of the first to demonstrate its use in cable service in 1975. Archer Taylor, *History Between Their Ears* (Denver, CO: The Cable Center, 2000), 222–23.

92. Jeff Hecht, *City of Light: The Story of Fiber Optics* (New York: Oxford University Press, 1999); an excellent history of fiber optics.

93. Ibid., 18.

94. Ibid., 93.

95. John Barrington, "Fiber Optics—Passkey Into Broadband Communications Era." In: Mary Louise Hollowell, ed. *The Cable/Broadband Communications Book, 1977–1978* (Washington, D.C.: Communications Press, Inc., 1977), 143.

96. Ibid.

97. "Buy While You Watch," *Business Week*, December 24, 1960, 62–63.

98. Taylor, 110.

99. "Rediffusion Unveils Pilot US CATV System," *Television Digest*, November 2, 1970, 2–3.

100. Linda K. Fuller, *Community Television in the United States* (Westport, CT: Greenwood Press, 1994), 2.

101. "New York Cables' Answer to Carson: Late-night Sex," *Broadcasting*, June 9, 1975, 48.

102. "Will the Mighty Inherit the CATV Earth?" *Broadcasting*, March 20, 1972, 22.

103. "Cable: The First Forty Years," *Broadcasting*, November 21, 1988, 35–49.

104. Turner's biographies offer slightly differing, but basically similar accounts of the incident, such as the following: Christian Williams, *Lead Follow or Get Out of the Way: The Story of Ted Turner*. (New York: Times Books, 1981), 152–153; Robert Goldberg and Gerald Jay Goldberg, *Citizen Turner: The Wild Rise of an American Tycoon*. (New York: Harcourt Brace & Co., 1995), 216–17; Porter Bibb, *Ted Turner: It Ain't As Easy As It Looks*. (Boulder: Johnson Books, 1997), 132–33.

105. Williams, 152, and Bibb, 133, although versions vary slightly. "Pike" was Sid Pike, Channel 17 General Manager.

106. Goldberg and Goldberg, 216.

107. Williams, 152–53.

108. Davis, 92.

109. Ibid.

110. "Dawn of Cable Programming," *Cable Television Business*, February 1, 1984, 64.

111. Goldberg and Goldberg, 159–160.

112. Goldberg and Goldberg, 61.

113. Williams, 98.

114. Southern Satellite Systems, 62 FCC 2d 153 (1976).

115. The four were Troy (Alabama) Cablevision, Newton (Kansas) Cable TV, Hampton Roads (Virginia) Cablevision, and Multi-Vue TV (Grand Island, NE).

116. Memorandum Opinion and Order, 68 FCC 2d 57 (1978).

117. Unless they altered the signal in some manner, the carriers were also immune from any copyright liability for retransmission of the programming. See, *Eastern Microwave, Inc. v. Doubleday Sports, Inc.*, 534 F. Supp. 533 (N.D.N.Y.), *rev'd*, 691 F.2d 125 (2d Cir. 1982); *WGN Continental Broadcasting Co., V. United Video, Inc.*, 523 F. Supp. 403 (N.D. Ill. 1981), *rev'd*, 693 F.2d 622 (7th Cir. 1982).

118. He had wanted those call letters originally, but they were being used at the time by a noncommercial FM station in Cambridge, MA.

119. FCC, Network Inquiry Special Staff, "Video Interconnection: Technology, Cost, Regulatory Policies," 1980, 68–73.

120. Biographies of Robertson include the following: John B. Donovan, *Pat Robertson: The Authorized Biography*. (New York: MacMillan Publishing Co. 1988); and Alec Foege, *The Empire That God Built: Inside Pat Robertson's Media Machine* (New York: John Wiley & Sons, Inc. 1996).

121. Donovan, 45.

122. Ibid., 116.

123. There are several books on the rise and fall of Jim and Tammy Bakker and PTL, including, Gary L. Tidwell, *Anatomy of a Fraud: Inside the Finances of the PTL Ministries* (New York: John Wiley & Sons, Inc. 1993).

124. Robert Rosencrans, Oral History, The Cable Center, Denver, CO.

125. Stephen Frantzich and John Sullivan, *The C-SPAN revolution* (Norman, OK: University of Oklahoma Press, 1996), 26. This work offers an excellent description of the development of C-SPAN and the discussion in this chapter draws heavily from the book.

126. Frantzich and Sullivan, 36.

127. Ibid.

128. Dennis Harp, "'Nickelodeon' Cable Programmers Handbook." In: Robert Picard, ed. 1993. *The Cable Networks Handbook* (Riverside, CA: Carpelan Publishing, 1993), 150.

129. Les Brown, "Children's Programming without Commercials," *New York Times*, March 4, 1979, E20.

130. The ATAS determined that only programming that reached 51 percent of the country would be eligible for EMMY competition; national cable penetration at the time was less than 20 percent.

131. The first ACE was presented to HBO for a Bette Midler special.

132. Barry Litman and Susanna Eun, "The Emerging Oligopoly of Pay Cable TV in the USA," *Telecommunications Policy* 5:1 (1981): 20–34.

133. *American TV & Communications Corp.*, 70 FCC 2d 2175, at 2186 (1978).

134. Ralph Baruch, Oral History, The Cable Center, Denver, CO.

135. Ibid.

136. Frank Cooper, Oral History, The Cable Center, Denver, CO.

137. Technology and Economics, Inc., "The Emergence of Pay Cable Television," 104

138. Ibid., 320.

139. Quoted in Michele Hilmes, *Hollywood and Broadcasting: From Radio to Cable* (Urbana, IL: University of Illinois Press, 1990), 176.

140. Tony Schwatrz, "Pay Cable is Fighting for Movies," *New York Times*, August 22, 1981, sec. C, 24.

141. In 1979, there were three domestic, commercial satellite providers: RCA with its Satcom series, Western Union with Westar, and Comsat with its Comstar satellites.

142. Galavision was programmed for cable and used Satcom I, whereas SIN continued to provide feeds for its broadcast affiliates using Westar.

143. Satcom II was used primarily for voice and data communications and was a heavy provider of satellite telecommunications facilities in Alaska. It also serviced broadcast network feeds.

CHAPTER 9

1. House Committee on Small Business. Subcommittee on SBA and SBIC Authority, *Cable Television Industry*, Hearings, 97th Cong., 1st sess., September 23, October 13, and November 4, 1981, 259.

2. "The Gold Rush of 1980," *Broadcasting*, March 31, 1980, 35–56.

3. Allan Sloan, "Bring Plenty of Money," *Forbes*, December 10, 1979, 49–54.

4. James Roman, *Cablemania* (Englewood Cliffs, NJ: Prentice Hall 1983).

5. "Cable: The First 40 Years," *Broadcasting*, November 21, 1988, 44.

6. House Committee on Energy and Commerce, Subcommittee on Telecommunications, Consumer Protection and Finance. *Telecommunications in Transition: The Status of Competition in the Telecommunications Industry*, Staff Report. 97th Cong., 1st sess., November 3, 1981, Comm. Print 97-V, 348–49.

7. It is worth noting at the same time, that cable operators were not that enthusiastic about servicing apartment buildings. Apart from the problems noted in Chapter 7 on wiring New York City, apartment dwellers are more transient than homeowners, creating a great deal more churn. Theft of signals and equipment is a greater

problem in apartments as is delinquent or nonpayment of the monthly bill. A June 1982 Supreme Court decision held that apartment owners could not prohibit cable companies access to a tenant, but operators could be charged a reasonable fee for access to the building. *Loretto v. Teleprompter Manhattan CATV Corp.*, 458 US 419 (1982).

8. "The Gold Rush," 36.

9. "ATC-Daniels Partnership takes Denver Franchise," *Broadcasting*, March 1, 1982, 34.

10. Sloan, 54.

11. Thomas Whiteside, "Upward and Onward with the Arts," *The New Yorker*, May, 27, 1985, 72.

12. *Affiliated Capital Corp. v. City of Houston*, 519 F. Supp. 991 (S.D. Tex. 1981); *rev'd*, 700 F.2d 226 (5th Cir. 1983); *en banc*, 735 F.2d. 1555 (5th Cir. 1984).

13. House Committee on Small Business, *Cable Television Industry*, 1981, 259.

14. Sloan, 51.

15. Fred Dawson, "The Hottest Show in Town," *Cablevision*, November 22, 1982, 219–60.

16. *Central Telecommunications, Inc. v. TCI Cablevision, Inc.*, 800 F.2d 711 (8th Cir. 1986), No. 85–1805WM Brief for the Appellee, Affidavit of Elmer Smalling.

17. *Central Telecommunications, Inc. v. TCI Cablevision, Inc.*, 800 F.2d 711 (8th Cir. 1986), *cert. denied*, 480 U.S. 910 (1987).

18. Whiteside, 101.

19. Ibid.

20. House Committee on Energy and Commerce, *Telecommunications in Transition*, 1981, 349.

21. Ibid.

22. *Television & Cable Factbook*, Services Volume. (Washington, D.C.: Warren Publishing, 1985, 1391).

23. House Committee on Interstate and Foreign Commerce, Subcommittee on Communications, *Cable Television Promise versus Regulatory Performance*, 94th Cong., 2d sess., January 1976, Subcomm. Print, 3–4.

24. Ernest Holsendolph, "Tougher Times for Cable TV," *New York Times*, July 11, 1982, F1.

25. "A Year of Decision—Fowler," *Broadcasting*, April 12, 1982, 34.

26. Although TV station ownership was further restricted, following debate in Congress, by a rule which stipulated that total audience reach could not exceed 25 percent of national audience. Memorandum Opinion and Order in re: Multiple Ownership of AM, FM and television Broadcast Stations, 100 FCC 2d 74, 1984.

27. TV Programming for Children, 96 FCC 2d 648 (1984).

28. Commercial TV Stations, 98 FCC 2d 1078–1079 (1984).

29. Notice of Proposed Rulemaking, Deregulation of Radio, 733 FCC 2d 457 (1979). Although this proceeding never came to fruition.

30. Notice of Proposed Rulemaking, 45 F.R. 72902 (1980).

31. Low Power Television Service, Notice of Proposed Rulemaking, 82 FCC 2d 47 (1980).

32. Low Power Television Broadcasting, 46 F.R. 26062 (1981).

33. Low Power Television Service, 51 RR 2d 476 (1982).

34. Subscription Television Rules, 44 F.R. 60091, Oct. 18, 1979.

35. In particular, the FCC dropped its "complement of four" rule, which confined STV to markets with at least four commercial stations. Third Report & Order in Subscription TV Service, 90 FCC 2d 341 (1982).

36. House Committee on Energy and Commerce, *Telecommunications in Transition*, 254.

37. F.C.C. Report and Order, Docket No. 14712, 27 F.R. 12372 (December 1962).

38. Memorandum Opinion and Order, 47 FCC 2d 957 (1970).

39. In Multipoint Distribution Service, Notice of Proposed Rulemaking, 34 F.C.C. 2d 719 (1972).

40. Multiple Point Distribution Service, Report and Order, 45 FCC 2d 616 (1974).

41. John Gwin, one of the architects of CSAE, was Cox Cable Vice President and in charge of the Microband arrangement at the time, so the satellite connection was likely not a coincidence.

42. Notice of Proposed Rulemaking, FCC Docket No. 80–112 (May 2, 1981).

43. Use of Private Microwave Frequencies, 86 FCC 2d 299 (1981).

44. Instructional Television Fixed Service (MDS Reallocation), 94 FCC 2d 1203 (July 15, 1983).

45. Reregulation of Receive-Only Domestic Earth Stations, 74 FCC 2d 205 (1979).

46. Report and Order, 60 FCC 2d 700 (1976).

47. The FCC allocated the Ku band from 1.2 to 12.7 GHz for broadcast satellite consistent with international agreements reached at the 1977 World Broadcasting Satellite Administrative Conference (WARC-BS). Report and Order, 90 FCC 2d 676 (1982).

48. Notice of Inquiry, 45 RR2d 727, 45 Fed. Reg. 72719 (Oct. 29, 1980).

49. Notice of Proposed Policy Statement and Rulemaking, 46 Fed. Reg. 30124 (1981).

50. Direct Broadcast Satellites, 90 FCC 2d 676 (1982). In fact, the Commission was bound by international agreement in the allocation of orbital slots for DBS and, by 1982, the governing international body, the International Telecommunications Union (ITU) had not yet assigned space for these services. The FCC explained that it had granted conditional permits to provide a better sense of U.S. needs in space when the ITU met in 1983 to assign DBS service. At that meeting, the Region 2 Regional Administrative Radio Conference, the United States received eight orbital slots for commercial DBS service in the 12.2- to 12.7-GHz band.

51. Satellite Television Corp., 91 FCC 2d 953 (1982).

52. Subscription TV company, Oak Industries, had proposed a similar low-power service using Anik satellites, but abandoned the plans early in their development.

53. Teletext Transmission, 53 RR 2d 1309 (1983).

54. National Cable Television Association. *A Cable Primer*. (Washington, D.C., 1981), 9.

55. "Teletext and Videotex," *Broadcasting*, December 10, 1984, 66–70.

56. State-supported deployment and operation continued in France, but the country's "Minitel" terminals were used primarily as electronic substitutes for the phone book and lost money annually.

57. Larry Pryor, "The Videotex Debacle," *American Journalism Review*, November 1994, 41.

58. *Sony Corp. of America v. Universal City Studios, Inc.* 464 U.S. 417 (1984).

59. Eugene Marlow and Eugene Secunda. *Shifting Time and Space: The Story of Videotape* (New York: Praeger, 1991), 132.

60. Thomas Schatz, "The New Hollywood." In: Jim Collins, Hilary Radner, and Ava Collins, eds. *Film Theory Goes to the Movies.* (New York: Routledge, 1993), 25.

61. Bruce Klopfenstein, "The Diffusion of the VCR in the United States." In: Mark Levy, ed. *The VCR Age: Home Video and Mass Communication* (Newbury Park: Sage Publications, 1989).

62. Bruce Klofenstein, "From Gadget to Necessity: The Diffusion of Remote Control Technology; and Dean M. Krugman and Roland T. Rust. "The Impact of Cable and VCR Penetration on Network Viewing: Assessing the Decade," *Journal of Advertising Research*, January/February 1993, 67–73.

63. Fred Henck and Bernard Sraussburg. *A Slippery Slope: The Long Road to the Breakup of AT&T* (New York: Greenwood Press, 1988). The authors (and Stausburg served 10 years as Chief of the FCC's Common Carrier Bureau) described Goeken as likable but somewhat erratic and eccentric.

64. In re: Microwave Communications, Inc. 18 FCC 2d 953 (1969).

65. Steve McColl, *The Deal of the Century: The Breakup of AT&T.* (New York: Atheneum, 1986).

66. First Report and Order, 29 FCC 2d 870 (1971).

67. Bell System Offerings 46 FCC 2d 413 (1974), *aff'd sub nom, Bell Tel. Co. v. FCC*, 503 F.2d 1250 3rd Cir. 1974), *cert. denied*, 422 U.S. 1026 (1975).

68. Extended litigation resulted in an MCI victory in the case and an initial jury award of $1.8 billion, a figure that was later reduced to more about $113 million (figuring the trebled damages associated with antitrust violations). In the end, MCI reach an out-of-court settlement for an undisclosed sum with AT&T and the RBOCs. It is also worth noting parenthetically that MCI subsequently hired a new senior vice president, freshly retired from the Federal Communications Commission, Kenneth Cox (who had voted with the majority in favor of MCI's petitions).

69. Second Computer Inquiry, 77 FCC 2d 384 (1980), *aff'd sub nom, Computer & Communications Industries Association v. FCC*, 693 F.2d 198 (D.C. Cir. 1982).

70. *U.S. v. AT&T*, 552 F. Supp. 131 (D.D.C. 1982), *aff'd*, 103 S.Ct. 1240 (1983).

71. Huffman, L. "Wheeler Calls AT&T Pact 'Deceptive,'" *Multichannel News*, March 1, 1982, 4.

72. Technology and Economics, Inc., "The Emergence of Pay Cable Television, Final Report." Prepared for the National Telecommunications and Information Administration, July 1980, 351.

73. 1981, 3.8 million (33.1% total new color sets); 1982, 5.1 million (44.4% total); 1983, 6.5 million (45.9%); 1984 9.6 million (58.1%), 1985, 11.4 million (66.6%; 1986 12.8 million (68.6%). *Television and Cable Factbook*, Cable & Services vol. 56, 1988, C-306.

74. Louise Benjamin, "At the Touch of a Button: A Brief History of Remote Control Devices." In: James R. Walker and Robert V. Bellamy Jr., *The Remote Control in the New Age of Television* (Westport, CT: Praeger, 1993), 15–22.

75. Klopfenstein, 33.

76. Thomas Baldwin and D. Stevens McVoy, *Cable Communication*, 2nd ed. (Englewood Cliffs, NJ: Prentice Hall, 1988), 228.

77. Until the partnerships were closed out, Intercable, the general manager, did not know how much money would be generated, hence, the "blind" nature of the pooled investment.

78. John Malone, Oral History, The Cable Center, Denver, CO.

79. "Move into CATV, ANPA's Smith Advises Dailies," *Advertising Age*, November 7, 1966, 2.

80. Ralph Baruch, Oral History, The Cable Center, Denver, CO.

81. George Mair, *Inside HBO*. (New York: Dodd, Mead, 1988), 77.

82. Kay Koplovitz, Oral History, The Cable Center, Denver, CO.

83. Satcom IIIR was the RCA replacement for the lost Satcom III. Comstar D-2 was operated by AT&T and GT&E, and Western Union owned Westar III.

84. Thomas Baldwin and D. Stevens McVoy. *CablecCommunication* (Englewood Cliffs, NJ: Prentice Hall, 1983), 112.

85. Brett Pulley, *The Billion Dollar BET*. (Hoboken, NJ: John Wiley & Sons, Inc. 2004); and Leonard Barchak, "Black Entertainment Television." In: Robert Picard, *The Cable Networks Handbook*. (Riverside, CA: Carpelan Publishing Co. 1993).

86. Pulley, 31–32.

87. John Malone, Oral History.

88. House Committee on Small Business. "*Cable Television Industry*."

89. "Bob Johnson: Making a BET on Black Programming," *Broadcasting*, October 4, 1982, 87.

90. Reese Schonfeld, *Me and Ted Against the World: The Unauthorized Story of the Founding of CNN*. (New York: Harper Collins, 2001), 14.

91. Said one ITNA board member, "Cable is the enemy. I will not let them kill me with my news and then piss on my grave." Ibid., 14.

92. Ibid., 15.

93. Porter Bibb, *Ted Turner: It Ain't As Easy As It Looks* (Boulder: Johnson Books, 1977), 157–158.

94. Goldberg and Goldberg, 227–28.

95. Schonfeld, 52.

96. Hank Whitmore, *CNN: The Inside Story* (Boston: Little Brown, 1990), 1.

97. Mary Louise Hollowell, *The Cable/Broadband Communications Books*, vol. 2, 1980–1981. (Washington, D.C.: Communications Press, Inc., 1982), xi.

98. In re: CBS, Inc., 87 FCC 2d 587 (1981).

99. Hurst, TX, with 16,671 subscribers, and Mansfield, TX, with 2,714 subscribers.

100. Notice of proposed rulemaking, 91 FCC 2d 76 (1982).

101. Raymond Joslin, Oral History, The Cable Center, Denver, CO.

102. David Waterman, "The Failure of Cultural Programming on Cable TV: An Economic Interpretation," *Journal of Communication* 36:3 (1986): 92–107.

103. "Who's Ahead in Making Case on De-regulation of Cable?" *Broadcasting*, October 27, 1975, 32.

104. Harold Fisher, "Financial News Network." In Picard.

105. Robert Ogles, "Music Television (MTV)." In Picard.

106. Patrick Parsons, "The Business of Popular Music," In: Kenneth Bindas, *America's Musical Pulse* (Westport, CN: Greenwood Press 1992). In addition, a

general recession made less money available for consumer luxury purchases, including recorded music, and illegal copying of tapes and records was widespread.

107. Ogles, 138.

108. Frank Batten with Jeffrey Cruikshank, *The Weather Channel: The Improbable Rise of a Media Phenomenon* (Boston: Harvard Business School Press, 2002); A general history.

109. John M. "The Weather Channel." In Picard, 213.

110. Joanne Gula and Coral Ohl. "The Disney Channel." In Picard, 79.

111. Robert M. Ogles and Herbert H. Howard. "The Nashville Network." In Picard.

112. "Ted Turner: A Seller's Remorse," *Broadcasting & Cable*, December 3, 2001, 9.

113. Batten with Cruikshank, 138–146.

114. *National Association of Broadcasters v. Copyright Royalty Tribunal*, 675 F.2d 367 (D.C. Cir, 1982).

115. *National Cable Television Association v. Copyright Royalty Tribunal*, 689 F.2d 1077 (D.C. Cir. 1982).

116. *National Cable Television Association v. Copyright Tribunal*, 724 F.2d 176 (D.C. Cir. 1983).

117. Rates varied based on several criteria including whether the station was a network affiliate, independent, or noncommercial.

118. The 3.75 percent fee increase applied only to signals added beyond the number permitted under the old FCC ceiling. The surcharge applied to initial signals permitted under the old FCC ceiling.

119. "Cable Begins Cutback," *Television Digest*, December 20, 1982.

120. "The New Order Passeth," *Broadcasting*, December 10, 1984, 54.

121. "The Communications Conscience of Senator Goldwater," *Broadcasting*, June 25, 1979, 97.

122. The subcommittee, however, was renamed the Subcommittee on Telecommunications, Consumer Protection, and Finance, and its responsibilities expanded.

123. Wheeler announced his resignation in August of 1984, taking a job as President and CEO of the start-up Nabu Network, a company created to deliver computer software to customers via satellite and cable.

124. Malone.

125. 47 USC 542 (b).

126. In addition, systems with thirty-six to fifty-four channels had to assign 10 percent of capacity for commercial leased access; systems with fifty-five or more activated channels had to set aside 15 percent of their capacity for leased access.

127. "Free at Last: Cable Gets Its Bill," *Broadcasting*, October 15, 1984, 38.

128. This was for home use only, and did not extend to hotels, apartment houses, taverns or, interestingly, fraternities.

129. The FCC's position was largely upheld by the courts in *American Civil Liberties Union v. FCC*, 823 F.2d 1554 (D.C. Cir 1987).

130. Even where a local system did not face effective competition, rate regulation was restricted to tiers that carried local TV signals, giving operators the ability to organize tiers in such a way as to avoid most of the impact of rate controls.

131. Although a 1983 FCC ruling thwarted an attempt by the Nevada Public Service Commission to establish rate regulation for cable in the state (Community Cable TV, Inc., 95 FCC 2d 1204 (1983).

132. Local Government Antitrust Act, Pub. L. No. 98–544.

CHAPTER 10

1. "DBS Tie Looks Likely— Turner," *Television Digest*, July 4, 1983.

2. The recently launched Discovery network also programmed live coverage. Its estimated subscriber base at the time was 6.7 million homes. All three broadcast networks, each monitoring the launch, switched to live coverage within five minutes of the explosion.

3. Porter Bibb, *Ted Turner: It Ain't As Easy as It Looks* (Boulder, CO: Johnson Books, 1997), 306.

4. CNN maintained its telephone line, for that period, at the consent of the government, but Iraqi officials eventually cut their direct communications, as well.

5. William Henry III., "History As It Happens," *Time*, January 6, 1992, 24.

6. Ibid.

7. Figures from *Broadcasting and Cable Yearbook* (New Providence, RI: R. R. Bowker, annual.)

8. Figures from, NCTA, *Cable Television Developments* (Washington, D.C.: National Cable Television Association, annual.)

9. Craig Kuhl, "Construction Surge Seen Booming into '89," *Cablevision*, September 26, 1988, 40. (Above-ground construction in 1987 was estimated at around $10,000 a mile; underground suburban construction, $15,000 a mile; and underground urban construction, $50,000 a miles.)

10. Figures from NCTA, *Cable Television Developments*.

11. Cable could have implemented digital or FM analog fiber systems, but it would have necessitated an additional and cost conversion process.

12. Larry Stark, "HFC: Interactive and Inexpensive," *Communications Technology*, October 1995, 37.

13. The Association of America's Public Television Stations., *A plain English guide to technology: Telecommunications trends and their impact on public television stations*, September. 1993, 5.

14. See also, Anne M. Hoag, "Measuring Regulatory Effects With Stock Market Evidence: Cable Stocks and the Cable Communications Policy Act of 1984," *Journal of Media Economics*, 15:4 (2002): 259–272; and Praeger, R. A. "The Effects of Deregulating Cable Television: Evidence from the Financial Markets," *Journal of Regulatory Economics*, 4 (1992): 347–363.

15. Figures from NCTA, *Cable Television Developments*.

16. See also, A. B. Jaffe and D. M. Kanter., "Market Power of Local Cable Television Franchises: Evidence from the Effects of Deregulation," *RAND Journal of Economics*, 21 (1990): 226–234.

17. Maxwell E. McCombs, "Mass Media in the Marketplace," *Journalism Monographs*, 1972, 24.

18. See, for example, William Wood and Sharon O'Hare, "Paying for the Video Revolution: Consumer Spending on the Mass Media," *Journal of Communication*, 41:1 (1991): 24–30; and Hugh Fullerton, "Technology Collides with Relative Constancy," *Journal of Media Economics*, 1988:75–84.

19. Ratings are the percentage of viewers out watching of the universe of people with TV sets; share is the percentage watching a particular program or network out of all those with their sets turned on at that time.

20. See, Dean Krugman and Roland Rust, "The Impact of Cable Penetration on Network Viewing," *Journal of Advertising Research*, 1987:9–13; and "The Impact of Cable and VCR Penetration on Network Viewing: Assessing the Decade," *Journal of Advertising Research*, 1993:67–73. Note also that figures vary by source. NCTA, *Cable Television Developments* reports combined network broadcast affiliates share at 53 percent in 1990.

21. NCTA, *Cable Television Developments*, Spring 1996, 5.

22. See also, Glasock, Jack. "Effect of Cable Television on Advertiser and Consumer Spending on Mass Media, 1978–1990," *Journalism Quarterly*, 70:3 (1993): 509–517.

23. Krugman and Rust, 1993, 69.

24. Richard Munro, Oral History, The Cable Center, Denver, CO.

25. *FCC V. Pacifica Foundation*, 438 U.S. 726 (1978).

26. *Home Box Office v. Wilkinson*, 531 F. Supp. 986 (D. Utah 1982). Although undaunted, the legislature passed a second, similar ordinance, which the courts struck down one more. *Community Television of Utah v. Wilkinson*, 611 F. Supp. 1099 (D. Utah 1985), *aff'd sub nom, Jones v. Wilkinson*, 800 F.2d 989 (10th Cir. 1986).

27. *Community Television of Utah, Inc. v. Roy City*, 555 F. Supp. 1164 (N.D. Utah 1982).

28. *Cruz v. Ferre*, 571 F. Supp. 125 (S.D. Fla. 1983), *aff'd*, 755 F.2d 1415 (11th Cir. 1985).

29. Although as noted in Chapter 6, some in the industry sought to redefine cable into a more active medium before the 1968 *Fortnightly* case,. much to the worry of NCTA lawyers.

30. *NBC v. U.S.*, 319 U.S. 190 (1943).

31. *Wheeling Antenna Co., v. United States*, 391 F.2d. 179, 183 (4th Cir. 1968). And in *Midwest Video I*. In 1972, the Supreme Court acknowledged that the scarcity doctrine did not apply to cable, but concluded that FCC regulation was nonetheless permissible based on cable's ancillary relationship to broadcasting. 406 U.S. at 669.

32. *Miami Herald Publishing Co., v. Tornillo*, 418 U.S. 214 (1974).

33. *Home Box Office v. FCC*, 567 F.2d 9 (D.C. Cir. 1977), *cert. denied*, 434 U.S. 829 (1977).

34. *Home Box Office*, 567 F.2d, 46.

35. *Midwest Video II*, 571 F.2d. 1025, 1056 (8th Cir. 1978).

36. *Black Hills Video Corp. v. FCC*, 399 F.2d 65 (8th Cir. 1968).

37. *Quincy Cable TV, Inc. v. FCC*, 768 F.2d 1434 (D.C. Cir. 1985), *cert. denied*, 476 U.S. 1169 (1986).

38. *Community Communications Corp. v. Boulder*, 485 F. Supp. 1035 (D. Colo. 1980), *rev'd*, 630 F.2d 704 (10th Cir. 1980) (*Boulder I*); *C.C.C. v. Boulder*, 496 F. Supp. 823 D. Colo. 1980), *rev'd*, 660 F.2d 1370 (10th Cir. 1980), *rev'd*, 455 U.S. 40 (1982), (*Boulder II*).

39. *C.C.C. v. Boulder*, 660 F.2d, 1377.

40. *City of Los Angeles v. Preferred Communications*, 476 U.S. 488 (1986).

41. Patrick Parsons, "In the Wake of *Preferred*: Waiting for Godot," *Mass Comm Review*, 16:1 (1989): 26–37.

42. See, for example, Benjamin Compaine, and Douglas Gomery, *Who Owns the Media.* (Mahwah, NJ: Lawrence Erlbaum Associates, 2000); Eli Noam, "Towards an Integrated Communications Market: Overcoming the Local Monopoly of Cable Television," *Federal Communications Law Journal,* 34:2 (1982): 209–257; Sylvia M. Chan-Olmstead and Barry Litman. "Antitrust and Horizontal Mergers in the Cable Industry," *Journal of Media Economics,* 1988: 3–28; David Waterman, "A New Look at Media Chains and Groups: 1977–1989," *Journal of Broadcasting & Electronic Media,* 35:2 (1991): 167–178; David Atkin, "Cable Exhibition in the USA," *Telecommunications Policy,* 18:4 (1994): 331–341; and Herbert Howard, *Ownership Trends in Cable Television: 1985* (Washington, D.C.: National Association of Broadcasters, Washington, D.C., September 1986).

43. See, Waterman.

44. Chan-Olmstead and Litman, 7.

45. Atkin, 337.

46. Chan-Olmstead and Litman, 7.

47. FCC Staff Report. "FCC Policy on Cable Ownership," November 1981, 111–12.

48. Report and Order in Docket no. 18891, 91 FCC 2d 46 (1982).

49. Charles Sammons died in 1988. James Leonard Reinsch, broadcast cable pioneer and retired head of Cox Broadcasting, passed away in 1991.

50. For a detailed description of the evolution and impact of the ownership changes at ABC, NBC, and CBS, see, Ken Auletta, *Three Blind Mice: How the TV Networks Lost Their Way* (New York: Vintage Books, 1992).

51. "ABC Is Being Sold For $3.5 Billion; 1st Network Sale," *The New York Times,* March 19, 1985, A1.

52. "DBS Tie Looks Likely— Turner."

53. Bibb, 233.

54. Lauren Rublin, "No Money Down. Will Ted Turner Buy CBS on the Cuff?" *Barron's,* April 22, 1985, 1, 16.

55. John Malone, Oral History, The Cable Center, Denver, CO.

56. "Cable: The First Forty Years," *Broadcasting,* November 21, 1988;35–49, 46.

57. Some speculated Wyman was seeking to prevent the company from falling completed into Tisch's hands.

58. Dan Rather, "From Murrow to Mediocrity," *The New York Times,* March 10, 1987, A27.

59. In February of 1983, Charles Bluhdorn, the Chairman of Paramount's corporate parent Gulf + Western, died following a long struggle with cancer. Diller had had good relations with Bluhdorn, but found that his style did not mesh with the new Gulf + Western Chairman, Martin Davis, setting the stage for his move to 20th Century-Fox.

60. Transfer of Broadcast Facilities, 52 RR 2d 1081 (1982).

61. Christopher Knowlton, "Want This Stock? It's Up 91,000%," *Fortune,* July 31, 1989, 97, 100.

62. Mark Robichaux, *Cable Cowboy: John Malone and the Rise of the Modern Cable Business* (Hoboken, NJ: John Wiley & Sons, Inc., 2002), 94.

63. For a detailed and lively description of the Time Warner merger, see Richard M. Clurman, *To the End of Time: The Seduction and Conquest of a Media Empire* (New York: Simon & Schuster, 1992.)

64. Ibid., 145.

65. Ibid., 151.

66. Paramount was owned by Gulf + Western and run by former press agent Martin Davis. Davis's predecessor CEO Charles Bluhdorn had built the company into a holding pen for scores of widely disparate businesses. On his ascension to corporate control in 1983, Davis pared back, selling most of the subsidiaries that did not involve entertainment, including the oil and gas assets, and, in 1989, just after making his bid for Time, formally changed the company name to Paramount Communications.

67. Clurman, 221.

68. Before the rule change, banks were restricted from lending to companies with debt-to-equity ratios 75 percent or more. The Comptroller lowered the figure to 50 percent and redefined various forms of debt and accounting with the effect of further limiting lending practices.

69. Robert O'Brien, "It was a Great Year for Stocks, Despite a Fourth-quarter Pounding," *Cablevision*, January 15, 1990, 58.

70. Thomas Southwick, *Distant Signals* (Overland Park, KS: Primedia Intertec, 1998), 273.

71. It dropped the threshold to 4 million at the introduction of its Peoplemeter technology.

72. "Cable: The First Forty Years," 46.

73. Donna Lampert, "Cable Television: Does Leased Access Mean Least Access?" *Cable Television Leased Access.* A Report of the Annenberg Washington Program, Communication Policy Studies, Northwestern University, 1991.

74. See, for example, Chipty Tasneem, "Horizontal Integration for Bargaining Power: Evidence from the Cable Television Industry," *Journal of Economic Management Strategy*, 1995;4:375.

75. U.S. Department of Commerce, National Telecommunications and Information Administration (NTIA), *Video Program Distribution and Cable Television: Current Policy Issues and Recommendations.* Anita Wallgren, Project Manager. NTIA Report 88–233 (June 1988), 80–82.

76. TNT signed a 4-year deal with National Basketball Association in late 1989; seventy-five games per year for total $275 million.

77. For an excellent history of the Discovery Channel, see Susan M. Strohm, "The Discovery Channel," In: Robert Picard, ed. *The Cable Networks Handbook* (Riverside, CA: Carpelan Publishing Co., 1993).

78. Malone.

79. Extended advertising time was not formally prohibited, but exceeding the guidelines triggered more stringent FCC review of license renewals, Amendment of Part O of the Commissions Rules, 43 FCC 2d 638,639 (1973).

80. Revised Programming and Commercialization Polices, 98 FCC 2d 1076, 1105 (1984).

81. Commercial TV Stations, 98 FCC 2d 1102 (1984).

82. Although shorter, 6- to 10-minute "Infomercials" were being run experimentally on the Columbus QUBEube system in the late 1970s." See, John Wicklein,

"Two-Way Cable: Much Promise, Some Concern." In: Mary Louise Hollowell, ed. *The Cable/Broadband Yearbook, Vol. 2, 1980–1981.* (Washington, D.C.: Communications Press, Inc., 1980), 190–207.

83. See, Radar Hayes and Herb Rotfeld., "Infomercials and Cable Network Programming," *Advancing the Consumer Interest,* 1:2 (1989): 17–22.

84. Following the CVN/QVC merger, former employees of CVN started a new shopping network, ValueVision International. Based in Minneapolis, it began operation in October 1991.

85. See, generally, Alec Foege, *The Empire That God Built* (New York: John Wiley & Sons, Inc., 1996).

86. Cited in John Taylor, "Pat Robertson's God, Inc.," *Esquire,* November 1994, 82.

87. Robertson and his son Tim bought about $180,000 worth of IFC stock, Class A shares at 3.3 cents a share, and Class B shares at 2.2 cents a share. Two years later the (Class B) share price was around $15 and the Robertson's investment was worth about $90 million.

88. Herbert Howard and Robert Ogles, "The Family Channel." In: Picard.

89. CBS had been a Cablevision partner in Rainbow through 1985 and 1987, but sold its interest back to Cablevision by 1987 when the broadcast company was going through its deep retrenchment.

90. Waterman, 175.

91. Benjamin Klein, "The Competitive Consequences of Vertical Integration in the Cable Industry," Paper for the National Cable Television Association, June 1989.

92. Cable scholar and economist Robert Crandall along with Harold Furchtgott-Roth concluded in a 1996 book for the conservative Brookings Institution that vertical integration favored cable network proliferation to the benefit of the consumer. Robert Crandall and Harold Furchtgott-Roth, *Cable TV: Regulation or Competition?* (Washington D.C.: The Brookings Institution, 1996). Media Economists Ahn and Litman suggested a curvilinear model "in which consumer welfare is enhanced as vertical integration increases up to an optimal threshold, after which consumer welfare decreases." Hoekyun Ahn and Barry Litman, "Vertical Integration and Consumer Welfare in the Cable Industry," *Journal of Broadcasting & Electronic Media,* 41 (1997): 453–477, 453. Chan-Olmsted and Litman could find no evidence that consumers were benefiting from the price breaks negotiated by the larger MSOs, Chan-Olmsted and Litman, 1988, 9.

93. Senate Committee on Commerce, Science and Transportation, Subcommittee on Communications, Hearings on Media Ownership, Diversity and Concentration. 101st Cong., 1st sess., June 14, 21, and 22, 1989, 609–610.

94. See, Report on Cable Competition and Rate Deregulation, FCC 90–276, 67 RR 2d 1771 1990, at para. 120–122.

95. "Jones Drops USA," *Cablevision,* October 10, 1988, 16.

96. For detailed reviews of the creation of Liberty Media Corp., see: Robichaux, 112–117; and L. J. Davis, *The Billionaire Shell Game* (New York: Doubleday, 1998), 139–143.

97. Robichaux, 113.

98. Ibid., 114.

99. Because the interest was held in Class B super-voting shares, he had a controlling 40 percent of the voting shares.

100. Robichaux, 114.

101. "Mr. Levin's Giant Step for Pay TV," *Broadcasting*, April 21, 1975, 16–18, 18.

102. "Spurt Begins in Pay Cable with Movies as Main Fare, Sports, Second," *Broadcasting*, August 18, 1975, 17–19, 18.

CHAPTER 11

1. Bill Daniels, *Daniels Letter* (a Bill Daniels industry newsletter), December 1986.

2. Under the provisions of the 1984 Cable Act defining effective competition, 96.5 percent of all cable systems were immune from rate regulation.

3. Cecilia Capuzzi, "Wall Street's Affair with the Wire," *Channels 1987 Field Guide*, 69.

4. Michael O. Wirth, "Cable's Economic Impact on Over-the-Air Broadcasting," *Journal of Media Economics*, 1990, 39–53,41.

5. Ibid., 41.

6. Figures from NCTA. *Cable Television Developments* (Washington, D.C:. National Cable Television Association, annual).

7. "Working at Wizardry," *Cable Television Business*, February 1, 1984, 100.

8. Senate Committee on Commerce, Science and Transportation, Subcommittee on Communications. Hearings on the Cable TV Consumer Protection Act of 1991, S. 12, 102d Cong., 2d sess., 1991, 8.

9. See, generally, Julia Dobrow, ed. *Social and Cultural Aspects of VCR Use*. (Hillsdale, NJ: Lawrence Erlbaum Associates, 1990); and Compaine, Benjamin, and Douglas Gomery. 2000. *Who Owns the Media*, 3rd ed. (Mahwah, NJ: Lawrence Erlbaum Associates, 2000;411–419.

10. Cary Dekmark and Ray Seggern. "Home Video." In: August Grant, ed. *Communication Technology Update*, 4th ed. (Boston: Focal Press, 1995), 274.

11. Richard Zacks, "Cassettes Rewrite Studios' Books," *Channels 1987 Field Guide*, 90.

12. NCTA. *Cable Television Developments*, 1990.

13. "Cable at the Crossroads in 1987," *Broadcasting*, January 5, 1987, 172.

14. "Wiring the Cities," *Cablevision*, November 22, 1982, 252.

15. See, Richard Katz, "Swinging for the Fences?" *Channels 1990 Field Guide*, 82.

16. Mark Frankel, "Forcing Open a New Window," *Channels 1987 Field Guide*, 82.

17. Ibid.

18. Richard Katz, "A Low-wire Act," *Channels 1990 Field Guide*, 92.

19. See Mark Banks, "Low Power Television.," in Grant, 118–124.

20. Estimate by the Wireless Cable Association. Quoted in: Andrew Servetas, "License to Thrive," *Channels 1990 Field Guide*, 93.

21. FCC. *Annual Assessment of the Status of Competition in Markets for the Delivery of Video*

22. *Programming, 6th Annual Report*, FCC (2000), 43.

23. The application was challenged, but survived in the courts. See: *Century Fed. v. FCC*, 846 F.2d 1479 (D.C. Cir 1988).

24. See, Memorandum Opinion and Order in General Telephone Co. of California, 4 FCC Rcd. 5693 (1989).

25. David Bollier, "There's Life after Scrambling." *Channels 1987 Field Guide*, 66.

26. Edward Horowitz, Oral History, The Cable Center, Denver, CO.

27. Bollier.

28. "Satellites," *Broadcasting*, July 20, 1987, 33–58.

29. FCC, *6th Annual Report*, 42.

30. "SMATV," *Channels 1987 Field Guide*, 70.

31. A total of eight orbital positions and 256 channels were available. The full-CONUS 61.5-degree W slot was less desirable than 101, 110, and 119 positions because of problems involving solar eclipse outages.

32. See, Continental Satellite Corp., 4 FCC Rcd. 6292 (1989). Other applicants receiving assignments in August, but ultimately failing to act on them, included: Anaheim-based Continental Satellite Corp.; Direct Broadcast Satellite Corp., owned by Kansas City Southern Railroad; and Directsat Corp., which had plans for religious programming.

33. Satellites," *Broadcasting*, 44.

34. Continental Satellite Corp. 4 FCC Rcd. 6292 (1989).

35. Satellite Syndicated Systems, Inc., 99 FCC 2d 1369 (1984).

36. Continental Satellite Corporation, 4 FCC Rcd 6292 (1989).

37. Advanced Communications Corp., 6 FCC Rcd. 2269 (1991).

38. Allard Sicco De Jong and Benjamin J. Bates, "Channel Diversity in Cable Television," *Journal of Broadcasting & Electronic Media* 35:2 (1991): 159–166.

39. National League of Cities. *Impact of the Cable Act on Franchising Authorities and Consumers*, September 18, 1987.

40. The five states included Maryland, New Hampshire, Ohio, Texas, and West Virginia.

41. Must-carry signals would have included stations within fifty miles of the cable system that drew at least a 2 percent weekly share in non-cable homes in the market. Systems with twenty or fewer channels were exempt from the requirement. Total carriage limits also applied to larger systems.

42. Report and Order, 1 FCC Rcd 864, modified in part 2 FCC Rcd. 3593 (1987).

43. *Century Communications v. FCC*, 835 F.2d 292 (D.C. Cir. 1987), *cert. denied*, 486 U.S. 1032 (1988).

44. Wirth.

45. FCC, Mass Media Bureau. Cable System Broadcast Signal Carriage Survey Report, September 1, 1988.

46. Program Exclusivity in the Cable and Broadcast Industries, 3 FCC Rcd. 5299, 64 RR 2d 1818 (1988). The new rules required cable operators to black out programming imported from any distant station, including superstations that duplicated locally broadcast syndication programming for which the local station held exclusive rights. The rules permitted, but did not require, broadcasters to negotiate for exclusive distribution rights. Cable operators did not have to protect (black out) programming from any broadcast signals generally available in their market, and systems with fewer than 1,000 subscribers are exempt. The rules required TV stations to give cable operator sixty days notice for protection of exclusive programming.

47. *United Video v. FCC*, 890 F.2d 1173 (D.C. Cir. 1989).

48. Notice of Inquiry, Compulsory Copyright License for Cable Retransmission, Gen. Docket No. 87–25, FCC Rcd. 87–66 (1987).

49. Notice of Inquiry, Telephone Company–Cable Television Cross-Ownership Rules, 2, FCC Rcd. 5092 (1987).

50. Further Notice of Inquiry and Proposed Rulemaking, 3 FCC RCD. 5849 (1988).

51. *United States. v. A.T.& T*, 552 F. Supp. 131 (1988).

52. *United States v. Western Electric Co.*, 900 F.2d 283 (D.C. Cir. 1990).

53. *United States v. Western Electric Co.*, 767 F. Supp. 308 (D.D.C. 1991).

54. The MFJ was formally rescinded in April of 1996, quietly vanishing after more than ten years of momentous and controversial life.

55. Broadcast networks also declined to offer their scrambled programming to dish owners outside the reach of an existing network market. The networks did not want to distribute backhaul feeds intended only for stations to dish owners. As for standard broadcast programming, from both networks and superstations, the Copyright Act of 1976 restricted distribution of those signals to the dish community. Tempo Enterprises indicated a willingness to provide WTBS to dish owners, but noted that the Copyright Act had to be amended first to extend cable's compulsory license to the distribution of broadcast signals directly to dish owners.

56. In Re: Competition, Rate Deregulation and the Commission's Policies Relating to the Provision of Cable Television Service ("Cable Report"), 5 Rcd. No. 16, 4962, July 1990; and see, Waterman, David. "Vertical Integration and Program Access in the Cable Television Industry," *Federal Communications Law Journal* 47:3 (1995): 511–535.

57. Cable Report, para. 114.

58. "Earth Station Battle Goes to the Hill," *Broadcasting*, March 10, 1986, 33–34.

59. Albert Gore, "Protecting Fair Access to Cable Satellite Programming," *Memphis State University Law Review* 15:3 (1985): 341–368.

60. The wind also was taken out of the legislative sails by the success of Tim Wirth to negotiate a small settlement between cable and TVRO interests. HBO and Showtime agreed, in response to the growing criticism and to Wirth's lobbying, to halt the controversial practice of rebating fees from dish owners to local cable operators

61. Senate Bill. 889, 100th Congress.

62. Senate Committee on the Judiciary, Subcommittee on Antitrust, Monopolies, and Business Rights. Competitive *Issues in the Cable Television Industry*. 100th Cong., 2 sess., March 17, 1988, S. Hearing, 100–818. Comm. Serial No. J-100–55, 470.

63. Ibid., 440.

64. House Committee On Energy and Commerce, Subcommittee on Telecommunications and Finance, *Cable Television*, 100th Cong., 2d sess., March 30, May 11, 1988, Comm. Serial No. 100–195.

65. U.S. Department of Commerce, National Telecommunications and Information Administration (NTIA), *Video Program Distribution and Cable Television: Current Policy Issues and Recommendations*, Anita Wallgren, Project Manager. NTIA Report 88–233, June 1988.

66. Although the NTIA report diverged from the OTP report in one aspect, stating that existing cable operators should not be separated from content control. See, 34, note 95.

67. 135 Cong. Record E740 (daily ed. March 13, 1989).

68. See, for example, Senate Committee on Commerce, Science and Transportation, Subcommittee on Communications, Hearings on S. 1880. *Cable TV Consumer Protection Act of 1989*, 101st Cong., 2d sess., March 29, 1990. Testimony of James Mooney, President, NCTA, 47–117, although similar testimony was presented at numerous congressional hearings during this period.

69. Ibid., 51.

70. Robert Ekelund and Robert Tollison, "The High Costs of Cable Regulation." Working paper, Center for Study of Public Choice, George Mason University, April 1992.

71. U.S. General Accounting Office (GAO), *Telecommunications: National Survey of Cable Television Rates and Services* (August, 1989).

72. "GAO on Cable," *Broadcasting*, August 7, 1989, 30.

73. Senate Committee on Commerce, Science and Transportation, Subcommittee on Communications., *Oversight of Cable TV*, 101st Cong., 1st sess., November 16 and 17, 1989, S. Hrg. 101–464.

74. "Danforth Throws the Book at Cable," *Broadcasting*, November 20, 1989, 29.

75. "Re-regulatory pendulum swings toward cable," *Broadcasting*, November 20, 1989, 27.

76. Senate Committee on Commerce, *Oversight of Cable TV*, 4.

77. Ibid., 290–292.

78. Robichaux, 107.

79. Senate Committee on Commerce, *Oversight of Cable TV*, 161.

80. Ibid., 162.

81. "On the Brink of War, NAB Tells Cable: It's Cash and Carry Time," *Broadcasting*, December 4, 1989, 35–37, p. 36.

82. "On the Brink of War," 36.

83. Pull back from Armageddon," *Broadcasting*, December 4, 1989, 130.

84. Cable Report, 5 FCC Rcd. 4962 (1990).

85. Ibid.

86. U.S. General Accounting Office (GAO), *Follow-Up National Survey of Cable Television Rates and Services*, GAO-RCED 90–199 (June 1990).

87. Ibid., 57, 61.

88. Reexamination of the Effective Competition Standard for the Regulation of Cable Television Basic Service Rtes, 6 FCC Rcd. 4545 (1991).

89. S. 12 and H.R. 4850.

90. Provisions of the 1992 Act relating to rate regulation subsequently survived court challenges. See *Time Warner Entertainment v. FCC*, 56 F.2d 151 (D.C. Cir. June 6, 1995).

91. The FCC promulgated rules applying this requirement in April 1993. See:, In re: Implementation of Sec. 12 and 19 of the Cable TV Consumer Protection and Competition Act of 1992, First Report and Order, 8 FCC Rcd. 3359 (1993).

92. While Congress adopted new rules for local carriage, it retained the general copyright framework for distant signals. It abolished the CRT in 1993. The process

had become too cumbersome and complicated. The CRT's functions were given to Library of Congress copyright office.

93. Eric Schmuckler, "Cable vs. Telephone: The War of the Wires," *Channels*, May/June 1984, 36–39, 36.

CHAPTER 12

1. "Seeing Over the Telephone," *The American Review of Reviews*, May 1927, 516–518.

2. Mark Robichaux, *Cable Cowboy: John Malone and the Rise of the Modern Cable Business* (Hoboken, NJ: John Wiley & Sons, Inc., 2002), 123.

3. Edmund Andrews, "A Cable Vision (Or Nightmare): 500 Channels," *The New York Times*, December 3, 1992, A1.

4. Ibid.

5. Ibid.

6. Patricia Aufderheide, *Communications Policy and the Public Interest: The Telecommunications Act of 1996* (New York: The Guilford Press, 1999), 39.

7. Also from a lexicography perspective because it began with a "T" and could be substituted for "television" in group titles without altering the existing acronym.

8. "Goodbye to All That," *Broadcasting & Cable*, May 15, 1995, 78.

9. The trend even hit equipment manufacturing. The old-line cable manufacturing firm General Instruments changed its corporate identity to "NextLevel," only to change back to G.I. in early1998.

10. Will Workman, "Cable Ratings Reach All-Time High as Broadcasters' Tailspin Continues," *Cable World*, May 27, 1996, 1.

11. Four weeks after unveiling the system at a San Francisco meeting of the Society of Motion Picture and Television Engineers (SMPTE), NHK partnered with Sony and CBS to put on an HDTV demonstration in Washington, D.C., as part of the CBS plan to create a DBS HDTV service.

12. Television Broadcasting Services; Advanced Television Technology, 52 Fed. Reg. 34, 259 (1987).

13. Advanced Television Systems, 65 RR 2d 295 (1988).

14. Advanced Television Systems. Third Report and Order, 71 RR 2d 375 (1992).

15. Fourth Report and Order, 11 FCC Rcd. 17,771 (1996).

16. In 1994, the Motion Picture Experts Group, established technical standards for compression. The first MPEG standard for video compression, called MPEG-1, permitted video to be transmitted at about 1.5 megabits per second (Mbps), roughly the quality of an analog VHS signal.

17. Eventually, different kinds of video information became amenable to different levels of compression. A TV news conference, for example, which involves little action or picture detail, can be subjected to high levels of compression because the need to continually refresh the data continually is low. A fast-moving basketball game requires more updating and, therefore, lower compression rates. Contemporary systems can sense the need for varying levels of compression and allocate bandwidth as necessary, creating variable or dynamic compression rates.

18. The name change was necessitated by a University of Illinois claim on the MOSAIC title.

19. Under the 1992 Act, a cable system faced effective competition only if: (1) less than 30 percent of the households in the franchise area subscribed to the service, and (2) a competing and comparable provider offered service to 50 percent of the homes in the franchise area and 15 percent subscribed, or (3) the franchising authority offered its own service and at least 50 percent of the homes in the area subscribed.

20. The FCC's benchmark was based on system rates in effect on September 30, 1992, and subsequent rate adjustment orders linked to rates in effect on that date, adjusted for inflation.

21. Implementation of sections of the Cable Television Consumer Protection and Competition Act of 1992, 8 FCC Rcd. 5631 (1993).

22. The Commission first set the implementation deadline for June 21, 1993, then delayed it until October 1, 1993, drawing a sharp complaint from Congress about foot-dragging and prompting the FCC to set a final start date of September 1.

23. Systems were permitted to challenge rate limitations by demonstrating that proposed increases were simple pass-through costs, but again the task of establishing the case was a daunting one.

24. FCC, Cable Services Bureau, "FCC Regulation Impact Survey Changes in Cable Television Rates between April 5, 1993 and September 1, 1993: Report and Summary," February 22, 1994.

25. Mark Robichaux, "TCI Offers Apology to FCC for Memo on Cable Rate Rises," *Wall Street Journal*, November 17, 1993, B1.

26. With the caveat that the increased rates remain below benchmark levels.

27. Reed E. Hundt, *You Say You Want a Revolution: A Story of Information Age Politics* (New Haven, CT: Yale University Press, 2000), 5.

28. Clinton did not act quickly filling the four seats that were vacant at the Commission when he took office. Commission Quello was acting chairman until the appointment of Hundt.

29. Hundt, 31.

30. Second Report on Reconsideration., Fourth Report & Order, 9 FCC Rcd. 4119 (1994). In the original formulation, the Commission combined the three specified criteria for effective competition into one unified variable in a regression equation. In the 1994 revision, the relative influence of the three measures was considered independently.

31. The weight of cable-related business became so heavy that, in 1994, the Commission created a new Cable Bureau to handle the workload. Previously, a Cable Bureau had existed between 1972 and 1981.

32. Robert W. Crandall and Harold Furchtgott-Roth, *Cable TV: Regulation or Competition?* (Washington, D.C.: The Brookings Institution, 1996), 41.

33. *Time Warner Entertainment v. FCC*, 56 F.3d 15 (D.C. Cir.), *cert. denied*, 116 S.Ct. 911 (1996).

34. Crandall and Furchtgott-Roth, 73.

35. NCTA. *Cable television developments*. (Washington, D.C.: National Cable and Telecommunications Association, annual).

36. In its initial ruling, the Supreme Court voted 5 to 4 that the First Amendment did not necessarily prohibit the government's action and sent the case back to the lower court for further review. *Turner Broadcasting, Inc. v. FCC*, 819 F. Supp 32 (D.D.C. 1993), *aff'd*, 512 U.S. 622 (1994). Both the lower court and the Supreme Court subsequently affirmed the economic theory behind the law and determined the

government had made its case. *Turner Broadcasting, Inc. v. FCC*, 910 F. Supp. 734 (1995), *aff'd*, 520 U.S. 180 (1997).

37. See, Richard Gershon and Bradley Egen, "Retransmission Consent, Cable Franchising, and Market Failure: A Case Study Analysis of WOOD-TV 8 Versus Cablevision of Michigan," *Journal of Media Economics* 12:3 (1999): 201–224.

38. Under the 1992 Act, broadcasters and cable operators were given until October 5, 1993, to negotiate retransmission consent fees (with subsequent renewals every three years).

39. Subsequent improvements in DSL capacity, with such the acronym s as HDSL (high bit rate DSL) and VDSL (very high bit rate DSL), boosted capacity to 50 Mbps, or more.

40. Anita Wallgren, "Video Program Distribution and Cable Distribution: Current Policy Issues and Recommendations," National Telecommunications and Information Agency. U.S. Department. of Commerce, June 1988.

41. Second Report and Order, 7 FCC Rcd. 5781 (1992).

42. Ibid.

43. *The Chesapeake and Potomac Telephone Company of Virginia v. United States*, 830 F. Supp. 909 (E.D. Va. 1993), *aff'd*, 42 F.3d 181 (4th Cir. 1994).

44. Appeals of the cases were made moot by repeal of the ban in the 1996 Telecommunications Act., discussed below.

45. Bob Hughes, Oral History, The Cable Center, Denver, CO.

46. Robichaux, 138.

47. At this time, the RBOCs were still banned from owning cable systems in their own service areas, so Bell Atlantic indicated it would divest non-compliant systems.

48. Senate Committee on the Judiciary, Hearings before the Subcommittee on Antitrust, Monopolies and Business Right., *Examining the Effects of Megamergers in the Telecommunications Industry*, 103rd Cong., 1st. sess., October 27, November 16, December 16, 1993, S. Hrg. 103–850, Serial No. J-103–33.

49. Primsestar entered into a similar agreement to end antitrust investigations initiated by attorneys general in forty-five states.

50. That same day Ergen filed suit against Advanced for breach of contract, a suit he eventually won, recouping $1 million he had invested in the partnership plus $230,000 in interest..

51. Advanced communications Corp., 6 FCC Rcd. 2269 (1991).

52. His 1991 extension was due to expire in December of 1994.

53. Advanced Communications, Inc. Memorandum Opinion and Order, 10 FCC Rcd. 13337 (1995).

54. Mark Robichaux, "Dishing it Out: Once a Laughing Stock, Direct-Broadcast TV Gives Cable a Scare." *Wall Street Journal*, November 7, 1996, A1.

55. Interactivity through an upstream path from the viewer's home is available, but in most architectures restricted to the 0- to 50-MHz band.

56. Johnnie Roberts and Laura Landro, "King of Cable: John Malone of TCI is Formidable Player in Bid for Paramount," *Wall Street Journal*, September 27, 1993, A1.

57. "USA Makes It Six in A Row," *Broadcasting & Cable*, January 1, 1996, 39.

58. In re: Evaluation of the Syndication and Financial Interest Rules, 56 Fed. Reg. 26242 (May 29, 1991).

59. *Schurz Communications v. FCC*, 982 F.2d 1043 (7th Cir. 1992).

60. *United States v. American Broadcasting Co., Inc.*, 1993–2 Trade Cas. 70,418 (C.D Cal, 1993).

61. In re: Evaluation of the Syndication and Financial Interest Rules, 8 FCC Rcd 3282 (1993). The remaining fin-syn restrictions expired in November 1995 and were not renewed.

62. CNN had been running news stories on young women and girls in Africa who, in a ritual practice, had their clitorises surgically removed. At a National Press Club speech in September 1994, Turner retold the tale of the "barbaric mutilation," adding, "Well, I'm angry, I'm being clitorized by Time Warner." See, Alec Klein, *Stealing Time: Steve Case, Jerry Levin, and Collapse of AOL Time Warner*. (New York: Simon & Schuster, 2003), 237.

63. U.S. West, which retained a 22.5 percent interest in TWE, attempted, unsuccessfully, to block the acquisition in court.

64. "The Top 200 Cable Operators," *Cablevision*, April 29, 1996, 105.

65. Chan-Olmsted, Sylvia. "Market Competition for Cable Television: Reexamining Its Horizontal Mergers and Industry Concentration," *Journal of Media Economics*, 1996 (9:2): 24–41.

66. Ibid., 32.

67. Second Report and Order, 8 FCC Rcd. 8565 (Oct. 23, 1993).

68. *Daniels Cablevision, Inc. v. U.S.*, 835 F. Supp. 1 (D.D.C. 1993).

69. *Time Warner Entertainment Co. v. United States*, 211 F 3d 1313 (D.C. Cir. 2000).

70. *Time Warner Entertainment Co. v. FCC*, 240 F.3d 1126 (D.C. Cir. 2001).

71. With the important exception of the FCC requirement issued in 2000 that AT&T divest itself of sufficient a number of systems to bring it below the 30 percent cap. When counting AT&T equity interest in TWE, the company reach extended to 42 percent of national subscribership.

72. Hundt, 9.

73. IITF. *National Information Infrastructure: Agenda for Action*. (Washington, D.C.: National Telecommunications and Information Administration, 1993).

74. Gore, Albert. U.S. Vice President's Remarks at the National Press Club. Washington, D.C., December 21, 1993; and Gore, Albert. U.S. Vice-President's Speech on the National Information Infrastructure. UCLA, Los Angeles, January 11, 1994.

75. Ralph Lee Smith, "The Wired Nation," *The Nation*, May 18, 1970, 602.

76. Gore. Remarks at the National Press Club. December 21, 1993.

77. Hundt, 7.

78. David Kline, "Infobahn Warrior," *Wired*, July 1994.

79. Hundt, 31.

80. Hundt, 56.

81. In re: Implementation of Sections of the Cable television Consumer Protection and Competition Act of 1992: Rate Regulation. 10 FCC Rcd. 1226 (1994). Technically, all tiers remained regulated, but the FCC offered assurances that it would not interfere with product and pricing decisions on "going-forward" services.

82. Thomas Southwick, *Distant Signals*. (Overland Park, KS: Primedia Intertec, 1998), 336.

83. For a thoughtful and detailed review of the Act and its gestation, see Aufderheide (ref 6).

84. S. 1822, the Communications Act of 1994.

85. LECS were prohibited from acquiring more than a 10 percent financial interest in any cable company operating in their service territory. Similarly, a cable operator could not obtain more than a 10 percent financial interest in any LEC providing telephone services in its franchise area.

86. Defined as MSOs with more than $250 million in annual income.

87. If that competitive service is available to at least 50 percent—and subscribed to by at least 15 percent—of the homes in the service area.

CHAPTER 13

1. See footnote no. 2

2. Various accounts are reported in the following: Mark Robichaux, *Cable Cowboy* (Hoboken, NJ: John Wiley & Sons, Inc, 2002), 208–213; Thomas Southwick, *Distant Signals* (Overland Park, KS: Primedia Intertec, 1998), 325–327; and Joseph N. DiStefano, *COMCASTed*. (Philadelphia: Camino Books, Inc., 2005), 114–115.

3. "The Handbook for the Competitive Market," *Cablevision Blue Book Vol. III,* 1996, 28.

4. James J. Lock, "The Internet as Mass Medium: The Media Enter the World of Cyberspace," *Feedback,* 1995(36):7–10.

5. See, Compaine, Benjamin, and Douglas Gomery. *Who Owns the Media* (Mahwah, NJ: Lawrence Erlbaum Associates, 2000), 438.

6. Estimates on Internet use varied widely for this period, but see generally, Compaine and Gomery. *Who Owns the Media,* 437–480; Edward O'Neill, Brian Lavoie, and Rick Bennett. "Trends in the Evolution of the Public Web," *D-Lib Magazine.* 2003 (9:4) at www.dlib.org. Useful information also is available from surveys conducted through the mid-1990s by the Georgia Institute of Technology's GVU Center.

7. See, Reed E. Hundt, *You Say You Want a Revolution: A Story of Information Age Politics.* (New Haven: Yale University Press, 2000), 224.

8. Robichaux, 171.

9. A two-year court battle was settled in late 1997.

10. Patrick Parsons and Robert Frieden. *The Cable and Satellite Television Industries.* (Boston: Allyn & Bacon, 1998), 140.

11. Although later he would be personally disappointed and very angry when U.S. West reneged on a promise to maintain Continental headquarters in Boston, where it had been based nearly from its inception. When MediaOne announced a decision to consolidate operations and move the cable headquarters to U.S. West headquarters in Denver, Hostetter told the press, "I did business until I was 61 years old on a handshake and my word. It took me until this to be undone by it. I was so angry I was cross-eyed." "Architects of an Industry." (Special supplement to *Cablevision, Multichannel News,* and *CED.* Celebrating the 1999 Cable Television Hall of Fame. Cahners Publishing, 1999, 14a.). He quit the company, as did several other senior executives who refused to make the move.

12. Rich Brown, "US West buys Continental," *Broadcasting & Cable,* March 4, 1996, 12, 13.

13. Kim Mitchell and Vince Vittore, "Hostetter: 'Companies of Size Will Survive'," *Cable World,* March 4, 1996, 1

14. Parsons and Frieden, 131.

15. K. C. Neel, "Bell Atlantic Boxes Gather Dust at Adelphia," *Cable World*, January 25, 1999, 12.

16. Long-time company leader, William McGowan, died from a heart attack in 1992.

17. "Is Cable Telephony Here Yet?" *Cable World*, March 25, 1996, 37.

18. National Cable Television Association. Telecommunications and Advanced Services Provided by the Cable Television Industry, Washington, D.C., April 1996.

19. In 1996, AT&T also broke itself into three separate firms in what was called a "tri-vestiture." Computers under the NCR logo were group into one company. Manufacturing, with the old Bell Laboratories and Western Electric groups, were placed under the new Lucent Technologies firm, and the new AT&T focused on telecommunications services.

20. In 1996, AT&T also spun off Lucent Technologies, the renamed division that had once been the historic Bell Laboratories, the research and development and equipment manufacturing arm of the old AT&T.

21. Hundt, 217.

22. See, DiStefano, 150.

23. DirecTV bought out the DBS interests of its long-time partner USSB for $1.3 billion in December 1998

24. Data Over Cable Service Interface Specifications, or DOCSIS.

25. NCTA, *Cable Television Developments 2001* (Washington, D.C.: National Cable and Telecommunications Association, 2001), 14.

26. Lewis, "Most Go On Line At Home."

27. *A.T.&T. v. City of Portland.* 43 F. Supp. 2d 1146, 1155 (D. Or. 1999), *rev'd*, 216 F.3d 871 (9th Cir. 2000).

28. H.R. 1686, the Internet Freedom Act would have permitted ISPs denied access to file antitrust claims.

29. NCTA, *Cable Television Developments 2001*, 14.

30. In 1998, the FCC issued rules requiring digital boxes be made available for sale to the public. Commercial availability of navigational devices, 13 FCC Rcd 14775 (1998).

31. Fox and Liberty also acquired 40 percent of Rainbow Media Holdings sports channels from Cablevision System for $850 million.

32. See generally, Collins, Scott. 2004. *Crazy Like a Fox: The Inside Story of How Fox News Beat CNN.* (New York: Portfolio, Penguin Books, 2004).

33. "Turner on Murdoch: I'll squish Him like a bug," *Broadcasting & Cable*, December. 4, 1995, 56.

34. Will Workman, "Will Smaller Nets Survive the Bidding Wars?" *Cable World*, June 3, 1996, 2.

35. "Is Ted Turner Nuts? You Decide." *New York Post*, October 21, 1996, 1.

36. *Time Warner Cable v. Bloomberg LP*, 1997 US App Lexis 18283 (2d Cir. 1997).

37. See, Digital Television Service, Sixth Report and Order, 12 FCC Rcd. 14588 (1997).

38. The broadcaster's digital bandwidth, in broad terms, could be used for a single HDTV signal or, alternatively, several standard digital TV signals.

39. Carriage of Digital Television Broadcast Signals, First Report and Order, 16 FCC Rcd. 2598 (2001).

40. Carriage of Digital Television Broadcast Signals, Second Report and Order, 20 FCC Rcd 4516 (2005).

41. *ACLU v. Reno*, 521 U.S. 844 (1997).

42. *United States v. Playboy Entertainment Group, Inc.*, 529 U.S. 803 (2000).

42. FCC., *Annual Assessment of the Status of Competition in the Market for the Delivery of Video.*

43. *Programming, 10th Annual Report* FCC 04–5 (2004), 6.

44. The "Video Competition and Consumer Choice Act of 1998" would have given municipalities the option of delaying the ends of rate controls under certain conditions.

45. John Higgins, "Meager Mergers, Unless AT&T's in," *Broadcasting & Cable*, February 19, 2001, 28.

46. See, Patrick Parsons, "Horizontal Integration in the Cable Television industry: History and Context," *Journal of Media Economics* 2003 16:1 (2003): 23–40.

47. See generally, DiStefano.

48. NCTA., *Cable Television Developments 2000.*

49. MCI-WorldCom emerged from bankruptcy in 2004, dropping the now-tainted WorldCom name and becoming MCI.

50. For a detailed and well-written account of the unraveling of AT&T, see, Leslie Cauley, *End of the Line: The Rise and Fall of AT&T* (New York: The Free Press, 2005).

51. Joshua Cho, "Adelphia Cool to Partnering with AT&T," *Cable World*, February 22, 1999, 10.

52. Rebecca Blumenstein, "Sweet Revenge: Bid for AT&T Cable," *The Wall Street Journal*, July 10, 2001, B1.

53. After the sale of some assets the final price tag for AT&T had been about $80 billion. Comcast purchased AT&T cable for about $5,000 a subscriber.

54. For a detailed description of the AOL-Time Warner merger and its aftermath, see: Alec Klein, *Stealing Time: Steve Case, Jerry Levin, and Collapse of AOL Time Warner.* (New York: Simon & Schuster, 2003).

55. Klein, 94.

56. Ibid., 115.

57. In addition to the business burdens, Levin suffered a tragic personal loss in 1997 when his 31-year-old son, Jonathon, was abducted, tortured, and killed in New York City. The killer had been after Jonathon's ATM card and PIN number.

58. Klein, 274.

59. Time Warner settled the resultant cases with the SEC and the Justice Department in late 2004 paying $60 million in penalties and agreeing to make corrections in its accounting practices.

60. Pittman left MTV in 1987, worked briefly for Time Warner and joined AOL in 1996.

61. Adelphia was not the only cable company hurt by accounting scandals. In April 2005, four former Charter executives were sentenced to up to 14 months in prisons after pleading guilty to inflating subscriber figures.

62. Andrew Ross Sorkin, "Fallen Founder of Adelphia Tries to Explain," *New York Times*, April 7, 2003, C-1.

63. The FCC approved the acquisition in December 2003.

64. As a condition of FCC approval of the merger, Murdoch promised not to impede cable access to Fox programming for four years. As a footnote to his program access commitment, Murdoch said it did not extend to affiliated companies, including the many programming interests of Liberty Media.

65. FCC. *Trends In Telephone Service* (2005), Table 7.1, p. 7–3

66. Vonage Holdings Corporation Petition for Declaratory Ruling Concerning an Order of the Minnesota Public Utilities Commission, 19 FCC Rcd. 22404 (2004).

67. Kagan World Media., *Cable TV Investor: Deals & Finance*, July 29, 2004, 8–9.

68. Lorne Manly and Ken Belson. "Calling Out the Cable Guy: Why the Phone Company Is Going Hollywood," *New York Times*, November 27, 2005, sec. 3, p. 1.

69. FCC. *Annual Assessment of the Status of Competition in the Market for the Delivery of Video Programming*, 11th Annual Report, FCC 05–13 (2005), 5.

70. Ibid.

71. On the equipment side, the Internet server company Cisco Systems, Inc. purchased the old-line cable supplier Scientific-Atlanta for $7 billion in 2005.

72. RCN originally stood for Revolutionary Cable Network. This was later changed to Residential Cable Network.

73. FCC, *11th Annual Report*, 17.

74. Ibid.

75. NCTA., 2006. *Industry overview 2006*. (Washington D.C.: National Cable and Telecommunications Association, 2006).

76. Ibid.

77. See, Heeter, Carrie, and Bradley Greenberg. 1988. *Cableviewing*. (Norwood, NJ: Ablex, 1988).

78. FCC. *11th Annual Report*, para. 81.

79. Second Report and Order, FCC 05–190 (2005).

80. Digital Television Transition and Public Safety Act of 2005, Public Law 109–171 (Feb. 8, 2006).

81. FCC. *11th Annual Report*, 28.

82. Ibid., 8–9.

83. Ibid., 6.

84. NCTA. *Industry Overview 2006*, 12.

85. FCC, *11th Annual Report*, 32.

86. DiStefano, 90.

Index

Patrick R. Parsons is Don Davis Professor of Ethics, College of Communications, Penn State University. He is the co-author (with Robert Frieden) of *The Cable and Satellite Television Industries*. He also is the author of *Cable Television and the First Amendment* and co-editor (with Steven Knowlton) of *The Journalist's Moral Compass*.